Continuous Casting

Continuous Casting

Special Issue Editor

Michael Vynnycky

MDPI • Basel • Beijing • Wuhan • Barcelona • Belgrade

MDPI

Special Issue Editor
Michael Vynnycky
University of Limerick,
Ireland

Editorial Office
MDPI
St. Alban-Anlage 66
4052 Basel, Switzerland

This is a reprint of articles from the Special Issue published online in the open access journal *Metals* (ISSN 2075-4701) from 2018 to 2019 (available at: https://www.mdpi.com/journal/metals/special_issues/continuous_casting)

For citation purposes, cite each article independently as indicated on the article page online and as indicated below:

LastName, A.A.; LastName, B.B.; LastName, C.C. Article Title. *Journal Name* **Year**, *Article Number, Page Range.*

ISBN 978-3-03921-321-4 (Pbk)
ISBN 978-3-03921-322-1 (PDF)

Contents

About the Special Issue Editor

Michael Vynnycky is a Professor of Applied Mathematics in the Department of Mathematics and Statistics at the University of Limerick in Limerick, Ireland. He holds a D. Phil. in Applied Mathematics from Oxford University in the U.K., and has previously held research posts at Tohoku National Industrial Research Institute in Sendai, Japan, at the KTH Royal Institute of Technology in Stockholm, Sweden, and at the University of São Paulo in São Carlos, Brazil. His primary research interests lie in the deterministic mathematical modelling of natural and industrial processes, involving the use of asymptotic and numerical methods. He is the author of around 110 peer-reviewed journal publications and over 20 reviewed papers in international conference proceedings.

![metals logo] *metals*

MDPI

Editorial

Continuous Casting

Michael Vynnycky [1,2]

[1] Department of Materials Science and Engineering, The Royal Institute of Technology (KTH), Brinellvägen 23, 100 44 Stockholm, Sweden; michaelv@kth.se

[2] Department of Mathematics and Statistics, University of Limerick, Limerick V94 T9PX, Ireland; michael.vynnycky@ul.ie; Tel.: +353-61-213199

Received: 29 May 2019; Accepted: 31 May 2019; Published: 3 June 2019

1. Introduction and Scope

Continuous casting is a process whereby molten metal is solidified into a semi-finished billet, bloom, or slab for subsequent rolling in finishing mills; it is the most frequently used process to cast not only steel, but also aluminum and copper alloys. Since its widespread introduction for steel in the 1950s, it has evolved to achieve improved yield, quality, productivity, and cost efficiency. It allows lower-cost production of metal sections with better quality, due to the inherently lower costs of continuous, standardized production of a product, as well as providing increased control over the process through automation. Nevertheless, challenges remain and new ones appear, as methods are sought to minimize casting defects and to cast alloys that could originally only be cast via other means. This Special Issue covers a wide scope in the research field of continuous casting.

2. Contributions

Fourteen research articles have been published in this Special Issue of *Metals*. Twelve of these [1–12] relate to the continuous casting of steel, a general schematic for which is shown in Figure 1. As is evident from this figure, the overall process consists of a ladle and a tundish through which molten steel passes, a cooling mould region where solidification starts and at which electromagnetic stirring (EMS) may be applied, secondary cooling regions where water is sprayed on the solidified steel, a so-called strand electromagnetic stirrer, a further region at which final EMS is applied, and withdrawal rollers, by which point the steel has completely solidified. In addition, Figure 2 shows which stage of the continuous casting process each of the articles has focused on.

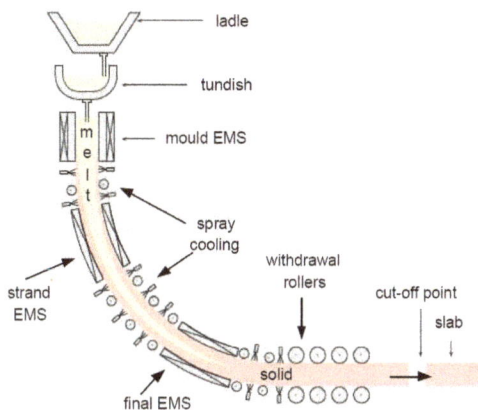

Figure 1. Schematic for the continuous casting of steel.

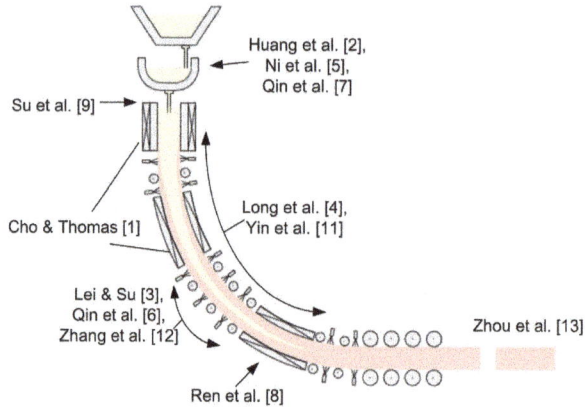

Figure 2. Schematic relating the articles in this special issue to the continuous casting of steel.

Commencing from the start of the process and working downwards [2,5,7], it is important to consider flow in the tundish. Huang et al. [2] use particle image velocimetry (PIV) and numerical simulation to investigate the flow characteristics for a two-strand tundish in continuous slab casting. On the other hand, Ni et al. [5] present a numerical study on the influence of a swirling flow tundish on multiphase flow and heat transfer in the mould, whereas Qin et al. [7] conduct a simulation study on the flow behavior of liquid steel in a tundish with annular argon blowing in the upper nozzle. Su et al. [9] use machine-learning techniques for mold-level prediction by means of variational mode decomposition and support vector regression (VMD–SVR), whereas Cho and Thomas [1] review the literature on electromagnetic forces in continuous casting of steel slabs. Yin et al. [10] consider modelling on inclusion motion and entrapment during full solidification in a curved billet caster, while Long et al. [4] develop a combined hybrid 3-D/2-D model for flow and solidification prediction during slab continuous casting. Qin et al. [6] perform an analysis of the influence of segmented rollers on slab bulge deformation, while Lei and Su [3] use machine learning in the research and application of a rolling gap prediction model. Zhang et al. [11] devise a laboratory experimental setup and consider heat transfer characteristics during secondary cooling, whereas Ren et al. [8] carry out numerical simulations of the electromagnetic field in round bloom continuous casting with final electromagnetic stirring. Zhou et al. [12] consider control of upstream austenite grain coarsening during the thin-slab cast direct-rolling (TSCDR) process, after complete solidification has occurred.

Aside from all of the above, Yang et al. [13] simulate crack initiation and propagation in the crystals of a continuously-cast beam blank, whereas Vynnycky [14] gives a review of applied mathematical modelling of continuous casting; this considers a hybrid of analytical and numerical modelling with an emphasis on the use of asymptotic techniques, and gives examples of problems not only in the continuous casting of steel, but also that of copper and aluminum alloys.

3. Conclusions and Outlook

A variety of topics have composed this Special Issue, presenting recent developments in continuous casting. Nevertheless, there are still many challenges to overcome in this research field and applications still need to be more widespread. As a Guest Editor, I hope that all of the scientific results in this Special Issue contribute to the advancement and future developments of research on continuous casting.

Finally, I would like to thank all reviewers for their invaluable efforts to improve the academic quality of published research in this Special Issue. I would also like to give special thanks to all staff at the Metals Editorial Office, especially to Toliver Guo, Assistant Editor, who managed and facilitated the publication process.

Conflicts of Interest: The author declares no conflicts of interest.

References

1. Cho, S.M.; Thomas, B.G. Electromagnetic forces in continuous casting of steel slabs. *Metals* **2019**, *7*, 471. [CrossRef]
2. Huang, J.; Yuan, Z.; Shi, S.; Wang, B.; Liu, C. Flow characteristics for two-strand tundish in continuous slab casting using PIV. *Metals* **2019**, *9*, 239. [CrossRef]
3. Lei, Z.; Su, W. Research and application of a rolling gap prediction model in continuous casting. *Metals* **2019**, *9*, 380. [CrossRef]
4. Long, M.; Chen, H.; Chen, D.; Yu, S.; Liang, B.; Duan, H. A combined hybrid 3-D/2-D model for flow and solidification prediction during slab continuous casting. *Metals* **2018**, *8*, 182. [CrossRef]
5. Ni, P.; Ersson, M.; Jonsson, L.T.I.; Zhang, T.; Jonsson, P.G. Numerical study on the influence of a swirling flow tundish on multiphase flow and heat transfer in mold. *Metals* **2018**, *8*, 368. [CrossRef]
6. Qin, Q.; Li, M.; Huang, J. Analysis of the influence of segmented rollers on slab bulge deformation. *Metals* **2019**, *9*, 231. [CrossRef]
7. Qin, X.; Cheng, C.; Li, Y.; Zhang, C.; Zhang, J.; Jin, Y. A simulation study on the flow behavior of liquid steel in tundish with annular argon blowing in the upper nozzle. *Metals* **2019**, *9*, 225. [CrossRef]
8. Ren, B.; Chen, D.; Xia, W.; Wang, H.; Han, Z. Numerical simulation of electromagnetic field in round bloom continuous casting with final electromagnetic stirring. *Metals* **2018**, *8*, 903. [CrossRef]
9. Su, W.; Lei, Z.; Yang, L.; Hu, Q. Mold-level prediction for continuous casting using VMD–SVR. *Metals* **2019**, *9*, 458. [CrossRef]
10. Yin, Y.; Zhang, J.; Dong, Q.; Li, Y. Modelling on inclusion motion and entrapment during the full solidification in curved billet caster. *Metals* **2018**, *8*, 320. [CrossRef]
11. Zhang, Y.; Wen, Z.; Zhao, Z.; Bi, C.; Guo, Y.; Huang, J. Laboratory experimental setup and research on heat transfer characteristics during secondary cooling in continuous casting. *Metals* **2019**, *9*, 61. [CrossRef]
12. Zhou, T.; O'Malley, R.J.; Zurob, H.S.; Subramanian, M.; Cho, S.H.; Zhang, P. Control of upstream austenite grain coarsening during the thin-slab cast direct-rolling (TSCDR) process. *Metals* **2019**, *9*, 158. [CrossRef]
13. Yang, G.; Zhu, L.; Chen, W.; Guo, G.; He, B. Simulation of crack initiation and propagation in the crystals of a beam blank. *Metals* **2018**, *8*, 905. [CrossRef]
14. Vynnycky, M. Applied mathematical modelling of continuous casting processes: A review. *Metals* **2018**, *8*, 928. [CrossRef]

metals

MDPI

Review

Electromagnetic Forces in Continuous Casting of Steel Slabs

Seong-Mook Cho [1] and Brian G. Thomas [1,2,*]

[1] Department of Mechanical Engineering, Colorado School of Mines, 1610 Illinois Street, Golden, CO 80401, USA; seongmookcho1@mines.edu

[2] Department of Mechanical Science and Engineering, University of Illinois at Urbana-Champaign, 1206 West Green Street, Urbana, IL 61801, USA

* Correspondence: bgthomas@mines.edu; Tel.: +1-303-273-3309

Received: 23 March 2019; Accepted: 20 April 2019; Published: 23 April 2019

Abstract: This paper reviews the current state of the art in the application of electromagnetic forces to control fluid flow to improve quality in continuous casting of steel slabs. Many product defects are controlled by flow-related phenomena in the mold region, such as slag entrapment due to excessive surface velocity and level fluctuations, meniscus hook defects due to insufficient transport of flow and superheat to the meniscus region, and particle entrapment into the solidification front, which depends on transverse flow across the dendritic interface. Fluid flow also affects heat transfer, solidification, and solute transport, which greatly affect grain structure and internal quality of final steel products. Various electromagnetic systems can affect flow, including static magnetic fields and traveling fields which actively accelerate, slow down, or stir the flow in the mold or strand regions. Optimal electromagnetic effects to control flow depends greatly on the caster geometry and other operating conditions. Previous works on how to operate electromagnetic systems to reduce defects are discussed based on results from plant experiments, validated computational models, and lab scale model experiments.

Keywords: magnetohydrodynamics; fluid flow; bubbles; inclusions; entrapment; entrainment; heat transfer; solidification; slab mold; continuous casting

---

1. Introduction

Continuous casting is the dominant process to manufacture steel, producing over 96% of steel in the world [1]. Thus, even small improvements to this process can have great impact. During solidification of molten steel, many complex phenomena arise, including multiphase fluid flow, particle transport and capture, heat transfer, solidification, and solute redistribution, which are strongly interrelated and can affect various surface and internal defects if process parameters are not controlled within optimal ranges for a given caster. In the mold region, where initial solidification occurs, fluid flow greatly affects steel surface quality, according to surface turbulence, flow instability, transport and capture of argon bubbles and inclusions, superheat transport, and meniscus solidification. In addition, steel internal quality depends greatly on heat transfer, solidification, and solute and dissolved gas transport, which affect segregation, porosity, and microstructure formation in the strand region of the process. These phenomena must be controlled within acceptable process windows to avoid defects and achieve ideal grain structure, and solute distribution.

Many efforts have been made to control fluid flow in the mold, which is responsible for many surface and internal defects in steel slab casting [2–4]. To control the mold flow pattern and reduce surface instability in the mold, the effects of nozzle geometry including port shape [5–7], size [5], angle [5,8–10], bottom design [5,11], and the flow control system [12] (stopper rod vs. slide gate) have been investigated. Other studies have investigated casting conditions including casting speed [13], argon gas injection [14–16], and nozzle submergence depth [17]. The effects of soft reduction on center segregation [18] and porosity [18,19] in the strand has been investigated to improve internal quality. Together with these process parameters, the application of electromagnetic (EM) forces is an attractive method to control phenomena related to fluid flow because the induced forces intrinsically adjust to molten steel flow variations, and field strength has the potential to be adjusted during operation. Thus, understanding and optimizing the effects of electromagnetic forces on the various continuous casting phenomena offer an important way to maintain quality and increase production for a given caster.

Tools to quantify the effects of electromagnetic forces on continuous casting phenomena include plant measurements, lab scale modeling with low melting temperature alloys, and computational modeling. In particular, computational models validated with plant data are a powerful methodology to understand the phenomena and to suggest practical strategies to optimize the operation of electromagnetic systems [20–23].

Various types of the electromagnetic systems have been developed and implemented into commercial slab casters to apply static and/or traveling magnetic fields to control fluid flow, particle transport and capture, heat transfer, and solidification during the continuous casting of steel slabs. Static magnetic fields have been widely applied as Electromagnetic Braking (EMBr) systems, including local, single-ruler, and double-ruler systems. These EMBr systems were invented to maintain a double-roll flow pattern [16] and to stabilize the fluid flow, especially at high casting speed. Local EMBr fields tend to slow down jet flow as it passes through a circular-shaped field region through the mold thickness [24], as shown in Figure 1a. Single-ruler EMBr (Figure 1b) produces a horizontal rectangular-shaped field across the entire mold width [25]. Placing the ruler above the nozzle tends to slow down surface flow and lessens surface turbulence [26]. Alternatively, positioning the ruler below the nozzle tends to deflect the jet upwards to accelerate surface flow [27–29] and also lessens mixing of the upper and lower zones, which is important during the casting of clad steel slabs [30]. Double-ruler EMBr (Figure 1c), also called Flow Control Mold (FC-Mold) [24], generates two horizontal static fields across the mold width, one above and one below the nozzle ports. Adjusting the relative strengths of the upper and lower fields enables more control of the flow field. To enable further adjustment of the static field, Multi-Mode Electromagnetic Brake (MM-EMB) has recently been developed, which aims to brake, dampen, and stabilize the flow in thin-slab casting molds [31]. Specifically, MM-EMB employs different combinations of five local static magnets: three aligned horizontally below the nozzle (one central and two near narrow faces) and two aligned above the nozzle (near narrow faces) [31].

Moving magnetic fields have been developed to control mold flow more actively than the EMBr systems. This is achieved using alternating current (AC) through a set of magnetics with increasing phase-shift, to achieve apparent motion of the magnetic field near each of the two wide faces of the mold. Four sets of magnets are installed two on each wide face, as shown in Figure 1d, and can generate three different moving fields: Electromagnetic Level Stabilizer (EMLS) which moves the fields toward the nozzle for slowing the jet; Electromagnetic Level Accelerator (EMLA), which moves the fields toward the narrow faces for accelerating the jet; and Electromagnetic Rotating Stirrer (EMRS) [32,33], also called Electromagnetic Stirring (EMS) in the Mold (M-EMS) [34,35], which move the fields in opposite directions across each wide face, for horizontally-rotating the flow around the perimeter of the mold. Alternatively, another moving field system, called Electromagnetic Casting (EMC), creates vertical rotating fields near the meniscus in the mold, to reduce oscillation mark depth and hook formation [36–38]. These fields can be set according to standard operating conditions, or adjusted continuously according to current conditions. Adjustments to the mode, moving (phase-shift) velocities, and field strengths can be accomplished manually by operators or automatically adjusted in real time, based on available plant conditions, potentially monitored with real-time sensors [32,39].

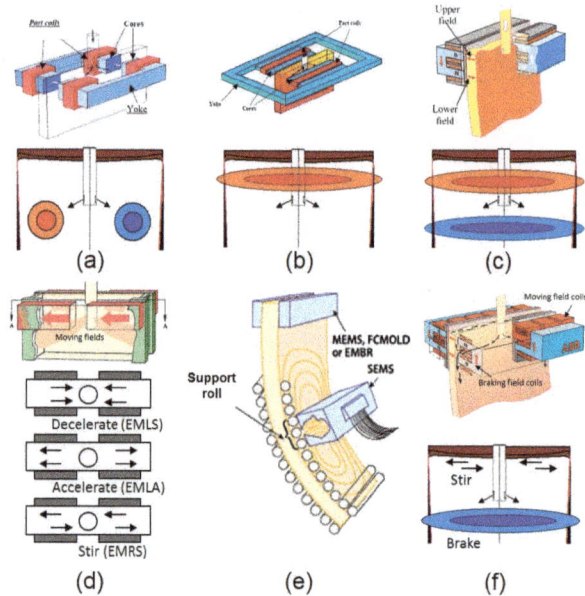

Figure 1. Types of electromagnetic systems showing hardware and field shape: (**a**) local Electromagnetic Braking (EMBr), (**b**) single-ruler EMBr, (**c**) double-ruler EMBr, (**d**) Electromagnetic Level Stabilizer (EMLS), Electromagnetic Level Accelerator (EMLA), and Electromagnetic Rotating Stirrer (EMRS) moving field systems, (**e**) Strand EMS (SEMS), and (**f**) combined fields system.

Below the mold, Strand EMS (SEMS) (Figure 1e), generates a horizontally traveling magnetic field towards one narrow face, by employing one or several box-type (behind the rolls) or in roll-type stirrers including magnets, on one or both of the strand wide faces. This slab-casting SEMS differs from S-EMS in bloom and billet casting, which can have either horizontal rotating magnetic fields around the strand perimeter as with M-EMS [33,40] or longitudinal moving fields that produce recirculating flow in the vertical plane through the strand thickness [41]. The slab-casting SEMS produces vertical recirculating flow regions across the strand width, both above and below the SEMS field region [33,40–47], which aims to control heat transfer and solidification/nucleation phenomena, to increase equiaxed grains and to reduce segregation and porosity [33,42,43].

Finally, a combined system, called FC 3rd generation system (FC3) [48], has recently been developed which combines traveling and static field systems together. An upper traveling field system similar to EMSR is applied above the nozzle ports designed to wash away particles from the meniscus region and make superheat more uniform [48]. A lower static field EMBr single-ruler system is applied below the nozzle ports, designed to lessen particle penetration deep into the mold cavity as shown in Figure 1f. Alternatively, EMC has been combined with single-ruler EMBr, to reduce both oscillation marks at the surface, and particle capture into the steel shell low in the strand [49]. With so much ability to customize the electromagnetic fields, there is a great need to understand how these systems affect fluid flow and steel quality. This has stimulated significant research over the past three decades, which is reviewed in this paper.

This paper first reviews the various research tools available to quantify the effects of electromagnetic forces on continuous steel slab casting, which include plant measurements, lab scale model experiments, and computational modeling. It then reviews current understanding of how each available electromagnetic field system affects important phenomena during slab casting, including fluid flow, surface instability, superheat transport, initial solidification, particle transport and capture, grain structure and internal quality, and steel composition distribution during the casting of steel slabs. Based on these findings, some practical strategies are offered on how to operate electromagnetic systems to reduce defect formation and to improve the quality of the steel product.

2. Tools to Quantify Electromagnetic Effects

To understand and optimize the use of electromagnetic forces to control fluid flow and the associated complex phenomena that affect steel quality in continuous casting, previous researchers have employed several different tools: plant measurements, lab scale modeling, and computational modeling. The plant measurement method is an essential tool to quantify fundamental phenomena in the real commercial process, to validate computational modeling predictions, and to test potential improvements in practice. However, plant measurements have many limitations, owing to the harsh environment of continuous casting and the difficulty of controlling conditions to conduct controlled experiments. Lab scale physical modeling using water is difficult because electromagnetic field effects cannot be accurately mimicked. Physical modeling with low melting alloys can provide important insights into the fundamental phenomena, owing to better flow visualization methods, and better control of the process parameters related to defect formation. However, this method has its own limitations, and it can be difficult to extrapolate the results of lab experiments to the real process. Thus, the best way to investigate how electromagnetic forces should be applied in continuous casting is to develop fundamental computational models of the phenomena, to validate them via both plant measurements and lab scale model experiments, to conduct parametric modeling studies to predict optimal operations based on fundamental understanding, and finally, to test the suggested improvements in the real caster, based on long-term plant measurements to gain reliable statistical evidence.

2.1. Plant Measurements of Fluid Flow Velocity

Fluid flow velocity must be accurately measured to quantify the average mold flow pattern, and velocity fluctuations, especially at the meniscus region where defects can be generated. Methods to measure the flow of molten steel focus on velocity near the top surface in the mold, and include strain gauge, paddle-rod, and nail dipping tests. Indirect measurements of subsurface velocities include electromagnetic Mold Flow Control (MFC) sensors and dendrite angle measurements.

2.1.1. Strain Gauge Rod Tests

Dipping a refractory rod equipped with a strain gauge through the slag steel interface into the molten steel can be used to measure velocity near the top surface in the mold in two ways. Firstly, the average velocity near the top surface depends on the measured torque, which can be related to the drag force applied by the steel moving past the rod, averaged over time and length of the rod [39,50]. The Submeniscus Velocity Control (SVC) device is an example commercial implementation of this method [51]. Secondly, a more sophisticated method is to measure the shedding frequency of the Kármán vortices forming behind the rod, based on the frequency of the time-varying deflections, [52–54] as shown in Figure 2a. The vortex shedding frequency increases linearly with the fluid velocity around the rod. This method has been used to measure surface velocity from 0.05 to 0.7 m/s, which covers the range of interest in slab casting [53]. In both methods, the refractory rod should not be affected by the magnetic field or the high temperature, so should be made from an insulating, thermal shock-resistant material such as a ZrO_2 coated Mo rod [39].

2.1.2. Paddle Rod Tests

Another method to measure flow velocity and direction near the mold top surface by dipping a refractory rod, into the top surface of the molten steel pool is the paddle rod test [55–57]. As shown in Figure 2b, one end of the rod is connected to a pivot and the other end is dipped into the molten steel pool. As the molten steel flow just near the surface impinges on and pushes the rod, it rotates around the pivot. The angle produced depends on a balance between the drag force exerted by the steel flow and the weight of the rod, which can be related to submeniscus velocity of the molten steel.

2.1.3. Nail Dipping Tests

Nail dipping tests are commonly used to quantify mold surface velocity and level due to its convenience and efficiency. This method was introduced to measure surface level profile and liquid slag layer thickness [58,59], and then extended to estimate surface velocity [60–64]. One or more stainless steel nails are immersed into the molten steel pool for a short time (~2–3 s) and removed. To quantify surface velocity, the shape of the solidified steel lump on the nail is measured shown in Figure 2c. The surface velocity is estimated according to [62]

$$U_{surface} = 0.624 \left(\varphi_{lump} \right)^{-0.696} \left(h_{lump} \right)^{0.567} \tag{1}$$

where φ_{lump} is the diameter of the lump solidified on the stainless steel nail and h_{lump} is the height difference built up by the flow. In addition, dipping several different nails at different times enables to quantifying transient variations of the surface velocity. The nail dipping test method shows good ability to detect surface velocity variations due to changes in casting speed, and a reasonable match with SVC measurements in a real caster [62], as shown in Figure 3.

2.1.4. Electromagnetic Mold Flow Control (MFC) Sensor Measurements

The electromagnetic MFC sensor, developed by AMEPA GmbH [65], consists of a permanent magnet and pair of highly sensitive current detectors mounted behind a copper mold plate as shown in Figure 2d for the wide face. The time delay for steel flow variations to travel between the two detectors is evaluated from the measured variations in the induced current, which is generated in proportion to the local velocity of the conducting molten steel traveling through the magnetic field [66]. The time-dependent spatially-averaged velocity near the solidification front in that region of the mold is then output knowing the distance between the two detectors. The method only works in regions where the steel generally flows across the solidification front from one sensor to the other, so the sensors should be positioned at reliable locations, such as near the meniscus between the Submerged Entry Nozzle (SEN) and Narrow Face (NF) in the Wide Face (WF) mold, or perhaps vertically in the NF mold [65,67].

2.1.5. Columnar Dendrite Angle Measurements

The tangential velocity of the molten steel across the solidification front can be estimated from the angle of the columnar dendrite growth direction [68–70]. As solute is washed away from the upstream side of the dendrites, they grow towards (into) the direction of the molten steel flow [69,70]. The angle of the columnar dendrites relative to the growth direction perpendicular to the strand surface increases in direct proportion to the liquid velocity, up to ~0.3 m/s [70], with further increases showing only minor effects. This indirect method enables velocity to be measured deep into the mold cavity and strand regions, where other measurement methods are not available due to the harsh environment. This method is costly as it requires microscopy on solidified slab samples [69] and may also need calibration.

2.1.6. Other Methods

Several other methods have great potential to measure velocity magnitude and/or direction of the liquid metal. One method, which has been demonstrated in laboratory tests with nonferrous alloys, is to immerse a steel sphere into the molten steel, and then measure the melting time with embedded wire(s) [71,72]. The melting rate increases with velocity, so the decrease in time to melt the sphere correlates with the flow velocity. In addition, other methods include photographic methods, reaction probes, tracer methods, electromagnetic probes, hot wire and hot film methods, dissolution methods, and fiber optic sensor measurements, which are all reviewed elsewhere [73].

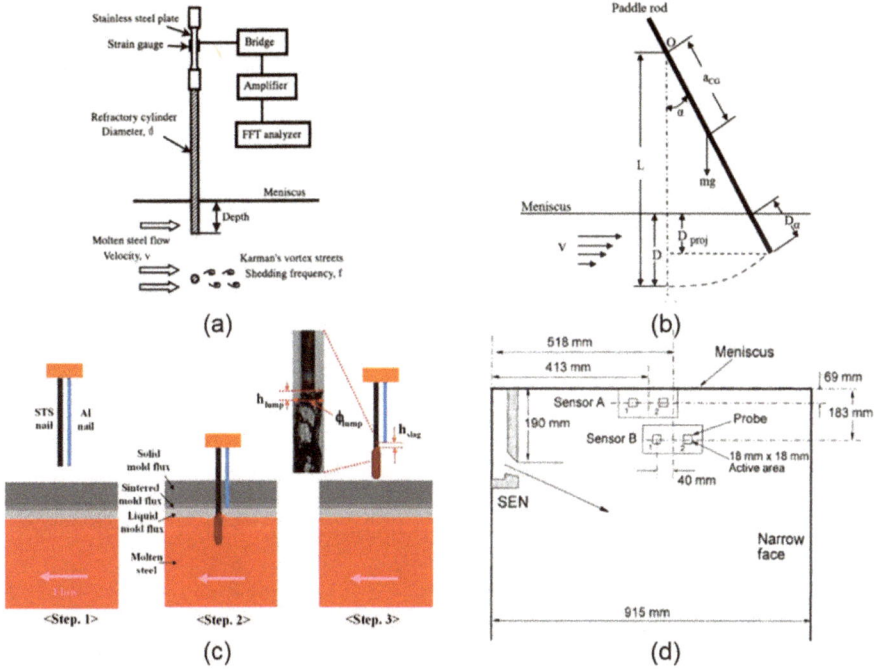

Figure 2. Plant flow–velocity measurements: (**a**) strain gauge test [52], (**b**) paddle rod test [55], (**c**) nail dipping method [64], and (**d**) electromagnetic mold flow control (MFC) test [66].

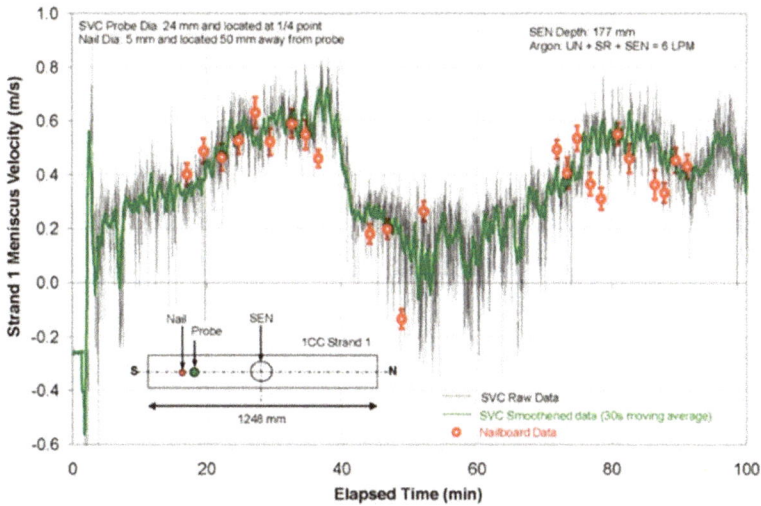

Figure 3. Comparison of surface velocity history between submeniscus velocity control (SVC) and nail dipping measurements [62].

2.2. Plant Measurements of Surface Level Profile and Fluctuations

Measurements of the surface level profile and its fluctuations are needed to understand transient phenomena related to surface defect formation. Especially, surface level fluctuations near the meniscus are known to cause slag entrapment during initial solidification [74]. Thus, it is important to accurately measure and control the surface level during continuous casting. Eddy current sensors are widely used to measure surface level for real-time control in the plant. Other methods to measure the surface level profile include nail board, sheet dipping, and oscillation mark measurements.

2.2.1. Eddy Current Sensor Measurements

In most casters, one eddy current sensor is located above the mold top surface to detect the transient surface level, which is sent to a flow controller (stopper rod or slide gate), to control the steel flow rate. This method is also the most common way to quantify level fluctuations. However, for best control of flow rate, it is best to measure the average liquid level, which is achieved by filtering the signal to remove high-frequency level variations produced by natural turbulence which cannot be controlled, and by placing the sensor above the most stable region (such as the quarter region in the mold with a double-roll flow pattern) [60]. Thus, this measurement underestimates the level fluctuations, especially near the narrow faces and SEN, which tend to be more severe. Moreover, level fluctuations in these regions are more important to quality problems such as deep meniscus hooks, owing to the lower molten steel temperature [56]. Thus, to measure the surface level profile, it is better to keep one senor at the most stable location (1/4 mold width) for the flow controller and to add other sensors near the narrow faces to monitor the surface level variations [75].

2.2.2. Nail Board Tests

In addition to surface velocity described in Section 2.1.3, nail dipping can also be used to measure the surface profile and its variations by using sets of "nail boards", which are made by attaching several nails to a wood or metal board [60,64]. After dipping and removing each board, an instantaneous surface profile is made by comparing the relative heights of the steel lumps solidified on the nails [60,64]. By dipping several nail boards, transient variations of the surface profile are quantified and the surface fluctuations are calculated [60,64]. Nail dipping can also be used to measure liquid mold–flux layer thickness by adding an aluminum wire beside each stainless steel nail [59,64]. Due to the aluminum melting temperature lower than the mold flux melting temperature, the height difference between the steel lump and the aluminum wire end, h_{slag} correlated with the liquid mold-flux layer thickness as shown in Figure 2c. The slag layer thickness can also be revealed by the location of the colored bands of scale that form on each nail [76].

2.2.3. Sheet Dipping Tests

Dipping a thin sheet of steel into the top surface is an alternative method to nail boards to determine the profile of the liquid mold flux/molten steel interface. Time averages and standard deviations of the transient interface profiles, revealed by serial dipping of the sheets, are useful to validate transient computational model predictions [77].

2.2.4. Oscillation Mark Measurements

Partial overflow and freezing of the molten steel over the meniscus produces an oscillation mark during each mold oscillation cycle. The oscillation mark appears as a small transverse depression in the surface of the solidified steel slab. Thus, each mark represents the instantaneous profile of the interface between the liquid mold flux and molten steel around the mold perimeter at the time it formed. Tracing and graphing a series of oscillation marks accurately reveals the transient liquid mold flux/molten steel interface profiles, such as shown in Figure 4a. These measured profiles can be used to validate computational models of time-dependent (Figure 4b) meniscus level profiles and their fluctuations [61].

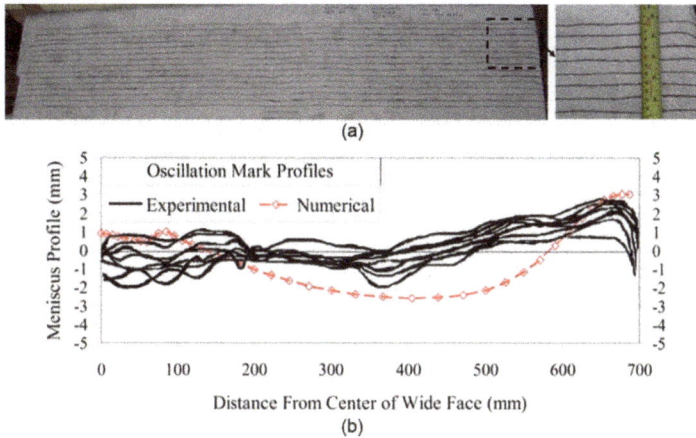

Figure 4. (a) Measurements and (b) model validation of oscillation mark profiles on steel slab surface [61].

2.3. Plant Measurements of Particle Capture

Particles, including argon bubbles, alumina, and slag inclusions, can be entrapped by the solidifying steel shell during continuous casting, and are greatly affected by the flow pattern, and EM effects. This results in surface and/or internal defects on the final steel products. Thus, it is important to quantify particle capture defects in as-cast steel slabs. Many different measurement methods are available, including ultrasonic testing, step milling, and other methods which are reviewed elsewhere [78].

2.3.1. Ultrasonic Testing (UT) Measurements

The locations of particles entrapped by the steel shell, can be measured in a steel slab in width, length, and depth directions, by using an ultrasonic detection system [79,80]. Recently, the dead zone, which is produced by reflected waves on the outer surface of the steel slab, has been reduced by using a V-shaped receiving probe and transmitting probe system with acoustic shielding between the probes [79]. This enables the detection of captured particles from 2 to 10 mm beneath the slab outer surface [79].

2.3.2. Step Milling Measurements

Step milling followed by microscopy, with automated surface scanning for particles, such as the ASPEX system [81], is a useful method to quantify both the location and size of particles captured into the solidifying steel shell [82]. Samples are cut from wide and narrow faces of an as-cast steel slab. The outer surface of each sample is milled away layer by layer, using an optical microscope to find and examine each particle observed on each exposed surface and recording the size and location of each particle. This procedure has been used to measure inclusions in slab samples in previous work [82,83], such as shown in Figure 5. Compared to the UT method, the step milling method is more difficult and expensive; but, step milling can quantify further details of the captured particles such as their shape and composition, in addition to their size and location.

2.3.3. Other Methods to Measure Particle Capture

Many other methods to measure size distribution, morphology, and composition of non-metallic oxide inclusions in the molten steel have been developed as reviewed elsewhere [78]. Direct methods include section method, volume method, extraction, and liquid evaluation methods. Section methods include Metallographic Microscope Observation (MMO), Image Analysis (IA), Sulfur Print, Scanning Electron Microscopy (SEM), Optical Emission Spectrometry with Pulse Discrimination Analysis (OES-PDA), Laser Microprobe Mass Spectrometry (LAMMS), X-ray Photoelectron Spectroscopy (XPS), and Auger Electron Spectroscopy (AES). Volume methods include Conventional Ultrasonic Scanning (CUS), Mannesmann Inclusion Detection by Analysis Surfboards (MIDAS), Scanning Acoustic Microscope (SAM), X-ray Detection, Slime (Electrolysis), Electron Beam (EB) melting, Cold Crucible (CC) melting, and Fractional Thermal Decomposition (FTD). Extraction methods include Coulter Counter Analysis, Photo Scattering Method, and Laser Diffraction Particle Size Analyzer (LDPSA). Inclusion evaluation methods in the liquid include Ultrasonic Techniques for Liquid System, Liquid Metal Cleanliness Analyzer (LIMCA), and Confocal Scanning Laser Microscope. In addition to these direct methods, indirect ways to estimate inclusion content include measurements of total oxygen, nitrogen pickup, dissolved aluminum loss, alumina pickup in the slag, and Submerged Entry Nozzle (SEN) clogging, which are easier and less costly [78].

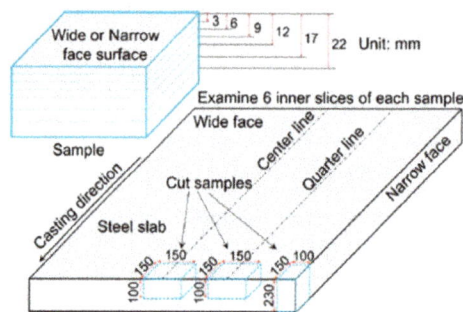

Figure 5. Step milling measurement of particles captured by the solidifying steel shell [82].

2.4. Lab Scale Modeling

Water modeling is a useful tool to simulate fluid flow phenomena of continuous steel casting due to kinematic viscosity of water similar with that of molten steel. However, the negligible electrical conductivity of water (See Table 1) prevents the study of magnetic forces that change the flow. Instead, physical devices to mimic the flow effects of electromagnetic forces have been applied to simulate stirring via two pairs of tuyeres that produce jets, and braking via a layer of beads suspended between screens that resist flow [84,85].

Table 1. Comparison of fluid properties.

Fluid	Melting Temperature (°C)	Density (kg/m^3)	Dynamic Viscosity (kg/m·s)	Electrical Conductivity (/Ω·m)	Thermal Conductivity (W/m·K)
Steel	1480–1510	7000	0.0067	714,000	35.0
Water	0	998.2	0.001	0.05	0.60
Mercury	−38.8	13,534	0.001555	1,020,000	8.3
$Sn_{60}Bi_{40}$	138–170	8250	0.0016	1,050,000	35
$Ga_{68}In_{20}Sn_{12}$	10.5	6360	0.0021	3,290,000	39

A more accurate method is to use low melting temperature alloys such as mercury, $Sn_{60}Bi_{40}$, or eutectic $Ga_{68}In_{20}Sn_{12}$ alloy, in physical models, which have the benefits of being conductive to simulate induced flow from electromagnetic fields, having high surface tension and nonwettability of molten steel to simulate multiphase flow, and being liquid at or near room temperature for easy measurements (Table 1) [86]. Lab scale modeling with these liquid metals can use Ultrasound Doppler Velocimetry (UDV) and Mutual Inductance Tomography (MIT) probes to map internal velocity fields, as applied to single-ruler EMBr fields in a slab casting mold [86–88], such as shown in Figure 6 [87].

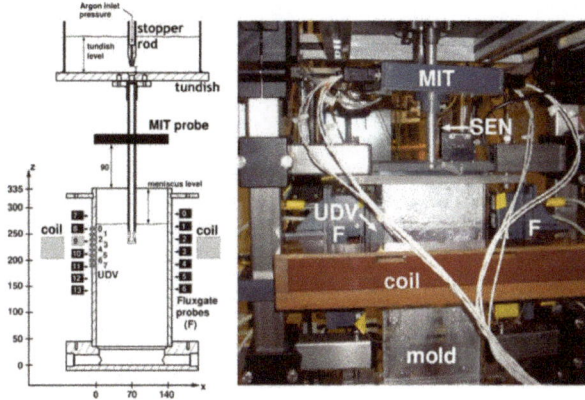

Figure 6. Low melting alloy ($Ga_{68}In_{20}Sn_{12}$) in 0.4-scale model of slab casting [87].

To maintain similarity between the low melting temperature alloy modeling and real caster conditions with electromagnetic forces, magnetohydrodynamics (MHD) dimensionless numbers, such as Hartmann number (Ha) and/or Stuart number (N), should be considered in addition to the standard fluid dynamics dimensionless numbers Froude number (Fr), Reynolds number (Re), and Weber number (We) for scaled physical models.

One way to obtain reasonable predictions of the real caster (R) from measurements in the lab scale model (M) when electromagnetic effects are present is to match both Fr and N. If the lab scale model stays fully turbulent, and multiphase flow is not dominant, then Re and We similarity is not critical [89]. First, Fr, which represents the ratio of inertial to gravitational forces, should be maintained the same in the scaled model and real caster, as follows

$$\frac{V_M}{\sqrt{gL_M}} = \frac{V_R}{\sqrt{gL_R}} \tag{2}$$

where V is a characteristic flow velocity, g is gravitational acceleration, and L is a characteristic length scale; "characteristic" refers to any pair of corresponding values in the model and real caster. This match can be achieved by choosing the casting speed in the model, $V_{Casting, M}$, by rearranging Equation (2):

$$V_{Casting, M} = V_{Casting, R} \sqrt{\frac{L_M}{L_R}} \tag{3}$$

Velocity anywhere in the real caster is then estimated by multiplying the measured velocity at the corresponding position in the scaled model by $\sqrt{\frac{L_M}{L_R}}$, according to Equation (2).

In addition, N, the ratio of electromagnetic to inertial forces, should be maintained constant, as follows

$$\left(\frac{\sigma_M}{\rho_M}\right)\frac{(B_M)^2 L_M}{V_M} = \left(\frac{\sigma_R}{\rho_R}\right)\frac{(B_R)^2 L_R}{V_R} \tag{4}$$

where σ is electric conductivity and ρ is density, as given in the property term. One way to achieve this match is to choose the magnetic field intensity in the model, B_M, as follows, which is found by rearranging Equation (4):

$$B_M = B_R \left(\frac{\rho_M}{\rho_R}\frac{\sigma_R}{\sigma_M}\right)^{0.5}\left(\frac{L_R}{L_M}\right)^{0.5}\left(\frac{V_M}{V_R}\right)^{0.5} \tag{5}$$

To match both Fr and N in the scaled model and the real caster, B_M can be chosen as follows

$$B_M = B_R \left(\frac{\rho_M}{\rho_R} \frac{\sigma_R}{\sigma_M} \right)^{0.5} \left(\frac{L_R}{L_M} \right)^{0.25} \tag{6}$$

which is found by replacing V_M/V_R in Equation (5) with $\sqrt{L_M/L_R}$, according to Fr in Equation (2). This approach was used successfully with a mercury model to investigate fluid flow velocities in a slab mold with static magnetic fields (local EMBr and single-ruler EMBr) during continuous casting [90].

Alternatively, it is possible to obtain reasonable predictions of the real caster from lab scale measurements on molten metal with electromagnetics, without matching Fr [27]. In this method, the casting speed in the model, $V_{Casting,\ M}$ is chosen to match N, by rearranging Equation (4) as follows

$$V_{Casting,\ M} = V_{Casting,\ R} \left(\frac{\rho_R}{\rho_M} \frac{\sigma_M}{\sigma_R} \right) \left(\frac{L_M}{L_R} \right) \left(\frac{B_M}{B_R} \right)^2 \tag{7}$$

With this method, velocity anywhere in the real caster is then predicted by multiplying the measured velocity at the corresponding position in the scaled model by $\left(\frac{\rho_R}{\rho_M} \frac{\sigma_M}{\sigma_R} \right) \left(\frac{L_M}{L_R} \right) \left(\frac{B_M}{B_R} \right)^2$, according to Equation (4). In addition, surface level in the real caster, l_R, is estimated from the measured surface level in the model, l_M, by applying the Froude number based scaling method, as follows [27]

$$l_R = l_M \left(\frac{L_R}{L_M} \right)^{0.5} \left(\frac{V_R}{V_M} \right) \tag{8}$$

Using this method, the flow pattern, velocities, and surface level predicted in a real caster matched well with both measurements and Large Eddy Simulation (LES) modeling of a scaled $Ga_{68}In_{20}Sn_{12}$ model [27]. In particular, the surface level in the real caster estimated with the Fr based scaling method (Equation (8)) showed much better agreement than a simple method of multiplying the measured surface level by the scale factor, L_R/L_M [27].

2.5. Computational Modeling: Magnetohydrodynamics (MHD) Models

Many studies of fluid flow-related phenomena in continuous steel slab casting have been conducted using three-dimensional Computational Fluid Dynamics (CFD) models, as reviewed previously [22,23]. These models solve the continuity equation for mass conservation and the Navier–Stokes equations for momentum conservation of incompressible Newtonian fluids, such as molten steel. To simulate other important phenomena, these flow equations are solved together with further coupled equations including turbulence equations, such as Reynolds-averaged Navier-Stokes (RANS)-based standard k-ε [91], Shear-Stress Transport (SST) k-ω [92], or LES-based subgrid-scale viscosity models such as the Wall-Adapting Local Eddy (WALE) model [93]; multiphase fluid flow equations [22] using the mixture, Eulerian, or Volume Of Fluid (VOF) methods; Lagrangian particle transport model such as the Discrete Phase Model (DPM); and advanced particle-capture criterion equations [82,94]. These equations are solved using classical finite-volume or finite difference methods, often with a commercial CFD package program, such as the ANSYS Fluent code [95], or an in-house code, such as the multi-GPU-based CUFLOW [96].

To consider effects of electromagnetic forces on fluid flow-related phenomena and Magnetohydrodynamics (MHD) models [97,98] are incorporated into the CFD models by adding a Lorentz force term to the momentum equation. Conducting fluid moving in applied magnetic fields induces current and the induced current and the magnetic fields generate Lorentz forces on the fluid flow. There are two methods including magnetic induction method and electrical potential method to calculate the Lorentz force term, as follows.

The total magnetic field, B consists of applied magnetic field, B_0, and induced magnetic field, b.

$$B = B_0 + b \qquad (9)$$

Next, to find B and the induced current density J, the magnetic induction method first finds B by solving the magnetic induction equation, derived by combining Ohm's law, Faraday's equation, and Ampère's law:

$$\frac{\partial B}{\partial t} = \nabla \times (u \times B) + \frac{1}{\mu\sigma}\nabla^2 B \qquad (10)$$

where t is time, u is fluid flow velocity, μ is magnetic permeability, and σ is electrical conductivity of the fluid. Knowing B, the induced current density Ampère's equation is then solved for J:

$$J = \frac{1}{\mu}\nabla \times B \qquad (11)$$

The other method to find J and B, the electric potential method, first assumes that B can be approximated by $B0$. This assumption is good for systems with high magnetic diffusion compared to magnetic induction, such as continuous casting, where the induced magnetic field b is small relative to the applied field, $B0$. With this electromagnetic condition, a Poisson equation for electric potential, φ can be derived from Ohm's law and conservation of charge, $\nabla \cdot J = 0$, as given by Equation (12).

$$\nabla^2 \varphi = \nabla \cdot (u \times B_0) \qquad (12)$$

Then, from the calculated φ, the induced current density is found from Equation (13).

$$J = \sigma(-\nabla\varphi + u \times B_0) \qquad (13)$$

Finally, in both methods, the Lorentz force, F_L is calculated from the calculated current density and magnetic field as follows

$$F_L = J \times B \qquad (14)$$

Note that the magnetic field is always perpendicular to the flow direction due to the right-hand rule. The force direction is perpendicular to both the induced current density and the magnetic field, according to the cross product in Equation (14). Together, this means that the force direction is 180°, or directly opposite to the flow direction, so is intrinsically a braking force. The actual effect on the flow is much more complicated, however. For a static magnetic field, mass conservation often makes the flow deflect around a region of local magnetic field braking, and as a result can produce faster flow in another region. For a time-varying field, the apparent movement of the field created by a consistent phase shift in a series of magnetics can create an apparent motor force that can almost match the flow direction in some cases.

Regardless of the computational method, model validation is essential. A classic test problem for MHD CFD involves turbulent flow (Re ~40,000) in a channel with a constant static magnetic field applied uniformly over a 304-mm-long rectangular region near its center. Lorentz forces in the channel domain are shown in Figure 7. Vectors of Lorentz force predicted with both the electric potential method [99] and the magnetic induction method [61] match.

In addition, model predictions of velocity profile across the channel cross section should match with the experimental measurements, such as shown in Figure 8 [61]. An M-shaped profile is observed with higher velocities near the channel walls and lower velocities towards the center region of the channel. This is because the Lorentz forces slow down the flow in the center region, while flow accelerates near the walls to conserve mass. Validation with a test problem such as this one demonstrates that a given MHD CFD model can enable accurate investigation of the effect of electromagnetics on flow in continuous steel casting.

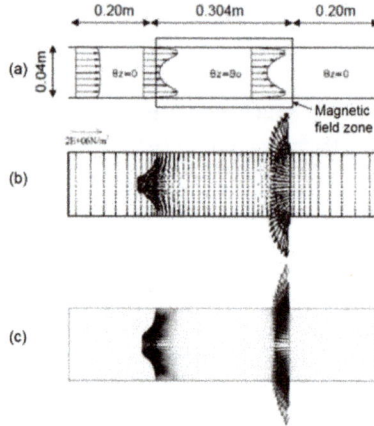

Figure 7. (a) Test channel domain of the magnetohydrodynamics (MHD) models and Lorentz force vectors calculated (b) with potential method [99] and (c) with magnetic induction method [61].

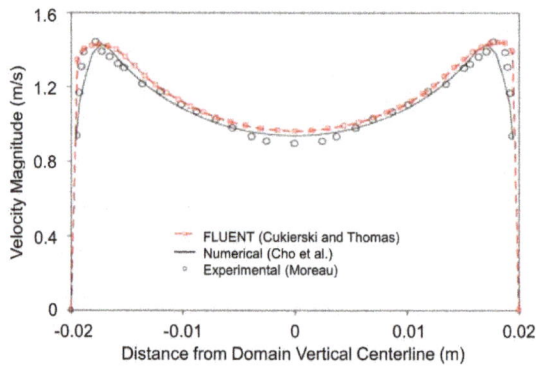

Figure 8. MHD model validation: comparison of velocity profiles in the test channel domain [61].

3. Electromagnetic Effects on Fluid Flow Pattern

The quality of continuously-cast steel slabs depends greatly on fluid flow phenomena including the flow pattern, which governs jet impingement, turbulence, surface velocity, and level variations in the mold region of the process. Excessive surface velocities, and the accompanying turbulence, large surface profile variations, and surface level fluctuations, can cause the entrainment of slag into the molten steel pool, which may lead to the entrapment of inclusions by the solidifying steel shell, and defects in the final product [74]. On the other hand, abnormally slow surface flow can result in low and nonuniform liquid temperature near the meniscus and problems associated with low superheat, which include insufficient slag melting and infiltration, meniscus freezing, and hook formation, leading to surface defects in the product [100,101]. Thus, process parameters must be controlled within optimum windows [16,32] of process operation, as shown in Figure 9 for data measured at NKK Corporation [16], to avoid these quality problems [2,16].

The flow pattern significantly affects many important phenomena in the slab mold and upper strand regions. Flow in the mold can be classified as having single-, double-, or unstable flow patterns [2,16]. Unstable flow often involves chaotic transient transitions between single- and double-roll, and should be avoided because it is the worst for steel quality, owing to severe surface instability during the flow transitions [2,16]. Electromagnetic control of the flow pattern aims to maintain a stable double-roll flow pattern that keeps surface velocity within an optimum range for best steel quality [16,32].

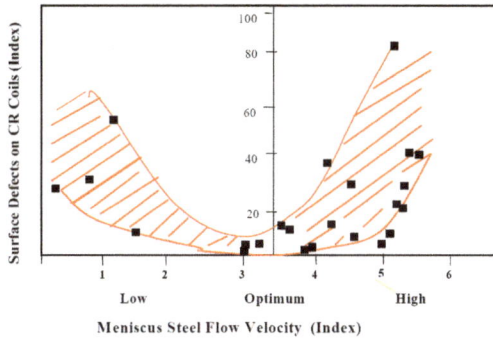

Figure 9. Slab/coil defects are minimum in an optimum window of steel meniscus velocities between weak and excessive double-roll flow conditions. NKK Corporation data; 700 to 1650 mm wide, 235-mm-thick slabs cast at 1.6 to 2.8 m/min [16].

3.1. Local EMBr

Local EMBr creates roughly circular-shaped static magnetic fields near the nozzle ports. The fields are designed to make the jet pass below the strongest core region of the magnetic field and deflected downward [24]. This results in deeper jet impingement on the narrow face, slower surface flow, and a flatter surface profile [61,102,103], as seen by comparing Figure 10a,b. These surface-flow braking effects become stronger with higher EMBr strength [102]. Note that excessive EMBr strength may produce too low surface velocity which could cause the meniscus freezing and hook formation associated with low surface temperature due to the stagnant surface flow.

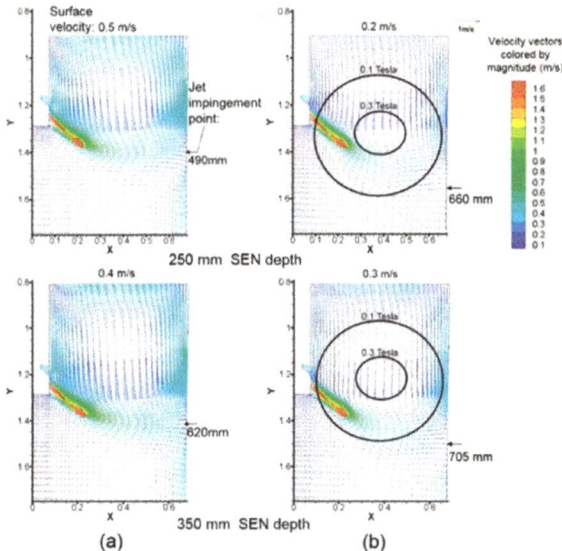

Figure 10. Mold flow pattern (**a**) without and (**b**) with local EMBr [61].

On the other hand, local EMBr causes the jet to deflect upward if the jet passes above the strongest region of the magnetic field [104]. This produces shallower jet impingement on the narrow face, higher surface velocity and larger variations in surface profile across the mold width [104]. Abnormally strong local EMBr fields may change the mold flow pattern from a double-roll to a single-roll pattern, making the jet directly go towards the surface without impinging first on the narrow faces [103]. This is usually detrimental to surface quality, and may also cause slag entrainment and other defects.

Thus, properly locating local EMBr field relative to the nozzle ports and optimizing the field strength is important to achieve surface flow conditions in a safe window of operation [105]. In addition to the EMBr magnetic field position, this also requires consideration of the SEN depth [61], nozzle port angle [90], and slab width [106] that strongly affect the jet behavior in the mold. As shown by comparing the two frames in Figure 10b, the EMBr effects on braking surface flow are weakened if the jet passes below the magnetic field region due to deeper SEN location, which leads to less deepening of the jet impingement point and less reduction in surface velocity. This trend is contrary to that without EMBr where surface velocity deceases with increasing SEN depth, as shown in Figure 10a. It is important to carefully monitor and control the SEN depth and field strength according to the casting conditions and the EMBr field location.

3.2. Single-Ruler EMBr

Single-ruler EMBr systems are designed to slow down the mold surface flow by applying a horizontal rectangular-shaped static field across the entire width of the slab mold region near the meniscus [25]. With increasing casting speed, surface flow tends to be stronger, so stronger electromagnetic braking is needed, such as in thin-slab casting at high casting speed [107,108]. This is useful to avoid excessive surface velocities and vortex formation [26], which can cause slag entrainment into the molten steel pool in the mold. Compared to local EMBr, single-ruler EMBr is more effective to brake the surface velocity in the mold because the magnetic field shape of single-ruler EMBr extends across the entire width, including the center and the narrow face regions [90]. However, excessive application of single-ruler EMBr near the meniscus can be detrimental by slowing surface velocities too much for the casting conditions. This can result in excessive cooling of the meniscus region and insufficient mixing of the surface slag layers, leading to meniscus freezing, where the accompanying subsurface hooks can capture particles including argon bubbles, alumina inclusions, and mold slag droplets into the solidifying steel shell, leading to surface defects.

On the other hand, placing the single ruler below the nozzle can increase surface velocity [27–29]. In this case, the field deflects the downward-flowing jet to become more horizontal, impinging higher on the narrow face, and causing more and faster flow up the narrow faces towards the meniscus. This can help by increasing surface velocity into the optimal range in some situations such as thick slab casting at low casting speed. Care should be taken to avoid locating the maximum of the magnetic field directly across the nozzle ports, as this produces flow instability resulting in severe jet wobbling and surface fluctuations, as discussed further in Section 4.1 [27–29]. The strong horizontal field also tends to lessen mixing between the upper and lower recirculation zones of the double-roll flow pattern. This is useful during casting of clad steels, where the field is positioned between two nozzles which deliver two different steel grades, as discussed later [30].

Vertical single-ruler EMBr, recently proposed, is designed to decrease surface velocity and surface profile variations by imposing two strong static ruler-shaped magnetic fields oriented vertically near the narrow faces [109–111]. The flow velocity up the narrow face is predicted to become slower and more uniform with vertical EMBr [110]. This is proposed to be useful for high-speed thin-slab mold casting, to avoid excessive surface flow problems. If the flow has a double-roll pattern with a downward jet, then the vertical ruler field is predicted to reduce surface velocity more than a horizontal single-ruler EMBr field [109,110], as shown in Figure 11.

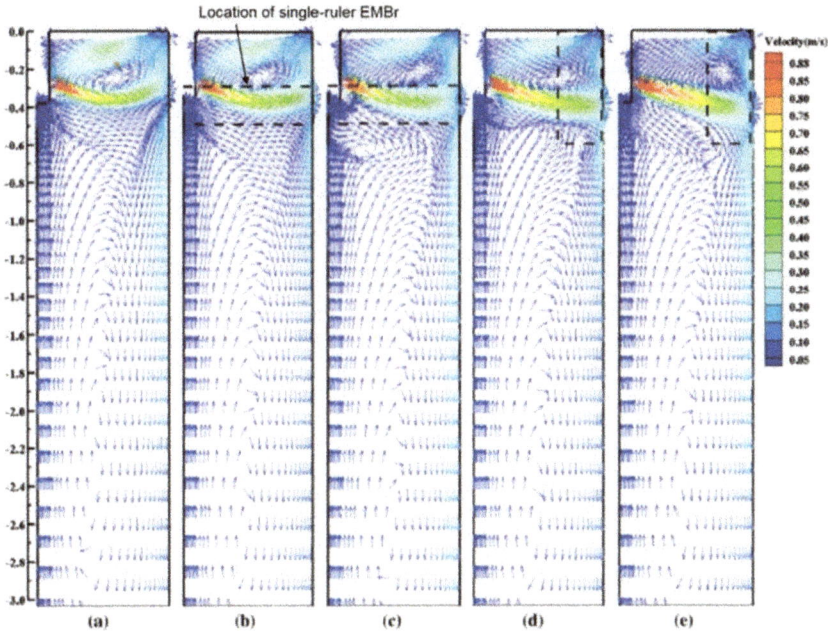

Figure 11. Mold flow pattern (**a**) with no magnetic field, (**b**) horizontal single-ruler EMBr 0.1 T, (**c**) 0.2 T, (**d**) vertical single-ruler EMBr 0.1 T, and (**e**) 0.2 T [109].

3.3. Double-Ruler EMBr

Double-ruler EMBr combines two horizontal rectangular-shaped static ruler magnetic fields across the mold width: one (upper ruler) above and the other (lower ruler) below the nozzle ports [24]. The upper ruler tends to decrease surface velocities and surface level variations across the mold width [17,29,112–114]. This ruler acts in several ways: stabilizing flow inside the nozzle, deflecting the jets exiting the ports downward, especially when the ruler is located just above the nozzle ports, which slows the flow up the narrow faces and finally by slowing the flow along the meniscus [29,112–114]. The lower ruler generally tends to lessen the jet penetration depth [17,79,114–119], which affects particle transport deep into the mold cavity, as discussed in Section 6.2.

Compared to local and single-ruler EMBr systems, the double-ruler EMBr has more flexibility, with independent control of the strength of two rulers offering better potential to optimize the flow pattern, by either increasing or decreasing the intensity of surface flows. However, nonoptimal choice of the ruler intensities has the potential to worsen quality problems. Excessive upper-ruler strength can make surface flow too slow, especially in low-speed operations, resulting in meniscus freezing and hook formation, which is detrimental to the slab surface quality. Alternatively, excessive lower ruler field strength could lead to excessive surface velocities, turbulence and associated defects, especially in high-speed operations.

The ruler-shaped field often tends to decrease in strength towards the narrow faces. If the strength of the lower ruler near the narrow faces is too small, then the effect of the upper ruler deflecting the jet downwards can increase flow down the narrow faces below the mold, resulting in deeper jet penetration and associated particle defects [113]. In addition, this problem may decrease flow towards the surface, resulting in slower surface flows and associated stagnation problems, even with a strong lower-ruler field and no upper field current. This is because the double-ruler still generates a magnetic field peak above the nozzle that can deflect the jet downward, which differs from the single-ruler EMBr field below the nozzle ports [17].

Finally, confining the jet between the upper and lower rulers tends to make the jet thinner, slightly faster, and with less vertical variations [113]. Thus, the locations and strengths of the double-ruler EMBr system should be adjusted to achieve optimal mold flow, keeping the surface flow velocity within the optimal range, and minimizing jet penetration deep into the mold cavity, in order to improve both surface and internal quality of steel slabs.

3.4. MM-EMB

By adjusting the strengths of the five local static fields, multi-mode EMB can be designed to change, and hopefully improve flow velocity and stability in several different ways [31,120]. With two strong local magnetic fields located near the meniscus near the narrow faces and an optional strong field just below the nozzle, MM-EMB in braking mode 1 or damping mode tends to slow down surface flow [31,120], and lessen surface level fluctuations, in order to avoid defects associated with excessive surface velocity, such slag entrainment.

Alternatively, applying two strong local magnetic fields below the nozzle near the narrow faces, MM-EMB in braking mode 2 can achieve similar flow behavior. This mode is similar to single-ruler EMBr positioned below the nozzle, except there is no field in the center of the mold below the nozzle. The observed decrease in surface flow is likely due to the field cores being positioned above the jets near impingement on the narrow faces. Care should be taken in positioning the fields, however, because if the two fields are located below the jets near narrow-face impingement, surface flow would be expected to increase. Thus, this mode likely experiences similar behavior and operation guidelines as the local-EMBr field discussed in Section 3.1. Other MM-EMB modes are possible, which deserve further investigation.

3.5. Moving Fields: EMLS, EMLA, EMRS, M-EMS, EMC, and SEMS

Moving electromagnetic fields are generated by passing alternating currents through a series of magnets, each having a different phase shift in order to create a traveling Lorentz force to actively drive the flow tangentially across the surfaces of the solidifying steel shell in the mold. The strength and direction of the force depends on the magnet orientation, the applied current, and the effective frequency of the phase shift. Three types of horizontally-moving magnetic fields near the nozzle ports are EMLA [16,32,33,40,51,56,121], EMLS [16,32,33,40,51,56,122], and EMRS [16,32,33,40,51] or M-EMS [123,124]. Alternatively, EMC applies vertically-moving magnetic fields near the meniscus [36–38]. Other types of moving field systems—SEMS and S-EMS—are applied to the strand below the mold [33,40–47].

When surface flow is too slow, EMLA is designed to accelerate the jets in order to increase the surface velocity, turbulence, and surface temperature. This is useful for situations, such as low-speed casting and wide molds, to prevent stagnant surface flow, meniscus freezing, and the corresponding surface defects [121]. Also, EMLA can be applied when argon gas flow rate is high, to transform a detrimental single-roll pattern to a better double-roll pattern [32].

Alternatively, when surface flow is too fast, EMLS can slow down the jets in order to decrease surface velocity, surface profile variations, level fluctuations, and associated quality problems such as slag entrainment [32]. However, excessive magnetic field strength may lead to a single-roll pattern in the mold when argon gas fraction is high [51,122].

Thirdly, EMRS or M-EMS can rotate the flow around the perimeter of the mold, which is expected to wash particles away from the dendritic solidification front, in order to lessen particle entrapment, especially near the meniscus region [123,124]. In addition, the mixing effect of the rotating surface flow is designed to make the temperature distribution in the liquid near the mold top surface and meniscus region more uniform, especially in the central region of the mold near the nozzle, lessening the associated problems of meniscus freezing and lowering hook depth [123,124]. This is an alternative way to previously discussed methods to increase surface flows, with the potential benefit of less detrimental level fluctuations, if great care is taken.

Fourthly, the vertically moving magnetic fields generated by EMC induce vertical rotating flows, consisting of very small upper and lower counter-rotating recirculation regions near the meniscus in the mold, where the magnets are located [36,38]. This lessens the tendency of molten steel to overflow the meniscus during oscillation mark and hook formation. Thus, EMC can decrease surface defects including uneven slab surface and particle capture during initial solidification [36]. More details are given in Section 5.3.

Finally, SEMS in slab casting creates large vertical recirculation regions below the mold, designed to control superheat, solidification, grain structure, and solute distribution [33,40–47]. The one-way horizontally-moving fields in SEMS create Lorentz forces which directly drive the molten steel across the strand towards one of the narrow faces. After impinging on that narrow face, the flow splits vertically upward and downward. With one magnetic field or an "in-roll" pair of moving field systems on each side of the strand, this results in two large, counter-rotating vertical recirculation regions across the entire strand width, which extend far above and below the SEMS field region [41,42,45–47]. With two sets of two pairs each of in-roll-type SEMS stirrers employed, SEMS can generate three flow recirculation zones across the strand, which induce stronger effects on mixing the flow, compared to single-pair in-roll SEMS [42]. The top surface flow in the mold is not affected much by SEMS, so the importance of this mixing flow is related mainly to effects on superheat distribution, temperature gradients at the solidification front, grain structure, porosity, and segregation in the strand, which are discussed later. All of these moving-field methods deserve further investigation to understand and optimize the behavior.

3.6. Combined Traveling and Static Fields

Static and traveling magnetic fields can be combined together to offer even more flexibility to tailor electromagnetic control of the flow pattern in the mold. The FC3 system upper EMS field is designed to act like EMRS to produce the rotating flow around the perimeter of the upper mold [55,84,85], as shown in Figure 12. This aims to reduce particle capture during initial solidification and to make molten steel temperature and superheat removal more uniform around the mold perimeter. The static field in the lower mold region of FC3 is designed to slow down the jet flow to control surface flow intensity, and to shorten the jet penetration deep below the mold [55,84,85], which is designed to lessen deep particle penetration and capture. The combined effects of these two fields have great potential to reduce both surface and internal defects if they are optimized.

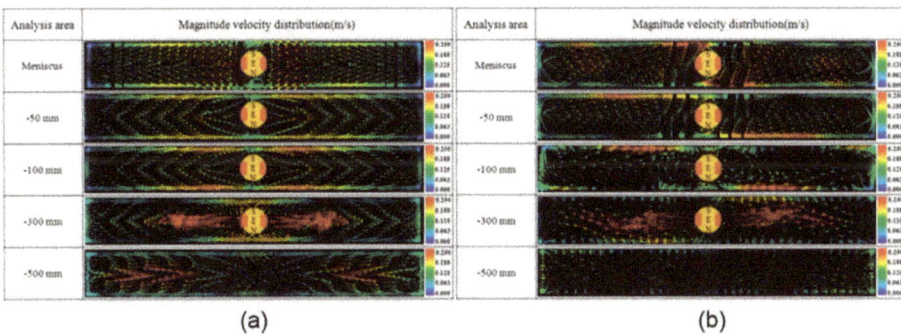

Figure 12. Flow patterns at various horizontal sections of the mold with (a) no magnetic field and (b) combined moving and static magnetic fields [55].

Another type of combined field system employs EMC near the meniscus on the narrow faces and a single-ruler static EMBr field below the nozzle ports [49]. The EMBr field is designed to brake the jet and control surface flow intensity. The moving field is designed to decrease meniscus level fluctuations to lessen slag entrapment and has been shown to greatly decrease oscillation mark depth, making a smooth slab surface with less hook formation [36].

4. Electromagnetic Flow Control and Surface Instability Effects

Surface instabilities associated with high surface velocity, velocity fluctuations, and severe level fluctuations are detrimental to surface quality of the final steel products. For many quality concerns, flow instabilities and transient events are often more important than the time-averaged flow pattern itself [16,32]. These instabilities can cause intermittent slag entrainment into the molten steel pool and slag entrapment by the solidifying steel shell, resulting in slag capture defects [74]. In addition, oscillation mark profiles become deeper and more uneven due to severe surface level fluctuations, especially near the meniscus. Electromagnetic forces affect both the flow pattern and the flow stability. Optimal application of electromagnetics to better control the surface flow and level can help to prevent these instabilities and the related quality problems.

4.1. Local and Single-Ruler EMBr

Local EMBr is designed to decrease velocity fluctuations of the jet passing below the strong magnetic field. However, the jet instability becomes severe when the local EMBr core is located below the jet, especially with excessive field strength [106]. This is because the jet deflects upwards to avoid the strong field region, and may disrupt the top surface, especially in wide slabs [106]. This further emphasizes the importance of optimizing the vertical orientation of the magnets relative to the SEN depth, according to the port angle and slab width, as discussed in Section 3.1.

The effects of single-ruler EMBr also depend strongly on the location of the ruler [27,28]. Locating a single-ruler EMBr above the nozzle decreases surface level fluctuations, resulting in a more stable surface. On the other hand, lowering the magnetic field below the nozzle ports produces shallower downward jet angles, higher surface velocities, lower turbulent kinetic energy at the surface, and better surface flow stability, especially at higher level fluctuation frequencies. In addition, vortex formation at the mold top surface, caused by biased surface flow due to nozzle misalignment between left and right sides in the mold, can be lessened with single-ruler EMBr together with optimized argon injection [125]. Applying the ruler across nozzle ports worsens the flow instability, however, producing strong unbalanced, asymmetric transient behavior (jet wobbling) and complex flow [28,29,126], as shown in Figure 13. Thus, the electromagnetic field should not be placed with its maximum directly across the nozzle ports, where it may aggravate unstable flow. When modeling these phenomena, it is important to note that the flow instability with realistic conducting walls is much less than with insulated walls [27,28,127,128], as shown in Figure 13. With single-ruler EMBr below the nozzle and electrically-conducting walls, the low-frequency oscillations of the jet flow are suppressed and results in a stable double-roll flow pattern with surface velocity within the window of safe operation, for the conditions simulated [28].

Vertical single-ruler EMBr is predicted to reduce both surface velocity and level fluctuations, especially near the meniscus [111]. The strong magnetic field oriented vertically along near the narrow faces tends to brake the upward flow produced after jet impingement on the narrow faces. Deepening SEN depth and lowering nozzle port angles (more downward) should further lessen surface fluctuations [111].

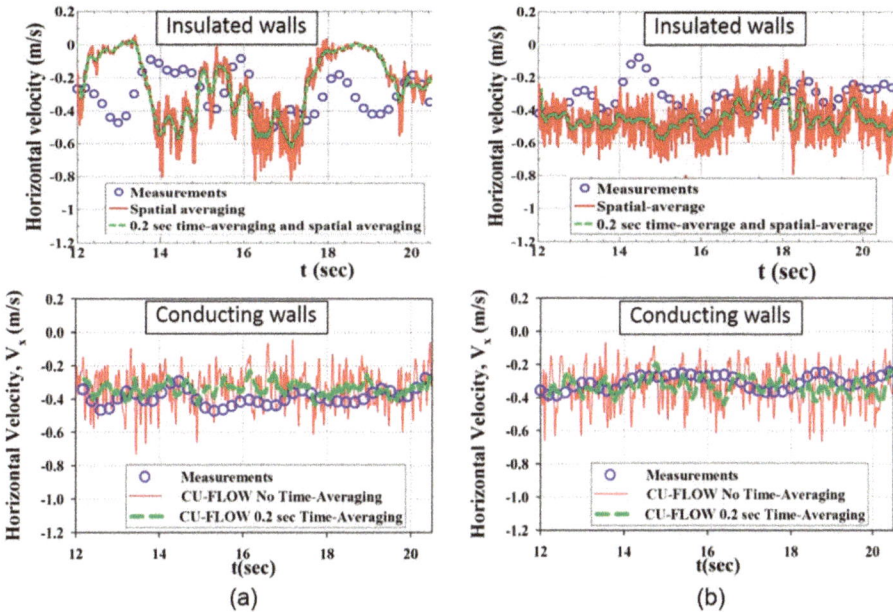

Figure 13. Transient history of jet velocity component (minus sign (−) indicates toward NF) in the mold with single-ruler EMBr locating at (**a**) 92 mm (near the nozzle ports) and (**b**) 121 mm (below the nozzle) below meniscus [28].

4.2. Double-Ruler EMBr

By surrounding the jets leaving the ports with static fields both above and below, double-ruler EMBr acts to stabilize the flow, deflecting chaotic turbulent variations in the jet direction back towards its designed path towards the narrow faces. Thus, asymmetrical flow and its associated variable surface defects tend to be reduced [79]. For example, surface level fluctuations caused by unbalanced flows in a slab mold are reduced with double-ruler EMBr [79]. Swirl flow inside the nozzle causes jet wobbling, especially with a slide-gate control system. This results in stronger variations in surface velocity and low-frequency level fluctuations [113], as shown in Figure 14a. The clockwise swirls, which have higher momentum than the counterclockwise swirls, due to shorter flowing path from the slide gate to the nozzle bottom on the right side in this orientation, produce higher surface velocity and surface level height. Thus, surface velocity and level fluctuation variations occur whenever the swirl flow direction changes in the nozzle bottom. Double-ruler EMBr reduces the surface velocity fluctuations due to the swirl direction changes as shown in Figure 14b, by dampening the momentum differences between swirl rotation directions. This is because the upper ruler field across the nozzle makes the velocity profile more uniform inside the nozzle [17,113], which then reduces jet wobbling and stabilizes the surface velocity and level across the mold width, as shown in Figures 15 and 16 [113]. This effect is also likely with single-ruler EMBr above the nozzle ports.

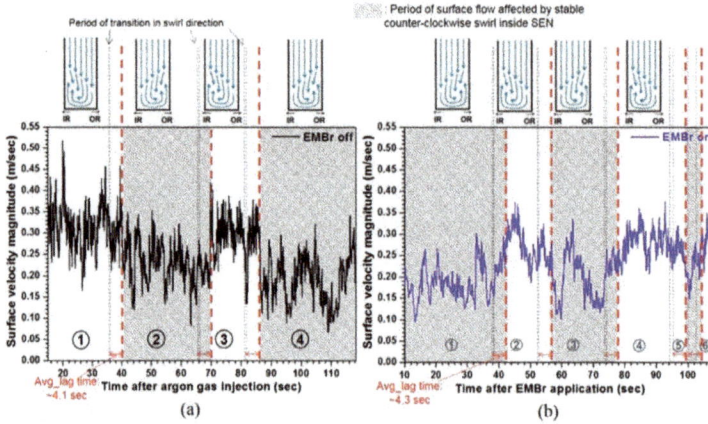

Figure 14. Surface velocity magnitude histories in the mold (**a**) without and (**b**) with double-ruler EMBr [113].

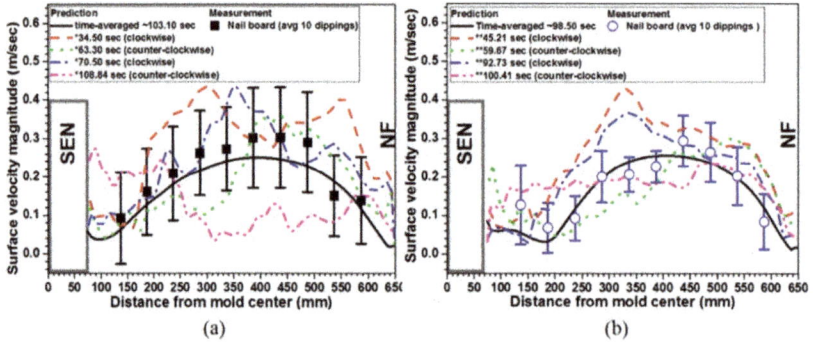

Figure 15. Surface velocity variations comparing LES modeling and measurements (**a**) without and (**b**) with double-ruler EMBr [113].

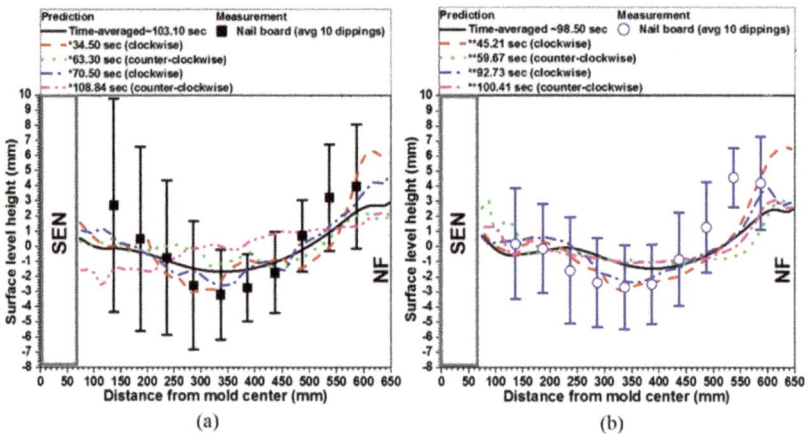

Figure 16. Surface level variations comparing LES modeling and measurements (**a**) without and (**b**) with double-ruler EMBr [113].

4.3. Moving Fields: EMLS, EMLA, and EMRS

Moving field effects on flow stability in the mold region have received less previous study [39,51,56,75], compared to that of static fields. Feedback control systems have been implemented in a few commercial operations with EMLS and EMLA, adjusting the magnetic field strength according to the current "F-value" [105], which is an estimate of surface flow strength based on SEN geometry, mold width, and casting speed [51,75]. This system is reported to maintain stable surface level in the mold [51,75]. For this to be effective, the EMLS and EMLA fields should be located just below the nozzle ports [51,75] and the flow pattern should be double-roll [16]. It is especially important to avoid unstable flow due to continuous transition between single-roll and double-roll patterns, as shown in Figure 17, as this causes severe surface level fluctuations [16]. In addition, this system needs to be improved to handle argon gas [32].

As an alternative control method, the magnetic field strength could be adjusted more actively according to local flow conditions. Two feedback control systems, based on two sensor measurements of instantaneous surface level, are expected to be better. Flow rate from the stopper rod or slide gate can be controlled to maintain a constant average liquid level, based on the first sensor, located at the quarter point. This location, found midway between the SEN and NF, is the most stable location in the mold relative to the highest amplitude wave motion [60,64,113]. The other sensor can be located near the narrow-face meniscus. Then, the magnetic field strength can be controlled to decrease the estimated surface level profile variations, as indicated by the difference between the two sensor measurements [24].

Finally, EMRS appears not increase surface level fluctuations, even the magnetic field is located near the mold surface [51].

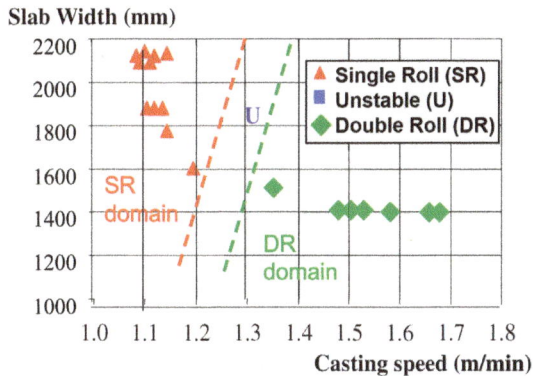

Figure 17. Unstable flow regime between single and double-roll flow patterns (4.4 t/min throughput) [16].

5. Electromagnetic Effects on Superheat Transport and Initial Solidification

Superheat is delivered with the molten steel flow towards top surface (mold flux layer), which affect initial solidification at the meniscus, and also deep into the mold cavity, which affects shell growth and potential thinning. Superheat transport depends on the mold flow pattern, which determines how the jet along with its superheat takes to reach different regions in the mold. The coldest regions tend to be found at the meniscus near the SEN and near the narrow faces. Stronger upward flow brings more superheat faster to these regions, resulting in higher meniscus temperatures. Optimal surface flow strength also mixes the liquid mold flux, which helps to melt the powder and encourage good gap infiltration and uniform lubrication. On the other hand, weaker upward flow results in a colder, stagnant meniscus, perhaps causing initial solidification problems including meniscus freezing, deeper oscillation marks and hooks, which can further increase particle capture, and slag infiltration problems.

Thus, through its effect on the flow pattern, electromagnetic forces can also help to control superheat distribution, initial solidification, and shell growth.

5.1. Static Magnetic Fields

As discussed in Sections 3.1 and 3.2, the location of static EMBr fields relative to the jet greatly affects the mold flow pattern. Thus, superheat transport is also influenced strongly by the magnetic field location [103,104,108,109,126,129,130]. Magnetic fields located below the jet deflect flow upwards towards the surface, leading to a hotter meniscus region [103,104], increasing meniscus temperature with increasing field strength, as shown in Figure 18 [104]. Optimal local EMBr fields can produce shallower oscillation marks [131].

On the other hand, magnetic fields above the jet deflect flow downwards, deeper into the mold cavity. The weaker upward flow transports less superheat to the mold upper mold, resulting in meniscus freezing, flow stagnation at the surface, and associated quality problems. In addition, local EMBr field can lessen jet intensity in the lower mold, resulting in a more uniform shell with less shell thinning at the jet impingement point [131].

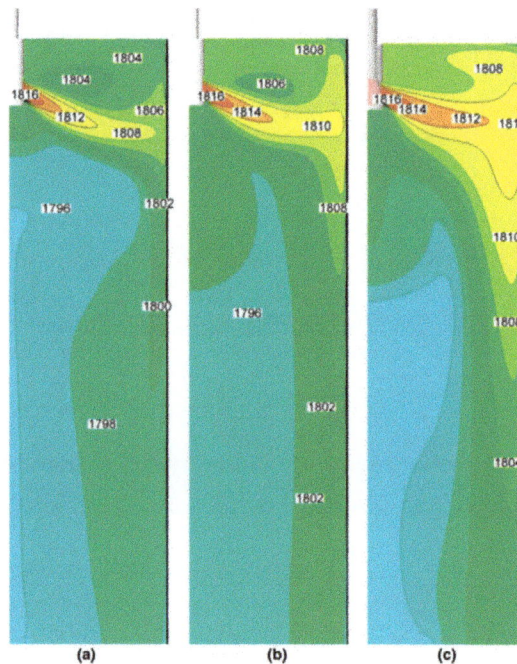

Figure 18. Temperature distribution in the mold with local EMBr field strength B_0 of (**a**) 0.0 T, (**b**) 0.2 T, and (**c**) 0.39 T [104].

Applying double-ruler EMBr has more flexibility to control the mold flow. Thus, superheat transport towards the stop surface is more adjustable by independent control of the field strengths of two rulers. More superheat is delivered towards the surface with stronger upward flows if the lower ruler is stronger and the upper ruler field is far above the nozzle ports [115,132]. This increases meniscus temperatures [115,132]. This can decrease the depths of oscillation marks and subsurface hooks, which are strongly related to low temperature distributions near the meniscus [100,101], as shown in Figure 19 [132]. Alternatively, abnormally high strength or low location of the upper ruler field can make the meniscus flow stagnant, causing associated surface defects.

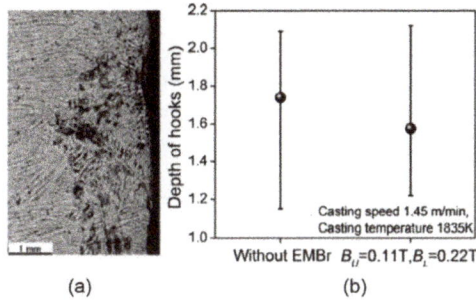

Figure 19. (a) Subsurface hooks and (b) the effect of double-ruler EMBr on hook depth in the steel slab [132].

5.2. Horizontally-moving Magnetic Fields: EMRS, M-EMS, and EMLA

The horizontally-rotating magnetic fields applied with EMRS or M-EMS produce rotating flow around the perimeter of the mold. This makes superheat transport and temperature distribution more uniform near the solidifying steel shell, especially where the meniscus tends to be cold, near the SEN and narrow faces [34]. Thus, depths of hooks and oscillation marks should be lessened. This also enables improvement of heat transfer uniformity through the slag layer in the steel shell/mold gap around the mold. Thus, the initial steel solidification around the mold is more uniform [34]. Uniformity of initial solidification can lessen defects such as longitudinal cracks, which initiate at local nonuniformities where stress can concentrate.

EMLA can increase temperature near the meniscus, by strengthening the upper flow along with delivering more superheat to the upper region of mold [32]. This is expected to have similar benefits of shallower oscillation marks and hooks.

5.3. Vertically Rotating Field near Meniscus: EMC

The moving EMC fields near the meniscus create two vertically counter-rotating recirculation zones near the meniscus [36,38]. The lower flow recirculation opposes molten steel trying to overflow the meniscus during each mold oscillation cycle. It also mixes the local superheat making temperature near the meniscus more uniform. These effects reduce oscillation mark depth and hook formation [36,38] as shown in Figure 20. This is expected to improve surface quality and to reduce particle capture defects.

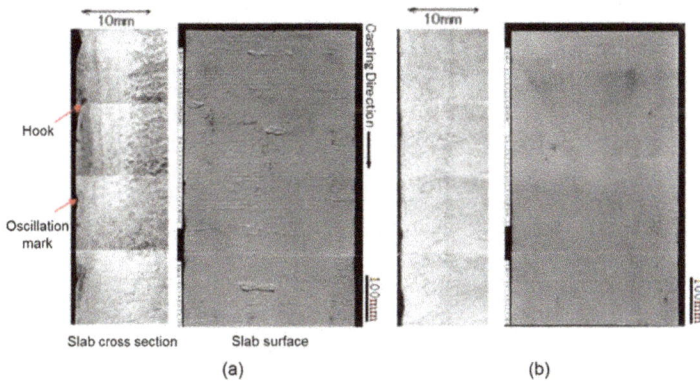

Figure 20. Oscillation marks and hooks in steel slabs (a) without and (b) with electromagnetic casting (EMC) [36].

6. Electromagnetic Effects on Particle Transport and Capture

Particles including bubbles, alumina, and slag inclusions may be transported with the flow and removed to the surface slag layer, or captured into the solidifying steel shell. Large particles are detrimental to product properties such as fatigue life, especially in high-strength steel. Bubbles are detrimental because they are usually coated with small inclusions. Captured large bubbles concentrate the inclusions into large clusters, so are detrimental to final steel product quality if they cannot be removed by further processes such as surface scarfing. Large particles are also detrimental by leading to delaminations, blisters, pencil pipes, and other defects during downstream operations [133,134].

Near the meniscus, excessive surface flow causes surface profile variations and level fluctuations, which can entrain slag as inclusions into the molten steel pool, and/or may directly entrap slag, inclusions, and bubbles into the steel shell at the meniscus [8,74]. In addition, single-roll surface flows towards narrow faces or cross-surface flows between wide faces may move particles trapped at the interface between the liquid mold slag and the molten steel pool, to the meniscus regions, where they may be captured. Alternatively, strong flow across (tangential to) the steel solidification front can wash particles away from the solidifying steel shell front, resulting in less particle capture [82,94].

Abnormally slow surface flow can lead to insufficient superheat transport to the meniscus, leading to deeper frozen hooks, as discussed in Section 5. This can cause more particle capture defects in the mold regions. Thus, the first objective of flow control is to maintain surface flows in a safe window of operation. Electromagnetic fields, combined with other casting conditions and nozzle geometry, can help to achieve this. The extra control possible with electromagnetics, however, may enable paying attention to particle transport and capture independent from surface flow.

Finally, electromagnetic fields can change the local velocity field adjacent to non-conducting particles, such as large argon bubbles, which can change the bubble shape and terminal rising velocity, relative to the flow field [135]. Thus, electromagnetic effects on the flow pattern and turbulence can significantly affect particle transport and the fraction and distributions of entrapped particles in several ways.

6.1. Local and Single-Ruler EMBr

As discussed in Sections 3–5, with local or single-ruler EMBr field located below the jets, the jets deflect upward towards the mold top surface, which could help argon bubbles and other particles carried with the flow to escape into the surface slag layer [104]. However, excessive upward flow may disrupt the top surface and allow more bubbles to accumulate beneath the slag layer, especially near the SEN, where they may be captured during initial solidification. In extreme cases, strong surface flows, such as caused by a poorly-designed local field, can push the liquid mold flux away, causing an open eye, where the molten steel is exposed to the powder and air [136]. This must be avoided because the associated reoxidation, contamination with carbon from the mold powder, bubble, and slag entrainment are all very detrimental to steel quality. Thus, the main objective of EMBr is to change the flow pattern to avoid excessive surface-directed flows to avoid particle entrapment and other defects associated with excessive surface flows.

On the other hand, if EMBr fields are located partly above the nozzle ports, so the jets are deflected downward, the flow may transport more particles deep into the strand below the mold. This will increase internal defects, as only a small fraction (lower than ~10%) of these particles are able to return to the top slag layer [137]. Thus, placing local or single-ruler EMBr fields at a proper location is important to control particle transport and capture in the mold, as expected from the effects on fluid flow pattern and surface instability, discussed in previous Sections 3.1, 3.2 and 4.1.

6.2. Double-Ruler EMBr

The upper ruler of double-ruler EMBr can help to avoid excessive surface flows and stabilizes the surface slag/molten steel interface in the mold, which tends to reduce the entrapment of inclusions due to level fluctuations [113]. These effects are similar to a single-ruler EMBr above the nozzle.

The lower ruler of double-ruler EMBr can help to lessen jet penetration and particle transport deep into the strand. As shown in Figure 21 [138], such an EMBr system causes fewer bubbles to be transported towards the narrow faces. The sizes and fractions of bubbles captured with these two flow patterns are shown in Figure 22 [138]. Small particles which contact the solidification front are easily entrapped between dendrites. On the other hand, large particles are only captured if the particle stays at the solidification front for long enough time to become surrounded by the growing shell front [82,94,138–140]. A simple criterion, which captures any particle touching the solidifying shell, overpredicts the capture of large particles, as shown in Figure 22a. Calculations with an advanced capture criterion [82,94,138] consider a potential balance of eight forces on the particle at the shell front, which include the drag from tangential velocities, which can wash away the particles from the solidification front, back into the main flow. This sophisticated capture model agrees better with plant measurements in Figure 22a, which show an average bubble diameter of ~0.1mm. Results also show that the average size of bubbles captured near the meniscus is slightly bigger than those captured deeper in the caster. Meniscus hooks, not included in the model, likely capture larger particles rising up beneath them. The change in the bubble transport with double-ruler EMBr tends to decrease the fraction of bubbles captured into the solidifying steel, as shown in Figure 22b, especially the 0.1mm size [82,138]. In addition, less deep hooks, due to more superheat towards the meniscus region with double-ruler EMBr, can reduce inclusion capture into the hooks [132] as shown in Figure 23.

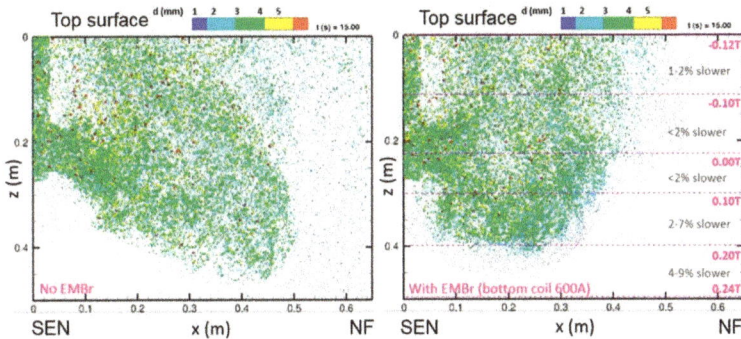

Figure 21. Bubble distributions in a slab caster centerplane without and with double-ruler EMBr [138].

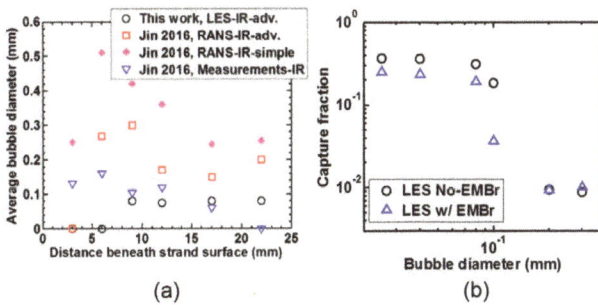

Figure 22. Captured bubble size distribution showing (**a**) the variation with distance beneath the strand surface and (**b**) the effect of double-ruler EMBr on capture fraction [138].

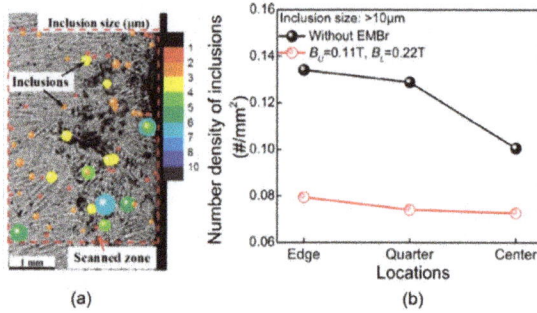

(a)   (b)

Figure 23. (**a**) Inclusions captured by hooks and (**b**) the effect of double-ruler EMBr on inclusion capture in the steel slab [132].

However, bubbles can penetrate deeper into the mold with double-ruler EMBr if the lower-ruler field strength is not enough to reduce the downward flow velocity near the narrow faces [113]. On the other hand, excessive strength of the upper ruler field can make the surface flow stagnant, leading to formation of more hooks which can entrap more slag inclusions into the shell. Thereby, optimizing the strength of both rulers is important to reduce the particle transport and capture [79,115,138], improving both surface and internal quality of the steel slabs.

6.3. Moving Magnetic Fields

Proper application of EMLS magnetic fields to stabilize surface velocity and surface level fluctuations may also help to reduce particle-related defects in the final steel product [32]. This is due to lessening the entrainment of slag inclusions and bubbles coated with inclusions and their capture into the solidifying steel shell, especially at the meniscus, as discussed in the previous section.

Horizontally-rotating magnetic fields induced by EMRS or M-EMS produce horizontally-rotating flow patterns around the perimeter of the mold. This flow pattern can wash away particles from the steel shell front in the mold [32,34,124,141–143], especially near the meniscus region, and from beneath subsurface hooks. This can significantly reduce defects related to the capture of bubbles near the surface (sometimes called pinholes or blowholes) [32,34], as shown in Figure 24, and other defects with large inclusions (slag, alumina [34], or calcium-alumina). In addition, more uniform surface temperature near the meniscus can lessen hook depth as discussed in Section 5. This in turn can reduce particle capture at the meniscus. Both the washing effect and the reduction of hooks are effective to reduce particle capture defects. Similar washing effects to lessen the capture of large inclusions might be involved with EMLA, and/or SEMS, but this needs further investigation, as previous work could not be found.

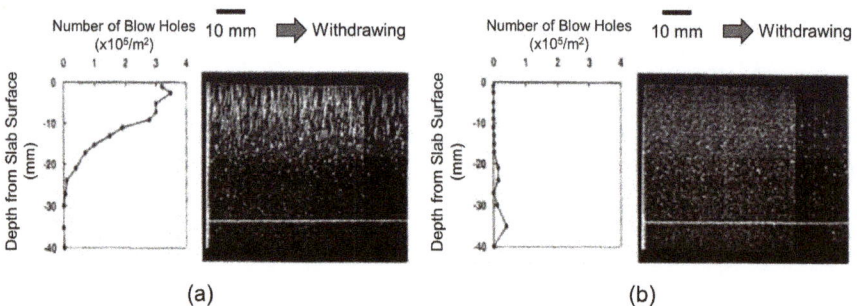

(a)   (b)

Figure 24. Subsurface-entrapped bubble defects in steel slabs (**a**) without and (**b**) with EMRS [32].

7. Electromagnetic Effects on Grain Structure and Internal Quality

Grain structure including the chill zone, columnar zone, Columnar-Equiaxed Transition (CET), and equiaxed zone is important for steel mechanical properties. Having more equiaxed grains in the slab correlates with less center segregation [144]. Many efforts have been made to increase the equiaxed zone size using electromagnetics. M-EMS can increase nuclei formation, resulting in smaller grains near the strand surface [34], perhaps due to the higher velocity flow across the solidification front melting off dendrite tips [145]. If more nuclei can survive and flow deep into the molten steel pool, this would be expected to lead to a larger fraction of equiaxed grains in the final product. Similarly, SEMS induces rotating flow below the mold, specifically in the vertical plane of the width and casting directions [33,40,42,45], which should decrease temperature gradients in the liquid, and has been shown to increase the size of the central equiaxed zone [33,40,42,45,146], as shown in Figure 25. In nonoriented electrical steel casting, the application of two sets of two pairs of in-roll SEMS magnetic fields at 400 A and 5 Hz was reported to show great improvement of grain structure, to over 60% equiaxed, as shown in Figure 26 [42].

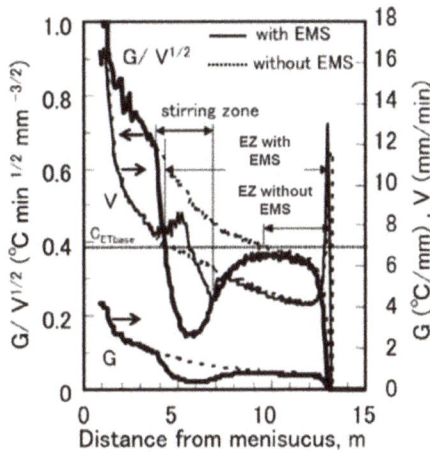

Figure 25. Effect of strand EMS on temperature gradient, dendrite tip velocity, and location of Equiaxed Zone (EZ) [146].

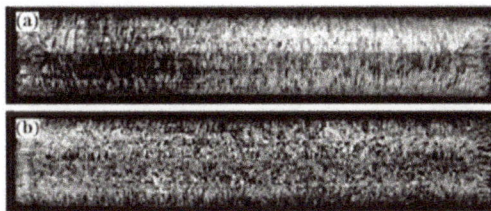

Figure 26. Microstructure of nonoriented electrical steel in slab horizontal cross-sections (a) without and (b) with SEMS rollers (54% equiaxed grains) [42].

In addition, strand EMS also affects segregation and centerline quality. Applying S-EMS together with soft reduction (Posco Heavy strAnd Reduction Process: PosHARP) has been reported to reduce center porosity and centerline segregation (abnormal high and varying solute concentration between the dendrites near final solidification) better than soft reduction alone [43], as shown in Figure 27.

Method	Etching image	Mn segregation degree (EPMA results)	Center porosity (mm³/g)
Soft reduction	Centerline →	Mn/Mn₀ = ~ 1.04	0.25 ~ 0.65
PosHARP	Centerline →	Mn/Mn₀ = ~ 0.97	0.09 ~ 0.20

Figure 27. Effects of SEMS with soft reduction on segregation and porosity defects [43].

However, white bands (low solute concentration) and/or dark bands (high solute concentration) in the steel microstructure are caused by tangential flow across the solidification front, due to nonuniform solute distribution in the strand region. These distinctive bands of macrosegregation can be seen in slabs cast with SEMS, always at the depth into the slab where the flow was induced [147], such as in Figure 26b. In addition, distinctive changes in the angle of the columnar dendrite growth directions can be seen in the cross-sections of slabs cast with SEMS, corresponding with the changing flow directions across the solidification front with distance down the strand. The effect of casting conditions and flow control methods with and without electromagnetic effects on segregation and grain structure needs more study to more accurately quantify the relation between fluid flow, macrosegregation, grain structure, and related steel quality.

8. Electromagnetic Control of Steel Composition Distribution: Clad Steel Casting

Another type of horizontal rectangular-shaped static magnetic field with a single-ruler, also called Level DC Magnetic Field (LMF) [30], is useful to manufacture clad steel slabs. This strong static field is employed across the slab mold in between the submergence depths of two nozzles, which deliver two different steel grades into the mold, as shown in Figure 28a. This magnetic field applied just below the mold exit produces two separated flow zones in the upper and lower pools, as shown in Figure 28b. This tends to lessen mixing of the two steel alloys, which consist of a stainless steel surface layer, which solidifies first, and a low carbon steel interior that solidifies later below the mold.

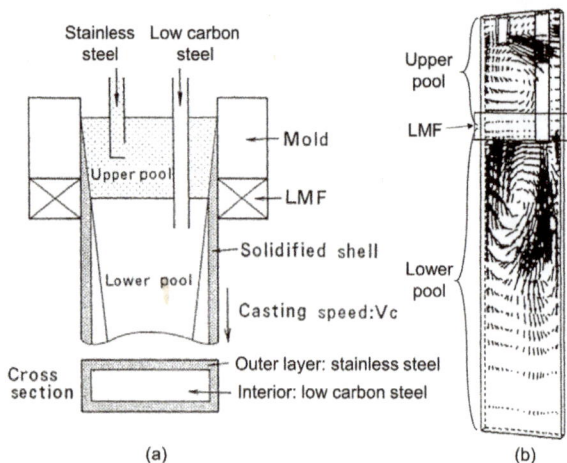

Figure 28. (a) Schematic of continuous casting and (b) flow patterns in the mold and strand region with Level DC Magnetic Field (LMF) for clad steel slabs [30].

Figure 29 shows the effects of LMF on microstructure and composition distribution in the clad steel slab. With LMF, the sharp gradient in nickel concentration at the interface between the stainless steel (outer surface layer) and the low carbon steel (interior) and the negligible nickel in the interior both indicate the effectiveness of LMF to prevent mixing of stainless steel into the interior.

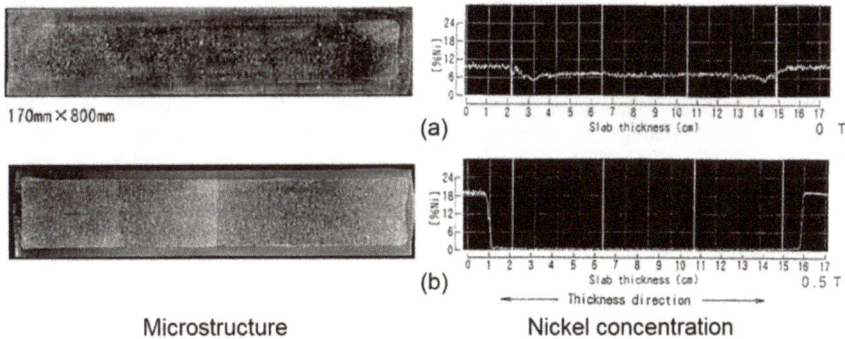

Figure 29. Comparison of microstructure and nickel distribution in the clad steel slabs (**a**) without and (**b**) with LMF [30].

9. Summary and Conclusions

This paper has reviewed the many different types of electromagnetic systems used in slab casting to affect fluid flow and related phenomena and the research tools that can be used to investigate and understand the phenomena that affect steel product quality. Some of these effects and practical strategies to operate these electromagnetic systems are summarized below.

- Combining several plant measurement methods is recommended to quantify the flow fluid and effects on quality, owing to the complexity of the continuous casting process with electromagnetics and the difficulty of making direct measurements.
- Computational modeling validated with plant measurements and lab experiments is the best way to quantify and understand the effects of electromagnetic forces on fluid flow, superheat transfer, solidification, particle transport and capture, grain structure, steel composition, and other phenomena and defects.
- Static magnetic fields (local, single-ruler, and double-ruler EMBr and EMB), moving magnetic fields (EMLS, EMLA, EMRS/M-EMS, EMC, and SEMS), and combined systems have been developed to affect the flow pattern and flow stability, aiming to control the intensity of surface flows in the mold to reduce various defects including surface defects, slag entrainment, inclusion entrapment, and deep oscillation marks, and/or to control internal cleanliness, grain structure, segregation, and porosity.
- EMBr, EMB, EMLS, and EMLA are designed to maintain a stable double-roll flow pattern which keeps surface velocity, profile, and level fluctuations within a safe operating window, which is most useful for higher casting speed operations, especially thin-slab casting.
- Placing static EMBr fields at a proper location relative to the flowing jets is critical to achieving the flow objectives. A strong magnetic field above the jet core tends to deflect the jets downward, and to slow surface velocity, which decreases variations in surface level and profile. In this case, care is needed to avoid over cooling the surface if the field is too strong.
- On the other hand, a strong magnetic field below the jet core tends to deflect the jets upwards, increasing surface velocities when casting at low speed and lessening deep penetration of inclusions. In this case, care is needed to avoid upward excessive surface flows, if the field is too strong.

- Locating the core of an EMBr magnetic field directly across the jets exiting the nozzle ports should be avoided to prevent unstable jet flow and associated defects.
- A static ruler EMBr field across the nozzle above the ports helps to stabilize flow inside the nozzle, with consequent improvement of flow stability in the mold.
- Maintaining proper ruler-EMBr field strength across the mold towards the narrow faces is important to reduce surface level fluctuations near the meniscus and jet penetration deep into the mold cavity.
- Moving magnetic fields in the mold (EMLS, EMLA, EMRS, and M-EMS) actively drive the flow, providing an alternative method to achieve flow objectives. These include: EMLS moving fields towards the SEN, aiming to lower surface velocity and turbulence; EMLA moving fields towards the narrow faces, aiming to increase surface velocity and turbulence; and EMRS (M-EMS) fields rotating around the perimeter of the mold surface, aiming (in part) to wash particles away from the solidifying steel shell to lessen particle capture.
- Superheat transport and initial solidification depend greatly on the mold flow pattern. Thus, adjusting the magnetic fields to deflect (static fields) or accelerate (EMLA) the jet upwards towards the top surface in the mold can reduce meniscus freezing, hook formation, and oscillation mark depth. Furthermore, rotating magnetic fields generated by EMRS (M-EMS), or EMC can make superheat and temperature near the meniscus more uniform.
- Lessening the jet impingement depth, with a uniform ruler EMBr field across the mold below the jet, can reduce particle capture deep into the solidifying steel shell. In addition, the washing effect generated by a rotating flow pattern with EMRS or M-EMS can reduce surface defects including particle capture during initial solidification at the meniscus including subsurface hooks.
- Below the mold, horizontally-moving fields towards one narrow face (SEMS) produces vertically rotating flows in the strand region, which mixes superheat, resulting in increased equiaxed grains, and less center segregation and porosity defects.
- Strong static magnetic fields can enable clad steel casting, by helping to separate two steel alloys without mixing, by generating two separate flow recirculation zones above and below the magnetic field.
- The application of combined fields, employing a traveling field either horizontally (via EMRS or M-EMS) or vertically (EMC) in the upper part of the mold and a static field (single-ruler EMBr) in the lower part of a mold, has great potential to reduce both surface and internal defects: The horizontally-moving field around the perimeter of the mold surface can wash away particles at the solidifying steel shell front and prevent their capture. At the same time, the static field prevents the jet flow go deep into the mold cavity, thereby reducing particle penetration, capture, and internal defects.
- The vertically moving EMC field near the meniscus can greatly reduce oscillation mark and hook depth.
- One of the greatest benefits of electromagnetics over conventional flow control devices, (such as port geometry) is the potential to adjust the field strength during operation according to the current flow conditions. Even better is the potential to adjust the magnetic field according to real-time feedback from in-mold sensors, such as multiple sensors of surface level, in order to maintain the intensity of surface flows in window of safe operation real time. More work is needed to implement this into practice.

Electromagnetic systems have been designed to prevent defect formation and to improve steel quality. However, all of the process geometry and conditions including nozzle port angle, SEN depth, casting speed, and argon gas flow rate need to be considered together to find optimal flow system operation conditions including electromagnetics for a specific caster, according to its needs regarding steel quality concerns.

Author Contributions: S.-M.C. and B.G.T. reviewed the references and wrote the article. Furthermore, B.G.T. as a corresponding author supervised the paper preparation.

Funding: Support from the Continuous Casting Center at Colorado School of Mines, the Continuous Casting Consortium at University of Illinois at Urbana-Champaign, and the National Science Foundation GOALI grant (Grant No. CMMI 18-08731) are gratefully acknowledged.

Acknowledgments: Provision of FLUENT licenses through the ANSYS Inc. academic partnership program is much appreciated.

Conflicts of Interest: The authors declare no conflicts of interest.

References

1. World Steel Association. Table 4. Production of Continuously Cast Steel. In *Steel Statistical Yearbook 2018*; World Steel Association: Brussels, Belgium, 2018; pp. 9–12.
2. Thomas, B.G. Chapter 14. Fluid Flow in the Mold. In *Making, Shaping and Treating of Steel*, 11th ed.; Cramb, A., Ed.; Casting Volume; AISE Steel Foundation: Pittsburgh, PA, USA, 2003; Volume 5, pp. 14.1–14.41.
3. Thomas, B.G. Chapter 15. Continuous Casting of Steel. In *Modeling for Casting and Solidification Processing*; Yu, O., Ed.; Marcel Dekker: New York, NY, USA, 2001; pp. 499–540.
4. Thomas, B.G. Modeling of Continuous Casting Defects Related to Mold Fluid Flow. *Iron Steel Technol.* **2006**, *3*, 128–143.
5. Najjar, F.M.; Thomas, B.G.; Hershey, D.E. Numerical Study of Steady Turbulent Flow through Bifurcated Nozzles in Continuous Casting. *Metall. Mater. Trans. B* **1995**, *26B*, 749–765. [CrossRef]
6. Calderon-Ramos, I.; Morales, R.D.; Salazar-Campoy, M. Modeling Flow Turbulence in a Continuous Casting Slab Mold Comparing the use of Two Bifurcated Nozzles with Square and Circular Ports. *Steel Res. Int.* **2015**, *86*, 1610–1621. [CrossRef]
7. Salazar-Campoy, M.M.; Morales, R.D.; Najera-Bastida, A.; Calderon-Ramos, I.; Cedillo-Hernandez, V.; Delgado-Pureco, J.C. A Physical Model to Study the Effects of Nozzle Design on Dispersed Two-Phase Flows in a Slab Mold Casting Ultra-Low-Carbon Steels. *Metall. Mater. Trans. B* **2018**, *49B*, 812–830. [CrossRef]
8. Cho, S.-M.; Thomas, B.G.; Kim, S.-H. Effect of Nozzle Port Angle on Transient Flow and Surface Slag Behavior during Continuous Steel slab casting. *Metall. Mater. Trans. B* **2018**, *50B*, 52–76. [CrossRef]
9. Thomas, B.G.; Mika, L.J.; Najjar, F.M. Simulation of Fluid Flow inside a Continuous Slab-Casting Machine. *Metall. Mater. Trans. B* **1990**, *21B*, 387–400. [CrossRef]
10. Cho, S.-M.; Thomas, B.G.; Lee, H.-J.; Kim, S.-H. Effect of Nozzle Port Angle on Mold Surface Flow in Steel Slab Casting. *Iron Steel Technol.* **2017**, *14*, 76–84.
11. Chaudhary, R.; Lee, G.-G.; Thomas, B.G.; Kim, S.-H. Transient Mold Fluid Flow with Well- and Mountain-Bottom Nozzles in Continuous Casting of Steel. *Metall. Mater. Trans. B* **2008**, *39B*, 870–884. [CrossRef]
12. Gursoy, K.A.; Yavuz, M.M. Effect of Flow Rate Controllers and their Opening Levels on Liquid Steel Flow in Continuous Casting Mold. *ISIJ Int.* **2016**, *56*, 554–563. [CrossRef]
13. Wang, Y.; Zhang, L. Transient Fluid Flow Phenomena during Continuous Casting: Part II—Cast Speed Change, Temperature Fluctuation, and Steel Grade Mixing. *ISIJ Int.* **2010**, *50*, 1783–1791. [CrossRef]
14. Bai, H.; Thomas, B.G. Turbulent Flow of Liquid Steel and Argon Bubbles in Slide gate Tundish Nozzles: Part II. Effect of Operation Conditions and Nozzle Design. *Metall. Mater. Trans. B* **2001**, *32B*, 269–284. [CrossRef]
15. Liu, Z.Q.; Qi, F.S.; Li, B.K.; Cheung, S.C.P. Modeling of bubble behaviors and size distribution in a slab continuous casting mold. *Int. J. Multiphase Flow* **2016**, *79*, 190–201. [CrossRef]
16. Dauby, P.H. Continuous casting: Make better steel and more of it! *Revue de Métallurgie* **2012**, *109*, 113–136. [CrossRef]
17. Jin, K.; Vanka, S.P.; Thomas, B.G. Large Eddy Simulations of the Effects of EMBr and SEN Submergence Depth on Turbulent Flow in the Mold Region of a Steel Caster. *Metall. Mater. Trans. B* **2017**, *48B*, 162–178. [CrossRef]
18. Jacobi, H.F. Investigation of Centreline Segregation and Centreline Porosity in CC-Slabs. *Steel Res. Int.* **2003**, *74*, 667–678. [CrossRef]

19. El-Bealy, M.O. Macrosegregation Quality Criteria and Mechanical Soft Reduction for Central Quality Problems in Continuous Casting of Steel. *Mater. Sci. Appl.* **2014**, *5*, 724–744. [CrossRef]

20. Thomas, B.G. Review on Modeling and Simulation of Continuous Casting. *Steel Res. Int.* **2018**, *89*, 1700312. [CrossRef]

21. Thomas, B.G. Chapter 5. Modeling of Continuous Casting. In *Making, Shaping and Treating of Steel*, 11th ed.; Cramb, A., Ed.; Casting Volume; AISE Steel Foundation: Pittsburgh, PA, USA, 2003; Volume 5, pp. 5.1–5.24.

22. Thomas, B.G.; Zhang, L. Mathematical Modeling of Fluid Flow in Continuous Casting. *ISIJ Int.* **2001**, *41*, 1181–1193. [CrossRef]

23. Yuan, Q.; Zhao, B.; Vanka, S.P.; Thomas, B.G. Study of Computational Issues in Simulation of Transient Flow in Continuous Casting. *Steel Res. Int.* **2005**, *76*, 33–43. [CrossRef]

24. Kollberg, S.; Lofgren, P.M.; Hanley, P. Improving Quality and Productivity in Thick Slab Casting by Direct Control of ElectroMagnetic Brake (EMBR). In Proceedings of the AISTech 2004, Nashville, TN, USA, 15–17 September 2004; pp. 977–984.

25. Zeze, M.; Harada, H.; Takeuchi, E.; Ishii, T. Application of DC Magnetic Field for the Control of Flow in the Continuous Casting Strand. In Proceedings of the 76th Steelmaking Conference, Dallas, TX, USA, 28–31 March 1993; pp. 267–272.

26. Qian, Z.-D.; Wu, Y.-L. Large Eddy Simulation of Turbulent Flow with the Effects of DC Magnetic Field and Vortex Brake Application in Continuous Casting. *ISIJ Int.* **2004**, *44*, 100–107. [CrossRef]

27. Singh, R.; Thomas, B.G.; Vanka, S.P. Effects of a Magnetic Field on Turbulent Flow in the Mold Region of a Steel Caster. *Metall. Mater. Trans. B* **2013**, *44B*, 1201–1221. [CrossRef]

28. Thomas, B.G.; Singh, R.; Vanka, S.P.; Timmel, K.; Eckert, S.; Gerbeth, G. Effect of Single-Ruler Electromagnetic Braking (EMBr) Location on Transient Flow in Continuous Casting. *J. Manuf. Sci. Prod.* **2015**, *15*, 93–104. [CrossRef]

29. Chaudhary, R.; Thomas, B.; Vanka, S. Effect of Electromagnetic Ruler Braking (EMBr) on Transient Turbulent Flow in Continuous Slab Casting using Large Eddy Simulations. *Metall. Mater. Trans. B* **2012**, *43*, 532–553. [CrossRef]

30. Harada, H.; Takeuchi, E.; Zeze, M.; Tanaka, H. MHD analysis in hydromagnetic casting process of clad steel slabs. *Appl. Math. Modell.* **1998**, *22*, 873–882. [CrossRef]

31. Kunstreich, S. Recent developments of electromagnetic actuators for continuous casting of long and flat products. *MILLENNIUM STEEL* **2014**, 57–63.

32. Kunstreich, S.; Dauby, P.H. Effect of liquid steel flow pattern on slab quality and the need for dynamic electromagnetic control in the mold. *Ironmak. Steelmak.* **2005**, *32*, 80–86. [CrossRef]

33. Kunstreich, S. Electromagnetic stirring for continuous casting-Part 2. *Rev. Met. Paris.* **2003**, *100*, 1043–1061. [CrossRef]

34. Nakashima, J.; Fukuda, J.; Kiyose, A.; Kawase, T.; Ohtani, Y.; Doki, M. *Improvement of Slab Surface Quality with In-mold Electromagnetic Stirring*; Nippon Steel and Sumitomo Metal Corporation: Tokyo, Japan, 2002; pp. 61–67.

35. Fujisaki, K. In-Mold Electromagnetic Stirring in Continuous Casting. *IEEE TRANSACTIONS ON INDUSTRY APPLICATIONS* **2001**, *37*, 1098–1104. [CrossRef]

36. Tani, M.; Zeze, M.; Toh, T.; Tsunenari, K.; Umetsu, K.; Hayashi, K.; Tanaka, K.; Fukunaga, S. Electromagnetic Casting Technique for Slab Casting. *Nippon Steel Technical Report* **2013**, *104*, 62–68. [CrossRef]

37. Cha, P.-R.; Hwang, Y.-S.; Nam, H.-S.; Chung, S.-H.; Yoo, J.-K. 3D Numerical Analysis on Electromagnetic and Fluid Dynamic Phenomena in a Soft Contact Electromagnetic Slab Caster. *ISIJ Int.* **1998**, *38*, 403–410. [CrossRef]

38. Toh, T.; Takeuchi, E.; Hojo, M.; Kawai, H.; Matsumura, S. Electromagnetic Control of Initial Solidification in Continuous Casting of Steel by Low Frequency Alternating Magnetic Field. *ISIJ Int.* **1997**, *37*, 1112–1119. [CrossRef]

39. Kubota, J.; Kubo, N.; Ishii, T.; Suzuki, M.; Aramaki, N.; Nishimachi, R. Steel Flow Control in Continuous Slab Caster Mold by Traveling Magnetic Field. *NKK TECHNICAL REVIEW* **2001**, *85*, 1–9.

40. Kunstreich, S. Electromagnetic stirring for continuous casting-Part 1. *Rev. Met. Paris.* **2003**, *100*, 395–408. [CrossRef]

41. Dubke, M.; Tacke, K.-H.; Spitzer, K.-H.; Schwerdtfeger, K. Flow fields in electromagnetic stirring of rectangular strands with linear inductors: Part I. theory and experiments with cold models. *Metall. Trans. B.* **1988**, *19B*, 581–593. [CrossRef]
42. Gong, J.; Liu, H.-p.; Wang, X.-h.; Bao, Y.-p. Numerical Simulation of Electromagnetic Field and Flow Pattern in a Continuous Slab Caster with in roll-type Strand Electromagnetic Stirring. *J. Iron Steel Res. Int.* **2015**, *22*, 414–422. [CrossRef]
43. Kim, G.H.; Kwon, S.H.; Won, Y.M.; Lee, C.H. Enhancement of slab internal quality by electromagnetic stirring of molten steel. In Proceedings of the 8th International Conference on Electromagnetic Processing of Materials (EPM 2015), Cannes, France, 12–16 October 2015.
44. Kunstreich, S. Strand electromagnetic stirring (S-EMS) for thick slab casters: Box-type or In-roll stirrers? *Millennium Steel* **2008**, 122–124.
45. El-Kaddah, N.; Natarajan, T.T. Electromagnetic Stirring of Steel: Effect of Stirring Design on Mixing in Horizontal Electromagnetic Stirring of Steel Slabs. In Proceedings of the Second International Conference on CFD in the Mineral and Process Industries, Melbourne, Australia, 6–8 December 1999; pp. 339–344.
46. Dubke, M.; Tacke, K.-H.; Spitzer, K.-H.; Schwerdtfeger, K. Flow fields in electromagnetic stirring of rectangular strands with linear inductors: Part II. Computation of flow fields in billets, blooms, and slabs of steel. *Metall. Trans. B.* **1988**, *19B*, 595–602. [CrossRef]
47. Lambert, V.; Galpin, J.-M.; Hackl, H.R.; Jacobson, N.P. New Strong Strand Stirrer Boosting Quality for Ferritic Stainless Steel. *Iron Steel Technol.* **2008**, *5*, 71–79.
48. Sedén, M.; Jacobson, N.; Lehman, A.; Eriksson, J.-E. Control of Flow Behavior by FC Mold G3 in Slab Casting Process. In Proceedings of the 8th European Continuous Casting Conference, Graz, Austria, 23–26 June 2014.
49. Qian, Z.-D.; Wu, Y.-L.; Li, B.-W.; He, J.-C. Numerical Analysis of the Influences of Operational Parameters on the Fluid Flow in Mold with Hybrid Magnetic Fields. *ISIJ Int.* **2002**, *42*, 1259–1265. [CrossRef]
50. Yavuz, M.M. The Effects of Electromagnetic Brake on Liquid Steel Flow in Thin Slab Caster. *Steel Res. Int.* **2011**, *82*, 809–818. [CrossRef]
51. Domgin, J.-F.; Anderhuber, M.; Doncker, M.D.; Paepe, A.D. Optimization of an Electromagnetic Technology in ArcelorMittal Gent for Improving Products Quality in Steel Industry. *J. Manuf. Sci. Prod.* **2015**, *15*, 105–117. [CrossRef]
52. Mizukami, H.; Hanao, M.; Hiraki, S.; Kawamoto, M.; Watanabe, T.; Hayashi, A.; Iguchi, M. Measurement of Meniscus Flow Velocity in High Speed Continuous Casting Mold. *Tetsu-to-Hagane* **2000**, *86*, 265–270. [CrossRef]
53. Iguchi, M.; Terauchi, Y. Karman Vortex Probe for the Detection of Molten Metal Surface Flow in Low Velocity Range. *ISIJ Int.* **2002**, *42*, 939–943. [CrossRef]
54. Iguchi, M.; Kawabata, H.; Ogura, T.; Hayashi, A.; Terauchi, Y. A New Probe for Directly Measuring Flow Velocity in a Continuous Casting Mold. *ISIJ Int.* **1996**, *36*, S190–S193. [CrossRef]
55. Han, S.-W.; Cho, H.-J.; Jin, S.-Y.; Sedén, M.; Lee, I.-B.; Sohn, I. Effects of Simultaneous Static and Traveling Magnetic Fields on the Molten Steel Flow in a Continuous Casting Mold. *Metall. Mater. Trans. B* **2018**, *49B*, 2757–2769. [CrossRef]
56. Kubota, J.; Kubo, N.; Suzuki, M.; Ishii, T.; Nishimachi, R.; Aramaki, N. Steel Flow Control with Travelling Magnetic Field for Slab Continuous Caster Mold. *Tetsu-to-Hagane* **2000**, *86*, 69–75. [CrossRef]
57. Jin, K.; Vanka, S.P.; Thomas, B.G.; Ruan, X. Large Eddy Simulations of the Effects of Double-Ruler Electromagnetic Braking and Nozzle Submergence Depth on Molten Steel Flow in A Commercial Continuous Casting Mold. In Proceedings of the TMS Annual Meeting, CFD Modeling and Simulation in Materials Processing Symposium 2016, Nashville, TN, USA, 14–18 March 2016; pp. 159–166.
58. Dauby, P.H.; Emling, W.H.; Sobolewski, R. Lubrication in the Mold: A Multiple Variable System. *Ironmaker Steelmaker* **1986**, *13*, 28–36.
59. Mcdavid, R.M.; Thomas, B.G. Flow and Thermal Behavior of the Top Surface Flux/Powder Layers in Continuous Casting Molds. *Metall. Mater. Trans. B* **1996**, *27B*, 672–685. [CrossRef]
60. Cho, S.-M.; Lee, H.-J.; Kim, S.-H.; Chaudhary, R.; Thomas, B.G.; Lee, D.-H.; Kim, Y.-J.; Choi, W.-R.; Kim, S.-K.; Kim, H.-S. Measurement of Transient Meniscus Flow in Steel Continuous Casters and Effect of Electromagnetic Braking. In Proceedings of the TMS Annual Meeting Symposium 2011, San Diego, CA, USA, 27 February–3 March 2011.

61. Cukierski, K.; Thomas, B.G. Flow Control with Local Electromagnetic Braking in Continuous Casting of Steel Slabs. *Metall. Mater. Trans. B* **2008**, *38B*, 94–107. [CrossRef]

62. Liu, R.; Thomas, B.G.; Sengupta, J.; Chung, S.D.; Trinh, M. Measurements of Molten Steel Surface Velocity and Effect of Stopper-rod Movement on Transient Multiphase Fluid Flow in Continuous Casting. *ISIJ Int.* **2014**, *54*, 2314–2323. [CrossRef]

63. Rietow, B.; Thomas, B.G. Using Nail Board Experiments to Quantify Surface Velocity in the CC Mold. In Proceedings of the AISTech 2008, Pittsburgh, PA, USA, 5–8 May 2018; pp. 1–11.

64. Cho, S.-M.; Kim, S.-H.; Thomas, B.G. Transient Fluid Flow during Steady Continuous Casting of Steel Slabs: Part I. Measurements and Modeling of Two-phase Flow. *ISIJ Int.* **2014**, *54*, 845–854. [CrossRef]

65. Köhler, K.U.; Andrzejewski, P.; Julius, E.; Haubrich, H. Steel Flow Velocity Measurement and Flow Pattern Monitoring in the Mould. In Proceedings of the 78th Steelmaking Conference, Nashville, TN, USA, 2–5 April 1995; pp. 445–449.

66. Thomas, B.G.; Yuan, Q.; Sivaraj Sivaramakrishnan, S.; Shi, T.; Vanka, S.P.; Assar, M.B. Comparison of Four Methods to Evaluate Fluid Velocities in a Continuous Slab Casting Mold. *ISIJ Int.* **2001**, *41*, 1262–1271. [CrossRef]

67. Assar, M.; Dauby, P.H.; Lawson, G.D. Opening the Black Box: PIV and MFC Measurements in a Continuous Caster Mold. In Proceedings of the 83rd Steelmaking Conference, Pittsburgh, PA, USA, 26–29 March 2000.

68. Okano, S.; Nishimura, T.; Ooi, H.; Chino, T. Relation between Large Inclusions and Growth Directions of Columnar Dendrites in Continuously Cast Slabs. *Tetsu-to-Hagane* **1975**, *61*, 2982–2990. [CrossRef]

69. Esaka, H.; Toh, T.; Harada, H.; Takeuchi, E.; Fujisaki, K. Deflection of Steel Dendrite Growing in the Fluid Flow Driven by Electromagnetic Stirrer. *Tetsu-to-Hagane* **2000**, *86*, 247–251. [CrossRef]

70. Wang, X.; Wang, S.; Zhang, L.; Sridhar, S.; Conejo, A.; Liu, X. Analysis on the Deflection Angle of Columnar Dendrites of Continuous Casting Steel Billets Under the Influence of Mold Electromagnetic Stirring. *Metall. Mater. Trans. A* **2016**, *47A*, 5496–5509. [CrossRef]

71. Melissari, B.; Argyropoulos, S.A. Measurement of Magnitude and Direction of Velocity in High-Temperature Liquid Metals. Part I: Mathematical Modeling. *Metall. Mater. Trans. B* **2005**, *36B*, 691–700. [CrossRef]

72. Melissari, B.; Argyropoulos, S.A. Measurement of Magnitude and Direction of Velocity in High-Temperature Liquid Metals. Part II: Experimental Measurements. *Metall. Mater. Trans. B* **2005**, *36B*, 639–649. [CrossRef]

73. Argyropoulos, S.A. Measuring velocity in high-temperature liquid metals: A review. *Scand. J. Metall.* **2000**, *30*, 273–285. [CrossRef]

74. Hibbeler, L.C.; Thomas, B.G. Mold Slag Entrainment Mechanisms in Continuous Casting Molds. *Iron Steel Technol.* **2013**, *10*, 121–134.

75. Kubota, J.; Okitoto, K.; Shirayama, A.; Murakati, H. Meniscus Flow Control in the Mold by Traveling Magnetic Field for High Speed Slab Caster. In Proceedings of the 74th Steelmaking Conference, Washington, DC, USA, 14–17 April 1991; pp. 233–241.

76. Akhtar, A.; Thomas, B.G.; Sengupta, J. Analysis of Nail Board Measurements of Liquid Slag Layer Depth. In Proceedings of the AISTech 2016, Pittsburgh, PA, USA, 16–19 May 2016; pp. 1427–1438.

77. Yuan, Q.; Thomas, B.G.; Vanka, S.P. Study of Transient Flow and Particle Transport in Continuous Steel Caster Molds: Part I. Fluid Flow. *Metall. Mater. Trans. B* **2004**, *35B*, 685–702. [CrossRef]

78. Zhang, L.; Thomas, B.G. State of the Art in Evaluation and Control of Steel Cleanliness. *ISIJ Int.* **2003**, *43*, 271–291. [CrossRef]

79. Furumai, K.; Matsui, Y.; Murai, T.; Miki, Y. Evaluation of Defect Distribution in Continuously-Cast Slabs by Using Ultrasonic Defect Detection System and Effect of Electromagnetic Brake on Decreasing Unbalanced Flow in Mold. *ISIJ Int.* **2015**, *55*, 2135–2141. [CrossRef]

80. Lee, G.-G.; Shin, H.-J.; Thomas, B.G.; Kim, S.-H. Asymmetric Multi-phase Fluid Flow and Particle Entrapment in a Continuous Casting Mold. In Proceedings of the AISTech 2008, Pittsburgh, PA, USA, 5–8 May 2008.

81. Ren, Y.; Wang, Y.; Li, S.; Zhang, L.; Zuo, X.; Lekakh, S.N.; Peaslee, K. Detection of Non-metallic Inclusions in Steel Continuous Casting Billets. *Metall. Mater. Trans. B* **2014**, *45B*, 1291–1303. [CrossRef]

82. Jin, K.; Thomas, B.G.; Ruan, X. Modeling and Measurements of Multiphase Flow and Bubble Entrapment in Steel Continuous Casting. *Metall. Mater. Trans. B* **2016**, *47B*, 548–565. [CrossRef]

83. Demmon, F.; Gass, R.; Yin, H. Step Milling as a Tool for Characterizing Defects in Slabs at ArcelorMittal. *Iron Steel Technol.* **2015**, *12*, 82–91.

84. Yang, H.; Tehranchi, F.; Eriksson, J.-E.; Song, J. Water Modeling of Stirring and Braking Processes in a Slab Caster Mold. In Proceedings of the AISTech 2010, Pittsburgh, PA, USA, 3–6 May 2010; pp. 135–146.

85. Yang, H.; Sedén, M.; Jacobson, N.; Eriksson, J.-E.; Hackl, H. Development of the Third-Generation FC Mold by Numerical and Water Model Simulations. In Proceedings of the AISTech 2012, Atlanta, GA, USA, 7–9 May 2012.

86. Timmel, K.; Kratzsch, C.; Asad, A.; Schurmann, D.; Schwarze, R.; Eckert, S. Experimental and Numerical Modeling of Fluid Flow Processes in Continuous Casting: Results from the LIMMCAST-Project. In Proceedings of the Final LIMTECH Colloquium and International Symposium on Liquid Metal Technologies, Dresden, Germany, 19–20 September 2017; IOP Conf. Series: Materials Science and Engineering. p. 012019.

87. Wondrak, T.; Eckert, S.; Gerbeth, G.; Klotsche, K.; Stefani, F.; Timmel, K.; Peyton, A.J.; Terzija, N.; Yin, W. Combined Electromagnetic Tomography for Determining Two-phase Flow Characteristics in the Submerged Entry Nozzle and in the Mold of a Continuous Casting Model. *Metall. Mater. Trans. B* **2011**, *42B*, 1201–1210. [CrossRef]

88. Timmel, K.; Eckert, S.; Gerbeth, G. Experimental Investigation of the Flow in a Continuous-Casting Mold under the Influence of a Transverse, Direct Current Magnetic Field. *Metall. Mater. Trans. B* **2011**, *42B*, 68–80. [CrossRef]

89. Chaudhary, R.; Rietow, B.T.; Thomas, B.G. Difference between Physical Water Models and Steel Continuous Caster: A Theoretical Evaluation. In Proceedings of the Materials Science and Technology 2009, Pittsburgh, PA, USA, 25–29 October 2009; pp. 1090–1101.

90. Harada, H.; Toh, T.; Ishii, T.; Kaneko, K.; Takeuchi, E. Effect of Magnetic Field Conditions on the Electromagnetic Braking Efficiency. *ISIJ Int.* **2001**, *41*, 1236–1244. [CrossRef]

91. Launder, B.E.; Spalding, D.B. *Lectures in Mathematical Models of Turbulence*; Academic Press: London, UK, 1972.

92. Menter, F.R. Two-Equation Eddy-Viscosity Turbulence Models for Engineering Applications. *AIAA J.* **1994**, *32*, 1598–1605. [CrossRef]

93. Nicoud, F.; Ducros, F. Subgrid-Scale Stress Modelling Based on the Square of the Velocity Gradient Tensor. *Flow Turbul. Combust.* **1999**, *62*, 183–200. [CrossRef]

94. Thomas, B.G.; Yuan, Q.; Mahmood, S.; Liu, R.; Chaudhary, R. Transport and Entrapment of Particles in Steel Continuous Casting. *Metall. Mater. Trans. B* **2014**, *45B*, 22–35. [CrossRef]

95. *ANSYS FLUENT 14.5-Theory Guide*; ANSYS. Inc.: Canonsburg, PA, USA, 2012.

96. Kumar, P.; Vanka, S.P. Effects of confinement on bubble dynamics in a square duct. *Int. J. Multiphase Flow* **2015**, *77*, 32–47. [CrossRef]

97. Moreau, R. *Magnetohydrodynamics*; Kluwer Academic Publishers: Dordrecht, The Netherlands, 1990; pp. 110–164.

98. Davidson, P.A. *Introduction to Magnetohydrodynamics*, 2nd ed.; Cambridge University Press: Cambridge, UK, 2017; pp. 27–184.

99. Cho, M.J.; Kim, I.C.; Kim, S.J.; Kim, J.K. 3-D Turbulent Heat & Fluid Flow Analysis with Solidification under Electro-Magnetic Field. *Trans. KSME* **1999**, *23B*, 1491–1502.

100. Sengupta, J.; Shin, H.-J.; Thomas, B.G.; Kim, S.-H. Micrograph Evidence of Meniscus Solidification and Sub-Surface Microstructure Evolution in Continuous-Cast Ultra-Low Carbon Steels. *Acta Mater.* **2006**, *54*, 1165–1173. [CrossRef]

101. Sengupta, J.; Thomas, B.G.; Shin, H.-J.; Lee, G.-G.; Kim, S.-H. A New Mechanism of Hook Formation during Continuous Casting of Ultra-Low-Carbon Steel Slabs. *Metall. Mater. Trans. A* **2006**, *37A*, 1597–1611. [CrossRef]

102. Takatani, K.; Nakai, K.; Kasai, N.; Watanabe, T.; Nakajima, H. Analysis of Heat Transfer and Fluid Flow in the Continuous Casting Mold with Electromagnetic Brake. *ISIJ Int.* **1989**, *29*, 1063–1068. [CrossRef]

103. Kim, D.-S.; Kim, W.-S.; Cho, K.-H. Numerical Simulation of the Coupled Turbulent Flow and Macroscopic Solidification in Continuous Casting with Electromagnetic Brake. *ISIJ Int.* **2000**, *40*, 670–676. [CrossRef]

104. Wang, Y.; Zhang, L. Fluid Flow-Related Transport Phenomena in Steel Slab Continuous Casting Strands under Electromagnetic Brake. *Metall. Mater. Trans. B* **2011**, *42B*, 1319–1351. [CrossRef]

105. Teshima, T.; Osame, M.; Okimoto, K.; Nimura, Y. Improvement of Surface Property of Steel at High Casting Speed. In Proceedings of the 71th Steelmaking Conference, Toronto, Canada, 17–20 April 1988; pp. 111–118.

106. Hwang, Y.-S.; Cha, P.-R.; Nam, H.-S.; Moon, K.-H.; Yoon, J.-K. Numerical Analysis of the Influences of Operational Parameters on the Fluid Flow and Meniscus Shape in Slab Caster with EMBR. *ISIJ Int.* **1997**, *37*. [CrossRef]

107. Tian, X.-Y.; Li, B.-W.; He, J.-C. Electromagnetic Brake Effects on the Funnel Shape Mold of a Thin Slab Caster Based on a New Type Magnet. *Metall. Mater. Trans. B* **2009**, *40B*, 596–604. [CrossRef]

108. Zhang, L.-S.; Zhang, X.-F.; Wang, B.; Liu, Q.; Hu, Z.-G. Numerical Analysis of the Influences of Operational Parameters on the Braking Effect of EMBr in a CSP Funnel- Type Mold. *Metall. Mater. Trans. B* **2014**, *45B*, 295–306. [CrossRef]

109. Xu, L.; Wang, E.; Karcher, C.; Deng, A.; Xu, X. Numerical Simulation of the Effects of Horizontal and Vertical EMBr on Jet Flow and Mold Level Fluctuation in Continuous Casting. *Metall. Mater. Trans. B* **2018**, *49B*, 2779–2793. [CrossRef]

110. Li, F.; Wang, E.; Feng, M.; Li, Z. Simulation Research of Flow Field in Continuous Casting Mold with Vertical Electromagnetic Brake. *ISIJ Int.* **2015**, *55*, 814–820. [CrossRef]

111. Li, Z.; Wang, E.; Zhang, L.; Xu, Y.; Deng, A. Influence of Vertical Electromagnetic Brake on the Steel/Slag Interface Behavior in a Slab Mold. *Metall. Mater. Trans. B* **2017**, *48B*, 2389–2402. [CrossRef]

112. Cho, S.-M.; Thomas, B.G.; Kim, S.-H. Transient Fluid Flow during Steady Continuous Casting of Steel Slabs: Part II. Effect of Double-Ruler Electro-Magnetic Braking. *ISIJ Int.* **2014**, *54*, 855–864. [CrossRef]

113. Cho, S.-M.; Thomas, B.G.; Kim, S.-H. Transient Two-Phase Flow in Slide-Gate Nozzle and Mold of Continuous Steel Slab Casting with and without Double-Ruler Electro-Magnetic Braking. *Metall. Mater. Trans. B* **2016**, *47B*, 3080–3098. [CrossRef]

114. Singh, R.; Thomas, B.G.; Vanka, S.P. Large Eddy Simulations of Double-Ruler Electromagnetic Field Effect on Transient Flow During Continuous Casting. *Metall. Mater. Trans. B* **2014**, *45B*, 1098–1115. [CrossRef]

115. Yu, H.; ZHU, M. Numerical Simulation of the Effects of Electromagnetic Brake and Argon Gas Injection on the Three-dimensional Multiphase Flow and Heat Transfer in Slab Continuous Casting Mold. *ISIJ Int.* **2008**, *48*, 584–591. [CrossRef]

116. Sarkar, S.; Singh, V.; Ajmani, S.K.; Singh, R.K.; Chanko, E.Z. Effect of Argon Injection in Meniscus Flow and Turbulence Intensity Distribution in Continuous Slab Casting Mold Under the Influence of Double Ruler Magnetic Field. *ISIJ Int.* **2018**, *58*, 68–77. [CrossRef]

117. Sarkar, S.; Singh, V.; Ajmani, S.K.; Ranjan, R.; Rajasekar, K. Effect of Double Ruler Magnetic Field in Controlling Meniscus Flow and Turbulence Intensity Distribution in Continuous Slab Casting Mold. *ISIJ Int.* **2016**, *56*, 2181–2190. [CrossRef]

118. Li, B.; Okane, T.; Umeda, T. Modeling of Molten Metal Flow in a Continuous Casting Process Considering the Effects of Argon Gas Injection and Static Magnetic-Field Application. *Metall. Mater. Trans. B* **2000**, *31B*, 1491–1503. [CrossRef]

119. Idogawa, A.; Sugizawa, M.; Takeuchi, S.; Sorimachi, K.; Fujii, T. Control of Molten Steel Flow in Continuous Casting Mold by Two Static Magnetic Fields Imposed on Whole Width. *Mater. Sci. Eng. A* **1993**, *A173*, 293–297. [CrossRef]

120. Kunstreich, S.; Gautreau, T.; Ren, J.Y.; Codutti, A.; Guastini, F.; Petronio, M. Development and Validation of Multi-Mode® EMB, a New Electromagnetic Brake for Thin Slab Casters. In Proceedings of the 8th European Continuous Casting Conference, Graz, Austria, 23–26 June 2016.

121. Kubo, N.; Kubota, J.; Suzuki, M.; Ishii, T. Molten Steel Flow Control under Electromagnetic Level Accelerator in Continuous Casting Mold. *ISIJ Int.* **2007**, *47*, 988–995. [CrossRef]

122. Kubo, N.; Ishii, T.; Kubota, J.; Ikagawa, T. Numerical Simulation of Molten Steel Flow under a Magnetic Field with Argon Gas Bubbling in a Continuous Casting Mold. *ISIJ Int.* **2004**, *44*, 556–564. [CrossRef]

123. Okazawa, K.; Toh, T.; Fukuda, J.; Kawase, T.; Toki, M. Fluid Flow in a Continuous Casting Mold Driven by Linear Induction Motors. *ISIJ Int.* **2001**, *41*, 851–858. [CrossRef]

124. Yin, Y.; Zhang, J.; Lei, S.; Dong, Q. Numerical Study on the Capture of Large Inclusion in Slab Continuous Casting with the Effect of In-mold Electromagnetic Stirring. *ISIJ Int.* **2017**, *57*, 2165–2174. [CrossRef]

125. Li, B.; Okane, T.; Umeda, T. Modeling of Biased Flow Phenomena Associated with the Effects of Static Magnetic-Field Application and Argon Gas Injection in Slab Continuous Casting of Steel. *Metall. Mater. Trans. B* **2001**, *32B*, 1053–1066. [CrossRef]

126. Moon, K.H.; Shin, H.K.; Kim, B.J.; Chung, J.Y.; Hwang, Y.S.; Yoon, J.K. Flow Control of Molten Steel by Electromagnetic Brake in the Continuous Casting Mold. *ISIJ Int.* **1996**, *36*, S201–S203. [CrossRef]

127. Liu, Z.; Vakhrushev, A.; Wu, M.; Karimi-Sibaki, E.; Kharicha, A.; Ludwig, A.; Li, B. Effect of an Electrically-Conducting Wall on Transient Magnetohydrodynamic Flow in a Continuous-Casting Mold with an Electromagnetic Brake. *Metals* **2018**, *8*, 609. [CrossRef]

128. Miao, X.; Timmel, K.; Lucas, D.; Ren, Z.; Eckert, S.; Gerbeth, G. Effect of an Electromagnetic Brake on the Turbulent Melt Flow in a Continuous-Casting Mold. *Metall. Mater. Trans. B* **2012**, *43B*, 954–972. [CrossRef]

129. Tian, X.-Y.; Zou, F.; Li, B.-W.; He, J.-C. Numerical Analysis of Coupled Fluid Flow, Heat Transfer and Macroscopic Solidification in the Thin Slab Funnel Shape Mold with a New Type EMBr. *Metall. Mater. Trans. B* **2010**, *41B*, 112–120. [CrossRef]

130. Hwang, J.Y.; Cho, M.J.; Thomas, B.G.; Cho, S.M. Numerical Simulation of Turbulent Steel CEM® Mold under High Mass Flow Condition. In Proceedings of the 9th International Symposium on Electromagnetic Processing of Materials (EPM2018), Hyogo, Japan, 14–18 October 2018; p. 012033.

131. Satou, Y.; Baba, N.; Kasai, N.; Mutou, A.; Hanao, M. Increase of Casting Speed of Hypo-peritectic Steel at Kashima No.3 Caster. In Proceedings of the AISTech 2009, St. Louis, MO, USA, 4–7 May 2009; pp. 663–671.

132. Wang, S.; Zhang, X.; Zhang, L.; Wang, Q. Influence of Electromagnetic Brake on Hook Growth and Inclusion Entrapment beneath the Surface of Low-Carbon Continuous Casting Slabs. *Steel Res. Int.* **2018**, *89*, 1800263. [CrossRef]

133. Zhang, L.; Aoki, J.; Thomas, B.G. Inclusion Removal by Bubble Flotation in a Continuous Casting Mold. *Metall. Mater. Trans. B* **2006**, *37B*, 361–379. [CrossRef]

134. Gass, R.; Knoepke, H.; Moscoe, J.; Shah, R.; Beck, J.; Dzierzawski, J.; Ponikvar, P.E. Conversion of Ispat Inland's No. 1 Slab Caster to Vertical Bending. In Proceedings of the ISSTech 2003 Conference, Indianapolis, Indiana, 27–30 April 2003; pp. 3–18.

135. Jin, K.; Kumar, P.; Vanka, S.P.; Thomas, B.G. Rise of an argon bubble in liquid steel in the presence of a transverse magnetic field. *Phys. Fluids* **2016**, *28*, 093301. [CrossRef]

136. Wang, Y.; Dong, A.; Zhang, L. Effect of Slide Gate and EMBr on the Transport of Inclusions and Bubbles in Slab Continuous Casting Strands. *Steel Res. Int.* **2011**, *82*, 428–439. [CrossRef]

137. Yuan, Q.; Thomas, B.G.; Vanka, S.P. Study of Transient Flow and Particle Transport in Continuous Steel Caster Molds: Part II. Particle Transport. *Metall. Mater. Trans. B* **2004**, *35B*, 703–714. [CrossRef]

138. Jin, K.; Vanka, S.P.; Thomas, B.G. Large Eddy Simulations of Electromagnetic Braking Effects on Argon Bubble Transport and Capture in a Steel Continuous Casting Mold. *Metall. Mater. Trans. B* **2018**, *49B*, 1360–1377. [CrossRef]

139. Shibata, H.; Yin, H.; Yoshinaga, S.; Emi, T.; Suzuki, M. In-situ Observation of Inclusions in Steel Melt Engulfment and Pushing by Advancing Melt/Solid of Nonmetallic Interface. *ISIJ Int.* **1998**, *38*, 149–156. [CrossRef]

140. Lee, S.-M.; Kim, S.-J.; Kang, Y.-B.; Lee, H.-G. Numerical Analysis of Surface Tension Gradient Effect on the Behavior of Gas Bubbles at the Solid/Liquid Interface of Steel. *ISIJ Int.* **2012**, *52*, 1730–1739. [CrossRef]

141. Pesteanu, O. The Washing Effect in Electromagnetic Rotational Stirrers for Continuous Casting. *ISIJ Int.* **2005**, *45*, 1073–1075. [CrossRef]

142. Pesteanu, O. Short Contribution to the Study of the Washing Effect in Electromagnetic Stirrers for Continuous Casting. *ISIJ Int.* **2003**, *43*, 1861–1862. [CrossRef]

143. Toh, T.; Hasegawa, H.; Harada, H. Evaluation of Multiphase Phenomena in Mold Pool under In-mold Electromagnetic Stirring in Steel Continuous Casting. *ISIJ Int.* **2001**, *41*, 1245–1251. [CrossRef]

144. Kittaka, S.; Fukuokaya, T.; Maruki, Y.; Kanki, T. *Nippon Steel Strand Electro-Magnetic Stirrer "S-EMS" for Slab Caster*; Nippon Steel: Tokyo, Japan, 2003; pp. 70–74.

145. Xu, Y.; Xu, X.-J.; Li, Z.; Wang, T.; Deng, A.-Y.; Wang, E.-G. Dendrite Growth Characteristics and Segregation Control of Bearing Steel Billet with Rotational Electromagnetic Stirring. *High Temp. Mater. Proc.* **2017**, *36*, 339–346. [CrossRef]

146. Shibata, H.; Itoyama, S.; Kishimoto, Y.; Takeuchi, S.; Sekiguchi, H. Prediction of Equiaxed Crystal Ratio in Continuously Cast Steel Slab by Simplified Columnar-to-Equiaxed Transition Model. *ISIJ Int.* **2006**, *46*, 921–930. [CrossRef]

147. Bridge, M.R.; Rogers, G.D. Structural Effects and Band Segregate Formation during the Electromagnetic Stirring of Strand-Cast Steel. *Metall. Trans. B.* **1984**, *15B*, 581–589. [CrossRef]

![metals logo] *metals*

MDPI

Article

Flow Characteristics for Two-Strand Tundish in Continuous Slab Casting Using PIV

Jun Huang *, Zhigang Yuan, Shaoyuan Shi, Baofeng Wang and Chi Liu *

School of Energy and Environment, Inner Mongolia University of Science and Technology, Baotou 014010, China; 18940046801@163.com (Z.Y.); shishaoyuan@126.com (S.S.); 13327185758@163.com (B.W.)
* Correspondence: hjun8420@imust.edu.cn (J.H.); liuchiyd@163.com (C.L.);
   Tel.: +86-472-595-1568 (J.H.); +86-131-9070-8932 (C.L.)

Received: 4 January 2019; Accepted: 14 February 2019; Published: 17 February 2019

Abstract: With the development of continuous casting technology, there has been an increase in the stringent requirements for the cleanliness and quality of steel being produced. The flow state of molten steel in tundish is the key to: Optimizing the residence time of molten steel in the tundish; homogenizing the temperature of molten steel; and removing inclusions by floatation. Hence, from theoretical and practical aspects, it is imperative to examine and analyze the flow field of molten steel in the tundish in order to ensure the desired molten steel flow. In this study, a two-strand tundish with 650 mm × 180 mm slab casting is considered as the subject for this research. According to the similarity theory, combined with the geometrical shape and dimension of the prototype tundish, a tundish model with a geometric similarity ratio of 2:3 is established in the laboratory. Digital particle image velocimetry (PIV) is employed to measure and examine the flow fields at different casting speeds for a tundish containing different flow control devices. The flow in the tundish is typically turbulent and also consists of a vortex motion; it exhibits both random and ordered characteristics. Results reveal that the presence of baffles with 15° holes can cause an upward-directed flow in the outlet section and give rise to a large circulation. When the casting speed is doubled, the overall velocity of the flow field and turbulent intensity increase, leading to an increase in the molten steel surface velocity.

Keywords: steel tundish; baffle; flow field; velocity; PIV

---

1. Introduction

The main function of the tundish is to act as a steel reservoir between the ladle and the mold, and in the case of multistrand casters, to distribute the liquid into the molds. In addition to being a reservoir of liquid steel, the tundish is more increasingly used as a metallurgical reactor vessel aimed at improving control of steel cleanness, temperature, and composition. Tundish metallurgy was proposed in the early 1980s as a special secondary refining technology and an important link in ensuring excellent steel quality during production from smelting and refining to the formation of slabs or billets [1]. In the past three decades, many experts and researchers have done a lot of work and published many important papers. A comprehensive review was given in the paper by Sahai [2] and the book by Sahai and Emi [3].

The basic physical phenomenon in the tundish involves the flow of molten steel. The metallurgical effect of the tundish is mainly achieved by the reasonable flow of molten steel in the tundish. The flow is intended to deliver the molten metal to the molds evenly and at a designed throughput rate and temperature with minimal contamination by, and maximize flotation of macro inclusions. Hence, the investigation of the flow phenomenon in the tundish is the foundation of tundish metallurgy.

Physical modeling has played a key role in tundish research [4–6]. In physical modeling, a low temperature aqueous analog, generally water, is used to represent molten metal in a tundish.

In particular, the kinematic viscosity of water is similar to that of molten steel. Water flow in a transparent model tundish can be used to observe melt flow physically taking place in an actual tundish. A full or reduced scale tundish model may be designed based on appropriate similarity criteria in which the flow of molten metal is simulated by the flow of water. As water flow in the model is a realistic representation of the actual tundish melt flow, it may be used to study the melt flow in a tundish.

According to the similarity principle for flow phenomena, the use of a water model to investigate the flow of molten steel in tundish is not only feasible, but also can accurately reflect the values and law of the actual flow of molten steel. However, in the actual hydraulics simulation of a tundish, it is difficult to completely maintain all forces to be equal between the model and the prototype; hence, different situations need to be considered. In a water simulation experiment using a reduced-scale model, it is imperative to use an approximate model method to ensure that the flow in the model is similar to that in the prototype.

Based on flow visualization and image processing technology, a digital image technique has been developed, which combines single point measurement technology (hotline, turbine) and flow visualization technology. Digital image measurement technology can not only achieve the overall structure and transient image field, but also obtain the velocity data of the whole flow field quantitatively. Some researchers [7–9] have measured the flow field in the tundish by the particle image velocimetry (PIV), thus that the overall structure and transient image of a planar flow field was obtained quantitatively.

Many attempts have been made to improve melt flow characteristics in existing tundish's by the installation of various flow control devices, such as weirs, dams, baffles with holes, and turbulence suppressors. The beneficial effects of various flow modification devices have been applied in actual industrial trials as well as physical and mathematical modeling studies [10,11]. Optimum placement of flow control devices has been found to result in an increase in the average residence time of fluid as well as an increase in the plug flow volume in the tundish. These flow control devices, properly installed, may create localized mixing in contained regions, which may help in inclusion agglomeration and hence, inclusion removal.

Due to the limited of view of the camera in PIV, developing an approach for measuring a water model with a length close to 5m is an important problem to be solved in this paper. In this study, the PIV flow measurement is carried out for a reduced-scale two-strand tundish model for slab production. The paper analyzes the flow characteristics in a two-strand tundish for continuous slab casting with different flow-control devices. The effect of eddies is discussed.

2. Experimental Object and Scheme

To transfer results of measurements from a model to the original tundish, apart from the geometric similarity, the fluid-dynamic similarity must be considered as well. For an isothermal water model experiment, geometrical similarity, and dynamic similarity between the model and the prototype are required. In order to maintain similarity in depth of liquid, the relationship between the velocities and hence the inlet flow rate of the fluid in the model and in the prototype is to be obtained by following either of Froude (*Fr*) similarity or Reynolds (*Re*) similarity. For the dynamic similarity, the *Re* number and Fr number in the model should be equivalent to those in the prototype. As steel flow in the tundish is gravity-driven, it is understood that flow inside the tundish is Froude criteria dominated. As the flow of liquid steel within the tundish is severely turbulent, the *Re* number of the model is in the same self-modelling region as the prototype, the *Re* number can meet the requirement naturally. Thus, most reduced scale-modeling studies are done mainly based on Froude similarity criteria [12].

In the experiments, a two-strand tundish of a stainless-steel slab continuous casting is used as the experimental object. The tundish model in the hydraulics experiment comprised clear glass with a 2:3 ratio to the prototype. The flow rate passing the long nozzle in the experiment is determined on the basis of the typical sectional dimension of the cast slab, i.e., 180 mm × 650 mm, and casting speed.

The typical casting speed is 1.2 m/min. The long nozzle diameter is 75 mm. The main dimensions of the water model are shown in Figure 1, all dimensions are in millimeters. The scale factor of 2:3 is defined as the ratio of lengths in the model and prototype systems to ensure geometric similarity.

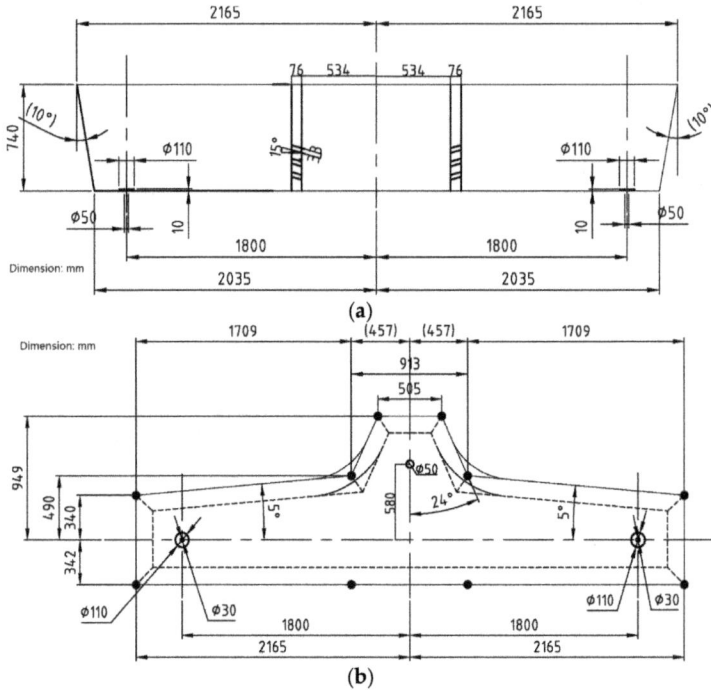

Figure 1. Two-strand tundish water model (**a**) shell diagram; (**b**) control device.

To satisfy the Froude similarity, the Froude number of the water model should be equal to that of the prototype:

$$Fr = \frac{v^2}{gl} = \frac{v'^2}{gl'} \tag{1}$$

The similarity ratio in velocity can be obtained (model velocity/prototype velocity):

$$\frac{v'}{v} = \sqrt{\frac{l'}{l}} = \lambda^{0.5} = \left(\frac{2}{3}\right)^{0.5} \tag{2}$$

The similarity ratio in the flowrate can be obtained (model flowrate /prototype flowrate):

$$\frac{Q'}{Q} = \frac{l'^2 \times v'}{l^2 \times v} = \lambda^{2.5} = \left(\frac{2}{3}\right)^{2.5} \tag{3}$$

where v and v' represent the velocity of the fluid in the tundish in the prototype and model respectively, m/s; g is the gravitational acceleration, m/s^2; l is the characteristic length, m; λ is the similarity ratio, Q and Q' are the flowrate in the tundish in the prototype and model respectively, m^3/h.

In the water simulation experiment, the quantity of poured steel is controlled by the flowrate at the long nozzle, meanwhile the liquid level is controlled by the level gauge in the tundish model. Table 1 summarizes the concrete experimental parameters of the tundish prototype and the water model.

Table 1. Experimental parameters of the tundish prototype and water model.

Object	Melt Depth in Tundish/mm	Volume/m^3	Long Nozzle Inserted Depth/mm	Flowrate/(m^3/h)
Prototype	900	6.00	250	8.42
Model	600	1.78	167	3.06

Fluid flow characteristics in the tundish with the incorporation of flow control devices is dependent on the optimum location and size. Numerical simulations have been used to optimize current control devices and installation locations. The optimized results are obtained by numerical simulation and must be validated by actual water simulation experiments. Scheme A is an optimization scheme, while the other three are in contrast to it. The installation location and size of these devices is shown in Figure 1. The flow control devices in the tundish mainly include the baffles with holes and the turbulence suppressor. Table 2 summarizes the experimental scheme. Experiments are carried out in the presence or absence of the flow control device and different casting speeds. The baffles with holes are placed in the middle of the tundish. The baffles are perpendicular to the walls of the tundish. Each baffle has six circular holes with a diameter of 33 mm and an upward inclination of 15°. To highlight the effect of the flow control device on the flow, the casting speed is doubled for comparing and manifesting the results obtained from the flow test.

Table 2. Experimental scheme and relevant processing parameters of tundish.

No.	Flow Control Device	Prototype Casting Speed/(m/min)
A	baffles with holes and suppressor	1.2
B	baffles with holes and suppressor	2.4
C	Suppressor only	1.2
D	Suppressor only	2.4

The model tundish is placed on the experimental platform, and water pipelines are connected. After setting the experimental parameters and waiting for 30 min for the flow to become steady, PIV flow measurement starts.

According to the flow path of the molten steel in the tundish, the measuring section is shown in Figure 2. The measurement section is sequentially scanned with the PIV. The photographs by PIV are processed by PIVview software, then the flow field diagrams from different locations are merged into the full section. PIVview is a compact program package for the evaluation of particle image velocimetry with the PIV-Groups of the German Aerospace Center. The software is be developed in close cooperation with the Institute of Aerodynamics and Flow Technology of the German Aerospace Center in Goettingen, Germany.

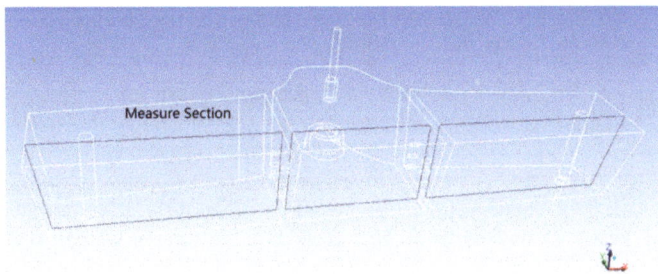

Figure 2. Schematic of the measuring section positions in the tundish.

3. PIV Measurement Process

Figure 3 shows the hardware structure diagram of the test system developed in this study for the flow field. A certain number of tracing particles are evenly dispersed in the flow field. The 2D flow field with an approximate thickness of 3 mm is illuminated by a sheet light source comprising a laser and its lens. Two high-resolution cameras are arranged along the vertical direction of the laser to form a camera array, which can record the trajectory of the tracing particles of a particular frequency under the illumination of the sheet light source. The time interval between the two pulses of the laser that illuminate the flow field is set to meet the camera's recording frequency and synchronize the camera's recording frequency with the laser's stroboscopic frequency using the synchronizer, which are similar to commercial PIV requirements in that two continuous frames can be continuously recorded in a short time period. The use of a multi-camera array can meet the measurement for a large sight field. The machine vision systems are employed related to the acquisition and processing of data during the image recognition and velocity measurements, enabling the system to solve low-speed measurement problems at a lower cost. The motion drive unit controlled by the servo motor can meet the requirements of segmentation measurement. Combined with the camera array setting, it can meet the measurement requirement for velocity in a wider range, which also provides the necessary hardware and software conditions for implementing the different scale measurement for the flow field in the tundish.

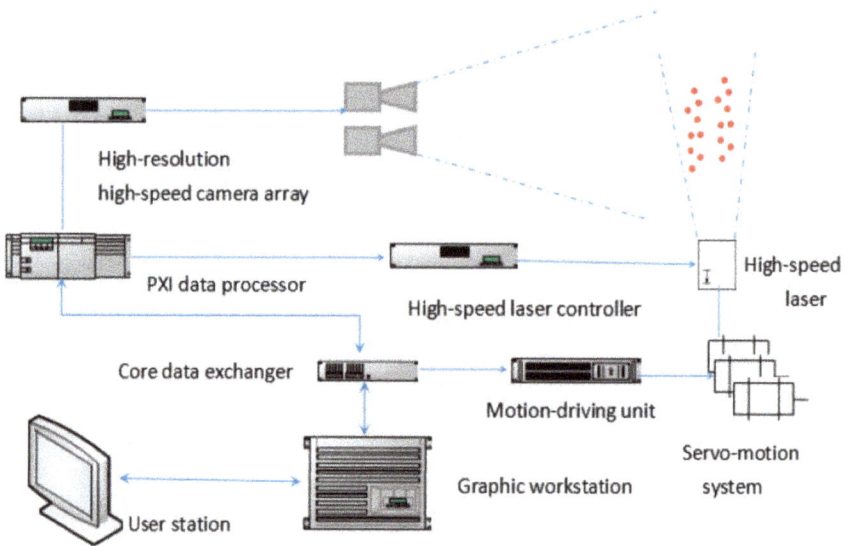

Figure 3. Diagram of the particle image velocimetry (PIV) system.

Owing to the low flow velocity in the tundish (average velocity < 1.0 m·s^{-1}), the continuous shooting mode of a digital camera with a fixed number of frames can meet the requirement to trace particles. The method does not require two consecutive frames in an extremely short time (5 µs) as it does by the PIV technology and only needs to ensure the synchronization between the continuous shooting speed of the digital camera and the laser stroboflash, hence, the measurement system can not only decrease the price, but also the array comprising multiple cameras can meet the requirements for obtaining a large field of view simultaneously. The image acquisition system for the flow field in the experimental tundish mainly comprises of an MGL-N-532 laser source, two Canon 5D MARK III cameras (Tokyo, Japan) and the background control and data processing software, which is shown in Figure 4. The laser emitted by the MGL-N-532 laser is the green laser with a wavelength of 532 nm and a shooting frequency of 30 Hz. Polystyrene beads with a density of 1.02 g/cm^3 and a diameter of 50 μm are tracing particles. Similar to a commercial PIV system, the displacement of tracing particles is obtained by the cross-correlation algorithm for the two images in the tested area, and the data are processed by the authorized PIVview software to display the two-dimensional velocity vector field within the areas in which velocity is measured [13].

Figure 4. The image acquisition system.

As the sectional flow field of the tundish water model exceeds the shooting range of the camera array, the sectional flow field of the entire tundish water model must be obtained by the image stitching method. Although the measuring moments for different areas are not the same, the stitched image can reflect the flow distribution characteristics in the cross section.

4. Experimental Results and Discussion

Figure 5 shows the sectional flow field of the tundish with PIV measurements and numerical simulation under schemes A. The position of the measured and simulated results is the same. The simulation shows only half. The position of two baffles with holes is indicated in the diagram. Of the black stripes that appear in the PIV measurement drawing, three are the supporting frames in the process of making the tundish model. The white stripes represent the results of different cameras at different times in Figure 5c. The results obtained from the different method are shown together.

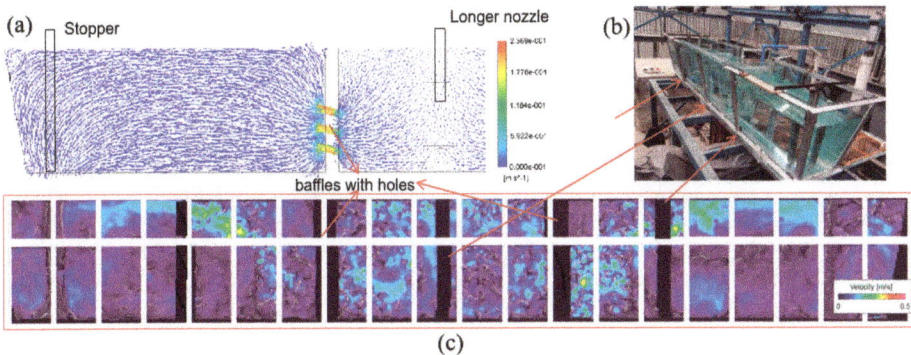

Figure 5. Flow measurement and simulation result of scheme A; (**a**) numerical simulation; (**b**) experimental setup; (**c**) PIV measurements.

Figure 5c shows a very detailed velocity field in which the flow is more active in the inlet section. The very marked large circulation is readily apparent from the plot. It can be seen that flow in the tundish is a local pulse, partially rendering a lot of mixing, but the overall flow shows the trend of a certain flow movement.

Figure 5a also shows the flow field for the numerical simulation results obtained by test scheme A. The commercial CFD package FLUENT with ANSYS 15.0 is used to model and solve the tundish fluid flow and it is compared with the available experiment results. The tundish with a long nozzle and exit is modelled. A non-uniform Cartesian mesh is used, with a power-law expansion of cells from the walls outwards. The finest cells are placed next to the jet-impact wall, since that is where maximum resolution is needed for the fast-moving boundary layers. A fine mesh is also used to cover the jet region. In order to test the grid independence of the results, it is necessary to test the sensitivity of the results on the number of mesh points used. A total of about 8,500,000 cells are used in the finest calculation. The mathematical model is based on the assumptions of continuum hypothesis, the standard two equation model, k-ε equation is used to model the turbulence. A steady-state incompressible solution is adopted, with the main dependent variables being the pressure, three velocity components and the two turbulent quantities. The SIMPLE (Semi-Implicit Method for Pressure Linked Equations) algorithm is used for the pressure-velocity coupling and QUICK (Quadratic Upstream Interpolation for Convective Kinetics) scheme is used for discretization of momentum, turbulent equations.

As can be observed from the measurement results, the flow complexity in the tundish is far from the simple recirculating flow that is obtained by the Reynolds-averaged Navier-Stokes (RANS) equations, plus an appropriate turbulence model, combining large circulation and small-scale eddies. To gain more knowledge about the transient turbulence process, which cannot be achieved via Reynolds-averaged equations solutions, large-eddy simulations (LES) of the tundish flow field are performed. Alkishriwi et al. [14,15] and Jha et al. [16,17] have carried out such simulations to investigate the turbulent flow structure and vortex dynamics. They have confirmed that the metallurgical effect of the tundish is mainly accomplished by the flow behavior. LES simulation of the flow field in a tundish is conducted to analyze the flow structure, which determines to a certain extent the steel quality. Many intricate flow details have not been observed by customary RANS approaches.

The motion of the liquid steel is generated by jets into the tundish and continuous casting mold. The flow regime is mostly turbulent, but some turbulence attenuation can occur far from the inlet. The characteristics of the flow in a tundish include jet spreading, jet impingement on the wall, wall jets, and an important decrease of turbulence intensity in the core region of the tundish far from the jet. Compared to the numerical simulation, physical simulation for the flow field provides more details on the evolution of not only velocity but also eddies. Mathematical modeling may provide a much more detailed picture of velocity, turbulence, and temperature fields as a function of location and time. In general, the numerical simulation shows the flow trend, while the water simulation work presents flow details. The two means complement each other and provide a comprehensive understanding of the flow.

Figures 6–8 shows the flow for schemes B, C, and D. C and D do not have baffles with holes. These results also show turbulent flow behavior in the tundish. Eddies can be clearly seen by measurement. Moreover, the evolution and interaction between eddies can be observed by the measurement of a large amount of transient data.

As can be seen from the results of the flow field in the tundish, the flow is typically turbulent and consists of eddies in motion. Turbulence exhibits random and orderly characteristics. One of its basic structures includes the eddies having various scales. Statistically, a large number of random small eddies form the background flow field, and large-scale eddies structures with quasi-ordered structures are statistically significant. The large eddies are limited and affected by flow boundary conditions and flow interfaces. The large eddies contain the turbulent kinetic energy, which is associated with the fluctuating velocity components. The small eddies, which are called the micro scale of turbulence, sometimes do not show up due to the limitation of image sampling setting and image resolution in

PIVview. According to turbulence theory, the turbulent kinetic energy is being dissipated in the smaller eddies. The small eddies may also play an important role in promoting the coalescence of inclusion particles [12].

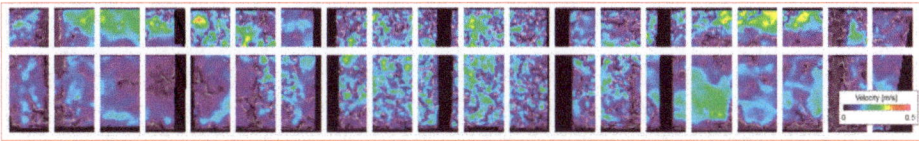

Figure 6. Flow measurement results obtained by scheme B.

Figure 7. Flow measurement results obtained by scheme C.

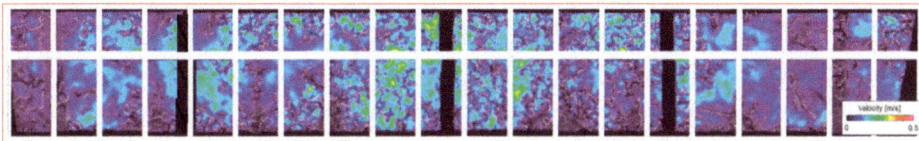

Figure 8. Flow measurement results obtained by scheme D.

The very important manifestation of turbulent flow is the presence of the eddies. The turbulent fluctuations will bring about a very effective mixing, mass (and also heat and momentum) can be readily transported by the eddies from one part of the fluid to the other [12]. The metallurgical effect of the tundish needs to be exerted via the transmission effect of these eddies.

The eddy is a flowing form, the vorticity is a physical quantity. The mathematical definition of vorticity is extremely clear. The curl of the flow field is defined as the vorticity:

$$\vec{\omega} = \nabla \times \vec{V} \tag{4}$$

where $\vec{V}$ is the velocity, m/s; $\vec{\omega}$ is the vorticity, 1/s; ∇ is the nabla operator. From the definition of vorticity, vorticity is caused by the velocity gradient in the flow field. The positions at which the velocity gradients are large exhibit high strain rates according to Newton's law of viscosity, the flow's viscosity depends on the strain rate. A causal relationship clearly exists between the vorticity and viscosity of the fluid. For homogeneous incompressible fluids, the vorticity is generated from the fluid-solid interface. Besides moving with the flow, the vorticity diffuses like heat. At the inlet section of the tundish, the impacting jet on the turbulence suppressor generates a large number of eddies. Owing to the fluid viscosity and flow control device, a large number of eddies are generated. For preventing the entrapped slag, the molten steel is required to flow out of the tundish outlet in a stable manner. With respect to the aspects of satisfying the requirements of process and the improvement on the metallurgical effect, the application of either a flow control device or a plug rod or increase in the distance between the inlet and outlet can avoid the interference of the inlet's strong vortex flow on the outlet.

For a casting speed of 1.2 m·s^{-1} the theoretical velocity in the outlet section can be calculated to be 0.001 m·s^{-1} according to the flow rate and sectional area of the tundish. At such low relative average velocity, it is difficult to produce relatively intense heat, mass, and momentum transfer. However, owing to the strong impacting jet action occurring near the long nozzle and the effect of the flow control device, the turbulence occurring in the tundish is clear, and a large number of eddies exist.

By the comparison between the presence and absence of the baffles with holes, the flow control device exerts a fundamental change in the flow within the tundish. In the absence of the flow control device, the turbulent areas concentrate near the long nozzle. According to the analysis related to the reason for the generation of the flow eddies in the tundish, owing to the presence of the flow control device inside the tundish and the number of fluid–solid interfaces increasing, the vorticity generating positions increase, intensifying the momentum exchange of the internal flow and making the overall flow easily uniform.

In the presence of the baffles with holes, the flow is in accordance with the requirement of the designed flow control device, especially in the flow direction and flow velocity. After flow passes through the inlet section, the flow is "thrown up" through the 15° holes to the outlet section surface and flows out from the outlet after sufficient exchange with the surface. Simultaneously, a large circulation is formed in the outlet section. The flow control devices, properly installed, may create localized mixing in contained regions, which may help in inclusion agglomeration and hence, their removal [18].

By doubling the casting speed, the shape of the flow does not significantly change. The overall velocity of the flow field and the turbulent intensity increase. In the presence of the flow control device, the surface velocity in water simulation can reach 0.3 m·s^{-1}. After similarity conversion, the velocity of molten steel reaches 0.37 m·s^{-1}. Such high speed may destroy the slag layer in the industrial production. Higher surface velocity could cause tundish slag entrainment at the slag/metal boundary due to turbulence arising and also cause refractory erosion near the nozzle or stopper, which requires the use of more expensive refractory practices.

By the millimeter-level spatial resolution of the tundish flow and the measurement and display at the millisecond-level time dimension scale, it is possible to achieve the precise understanding of the flow in the tundish. After combining the reality of the eddy structure and the evolution of eddies, an extremely comprehensive and precise understanding on the flow within the tundish can be achieved. These results provide a good precondition for examining the metallurgical effects of other tundishes on the basis of understanding the flow.

5. Conclusions

(1) Based on camera array scanning, reduced-scale two-strand water tundish flow fields are obtained at different casting speeds using different flow control devices. Flow measurement results show richer flow-field details than RANS simulation.

(2) There is a typical turbulent flow as well as vortex motions simultaneously in the tundish. The flow-control device and the boundary have an effect on the generation and dissipation of eddies, which has an important influence on the metallurgical effect of the tundish.

(3) Baffles with 15° holes can cause an upward-directed flow in the outlet section and form a large circulation. Eddies are generated and the flow in the tundish tends to be more uniform. As the casting speed is doubled, the overall velocity of the flow field and the turbulent intensity increase, resulting in a molten steel surface velocity of up to 0.37 m/s.

Author Contributions: Conceptualization, J.H. and B.W.; methodology, J.H.; formal analysis, Z.Y.; investigation, S.S. and J.H.; resources, J.H.; data curation, S.S. and C.L.; writing—original draft preparation, B.W.; writing—review and editing, J.H.; visualization, C.L. and Z.Y.; supervision, Z.Y.; project administration, J.H.; funding acquisition, J.H. and B.W.

Funding: The work was supported by the China National Heavy Machinery Research Institute Co., Ltd. and Natural Science Foundation of Inner Mongolia under Grant 2017MSLH0534.

Acknowledgments: The author would like to thanks Baotou Lianfang High Tech Co., Ltd. for their technical support.

Conflicts of Interest: The authors declare no conflict of interest.

Metals **2019**, *9*, 239

References

1. Mazumdar, D. Tundish metallurgy: Towards increased productivity and clean steel. *Trans. Indian Inst. Met.* **2013**, *66*, 597–610. [CrossRef]
2. Sahai, Y. Tundish technology for casting clean steel: A review. *Metall. Mater. Trans. B-Process Metall. Mater. Process. Sci.* **2016**, *47*, 2095–2106. [CrossRef]
3. Sahai, Y.; Emi, T. *Tundish Technology for Clean Steel Production*; World Scientific: Hackensack, NJ, USA, 2008; pp. 1–14.
4. Ramos-Banderas, A.; Sanchez-Perez, R.; Morales, R.D.; Palafox-Ramos, J.; Demedices-Garcia, L.; Diaz-Cruz, M. Mathematical simulation and physical modeling of unsteady fluid flows in a water model of a slab mold. *Metall. Mater. Trans. B-Process Metall. Mater. Process. Sci.* **2004**, *35*, 449–460. [CrossRef]
5. Sahai, Y. Advances in tundish technology for quality improvements of cast steel. *J. Iron Steel Res. Int.* **2008**, *15*, 643–652.
6. Braun, A.; Warzecha, M.; Pfeifer, H. Numerical and physical modeling of steel flow in a two-strand tundish for different casting conditions. *Metall. Mater. Trans. B-Process Metall. Mater. Process. Sci.* **2010**, *41*, 549–559. [CrossRef]
7. Odenthal, H.J.; Bolling, R.; Pfeifer, H. Numerical and physical simulation of tundish fluid flow phenomena. *Steel Res. Int.* **2003**, *74*, 44–55. [CrossRef]
8. Giurgea, C.; Bode, F.; Nascutiu, L.; Dudescu, C. Considerations regarding the optically transparent rigid model for piv investigations. A case study. Part 2: Notes on the failure of the model. *Energy Procedia* **2016**, *85*, 235–243. [CrossRef]
9. Cwudziński, A. Piv method and numerical computation for prediction of liquid steel flow structure in tundish. *Arch. Metall. Mater.* **2015**, *60*, 11–17. [CrossRef]
10. Merder, T. Influence of design parameters of tundish and technological parameters of steel continuous casting on the hydrodynamics of the liquid steel flow. *Metalurgija* **2014**, *53*, 443–446.
11. Cwudzinski, A. Numerical and physical modeling of liquid steel active flow in tundish with subflux turbulence controller and dam. *Steel Res. Int.* **2014**, *85*, 902–917. [CrossRef]
12. Szekely, J.; Ilegbusi, O.J. *The Physical and Mathematical Modeling of Tundish Operations*; Springer: New York, NY, USA, 1989; pp. 1–52.
13. Huang, J.; Zhang, Y.; Zhang, Y.; Zhang, Y.; Ye, X.; Wang, B. Study of flow characteristics of tundish based on digital image velocimetry technique. *Metall. Mater. Trans. B* **2016**, *47*, 3144–3157. [CrossRef]
14. Alkishriwi, N.; Meinke, M.; Schroder, W. A large-eddy simulation method for low Mach number flows using preconditioning and multigrid. *Comput. Fluids* **2006**, *35*, 1126–1136. [CrossRef]
15. Alkishriwi, N.; Meinke, M.; Schroder, W.; Braun, A.; Pfeifer, H. Large-eddy simulations and particle-image velocimetry measurements of tundish flow. *Steel Res. Int.* **2006**, *77*, 565–575. [CrossRef]
16. Jha, P.K.; Ranjan, R.; Mondal, S.S.; Dash, S.K. Mixing in a tundish and a choice of turbulence model for its prediction. *Int. J. Numer. Methods Heat Fluid Flow* **2003**, *13*, 964–996. [CrossRef]
17. Jha, P.K.; Rao, P.S.; Dewan, A. Effect of height and position of dams on inclusion removal in a six strand tundish. *ISIJ Int.* **2008**, *48*, 154–160. [CrossRef]
18. Warzecha, M.; Merder, T.; Warzecha, P. Investigation of the flow structure in the tundish with the use of RANS and LES methods. *Arch. Metall. Mater.* **2015**, *60*, 215–220. [CrossRef]

metals

Article

Research and Application of a Rolling Gap Prediction Model in Continuous Casting

Zhufeng Lei * and Wenbin Su

School of Mechanical Engineering, Xi'an Jiaotong University, 28 West Xianning Road, Xi'an 710049, China;
wbsu@mail.xjtu.edu.cn
* Correspondence: leizhufeng@stu.xjtu.edu.cn; Tel.: +86-029-8266-5304

Received: 6 March 2019; Accepted: 23 March 2019; Published: 25 March 2019

Abstract: Control of the roll gap of the caster segment is one of the key parameters for ensuring the quality of a slab in continuous casting. In order to improve the precision and timeliness of the roll gap value control, we proposed a rolling gap value prediction (RGVP) method based on the continuous casting process parameters. The process parameters collected from the continuous casting production site were first dimension-reduced using principal component analysis (PCA); 15 process parameters were chosen for reduction. Second, a support vector machine (SVM) model using particle swarm optimization (PSO) was proposed to optimize the parameters and perform roll gap prediction. The experimental results and practical application of the models has indicated that the method proposed in this paper provides a new approach for the prediction of roll gap value.

Keywords: multi-source information fusion; data stream; continuous casting; roll gap value; prediction; global optimization; support vector regression

---

1. Introduction

Motivation

High quality continuous casting technology has become the most internationally competitive core technology in the modern steel industry [1–4]. Due to the complexity of the continuous casting process, there are many factors that can affect the quality of continuous casting. Among them, the roll gap of the caster segment is a key parameter. Calculation of the roll gap remains an important problem in continuous casting production. Establishing a dynamic adaptive predictive model for the caster segment in continuous casting, and real-time adaptive adjustment of the roll gap according to actual working conditions, is theoretically and practically valuable for improving the quality of the slab.

The continuous casting process involves several steps as shown in Figure 1. First, the molten steel enters the mold from the tundish and a certain thickness of the shell solidifies. The slab, with a shell of a certain thickness, then enters the caster segment from the mold. Secondary cooling is then performed until the slab is completely solidified. Next, soft reduction of the slab is performed in the caster segment by adjusting the distance between the upper and lower rollers; this adjusts the internal crystal arrangement of the slab and improves the internal quality of the slab.

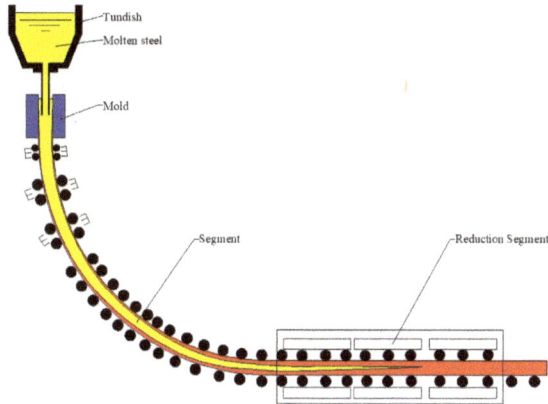

Figure 1. Continuous casting process.

There are many published studies on the prediction of continuous casting process parameters. Lait [5] developed a one-dimensional finite-difference model to calculate the temperature field and pool profile of continuously cast steel. The result obtained was reasonable for low-carbon billets over most of the mold region. Rappaz [6] described how the microscopic models of microstructure formation could be coupled to macroscopic heat flow calculations in order to predict microstructural features at the scale of the whole process. Choudhary et al. [7] developed a steady-state three-dimensional heat flow model based on the concept of artificial effective thermal conductivity. The model can be applied to various geometrical shapes of relevance to continuous casting of steel. Koric et al. [8] created an accurate multi-physics model of metal solidification at the continuum level; this model comprised of separate three-dimensional models for the thermomechanical behavior of the solidifying shell, turbulent fluid flow in the liquid pool, and thermal distortion of the mold. The model was applied to simulate continuous casting of steel. Numerical modeling is still the primary design tool used for continuous casting studies.

Artificial intelligence (AI) is a branch of the computer science discipline. AI is considered one of three cutting-edge technologies since the 1970s (those being, space technology, energy technology, and artificial intelligence). It is also considered to be one of three cutting-edge technologies of the 21st century (those being, genetic engineering, nanoscience, and artificial intelligence). AI has developed rapidly over the past three decades, has been widely used in many subject areas, and has achieved important results. AI has gradually become an independent branch of study [9]. In recent years, AI methods have been widely used in the field of intelligent manufacturing. Yang et al. [10] proposed a framework and several general guidelines for implementing big data analytics in a high-performance computing environment. AI methods have also been introduced to the field of continuous casting. Hore et al. [11] developed a model based on adaptive neural network formalism coupled with a fuzzy inference system to predict the mechanical properties of hot-rolled TRIP steel. The present model provides a predictive platform for possible application of these AI-based tools for automation, real-time process control, and operator guidance in plant operation. Liu and Gao [12] established a method for online prediction of the silicon content in blast furnace ironmaking processes. The superiority of the proposed method was demonstrated and compared with other soft sensors in terms of online prediction of the silicon content in an industrial blast furnace in China. Mahmoodkhani et al. [13] considered the friction coefficient as an input parameter in the neural network; it was optimized using an iterative method employing an equation that related the friction coefficient to the rolling force in order to rapidly predict the roll force during skin pass rolling of 980DP and 1180CP high strength steels. Tiensuu et al. [14] used statistical models to improve the dimensional accuracy of a steel plate by updating the selection of parameters for slab design. Zhang et al. [15] developed a model to predict the critical point of interfacial instability of liquid-liquid stratified flow based on the Kelvin-Helmholtz instability. The results of the water model indicated that the prediction model was correct.

Particle swarm optimization (PSO), described by Eberhart and Kennedy in 1995, is a stochastic optimization technique based on population [16]. The particle swarm algorithm mimics the clustering behavior of insects, herds, flocks, and fish groups. These groups search for food in a cooperative way. Each member of the group changes its search mode by learning from its own experience and the experience of other members of the group. Valvano et al. [17] introduced a novel decline PSO procedure. This method was used to select the optimal parameters for sound control. Similarly, PSO was used to select the optimal parameters of the support vector machine (SVM) kernel function in this paper.

The remainder of this paper is organized as follows: The data source is introduced in Section 2. The process parameters collected at the continuous casting production site were dimension reduced by PCA, as described in Section 3. The PSO-SVM is introduced in Section 4. In Section 5, the proposed model is applied to a dataset and the results are analyzed. In Section 6, the model is applied to industrial production and compared with products produced without the model. Finally, a conclusion is presented in Section 7.

2. Continuous Casting Process Parameters

2.1. Acquisition of Continuous Casting Process Parameters

In this paper, a Chinese steel company was selected as the research object; data on the continuous casting production line was collected online from this site. The continuous casting machine under study had more than 6000 distributed sensors to record most of the process state, including the casting speed, the amount of water in each cold zone, the type of steel, and the casting temperature. Figure 2 depicts the topological structure of the continuous casting production data acquisition system. The main core equipment of the continuous casting production data acquisition system included a basic automation level Programmable Logic Controller (PLC), a man-machine interface server, a monitoring operation station, a process control-level computer database, a model application server, and a terminal client. The system collected data through the PLC controller and monitoring operation station; the data then entered the computer database. Data were analyzed and the model was updated and corrected in the model application server. Data was displayed to the operator through the terminal client; process adjustment and optimization occurred through the PLC.

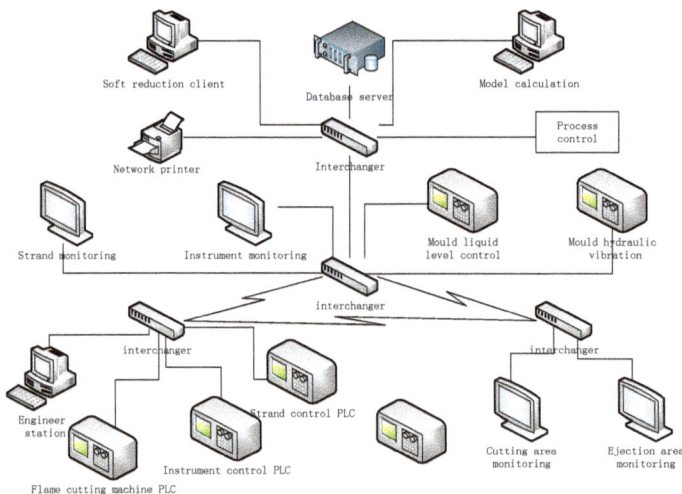

Figure 2. Topological structure of the continuous casting production data acquisition system.

In this paper, the process parameter data from the continuous casting production line were obtained with the help of field engineers; the data dimension was 6153. The process parameters included the tundish, ladle, mold, and caster segment, which were closely correlated with the roll gap value obtained; the closely related data dimension was 1020. These data were correlated with the product quality. The collection time was 1 h, the sensor recorded every 0.02 s, and the data volume was 180,000.

A correlation analysis of these process parameters was performed; the parameters that were more relevant were classified as a category. The coefficient of the association between two parameters could be obtained by:

$$r = \frac{\sum\limits_{i=1}^{n} (x_i - \overline{x})(y_i - \overline{y})}{\sqrt{\sum\limits_{i=1}^{n} (x_i - \overline{x})^2 \cdot \sum\limits_{i=1}^{n} (y_i - \overline{y})^2}} \tag{1}$$

where r is a correlation coefficient reflecting the relationship between two variables and the related direction of this relationship, x_i is a data point in data set X, $\overline{x}$ is the mean of X, y_i is a data point in data set Y, and $\overline{y}$ is the mean of Y.

The analyses revealed that the correlation coefficients between each process parameter were more than 0.5; many correlation coefficients were even greater than 0.8. As a result, 15 types of continuous casting process parameters were chosen, as listed in Table 1.

Table 1. Process parameter names and abbreviations.

Name	Process Parameter	Name	Process Parameter
GSWD	Temperature of tundish molten steel	KMLL	Water flow of mold width surface
GRD	Overheating of molten steel	KMWD	Outlet temperature of mold width surface
LS	Pulling rate	ZMYL	Water pressure of mold narrow surface
ZDPL	Vibration frequency of mold	ZMLL	Water flow of mold narrow surface
ZDFZ	Vibration amplitude of mold	ZMWD	Outlet temperature of mold narrow surface
CDGYL	Average pressure of 18 drive rollers	ELSLL	Average flow of 18 second cold water loops
JJQYW	Mold liquid level	ELSYL	Average pressure of 18 second cold water loops
KMYL	Water pressure of mold width surface	-	-

2.2. Data Pre-Processing

The Pauta criterion was used to detect outliers in large monitoring data sets [18]. In this paper, the Pauta criterion was used to detect outliers in the data. The detected outliers were changed to values nearby so as not to damage the sequence of the data.

Assuming that all data were measured with the same precision and in order to obtain $x_1, x_2, \ldots, x_n$, the arithmetic mean x and the residual error $v_i = x_i - x$ ($i = 1, 2, \ldots, n$) were calculated. In addition, the standard error σ was calculated using the Bessel formula. If a residual error v_b of a measured value x_b satisfied Equation (2):

$$|v_b| = |x_b - x| > 3\sigma \tag{2}$$

then x_b was an outlier, which contained a larger error, and thus, this value was replaced with the next value adjacent to it.

3. Dimension Reduction of Streaming Data from Multi-Source Information

3.1. Standardization of Continuous Casting Process Parameters

The continuous casting process is a complex continuous phase transition process. There are many links which affect the quality of the casting billet and there are many collected production process parameters. Therefore, the analysis of the data is particularly important. Analysis revealed that each process parameter could be a dimension of the data samples. The predicted roll gap of data was seen as a label, a group of N-dimensional process parameters was seen as an input vector x_i, $x_i \in \mathfrak{R}^{N \times 1}$, and another adjacent group of process parameters was seen as the second set of the input vector x_{i+1}.

The unit and magnitude of the input parameters were different. Thus, the data were standardized in order to produce data in the same range for analysis under the condition of mutual equality. A linear transformation of raw data was performed to standardize the data; this mapped the result to [0, 1], according to:

$$x_{im*} = \frac{x_{im} - x_{min}}{x_{max} - x_{min}} \tag{3}$$

where x_{im} was the m-dimensional process parameters of any group of input vector, x_{im*} was the standardization of x_{im}, x_{max} was the maximum of the sample data, and x_{min} was the minimum of the sample data.

3.2. Dimension Reduction of Continuous Casting Process Parameters

In this paper, in order to obtain a highly responsive roll gap value prediction model for the continuous casting caster segment, dimension reduction of the continuous casting process parameters was considered; this would reduce the operation time and the complexity of the algorithm.

The principal component contribution rate method was used to determine the number of parameters [19]. The variance contribution ratio and summation variance contribution ratio of each parameter is shown in Table 2.

Table 2. Variance contribution ratio of each parameter.

NO.	Feature	Variance Contribution Ratio (%)	Summation Variance Contribution Ratio (%)
1	4.79	36.97	36.97
2	3.43	32.85	69.82
3	1.76	14.74	84.56
4	1.24	5.25	89.81
5	1.01	4.74	93.55
6	0.903	3.02	96.57
7	0.605	1.92	98.49
8	0.455	0.43	98.92
9	0.350	0.32	99.24
10	0.203	0.21	99.45
11	0.113	0.15	99.6
12	0.058	0.12	99.72
13	0.044	0.1	99.82
14	0.030	0.09	99.91
15	0.008	0.09	100.00

Table 2 shows that the cumulative variance contribution ratio of the first six parameters was greater than 95%. Thus, the raw data could be fully captured with six parameters.

Figure 3 shows the classification results of principal component analysis (PCA), each color represents a principal component, there are six colors in Figure 3, representing six principal components. PCA demonstrated the effect of the dimensionality specification. The PCA results demonstrated that the data was reduced from 15 dimensions to 6.

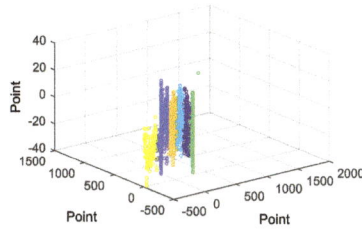

Figure 3. Dimensionality reduction results of Principal Component Analysis (PCA).

4. Establishing a Roll Gap Value Prediction Model from Multi-Source Information

4.1. PSO-SVM Model

SVM can not only solve the classification problem, but can also solve the regression problem; the basic model is the largest linear classifier defined in the feature space. SVM aims to achieve a distinction between samples by constructing a hyperplane for classification so that the sorting interval between the samples is maximized and the sample to the hyperplane distance is minimized.

Set a training data set for a feature space $D = \{(x_1, y_1), (x_2, y_2), \dots, (x_m, y_m)\}$, $x_i \in \chi = \Re^n$, $y_i \in y = \{+1, -1\}$, $i = 1, 2, \dots, N$, where x_i is the i-th feature vector, y_i is the class tag of x_i.

The corresponding equation of the classification hyperplane was

$$h(x) = w \cdot x + b \tag{4}$$

where x was the input vector, w was the weight, and b was the offset.

The classification decision function was

$$\text{Sign}\,(h(x)) \tag{5}$$

$$\begin{cases} h(x) > 0, y_i = 1 \\ h(x) < 0, y_i = -1 \end{cases} \tag{6}$$

The support vector machine was implemented to find the w and b when the interval between the separation hyperplane and the nearest sample point was maximized. When the training set was linearly separable, the sample points belonging to different classes could be separated by one or several straight lines with the largest interval. The maximum interval was solved by the following formula:

$$\max \gamma_i = y_i \left(\frac{w}{\|w\|} \cdot x_i + \frac{b}{\|w\|} \right) \tag{7}$$

$$s.t. y_i \left(\frac{w}{\|w\|} \cdot x_i + \frac{b}{\|w\|} \right) \geq \gamma, i = 1, 2, \dots, N \tag{8}$$

where γ is the geometric interval.

Thus, we could obtain the linear separable support vector machine optimization problem.

$$\min_{w,b} \frac{1}{2} \|w\|^2 \tag{9}$$

$$s.t. y_i (w \cdot x_i + b) - 1 \geq 0, i = 1, 2, \dots, N \tag{10}$$

In the actual data set, there were many specific points, making the data set linearly inseparable; in order to solve this problem, we introduced a slack variable for each sample point $\xi_i \geq 0$, so that

$$y_i (w \cdot x_i + b) \geq 1 - \xi_i \tag{11}$$

for each slack variable ξ_i, pay a price ξ_i, and the optimization problem becomes

$$\frac{1}{2}\|w\|^2 + C\sum_{i=1}^{N}\xi_i \qquad (12)$$

where $C > 0$ is the penalty factor.

Most of the data in the actual data were linearly inseparable. Therefore, these data could be mapped to a high-dimensional feature space through non-linear mapping, letting the non-linear problem be transformed into a linear problem. The linear indivisible problem was transformed into a linear separable problem.

Introduce kernel functions:

$$K(x_i, x_j) = \varphi(x_i) \cdot \varphi(x_j) \qquad (13)$$

where the value of the kernel equaled the inner product of two vectors, x_i and x_j.

At this point, we obtained

$$W(\alpha) = \frac{1}{2}\sum_{i=1}^{N}\sum_{j=1}^{N}\alpha_i\alpha_j y_i y_j K(x_i, x_j) - \sum_{i=1}^{N}\alpha_i \qquad (14)$$

where $\alpha_i \geq 0$, $i = 1, 2, \ldots, N$ was the Lagrangian multiplier and N was the number of samples.

In this paper, the radial basis function (RBF) was chosen as the SVR kernel function, and the expression was

$$K(x_i, x) = \exp(\frac{-\|x_i - x\|^2}{2g^2}) \qquad (15)$$

where g was the kernel function coefficient.

At this point, the classification function became

$$f(x) = \text{sign}[\sum_{i=1}^{N}\alpha_i y_i \exp(\frac{-\|x_i - x\|^2}{2g^2}) + b] \qquad (16)$$

In the SVM model, training data on the cost function and the constraint condition were known. Only the penalty factor C and the kernel function parameter g could be adjusted. When the input sample points are wrongly divided, the impact of this error can be adjusted by C; this highlights the important effect of the misclassification of sample points. The kernel function parameter g represents the kernel function parameter γ, k in the g represents the number of attributes in the input data. Thus, the hit rate of the roll gap value prediction model is governed by these two parameters [20–29].

PSO, also referred to as the particle swarm optimization algorithm or bird flock foraging algorithm, is a type of evolutionary algorithm (EA). Starting from the random solution, the optimal solution is found through iteration, and the quality of the solution is evaluated through fitness. However, the PSO algorithm rule is simpler. It searches for the global optimum by following the current optimal value. This algorithm is easy to implement, has high-precision and fast-convergence, and is suitable for solving practical problems.

PSO was used to optimize the penalty factor parameter C and kernel function parameter g in the SVM. Cross-validation of the prediction results were performed to obtain the optimal C and g so as to optimize the SVM prediction results.

Assuming that C and g are in a D-dimension target-searching space, a group was composed of m particles. The position of the i-th particle represented vector $c_i = (c_{i1}, c_{i2}, \ldots, c_{iD})$, $i = 1, 2, \ldots, m$, whose speed was also a D-dimension vector, $g_i = (g_{i1}, g_{i2}, \ldots, g_{iD})$. The optimal location that the i-th particle had searched so far was $P_i = (P_{i1}, P_{i2}, \ldots, P_{iD})$. The optimal position in which the whole particle swarm had searched was $P_k = (P_{k1}, P_{k2}, \ldots, P_{kD})$. The particle updating equation was as follows:

$$g_{id}(t+1) = g_{id}(t) + h_1 r_1 (P_{id} - c_{id}(t)) + h_2 r_2 (P_{kd} - c_{id}(t)) \tag{17}$$

$$c_{id}(t+1) = c_{id}(t) + g_{id}(t+1) \tag{18}$$

while $g_{id} > G_{max}$, $g_{id} = G_{max}$;
while $g_{id} < -G_{max}$, $g_{id} = -G_{max}$.

In the formula, $i = 1, 2, \ldots, m$; $d = 1, 2, \ldots, D$. h_1 and h_2 were non-negative constants, r_1 and r_2 were uniform distribution random numbers within the range of [0, 1], $c_{id}(t)$ was the current position of the i-th particle, t is the current moment, P_{id} was the optimal location that the i-th particle had searched so far, and g_{id} was the current speed of the i-th particle. $g_{id} \in [-G_{max}, G_{max}]$, G_{max}, the maximum limit speed, was a negative number.

4.2. Establishing the PSO-Roll Gap Value Prediction

This section reports the process for establishing the PSO-SVM model. The SVM algorithm was used to establish the roll gap value prediction (RGVP) model of streaming data from multi-source information, as shown in Figures 4 and 5. Six types of continuous casting process parameters comprised the input data, $k(x_1, x)$; $k(x_2, x)$; $\ldots$; $k(x_N, x)$ were the kernel functions of the SVM, and the roll gap value was the output data.

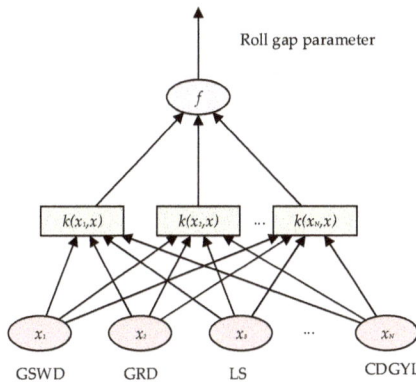

Figure 4. Roll gap value prediction model of streaming data from multi-source information. f is the output of the model.

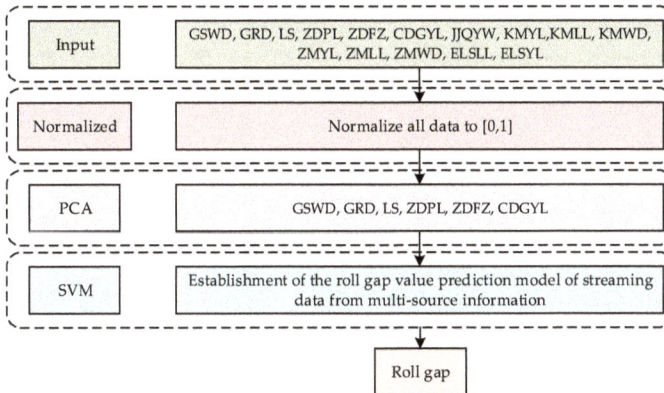

Figure 5. Common computational architecture of roll gap value prediction model.

In order to enhance the accuracy of the roll gap value prediction model, PSO was used to optimize the parameters of the PSO-RGVP.

The specific steps of the PSO-RGVP were as follows:

Step 1: Pre-processing of process parameter data to obtain 15 types of process parameters.

Step 2: Standardization of process parameter data and feature reduction to obtain six types of parameters.

Step 3: Determination of the scope of C and g by PSO.

Step 4: Testing of model parameters using the method of cross validation to obtain the optimal C and g.

Step 5: Establishment of the roll gap value prediction model of streaming data from multi-source information.

Step6: Prediction of the results.

5. Experiments and Results

This section presents the results of the PSO-RGVP model and compares these results to the traditional numerical heat transfer metallurgical model; MATLAB was used to perform the experiment. This section is divided into two parts. The first was the establishment and training of the model using historical data; the second was the prediction of the test sample. Two thousand sets of process parameter data from continuous casting were collected to establish and train the model. In addition, another 500 sets of data were collected to carry out the forecast test. The experiment results are shown in Table 3.

Table 3. Results of establishment and training of the model experiment.

Parameter Optimization Results	Training Time	Mean Square Deviation
$C = 0.1; g = 0.1$	304.8 s	97.5%

Figure 6 presents the parameter optimization results of the PSO-RGVP. When the termination generation was 100 and the population number was 20 in the PSO-RGVP, it could be concluded that the optimal penalty factor $C = 0.1$ and $g = 0.1$.

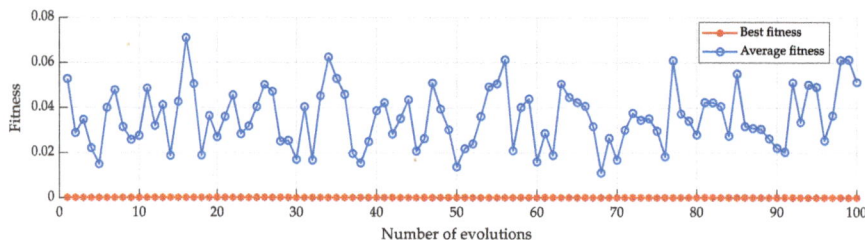

Figure 6. Parameter optimization results of PSO.

After predicting 500 sets of process parameter data that were collected with the PSO-RGVP, the predicted roll gap value was compared with the actual roll gap value, as shown in Figure 7. Relative error of prediction of the PSO-RGVP is shown in Figure 8.

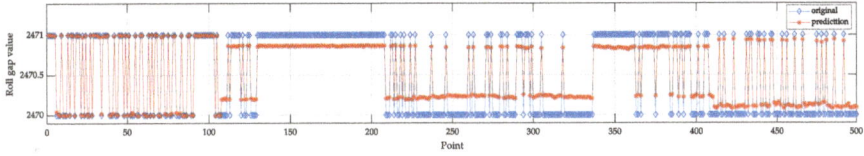

Figure 7. Prediction results of PSO-RGVP.

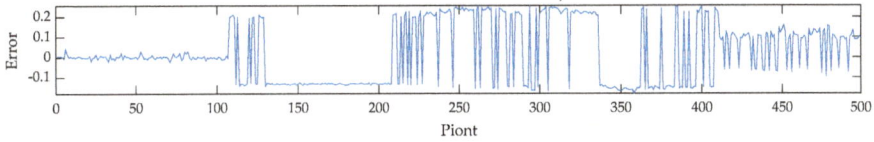

Figure 8. Relative error of prediction of the PSO-RGVP.

The PSO-RGVP is a new way to predict the value of roll gap. The maximum relative error between the predictive value and the actual value of the method proposed in this paper was 2.5%, and the prediction accuracy was 97.5%.

6. Industrial Application

This section reports the actual application of the PSO-RGVP. Modular design was used in the system; the modular structure is shown below in Figure 9.

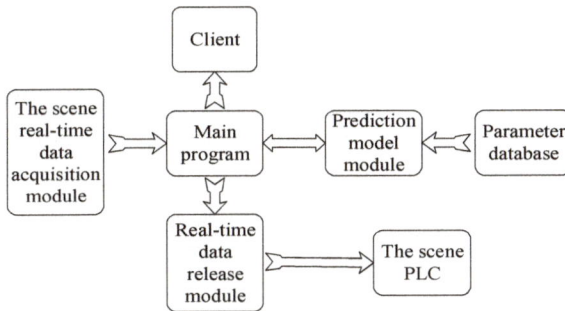

Figure 9. System modular structure diagram.

The test system had two main modules: The prediction model module and the main application program. Data interaction in the prediction model was achieved with the main application program; the program was structured and clearly defined in Figure 9.

A brief operational flow chart of the test system is shown in the following diagram (Figure 10).

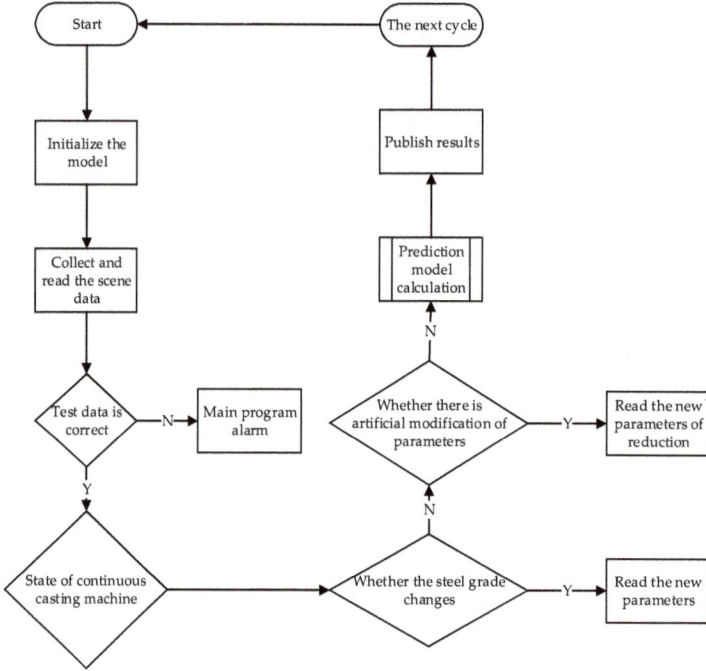

Figure 10. Test system calculation process.

As shown in Figure 10, after initializing the model, the system first acquired the real-time process parameters and filtered for correctness of the process parameters. Through the identification of real-time data, the system obtained the status of the continuous casting machine at the time and decided whether to carry out the roll gap adjustment strategy. Through the identification of the type of steel, the system read the total rolling reduction, the rolling interval, the rolling distribution, etc. The prediction model was used to obtain the roll gap prediction value, which was then released.

The prediction data was imported into the test system for trial production, then casting was performed and the quality of the slab was observed.

In order to verify the application of the model, the center segregation and center porosity of the slab, before and after use of the model, were analyzed by macroscopic examination. In particular, the continuous casting machine was compared before and after use of the model for the production of steel for Q235; the section size was 230 × 1350, the cast speed was 1.30 m/min, and the total reduction was 4 mm. The results of macroscopic examination were determined according to Chinese metallurgical standards. The results are reported in Table 4 and in Figures 11 and 12.

Table 4. Comparison of quality rating.

Serial Number	Section Size (mm × mm)	Centre Segregation	Centre Porosity	Intermediate Cracks	Triangle Area Cracks
Without proposed model	230 × 1350	B1.0	2.0	1.5	1.5
Proposed model	230 × 1350	C0.5	1.0	1.0	0.5

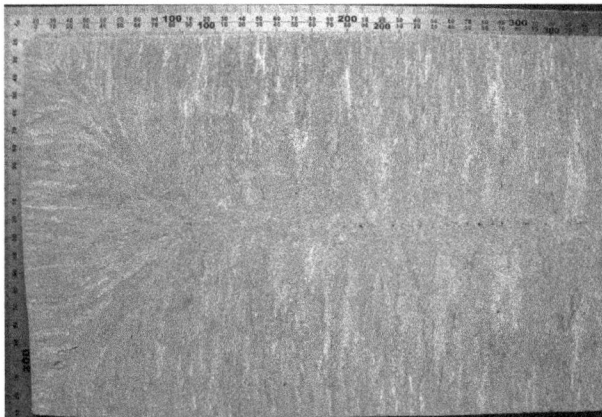

Figure 11. Macrosegregation examination of results of the proposed model.

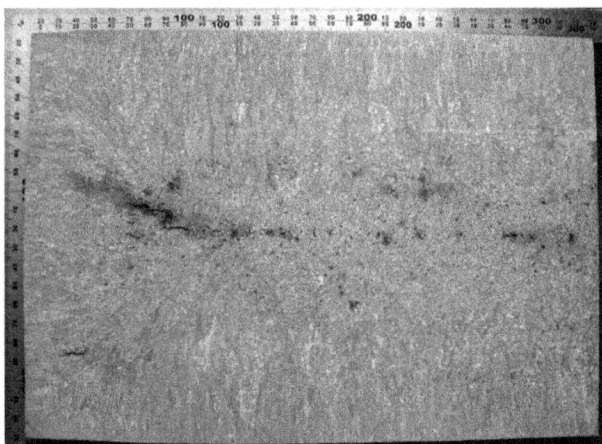

Figure 12. Macrosegregation examination of the results of the traditional method.

Figures 11 and 12 show the macrosegregation examination of the two methods: The proposed model and the traditional method. The macrosegregation sample was cut from the continuous casting slab. After polishing and pickling the surface, it was photographed under a low-power microscope. This was the main approach used to check the internal quality of a slab. It was obvious that under the same production conditions, the slab produced under the proposed model only had very few intermediate cracks; the quality rating of center segregation was C0.5; center porosity was 1.0; intermediate cracks was 1.0; triangle area cracks was 0.5. According to the quality of the intermediate cracks produced by the proposed model, the center segregation and center porosity in the slab were greatly reduced. After using the proposed model, the center segregation and center porosity in the slab were greatly reduced. In addition, the proposed model greatly reduced the labor intensity and maintenance time and improved the maintenance efficiency for production.

The results indicate that the roll gap prediction model proposed in this paper has a short computational time, can accurately predict the roll gap, and allows for real-time prediction and adjustment during production. The findings from the industrial application of this model demonstrate its accuracy. Thus, the roll gap value prediction efficiency is greatly increased in the proposed model. In this study, 2000 sets of data were collected to establish and train the model. In addition, another 500 sets of data were collected to carry out the forecast test. The prediction accuracy was 97.5%. This showed that the PSO-RGVP was superior for prediction of the roll gap value.

7. Conclusions

In this paper, a new prediction approach, PSO-RGVP, was proposed. Multi-source information process parameters from the continuous casting process were excavated and analyzed, and PSO was used for global optimization of an adaptive prediction model for the roll gap value of the caster segment; SVM was used for the prediction of the roll gap. This method takes into account the mutual influence and restriction of the multi-source information process parameters in the actual production process; the model parameters were obtained quickly and accurately, and adaptive prediction of the roll gap value was achieved. The experimental results confirmed the efficiency of the PSO-RGVP, because actual data was used. PSO-RGVP provides a new approach for the prediction of the roll gap value.

Author Contributions: W.S. conceived and designed the research, Z.L. performed the experiment and wrote the manuscript.

Funding: This work was financially supported by the National Natural Science Foundation of China (NO. 51575429).

Acknowledgments: Q.G., X.L., H.Z., B.H., and Y.Z. are acknowledged for their valuable technical support.

Conflicts of Interest: The authors declare no conflict of interest.

References

1. Ataka, M. Rolling technology and theory for the last 100 years: The contribution of theory to innovation in strip rolling technology. *ISIJ Inter.* **2015**, *55*, 89–102. [CrossRef]
2. Ge, S.; Isac, M.; Guthrie, R.I.L. Progress of strip casting technology for steel; historical developments. *ISIJ Inter.* **2012**, *52*, 2109–2122. [CrossRef]
3. Tacke, K.H.; Schwinn, V. Recent developments on heavy plate steels. *Stahl Eisen* **2005**, *125*, 55.
4. Wolf, M.M. History of Continuous Casting. In Proceedings of the 75th Steelmaking Conference, Toronto, ON, Canada, 5–8 April 1992; pp. 47–101.
5. Lait, J. Mathematical modelling of heat flow in the continuous casting of steel. *Ironmak. Steelmak.* **1974**, *1*, 90–97.
6. Rappaz, M. Modeling of microstructure formation in solidification processes. *Int. Mater. Rev.* **1989**, *34*, 93–123. [CrossRef]
7. Choudhary, S.K.; Ganguly, S. Morphology and segregation in continuously cast high carbon steel billets. *ISIJ Inter.* **2007**, *47*, 1759–1766. [CrossRef]
8. Koric, S.; Hibbeler, L.C.; Liu, R.; Thomas, B.G. Multiphysics model of metal solidification on the continuum level. *Numer. Heat Transfer, Part B-Fundam.* **2010**, *58*, 371–392. [CrossRef]
9. Nilsson, N.J. *Principles of Artificial Intelligence*; Morgan Kaufmann Publishers: Burlington, MA, USA, 2014.
10. Yang, Y.; Cai, Y.D.; Lu, Q.; Zhang, Y.; Koric, S.; Shao, C. High-Performance Computing Based Big Data Analytics for Smart Manufacturing. In Proceedings of the ASME 2018, the 13th International Manufacturing Science and Engineering Conference, College Station, TX, USA, 18–22 June 2018.
11. Hore, S.; Das, S.K.; Banerjee, S.; Mukherjee, S. An adaptive neuro-fuzzy inference system-based modelling to predict mechanical properties of hot-rolled TRIP steel. *Ironmak. Steelmak.* **2016**, *44*, 656–665. [CrossRef]
12. Liu, Y.; Gao, Z. Enhanced just-in-time modelling for online quality prediction in BF ironmaking. *Ironmak. Steelmak.* **2015**, *42*, 321–330. [CrossRef]
13. Mahmoodkhani, Y.; Wells, M.A.; Song, G. Prediction of roll force in skin pass rolling using numerical and artificial neural network methods. *Ironmak. Steelmak.* **2016**, *44*, 281–286. [CrossRef]
14. Tiensuu, H.; Tamminen, S.; Pikkuaho, A.; Röning, J. Improving the yield of steel plates by updating the slab design with statistical models. *Ironmak. Steelmak.* **2016**, *44*, 577–586. [CrossRef]
15. Zhang, L.; Li, Y.; Wang, Q.; Yan, C. Prediction model for steel/slag interfacial instability in continuous casting process. *Ironmak. Steelmak.* **2015**, *42*, 705–713. [CrossRef]

16. Eberhart, R.; Kennedy, J. A New Optimizer Using Particle Swarm Theory. In Proceedings of the MHS'95, the Sixth International Symposium on Micro Machine and Human Science, Nagoya, Japan, 4–6 October 1995; pp. 39–43. [CrossRef]

17. Valvano, S.; Orlando, C.; Alaimo, A. Design of a noise reduction passive control system based on viscoelastic multilayered plate using PDSO. *Mech. Syst. Sig. Process.* **2019**, *123*, 153–173. [CrossRef]

18. Zhang, M.; Yuan, H. The Pauta criterion and rejecting the abnormal value. *J. Zhengzhou Univ. Tech.* **1997**, *1*, 84–88.

19. Martinez, A.M.; Kak, A.C. PCA versus LDA. *IEEE Trans. Pattern Anal. Mach. Intell.* **2001**, *23*, 228–233. [CrossRef]

20. Cervantes, J.; Garcia-Lamont, F.; Rodriguez-Mazahua, L.; Lopez, A.; Ruiz-Castilla, J.; Trueba, A. PSO-based method for SVM classification on skewed data sets. *Neurocomputing* **2017**, *228*, 187–197. [CrossRef]

21. Chiang, J.H.; Hao, P.Y. A new kernel-based fuzzy clustering approach: Support vector clustering with cell growing. *IEEE Trans. Fuzzy Syst.* **2003**, *11*, 518–527. [CrossRef]

22. Fang, Y.; Hu, C.; Liu, L.; Zhang, X. Breakout prediction classifier for continuous casting based on active learning GA-SVM. *China Mech. Eng.* **2016**, *27*, 1609–1614. [CrossRef]

23. Gaudioso, M.; Gorgone, E.; Labbe, M.; Rodriguez-Chia, A.M. Lagrangian relaxation for SVM feature selection. *Comput. Oper. Res.* **2017**, *87*, 137–145. [CrossRef]

24. Zhang, G.Z.; Sun, J. Application of fuzzy control on the caster segment's gap control of slab continuous casting machine. *Electr. Drive* **2009**, *39*, 51–53.

25. Jain, A.K.; Duin, R.P.W.; Mao, J. Statistical pattern recognition: A review. *IEEE Trans. Pattern Anal. Mach. Intell.* **2000**, *22*, 4–37. [CrossRef]

26. Muscat, R.; Mahfouf, M.; Zughrat, A.; Yang, Y.Y.; Thornton, S.; Khondabi, A.V.; Sortanos, S. Hierarchical fuzzy support vector machine (SVM) for rail data classification. *IFAC Papersonline* **2014**, *47*, 10652–10657. [CrossRef]

27. Refan, M.H.; Dameshghi, A.; Kamarzarrin, M. Improving RTDGPS accuracy using hybrid PSOSVM prediction model. *Aerosp. Sci. Technol.* **2014**, *37*, 55–69. [CrossRef]

28. Wang, J.-S.; Chiang, J.-C. A cluster validity measure with Outlier detection for support vector clustering. *IEEE Trans. Syst. Man Cybern. Part B Cybern.* **2008**, *38*, 78–89. [CrossRef] [PubMed]

29. Wei, J.; Zhang, R.; Yu, Z.; Hu, R.; Tang, J.; Gui, C.; Yuan, Y. A BPSO-SVM algorithm based on memory renewal and enhanced mutation mechanisms for feature selection. *Appl. Soft Comput.* **2017**, *58*, 176–192. [CrossRef]

metals

MDPI

Article

A Combined Hybrid 3-D/2-D Model for Flow and Solidification Prediction during Slab Continuous Casting

Mujun Long *, Huabiao Chen, Dengfu Chen *, Sheng Yu, Bin Liang and Huamei Duan

College of Materials Science and Engineering, Chongqing University, Chongqing 400044, China; chenhuabiao@cqu.edu.cn (H.C.); jscquys@163.com (S.Y.); binliang@cqu.edu.cn (B.L.); duanhuamei@cqu.edu.cn (H.D.)
* Correspondence: longmujun@cqu.edu.cn (M.L.); chendfu@cqu.edu.cn (D.C.);
  Tel.: +86-23-6510-2467 (M.L. & D.C.)

Received: 11 December 2017; Accepted: 8 March 2018; Published: 14 March 2018

Abstract: A combined hybrid 3-D/2-D simulation model was developed to investigate the flow and solidification phenomena in turbulent flow and laminar flow regions during slab continuous casting (CC). The 3-D coupling model and 2-D slicing model were applied to the turbulent flow and laminar flow regions, respectively. In the simulation model, the uneven distribution of cooling water in the width direction of the strand was taken into account according to the nozzle collocation of secondary cooling zones. The results from the 3-D turbulent flow region show that the impact effect of the molten steel jet on the formation of a solidification shell is significant. The impact point is 457 mm below the meniscus, and the plug flow is formed 2442 mm below the meniscus. In the laminar flow region, grid independence tests indicate that the grids with a cell size of 10×10 mm^2 are sufficient in simulations to attain the precise temperature distribution and solidification profile. The liquid core of the strand is not entirely uniform, and the solidification profile agrees well with the integrated distribution of cooling water in secondary cooling zones. The final solidification points are at a position of 400–500 mm in the width direction and are 17.66 m away from the meniscus.

Keywords: slab continuous casting; hybrid simulation model; uneven secondary cooling

1. Introduction

Continuous casting (CC) technology has become the primary method of producing steel strands in the steelmaking industry. During the CC process, the molten steel is continuously fed into the water-cooled mold through a submerged entry nozzle (SEN) and a solidified shell of sufficient thickness is formed. Subsequently, the strand is pulled into the secondary cooling zones and cooled by water spray or air-mist spray in order to solidify completely. The strand quality, particularly regarding surface and inner cracks, is closely related to the turbulent flow and the heat transfer during the solidification involved in a CC process [1–5]. This is particularly true given that, in the slab CC process, the solidification profile of the slab transverse section—which closely relates to the integrated distribution of cooling water in secondary cooling zones—is not entirely uniform [6–8]. This effect on centerline segregation is significant [7,9,10].

To date, both a two-dimensional (2-D) slicing model and three-dimensional (3-D) coupling model are widely used to predict the flow and solidification phenomena during the CC process. Owing to the high calculation efficiency, the 2-D slicing model is widely used to predict the temperature distribution and solidification profile during the CC process [7,10–12]. The calculation efficiency of the 2-D slicing model is improved by using an effective thermal conductivity concept to indirectly take the flow effect into account. However, the 2-D slicing model assumes that the solidification shell is a heat-transfer slice which moves from the meniscus to the solidification end. Evidently, the effect of turbulent flow on temperature distribution and the solidification shell is not directly considered in the 2-D slicing model. In fact, especially for slab the CC process, the molten steel jet which impacts the narrow face has a significant effect on the temperature distribution and solidification shell [13]. Hence, to consider the effect of turbulent flow, the 3-D coupling model has been applied during the CC process [13–17]. Compared with the 2-D slicing model, the computational domain is much larger. Thereby, the amount of calculation is very large. Furthermore, the integrated distribution of cooling water in the width directions of secondary cooling zones is not considered in the previous 3-D calculations [13,17]. This simple treatment of the heat transfer condition would evidently affect the accurate prediction of the solidification profile. Recently, Sun et al. [18] proposed a method to divide the computational domain of the CC bloom to improve the calculation efficiency. However, in that simulation model, the grid independence tests were not carried out and the heat transfer coefficient was assumed to be constant around the strand transverse section. Moreover, compared with the bloom, the heat transfer boundary of the CC slab is more complex. In the width direction, the distribution of the cooling water is not entirely uniform. Thus far, the combined hybrid 3-D/2-D numerical model has not been used in the simulation of a slab CC process.

In the present work, to simultaneously consider the effect of turbulent flow and improve calculation efficiency, a combined hybrid 3-D/2-D numerical model was established and used to explore the transport phenomena during the slab CC process. The 3-D coupling model and 2-D slicing model were adopted in the turbulent flow region and laminar flow region, respectively. The effects of flow in the turbulent region on the temperature distribution and the location of the final solidification point are considered by the data of the interface in the 3-D turbulent flow region. These are transmitted to the 2-D slice as the initial conditions. The grid independence tests were carried out to find a suitable mesh cell size. This was then adopted in the simulation of the laminar flow region. In the present model, the distribution of cooling water in the width direction was taken into account.

2. Model Description

2.1. Mathematical Formulation

The continuity equation is

$$\frac{\partial \rho}{\partial t} + \frac{\partial (\rho u_i)}{\partial x_i} = 0 \tag{1}$$

The momentum equation is

$$\frac{\partial (\rho u_i)}{\partial t} + \frac{\partial (\rho u_i u_j)}{\partial x_j} = -\frac{\partial P}{\partial x_i} + \frac{\partial}{\partial x_i}\left[\mu_{\text{eff}}\left(\frac{\partial u_i}{\partial x_j} + \frac{\partial u_j}{\partial x_i}\right)\right] + \rho g_i + S_{i,\text{mom}} \tag{2}$$

$$\mu_{\text{eff}} = \mu + \mu_t = \mu + c_\mu \rho \frac{k^2}{\varepsilon} \tag{3}$$

and

$$S_{i,\text{mom}} = \frac{(1 - f_\text{L})^2}{(f_\text{L}^3 + 0.01)} A_{\text{mushy}}\left(u_i - v_{i,p}\right) \tag{4}$$

$$f_\text{L} = \begin{cases} 0 & T < T_{\text{Solidus}} \\ \frac{T - T_{\text{solidus}}}{T_{\text{Liquidus}} - T_{\text{Solidus}}} & T_{\text{Solidus}} < T < T_{\text{Liquidus}} \\ 1 & T > T_{\text{Liquidus}} \end{cases} \tag{5}$$

where ρ, t, u_i, u_j, μ_{eff}, P, g_i, f_L, A_{mushy}, $v_{i,p}$, T, T_{Solidus}, and T_{Liquidus} are molten steel density (kg/m³), time (s), i-component of velocity (m/s), j-component of velocity (m/s), effective viscosity (kg·m⁻¹·s⁻¹), pressure (N/m²), i-component of acceleration due to gravity (m/s²), liquid fraction, morphology constant, i-component of casting speed (m/s), temperature (K), solidus temperature (K), and liquidus temperature (K), respectively. The value of A_{mushy} is usually between 10^5 and 10^8 in the numerical modelling of CC processes. In addition, the higher the value of A_{mushy}, the steeper the transition of the velocity of the material to zero as it solidifies. Values between 10^4 and 10^7 are recommended for most computations (based on the user's guide of Ansys Fluent 14.0). Based on these two things, the value of A_{mushy} was set to 10^7 in the present work.

Standard k-ε equations are

$$\frac{\partial(\rho k)}{\partial t} + \frac{\partial(\rho k u_i)}{\partial x_i} = \frac{\partial}{\partial x_i}\left(\frac{\mu_t}{\sigma_k}\frac{\partial k}{\partial x_j}\right) + G - \rho\varepsilon + S_k \tag{6}$$

$$\frac{\partial(\rho\varepsilon)}{\partial t} + \frac{\partial(\rho\varepsilon u_i)}{\partial x_i} = \frac{\partial}{\partial x_i}\left(\frac{\mu_t}{\sigma_\varepsilon}\frac{\partial\varepsilon}{\partial x_j}\right) + c_1\frac{\varepsilon}{k}G - c_2\frac{\varepsilon^2}{k}\rho + S_\varepsilon \tag{7}$$

where

$$G = \mu_t\frac{\partial u_i}{\partial x_j}\left(\frac{\partial u_i}{\partial x_j} + \frac{\partial u_j}{\partial x_i}\right) \tag{8}$$

$$S_k = \frac{(1-f_L)^2}{(f_L^3 + 0.01)}A_{mushy}k \tag{9}$$

$$S_\varepsilon = \frac{(1-f_L)^2}{(f_L^3 + 0.01)}A_{mushy}\varepsilon \tag{10}$$

where G, k, ε, μ, and μ_t are the generation of turbulence kinetic energy (kg·m^{-1}·s^{-3}), turbulent kinetic energy (m^2/s^2), dissipation rate of turbulence energy (m^2/s^3), molecular viscosity (kg·m^{-1}·s^{-1}), and turbulent viscosity (kg·m^{-1}·s^{-1}), respectively. The standard values of σ_k, σ_ε, c_1, c_2, and c_μ recommended by Launder and Spalding are 1.0, 1.3, 1.44, 1.92, and 0.09, respectively.

The energy equation is

$$\frac{\partial(\rho H)}{\partial t} + \frac{\partial(\rho u_i H)}{\partial x_i} = \frac{\partial}{\partial x_i}\left[\left(\lambda + \frac{c_p\mu_t}{Pr_t}\right)\frac{\partial T}{\partial x_i}\right] \tag{11}$$

where

$$H = h + \Delta H = h_{ref} + \int_{T_{ref}}^{T} c_p dT + f_L L \tag{12}$$

where H, Pr_t, h, L, λ, and c_p, are enthalpy (J/kg), turbulent Prandtl number, sensible enthalpy (J/kg), pure solvent melting heat (J/kg), thermal conductivity (W·m^{-1}·K^{-1}), and specific heat (J·kg^{-1}·K^{-1}), respectively. The value of Pr_t was set to 0.85 (based on the theory guide of Ansys Fluent 14.0). More details about the mathematical formulation are available in reference [14].

2.2. Computational Domain

According to the geometric symmetry, the strand was assumed to be ideally symmetrical and a quarter of the strand was selected to be calculated. The geometry and working parameters of the CC process are listed in Table 1. The whole 3-D computational domain (1530 × 190 × 20,362 mm^3) is very large. If the coupled model is adopted in it directly, the cost of computation will be very large. To cut down the great amount of calculation and improve the calculation efficiency, the computational domain is suitably divided into two parts, the 3-D turbulence region and the 2-D laminar flow region. The division method is illustrated in Figure 1. When the z-component velocity of molten steel is equal to the casting speed, the plug flow is formed. This means that the local molten steel and shell are in a relatively static status. That is, the first transverse section of the strand—where z-component velocities equal to the casting speed—is the interface between the turbulent flow region and laminar flow region. The data of the interface in the 3-D turbulent flow region are transmitted to the initial 2-D slice in the laminar flow region using a coordinate interpolation algorithm.

Table 1. The geometry and working parameters of the CC (continuous casting) process.

Parameters	Values	Secondary Cooling	Length (mm)	Cooling Water Flow Rate (L/min)
Mold section	1530 × 190 mm²	Zone 1	405	155
Mold length	800 mm	Zone 2	555	84
Inside size of SEN	86 × 45 mm²	Zone 3	800	54
Outside size of SEN	141 × 100 mm²	Zone 4	1730	65
Port size of SEN	45 × 73 mm²	Zone 5	1927	52
Port angle	−15 degrees	Zone 6	3854	86
Casting speed	1.2 m/min	Zone 7	5806	86
Casting temperature	1811 K	Zone 8	4485	59
Steel grade	Q345	-	-	-

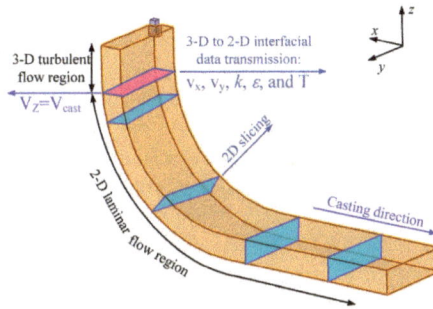

Figure 1. Illustration of 3-D turbulent flow and 2-D laminar flow regions during the CC (continuous casting) process.

In order to find out the interface between the 3-D turbulent flow region and the 2-D laminar region, a computational domain with a size of 1530 × 190 × 3250 mm³ was built and calculated. The position of this interface can be obtained by analyzing the z-component velocity variation of molten steel at the central symmetry plane of strand narrow face. The meshes created by ANSYS ICEM CFD 14.0 and adopted in 3-D turbulent flow region and 2-D laminar flow region are shown in Figure 2. The number of elements for the meshes of the 3-D part and 2-D part are 517,793 and 770, respectively.

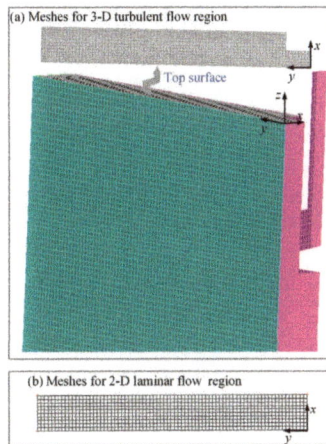

Figure 2. Meshes adopted in the (**a**) 3-D turbulent flow region and (**b**) 2-D laminar flow region.

2.3. Boundary Conditions and Physical Properties

The inlet velocity of SEN was calculated based on mass conservation. The k and ε values were estimated using the semi-empirical relations. The values of these parameters are v_{in} = 1.502 m/s, k = 0.003 m^2/s^2, ε = 0.0134 m^2/s^3, and T = 1811 K. The boundary condition of the outlet in the 3-D part was set as 'outflow'. The wall condition was employed on the meniscus without heat transfer.

The wall condition was employed on the strand surface. The heat flux of the mold surface was calculated using Equation (13) [19].

$$q_{\mathrm{m}} = 2,680,000 - 276,000 \times \sqrt{\frac{60L_z}{v_c}} \tag{13}$$

where L_z and v_c are the distance from the meniscus (m) and the casting speed (m/min), respectively.

In the secondary cooling zone, the cooling types include four aspects: water spray cooling, radiation cooling, water evaporation cooling, and roll contact cooling. In the present work, four aspects of cooling types have been considered. First, we determined the heat conditions for different cooling types with the same method described in our publications [7,9]. The heat flux for the water spray cooling region was calculated using Equation (14). More detailed descriptions about the other three cooling types in the secondary cooling zone are available in our publications [7,9]. Then, to simplify this heat transfer process, we use the integrated heat transfer coefficient (in which the four cooling types have been taken into account) to describe the heat transfer process.

$$q_{\mathrm{s}} = h(T_{\mathrm{w}} - T_0) = \left(2950.190 \times T_{\mathrm{w}}^{-0.235} \times w^{0.805}\right)(T_{\mathrm{w}} - T_0) \tag{14}$$

where w, T_{w}, and T_0 are the spray water impingement density (L·m^{-2}·s^{-1}), slab surface, and water temperature, respectively.

Due to the existence of spray water overlap, the distribution of spray water for each row of nozzles along the width direction should be identified based on the nozzle collocation, as shown in Figure 3a. More details about this technique are available in our publication [7]. According to the nozzle collocation of secondary cooling zones, the integrated distribution of cooling water from Zone 1 to Zone 8 is obtained, as shown in Figure 3b. In order to deal with the heat transfer condition conveniently, the strand surface in the width direction is divided into 26 pieces (sections No. 1 to No. 25 being 30 mm wide and section No. 26 being 15 mm wide) for heat transfer. The cooling water sprayed onto each piece is determined based on the nozzle collocation of each cooling zone. The cooling water on each piece is supposed to be uniformly distributed.

The physical properties of steel Q345 used in the calculation are listed in Table 2.

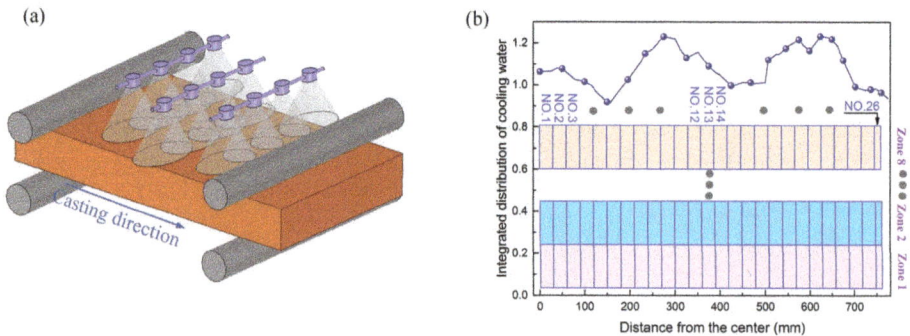

Figure 3. (a) Schematic of nozzle collocation; (b) integrated distribution of cooling water from Zone 1 to Zone 8.

Table 2. Physical properties of steel Q345.

Physical Properties	Values	Physical Properties	Values
Density, kg/m^3	7330	Viscosity, kg·(m·s)$^{-1}$	0.0062
Specific heat, J (kg·K)$^{-1}$	$319.59 + 0.1934 \times T$ (K)	Thermal conductivity, W (m·K)$^{-1}$	$57.524 - 0.0164 \times T$ (K)
Liquidus, K	1786	Solidus, K	1715
Latent heat, J/kg	255,500	-	-

2.4. Solution Procedure

The coupled calculation model was solved using the SIMPLE method with ANSYS Fluent 14.0. For both the 3-D turbulent flow region and the 2-D laminar flow region, the time step size was set to 0.05 s. Based on the relations between the length of model and casting speed, the total number of time steps for 3-D turbulent flow region and 2-D laminar flow region were 3500 and 17,802, respectively. The calculation was performed on a computer with a 3.50-GHz Intel Core i7-3770k processor and 16.0-GB RAM. For each time step, the tolerances of continuity, x-velocity, y-velocity, z-velocity, k, ε, and energy were set as 0.001, 0.001, 0.001, 0.001, 0.001, 0.001, and 1×10^{-6}, respectively.

3. Results and Discussion

3.1. Flow and Solidification Phenomena in the Turbulent Flow Region

Figure 4a shows the melt flow pattern in the turbulent flow region. It is seen that the molten steel jet flows through the port of SEN and impacts on the narrow face of the mold. The molten steel jet is divided into two parts at the impact point and forms upper and lower recirculation vortexes. The fresh molten steel with higher temperature is brought to the impact point, where the solidification shell is remelting. The remelting phenomenon of the solidification shell is shown in Figure 4b. It is obvious that the impact effect of the molten steel jet on the formation of the solidification shell near the impact point is remarkable.

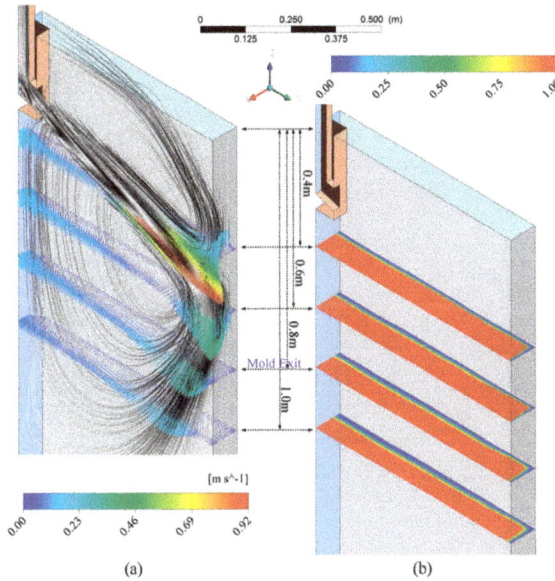

Figure 4. (a) Flow field in the turbulent flow region; (b) liquid fraction on the different transverse sections.

The impact point of the jet on the narrow face is obtained from Figure 5a,b. At the impact point, the z-component of velocity is equal to zero and the velocity magnitude reaches a valley value. In the present case, the impact point is at the position 457 mm below the meniscus. In order to find the interface between the turbulent flow region and the laminar flow region, the z-component of velocities of characteristic lines at the central symmetry plane of the strand narrow face were investigated. From Figure 5c, it can be seen that the z-component of the molten steel velocity is equal to the casting speed when arriving at the location 2442 mm below the meniscus, where it is roughly placed at the end of secondary cooling Zone 3.

To deal with the heat transfer condition for each secondary cooling zone separately, the transverse section at the end of secondary cooling Zone 3 (2560 mm below meniscus) is selected as the interface between the turbulent flow region and laminar flow region. Therefore, the 2-D slicing model is adopted in the laminar flow region from Zone 4 to Zone 8.

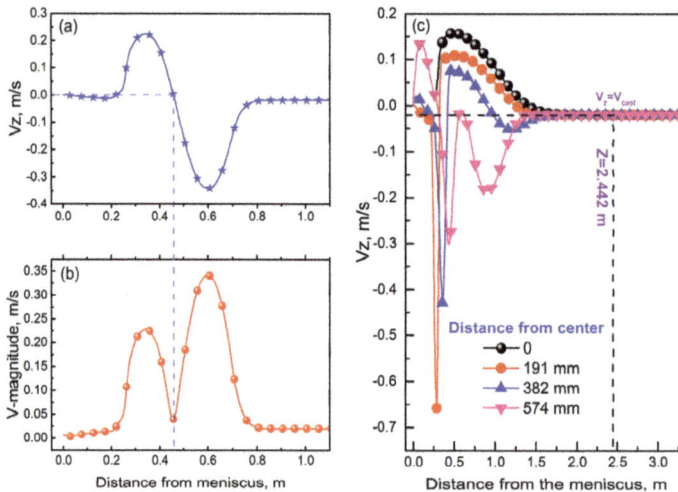

Figure 5. (**a**) z-component of velocity and (**b**) velocity magnitude variation of the line with 10 mm from the narrow face at the central symmetry plane. (**c**) z-component of the velocity variation of different lines at the central symmetry plane of the strand narrow face.

3.2. Grid Independence Tests for the Laminar Flow Region

To begin with, grid sizes for the laminar flow region have been carefully chosen to ensure the grid independence solution. As seen in the results of the grid independence tests shown in Figure 6, relatively larger differences exist in the solidification profile and temperature extreme in the transverse section (17.56 m from the meniscus) with different cell sizes. It is obvious that the cell size 25×25 mm^2, as shown in Figure 6a, is too large for this case to precisely predict the solidification profile and temperature distribution. The solidification profile does not correspond with the integrated distribution of cooling water in secondary cooling zones (shown in Figure 3). Also, the lowest temperature is higher than the other cases. In comparing Figure 6b,c, where the cell size decreases from 15×15 mm^2 to 10×10 mm^2, there is a similar solidification profile which is corresponding with the integrated distribution of cooling water in the width direction of strand. The maximum and minimum temperatures in the case where the cell size is 15×15 mm^2 are very close to those in the case where the cell size is 10×10 mm^2. When the cell size reduces from 10×10 mm^2 to 5×5 mm^2, as shown in Figure 6d, the temperature extreme of 1181 K to 1733 K shows little change compared with that using grids with a cell size of 10×10 mm^2. Thereby, in the present simulation, the grids with cell size 10×10 mm^2 are sufficient to obtain the precise temperature distribution and solidification profile.

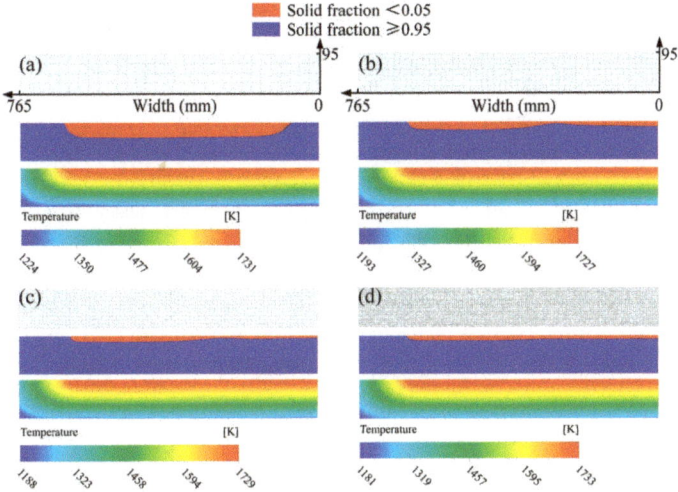

Figure 6. Solidification profile and temperature field in the transverse section (17.56 m from the meniscus) with different cell sizes: (**a**) 25 × 25 mm², (**b**) 15 × 15 mm², (**c**) 10 × 10 mm², (**d**) 5 × 5 mm².

3.3. Temperature Field and Solidification Profile

The slab center and wide surface center temperature along the casting direction predicted by the present model are compared to that predicted by our previous model, which is proven and widely employed in temperature prediction during the CC process. As the results show in Figure 7, the temperature predicted by the present model agrees well with that of the models by Long et al. [7] and Long and Chen [9]. The slab center temperature, as shown in Figure 7a, decreases slowly before complete solidification. However, after complete solidification, the slab center temperature decreases sharply. This is because there is no latent heat to maintain the temperature when the local region solidifies completely. The wide surface center temperature is shown in Figure 7b. Due to the strong cooling of the mold, the surface temperature decreases radically, and then recalesces when the strand is pulled out of the mold. In the secondary cooling zone, the surface temperature decreases gradually.

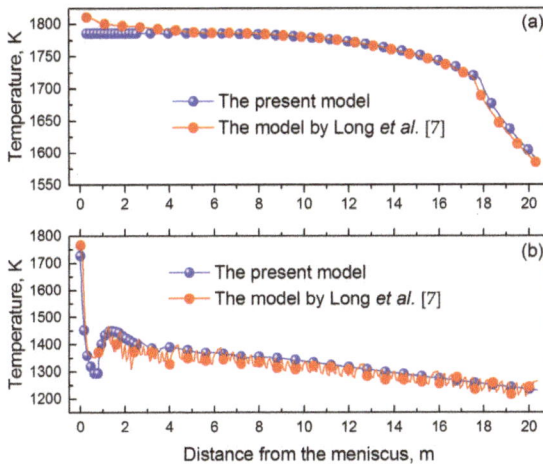

Figure 7. (**a**) Slab center and (**b**) wide surface center temperature along the casting direction.

The temperature distribution in transverse sections with different distances from the meniscus (as shown in Figure 8a) have a similar profile. The temperature decreases as the distance from the meniscus increases. Figure 8b shows the solidification profile evolution process during continuous casting. It is obvious that the liquid core of the strand is not entirely uniform. At the local region of the strand in the width direction, from roughly 400 to 500 mm, the molten steel solidifies completely owing to the minimum amount of the cooling water sprayed on it in the secondary cooling zone. The final solidification points (another one in the other half of the geometry) are 17.66 m away from the meniscus. Moreover, at the position about 150 mm in width direction, the molten steel solidifies completely a little earlier than that at the region from 400 to 500 mm. The reason is that the amount of cooling water at the former position is a little more than that at the region from 400 to 500 mm. This results from the valley of the distribution curve being very narrow at the position of about 150 mm, although the proportion of cooling water is slightly lower than that at the region from 400 to 500 mm. The solidification profile agrees well with the integrated distribution of cooling water in the secondary cooling zones.

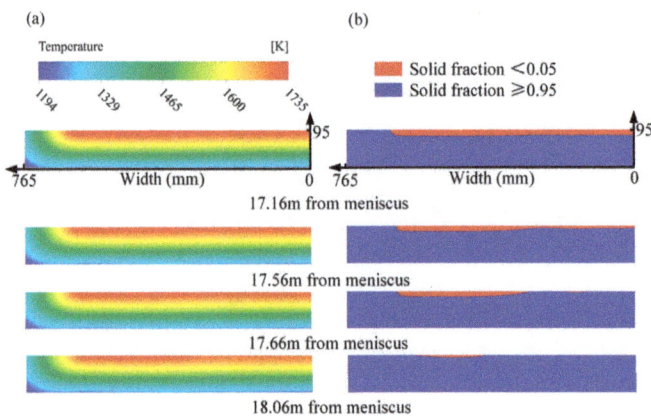

Figure 8. (**a**) Temperature field and (**b**) solidification profile in transverse sections with different distances from the meniscus.

4. Conclusions

A combined hybrid 3-D/2-D simulation model with a high calculation efficiency has been developed to investigate the flow and solidification behaviors during the slab CC process. The main conclusions are summarized as follows:

(1) The impact effect caused by the molten steel jet on the formation of a solidification shell is significant. The impact point is at the position 457 mm below the meniscus, and the plug flow is formed 2442 mm below the meniscus.

(2) For the simulation of the laminar flow region, the grids with a cell size of 10×10 mm^2 are sufficient to attain a precise temperature distribution and solidification profile.

(3) The solidification profile of the strand is not entirely uniform. The final solidification points, roughly being at the position from 400–500 mm in the width direction, are 17.66 m away from the meniscus.

Acknowledgments: The authors would like to thank the Natural Science Foundation of China (NSFC) for financial support (project No. 51504048, 51374260, 51611130062).

Author Contributions: Huabiao Chen, Mujun Long, and Dengfu Chen conceived the simulation model and the validation method. Huabiao Chen, Sheng Yu, Bin Liang, and Huamei Duan performed the calculations. All of the authors analyzed and discussed the data. Huabiao Chen wrote the paper.

Conflicts of Interest: The authors declare no conflict interest.

References

1. Ma, J.C.; Xie, Z.; Ci, Y.; Jia, G.L. Simulation and application of dynamic heat transfer model for improvement of continuous casting process. *Mater. Sci. Technol.* **2009**, *25*, 636–639. [CrossRef]
2. Zhao, Y.; Chen, D.F.; Long, M.J.; Shen, J.L.; Qin, R.S. Two-dimensional heat transfer model for secondary cooling of continuously cast beam blanks. *Ironmak. Steelmak.* **2014**, *41*, 377–386. [CrossRef]
3. Ma, J.C.; Lu, C.S.; Yan, Y.T.; Chen, L.Y. Design and application of dynamic secondary cooling control based on real time heat transfer model for continuous casting. *Int. J. Cast Met. Res.* **2014**, *27*, 135–140. [CrossRef]
4. He, D.F.; Chang, S.; Wang, H.B. Controlling transverse cracks of slab based on edge control technology. *J. Iron Steel Res. Int.* **2015**, *22*, 42–47. [CrossRef]
5. Chen, S.D.; Hu, Z.F.; Yuan, Y.Y.; Luo, Y.Z. Study on intermediate crack in continuous casting slab of medium carbon steel. *J. Iron Steel Res. Int.* **2011**, *18*, 383–388.
6. Long, M.J.; Chen, D.F.; Wang, Q.X.; Luo, D.H.; Han, Z.W.; Liu, Q.; Gao, W.X. Determination of CC slab solidification using nail shooting technique. *Ironmak. Steelmak.* **2012**, *39*, 370–377. [CrossRef]
7. Long, M.J.; Dong, Z.H.; Chen, D.F.; Liao, Q.; Ma, Y.G. Effect of uneven solidification on the quality of continuous casting slab. *Int. J. Mater. Prod. Technol.* **2013**, *47*, 216–232. [CrossRef]
8. Shen, H.F.; Hardin, R.A.; MacKenzie, R.; Beckermann, C. Simulation using realistic spray cooling for the continuous casting of multi-component steel. *J. Mater. Sci. Technol.* **2002**, *18*, 311–314.
9. Long, M.J.; Chen, D.F. Study on mitigating center macro-segregation during steel continuous casting process. *Steel Res. Int.* **2011**, *82*, 847–856. [CrossRef]
10. Ji, C.; Luo, S.; Zhu, M.; Sahai, Y. Uneven solidification during wide-thick slab continuous casting process and its influence on soft reduction zone. *ISIJ Int.* **2014**, *54*, 103–111. [CrossRef]
11. Xie, X.; Chen, D.F.; Long, H.J.; Long, M.J.; Lv, K. Mathematical modeling of heat transfer in mold copper coupled with cooling water during the slab continuous casting process. *Metall. Mater. Trans. B* **2014**, *45*, 2442–2452. [CrossRef]
12. Hardin, R.A.; Liu, K.; Kapoor, A.; Beckermann, C. A transient simulation and dynamic spray cooling control model for continuous steel casting. *Metall. Mater. Transf. B* **2003**, *34*, 297–306. [CrossRef]
13. Wang, Q.Q.; Zhang, L.F. Influence of FC-mold on the full solidification of continuous casting slab. *JOM* **2016**, *68*, 2170–2179. [CrossRef]
14. Shamsi, M.R.R.I.; Ajmani, S.K. Three dimensional turbulent fluid flow and heat transfer mathematical model for the analysis of a continuous slab caster. *ISIJ Int.* **2007**, *47*, 433–442. [CrossRef]
15. Shamsi, M.R.R.I.; Ajmani, S.K. Analysis of mould, spray and radiation zones of continuous billet caster by three-dimensional mathematical model based on a turbulent fluid flow. *Steel Res. Int.* **2010**, *81*, 132–141. [CrossRef]
16. Yang, J.W.; Du, Y.P.; Shi, R.; Cui, X.C. Fluid flow and solidification simulation in beam blank continuous casting process with 3d coupled model. *Iron Steel Res. Int.* **2006**, *13*, 17–21. [CrossRef]
17. Seyedein, S.H.; Hasan, M. A three-dimensional simulation of coupled turbulent flow and macroscopic solidification heat transfer for continuous slab casters. *Int. J. Heat Mass Transf.* **1997**, *40*, 4405–4423. [CrossRef]
18. Sun, H.B.; Zhang, J.Q. Study on the macrosegregation behavior for the bloom continuous casting: Model development and validation. *Metall. Mater. Transf. B* **2014**, *45*, 1133–1149. [CrossRef]
19. Cai, K.K. *Continuous Casting Mold*; Metallurgical Industry Press: Beijing, China, 2008; p. 6, ISBN 978-7-5024-4635-2.

![metals logo] *metals*

MDPI

Article

Numerical Study on the Influence of a Swirling Flow Tundish on Multiphase Flow and Heat Transfer in Mold

Peiyuan Ni [1,2,*], Mikael Ersson [3], Lage Tord Ingemar Jonsson [3], Ting-an Zhang [1] and Pär Göran JÖNSSON [3]

[1] Key Laboratory of Ecological Metallurgy of Multi-metal Intergrown Ores of Education Ministry, School of Metallurgy, Northeastern University, Shenyang 110819, China; zta2000@163.net
[2] Department of Materials and Manufacturing Science, Graduate School of Engineering, Osaka University, 2-1 Yamadaoka, Suita, Osaka 565-0871, Japan
[3] Department of Materials Science and Engineering, KTH Royal Institute of Technology, SE-100 44 Stockholm, Sweden; bergsman@kth.se (M.E.); lage@kth.se (L.T.I.J.); parj@kth.se (P.G.J.)
* Correspondence: peiyuan_ni@163.com; Tel.: +86-024-83686283

Received: 30 April 2018; Accepted: 18 May 2018; Published: 21 May 2018

Abstract: The effect of a new cylindrical swirling flow tundish design on the multiphase flow and heat transfer in a mold was studied. The RSM (Reynolds stress model) and the VOF (volume of fluid) model were used to solve the steel and slag flow phenomena. The effect of the swirling flow tundish design on the temperature distribution and inclusion motion was also studied. The results show that the new tundish design significantly changed the flow behavior in the mold, compared to a conventional tundish casting. Specifically, the deep impingement jet from the SEN (Submerged Entry Nozzle) outlet disappeared in the mold, and steel with a high temperature moved towards the solidified shell due to the swirling flow effect. Steel flow velocity in the top of the mold was increased. A large velocity in the vicinity of the solidified shell was obtained. Furthermore, the risk of the slag entrainment in the mold was also estimated. With the swirling flow tundish casting, the temperature distribution became more uniform, and the dissipation of the steel superheat was accelerated. In addition, inclusion trajectories in the mold also changed, which tend to stay at the top of the mold for a time. A future study is still required to further optimize the steel flow in mold.

Keywords: swirling flow tundish; multiphase flow; heat transfer; mold; continuous casting

1. Introduction

The mold is the final stage during the continuous casting process of steel, where the solidification of the molten steel occurs. Multiphase flow, heat and mass transfer, slag entrainment, inclusion and bubble entrapment, inclusion removal, and solidification are very important multiphysics concerns in the continuous casting process. This is due to the fact that these issues can significantly influence the quality of the semifinal steel product. As a matter of the first importance, a desirable steel flow in mold is wanted, since the other physical phenomena are directly affected by the steel flow inside the mold.

Direct investigations of the flow phenomena in a mold face significant challenges, due to the high temperature and high cost. Therefore, as an initial step to further improve the steel flow performance in a mold, numerical and physical modeling has become a common way to study the multiphase flow phenomena under various conditions. Specifically, some factors that may affect the mold flow, such as the SEN (Submerged Entry Nozzle) type (straight or bifurcated) [1,2], SEN port design (shape, angle, thickness) [3–10], argon bubbles [11–23], SEN immersion depth [3,4,6,24], nozzle clogging [25], mold flow modifier [26], EMBr (electromagnetic braking) [13,14,20–24,27–30] and M-EMS (mold electromagnetic stirring) [8–10,31], have been vastly investigated. Further improvements of the steel flow performance simply based on the parameter optimization become difficult. Therefore, in recent years, EMBr and M-EMS are widely applied to improve the steel flow performance in mold. It was found that EMBr can reduce the flow variation [23], suppress the flow velocity [20,30], increase the temperature near meniscus [14,29], decrease the temperature difference in mold [20,29], and reduce the impingement intensity near the narrow wall [29]. The use of M-EMS was found to contribute to a uniform temperature distribution [10,31], a large floating up rate of inclusions [31,32], a homogeneous solute distribution [10], a uniform solidified shell [10], and a high quality of the steel product [32]. However, their application relies on costly equipment, and also requires the consumption of electricity. Furthermore, it is sometimes difficult to realize a good flow pattern in a mold, since the original upstream flows from the SEN ports are unknown, due to flow fluctuations or biased flows. In addition, the effects of EMBr are directly related to some factors, such as the intensity of the magnetic field, the reciprocal position between the magnetic field and the acting region, the casting speed, the SEN depth, and so on [24,30]. Also, for the M-EMS case, the meniscus velocity magnitude and the level fluctuation height were found roughly linearly proportional to the applied current [33]. Therefore, over-stirring or insufficient stirring should be avoided, which sometimes is difficult, due to the transient steel flow in a real casting situation. Furthermore, it takes some time for the M-EMS to change the steel flow from a single port SEN in a billet or bloom casting, due to the high momentum of the impingement jet flow going deep into the mold. In summary, the performances of EMBr and M-EMS highly depend on the SEN port flow situation and application parameters. This leads to some uncertainties of their performances in applications.

An alternative way to optimize the mold flow is by a root measure to control the SEN outlet flow. This is realized by using a swirling flow SEN, which aims to produce a rotational flow component to optimize the SEN port flows, and afterwards, optimize the steel flow in a mold. The swirling flow SEN and its influence on the mold flow have been vastly studied [34–46]. It was found that the heat and mass transfer near the meniscus can be remarkably activated [34,38,40,42], and a uniform velocity distribution can be obtained within a short distance from the SEN outlet [34,38,40]. Furthermore, the penetration depth of the SEN outlet flow is remarkably decreased in a billet mold [34,42]. Industrial trial results [39] show that the swirling flow SEN effectively improved the steel product quality and reduced the clogging problem of the SEN side ports. Therefore, the swirling flow SEN has advantages in the continuous casting process.

In the past, several methods were studied both experimentally and numerically to produce a swirling flow inside a SEN. Specifically, a swirl blade method was investigated in many studies, where a swirling flow was produced by installing a swirl blade inside the SEN. It is a cost-saving method, and has been proved by plant trials [39] that it can improve the steel product quality. However, the lifespan of the swirl blade and the inclusion attachment on its surface, which may lead to nozzle clogging, restrict its application for longer casting times. Some studies have also been carried out to investigate the electromagnetic stirring method [43–46]. The swirling flow is obtained by installing the electromagnetic stirring equipment surrounding the SEN. Therefore, it is associated with an equipment cost and an electricity cost, which increases the steel product cost. Recently, Ni et al. [47–49] proposed a new method to produce a swirling flow in a SEN simply by using a cylindrical tundish design. It is a simple and cost-saving method to realize a swirling flow in the SEN. Furthermore, its effectiveness has been confirmed both by water model experiments, and also by numerical simulations [49]. However, the steel flow characteristics in the mold with this new tundish design remains to be studied.

Previous studies about the influence of a swirling flow SEN on the mold flow commonly ignored the top slag phase in the mold. The influence of a swirling flow SEN on the steel-slag interface phenomena and the steel flow in the vicinity of the solidified shell should be further studied. Moreover, swirling steel flows, produced from M-EMS and the swirling SEN, were mainly investigated by using k-ε type of turbulence models. However, high-intensity swirling flows normally have anisotropic turbulent fluctuations, and sometimes, a vortex core precession exists in this kind of flows [50]. A RSM

(Reynolds Stress Model) which directly solves the anisotropic turbulent fluctuations shows better performance, in general, compared to the RANS eddy-viscosity models [49,51–53]. Jakirlic et al. [53] also found that a good ability to capture the stress anisotropy in the near-wall region is very important to reproduce these types of flow. Therefore, the swirling steel flow in a cylindrical tundish design has previously been solved by using RSM coupled with the Stress-Omega submodel, where the turbulent boundary layer was also resolved with a very fine grid ($y^+ < 1$) [49]. In this study, the characteristics of the multiphase flow and heat transfer in a billet mold were studied during the casting process by using the new swirling flow tundish design. The swirling flow velocity profile on a cross section of the SEN obtained in the previous study [49] was used as the inlet flow condition for the mold flow solution to save the computational time. The RSM coupled with the Stress-Omega submodel was thereafter used to solve the flow in the mold, with a very fine grid near the solidified shell of y^+ value around 1. The VOF (volume of fluid) method was used to capture the steel-slag interface. The energy equation was solved to study the temperature distribution in the mold, and a Lagrangian particle tracking scheme was used to study the motions of non-metallic inclusions in the mold. The fluid flow, steel-slag interface fluctuation, temperature distribution, and inclusion motion in the mold were investigated. In addition, to show the change of the multiphysics in the mold, these characteristics were compared to those in a conventional tundish casting without a swirling flow effect.

2. Model Description

2.1. Model Assumptions

A three-dimensional mathematical model has been developed to describe the steel-slag-inclusion three-phase flow, and the temperature distribution in the mold. The model is based on the following assumptions:

1. Steel and slag behave as an incompressible Newtonian fluids;
2. Solidification in the mold is not considered;
3. A constant molecular viscosity for steel and slag was assumed. This is due to that the maximum temperature difference in the mold is only 30 K between 1788 K and 1818 K. The viscosity change in this temperature range is not significant, and this can be seen from a previous study [10];
4. A constant steel and slag density was used. The temperature influence on the steel density change was accounted for in the source term of the momentum equation;
5. The SEN wall was assumed to be a smooth wall;
6. Inclusions were assumed to be spherical.

2.2. Transport Equations

The conservation of a general variable ϕ within a finite control volume can be expressed as a balance among the various processes, which tends to increase or decrease the variable values. The conservation equations, e.g., continuity, volume fraction, momentum, turbulence equations, and energy equation can be expressed by the following general equation [54]:

$$\frac{\partial}{\partial t}(\rho\phi) + \frac{\partial}{\partial x_i}(\rho\phi u_i) = \frac{\partial}{\partial x_i}\left(\Gamma_\phi \frac{\partial\phi}{\partial x_i}\right) + S_\phi, \tag{1}$$

where the first term on the left-hand side is the instantaneous change of ϕ with time, the second term on the left-hand side represents the transport due to convection, the first term on the right-hand side expresses the transport due to diffusion where Γ_ϕ is the diffusion coefficient with different values for different turbulence models, or the effective thermal conductivity. Furthermore, the second term on the right-hand side is the source term.

2.3. Interface Tracking

In order to investigate the steel-slag interface fluctuation, the steel-slag interface must be properly tracked. This is done by employing the VOF model [55], where a volume fraction equation for the slag phase was solved. The sum of the slag phase fraction α_{slag} and the steel phase fraction α_{steel} is equal to 1. In addition, one set of momentum and energy equation was solved to obtain the predicted flow field in the mold. The mixed material properties in the grid cell, where the interface exists, are required by the momentum equation and can be calculated by the following equations:

$$\rho_{mix} = \alpha_{steel}\rho_{steel} + \alpha_{slag}\rho_{slag}, \tag{2}$$

$$\mu_{mix} = \alpha_{steel}\mu_{steel} + \alpha_{slag}\mu_{slag}. \tag{3}$$

2.4. Turbulence Modeling

As previously mentioned, one important concern about modeling the swirling flow is the anisotropic turbulent properties which commonly exist in high intensity swirling flows. Here, the RSM model [55–57] combined with the Stress-Omega submodel [55,58] was used to simulate the steel flow. The Stress-Omega submodel is good for modeling flows over the curved surfaces and swirling flows [55]. A near-wall treatment is automatically used to perform blending between the viscous sublayer and the logarithmic region [55]. In RSM model, the Reynolds stress terms emerging from the Reynolds averaging of Navier-Stokes equations are directly solved by resolving their transport equations to account for the possible anisotropic fluctuation in a swirling flow. In order to save the computational time, the realizable k-ε turbulence model [59], coupled with the enhanced wall treatment model [55], was first used to produce an initial flow field. Then, with this flow initialization, the RSM model calculation was carried out until a fully developed flow was obtained.

2.5. Heat Transfer

The temperature distribution in the mold was obtained by solving the following energy equation [55]:

$$\frac{\partial}{\partial t}(\rho E) + \frac{\partial}{\partial x_i}(u_i(\rho E + p)) = \frac{\partial}{\partial x_i}\left((k + \frac{c_p\mu_t}{Pr_t})\frac{\partial T}{\partial x_i}\right), \tag{4}$$

where E is energy in the unit of J, k is the thermal conductivity with the unit of W/(m·K), c_p is the specific heat capacity in J/(kg·K), μ_t is the turbulent viscosity, Pr_t is the turbulent Prandtl Number, ρ is fluid density in kg/m³, p is pressure in Pa, and T is temperature in K. The steel density change and subsequent natural convection due to temperature variance was accounted for by the Boussinesq model [60]. This model treats density as a constant value in all solved equations, except for the buoyancy in the momentum equation (it is normally put in the source term) as follows:

$$(\rho - \rho_o)g \approx -\rho_o\beta(T - T_o)g, \tag{5}$$

where ρ_o is the (constant) density of the liquid steel with the unit of kg/m³, T_o is the operating temperature in K, and β is the thermal expansion coefficient of the liquid steel. The thermal properties of the fluids and some parameters are shown in Table 1.

Table 1. Thermal properties of the steel and slag.

Parameters	Symbols	Steel	Slag
Density, kg/m^3	ρ_0	7000	2600
Viscosity, kg/(m·s)	μ	0.0064	0.09
Thermal conductivity, W/(m·K)	k	35	1.1
Specific heat, J/(kg·K)	c_p	628	1200
Thermal expansion coefficient, 1/K	β	10^{-4}	-
Interfacial tension, N/m	σ	1.6	
Operating temperature, K	T_0	1788	
Turbulent Prandtl number	Pr_t	0.85	

2.6. Lagrangian Particle Tracking Model

The inclusion velocity u_p was obtained by solving the following momentum equation, which has been introduced in detail in a previous study [61]:

$$\frac{du_p}{dt} = \frac{(u - u_p)}{\tau_r} + g\left(1 - \frac{\rho_f}{\rho_p}\right) + \frac{1}{2}\frac{\rho_f}{\rho_p}\left(u_p \nabla u - \frac{du_p}{dt}\right) + \frac{\rho_f}{\rho_p}u_p \nabla u + \frac{2\eta v^{\frac{1}{2}}\rho_f S_{ij}}{\rho_p d_p (S_{lk} S_{kl})^{\frac{1}{4}}}(u - u_p), \quad (6)$$

where, on the right-hand side, the first term is the drag force, the second term is the force per unit inclusion mass due to gravity and buoyancy, the third term is the virtual mass force, the fourth term is the pressure gradient force, and the fifth term is the Saffman's lift force [62,63]. Furthermore, u is the continuous-phase velocity, v is the kinematic viscosity of the fluid, d_p is the diameter of an inclusion, ρ_f and ρ_p are the densities of the fluid and the inclusion, respectively. Furthermore, S_{ij} and S_{lk} are the deformation tensor, and η is a constant which is equal to 2.59 [63].

2.7. Boundary Conditions

The velocity profile on the cross section of the cylindrical tundish SEN, which has been solved in a previous study [49], was used as the inlet boundary condition for the current simulation of the mold flow. Figure 1a,b show the location of the cross section on the cylindrical tundish SEN and the steel flow characteristics [49]. The cross section is located at 0.4 m below the tundish bottom with the total SEN length of 0.65 m. It can be seen that the maximum tangential velocity is around 2.5 m/s. The swirling number is defined by using the mean tangential velocity, W, and the mean vertical velocity, V, on the cross section with the ratio of $2W/3V$, and it is 1.24 on this cross section [49]. The inlet boundary condition for the conventional tundish casting is a uniform velocity distribution at the SEN cross section, with the steel flow velocity of -1.1 m/s in Z-direction, which corresponds to the same casting speed as the swirling flow tundish casting.

The calculation domain is shown in Figure 1c. A non-slip boundary condition was imposed on the SEN wall. For the top surface of the mold, a zero-shear slip wall boundary condition was used. For the mold wall, a moving wall boundary condition with the velocity of -0.013 m/s in Z direction was used to account for the movement of the solidified shell in a real casting process. The fully developed flow condition is adopted at the mold outlet, where the normal gradients of all variables are set to zero. For the heat transfer boundary condition, a constant steel temperature of 1818 K was used at the inlet. A constant temperature of 1788 K was imposed on the solidified shell. An adiabatic condition was used both at the SEN wall and at the free surface. In addition, a "reflect" wall boundary condition was used for the inclusion tracking, and an "escape" boundary condition was used at the bottom outlet of the mold.

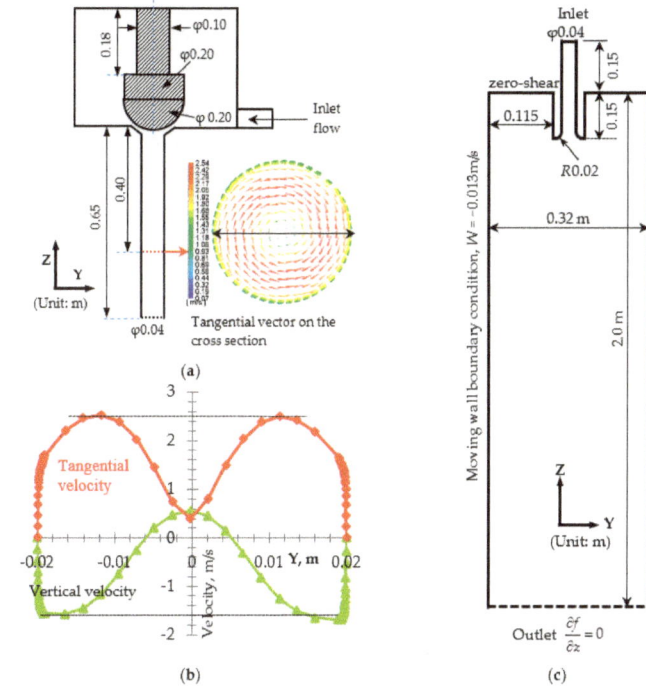

Figure 1. (a) Location of the cross section used as the inlet boundary condition of the mold simulation and the tangential velocity distribution on it, (b) velocity distribution along the line on the cross section (negative values for vertical velocity means in gravity direction), and (c) calculation domain in mold.

2.8. Solution Method

The multiphase flow and temperature distribution in the billet mold was solved by using the commercial software ANSYS FLUENT 18.0®. The numerical simulations were carried out based on 2.2 million grid cells to guarantee the grid-independent solution. A very fine grid was used in the near-wall region, with the y^+ value of the first grid layer around 1. The PISO scheme was used for the pressure-velocity coupling. Furthermore, the PRESTO method was adopted to discretize the pressure. The governing equations were discretized using a second order upwind scheme. The convergence criteria were as follows: the residuals of all dependent variables were smaller than 1×10^{-3} at each time step.

3. Results and Discussion

The multiphase flow and heat transfer in the mold both with a swirling flow tundish casting and also with a conventional tundish casting were firstly solved by the realizable k-ε model with an enhanced wall treatment for the first 75 s. After that, this solution was used as an initial condition for the RSM model calculation to 125 s for a developed flow field. The multiphysics in the mold from a conventional tundish casting and from a swirling flow tundish casting were analyzed and compared in the following.

3.1. Steel Flow Paths

Figure 2 shows the steel flow path in the mold both from a conventional tundish and from a swirling flow tundish design. It can be observed that a completely different flow pattern in the mold was observed. With a conventional tundish casting in Figure 2a, the SEN outlet flow goes deep into the mold to the depth of around 1.5 m. This results from a large vertical momentum of the steel flow. As the steel flows downwards, the jet is entraining the surrounding fluid due to the friction. This dissipates the jet momentum and also increases the jet width. Meanwhile, the pressure in the region near the downward jet flow decreases. Therefore, the steel surrounding the impingement jet moves towards it. This leads to a vertical rotational flow in the middle region of the mold as shown in Figure 2a indicated by the red color arrows. The upwards steel flow in this rotational movement further leads to a weak rotational flow in the upper part of the mold near the meniscus. However, with a swirling flow tundish design in Figure 2b, the SEN outlet flow moves towards the solidified shell rather than goes deeply into the mold. This is due to the rotational steel flow momentum. After the steel stream reaches the solidified shell, a part of the steel flows downwards along the solidified shell with a horizontal rotational flow momentum, which can be seen from the red arrow in Figure 2b. Another part of the steel moves upwards and towards the meniscus. Therefore, the steel flow pattern undergoes a significant change compared to a conventional tundish casting. The deep impingement jet into the mold observed in Figure 2a disappeared with the use of a swirling flow tundish. This is also one of the advantages for the use of a swirling flow SEN compared to M-EMS. Due to the high steel flow inertia from the one-port SEN, the impingement jet was actually not significantly changed by the M-EMS, which shows a high downwards steel flow velocity in the center of the mold. This can be seen from some previous studies [32,33]. Therefore, in some cases, the side-port SEN was investigated in bloom castings in combination with M-EMS [8,10], which can change the high temperature SEN port flow towards the solidification front, rather than moving deeply into the mold. However, side ports are the sensitive region for nozzle clogging, and they deliver the steel into the mold from a certain direction depending on the SEN port direction, rather than along the periphery of the SEN, which is in 360° in a swirling flow SEN. Therefore, the current swirling flow tundish SEN can deliver high temperature steel uniformly distributed towards the solidified shell. Furthermore, this also avoids the strong attack by the high temperature steel on some locations of the solidified shell which is in the case with side ports.

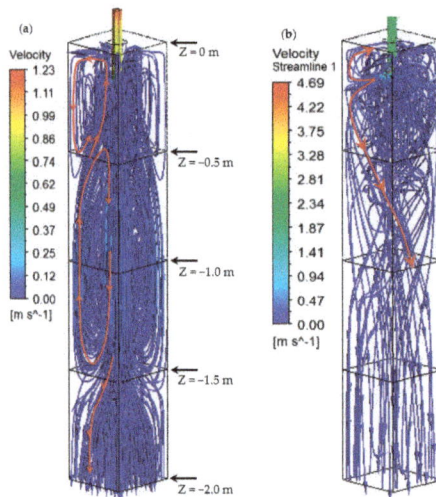

Figure 2. Comparison of steel flow paths in mold. (**a**) Casting with a conventional tundish and (**b**) casting with a swirling flow tundish.

3.2. Steel Flow Velocity

Figure 3 shows the steel flow velocity on the vertical planes at the middle of the mold. The locations of these vertical planes can be seen from the top view sketch of the mold in Figure 3. For all the vertical planes passing the mold center, the diagonal vertical plane is the largest vertical plane, and the plane perpendicular to the solid shell middle is the smallest vertical plane. Therefore, these planes are selected to show the flow characteristics. It can be seen in Figure 3a that the steel jet keeps a high velocity in the center of the mold, even for a large depth. At the top of the mold near the meniscus, the steel flow is very weak. However, with the swirling flow tundish as shown in Figure 3b, the high velocity region was located at the top region of the mold. This is expected to improve the heat transfer near the meniscus and the dissipation of the steel superheat. Furthermore, steel moves downwards at the region near the solidified shell, and it flows upwards in the center of the mold. This may be helpful to improve the mixing towards a homogeneous state in the mold. Furthermore, the velocity magnitude of the SEN outlet flow rapidly decreases to smaller than 0.4 m/s. As previously mentioned, there was no main impingement jet deep into the mold, which is different from the case of a conventional tundish casting with a straight SEN.

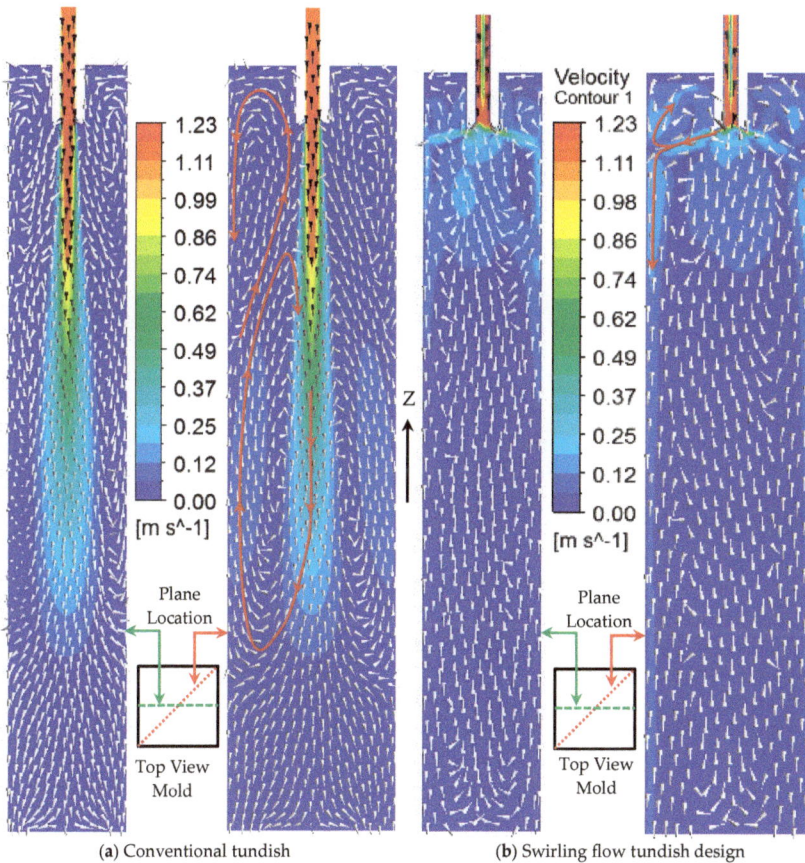

Figure 3. Steel flow velocity in the vertical middle plane of the mold. (**a**) Conventional tundish casting and (**b**) Swirling flow tundish design casting (arrows are the steel flow directions).

Figure 4 shows a comparison of the vertical velocity distributions along lines at different mold depths. It can be see that the vertical velocity magnitude is much smaller in the mold center when a swirling flow tundish design was used. This has been clearly shown in Figure 3b. However, at the locations close to the solidified shell, its velocity is larger than that with a conventional tundish casting. This can be seen from the enlarged part in Figure 4. A large velocity near the solidified shell is helpful to shear off the dendrites from the solidification interface and promotes the nucleate, which results in an enhancement of the transition from a columnar to an equiaxed solidification [33]. For the casting with a conventional tundish, the vertical velocity of the steel flow is still around 0.1 m/s at a depth of 1.5 m in the mold center. This kind of flow pattern, with a strong downwards flow stream, is not good for some issues in a steel continuous casting process, such as the inclusion removal and the dissipation of steel superheat. Therefore, it is clear that the swirling flow tundish design is helpful for the optimization of the mold flow.

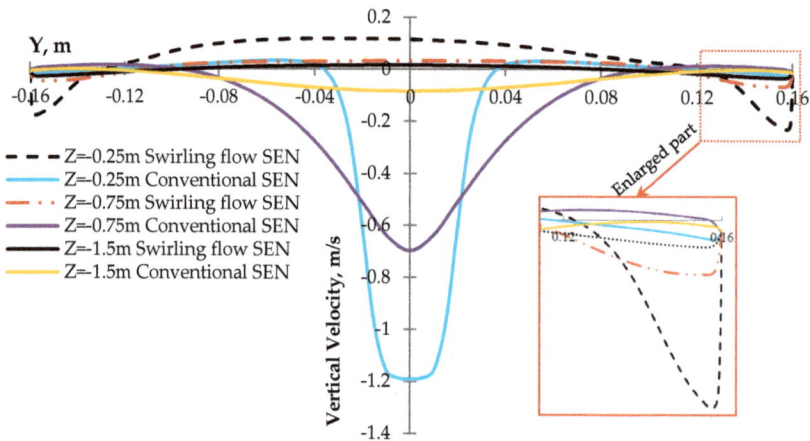

Figure 4. Vertical steel flow velocity along horizontal lines in different mold depths.

Another important characteristic about the swirling flow tundish casting is the swirling steel flow on the cross sections of the mold. Figure 5 shows the tangential velocities of the steel flow on two cross sections in different mold depths. It can be seen, in Figure 5b, that the maximum tangential velocity can reach around 0.076 m/s near the solidified shell on the cross section with a mold depth of 0.5 m. In addition, at a mold depth of 1.5 m, the maximum tangential velocity decreases to around 0.005 m/s, while the velocity distribution becomes more axisymmetric. This means that the swirling flow becomes more uniform after moving from the depth of 0.5 m to 1.5 m. Furthermore, the high tangential velocity region is still located near the solidified shell. Figure 6 shows the magnitude of the tangential velocity along different horizontal lines in different mold depths. The locations of the horizontal lines are shown in Figure 5b. It can be seen that the maximum tangential velocity gradually decreases when the steel moves downwards. This is similar as that in the mold with M-EMS, where the rotational velocity becomes smaller with an increased distance away from the stirrer midplane [33]. A large velocity gradient exists near the solidified shell, which can also be seen in Figure 4 for the vertical velocity component. In the mold with a conventional tundish, no obvious swirling flow was observed, and the steel flow velocity near the solidified shell is much smaller compared to the swirling flow tundish case. Therefore, both the tangential velocity magnitude and the axial velocity magnitude near the solidified shell are larger for the case with the swirling flow tundish casting than that with a conventional tundish casting.

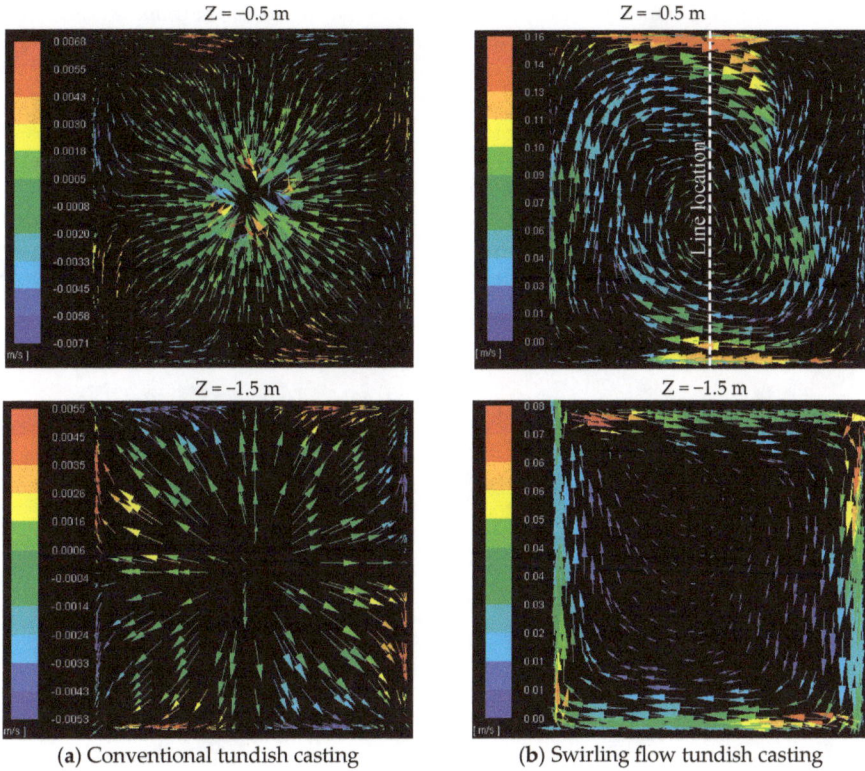

Figure 5. Tangential velocities on different cross sections of the mold. (**a**) Conventional tundish casting and (**b**) swirling flow tundish casting.

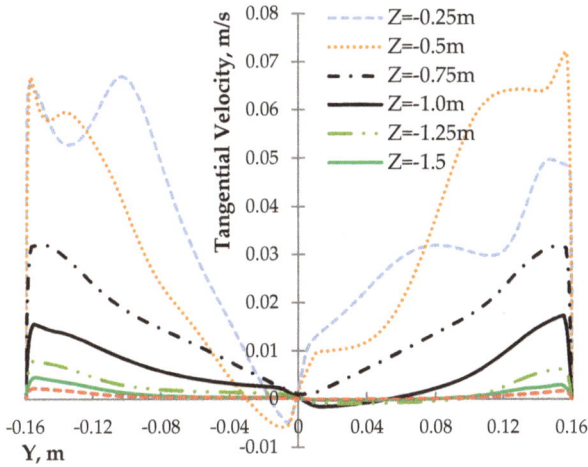

Figure 6. Tangential velocity distribution along different horizontal lines in different mold depths.

Due to the change of the steel flow pattern in mold, shear stresses on the solidified shell become different. Figure 7 shows the shear stress on the solidified shell for the first 0.5 m shell from the meniscus. It can be seen that the values of shear stresses are mostly smaller than 10 Pa both for the conventional tundish case and also for the swirling flow tundish case. Shear stress is proportional to the velocity gradient. Therefore, a large shear stress represents a large velocity gradient near the solidified shell. Due to that, a rotational flow exists in the mold as shown in Figure 5b, the shear stress values for the swirling flow tundish case are mostly larger than that with a conventional tundish. In addition, shear stresses at the locations where the SEN outlet flow hits the solidified shell are not very large, as shown in Figure 7b, with the values of around 6 Pa. This means that there is no strong flow stream towards the solidified shell, due to the uniformly spreading of the steel flow from the SEN outlet.

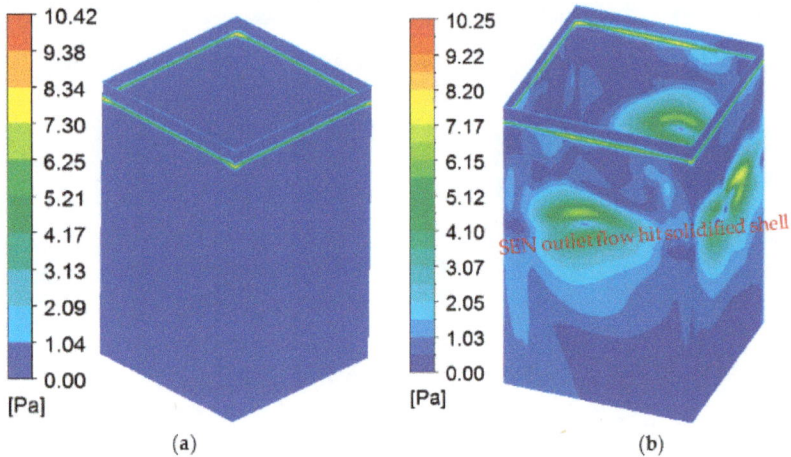

Figure 7. Shear stress on the solidified shell. (**a**) Conventional tundish casting and (**b**) swirling flow tundish casting.

3.3. Turbulence Properties

Due to the flow changes as presented above, the characteristics of the turbulence properties in the mold are also significantly changed, as shown in Figure 8. It can be seen that with a conventional tundish casting, the turbulence properties show jet flow characteristics. A high turbulent kinetic energy and Reynolds stress values exist in the region near the impingement jet, due to the shear between the jet and its surrounding steel. However, at the top of the mold, the magnitudes of the turbulence properties are very small. In addition, the turbulence properties show anisotropic characteristics. The Reynolds stress is large in the vertical direction, namely, the value of $\overline{w'w'}$ in the conventional tundish casting. With a swirling flow tundish casting, turbulence properties are sharply dissipated in the top region of the mold. This is due to the change of the steel flow pattern as previously shown in Figure 3b. Deeper in the mold, smaller values of both turbulent kinetic energy and Reynolds stresses were observed. Due to the large value of turbulent properties and the flow characteristic change with the swirling flow tundish casting, the heat transfer in the top of the mold was expected to be enhanced.

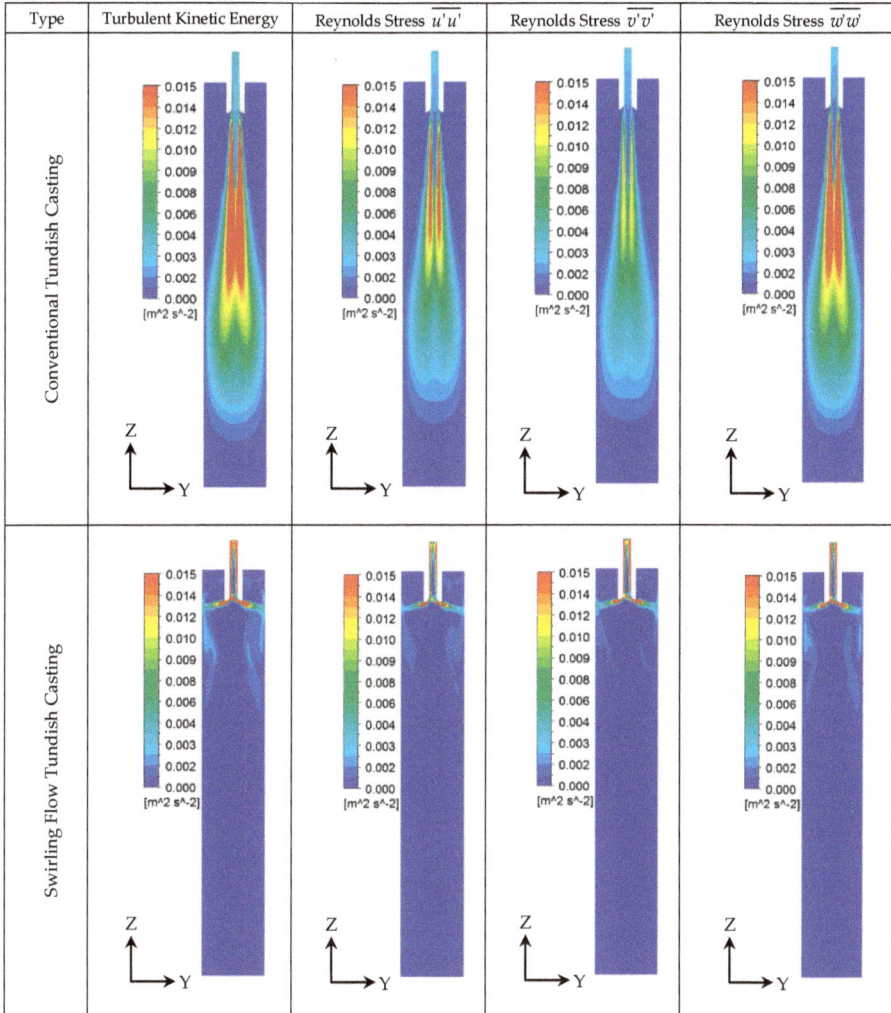

Figure 8. Turbulence properties on the YZ middle plane of the mold.

3.4. Steel/Slag Interface Phenomena

Steel/slag interface phenomena in the mold are very important during the continuous casting process. The reason is that the slag entrainment into the steel may lead to the formation of non-metallic inclusions. Therefore, a large steel/slag interface fluctuation is unwanted in the continuous casting process. Figure 9 shows the steel/slag interface and the flow pattern in the steel and slag. It can be seen that the thickness of the steel/slag interface region is larger when a swirling flow tundish was used, compared a conventional tundish. This can easily be seen from Figure 9b where the isosurface of the slag with the density of 2601 kg/m^3 was plotted (the pure slag has the density of 2600 kg/m^3). When a conventional tundish SEN was used, steel with a large momentum moves deeply into the mold, and a calm steel flow region at the top part of the mold was obtained. This can be seen from Figure 3a. However, the steel flow is activated in the top of the mold when the swirling flow tundish was used. This leads to a large level fluctuation, while it helps the heat transfer near the meniscus. In the mold with a swirling flow tundish casting, as shown in Figure 9, the steel flowing towards the solidified shell divided into an upwards and a downwards flow. The upwards flow shown in Figure 9b directly moves towards the steel/slag interface with the velocity of around 0.2 m/s, as shown in Figure 3b. Due to the small immersion depth of the SEN in steel with the value of around 12 cm, this velocity is still high when it reaches the steel/slag interface. This should cause a large steel/slag interface fluctuation. However, a flat interface is generally observed in this study. In the case with M-EMS, the level fluctuation was also found to be increased [10,33,64]. The meniscus surface has a swirl flow and the meniscus level rises near the bloom strand wall and sinks around the SEN wall, which shows an inclined steel/slag interface [33]. Sometimes, a vortex formation near the SEN wall was found with M-EMS [64]. Therefore, the mold level fluctuation should be considered to make it as low as possible, both for M-EMS applications and for the use of swirling flow SEN.

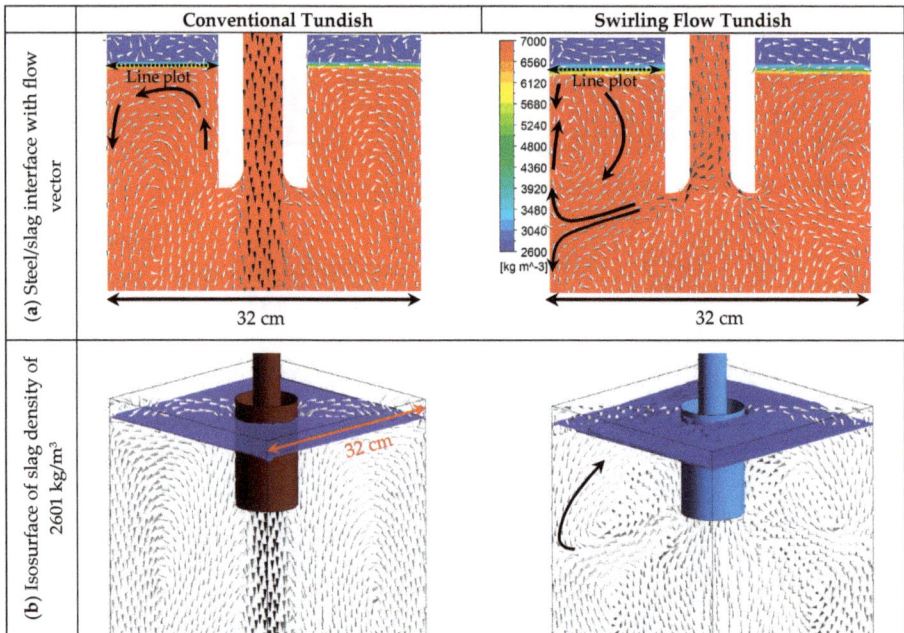

Figure 9. Steel/slag interface with steel flow vectors.

Figure 10 shows the distributions of the velocity magnitude and turbulent kinetic energy along the steel/slag interface. The location of the line plot is shown in Figure 9. It can be seen that the velocity magnitude with the swirling flow tundish case is around 3 times that of a conventional tundish, with the maximum value of 0.05 m/s. However, this maximum value is still smaller than that found in a mold with M-EMS application, where the values can reach around 0.2 to 0.3 m/s [33,64]. The turbulent kinetic energy is close to zero with a conventional tundish casting, and the maximum value is around 0.001 m^2/s^2 for the swirling flow tundish case. This means that the turbulence intensity is slightly increased when the swirling flow tundish design was used.

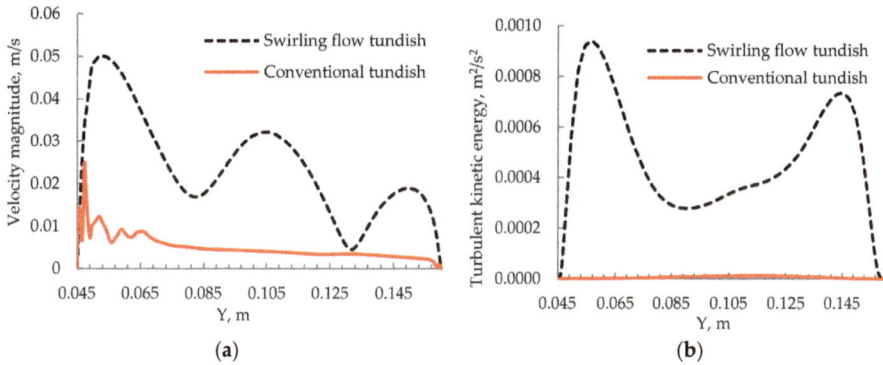

Figure 10. (a) Velocity magnitude at the steel/slag interface and (b) turbulent kinetic energy at the steel/slag interface.

According to a previous research [65], the slag entrainment into liquid steel may occur when the Weber number is greater than 12.3. The Weber number can be defined as

$$\mathrm{We} = \frac{u_1^2 \rho_1}{\sqrt{\sigma g(\rho_1 - \rho_s)}},$$ (7)

where u_1 is the radial steel velocity, g is gravitational acceleration, and σ is the interfacial tension between steel and slag. A slag density value of 2600 kg/m^3 was used, and the value of interfacial tension between the steel and the slag was set to 1.16 N/m [66]. The maximum total velocity at the steel-slag interface in the mold, 0.05 m/s, was used to calculate the Weber number. The calculated maximum Weber number is around 0.8 for the case with the swirling flow tundish casting. Therefore, the Weber number is still much smaller than 12.3, which means a small risk for the slag entrainment. Due to the swirling flow effect, the SEN outlet flow spreads towards the mold wall, rather than goes deep into the mold, as in a conventional tundish casting with a one-port SEN. Therefore, it is possible to increase the immersion depth of the SEN as well, where the value is 15 cm in the current study. This may help to decrease the fluctuations, as well as to further reduce the risk of the slag entrainment. In reality, the slag entrainment issue should be experimentally investigated in the future when a slag is used to protect the steel from the air reoxidation.

3.5. Temperature Distribution

Steel temperature in the mold is very important, since it significantly influences the solidification structure, which in turn determines the product quality. One important issue regarding the steel temperature is the removal of the steel superheat in the mold. Furthermore, a uniform steel temperature in the mold is also important, in order to obtain a uniform solidified shell. The influence of the swirling flow tundish casting on the temperature field in the mold was investigated, where the natural convection in the mold was also considered.

Figure 11 shows the temperature distribution in the mold. It can be seen that the swirling flow tundish design significantly change the temperature distribution in the mold. Specifically, in the mold with a conventional tundish SEN, the steel flow jet with a high temperature directly goes deeply into the mold. This leads to a high temperature region which is located deep in the center of the mold, as shown in Figure 11a. However, the temperature is low in the mold top. Therefore, the temperature field is not uniform in the mold, and it is not good for the removal of the steel superheat. In the top part of the mold, the density of steel in the region near the solidified shell is high, due to the low temperature. Therefore, steel tends to move downwards near the solidified shell. Furthermore, as stated before, the solidified steel shell moves downwards at a speed of 0.013 m/s in order to simulate the movement of the steel shell in a real casting. In addition, the weak rotational flow in the top of the mold also leads to the steel flowing downwards near the solidified shell. These factors lead to a downwards movement of steel with a low temperature as shown in Figure 11a. In the lower part of the mold, the rotational flow leads to an upwards steel flow near the solidified shell. This upwards flow comes from the main flow jet, and thus, has a high temperature. Finally, a low temperature region was formed in the mold at a depth of around 0.5 m, as shown in Figure 11a, where the low temperature downwards flow meets the high temperature upwards flow.

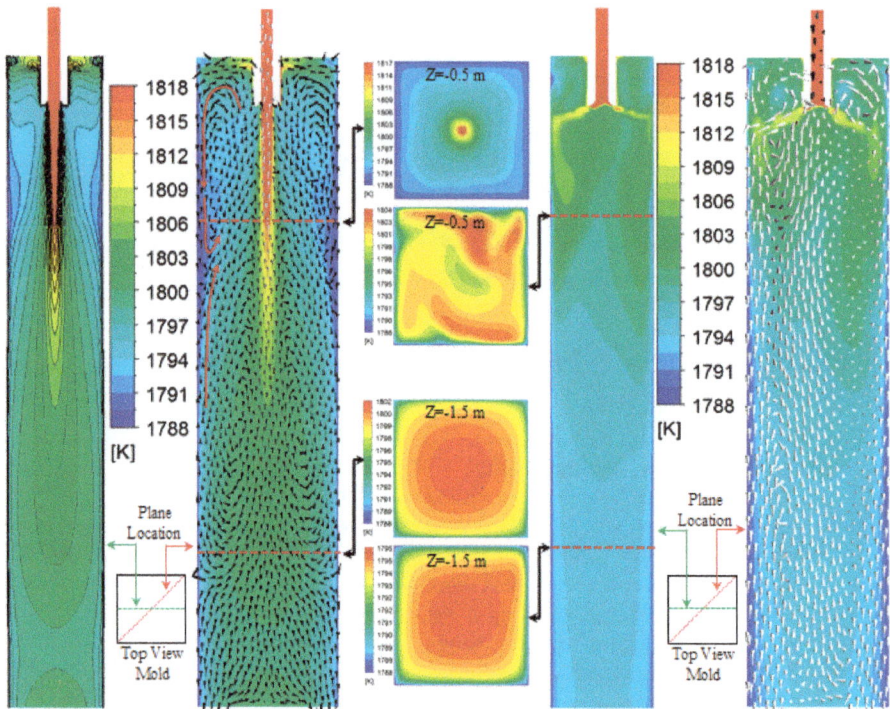

Figure 11. Temperature distribution in mold. (**a**) Conventional tundish casting and (**b**) swirling flow tundish design casting.

With a swirling flow tundish design, the temperature field in the mold changes a lot, as shown in Figure 11b. The high temperature impingement jet disappears, and the temperature field becomes more uniform in the mold. The maximum temperature gradually decreases, from the mold top to the bottom. Due to the swirling flow effect, steel with a high temperature changes the flow direction towards the solidified shell after it moves out from the SEN outlet. It increases the temperature near the solidified shell, as well as the temperature gradient there. Furthermore, this enhanced the removal of the steel superheat, and the core temperature of the billet was dramatically reduced. On the cross section at a depth of 0.5 m in the mold, the maximum temperature for the conventional tundish case and the swirling flow tundish case is 1817 and 1804 K, respectively. These values decrease to 1802 K for the conventional tundish casting and to 1795 K for the swirling flow tundish casting at the mold depth of 1.5 m. Therefore, the swirling flow improves the steel superheat removal and leads to a more uniform temperature field in the mold. This is good for the quality of the steel solidification structure with the formation of equiaxed grains. A previous study with the application of M-EMS shows that the high temperature region was located at the mold center [32]. This is due to the impingement jet from a straight SEN having a large inertia, and that M-EMS can only reduce the impingement depth, rather than completely remove the impingement jet. This may be the reason that the design of four horizontal side ports, with the port located in the tangential direction of the SEN circumference, was found to further improve the bloom quality cast in a M-EMS mold, compared to that with a straight SEN, due to the improvement in superheat dissipation [8].

3.6. Inclusion Behavior in Mold

In order to understand the behavior of inclusions, 30 inclusions of 1 μm, 10 μm, and 100 μm, released from the inlet, were individually tracked to see their behaviors in the mold. Figure 12 shows the trajectories of different sizes of inclusions in the mold. It can be seen that the steel flow pattern in the mold has a significant influence on the inclusion motions in the mold. Due to the impingement jet going deep into the mold, inclusions that follow the steel flow can reach a large mold depth in the conventional tundish casting. A few of them stay on the top part of the mold. This is true, even for large inclusions with a diameter of 100 μm, on which a large buoyancy force acts upon compared to small inclusions. Therefore, the steel flow pattern with a conventional tundish SEN is not beneficial for the removal of non-metallic inclusions. In the mold with a swirling flow tundish casting, some inclusions stay on the top of the mold for a time. This is due to the steel flow pattern change as previously shown in Figure 3. This may provide the chances for some inclusions to be removed. However, the buoyancy force is still not large enough to keep the large size inclusions at the top of the mold. In this study, a reflect wall boundary condition was used to show the moving path of inclusions. To realize a dynamic simulation on the inclusion removal into the slag, and the inclusion attachment on the wall, an interface capture model is required, and this is left for a future study.

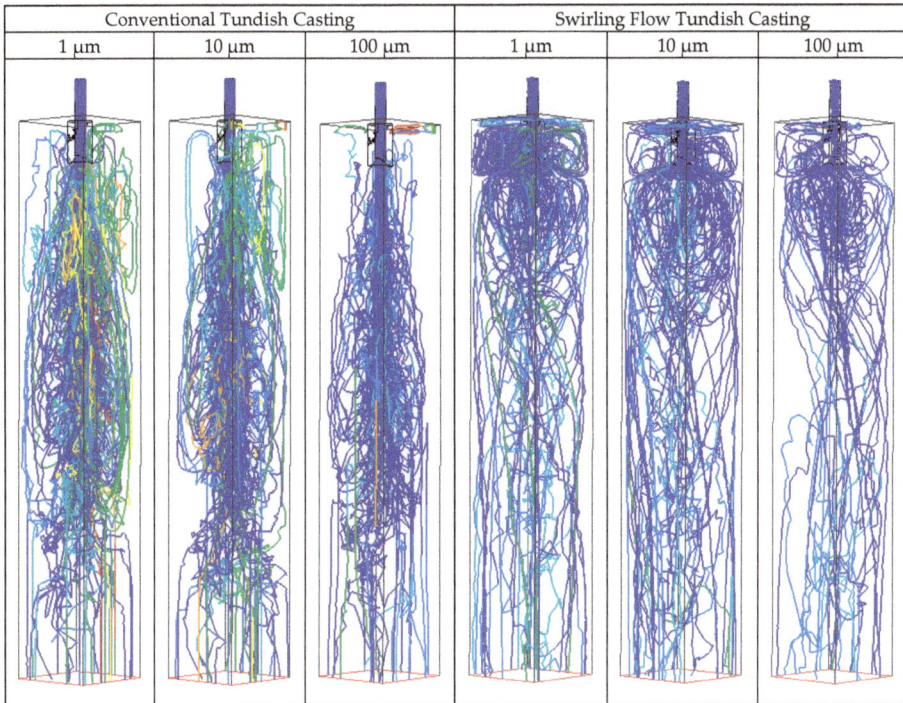

Figure 12. Trajectories of different size inclusions in the mold.

4. Concluding Discussion

A new cylindrical tundish design that produces a swirling flow in the SEN by using the steel flow potential has been investigated both by water model experiments and numerical simulations [47–49,61,67]. This study was the first to try to investigate the influence of such swirling flow tundish design on the steel flow, heat transfer, inclusion motion, and steel/slag interface stability in the mold. Previously, the effect of a swirling flow SEN on the steel flow in molds has been studied [34,38,40–42,46]. However, issues such as the steel/slag interface fluctuation, steel velocity in the vicinity of the solidified shell, and the field properties in the deep mold, are still not well investigated. In addition, the density change due to temperature variance was included in the model to consider the natural convection effect on the steel flow. In reality, there is a mushy region near the solidified shell, where the liquid volume fraction and the liquid viscosity change with the distance from the completely solidified shell; this is ignored in this study. Under this assumption, the multiphysics in the mold from a conventional tundish casting, and from the swirling flow tundish casting, are compared to show the influence of the swirling flow tundish design on the mold flow. However, a further study is still required to include the influence of the fluid property changes in the mushy region on the steel flow.

Swirling flow normally shows a certain swirl frequency [67,68], which means that the transient flow characteristics may not be axisymmetric. Therefore, a transient solver is recommended to solve this kind of flow phenomena. In this study, the plotted results are instantaneous flow field properties at 125 s, rather than the averaged flow properties in a time interval which is larger than the characteristic time of the swirling flow. This aims to give a clear observation on the swirling flow characteristics. The flow vectors in Figure 5 show that the flow is not axisymmetric. This is due to that the swirling flow in the SEN is not completely axisymmetric, as shown in Figure 1. However, the flow symmetry can be further improved by the cylindrical tundish design, such as using two tangential inlets in the cylindrical tundish in Figure 1a, which was investigated in a previous study [67]. For the study on M-EMS application, it was also found that the distribution of the electromagnetic force is not uniform in space [33]. The experimental measurements reveal that an axisymmetric flow cannot be maintained for situations with the simultaneous occurrence of the SEN jet flow and an electromagnetic stirring [64]. Therefore, efforts are still required in order to improve the flow uniformity.

The swirling flow intensity in the SEN is influenced by the cylindrical tundish design. Important design factors include the diameter of the cylinder and the steel flow velocity at the tangential inlet of the cylinder, as well as the tundish inlet area [47,49]. With a proper cylindrical tundish parameter, a certain intensity of the swirling flow in the SEN can be obtained. This should be determined by the desired flow behavior in the mold, such as the control of the mold level fluctuation. In addition, the swirling flow tundish casting can be simply realized either by connecting a small cylindrical tundish to a conventional tundish [47,48], or by installing a ceramic cylinder inside a conventional tundish [67]. Therefore, the functions of a conventional tundish, such as the inclusion removal, are not destroyed.

With a swirling flow in SEN, the steel flow has both a tangential momentum and a vertical momentum. These two momentum components influence the spreading angle of the SEN outlet flow. In addition, the steel flow in the mold is also affected by the geometry of the SEN outlet. Previous studies [8,38,40] regarding the swirling flow SEN show that a divergent nozzle can lead to the SEN outlet flow spreading more widely. In the current study, a divergent nozzle was used with only a small expansion in diameter at the SEN outlet, and it seemed to have almost no effect on the mold flow in a conventional tundish casting, where a straight impingement jet flow was observed. However, a wide spreading was observed when it was used together with the swirling flow tundish design. The spreading angle of the steel flow from the SEN is shown to be of great influence on the heat and fluid flow in the mold [38,40]. However, regarding the formation mechanism of the spreading angle, it is not well explained. Previous investigations on divergent nozzles were carried out in a swirling flow, which was produced by installing a swirl blade inside a SEN [38,40]. The mean value of tangential velocity in the SEN is around 1.72 m/s, and the averaged axial velocity is 2 m/s. A significant influence of the divergent nozzle angle on the steel flow spreading was found. However, a large spreading of the steel flow, as that shown in Figure 3b, was also found by Ying et al. [46], with a straight nozzle used in their study, where the swirling flow was produced by installing an electromagnetic stirrer surrounding the SEN. Furthermore, the spreading angle of the steel flow was found to change with the power input, which relates to the swirling flow intensity. This means that the divergent nozzle is not the necessary factor for the steel flow spreading. Figure 13 shows the SEN outlet flow direction in the mold. Theoretically, the spreading angle of the steel flow should be determined by the ratio of the tangential velocity to the axial velocity at the moment that steel leaves the SEN. This ratio in the current study is around 1.47, which corresponds to the spreading angle α of around 56° in Figure 13a. Figure 13b shows that the spreading angle is larger than this value. This is due to the induced upwards flow by the swirling flow tundish SEN in the mold center, which shifts the SEN outlet flow upwards. Since the instantaneous flow is not axisymmetric, the left side upwards flow is stronger than that on the right side. This may create a larger spreading angle on the left side than that on the right side. In addition, the steel tends to flow along the inner wall of the divergent nozzle, due to its inertia. Therefore, the swirling flow will undergo an expansion inside a divergent nozzle, due to the expanding nozzle diameter. This expansion further reduces the pressure in the swirling flow center and increases the low-pressure area. Therefore, steel near the divergent nozzle outlet will flow into the swirling flow center, creating an upwards steel flow. This may increase the shift of the steel flow angle compared to a straight nozzle, which has been observed in a previous study [38,40]. Finally, the swirling flow leads to a horizontal movement of the steel. The shortest distance for the steel reaching the solidified shell is shown in Figure 13c. During the way to the solidified shell, the velocity of the steel flow will decrease due to the momentum dissipation. Therefore, the magnitude of the steel flow velocity near the solidified shell should be greatly influenced by this distance. This distance in the current study is different, as shown in Figure 13c. This is also the reason that a non-uniform wall shear stress in Figure 7b was observed. Furthermore, the uniformity of shear stresses is expected to be improved in a round billet mold, since the distance from the SEN to the solidified shell will be equal in all directions. In addition, the velocity magnitude near the solidified shell might be increased by using a SEN with a large gradual expansion in diameter. With this kind of SEN, the swirling flow will have a gradually expanding and developing process, instead of a sudden expansion from a small SEN diameter to the large mold cross section. This may be helpful for the optimization of the mold flow, and it is left for a future study.

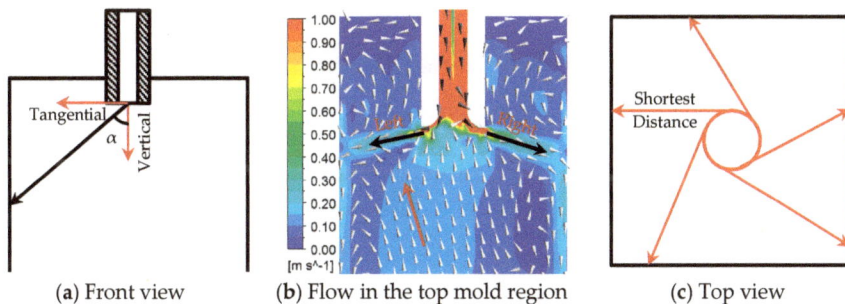

| (a) Front view | (b) Flow in the top mold region | (c) Top view |

Figure 13. Steel flow direction and its collision location with solidified shell. (a) Front view schematic, (b) flow in the top mold region, and (c) top view schematic.

5. Conclusions

Multiphase flow and heat transfer in a mold with a new cylindrical tundish design during the continuous casting process were investigated by using numerical simulations. Steel and slag flow, heat transfer, and inclusion motion in the mold were analyzed. The main conclusions were the following:

1. The new cylindrical tundish design for swirling flow casting significantly changed the flow behavior in the mold. The deep impingement jet in the mold disappeared, and the steel flow moved towards the solidified shell, due to the swirling flow effect. A large velocity in the vicinity of the solidified shell was obtained.
2. The steel flow velocity in the top part of the mold was increased. The calculated Weber number was round 0.8, which indicates a small risk for the slag entrainment.
3. With the swirling flow tundish casting, the temperature distribution became more uniform, and the dissipation of the steel superheat was accelerated. Furthermore, due to the high temperature steel directly flowing to the solidified shell, the temperature near the solidified shell was increased. A high temperature region was found at the top part of the mold, rather than in the deep center of the mold in a conventional tundish casting.
4. Inclusion trajectories in the mold change a lot, due to the change of the SEN outlet flow pattern. Instead of moving deeply into the mold following the impingement jet, some inclusions tended to stay for a time at the top part of the mold. This may be helpful for their removal.

Author Contributions: P.N. and L.T.I.J. designed the paper; P.N. and M.E. did the numerical simulation; all the authors analyzed and discussed the results; P.N. wrote the paper; M.E., L.T.I.J., T.Z. and P.G.J. revised the paper.

Funding: National Natural Science Foundation of China (Grant No. 51704062).

Acknowledgments: The authors want to thank the National Natural Science Foundation of China (Grant No. 51704062) for the support on this work.

Conflicts of Interest: The authors declare no conflict of interest.

References

1. Szekely, J.; Yadoya, R.T. The physical and mathematical modelling of the flow field in the mold region of continuous casting systems. Part II. The mathematical representation of the turbulence flow field. *Metall. Mater. Trans.* **1973**, *4*, 1379–1388. [CrossRef]
2. Xu, M.; Zhu, M. Transport phenomena in a Beam-Blank continuous casting mold with two types of submerged entry nozzle. *ISIJ Int.* **2015**, *55*, 791–798. [CrossRef]
3. Thomas, B.G.; Mika, L.J.; Najjar, F.M. Simulation of fluid flow inside a continuous slab-casting machine. *Metall. Mater. Trans. B* **1990**, *21*, 387–400. [CrossRef]
4. Calderon-Ramos, I.; Morales, R.D.; Garcia-Hernandez, S.; Ceballos-Huerta, A. Effects of immersion depth on flow turbulence of liquid steel in a slab mold using a nozzle with upward angle rectangular ports. *ISIJ Int.* **2014**, *54*, 1797–1806. [CrossRef]
5. Calderon-Ramos, I.; Morales, R.D.; Salazar-Campoy, M. Modeling flow turbulence in a continuous casting slab mold comparing the use of two bifurcated nozzles with square and circular ports. *Steel Res. Int.* **2015**, *86*, 1610–1621. [CrossRef]
6. Calderon-Ramos, I.; Morales, R.D. The role of submerged entry nozzle port shape on fluid flow turbulence in a slab mold. *Metall. Mater. Trans. B* **2015**, *46*, 1314–1325. [CrossRef]
7. Salazar-Campoy, M.; Morales, R.D.; Najera-Bastida, A.; Cedillo-Hernandez, V.; Delgado-Pureco, J.C. A physical model to study the effects of nozzle design on dense two-phase flows in a slab mold casting Ultra-Low carbon steels. *Metall. Mater. Trans. B* **2017**, *48*, 1376–1389. [CrossRef]
8. Sun, H.; Zhang, J. Macrosegregation improvement by swirling flow nozzle for bloom continuous castings. *Metall. Mater. Trans. B* **2014**, *45*, 936–946. [CrossRef]
9. Sun, H.; Li, L. Application of swirling flow nozzle and investigation of superheat dissipation casting for bloom continuous casing. *Ironmak. Steelmak.* **2016**, *43*, 228–233. [CrossRef]
10. Fang, Q.; Ni, H.; Zhang, H.; Wang, B.; Lv, Z. The effects of a submerged entry nozzle on flow and initial solidification in a continuous casting bloom mold with electromagnetic stirring. *Metals* **2017**, *7*, 146. [CrossRef]
11. Thomas, B.G.; Dennisov, A.; Bai, H. Behavior of argon bubbles during continuous casting of steel. In Proceedings of the ISS 80th Steelmaking Conference, Chicago, IL, USA, 13–16 April 1997; pp. 375–384.

12. Thomas, B.G.; Huang, X.; Sussman, R.C. Simulation of argon gas flow effects in a continuous slab caster. *Metall. Mater. Trans. B* **1994**, *25*, 527–547. [CrossRef]
13. Li, B.; Okane, T.; Umeda, T. Modeling of biased flow phenomena associated with the effects of static magnetic-field application and argon gas injection in slab continuous casting of steel. *Metall. Mater. Trans. B* **2001**, *32*, 1053–1066. [CrossRef]
14. Wang, Y.; Zhang, L. Fluid flow-related transport phenomena in steel slab continuous casting strands under electromagnetic brake. *Metall. Mater. Trans. B* **2011**, *42*, 1319–1351. [CrossRef]
15. Liu, Z.; Li, B.; Jiang, M. Transient asymmetric flow and bubble transport inside a slab continuous casting mold. *Metall. Mater. Trans. B* **2014**, *45*, 675–697. [CrossRef]
16. Liu, Z.; Li, L.; Qi, F.; Li, B.; Jiang, M.; Tsukihashi, F. Population balance modeling of polydispersed bubbly flow in continuous casting using multiple-size-group approach. *Metall. Mater. Trans. B* **2015**, *46*, 406–420. [CrossRef]
17. Liu, Z.; Sun, Z.; Li, B. Modeling of quasi-four-phase flow in continuous casting mold using hybrid Eulerian and Lagrangian approach. *Metall. Mater. Trans. B* **2017**, *48*, 1248–1267. [CrossRef]
18. Liu, Z.; Li, B. Large-Eddy simulation of transient horizontal gas–liquid flow in continuous casting using dynamic subgrid-scale model. *Metall. Mater. Trans. B* **2017**, *48*, 1833–1849. [CrossRef]
19. Pfeiler, C.; Wu, M.; Ludwig, A. Influence of argon gas bubbles and non-metallic inclusions on the flow behavior in steel continuous casting. *Mater. Sci. Eng. A* **2005**, *413–414*, 115–120. [CrossRef]
20. Yu, H.; Zhu, M. Numerical simulation of the effects of electromagnetic brake and argon gas injection on the three-dimensional multiphase flow and heat transfer in slab continuous casting mold. *ISIJ Int.* **2008**, *48*, 584–591. [CrossRef]
21. Cho, S.; Kim, S.; Thomas, B.G. Transient fluid flow during steady continuous casting of steel slabs: Part I. measurements and modeling of two-phase flow. *ISIJ Int.* **2014**, *54*, 845–854. [CrossRef]
22. Jin, K.; Thomas, B.G.; Ruan, X. Modeling and measurements of multiphase flow and bubble entrapment in steel continuous casting. *Metall. Mater. Trans. B* **2016**, *47*, 548–565. [CrossRef]
23. Cho, S.; Thomas, B.G.; Kim, S. Transient two-phase flow in slide-gate nozzle and mold of continuous steel slab casting with and without double-ruler electro-magnetic braking. *Metall. Mater. Trans. B* **2016**, *47*, 3080–3098. [CrossRef]
24. Cukierski, K.; Thomas, B.G. Flow control with local electromagnetic braking in continuous casting of steel slabs. *Metall. Mater. Trans. B* **2008**, *39*, 94–107. [CrossRef]
25. Zhang, L.; Wang, Y.; Zuo, X. Flow transport and inclusion motion in steel continuous-casting mold under submerged entry nozzle clogging condition. *Metall. Mater. Trans. B* **2008**, *39*, 534–550. [CrossRef]
26. Gonzalez-Trejo, J.; Real-Ramirez, C.A.; Miranda-Tello, R.; Rivera-Perez, F.; Cervantes-De-La-Torre, F. Numerical and physical parametric analysis of a SEN with flow conditioners in slag continuous casting mold. *Arch. Metall. Mater.* **2017**, *62*, 927–946. [CrossRef]
27. Miao, X.; Timmel, K.; Lucas, D.; Ren, Z.; Eckert, S.; Gerbeth, G. Effect of an electromagnetic brake on the turbulent melt flow in a continuous-casting mold. *Metall. Mater. Trans. B* **2012**, *43*, 954–972. [CrossRef]
28. Liu, Z.; Li, L.; Li, B. Large eddy simulation of transient flow and inclusions transport in continuous casting mold under different electromagnetic brakes. *JOM* **2016**, *68*, 2180–2190. [CrossRef]
29. Ha, M.Y.; Lee, H.G.; Seong, S.H. Numerical simulation of three-dimensional flow, heat transfer, and solidification of steel in continuous casting mold with electromagnetic brake. *J. Mater. Process. Technol.* **2003**, *133*, 322–339. [CrossRef]
30. Yu, H.; Wang, B.; Li, H.; Li, J. Influence of electromagnetic brake on flow field of liquid steel in the slab continuous casting mold. *J. Mater. Process. Technol.* **2008**, *202*, 179–187.
31. Yu, H.; Zhu, M. Three-dimensional magnetohydrodynamic calculation for coupling multiphase flow in round billet continuous casting mold with electromagnetic stirring. *IEEE Trans. Magn.* **2010**, *46*, 82–86.
32. Yu, H.; Zhu, M. Influence of electromagnetic stirring on transport phenomena in round billet continuous casting mould and macrostructure of high carbon steel billet. *Ironmak. Steelmak.* **2012**, *39*, 574–584. [CrossRef]
33. Liu, H.; Xu, M.; Qiu, S.; Zhang, H. Numerical simulation of fluid flow in a round bloom mold with In-Mold rotary electromagnetic stirring. *Metall. Mater. Trans. B* **2012**, *43*, 1657–1675. [CrossRef]
34. Yokoya, S.; Takagi, S.; Iguchi, M.; Asako, Y.; Westoff, R.; Hara, S. Swirling effect in immersion nozzle on flow and heat transport in billet continuous casting mold. *ISIJ Int.* **1998**, *38*, 827–833. [CrossRef]

35. Yokoya, S.; Takagi, S.; Iguchi, M.; Marukawa, K.; Yasugair, W.; Hara, S. Development of swirling flow generator in immersion nozzle. *ISIJ Int.* **2000**, *40*, 584–588. [CrossRef]
36. Yokoya, S.; Takagi, S.; Kaneko, M.; Iguchi, M.; Marukawa, K.; Hara, S. Swirling flow effect in off-center immersion nozzle on bulk flow in billet continuous casting mold. *ISIJ Int.* **2001**, *41*, 1215–1220. [CrossRef]
37. Yokoya, S.; Takagi, S.; Ootani, S.; Iguchi, M.; Marukawa, K.; Hara, S. Swirling flow effect in submerged entry nozzle on bulk flow in high throughput slab continuous casting mold. *ISIJ Int.* **2001**, *41*, 1208–1214. [CrossRef]
38. Yokoya, S.; Jönsson, P.G.; Sasaki, K.; Tada, K.; Takagi, S.; Iguchi, M. The effect of swirl flow in an immersion nozzle on the heat and fluid flow in a billet continuous casting mold. *Scan. J. Metall.* **2004**, *33*, 22–28. [CrossRef]
39. Tsukaguchi, Y.; Hayashi, H.; Kurimoto, H.; Yokoya, S.; Marukawa, K.; Tanaka, T. Development of swirling-flow submerged entry nozzles for slab casting. *ISIJ Int.* **2010**, *50*, 721–729. [CrossRef]
40. Kholmatov, S.; Takagi, S.; Jonsson, L.; Jönsson, P.; Yokoya, S. Development of flow field and temperature distribution during changing divergent angle of the nozzle when using swirl flow in a square continuous casting billet mould. *ISIJ Int.* **2007**, *47*, 80–87. [CrossRef]
41. Kholmatov, S.; Takagi, S.; Jönsson, P.; Jonsson, L.; Yokoya, S. Influence of aspect ratio on fluid flow and heat transfer in mould when using swirl flow during casting. *Steel Res. Int.* **2008**, *79*, 698–707. [CrossRef]
42. Kholmatov, S.; Takagi, S.; Jonsson, L.; Jönsson, P.; Yokoya, S. Effect of nozzle angle on flow field and temperature distribution in a billet mould when using swirl flow. *Steel Res. Int.* **2008**, *79*, 31–39. [CrossRef]
43. Geng, D.; Lei, H.; He, J.; Liu, H. Effect of electromagnetic swirling flow in slide-gate SEN on flow field in square billet continuous casting mold. *Acta Metall. Sin. (Engl. Lett.)* **2012**, *25*, 347–356.
44. Wondrak, Th.; Eckert, S.; Galindo, V.; Gerbeth, G.; Stefani, F.; Timmel, K.; Peyton, A.J.; Yin, W.; Riaz, S. Liquid metal experiments with swirling flow submerged entry nozzle. *Ironmak. Steelmak.* **2012**, *39*, 1–9. [CrossRef]
45. Li, D.; Su, Z.; Chen, J.; Wang, Q.; Yang, Y.; Nakajima, K.; Marukaw, K.; He, J. Effects of electromagnetic swirling flow in submerged entry nozzle on square billet continuous casting of steel process. *ISIJ Int.* **2013**, *53*, 1187–1194. [CrossRef]
46. Yang, Y.; Jönsson, P.G.; Ersson, M.; Su, Z.; He, J.; Nakajima, K. The Influence of swirl flow on the flow field, temperature field and inclusion behavior when using a half type electromagnetic swirl flow generator in a submerged entry and mold. *Steel Res. Int.* **2015**, *86*, 1312–1327. [CrossRef]
47. Ni, P.; Jonsson, L.; Ersson, M.; Jönsson, P. A new tundish design to produce a swirling flow in the SEN during continuous casting of steel. *Steel Res. Int.* **2016**, *87*, 1356–1365. [CrossRef]
48. Ni, P.; Jonsson, L.; Ersson, M.; Jönsson, P. Non-Metallic inclusion behaviors in a new tundish and SEN design using a swirling flow during continuous casting of steel. *Steel Res. Int.* **2017**, *88*, 1600155. [CrossRef]
49. Ni, P.; Wang, D.; Jonsson, L.; Ersson, M.; Zhang, T.; Jönsson, P. Numerical and physical study on a cylindrical tundish design to produce a swirling flow in the SEN during continuous casting of steel. *Metall. Mater. Trans. B* **2017**, *48*, 2695–2706. [CrossRef]
50. Wang, S.; Yang, V.; Hsiao, G.; Hsieh, S.; Mongia, H.C. Large-eddy simulation of gas-turbine swirl injector flow dynamics. *J. Fluid Mech.* **2007**, *583*, 99–122. [CrossRef]
51. Weber, R.; Visser, B.M.; Boysan, F. Assessment of turbulence modeling for engineering predictions of swirling vortices in the near burner zone. *Int. J. Heat Fluid Flow* **1990**, *11*, 225–235. [CrossRef]
52. Hoekstra, A.; Derksen, J.; Van Den Akker, H. An experimental and numerical study on turbulent swirling flow in gas cyclones. *Chem. Eng. Sci.* **1999**, *54*, 2055–2065. [CrossRef]
53. Jakirlic, S.; Hanjalic, K.; Tropea, C. Modeling rotating and swirling turbulent flows: A perpetual challenge. *AIAA J.* **2002**, *40*, 1984–1996. [CrossRef]
54. Patankar, S.V. *Numerical Heat Transfer and Fluid Flow*; Hemispere Publishing Corp.: New York, NY, USA, 1980.
55. ANSYS. *Fluent Theory Guide*; Release 18.0; ANSYS: Canonsburg, PA, USA, 2017.
56. Versteeg, H.K.; Malalasekera, W. *An Introduction to Computational Fluid Dynamics: The Finite Volume Method*, 2nd ed.; Pearson Education Limited: London, UK, 2007; p. 80.
57. Launder, B.E.; Reece, G.J.; Rodi, W. Progress in the development of a Reynolds-stress turbulence closure. *J. Fluid Mech.* **1975**, *68*, 537–566. [CrossRef]
58. Lien, F.S.; Leschziner, M.A. Assessment of turbulence-transport models including non-linear RNG eddy-viscosity formulation and second-moment closure for flow over a backward-facing step. *Comput. Fluids* **1994**, *23*, 983–1004. [CrossRef]

59. Shih, T.-H.; Liou, W.W.; Shabbir, A.; Yang, Z.; Zhu, J. A new k-ε eddy viscosity model for high Reynolds number turbulent flows. *Comput. Fluids* **1995**, *24*, 227–238. [CrossRef]

60. ANSYS. *Fluent User's Guide*; Release 18.0; ANSYS: Canonsburg, PA, USA, 2017.

61. Ni, P.; Ersson, M.; Jonsson, L.; Jönsson, P. A study on the nonmetallic inclusion motions in a swirling flow submerged entry nozzle in a new cylindrical Tundish design. *Metall. Mater. Trans. B* **2018**, *49*, 723–736. [CrossRef]

62. Saffman, P.G. The lift on a small sphere in a slow shear flow. *J. Fluid Mech.* **1965**, *22*, 385–400. [CrossRef]

63. Morsi, S.A.; Alexander, A.J. An investigation of particle trajectories in two-phase flow systems. *J. Fluid Mech.* **1972**, *55*, 193–208. [CrossRef]

64. Willers, B.; Barna, M.; Reiter, J.; Eckert, S. Experimental investigations of rotary electromagnetic mould stirring in continuous casting using a cold liquid metal model. *ISIJ Int.* **2017**, *57*, 468–477. [CrossRef]

65. Jonsson, L.; Jönsson, P. Modeling of fluid flow conditions around the slag/metal interface in a gas-stirred ladle. *ISIJ Int.* **1996**, *36*, 1127–1134. [CrossRef]

66. Shannon, G.N.; Sridhar, S. Film-drainage, separation and dissolution of Al_2O_3 inclusions at steel/interfaces. *High Temp. Mater. Process.* **2005**, *24*, 111–124. [CrossRef]

67. Ni, P.; Ersson, M.; Jonsson, L.; Jönsson, P.G. Application of a swirling flow producer in a conventional tundish during continuous casting of steel. *ISIJ Int.* **2017**, *57*, 2175–2184. [CrossRef]

68. Bai, H.; Ersson, M.; Jönsson, P. Experimental validation and numerical analysis of the swirling flow in a submerged entry nozzle and mold by using a reverse turboswirl in a billet continuous casting process. *Steel Res. Int.* **2017**, *88*, 1600399. [CrossRef]

metals

MDPI

Article

Analysis of the Influence of Segmented Rollers on Slab Bulge Deformation

Qin Qin *, Ming Li and Jianlin Huang

School of Mechanical Engineering, University of Science and Technology Beijing, Beijing 100083, China;
18567937527@163.com (M.L.); S20160419@xs.ustb.edu.cn (J.H.)
* Correspondence: qinqin@me.ustb.edu.cn; Tel.: +86-10-6233-4106

Received: 26 December 2018; Accepted: 12 February 2019; Published: 14 February 2019

Abstract: The bulge deformation of the continuous casting slab must be controlled in order to improve the slab quality. In this study, a coupled three-dimensional thermomechanical model is suggested based on dynamic contact between the slab and the rollers, so as to investigate the influence of the rollers in reducing slab bulge deformation. Moreover, the rigid casting rollers in this model are replaced by elastic casting rollers in order to improve the calculation accuracy. Further, the influence of two-segment and three-segment rollers on the slab bulge deformation is systematically studied. The results indicate that the bulge deformation of the slab increased by 74.3% when elastic casting rollers were adopted instead of rigid casting rollers. This deformation was reduced by 29.7% when three-segment rollers were used instead of two-segment rollers. Moreover, the influence of the roller spacing and the roller diameter of the segmented roller on the deformation was studied in detail. In order to achieve the purpose of controlling the bulge deformation, improved segmented roller spacing and diameter were proposed, leading to a 75.4% reduction in the bulge deformation.

Keywords: continuous casting; bulge deformation; thermomechanical coupling; segmented roller; finite element analysis

1. Introduction

It is important to effectively control bulge deformation to improve the quality of continuous casting slabs. Thermal creep is one of the main factors that can cause bulge deformation in slabs, as the temperature distribution of the slab on both the wide and narrow sides affects bulge deformation. It is necessary to obtain the temperature field of a continuous casting slab during the solidification process. Some scholars have developed a two-dimensional solidification model to obtain the temperature field distribution of the slab during continuous casting [1,2]. However, they neglected the temperature field distribution of the wide sides. The two-dimensional solidification model was replaced by a three-dimensional solidification model to analyze the slab temperature field because the temperature distribution of the slab on both the wide and narrow sides could be considered in the three-dimensional model [3,4], which could better match the actual solidification process in continuous casting. This deformation calculation is very complicated. Some scholars used the theoretical analysis method [5–8] and two-dimensional thermomechanical coupling models to calculate bulge deformation [9,10]. However, the calculation accuracy of the theoretical analysis method could not be guaranteed because the bulge deformation and temperature fluctuations on the wide side of the slab in the two-dimensional model were neglected. Therefore, three-dimensional models of slab bulge deformation were developed to improve the calculation accuracy. A three-dimensional elastic–plastic and creep model was developed to calculate the slab bulge deformation on the wide side [11–13]. This model considered static contact between the continuous casting slab and the rollers. However, this model neglected the movement of the slab under the casting rollers.

In order to consider the dynamic contact between the slab and the rollers, Qin et al. suggested a 3D thermomechanical coupling model based on this dynamic contact between the slab and the rollers to compare with the 2D bulge deformation model to calculate bulge deformation [14]. Thus, Liu et al. proposed a 3D finite element viscoelastic creep model to study slab bulge deformation and the influence of temperature field distribution on this deformation [15]. However, this model neglected the slab bulge deformation on the narrow side. The slab bulge deformation on the wide side and that on the narrow side are actually mutually influential, and bulge deformation on the wide and narrow sides has been observed by some scholars [16–18]. Thus, 3D thermomechanical coupling models that included the dynamic contacts between the slab and the rollers were established to study the slab bulge deformation of the narrow side [19,20]. The 3D finite element models mentioned above mainly focused on the deformation mechanism of bulge deformation and the influence of process and structural parameters on this deformation. However, control measures of bulge deformation were hardly used in these models.

Qin et al. built a three-dimensional thermomechanical coupling model to study the temperature and bulge deformation distributions of a slab during the casting process, and the bulge deformation of this model was compared with that of the 2D bulging model [13,18]. The simulation results in the 3D bulging models were further compared with measured data from the actual production process to explore the accuracy of the simulation results in the three-dimensional bulge deformation models [17,20]. This comparison made it possible to use the finite element method to explore a method for controlling slab bulge deformation, and the fixed-gap and variable-diameter methods were suggested to reduce bulge deformation [21–24].

In the above-mentioned models, all casting rollers were assumed to be rigid, in order to save computing time, and the influence of casting roller deformation on slab bulge deformation was neglected. In fact, these casting rollers also deform during continuous casting. Therefore, this simplification would lead to an inaccurate reduction of bulge deformation values for a continuous casting slab. Moreover, segmented rollers have been widely adopted in the actual production process to control for slab bulge deformation [25–27]. The reason for this is that the stiffness of segmented rollers is much greater than that of solid rollers [28,29]. However, the structural parameters of segmented rollers have not been systematically studied for controlling slab bulge deformation.

This paper aims to discover the influence of segmented rollers on reducing slab bulge deformation and investigate the relationship between the stiffness of the segmented rollers and slab bulge deformation. A 3D thermomechanical coupling model based on the dynamic contact between the slab and the elastic casting rollers was suggested, and the structural parameters of the segmented rollers are investigated for controlling bulge deformation.

2. 3D Solidification Model Description

The temperature distribution is the foundation of this bulging analysis because the thickness of the solidified slab and the material property parameters depend on the temperature field in the slab. Therefore, the first step was to obtain the temperature field by using a solidification model.

2.1. 3D Solidification Finite Element Model

The cooling process of the continuous casting slab was a three-dimensional transient heat transfer process with heat conduction, thermal convection, and thermal radiation accompanied by phase transition. The heat transfer differential equation can be expressed as [30]:

$$\rho c_p \frac{\partial T}{\partial t} = \frac{\partial}{\partial x}\left(\lambda \frac{\partial T}{\partial x}\right) + \frac{\partial}{\partial y}\left(\lambda \frac{\partial T}{\partial y}\right) + \frac{\partial}{\partial z}\left(\lambda \frac{\partial T}{\partial z}\right) + q \tag{1}$$

The axial position z along the liquid level of crystallizer is related to casting speed u and time t. The relation function is $z = ut$. The relation function is substituted into Equation (1) and Equation (2) is shown as follows:

$$\rho c_p u \frac{\partial T}{\partial z} = \frac{\partial}{\partial x}\left(\lambda \frac{\partial T}{\partial x}\right) + \frac{\partial}{\partial y}\left(\lambda \frac{\partial T}{\partial y}\right) + \frac{\partial}{\partial z}\left(\lambda \frac{\partial T}{\partial z}\right) + q \tag{2}$$

where

ρ is the density $(kg \cdot m^{-3})$,

c_p is the specific heat capacity under constant pressure $(J \cdot (kg \cdot K)^{-1})$,

T is the temperature (K),

t is time (s),

q is the internal heat source $(W \cdot m^{-3})$,

λ is the thermal conductivity $(W \cdot (m \cdot K)^{-1})$, and

u is the casting speed, $(m \cdot min^{-1})$.

There is a solid phase zone, a liquid phase zone, and a two-phase zone in the continuous casting process, and Equation (1) can be treated as follows:

(1) The solid phase and liquid phase can be calculated as:

$$\rho c_p \frac{\partial T}{\partial t} = \frac{\partial}{\partial x}\left(\lambda \frac{\partial T}{\partial x}\right) + \frac{\partial}{\partial y}\left(\lambda \frac{\partial T}{\partial y}\right) + \frac{\partial}{\partial z}\left(\lambda \frac{\partial T}{\partial z}\right) \tag{3}$$

The heat transfer differential equation of the slab in the solid phase zone and the liquid phase zone is exactly the same Equation (3). The corresponding thermal conductivity can be obtained in the solid phase zone and the liquid phase zone separately.

For the thermal conductivity λ_L of the liquid phase zone, the forced convection heat transfer was caused due to the flow of the molten steel, which accelerated the elimination of the degree of superheat. In 1967, Mizikar [31] introduced the effective thermal conductivity for the first time to take the effects of convective heat transfer into account. The method was to treat the liquid phase zone of the slab into a "quasi-solid" and convert the convective thermal conductivity of the molten steel into an effective thermal conductivity, which was equivalent to n times compared with the thermal conductivity of the static molten steel. The value of n was generally from 2 to 7. Therefore, this method was used to calculate the thermal conductivity of the liquid phase zone in the paper and the value of n was 5. The relationship between the thermal conductivity of the solid phase zone and the thermal conductivity of the liquid phase zone was as follows:

$$\lambda_L = 5\lambda_S \tag{4}$$

where

λ_L is the thermal conductivity of the solid phase zone $(W \cdot (m \cdot K)^{-1})$, and

λ_S is the thermal conductivity of the liquid phase zone $(W \cdot (m \cdot K)^{-1})$.

(2) Since the two-phase zone has the latent heat of solidification, the internal heat source q in Equation (1) must be taken into consideration. In this study, the equivalent specific heat method was used to solve the latent heat of solidification and q is calculated as:

$$q = -\rho_{ls} \frac{\Delta H_f}{T_L - T_S} \frac{\partial T}{\partial t} \tag{5}$$

where

ΔH_f is latent heat of solidification under the action of various metals $(kJ \cdot kg^{-1})$,

T_L is the liquidus temperature (*K*), and

T_S is the solidus temperature (*K*).

Equation (5) was introduced in the solidification heat transfer differential equation of the two-phase zone and Equation (1) was rearranged as:

$$\rho_{ls} c_{eff} \frac{\partial T}{\partial t} = \frac{\partial}{\partial x}\left(\lambda_{ls}\frac{\partial T}{\partial x}\right) + \frac{\partial}{\partial y}\left(\lambda_{ls}\frac{\partial T}{\partial y}\right) + \frac{\partial}{\partial z}\left(\lambda_{ls}\frac{\partial T}{\partial z}\right) \tag{6}$$

where c_{eff} is the equivalent heat capacity ($J \cdot (kg \cdot K)^{-1}$) and is calculated as:

$$c_{eff} = c_p + \frac{\Delta H_f}{T_L - T_S} \tag{7}$$

λ_{ls} is the thermal conductivity of the two-phase zone ($W \cdot (m \cdot K)^{-1}$), and is calculated as:

$$\lambda_{ls} = \lambda_s + \frac{\lambda_L - \lambda_S}{T_L - T_S}(T - T_S) \tag{8}$$

ρ_{ls} is the density of the two-phase zone ($kg \cdot m^{-3}$).

A half three-dimensional solidification finite element model was established due to the symmetry of the slab, and had dimensions of 1200 mm × 2000 mm × 250 mm. DC3D8 is an eight-node linear heat transfer hexahedral element in the ABAQUS software (6.14, Dassault Systèmes Simulia Corp., Providence, RI, USA) and was used to mesh the solidification model to perform thermal simulation analysis. The size of the element was 25 mm (length) × 25 mm (width) × 10 mm (height), and 46,080 heat transfer elements and 51,597 nodes were included in the model. The bulge deformation at the end of the foot roller section was investigated because the bulge deformation of the slab is serious when the slab leaves the foot roller section. Therefore, segmented rollers have been widely adopted to control this deformation. Then, the temperature field at the end of the foot roller section was studied for the analysis. The location of the segmented rollers and the final solidification point are listed in Figure 1.

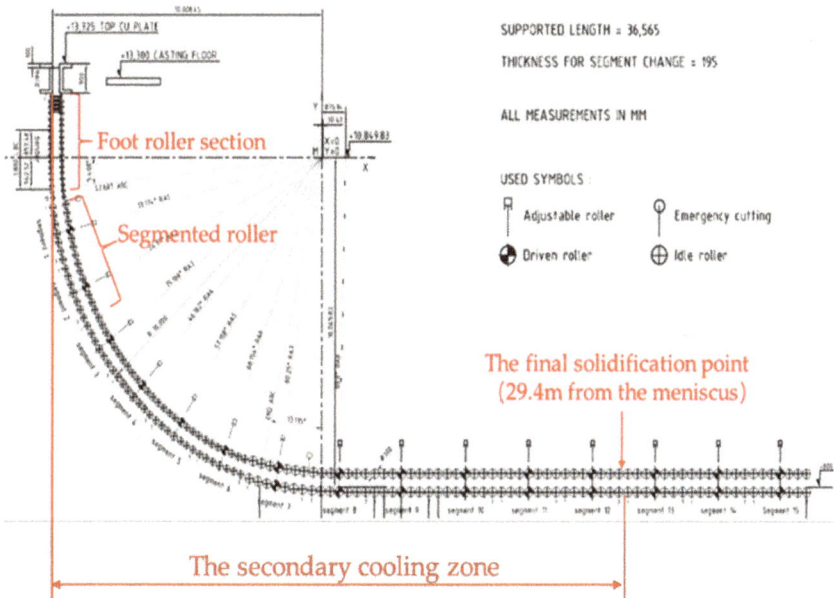

Figure 1. Slab caster roller map.

2.2. Physical Property Parameters

The material of the slab analyzed in this paper was Q235, and the material that was measured in Xiangtan Iron & Steel Co., Ltd. of Hunan Valin was AH36. The compositions of Q235 and AH36 are shown in Tables 1 and 2, respectively.

Table 1. Chemical composition and content of Q235.

Chemical Element	C	Si	Mn	P	S
Content (%)	0.18	0.20	0.40	≤0.025	≤0.022

Table 2. Chemical composition and content of AH36.

Chemical Element	C	Si	Mn	P	S	Al	Cr	Cu	Ni
Content (%)	0.157	0.2489	1.4143	0.0162	0.0044	0.0289	0.0375	0.0284	0.0177

According to the actual production process of an arc continuous casting machine, the slab geometry of the model and physical property parameters were quoted from the references [31–34] and they are listed in Table 3 and shown in Figure 2. For this study, the cooling water heat transfer, the slab surface radiation, and the heat transfer between the rollers and the slab were defined as the equivalent convection coefficient; the convection heat transfer of molten steel was expressed by effective thermal conductivity; and the latent heat of solidification was calculated by using the equivalent specific heat method.

Table 3. Simulation constants for solidification analysis.

Parameters	Values
Mold width	1200 mm
Half slab thickness	125 mm
Mold length	2000 mm
Casting speed	$1.5 \, m \cdot min^{-1}$
Water temperature	303 K
Inlet temperature	1808 K
Liquidus temperature	1793 K
Solidus temperature	1732 K
Specific heat capacity of the solid phase zone (c_S)	$706 \, J \cdot (kg \cdot K)^{-1}$
Specific heat capacity of the liquid phase zone (c_L)	$825 \, J \cdot (kg \cdot K)^{-1}$
Thermal conductivity of the solid phase zone (λ_S)	$34.83 \, W \cdot (m \cdot K)^{-1}$
Thermal conductivity of the liquid phase zone (λ_L)	$165 \, W \cdot (m \cdot K)^{-1}$
Latent heat of solidification under various metals (ΔH_f)	$284 \, kJ \cdot kg^{-1}$
Density	$7400 \, kg \cdot m^{-3}$

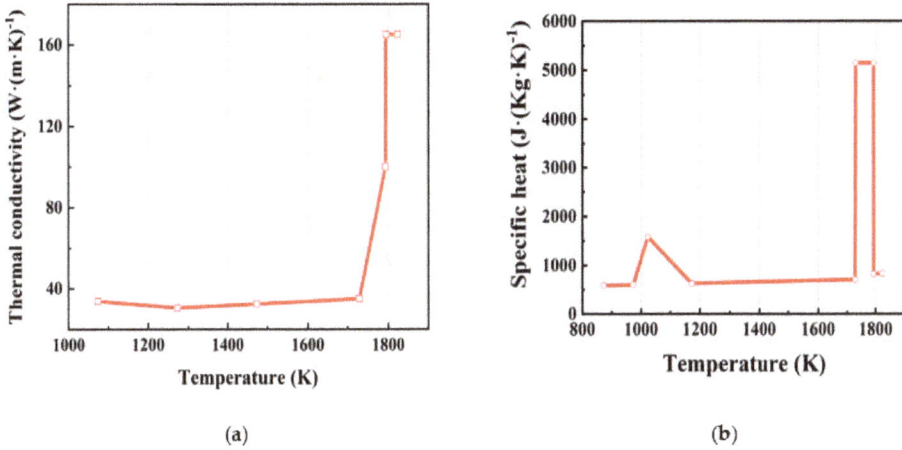

Figure 2. (a) Thermal conductivity of Q235 steel; (b) the specific heat of Q235 steel.

2.3. Boundary Conditions

(1) The comprehensive heat transfer coefficient was utilized to express the heat transfer process of the model, and the calculation expressions are as follows:

(a) The heat flux in the crystallizer was defined as follows:

$$q = 2680000 - b\sqrt{\frac{L}{v}} \tag{9}$$

where

b is determined by the actual heat balance calculation,
L is the crystallizer length (m), and
v is the casting speed ($m \cdot min^{-1}$).

(b) The heat flux in the secondary cooling zone was determined as follows, and the secondary cooling zone was from the end of the crystallizer to the final solidification point:

$$\Phi = h(T_b - T_w) \tag{10}$$

where

Φ is the heat flux ($W \cdot cm^{-2}$),
h is the heat transfer coefficient ($W \cdot (m^2 \cdot K)^{-1}$),
T_b is the slab surface temperature (K), and
T_w is the cooling water temperature (K).

(2) The initial temperature of the slab was uniformed as the inlet temperature of 1808 K.

2.4. Results and Discussion

The temperature and thickness distributions of the slab on the wide and narrow sides are shown in Figure 3. In this study, the simulation results were validated based on the actual slab temperature measurement results of the AH36 steel continuous casting process. The comparison of the measured and simulated temperatures in the center of the narrow face are exhibited in Figure 3b. As can be seen from this figure, the simulation temperature was close to the test temperature, and the maximum relative error was less than 5%. Thus, the simulation result that agreed well with the measured data could be used as the basis temperature field for bulge deformation analysis.

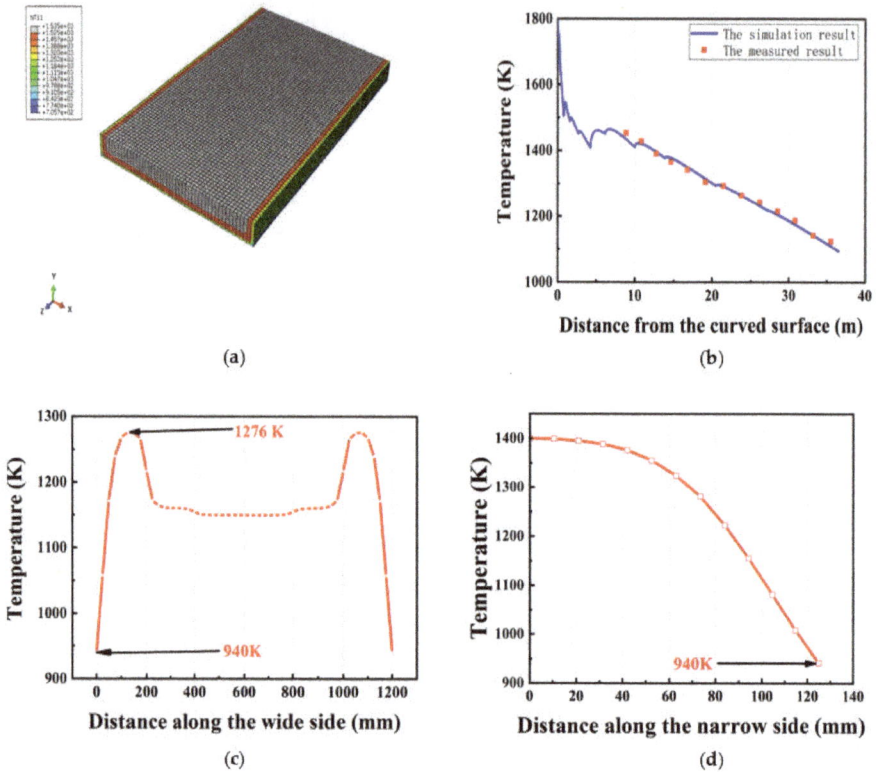

(a)

(b)

(c)

(d)

Figure 3. (a) Temperature field of the model. (b) The comparison of the measured and simulated temperature in the center of the narrow face. (c) Temperature on the wide side. (d) Temperature on the narrow side.

The temperature distributions on the wide and narrow sides are shown in Figure 3c,d. At the end of the foot roller section of the continuous slab caster, the highest temperature of the slab was 1276 K and the lowest temperature was 940 K on the wide side. The difference in temperature was 336 K, thus dropping by 26.3%. The highest temperature of the slab was 1400 K and the lowest temperature was 940 K on the narrow side. The difference in temperature was 460 K, a decrease of 32.9%. The thickness of the solidified slab at this segment was about 42.5 mm on the wide side, and the thickness of the narrow side was about 41 mm. Due to the uneven heat transfer, the temperature field of the slab was unevenly distributed overall. The lowest temperature of the slab appeared in the corner area owing to the bidirectional heat transfer characteristic.

3. The Establishment of the Bulge Deformation Model

The three-dimensional thermoelastic–plastic and creep coupling models were proposed to investigate slab bulge deformation after temperature distribution and the thickness of the solidified shell had been acquired. The relationship between the solidification heat transfer model and the slab bulge deformation model is illustrated in Figure 4. Q235 steel served as the analytical material of the casting slab. The casting roller material was 40CrMo.

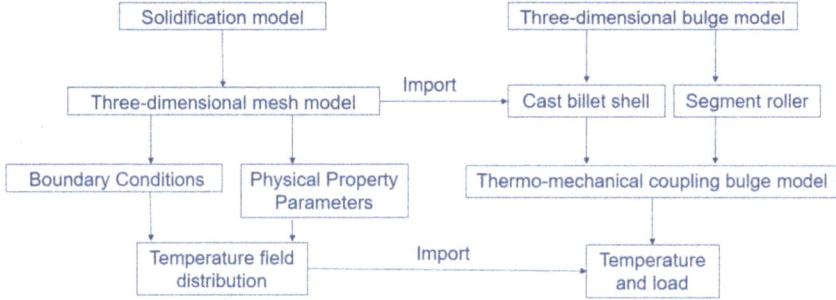

Figure 4. The relationship between the solidification model and the slab deformation model.

3.1. Physical Property Parameters

In the high-temperature continuous casting process, the bulge deformation of the slab is not only determined by the thermal process, as the high-temperature mechanical properties of the material also have an important influence on the deformation behavior of the slab. In order to accurately analyze the deformation of the casting slab, it was necessary to clarify the high-temperature mechanical properties of Q235 steel. Physical property parameters of the slab were quoted from the references [35,36] and are shown in Figure 5. According to the investigation, the material parameters of 40CrMo at the temperature of 523 K were suitable for the bulge deformation model analysis. The elastic modulus was 189.46 MPa, the Poisson ratio was 0.28, and the density was 7850 kg·m^{-3}.

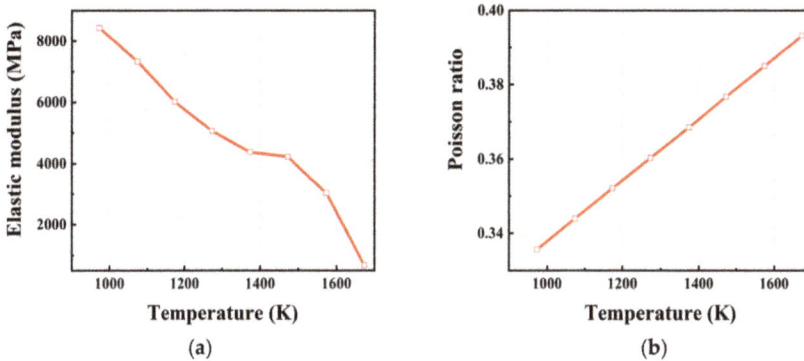

Figure 5. (a) Elastic modulus of Q235 steel; (b) Poisson ratio of Q235 steel.

3.2. Geometric Model with Various Segmented Casting Rollers

The geometric parameters of the slab and casting rollers used in the model are listed in Table 4. The three-dimensional geometric models of the interaction between the slab and the various casting rollers are shown in Figure 6, including rigid solid rollers, elastic solid rollers, elastic two-segment rollers, and elastic three-segment rollers. C3D8R was the eight-node linear three-dimensional stress hexahedron reduction integral element which was used to build the models. The casting slab was composed of 20,400 elements (C3D8R) and 40,581 nodes. In the continuous casting process, it was assumed that the casting rollers were linearly distributed, ignoring the casting blank curvature. The liquid core of the slab was removed and simplified as a cavity, and the static pressure of the molten steel in the slab was transformed into a uniform pressure on the inner surface of the slab. During the analysis, the thickness of the casting slab and the distribution of the temperature field did not change with the operation of the slab.

Table 4. Simulation constants for bulge deformations.

Parameters	Values
Thickness of the solidified slab on the wide side	42.5 mm
Thickness of the solidified slab on the narrow side	41 mm
Number of rollers	15
Roller spacing	300 mm
Radius of roller	115 mm
Length of roller	1240 mm
Casting speed	$1.5\ \mathrm{m\cdot min^{-1}}$

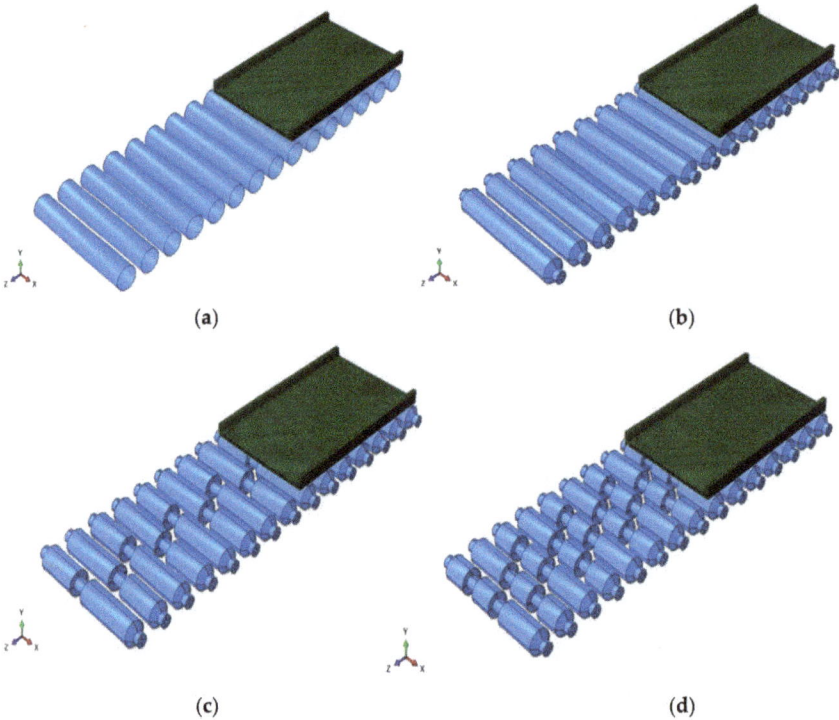

(a)

(b)

(c)

(d)

Figure 6. (a) Bulging model with rigid solid rollers. (b) Bulging model with elastic solid rollers. (c) Bulging model with two-segment rollers. (d) Bulging model with three-segment rollers.

3.3. Boundary Conditions and Contact Definition

(1) The uniform pressure was 369,634 Pa, which was to be the static pressure of the molten steel in the slab.

(2) The symmetrical displacement constraint was loaded on the slab because of the symmetry of the structure.

(3) When the longitudinal length of the slab was chosen as 2000 mm, the influence of the slab boundary on the internal structure could be neglected according to Saint-Venant's principle, and the calculation accuracy and efficiency could be guaranteed.

3.4. Creep Model

It was found that the time-hardening model could describe the creep behavior of the metal with a carbon content of 0.18% and a secondary cooling temperature range of 1173 K~T_s [37]. This could satisfy the requirements of this paper. Therefore, the time-hardening model was used as the creep model of the casting slab material. The constitutive equation of the time-hardening model is shown as Equation (7):

$$\dot{\varepsilon}_p = C \exp\left(\frac{-Q}{T}\right)\sigma^N t^m \tag{11}$$

where

$C = 0.3091 + 0.2090(pctC) + 0.1773(pctC)^2$,
$Q = 17160$,
$N = 6.365 - 4.521 \times 10^{-3}T + 1.439 \times 10^{-6}T^2$,
$m = -1.362 + 5.761 \times 10^{-4}T + 1.982 \times 10^{-8}T^2$,
$\dot{\varepsilon}_p$ is the creep strain rate (s^{-1}),
C is the influence parameter of carbon content $(MPa^{-n}s^{-m-1})$,
σ is the stress (MPa),
Q is the deformation energy constant (K),
T is the temperature (K),
t is the time (s),
m is the temperature-dependent time influence index, and
N is the temperature-dependent comprehensive stress influence index.

3.5. Predefined Temperature Field

The slab temperature distribution at the end of the foot roller section was introduced into the bulging analysis as a predefined field and kept invariable in the calculation process of bulge deformation.

4. Comparisons of Slab Bulge Deformations with Different Segment Rollers

The bulge deformation of the slab was obtained by establishing the three-dimensional finite element model of the coupling interaction between the various segmented rollers and the slab. The results calculated by ABAQUS were used to analyze the influence of the segmented rollers on slab bulge deformation.

4.1. Bulge Deformation on the Wide Side

The simulation results of slab bulge deformation along the wide side are shown in Figure 7. The influence of the segmented rollers was numerically analyzed by a comparison between the average values of the bulge deformation. As can be seen in Figure 7, the following conclusions can be drawn:

(1) In the bulge deformation model with the three-segment rollers, the deformation rapidly increased to 0.90 mm from the edge to the distance of 270 mm in the direction of the wide surface. The deformation of the slab formed into a bulging platform from the distance of 270 mm to the slab center (600 mm) in the wide direction. The value of the bulging platform was 0.90 mm. The reason for this was that the edge of the solidified slab had formed a thick shell of great stiffness, and the deformation was small. The thickness of the shell decreased with the increasing distance, resulting in the decrease of slab stiffness. The bulge deformation of the slab rapidly increased under the hydrostatic pressure. Then, the platform appeared with the uniform thickness and uniform static pressure of the slab. The deformation of the slab at the contact position with the roller sectional area was greater than the average value of the platform. This was because the fixed area of the slab intermittently passed through the sectional area of the segmented rollers, and the positive creep was more effective than the reverse creep in the fixed area. In the continuous casting process, the positive creep makes the solidified shell convex outwards, and the reverse creep makes the solidified shell recess.

(2) The minimum bulge deformation was 0.82 mm in the model with rigid solid rollers. The maximum bulge deformation of 1.43 mm appeared in the model with elastic solid rollers. The difference between the deformation of the two models was about 0.61 mm, and the increase remained around 74.3%. The results showed that the stiffness of the casting rollers had a great influence on slab bulge deformation. However, the casting rollers cannot be absolutely rigid rollers in the actual production process. Thus, the elastic rollers were more suitable than the rigid rollers to simulate the bulge deformation of the casting slab in the actual production process, as a more accurate result could be obtained.

(3) The two-segment and three-segment rollers were defined by using elasticity. The bulge deformation of the casting slab in the two-segment roller model was reduced to 1.28 mm in comparison with the elastic solid rollers model. The difference in bulge deformation was 0.15 mm, meaning there was a decrease of 10.5% between the two models, while the bulge deformation of the slab with the three-segment rollers reached 0.90 mm. The bulge deformation difference was 0.53 mm, dropping by 37.1%. The result indicated that the bulge deformation with the three-segment rollers was closer to that of the rigid solid rollers. This means that the bulge deformation of the casting slab could be effectively controlled by using segmented rollers, and the influence of the three-segment rollers was more effective than that of the two-segment rollers.

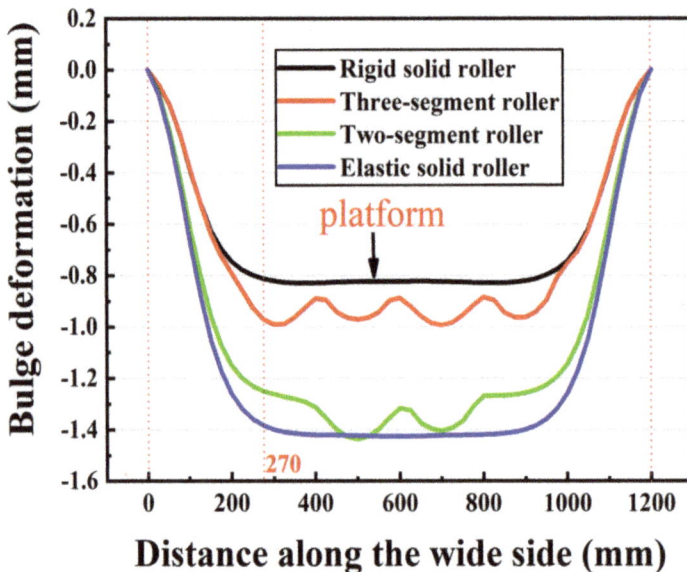

Figure 7. Bulge deformation of the slab on the wide side with different casting rollers.

4.2. Bulge Deformation on the Narrow Side

The simulation result of the slab bulge deformation along the narrow side is displayed in Figure 8.

The horizontal bulge deformation of the slab's narrow side was uneven, and the bulge deformation of the narrow side with the three-segment rollers was relatively close to that of the rigid rollers. The bulge deformation on the narrow side with the three-segment rollers reduced from 0.38 to 0.23 mm, a decrease of 39.5%. The bulge deformation on the narrow side with the two-segment rollers reduced from 0.39 to −0.008 mm, which was down by 102%. The bulge deformation on the narrow side with the rigid solid rollers reduced from 0.39 to 0.25 mm, dropping by 35.9%. The bulge deformation on the narrow side with the elastic solid rollers reduced from 0.37 to −0.08 mm, a decrease of 121.6%.

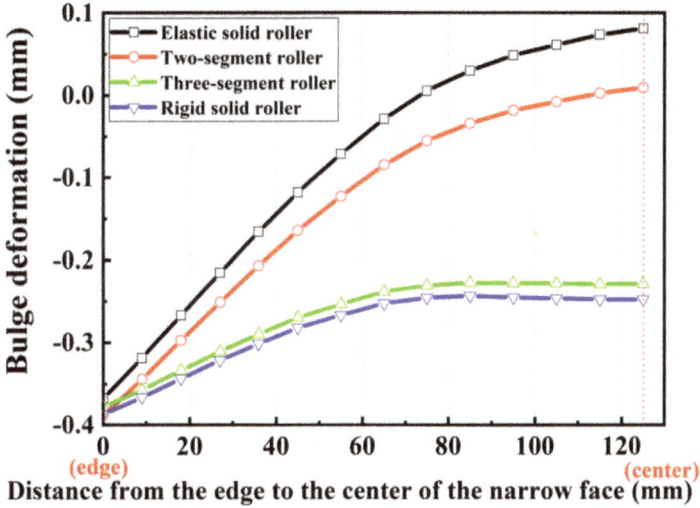

Figure 8. Bulge deformation of the slab on the narrow side with different casting rollers.

4.3. Analysis of Segmented Roller's Stiffness

The stiffness of the casting roller was an important factor affecting slab bulge deformation. In the bulging models, the vertical displacement of the elastic casting roller had occurred, which was selected to measure the stiffness of the casting roller. The results are shown in Figure 9. The vertical displacement of the elastic solid roller was a reverse parabola distribution. The maximum value of 0.14 mm appeared in the center of the solid roller. The rigid roller was not deformed, so the vertical displacement was 0 mm. The vertical displacement of the two-segment roller was evenly distributed around 0.004 mm. The displacement of the three-segment roller in the vertical direction was generally distributed around 0.0015 mm. The average vertical displacement of the roller was selected for comparison to intuitively analyze the stiffness of the casting rollers. The average vertical displacement of the elastic solid roller was 0.07 mm, the average vertical displacement of the two-segment roller was 0.004 mm, and that of the three-segment roller was 0.0015 mm. The stiffness of the two-segment roller was 17.5 times greater than that of the solid roller, and the stiffness of the three-segment roller was 46.7 times greater than that of the solid roller. As the number of sections increased, the deflection of the segment roller during the continuous casting process was smaller. This indicated that the stiffness of the segmented roller was effectively improved. Combined with the previous analysis of casting slab deformation, the conclusion could be drawn that the stiffness of the segmented roller could be better improved than that of the solid casting roller, and the bulge deformation of the slab could also be excellently reduced.

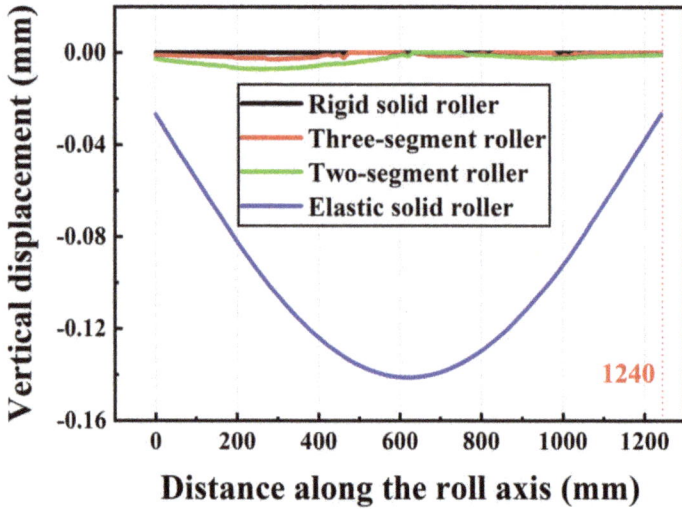

Figure 9. The vertical displacement of different casting rollers.

5. Influences of Roller Spacing and Roller Diameter on Slab Bulge Deformation

5.1. Establishment of Deformation Model with Different Roller Spacings and Diameters

In the continuous casting process, bulge deformation must be controlled to ensure the quality of the slab. There are many factors involved in slab bulge deformation in continuous casting production, among which the main parameters include casting speed, static pressure of molten steel, shell thickness, roller spacing, roller diameter, and surface temperature of the slab. The main purpose of this section is to explore and analyze the influence of the segmented roller structure parameters on bulge deformation. Under a constant casting speed condition (1.5 m·min^{-1}), the bulge deformation was calculated and analyzed for the casting slab. The specific simulation structural parameters are shown in Table 5. The influence of the casting process parameters was numerically analyzed by a comparison between the average values of the bulge deformation.

Table 5. Simulation parameters of bulging models.

Roller Spacing	Rigid Roller	Elastic Roller	Two-Segment Roller	Three-Segment Roller
	Roller Diameter			
300 mm	230 mm	230 mm	230 mm	230 mm
	250 mm	250 mm	250 mm	250 mm
	270 mm	270 mm	270 mm	270 mm
350 mm	230 mm	230 mm	230 mm	230 mm
400 mm	230 mm	230 mm	230 mm	230 mm

5.2. Results and Conclusions

5.2.1. Influences of Roller Spacing on Bulge Deformation

Roller spacing was adjusted by changing the gap between the rollers. Under the condition of constant casting speed (1.5 m·min^{-1}), the bulge deformation was calculated with 350 and 400 mm as the roller spacings. The average values of the bulge deformation were compared with that of a 300-mm roller spacing. The diameter of the roller was 230 mm. The results of slab bulge deformation on the wide side, shown in Figure 10 and Table 6, were established by the simulation model with different roller spacings. It could be concluded that the approximate distribution of the slab bulge deformation on the wide side would not change with a 50 mm increase of the roller spacing, and the improvement of the three-segment roller for bulge deformation was most obvious. However, the bulge deformation of the slab under the same continuous casting segmented roller increased greatly with the increase of the roller spacing. The maximum increment of bulging deformation was 67.1% under the rigid roller with the roller spacing changing from 300 to 350 mm. The maximum increment of bulge deformation under the elastic roller was 47.4%, with the roller spacing changing from 350 to 400 mm. The maximum increment of bulge deformation under the two-segment roller was 37.5%, with the roller spacing changing from 300 to 350 mm. The maximum increment of bulge deformation under the three-segment roller was 60.0%, with the roller spacing changing from 300 to 350 mm. The bulging increase was about the same for the rigid roller and the three-segment roller. This also proved that the effect of the three-segment roller was closest to that of the rigid roller on slab bulge deformation. The minimum increment was 32.9% under the elastic solid roller with the roller spacing changing from 300 to 350 mm.

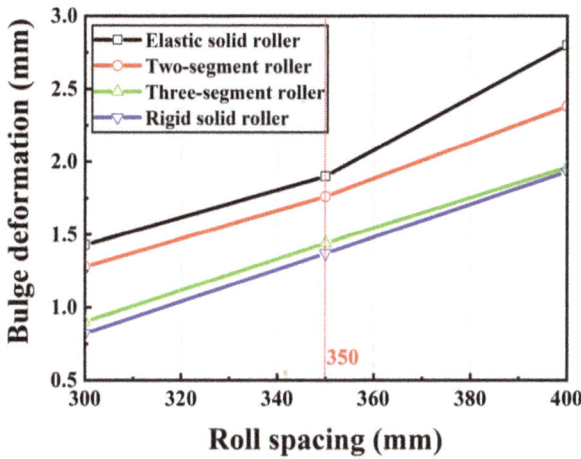

Figure 10. The bulge deformation on the wide side with different roller spacings.

Table 6. Bulge deformation on the wide side with different roller spacings.

Roller Spacing	Rigid Roller	Elastic Roller	Two-Segment Roller	Three-Segment Roller
300 mm	0.82 mm	1.43 mm	1.28 mm	0.90 mm
350 mm	1.37 mm	1.90 mm	1.76 mm	1.44 mm
400 mm	1.93 mm	2.80 mm	2.38 mm	1.96 mm

An analysis of variance (Table 6) showed that the value of variance ratio was 158.45 under different roller spacings, which was larger than the critical value of variance ratio (the critical value was 5.14). The value of variance ratio was 35.38 under different segmented rollers, which was larger than the critical value of variance ratio (the critical value was 4.76). This indicated that different roller spacings and segmented rollers all have a significant effect on slab bulge deformation. Generally, the slab bulge deformation on the wide and narrow sides increased with the increasing roller spacing. This was because the area of the slab between the two rollers increased with the increasing roller spacing. The effect of static pressure increased in the solidified shell, which indirectly reduced its stiffness. As a result, the value of slab bulge deformation became larger as the roller spacing increased. This indicated that roller spacing also played an important role in bulge deformation of the casting slab.

5.2.2. Influences of Roller Diameter on Bulge Deformation

The influence of roller diameter was observed by using the deformation models under various roller diameters and a constant roller spacing. The roller spacing was defined as 300 mm. The roller diameters of the bulging model are shown in Table 5. The results of bulge deformation on the wide side under different roller diameters are listed in Table 7. It was found that the slab bulge deformation under the rigid rollers was almost unchanged with different roller diameters. This was because the rigid roller was established by a rigid body that did not reform in the continuous casting process, and the changing roller diameters had no influence on slab bulge deformation. The average deformation of the slab on the wide side decreased by about 0.08 mm when the elastic solid roller diameter increased by 20 mm. The influence of changing roller diameters on bulging deformation was relatively small due to the low stiffness of the elastic solid roller. However, the slab bulge deformation was effectively influenced by changing the segmented roller diameter by 20 mm, including the two-segment and three-segment rollers. The maximum reduction of bulging deformation was 0.24 mm, with the two-segment roller diameter changing from 250 to 270 mm, and the decrease remained around 21.8%. The minimum reduction was 0.1 mm with the three-segment roller diameter changing from 230 to 250 mm, and the decrease remained around 11.1%. This was because the elastic roller was deformed in the model, which resulted in a larger contact area between the slab and the roller as the roller diameter increased. In other words, the rigidity of the solidified shell was indirectly increased. The bulge deformation of the continuous casting slab was effectively controlled with the improved stiffness of the solidified shell and the casting roller. An analysis of variance (Table 7) showed that the value of variance ratio was 6.30 under different roller diameters, which was larger than the critical value of variance ratio (the critical value was 5.14). The roller diameter had a significant effect on slab bulging deformation.

Table 7. Bulge deformation on the wide side with different roller diameters.

Roller Diameter	Rigid Solid Roller	Elastic Solid Roller	Elastic Two-Segment Roller	Elastic Three-Segment Roller
230 mm	0.82 mm	1.43 mm	1.28 mm	0.90 mm
250 mm	0.80 mm	1.37 mm	1.10 mm	0.80 mm
270 mm	0.80 mm	1.25 mm	0.86 mm	0.69 mm

In general, the slab bulge deformation increased with the increased roller spacing and decreased with the increased roller diameter. The comparisons of average values of bulge deformation along the wide side are shown in Figure 11. When the roller spacing was 400 mm and the roller diameter was 230 mm, the maximum average value of bulge deformation along the wide side was 2.80 mm with the elastic solid rollers. When the roller spacing was 300 mm and the roller diameter was 270 mm, the minimum average value of bulge deformation along the wide side was 0.69 mm with the three-segment rollers; the difference was of 2.11 mm in comparison with the previous one. The bulge deformation was reduced by 75.4%. The optimized structural parameters could effectively control slab bulge deformation.

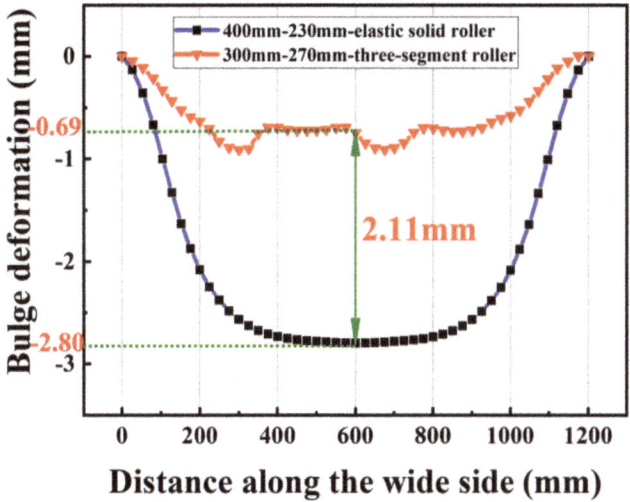

Figure 11. Comparisons of slab bulge deformation with different segment roller structure parameters.

6. Discussion

Slab bulge deformation is an important factor that affects the quality of the slab. Most researchers have focused on the deformation mechanism of the slab and rarely explored an effective way to control slab bulge deformation. Also, rigid casting rollers were most commonly used in the slab bulging models in order to simplify the calculations of slab bulge deformation. Although this simplification could greatly reduce the model's calculation time, the influence of roller deformation on slab deformation was neglected. In this paper, the influence of segmented rollers on slab bulge deformation was explored, and the structural parameters of segmented rollers were optimized to effectively control the bulge deformation of the casting slab. Also, elastic casting rollers were introduced to replace the traditional rigid casting rollers in the bulging model for simulation analysis to obtain more ideal results. The results showed that elastic casting rollers were more suitable for analyzing slab bulge deformation than rigid casting rollers. Then, by comparing the slab bulge deformation under the segmented and solid rollers separately, it was found that the stiffness of the segmented rollers was greater than that of the solid rollers, so that the bulge deformation of the slab could be effectively controlled. Finally, the roller spacing and diameter of the segmented rollers were further studied. It was found that reducing the segmented roller spacing and increasing the roller diameter could effectively control the bulge deformation of the slab. The research results in this paper could provide an effective theoretical basis for controlling slab bulge deformation in actual production.

7. Conclusions

A 3D finite element thermomechanical coupling model between segmented rollers and a casting slab was established to investigate the influence of the segmented rollers on slab bulge deformation. The traditional rigid continuous casting rollers were replaced by elastic continuous casting rollers in these models, which is an advantage that this model has over traditional models. The main conclusions are as follows:

(1) The slab bulge deformation under the elastic rollers increased by around 74.3% compared with that of traditional rigid rollers. The results indicated that the stiffness of the casting rollers would have a great influence on slab bulge deformation and the elastic rollers were more suitable for the slab bulge deformation analysis.

(2) The slab bulge deformation was 1.43 mm under the elastic solid rollers and was reduced by 10.5% under the two-segment rollers compared with elastic solid rollers, while the slab bulge deformation dropped by 37.1% under the three-segment rollers.

(3) The results showed that the bulge deformation could be controlled by reducing the roller spacing or increasing the roller diameter. The slab bulge deformation on the wide side increased when the roller spacing increased by 50 mm, with the maximum increase of 67.1% and the minimum increase of 32.9%. The average value of the slab bulge deformation was reduced by about 15% when the roller diameter was increased by 20 mm.

(4) The bulge deformation on the wide side was reduced by 75.4% when the roller diameter and spacing of the segmented rollers were optimized. This result indicates that the optimized structural parameters of casting segmented rollers could effectively control slab bulge deformation.

Author Contributions: Conceptualization, Q.Q.; methodology, Q.Q.; formal analysis, Q.Q.; investigation, M.L. and J.H.; writing—original draft preparation, M.L.; writing—review and editing, Q.Q.; visualization, M.L.; supervision, Q.Q.; funding acquisition, Q.Q.

Funding: This research was funded by the National Natural Science Foundation of China, grant number: 51375041.

Conflicts of Interest: The authors declare no conflict of interest.

References

1. Janik, M.; Dyja, H.; Berski, S.; Banaszek, G. Two-dimensional thermomechanical analysis of continuous casting process. *J. Mater. Process. Technol.* **2004**, *153*, 578–582. [CrossRef]
2. Ji, C.; Luo, S.; Zhu, M.; Sahai, Y. Uneven solidification during wide-thick slab continuous casting process and its influence on soft reduction zone. *ISIJ Int.* **2014**, *54*, 103–111. [CrossRef]
3. Dong, Q.; Zhang, J.; Yin, Y.; Wang, B. Three-dimensional numerical modeling of macrosegregation in continuously cast billets. *Metals* **2017**, *7*, 209. [CrossRef]
4. Long, M.; Chen, H.; Chen, D.; Yu, S.; Liang, B.; Duan, H. A combined hybrid 3-D/2-D model for flow and solidification prediction during slab continuous casting. *Metals* **2018**, *8*, 182. [CrossRef]
5. Yoshii, A.; Kihara, S. Analysis of bulging in continuously cast slabs by bending theory of continuous beam. *Trans. Iron Steel Inst. Jpn.* **1986**, *26*, 891–894. [CrossRef]
6. Liu, W.; Dai, Y. Study on the deformation of the liquid core continuous casting shell. *Heavy Mach.* **1994**, 10–16. [CrossRef]
7. Sun, J.; Sheng, Y.; Zhang, X. Analysis of bulging deformation and stress in continuous cast slabs. *J. Iron Steel Res.* **1996**, 11–15. [CrossRef]
8. Xu, R. Behavior analysis of bulging deformation in slab casting process. *Heavy Mach.* **2012**, 17–21. [CrossRef]
9. Toishi, K.; Miki, Y. Generation mechanism of unsteady bulging in continuous casting-2-fem simulation for generation mechanism of unsteady bulging. *ISIJ Int.* **2016**, *56*, 1764–1769. [CrossRef]
10. Ha, J.; Cho, J.; Lee, B.; Ha, M. Numerical analysis of secondary cooling and bulging in the continuous casting of slabs. *J. Mater. Process. Technol.* **2001**, *113*, 257–261. [CrossRef]
11. Zhang, Y. Analysis of solidification heat transfer and bulging in the cast slab. Master's Thesis, Yanshan University, Hebei, China, 2011.
12. Okamura, K.; Kawashima, H. Three-dimensional elasto-plastic and creep analysis of bulging in continuously cast slabs. *Tetsu-to-Hagane* **1989**, *75*, 1905–1912. [CrossRef]
13. Ning, Z.; Wu, D.; Qin, Q.; Zang, Y. Three-dimensional emulation research on the bulging deformation during continuously casting slab. *Metall. Equip.* **2007**, *2*.
14. Qin, Q.; Shang, S.; Wu, D.; Zang, Y. Comparative analysis of bulge deformation between 2D and 3D finite element models. *Adv. Mech. Eng.* **2014**, *6*, 942719. [CrossRef]
15. Liu, H.; Zhang, X.; Qian, L. 3D Finite element calculation of creep bulging system development and application. *Contin. Cast.* **2015**, *40*, 54–58.
16. Camporredondo, J.E.; Acosta, F.A.; Castillejos, A.H.; Gutierrez, E.P.; De la Gonzalez, R. Analysis of thin-slab casting by the compact-strip process: Part II. Effect of operating and design parameters on solidification and bulging. *Metall. Mater. Trans. B* **2004**, *35*, 561–573. [CrossRef]

17. Zhang, J.; Shen, H.F.; Huang, T.Y. Finite element thermal-mechanical coupled analysis of strand bulging deformation in continuous casting. *Adv. Mater. Res.* **2011**, *154–155*, 1456–1461. [CrossRef]

18. Qin, Q.; Yang, Z. Finite element simulation of bulge deformation for slab continuous casting. *Int. J. Adv. Manuf. Technol.* **2017**, *93*, 4357–4370. [CrossRef]

19. Triolet, N.; Bobadilla, M.; Bellet, M.; Avedian, L.; Mabelly, P. A thermomechanical modelling of continuous casting to master steel slabs internal soundness and surface quality. *Rev. De Métallurgie–Int. J. Metall.* **2005**, *102*, 343–353. [CrossRef]

20. Fu, J.-X.; Hwang, W.-S. Numerical simulation of slab broadening in continuous casting of steel. In *Numerical Simulation-From Theory to Industry*; InTech: Rijeka, Croatia, 2012.

21. Li, B. Study about the High-temperature creep property of q460e steel and its bulging in continuous casting process. Master's Thesis, Yan Shan University, Hebei, China, 2015.

22. Han, P.; Ren, T.; Jin, X. Influence of roll misalignment on bulging of continuous casting slab. *Iron Steel* **2016**, *51*, 53–58.

23. Ohno, H.; Miki, Y.; Nishizawa, Y. Generation mechanism of unsteady bulging in continuous casting-1-development of method for measurement of unsteady bulging in continuous casting. *ISIJ Int.* **2016**, *56*, 1758–1763. [CrossRef]

24. Verma, R.; Girase, N. Comparison of different caster designs based on bulging, bending and misalignment strains in solidifying strand. *Ironmak. Steelmak.* **2006**, *33*, 471–476. [CrossRef]

25. Yu, Y.; Duan, L.; Cui, X. Brief introduction of slab caster segment rollers. *Metall. Equip.* **2016**, 45–48. [CrossRef]

26. Li, S.; Shen, B. Roller type selection of sector of WISCO three steelmaking continuous casting machine. *Metall. Equip.* **1992**, *14*, 51–52.

27. Shi, J. The advantages and theoretical argument of caster segment sub-section roller. In Proceedings of the 8th (2011) China Steel Annual Meeting, Beijing, China, 26 October 2011; p. 6.

28. Liu, W.; Wang, W. Analysis on the stiffness of segmented baeking guide roll of slab continuous caster. *Bao Steel Technol.* **1994**, *12*, 58–63.

29. Liu, W.; Wang, W. The influence of structure and classified number of backup guide rolls in CCM on the deformation of slab bulge. *Shanghai Metals* **1995**, *17*, 25–30.

30. Qin, Q.; Wu, D. *Thermal and Mechanical Behavior of Continuous Casting Equipment*; Metallurgical Industry Press: Beijing, China, 2013.

31. Mizikar, E.A. Mathematical heat transfer model for solidification of continuously cast steel slabs. *Trans. Metall. Soc. AIME* **1967**, *239*, 1747–1753.

32. Cheng, J. *Continuous Casting Steel Manual*; Metallurgical Industry Press: Beijing, China, 1991.

33. Feng, K.; Chen, D.; Xu, C.; Wen, L.; Dong, L. Effect of main thermo-physical parameters of steel Q 235 on accuracy of concasting transport model. *Special Steel* **2004**, *25*, 28–31.

34. Mills, K.C.; Su, Y.; Li, Z.; Brooks, R.F. Equations for the calculation of the thermo-physical properties of stainless steel. *ISIJ Int.* **2004**, *44*, 1661–1668. [CrossRef]

35. Fu, J.; Li, J.; Wang, C.; Zhu, J. Research of Young's modulus of elasticity of steel Q235. *Mater. Rev.* **2009**, *23*, 68–70.

36. Uehara, M. Mathematical modelling of the unbending of continuously cast steel slabs. Master's Thesis, University of British Columbia, Vancouver, BC, Canada, 1983.

37. Kozlowski, P.F.; Thomas, B.G.; Azzi, J.A.; Wang, H. Simple constitutive equations for steel at high temperature. *Metall. Trans. A* **1992**, *23*, 903. [CrossRef]

metals

MDPI

Article

A Simulation Study on the Flow Behavior of Liquid Steel in Tundish with Annular Argon Blowing in the Upper Nozzle

Xufeng Qin [1,2], Changgui Cheng [1,2,*], Yang Li [1,2], Chunming Zhang [1,2], Jinlei Zhang [1,2] and Yan Jin [1,2]

1   The State Key Laboratory of Refractories and Metallurgy, Wuhan University of Science and Technology, Wuhan 430081, China; qinxufeng@wust.edu.cn (X.Q.); liyang@wust.edu.cn (Y.L.); Zhangspringming@hotmail.com (C.Z.); Jinleiazhang@outlook.com (J.Z.); jinyan@wust.edu.cn (Y.J.)
2   Hubei Provincial Key Laboratory for New Processes of Ironmaking and Steelmaking, Wuhan University of Science and Technology, Wuhan 430081, China
*   Correspondence: ccghlx@wust.edu.cn; Tel.: +86-027-68862651

Received: 27 December 2018; Accepted: 12 February 2019; Published: 13 February 2019

Abstract: A three-dimensional mathematical model of gas−liquid two-phase flow has been established to study the flow behavior of liquid steel in the tundish. The effect of the argon flow rate and casting speed on the flow behavior of liquid steel, as well as the migration behavior of argon bubbles, was investigated. The results from the mathematical model were found to be consistent with those from the tundish water model. There were some swirl flows around the stopper when the annular argon blowing process was adopted; the flow of liquid steel near the liquid surface was active around the stopper. With increased argon flow rate, the vortex range and intensity around the stopper gradually increased, and the vertical flow velocity of the liquid steel in the vicinity of the stopper increased; the argon volume flow in the tundish and mold all increased. With increased casting speed, the vortex range and intensity around the stopper gradually decreased, the peak value of vertical flow velocity of liquid steel at the vicinity of the stopper decreased, and the distribution and ratio of argon volume flow between the tundish and the mold decreased. To avoid slag entrapment and purify the liquid steel, the argon flow rate should not be more than 3 L·min^{-1}. These results provide a theoretical basis to optimize the parameters of the annular argon blowing at the upper nozzle and improve the slab quality.

Keywords: annular argon blowing; upper nozzle; flow behavior; argon gas distribution; tundish

1. Introduction

The tundish is a transitional container connecting the ladle and mold. The tundish makes the liquid steel composition and temperature uniform. It also distributes liquid steel and, more importantly, facilitates the removal of inclusions and then purifies the liquid steel. Many techniques have been adopted in the tundish to remove inclusions such as retaining walls and dams, diversion walls, tundish filtering, and electromagnetic stirring [1–4]. Moreover, the argon blowing in the tundish could effectively reduce the amount of the inclusion and purify the liquid steel. The mechanism is the injection of argon gas into the liquid steel in the tundish to form bubbles. Non-metallic inclusions could then be transported to the liquid surface in the tundish for removal.

The argon blowing patterns in the tundish mainly include the long shroud, the bottom permeable brick, the stopper, and the upper nozzle. Studies [5–7] of argon blowing through the long shroud demonstrated that the micro-bubbles generated by argon blowing could improve the removal rate of inclusions in the tundish, but these micro-bubbles had a short residence time in the tundish. Thus, the effect of removing small inclusions is less obvious.

Argon blowing through the bottom-permeable brick could effectively improve the flow pattern of liquid steel and promote the flotation and removal of fine inclusions [8–13]. The movement route of the liquid steel was more tortuous and closer to the liquid surface. This prolonged the residence time of liquid steel, improved the mixing degree of the liquid steel, and reduced the dead zone volume in the tundish. Argon blowing on the bottom-permeable brick could only stir and clean the liquid steel above the permeable brick, but this could not effectively clean the liquid steel passing through the gap between the tundish slope wall and the gas curtain. Most argon bubbles entered the tundish upper nozzle and the submerged entry nozzle (SEN) when the argon was blown by the stopper [14]. This step could clean the inner wall of the nozzle and reduce the adhesion of inclusions on the inner face of the nozzle. Moreover, some of the larger argon bubbles floated directly in the tundish and interacted with inclusions in the liquid steel during the floating process. This helped reduce the content of the inclusions in the strand [15].

The bubbles generated by the dispersed permeable portion of the upper nozzle could form a stable and continuous argon gas film between the inner wall of the nozzle and liquid steel [16–18]. This could effectively suppress the accumulation of inclusions such as Al_2O_3 on the inner wall of the nozzle and reduce the risk of nozzle clogging. Concurrently, argon bubbles generated from the ruptured gas film could wash the inclusions deposited on the inner wall of the nozzle [19].

Smirnov et al. [20–23] studied the argon-blowing process through a gas-permeable ceramic rod embedded in the nozzle pocket brick. Here, the argon bubbles raised around the stopper and formed an annular gas curtain barrier in the tundish. Their work showed that the technique could reduce the adhesion of inclusions on the inner face of the nozzle and prevent the nozzle from clogging. The loss of refractory material was significantly reduced versus argon blowing on the bottom-permeable brick. However, the void region between the permeable ceramic rods could make the liquid steel entering the tundish nozzle insufficiently clean. This would weaken the ability to remove inclusions in the liquid steel.

The preceding studies are significant for effectively controlling the liquid steel flow by argon blowing—this purifies the liquid steel in the tundish and prevents the nozzle from clogging. Here, we propose to use an annular permeable brick with a certain width set in a pocket brick on the outside of the upper nozzle to form a relatively complete annular gas curtain around the stopper. This is based on the work of Smirnov et al. and will improve the effect of controlling fluid flow in the tundish. The rising argon bubbles may promote the removal of inclusions in liquid steel of the tundish. Concurrently, the argon bubbles partially entering into the nozzle can realize the function of argon blowing with the stopper or upper nozzle to prevent the nozzle from clogging.

In this paper, we describe a three-dimensional mathematical model for the annular argon blowing at the upper nozzle in the tundish based on the actual process conditions of a continuous slab-casting tundish in a steel plant. The discrete phase model (DPM) was used to simulate the argon blowing process to analyze the effects of different argon flow rates and casting speed on the flow behavior of liquid steel and the migration behavior of argon bubbles in the tundish. In addition, the flow behavior of liquid steel with the annular argon blowing at upper nozzle was analyzed and compared with the established water model of the tundish. The results can be leveraged as a theoretical basis for the optimization of the annular argon blowing at the upper nozzle and the improvement of slab quality.

2. Model Description

2.1. Model Assumption

(1) The effect of liquid slag on the flow behavior of liquid steel in the tundish is neglected.

(2) The flow of liquid steel is a transient incompressible flow, and the physical properties of the liquid steel such as the density and viscosity are constant.

(3) Argon bubbles are regarded as rigid spheres, the bubble size does not change during the ascent, and the bubble diameters are distributed by Rosin−Rammler statistics, which were obtained by the water model experiments of tundish.

(4) The transport of tracer in the tundish is an unsteady mass transfer process.

2.2. Governing Equations

The flow of the liquid steel in the tundish is a three-dimensional transient incompressible flow and mass transfer process, which satisfies the basic physical laws of mass, momentum conservation. The continuity equation and the momentum equation are described as follows.

Continuity equation:

$$\frac{\partial \rho}{\partial t} + \frac{\partial (\rho u_i)}{\partial x_i} = 0 \tag{1}$$

Momentum equation (N–S):

$$\frac{\partial (\rho u_i)}{\partial t} + \frac{\partial (\rho u_i u_j)}{\partial x_j} = -\frac{\partial P}{\partial x_i} + \frac{\partial}{\partial x_j}\left[\mu_{eff}\left(\frac{\partial u_i}{\partial x_j} + \frac{\partial u_j}{\partial x_i}\right)\right] + \rho g + F_g \tag{2}$$

where ρ is the fluid density, in kg·m^{-3}; u_i and u_j are the velocity vectors, in m·s^{-1}, i and j each represent the three coordinate directions (x, y, and z), and repeated indices imply summation; P is the pressure, in Pa; μ_{eff} is the turbulent effective viscosity coefficient, in Pa·s; g is the gravitational acceleration, in m·s^{-2}. F_g is a momentum source term, which accounts for the presence of argon bubbles, in N·m^{-3}. Here, the standard k-ε turbulence equations were used in the mathematical model. The governing equations describing turbulent kinetic energy (k) and the dissipation rate of turbulence energy (ε) are, respectively:

$$\frac{\partial}{\partial t}(\rho k) + \frac{\partial}{\partial x_i}\left(\rho u_i k - \frac{\mu_{eff}}{\sigma_k}\frac{\partial k}{\partial x_i}\right) = G - \rho\varepsilon \tag{3}$$

$$\frac{\partial}{\partial t}(\rho\varepsilon) + \frac{\partial}{\partial x_i}\left(\rho u_i \varepsilon - \frac{\mu_{eff}}{\sigma_\varepsilon}\frac{\partial k}{\partial x_i}\right) = \frac{1}{k}\left(C_1 G - C_2 \rho\varepsilon^2\right) \tag{4}$$

$$G = \mu_t\frac{\partial u_i}{\partial x_i}\left(\frac{\partial u_i}{\partial x_i} + \frac{\partial u_i}{\partial x_j}\right) \tag{5}$$

$$\mu_{eff} = \mu_0 + \mu_t = \mu_0 + \rho C_\mu\frac{k^2}{\varepsilon} \tag{6}$$

where μ_0 is the dynamic viscosity, in Pa·s; μ_t is the turbulent viscosity, in Pa·s; k is the turbulent kinetic energy of the fluid, in m^2·s^{-2}; ε is the turbulent energy dissipation rate, in m^2·s^{-3}. Terms C_1, C_2, C_μ, σ_k, and σ_ε are empirical constants. The recommended values [11] of Launder and Spalding are C_1 = 1.42, C_2 = 1.92, C_μ = 0.09, σ_k = 1.0, and σ_ε = 1.0.

The trajectories and distributions of the argon bubbles are simulated using the discrete phase model (DPM). An equation for argon bubble velocity is obtained considering the drag force, buoyancy force, and virtual mass force exerted by the fluid on bubbles:

$$\frac{du_g}{dt} = \frac{18\mu}{\rho_g d_g^2}\cdot\frac{C_D Re_g}{24}(u_i - u_g) + \frac{\pi d_g^3}{6}(\rho_g - \rho)g + \frac{1}{2}\frac{\rho}{\rho_g}\frac{d}{dt}(u_i - u_g) \tag{7}$$

where u_g is the bubble velocity, in m·s^{-1}; μ is the molecular viscosity of the fluid, in Pa·s; ρ_g is the argon density, in kg·m^{-3}; d_g is the bubble diameter, in m; Re_g is the relative Reynolds number of bubbles; and C_D is the drag coefficient [24], which is a function of Re_g:

$$C_D = \begin{cases} \frac{24}{Re_g}\left(1 + \frac{1}{6}Re_g^{\frac{2}{3}}\right) & \text{if} Re_g < 1000 \\ 0.44 & \text{if} Re_g \geq 1000 \end{cases} \tag{8}$$

$$\mathrm{Re_g} = \frac{\rho d_g |u_g - u_i|}{\mu} \tag{9}$$

The momentum transfer from the discrete phase towards the melt is computed by examining their momentum change as [11,25]:

$$F_g = \sum_j^N (\frac{3\mu_0 C_D \mathrm{Re_g}}{4\rho_g d_g^2}(u_{gj} - u_i))m_p \Delta t \tag{10}$$

where N is the number of bubbles in a computational cell, which can be determined by the particle trajectory unsteady tracking method in Fluent software; u_{gj} is the velocity of bubble in a computational cell, in m·s^{-1}, m_p is the mass flow rate of argon bubbles, in kg·s^{-1}, which equals the argon density multiplied by the argon flow rate; Δt is the time step, in s, its value is 0.005 s.

In Equation (7), d_g adopts a Rosin-Rammler distribution, and the different bubble size range is divided into discrete size groups as shown in Equation (11).

$$Y_d = e^{-(d_g/\bar{d})^n} \tag{11}$$

where Y_d is the mass fraction with bubble diameter greater than d_g; $\bar{d}$ is the average diameter of bubbles, in m; and n is the distribution index. The mass flow with bubble diameter greater than d_g equals the mass flow rate of argon bubbles multiplied by the time step and the mass fraction with bubble diameter greater than d_g, the mass flow of different diameter range can be determined, and then the number of argon bubbles of different diameter range entering into tundish from the annular argon blowing brick can be determined.

In order to determine the residence time of liquid steel [26], the time evolution of tracer concentration C in the tundish was described by Equation (12):

$$\frac{\partial}{\partial t}(\rho C) + \frac{\partial}{\partial x_i}(\rho u_i C) = \frac{\partial}{\partial x_i}(\rho D_{eff} \frac{\partial C}{\partial x_i}) \tag{12}$$

$$D_{eff} = D_0 + \frac{\mu_{eff}}{\rho Sc_t} \tag{13}$$

where C is the concentration of the tracer, in kg·m^3; D_{eff} is the effective diffusion coefficient, in m^2·s^{-1}; D_0 is the molecular diffusion coefficient in m^2·s^{-1}, and its value is 0; and Sc_t is the turbulent Schmidt number and its value is 0.7.

2.3. Boundary Conditions

(1) The model inlet of tundish was set as the velocity-inlet, and the entry velocity was calculated based on the mass conservation principle according to the section size of the strand, the casting speed, and the inner diameter of the long shroud, namely, the velocity-inlet is equal to that the cross-sectional area of strand is multiplied by the casting speed and divided by the cross section area of the long shroud. The inlet value of k, the turbulence kinetic energy, and the ε, the rate of turbulence energy dissipation, were estimated from the following relations:

$$k = 0.01 u_{in}^2 \tag{14}$$

$$\varepsilon = \frac{k^{1.5}}{0.5 D_{in}} \tag{15}$$

where u_{in} is the inlet velocity, in m·s^{-1}, and D_{in} is the diameter of the inlet, in m. The value of D_{in} is 0.07 m. When the casting speed was controlled to be 1.05 m·min^{-1}, 1.2 m·min^{-1}, 1.35 m·min^{-1} and 1.5 m·min^{-1}, the value of u_{in} was 1.965 m·s^{-1}, 2.246 m·s^{-1}, 2.527 m·s^{-1} and 2.808 m·s^{-1}, respectively.

(2) The outlet of the liquid steel was set as the pressure-outlet according to the immersion depth of the SEN; the model outlet was the escape outlet for the argon bubbles.

(3) The surface of the molten pool was set as a free-surface [27], the normal velocity component and normal gradients of all other variables were assumed to be zero, and the bubbles were trapped at the liquid surface.

(4) Owing to the bilateral symmetry, only a half of the tundish is considered in the calculation in order to lower the computation cost. On symmetry plane, the boundary condition for velocity field is a zero normal component and zero gradient of tangential velocity component.

(5) The wall of the tundish was modeled as a no-slip wall boundary condition; the region near the wall was treated with a standard wall function [28].

2.4. Numerical Method

A three-dimensional mathematical model was established according to the real size of the industrial tundish. The schematic diagram of the tundish vertical view is shown in Figure 1. The half-tundish was taken as the computational domain considering the symmetry of the tundish. A schematic diagram of annular argon blowing at the upper nozzle in the tundish is shown in Figure 2. The physical properties and operational parameters are shown in Table 1.

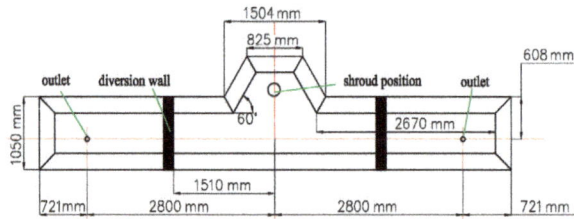

Figure 1. Schematic diagram of tundish vertical view.

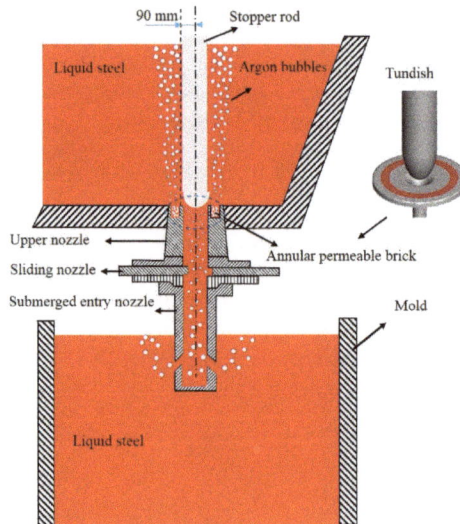

Figure 2. Schematic diagram of annular argon blowing at the tundish upper nozzle.

Table 1. Physical properties and process parameters.

Parameters	Value
Liquid steel density (kg·m^{-3})	7000
Liquid steel viscosity (Pa·s)	0.0065
Argon gas density (kg·m^{-3})	0.27
Working liquid surface height of tundish (mm)	960
Inner and outer diameter of the shroud (mm)	70/120
Immersion depth of the shroud (mm)	250
Diameter of the stopper (mm)	127
Inner diameter of the upper nozzle (mm)	50
Sectional dimensions of the slab (mm × mm)	1235 × 175

The coordinate axis X was parallel to the intersection line of the tundish front wall and the bottom wall, and the Y direction was perpendicular to the X direction. A schematic diagram of the computational domain coordinate system is shown in Figure 3, and the flow field in XZ plane and YZ plane will be shown for analyzing the effect of the argon bubbles on the flow of liquid steel.

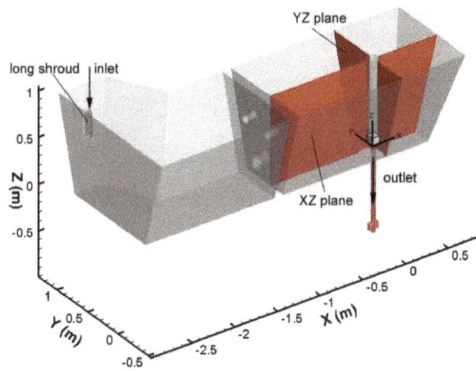

Figure 3. Schematic diagram of regional coordinate system of half tundish in the model calculation.

A mathematical model was developed using the finite-volume-based program ANSYS-FLUENT software (16.0, Ansys Inc., Canonsburg, PA, USA) based on these assumptions including the governing equations and boundary conditions described above. The velocity of liquid steel at the inlet and outlet of tundish is larger, and the flow behavior in the vicinity of stopper should be paid more attention to. So, local grid refinement was applied to simulate the behavior of blowing argon in the tundish; the meshes of the FLUENT computational domain included 6,700,000 unstructured grids using ANSYS-ICEM software (16.0, Ansys Inc.). The mesh size is 2 mm in the argon bubble action zone, and 8 mm far from the stopper. The SIMPLEC algorithm was applied to the velocity−pressure coupling. The second upwind order scheme was employed for discretization of momentum, k, and ε equations. The calculation was considered to be converged when the normalized residuals of all variables were smaller than 10^{-4}, when the flow field could achieve a relatively stable state. To reduce the simulation time and ensure the balance of equations at a discrete point of time, a fixed time step of 0.005 s was used in the time-dependent solution.

Considering the interaction between the continuous phase and the discrete phase, the stable flow field of the argon blowing in the tundish was obtained by calculating 15 s. Based on the stable flow field of gas–liquid two-phase, the tracer with the same properties as the liquid steel was added to the inlet of the stable flow field of the tundish for 1 s. The mixed flow of the tracer and the regional fluid were calculated by the transient mode simulation for 3000 s to obtain the residence time distribution (RTD) curve. The flow characteristics of the liquid steel in the tundish under different conditions were obtained by analyzing the RTD curve using the modified model proposed by Hong [29]. Next, we compared the vertical velocity of the liquid steel in the vicinity of stopper to analyze the influence of different process parameters on the flow of liquid steel. This was done 90 mm from the center of the stopper. The argon gas distribution between the tundish and mold was obtained by counting the number of bubbles entering the nozzle and the tundish in one second. The experimental outline is shown in Table 2. Combined with the results of water model experiments of tundish, the parameters of Rosin–Rammler distribution used in the calculation is shown in Table 3, then the mass fraction with bubble diameter greater than d_g can be determined by using of Equation (11).

Table 2. Mathematical simulation scheme.

Case	Casting Speed (m·min^{-1})	Argon Flow Rate (L·min^{-1})	Inner and Outer Diameter of Annular Permeable Brick (mm)
Case 1	1.35	2, 3, 4, 5	220/280
Case 2	1.05, 1.20, 1.35, 1.50	3	220/280

Table 3. Parameters of Rosin–Rammler distribution used in calculation.

Argon Flow Rate (L·min^{-1})	Minimum Bubble Diameter (mm)	Maximum Bubble Diameter (mm)	Average Bubble Diameter (mm)	Distribution Index
2	0.6	2.85	1.6	2.82
3	0.65	2.90	1.8	4.61
4	0.7	2.95	2.0	5.08
5	0.80	3.0	2.2	7.96

3. Comparison of Flow Behavior in Mathematical Model and Water Model

The flow of liquid steel in the tundish is mainly affected by the viscous force, gravity, and inertial force. The Froude number was chosen to ensure the motion similarity between the prototype and the model. A water model of the tundish with a 1:2 scale was made to simulate the argon blowing through the annular permeable brick in this work. When the argon flow rate was 3 L·min^{-1} and the casting speed was 1.35 m·min^{-1}, the inner and the outer diameters of the annular permeable brick were 220 mm and 280 mm, respectively. The distribution of argon bubbles in the water model and numerical simulation is shown in Figure 4, and the diffusion of the tracer in the tundish at the different times is shown in Figure 5, the left-hand diagram in Figure 5 shows the calculated flow of liquid steel in cross-section which is through the center of the lower diversion hole and the right side of the retaining wall.

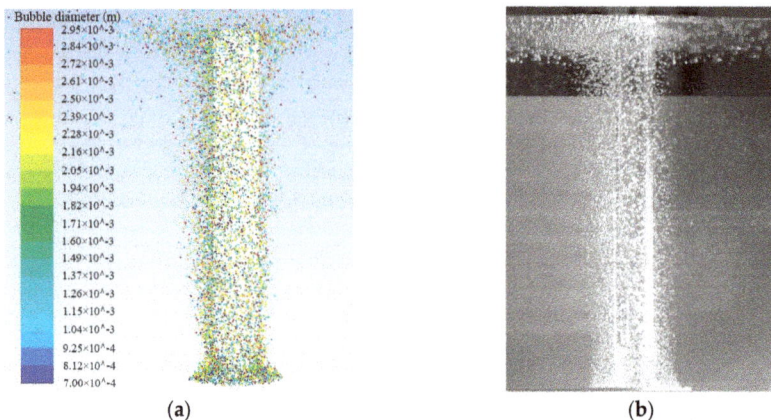

(a)   (b)

Figure 4. Distribution of argon bubbles in the tundish: (a) numerical simulation and (b) water model.

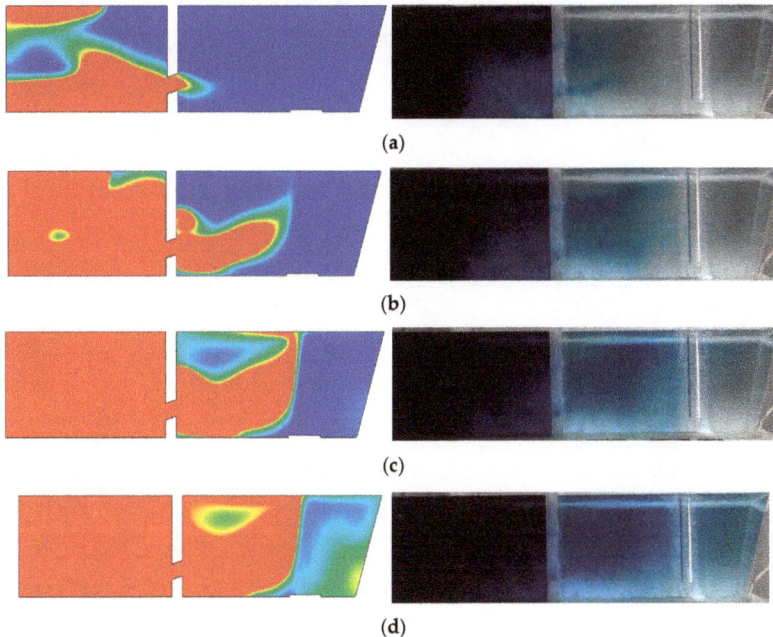

Figure 5. Flow field of the tracer in the half-tundish at different times of numerical model (left-hand diagram) and experiment (right-hand diagram): (**a**) 10 s, (**b**) 20 s, (**c**) 30 s, and (**d**) 40 s.

Figure 4 shows that the argon bubbles are asymmetrical on both sides of the stopper. The number of argon bubbles on the left side of the stopper is significantly higher than that on the right side of the stopper due to the action of the liquid steel flow.

Figure 5 demonstrates that the liquid steel departs from the diversion hole at the vicinity of the stopper and then flows upward to the liquid surface with the floating argon bubbles. The left side in Figure 5 is the flow field obtained by the mathematical modelling, right side is the flow field in water model of tundish. There are multi-swirl flow zones around the stopper. The diversion hole is not in the same XZ plane. It is part of the liquid steel that migrates from the back area of the stopper to the right wall of the tundish. There is weakened swirl flow to the right of the stopper. The distribution behavior of the argon bubbles calculated by the mathematical model is similar to that in the water model. Thus, the flow field of the liquid steel calculated by the numerical simulation is similar to that in the water model experiment.

4. Results and Discussion

4.1. Typical Flow Behavior of Liquid Steel in the Tundish with Annular Argon Blowing in the Upper Nozzle

When the casting speed was controlled at 1.35 m·min^{-1}, the inner and outer diameter of the annular permeable brick were 220 mm and 280 mm, respectively. The velocity streamlines of liquid steel in the tundish without the argon blowing and with the argon blowing at 3 L·min^{-1} are shown in Figure 6.

Figure 6a demonstrates that the liquid steel entered into the tundish from the shroud, and spreads after impinging on the turbulence controller. The liquid steel then passes through the diversion holes of the retaining wall and migrates obliquely upward. For the viscous resistance of the liquid steel and the obstruction of the stopper, the velocity at the near-surface of the liquid steel around the stopper is reduced gradually when the argon blowing is not adopted. Thus, the laminar flow in the liquid surface is weak, and there is no rising flow around the stopper.

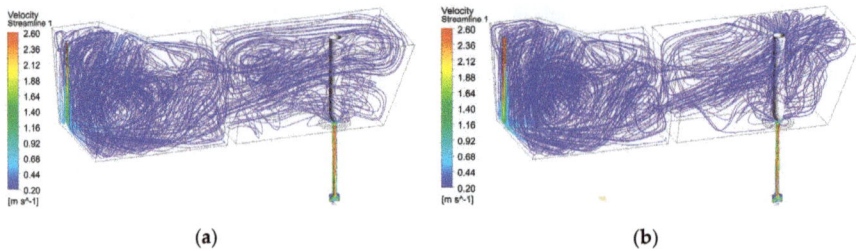

(a)   (b)

Figure 6. Velocity streamline of the liquid steel in the tundish: (**a**) without argon blowing and (**b**) with argon blowing.

When argon is blown through the annular permeable brick, the streamline of the liquid steel in the pouring area is obviously changed versus the pouring area without argon. Figure 6b demonstrates that the liquid steel is driven by the floating bubbles. It moves up to the liquid surface and diffuses to the surrounding area of the stopper. The liquid steel then moves to the vicinity of the tundish wall and flows downward; some swirl flows near the stopper are formed, which can make the level flow of liquid steel around the stopper active. The flow path of the liquid steel is extended, and the short circuit flow is significantly decreased in the tundish.

The flow characteristics of the liquid steel in the tundish with and without argon blowing by the mathematical calculation are shown in Table 4. Table 4 shows that the average residence time and the volume fraction of plug flow of liquid steel in the tundish increase with blowing argon versus no argon blowing. Thus, the annular argon blowing is beneficial to the floating and removal of inclusions in liquid steel.

Table 4. Flow characteristics of the liquid steel in the tundish with and without blowing argon.

Process Condition	Average Residence Time (s)	Volume Fraction of Plug Flow (%)	Volume Fraction of Dead Zone (%)	Volume Fraction of Mixed Flow (%)
without argon	592.12	30.91	12.75	56.34
with argon	593.56	31.59	12.53	55.88

4.2. Effect of Flow Rate of the Argon Blowing on the Flow Behavior of Liquid Steel in the Tundish

Figure 7 shows the flow behavior of the liquid steel along the XZ plane and the XY plane around the stopper in the tundish, the width of the XZ plane is only the distance between the diversion wall centerline and the right side wall of the tundish. The casting speed was 1.2 m·min^{-1}, and the inner and outer diameters of the annular permeable area in the nozzle pocket brick were 220 mm and 280 mm, respectively. The argon flow rate was 2 L·min^{-1}, 3 L·min^{-1}, 4 L·min^{-1}, and 5 L·min^{-1}. Under the same conditions, the vertical velocity of the liquid steel in the vicinity of the stopper is shown in Figure 8. The argon flow rate of 0 in Figure 8 indicates that the argon flow was absent.

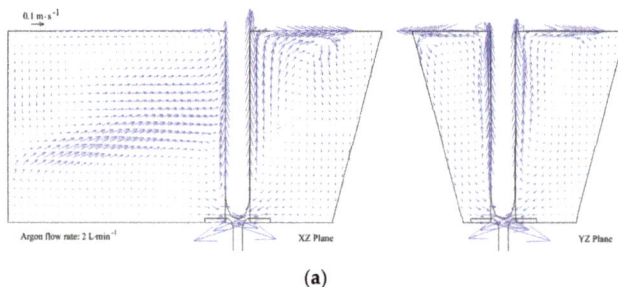

(a)

Figure 7. *Cont.*

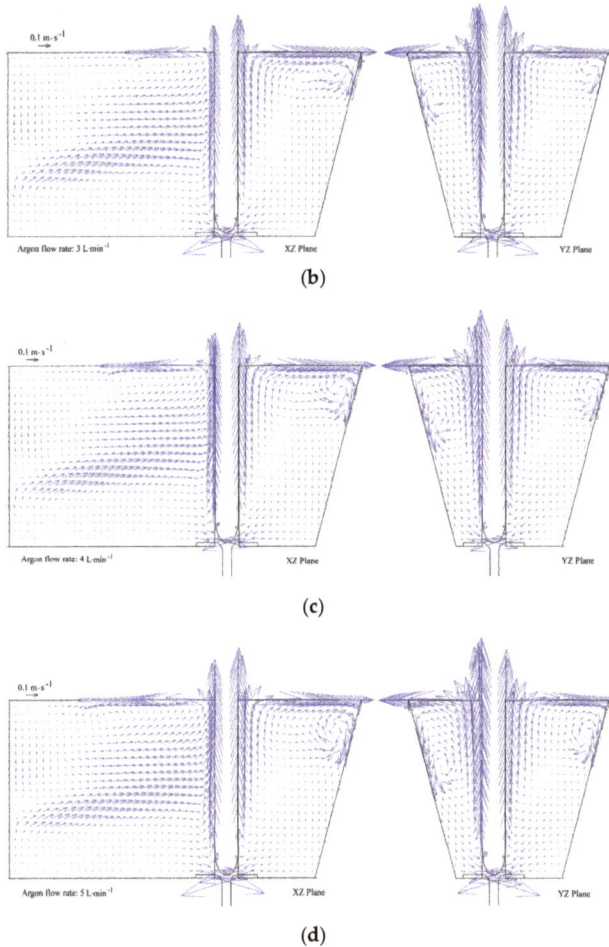

(b)

(c)

(d)

Figure 7. Flow behavior of liquid steel in the tundish under different argon flow rates: (**a**) 2 L·min⁻¹, (**b**) 3 L·min⁻¹, (**c**) 4 L·min⁻¹, and (**d**) 5 L·min⁻¹.

The vortex range and intensity around the stopper increase gradually with increased argon flowrates (Figure 7). The vortex center of liquid steel in the front, rear, and right sides of the stopper move towards the tundish wall, and then move downward along the tundish wall. The vortex center of liquid steel in the left side of the stopper moves towards the diversion wall. These flow behaviors are related to the coupling effect of the upflow of liquid steel near the stopper and the normal flow from the diversion hole. The flow intensity of the liquid steel near the tundish wall and diversion wall then increase, which decreases the dead zone of liquid steel in the tundish.

The near surface flow of liquid steel can promote floating of inclusions in the tundish, and then the increased flow rate of argon may improve the purity of liquid steel. However, an excessive surface flow velocity of liquid steel may cause slag entrapment. When the argon flowrate was 2 L·min⁻¹, 3 L·min⁻¹, 4 L·min⁻¹, and 5 L·min⁻¹, the peak value of the surface flow velocity in the tundish was 0.38 m·s⁻¹, 0.48 m·s⁻¹, 0.52 m·s⁻¹, 0.57 m·s⁻¹, respectively. Slag entrapment may occur at surface flow velocities over 0.45 m·s⁻¹ [30]. The argon flow rate should therefore not be larger than 3 L·min⁻¹.

As the distance from the upper nozzle surface increases, Figure 8 demonstrates that the vertical flow velocity of the liquid steel in the vicinity of the stopper first increases and then decreases.

It approaches 0 at the liquid surface in the tundish with blowing argon. These are related to the vortex flow around the stopper when the liquid steel arrives at this level. The vertical flow velocity is 0 while the horizontal flow velocity of liquid steel is the largest. In the mathematical model, the liquid surface was set to the free liquid surface in which the velocity variable at the normal direction is zero. In practical operation, the liquid surface of the tundish is covered with a slag layer. The floating argon bubbles around the stopper drive the liquid steel upward. The liquid steel from the left diversion hole impinges on the back side of the stopper, and some argon bubbles were carried by the liquid steel to the right side of the stopper as the argon bubbles floated. The vertical flow velocity is then higher than that in the left-hand of the stopper. Moreover, the distance between the front wall of the tundish and stopper is smaller, the floating argon bubbles cannot easily be dispersed. The driving effect of the floating bubbles then increases, and the vertical flow velocity of the liquid steel in the front vicinity of the stopper is higher.

There is no upflow of liquid steel without argon blowing. The liquid steel in the vicinity of the stopper flowed directly into the upper nozzle, and the vertical flow velocity in Figure 8 is negative. With increased argon flow rate, the driving effect of argon bubbles increases, and the vertical flow velocity of the liquid steel in the vicinity of the stopper increases. The peak value of vertical flow velocity of liquid steel increases with increased argon flow rate, and the peak value position also increases. The difference in peak value and peak value position with vertical flow velocity in different orientations is related to the coupling effect of the upflow of liquid steel near the stopper and the normal flow from the diversion hole.

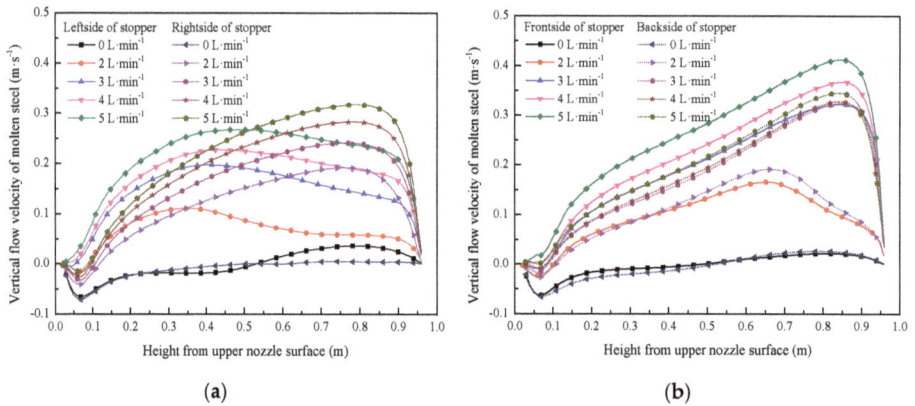

(a)   (b)

Figure 8. Vertical flow velocity of liquid steel in the vicinity of the stopper under different flow rates of argon: (**a**) left side and right side of the stopper; (**b**) front side and back sides of the stopper.

Figure 9 shows the volume flow distribution and ratio of argon in the tundish and mold under different argon flow rates. With increased argon flow rates, more argon bubbles may be released from the annular permeable brick. The number of argon bubbles entering the tundish and SEN then increases. The argon bubbles entering into the tundish nozzle can prevent the nozzle from clogging, and the high flow rate of argon may be appropriate. The bubble dimension increases with increasing argon flow rate. The larger bubbles float more easily, and the ratio of volume flow of argon between the tundish and mold then increases with increasing argon flow rate.

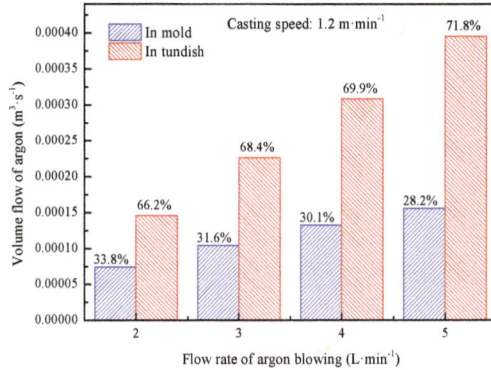

Figure 9. Volume flow distribution of argon under different argon flow rate.

4.3. Effect of Casting Speed on the Flow Behavior of Liquid Steel in the Tundish

When the flow rate of argon blowing was 3 L·min^{-1}, the inner and outer diameter of the annular permeable area in nozzle pocket brick were fixed at 220 mm and 280 mm, respectively, and the casting speed was 1.05 m·min^{-1}, 1.20 m·min^{-1}, 1.35 m·min^{-1} and 1.50 m·min^{-1}. The flow behavior of the molten steel in the XZ plane and the XY plane at different casting speeds is shown in Figure 10. The vertical flow velocity of liquid steel in the vicinity of the stopper at different casting speeds is shown in Figure 11.

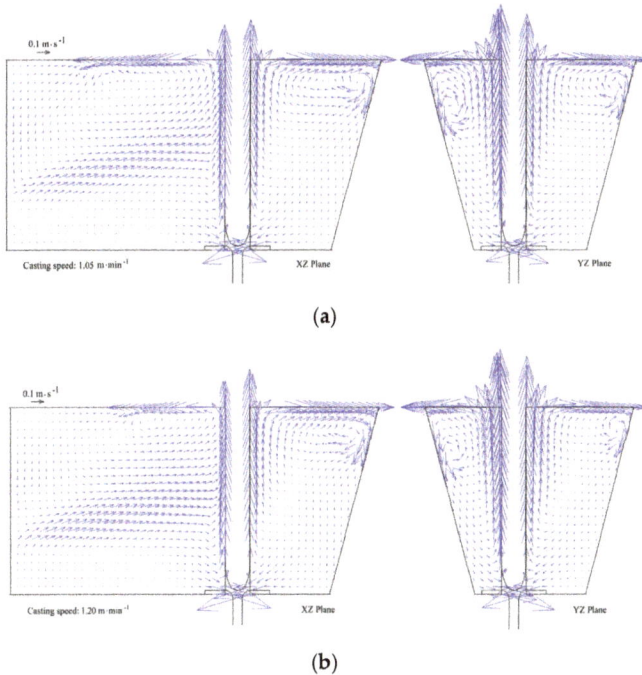

(a)

(b)

Figure 10. *Cont.*

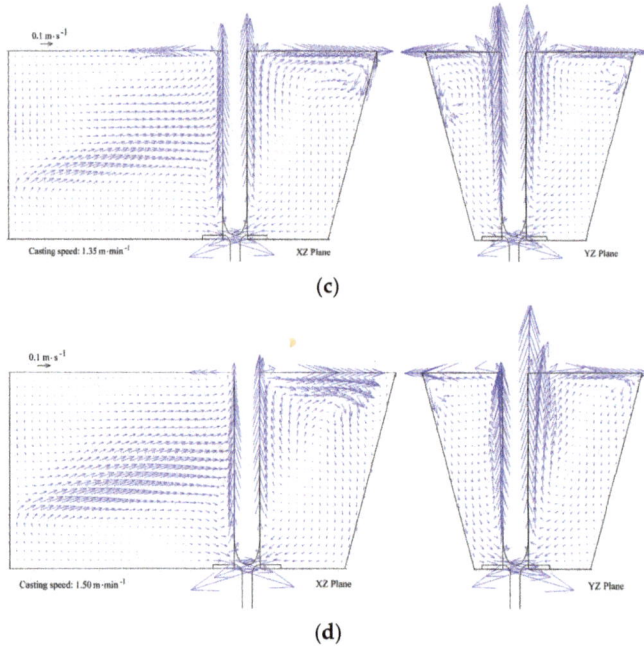

Figure 10. Flow behavior of liquid steel in the tundish under different casting speeds: (**a**) 1.05 m·min^{-1}, (**b**) 1.20 m·min^{-1}, (**c**) 1.35 m·min^{-1}, and (**d**) 1.50 m·min^{-1}.

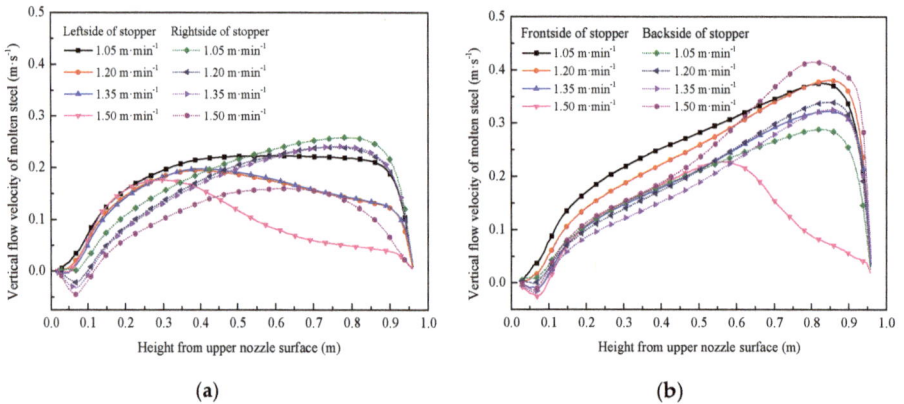

Figure 11. Vertical velocity of liquid steel in near area of stopper at different casting speeds: (**a**) left and right sides of the stopper; (**b**) front and back sides of the stopper.

Figure 10 shows that the vortex range and intensity around the stopper decrease gradually with increased casting speed. The vortex center of the liquid steel in different orientations moves towards the stopper. A higher casting speed leads to more bubbles entering the nozzle. The floating effect of the argon bubbles around the stopper weakens.

When the casting speed was 1.05, 1.20, 1.35, and 1.50 m·min^{-1}, the flow velocity peak of the liquid surface is 0.46, 0.50, 0.48, and 0.59 m·s^{-1}, respectively. The high casting speed can weaken the floating effect of the argon bubbles. The flow intensity near the back side of the stopper is high. This is related to the high flow velocity of liquid steel in the diversion hole; thus, the flow velocity peak at the liquid surface is high when the casting speed is 1.50 m·min^{-1}. This may lead to a slag entrapment problem, which can be eliminated by increasing the diameter of the diversion hole under the high casting speeds.

The peak value of the vertical flow velocity of liquid steel decreases with increased casting speed (Figure 11). This is seen at the left, right, and front side of the stopper; the height of peak value position also decreases. While the peak value of the vertical flow velocity of liquid steel at the back side of the stopper increases, the peak value position also moves up—this is related to the high flow velocity of liquid steel in the diversion hole, and can lead to slag entrapment. Thus, a high casting speed is disadvantageous for removal of inclusions.

Figure 12 shows the volume flow distribution and ratio of argon in the tundish and mold under different casting speeds. More argon bubbles were brought into the tundish nozzle with increased casting speed. The ratio of volume flow of argon between the tundish and mold decreases. When the casting speed is controlled to 1.05 m·min^{-1}, the argon volume fraction in the mold is minimized accounting for 15.9% of the total volume of blowing argon. This is 38.6% when the casting speed is controlled to 1.5 m·min^{-1}. The lower casting speed can promote the removal of nonmetallic inclusions in the tundish for a high floating effect around the stopper. This increases the residence time of the liquid steel. The high casting speed can catch more bubbles into the nozzle, which is beneficial to prevent the nozzle from clogging. It is important to regulate the floating bubbles around the stopper and the argon volume entering the nozzle.

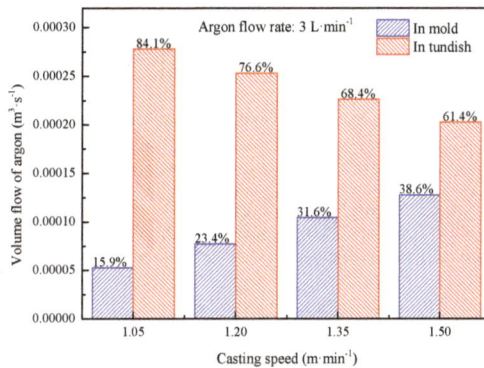

Figure 12. Volume flow distribution of argon under different casting speeds.

5. Conclusions

(1) There was some swirl flow of liquid steel around the stopper when the annular argon blowing at the upper nozzle was adopted. The liquid surface flow of liquid steel around the stopper is active, and the average residence time of liquid steel in the tundish slightly increases.

(2) With increased argon flow, the vortex range and intensity around the stopper gradually increase, and the vertical flow velocity of the liquid steel in the vicinity of the stopper increases. The argon volume flows in the tundish and the mold both increase.

(3) With increased casting speed, the vortex range and intensity around the stopper gradually decrease. The peak value of the vertical flow velocity of liquid steel at the vicinity of the stopper decreases, and the distribution and ratio of argon volume flow between the tundish and the mold decrease.

Author Contributions: X.Q. and C.C. conceived and designed the study; X.Q. and Y.L. conducted the experiment; X.Q. analyzed the experimental data and wrote the manuscript with the advice of C.C., C.Z., J.Z., and Y.J.

Funding: This research was funded by the National Nature Science Foundation of China (No. 51874215 and No. 51504172).

Acknowledgments: The authors would like to acknowledge the financial support from National Science Foundation of China (Grant No. 51874215 and No. 51504172).

Conflicts of Interest: The authors declare no conflicts of interest.

References

1. Moumtez, B.; Bellaouar, A.; Talbi, K. Numerical investigation of the fluid flow in continuous casting tundish using analysis of RTD curves. *J. Iron Steel Res. Int.* **2009**, *16*, 22–29.
2. Ding, N.; Bao, Y.P.; Sun, Q.S. Optimization of flow control devices in a single-strand slab continuous casting tundish. *Int. J. Miner. Metall. Mater.* **2011**, *18*, 292–296. [CrossRef]
3. Jin, Y.; Dong, X.S.; Yang, F. Removal mechanism of microscale non-metallic inclusions in a tundish with multi-hole-double-baffles. *Metals* **2018**, *8*, 611. [CrossRef]
4. Tripathi, A. Numerical investigation of electro-magnetic flow control phenomenon in a tundish. *ISIJ Int.* **2012**, *52*, 447–456. [CrossRef]
5. Chang, S.; Cao, X.; Zou, Z. Micro-bubble formation under non-wetting conditions in a full-scale water model of a ladle shroud Tundish system. *ISIJ Int.* **2018**, *58*, 60–67. [CrossRef]
6. Chatterjee, S.; Li, D.; Chattopadhyay, K. Modeling of liquid steel/slag/argon gas multiphase flow during Tundish open eye formation in a two-strand Tundish. *Metall. Trans. B* **2018**, *49*, 756–766. [CrossRef]
7. Chattopadhyay, K.; Hasan, M.; Isac, M. Physical and mathematical modeling of inert gas-shrouded ladle nozzles and their role on slag behavior and fluid flow patterns in a delta-shaped, four-strand tundish. *Metall. Trans. B* **2010**, *41*, 225–233. [CrossRef]
8. Cwudziński, A. Numerical and physical modeling of liquid steel flow structure for one strand tundish with modern system of argon injection. *Steel Res. Int.* **2017**, *88*, 1–14. [CrossRef]
9. Cwudziński, A. Numerical and physical modeling of liquid steel behaviour in one strand tundish with gas permeable barrier. *Arch. Metall. Mater* **2018**, *63*, 589–596.
10. Jin, Y.L.; Shen, C.; Zhang, J.P. Study on process technology of gas bubbling curtain substituting dam in the tundish of slab casting. *J. Iron Steel Res. Int.* **2011**, *18*, 256–262.
11. Chen, D.; Xie, X.; Long, M. Hydraulics and mathematics simulation on the weir and gas curtain in tundish of ultrathick slab continuous casting. *Metall. Trans. B* **2014**, *45*, 392–398. [CrossRef]
12. Vargas-Zamora, A.; Palafox-Ramos, J.; Morales, R.D. Inertial and buoyancy driven water flows under gas bubbling and thermal stratification conditions in a tundish model. *Metall. Trans. B* **2004**, *35*, 247–257. [CrossRef]
13. Zhong, L.; Li, L.; Wang, B. Water modelling experiments of argon bubbling curtain in a slab continuous casting tundish. *Steel Res. Int.* **2006**, *77*, 103–106. [CrossRef]
14. Wang, J.; Zhu, M.Y.; Zhou, H.B. Fluid flow and interfacial phenomenon of slag and metal in continuous casting tundish with argon blowing. *J. Iron Steel Res. Int.* **2008**, *15*, 26–31. [CrossRef]
15. Terzija, N.; Yin, W.; Gerbeth, G. Electromagnetic inspection of a two-phase flow of GaInSn and argon. *Flow Meas. Instrum.* **2011**, *22*, 10–16. [CrossRef]
16. Bai, H.; Thomas, B.G. Turbulent flow of liquid steel and argon bubbles in slide-gate tundish nozzles: Part 1. Model development and validation. *Metall. Trans. B* **2001**, *32*, 253–267. [CrossRef]
17. Bai, H.; Thomas, B.G. Turbulent flow of liquid steel and argon bubbles in slide-gate tundish nozzles: Part II. Effect of operation conditions and nozzle design. *Metall. Trans. B* **2001**, *32*, 269–284. [CrossRef]
18. Mohammadi-Ghaleni, M.; Zaeem, M.A.; Smith, J.D. Computational fluid dynamics study of molten steel flow patterns and particle–wall interactions inside a slide-gate nozzle by a hybrid turbulent model. *Metall. Trans. B* **2016**, *47*, 3056–3065. [CrossRef]
19. Li, Y.; Cheng, C.G.; Yang, M.L. Behavior characteristics of argon bubbles on inner surface of upper tundish nozzle during argon blowing process. *Metals* **2018**, *8*, 590. [CrossRef]
20. Smirnov, A.N.; Efimova, V.G.; Kravchenko, A.V. Design of a permeable annular refractory injection block for the tundish refining of steel. *Refract. Ind. Ceram.* **2014**, *55*, 173–178. [CrossRef]
21. Smirnov, A.N.; Efimova, V.G.; Kravchenko, A.V. Flotation of nonmetallic inclusions during argon injection into the tundish of a continuous-casting machine. Part 1. *Steel Transl.* **2013**, *43*, 673–677. [CrossRef]
22. Smirnov, A.N.; Efimova, V.G.; Kravchenko, A.V. Flotation of nonmetallic inclusions during argon injection into the tundish of a continuous-casting machine. Part 2. *Steel Transl.* **2014**, *44*, 11–16. [CrossRef]

23. Smirnov, A.N.; Efimova, V.G.; Kravchenko, A.V. Flotation of nonmetallic inclusions during argon injection into the tundish of a continuous-casting machine. Part 3. *Steel Transl.* **2014**, *44*, 180–185. [CrossRef]
24. Raghavendra, K.; Sarkar, S.; Ajmani, S.K.; Denys, M.B.; Singh, M.K. Mathematical modelling of single and multi-strand tundish for inclusion analysis. *Appl. Math. Modell.* **2013**, *37*, 6284–6300. [CrossRef]
25. Pfeiler, C.; Wu, M.; Ludwig, A. Influence of argon gas bubbles and non-metallic inclusions on the flow behavior in steel continuous casting. *Mater. Sci. Eng. A* **2005**, *413–414*, 115–120. [CrossRef]
26. Merder, T.; Warzecha, M. Optimization of a six-strand continuous casting tundish: Industrial measurements and numerical investigation of the tundish modifications. *Metall. Trans. B* **2012**, *43*, 856–868. [CrossRef]
27. Ramirez, O.S.D.; Torres-Alonso, E.; Banderas, J.Á.R. Thermal and fluid-dynamic optimization of a five strand asymmetric delta shaped billet caster tundish. *Steel Res. Int.* **2018**, *89*, 1700428. [CrossRef]
28. Zhang, H.; Luo, R.H.; Fang, Q. Numerical simulation of transient multiphase flow in a five-strand bloom tundish during lable change. *Metals* **2018**, *8*, 146. [CrossRef]
29. Lei, H. New insight into combined model and revised model for RTD curves in a multi-strand tundish. *Metall. Trans. B* **2015**, *46*, 2408–2413. [CrossRef]
30. Tang, D.T.; Li, Y.L.; Tian, Z.H. Numerical and physical simulation of bottom blowing gas rate effect on liquid steel fluidity feature in tundish. *Shanghai Met.* **2012**, *34*, 49–52. (In Chinese)

metals

MDPI

Article

Numerical Simulation of Electromagnetic Field in Round Bloom Continuous Casting with Final Electromagnetic Stirring

Bingzhi Ren [1,2], Dengfu Chen [2,*], Wentang Xia [1], Hongdan Wang [1,*] and Zhiwei Han [3]

1   School of Metallurgical and Materials Engineering, Chongqing University of Science & Technology, Chongqing 401331, China; renbingzhi@cqust.edu.cn (B.R.); wentangx@163.com (W.X.)
2   College of Materials Science and Engineering, Chongqing University, Chongqing 400044, China
3   Department of Continuous Casting, CISDI Engineering Co. Ltd., Chongqing 400013, China; zhiwei.han@cisdi.com.cn
*   Correspondence: chendfu@cqu.edu.cn (D.C.); wanghongdan@cqust.edu.cn (H.W.); Tel.: +86-023-6510-2467 (D.C.); +86-023-6502-3706 (H.W.)

Received: 25 September 2018; Accepted: 1 November 2018; Published: 5 November 2018

Abstract: A 3D mathematical model was developed to simulate the electromagnetic field in Φ600 mm round bloom continuous casting with final electromagnetic stirring (F-EMS), and the model was verified using measured data for the magnetic flux density in the stirrer centre. The distribution of electromagnetic force and the influence of current intensity and frequency were investigated. The results show that the Joule heat generated by F-EMS is very small and its influence on secondary cooling heat transfer in the stirring zone can be ignored. With an increase in current frequency, the electromagnetic force density at R/2 and R/3 of the Φ600 mm round bloom first increases and then decreases, reaching a maximum at 10 Hz.

Keywords: numerical simulation; round bloom; continuous casting; final electromagnetic stirring; electromagnetic field

1. Introduction

Continuous casting is a process generally used in steel production to make the molten steel solidified into a semi-finished billet, bloom, or slab for subsequent rolling in the finishing mills. Electromagnetic stirring (EMS) technology is widely used in the continuous casting production of steel. This technology utilises electromagnetic induction to provide a non-contact electromagnetic force to enhance the molten steel flow, heat transfer, and mass transfer and to promote columnar to equiaxed transition, thereby rectifying internal defects such as central segregation and shrinkage cavities [1]. According to different installation positions, three types of EMS exist in continuous casting steel. One is situated at the caster mould, referred to as mould electromagnetic stirring (M-EMS); another is situated along the strand in the secondary cooling zone, called a strand electromagnetic stirring (S-EMS); the last one is situated near the solidification end of the strand, known as final electromagnetic stirring (F-EMS) [2]. M-EMS is used in almost all billet/bloom casters. However, for high-carbon steel or large-section strands, the use of M-EMS cannot completely improve the internal quality of the strand, and F-EMS is generally required [3], which can more effectively improve the central porosity and V-shaped segregation of the strand [4,5]. Therefore, detailed analysis of the electromagnetic field, fluid flow, and solidification behaviour of the strand is very important for evaluating the impact of F-EMS on various metallurgical operations in the continuous casting process, and the electromagnetic field is the primary problem.

In 1986, Spitzer et al. deduced the analytical formula for the 2D electromagnetic force in an infinitely long stirring system [6]. Recently, Vynnycky [7] derived another form of analytical solution for the same issue with Ref. [6] when the tangential magnetic flux density was employed as the boundary condition. Trindade et al. [8], Yu et al. [9], and Liu et al. [10] studied the electromagnetic field of M-EMS, highlighting the obvious shielding effect of mould copper on the electromagnetic field, while that of F-EMS is different. Jiang et al. [11] and Sun et al. [3] focused mainly on fluid flow, solidification, and mass transfer of the continuous casting strand with F-EMS and rarely discussed the electromagnetic field. Therefore, this study aims to investigate the electromagnetic field in Φ600 mm round bloom continuous casting with F-EMS by numerical simulation, as well as examining the distribution features of the electromagnetic force and the influence of current intensity and frequency.

2. Mathematical Model

The continuous caster studied in this paper is produced by a domestic steel mill (Sunan Heavy Industry, Suzhou, China). Figure 1 is a structural diagram of the continuously cast round bloom with F-EMS. The origin of the coordinate is at the centre of the stirrer. The round bloom has a diameter of 600 mm and the stirrer core has a height of 300 mm.

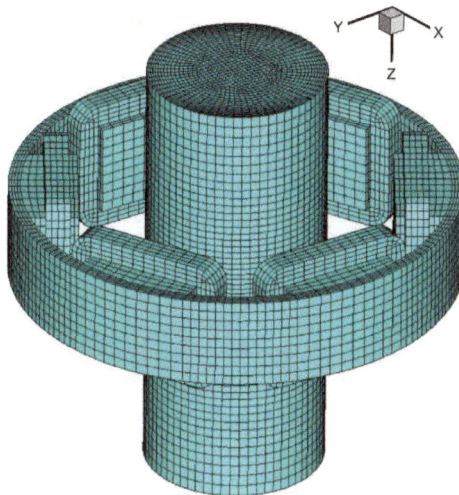

Figure 1. Schematic illustration of final electromagnetic stirring (F-EMS) for round bloom continuous casting. The air cylinder is not shown.

2.1. Basic Assumption

The fluid flow in a continuous casting round bloom with F-EMS can be described by Maxwell's equations and the Navier–Stokes equations, which are coupled by the electromagnetic force density and the flow velocity of the molten steel. In view of the complexity of the F-EMS process, several assumptions must be made in order to simplify the calculations.

1. In the F-EMS process, the magnetic Reynolds number (about 0.1) is far less than the shielding parameter (about 10) [10,12] and the angular frequency of the current intensity is much larger than the angular velocity of the molten steel; thus, the influence of the molten steel flow on the electromagnetic field is negligible [13]. Therefore, this problem can be decomposed into an electromagnetic problem and a fluid flow problem to be solved separately. In the present paper, only the electromagnetic field was calculated without regard to the flow field and solidification.
2. Since the current frequency of F-EMS is generally in the range of 4 Hz to 10 Hz, which belongs to the magnetic quasi-static field, the displacement current is ignored [8].

3. Both the outer shell and the cooling water jacket of the stirrer are approximated as an air zone for the benefit of the calculations.
4. The curve of the strand is negligible.

2.2. Control Equations

At low current frequencies, the time-varying electromagnetic field is assumed to be a magnetic quasi-static field, and Maxwell's equations can be simplified as follows [8]:

$$\nabla \cdot \boldsymbol{B} = 0 \tag{1}$$

$$\nabla \times \boldsymbol{B} = \mu_r \mu_0 \boldsymbol{J} \tag{2}$$

$$\nabla \times \boldsymbol{E} = -\frac{\partial \boldsymbol{B}}{\partial t} \tag{3}$$

$$\boldsymbol{J} = \sigma \boldsymbol{E} \tag{4}$$

where $\boldsymbol{B}$ is the magnetic flux density, $\boldsymbol{J}$ is the induced current density in the strand, μ_0 is the vacuum permeability, μ_r is the relative permeability, $\boldsymbol{E}$ is the electric field intensity, σ is the electrical conductivity, and t is the time.

The electromagnetic force density at each element is calculated by harmonic analysis using Equation (5) [8]:

$$\boldsymbol{F}_{time} = \boldsymbol{J} \times \boldsymbol{B} \tag{5}$$

The expression for the Joule heat power density generated by the induced current is calculated from Equation (6) [14]:

$$P_{time} = |J|^2 / \sigma \tag{6}$$

The time in Equations (5) and (6) refers to the real and imaginary parts of the variable at $0°$ and $90°$, respectively.

Since F-EMS uses a sinusoidal current to generate a harmonic electromagnetic field, the magnetic flux density, electromagnetic force density and Joule heat power density change with time. However, at low current frequencies, both the shielding parameter and the interaction parameter are small, and it is feasible to use the time-averaged electromagnetic force density and Joule heat power density [15]. Hence, the time-averaged values of the two variables are calculated from Equations (7) and (8) in the post-processing to investigate the distribution characteristics [8]:

$$F_{em} = (F_0 + F_{90})/2 \tag{7}$$

$$P_{ave} = (P_0 + P_{90})/2 \tag{8}$$

Because the time-averaged value of the magnetic flux density is 0, the time-averaged value of its modulus is transiently calculated by using Equation (9):

$$B_{mag} = f \int_{t_0}^{t_0 + 1/f} |B| dt \tag{9}$$

where f is the current frequency.

2.3. Boundary Conditions

An air cylinder with a radius of 0.9 m and a height of 2.2 m was established around the whole stirring system to calculate the electromagnetic field. The tangential magnetic condition ($\partial \Phi / \partial n$) = 0 was applied at the outer surface of the air cylinder, where Φ is the reduced scalar potential and n is the normal unit vector to the surface [8,10]. The three-phase coil excitation was applied with source current intensity (*I*) and frequency.

2.4. Physical Properties

Before the Φ600 mm round bloom is completely solidified, its temperature is above 1173 K [16], which is greater than the Curie point of the steel (about 1023 K) [17]. Therefore, the strand is a paramagnetic material. In the calculation, it is considered that both the strand and the iron core of the stirrer are isotropic materials and that their relative permeability is constant. The relevant physical properties are as follows: the vacuum permeability is 1.257×10^{-6} H·m^{-1}, the relative permeability of the strand and air is 1.0, the relative permeability of the iron core is 1000, and the electrical conductivity of the strand is 7.14×10^5 S·m^{-1}.

2.5. Solution

When calculating the electromagnetic field, the length of the strand is 1.8 m. ANSYS Emag software (ANSYS 14.0, ANSYS Inc., Canonsburg, PA, USA) was used to solve Maxwell's equations by the finite element method (FEM). We employed three different meshes to verify the independence of the grid, which included 220,000, 270,000, and 350,000 elements, respectively. The comparison between simulation results and experimental data indicated that the errors of the three meshes are 7.2%, 4.6%, and 4.0%, respectively. Hence, a mesh of 270,000 elements was selected to obtain the mesh-independent results. The distribution of the electromagnetic field in the strand was obtained and the time-averaged electromagnetic force density was extracted. Except for the transient analysis of B_{mag}, other variables were calculated using harmonic analysis.

3. Results and Discussion

In order to validate the mathematical model of the electromagnetic field, the magnetic flux density in the stirrer was measured along the axial direction without the strand and compared with the numerical simulation results, as shown in Figure 2. The measurement of the magnetic flux density was carried out in a Φ600 mm round bloom continuous caster at a domestic steel mill by using a portable Gauss meter (Bell, FW5180). The calculated values agree well with the measured data, which supports the correctness of the mathematical model and the solution method for the electromagnetic field. Figure 2 also demonstrates that the magnetic flux density in the stirrer with the strand is significantly lower than that without the strand, and the maximum value decreases from 55.4 mT calculated without the strand to 51.1 mT with the strand. This is mainly due to the 'skin effect': the strand can shield the electromagnetic field, resulting in a decrease in magnetic flux density.

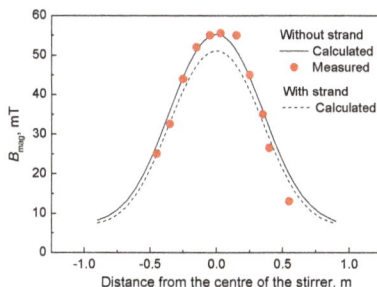

Figure 2. Profile of the magnetic flux density along the axial direction for $I = 350$ A and $f = 4$ Hz.

Figure 3 shows the distribution of the magnetic flux density in both the cross-section and the longitudinal section of the Φ600 mm round bloom at 400 A and 4 Hz. The magnetic flux density in the cross-section is roughly circularly symmetrical, and it is greatest at the edge and gradually attenuates towards the centre, as shown in Figure 3a. In the longitudinal section, the maximum value of the magnetic flux density appears in the middle of the stirrer and gradually weakens towards both ends, as shown in Figure 3b.

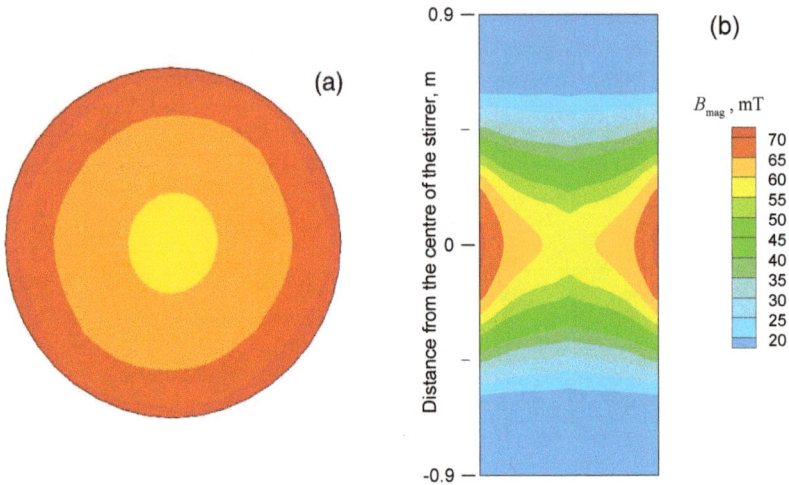

Figure 3. Contour plot of the magnetic flux density in the (a) cross-section and (b) longitudinal section for I = 400 A and f = 4 Hz.

Figure 4 displays the distribution of the time-averaged Joule heat power density in both the cross-section and the longitudinal section of the Φ600 mm round bloom at 400 A and 4 Hz. The characteristic of the Joule heat power density distribution is very similar to that of the magnetic flux density. It can also be seen from Figure 4a that the distribution of Joule heat power density has a hexagonal symmetry at the edge of the strand. This is because the iron core of the stirrer has six protruding tooth parts (see Figure 1), and the Joule heat power density at the edge of the strand near to them is large. Through the integral calculation, it is found that the Joule heat generated by F-EMS in the strand is 5.3 kW, and the quantity of the secondary cooling heat transfer (P) in this region is about 201 kW, which can be approximately calculated using Equation (10) [18]. Therefore, the Joule heat is small and negligible compared with the secondary cooling heat transfer in the stirring zone.

$$P = \varepsilon\delta\left(T_S^4 - T_0^4\right)CH/1000 \tag{10}$$

where ε is the emissivity of the steel (0.85 in the present study), δ is the Stefan–Boltzmann constant (5.67×10^{-8} W·m^{-2}·K^{-4}), T_s is the surface temperature of the round bloom (about 1223 K), T_0 is the environmental temperature (about 373 K), C is the circumference of the round bloom, and H is the height of the stirring zone, which is 1 m.

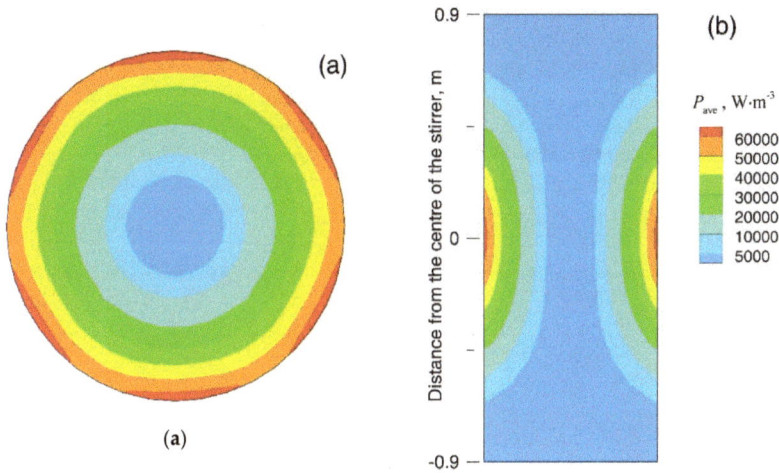

Figure 4. Contour plot of the time-averaged Joule heat power density in the (**a**) cross-section and (**b**) longitudinal section for $I = 400$ A and $f = 4$ Hz.

Figure 5a demonstrates the vector distribution of the time-averaged electromagnetic force density in the cross-section of the Φ600 mm round bloom. The electromagnetic force density is generally distributed circumferentially, thereby causing the rotation of the molten steel. It is greatest at the edge and decreases significantly towards the centre. Figure 5b shows the contour plot of the electromagnetic force density in the longitudinal section, which has a distribution similar to the magnetic flux density as well.

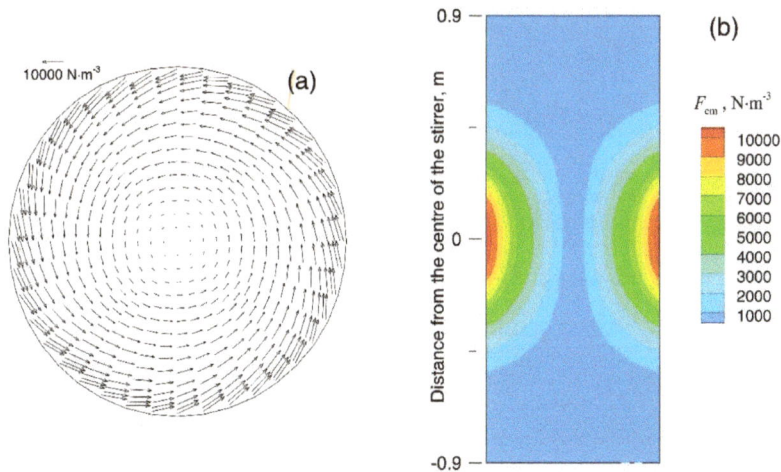

Figure 5. Distribution of the time-averaged electromagnetic force density in the (**a**) cross-section and (**b**) longitudinal section for $I = 400$ A and $f = 4$ Hz.

Figure 6 compares the electromagnetic force density calculated by FEM with that obtained from the analytical formula of the infinite-length stirring system deduced by Spitzer et al. [6]. Among the three components of the electromagnetic force density, the tangential electromagnetic force density is the largest, which is also the driving force of molten steel rotation. The tangential electromagnetic force density calculated by FEM is greater than that calculated by the analytical formula, especially at the edge of the strand. However, in the effective area of F-EMS on the strand ($r < R/2$), the difference between the two results is not large. In addition, the difference between the radial electromagnetic force densities calculated by FEM and by the analytical formula is also not large. Finally, the axial electromagnetic force density calculated by FEM is close to zero and negligible. In short, since the electromagnetic force generated by F-EMS usually acts in a region smaller than R/2 in a round bloom, it can be approximately calculated by the above analytical formula for the Φ600 mm round bloom.

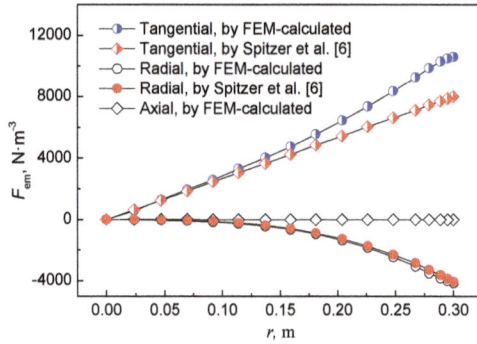

Figure 6. Comparison of FEM-calculated electromagnetic force density with that of the analytical solution by Spitzer et al. [6] in the middle plane of the stirrer for I = 400 A and f = 4 Hz.

Figure 7 illustrates the variation in the electromagnetic force density in the radial direction at different current frequencies. It can be observed that when the frequency is small, the electromagnetic force density has an approximately linear relationship with the radius, which is consistent with the analytical formula [6]. However, when the frequency is greater than 8 Hz, the electromagnetic force density is no longer linear with the radius due to the 'skin effect'.

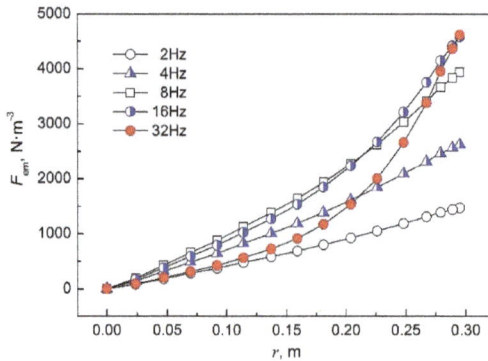

Figure 7. Electromagnetic force density profile as a function of radial distance at the middle plane of the stirrer for various current frequencies and I = 200 A.

Figure 8 shows the effect of the current frequency on the magnetic flux density and induced current density. It is found that different variations appear at different positions of the Φ600 mm round bloom. At the edge of the strand (r = R), the frequency has little effect on the magnetic flux density, while the induced current density increases significantly with increasing frequency. At r = R/2 of the strand, an increase in frequency leads to a decrease in magnetic flux density, and the induced current density first increases and then decreases, and the maximum value appears at 14 Hz. Since the electromagnetic force density is the cross-product of the magnetic flux density and the induced current density, it can be inferred from Figure 8 that there is a peak in the electromagnetic force density as a function of the current frequency.

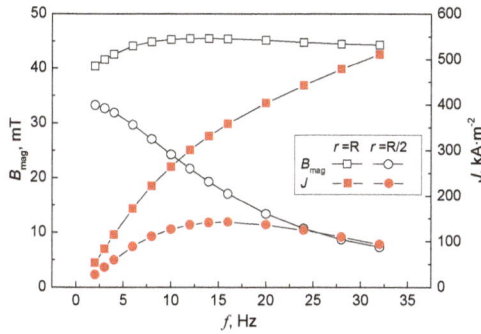

Figure 8. Effect of current frequency on the magnetic flux density and the induced current density at the middle plane of the stirrer for $I = 200$ A.

Figure 9 reveals the relationship between the electromagnetic force density and the current frequency at different radii of the $\Phi600$ mm round bloom. At the same current intensity, the electromagnetic force density at the edge of the strand ($r = R$) gradually increases with increasing frequency and remains basically constant until the frequency is greater than 16 Hz. However, the electromagnetic force density at $r = R/2$ and $R/3$ of the strand reaches a maximum at a frequency of 10 Hz and then gradually decreases. Since the radius of the melt core is generally less than $R/2$ at the position where the strand is installed with F-EMS, the F-EMS of the $\Phi600$ mm round bloom has the highest stirring efficiency at a frequency of 10 Hz.

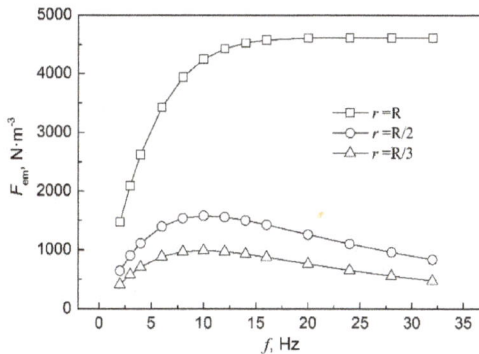

Figure 9. Effect of current frequency on the electromagnetic force density at different radii for $I = 200$ A.

Figure 10 shows the effect of current intensity on the magnetic flux density and the electromagnetic force density. The former increases linearly with an increase in current intensity, while the latter has a quadratic relationship with the current intensity, which is consistent with the analytical formula deduced by Spitzer et al. [6].

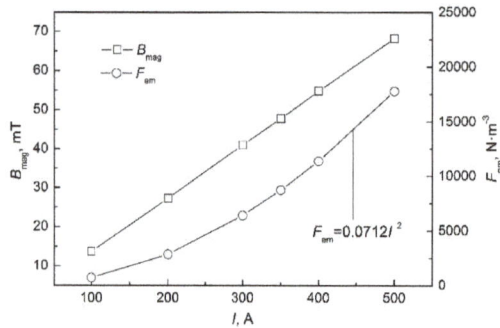

Figure 10. Effect of current intensity on the magnetic flux density and the electromagnetic force density at the middle plane of the stirrer for f = 4 Hz.

4. Conclusions

The three-dimensional electromagnetic field in a continuous casting round bloom with F-EMS was studied numerically, and the influence of current intensity and frequency on the electromagnetic field was discussed. The following conclusions were obtained:

1. The Joule heat generated by F-EMS in a continuous casting strand is very small, and its influence on the secondary cooling heat transfer in the stirring zone can be ignored.
2. Since the electromagnetic force generated by F-EMS usually provides stirring action in a region less than R/2 of a round bloom, it is feasible to calculate it approximately using the analytical formula for the infinite-length stirring system.
3. With increase in current frequency, the electromagnetic force density at R/2 and R/3 of the strand first increases and then decreases, and it reaches a maximum at 10 Hz. The results of the calculations show that the optimal current frequency of F-EMS is 10 Hz in continuous casting of a Φ600 mm round bloom.
4. The magnetic flux density increases linearly with current intensity, while the electromagnetic force density has a quadratic relationship with current intensity.

Author Contributions: B.R., W.X., and Z.H. validated the model. B.R. performed the calculation. B.R., D.C., and H.W. analyzed and discussed the data. B.R. wrote the paper. D.C. and H.W. reviewed and edited the paper. B.R., W.X., and H.W. acquired the funding.

Funding: This research was funded by the National Natural Science Foundation of China, grant number 51674057; the Chongqing Research Program of Basic Research and Frontier Technology, grant numbers cstc2016jcyjA0142 and cstc2017jcyjAX0236; the Scientific and Technological Research Program of Chongqing Municipal Education Commission, grant numbers KJ1601326 and KJ1713343; and the Research Foundation of Chongqing University of Science & Technology, grant number CK2016B19.

Conflicts of Interest: The authors declare no conflict of interest.

References

1. Barna, M.; Javurek, M.; Reiter, J.; Watzinger, J.; Kaufmann, B.; Kirschen, M. Continuous casting of round bloom strands with mould-electromagnetic stirring numerical simulations with a full coupling method. *World Iron Steel* **2012**, *12*, 29–33.
2. Hanley, P.J.; Kollberg, S.G. Electromagnetic methods for continuous casting. In *The Making, Shaping and Treating of Steel*, 11th ed.; Cramb, A.W., Ed.; Association for Iron & Steel Technology: Pittsburgh, PA, USA, 2003; pp. 287–297, ISBN 978-0-930767-04-4.
3. Sun, H.; Li, L.; Cheng, X.; Qiu, W.; Liu, Z.; Zeng, L. Reduction in macrosegregation on 380 mm × 490 mm bloom caster equipped combination M+F-EMS by optimising casting speed. *Ironmak. Steelmak.* **2015**, *42*, 439–449. [CrossRef]

4.   Mizukami, H.; Komatsu, M.; Kitagawa, T.; Kawakami, K. Effect of electromagnetic stirring at the final stage of solidification of continuously cast strand. *Tetsu-to-Hagane* **1984**, *70*, 194–200. [CrossRef]

5.   Xiao, C.; Zhang, J.; Luo, Y.; Wei, X.; Wu, L.; Wang, S. Control of macrosegregation behavior by applying final electromagnetic stirring for continuously cast high carbon steel billet. *J. Iron Steel Res. Int.* **2013**, *20*, 13–20. [CrossRef]

6.   Spitzer, K.H.; Dubke, M.; Schwerdtfeger, K. Rotational electromagnetic stirring in continuous casting of round strands. *Metall. Trans. B* **1986**, *17*, 119–131. [CrossRef]

7.   Vynnycky, M. On an anomaly in the modeling of electromagnetic stirring in continuous casting. *Metall. Mater. Trans. B* **2018**, *49*, 399–410. [CrossRef]

8.   Trindade, L.B.; Vilela, A.C.F.; Filho, Á.F.F.; Vilhena, M.T.M.B.; Soares, R.B. Numerical model of electromagnetic stirring for continuous casting billets. *IEEE Trans. Magn.* **2002**, *38*, 3658–3660. [CrossRef]

9.   Yu, H.Q.; Zhu, M.Y. Influence of electromagnetic stirring on transport phenomena in round billet continuous casting mould and macrostructure of high carbon steel billet. *Ironmak. Steelmak.* **2012**, *39*, 574–584. [CrossRef]

10.   Liu, H.; Xu, M.; Qiu, S.; Zhang, H. Numerical simulation of fluid flow in a round bloom mold with in-mold rotary electromagnetic stirring. *Metall. Mater. Trans. B* **2012**, *43*, 1657–1675. [CrossRef]

11.   Jiang, D.; Zhu, M. Center segregation with final electromagnetic stirring in billet continuous casting process. *Metall. Mater. Trans. B* **2017**, *48*, 444–455. [CrossRef]

12.   Marioni, L. *Computational Modelling and Electromagnetic-CFD Coupling Incasting Processes*; PSL Research University: Paris, France, 17 November 2017.

13.   Davidson, P.A.; Hunt, J.C.R. Swirling recirculating flow in a liquid-metal column generated by a rotating magnetic field. *J. Fluid Mech.* **1987**, *185*, 67–106. [CrossRef]

14.   Ren, B.Z.; Chen, D.F.; Wang, H.D.; Long, M.J.; Han, Z.W. Numerical simulation of fluid flow and solidification in bloom continuous casting mould with electromagnetic stirring. *Ironmak. Steelmak.* **2015**, *42*, 401–408. [CrossRef]

15.   Barna, M.; Javurek, M.; Reiter, J.; Lechner, M. Numerical simulations of mould electromagnetic stirring for round bloom strands. *Berg Huettenmaenn. Monatsh.* **2009**, *154*, 518–522. [CrossRef]

16.   Mao, B.; Ren, B.; Han, Z.; Cao, J.; Feng, K. Numerical simulation for heat transfer during solidification of round bloom continuous casting. *Ind. Heat.* **2012**, *41*, 50–53. [CrossRef]

17.   Guo, G.; Ma, H.; Zhang, J.; Huang, L. The relationship of thermophysical properties to microstructure of high-carbon steel. *Phys. Test. Chem. Anal.* **2006**, *42*, 167–170.

18.   Long, M.; Dong, Z.; Sheng, J.; Chen, D.; Chen, C. Universal secondary cooling structure for round blooms continuous casting of steels in various diameters. *Steel Res. Int.* **2015**, *86*, 154–162. [CrossRef]

![metals logo] *metals*

MDPI

Article

Mold-Level Prediction for Continuous Casting Using VMD–SVR

Wenbin Su [1], Zhufeng Lei [1,*], Ladao Yang [2] and Qiao Hu [1]

[1] School of Mechanical Engineering, Xi'an Jiaotong University, 28 West Xianning Road, Xi'an 710049, China; wbsu@mail.xjtu.edu.cn (W.S.); hqxjtu@xjtu.edu.cn (Q.H.)
[2] China National Heavy Machinery Research Institute Co., Ltd., 109 Dongyuan Road, Xi'an 710016, China; yld552008@163.com
* Correspondence: leizhufeng@stu.xjtu.edu.cn; Tel.: +86-177-4249-9272

Received: 5 March 2019; Accepted: 17 April 2019; Published: 18 April 2019

Abstract: In the continuous-casting process, mold-level control is one of the most important factors that ensures the quality of high-efficiency continuous casting slabs. In traditional mold-level prediction control, the mold-level prediction accuracy is low, and the calculation cost is high. In order to improve the prediction accuracy for mold-level prediction, an adaptive hybrid prediction algorithm is proposed. This new algorithm is the combination of empirical mode decomposition (EMD), variational mode decomposition (VMD), and support vector regression (SVR), and it effectively overcomes the impact of noise on the original signal. Firstly, the intrinsic mode functions (IMFs) of the mold-level signal are obtained by the adaptive EMD, and the key parameter of the VMD is obtained by the correlation analysis between the IMFs. VMD is performed based on the key parameter to obtain several IMFs, and the noise IMFs are denoised by wavelet threshold denoising (WTD). Then, SVR is used to predict each denoised component to obtain the predicted IMF. Finally, the predicted mold-level signal is reconstructed by the predicted IMFs. In addition, compared with WTD–SVR and EMD–SVR, VMD–SVR has a competitive advantage against the above three methods in terms of robustness. This new method provides a new idea for mold-level prediction.

Keywords: variational mode decomposition; empirical mode decomposition; support vector regression; mold level; continuous casting

---

1. Introduction

In the modern steel industry, high-efficiency continuous casting technology has become the most internationally competitive key technology [1]. The continuous casting process is a complex and continuous phase change process. Many factors affect the quality of slabs. The research into the key technology in the high-quality steel continuous-casting process is mainly focused on mold-level precision, as well as the segment and secondary cooling dynamic control [2].

At present, mold-level control is mainly based on the principle of predictive control, which combines prediction and control to improve the timeliness of prediction, but affects its accuracy. In view of the large mold-level disturbance, Guo et al. [3] used the prediction method in mold-level control. Aiming at the nonlinear characteristics of mold-level data, Tong et al. [4] carried out a constrained generalized prediction method based on the genetic algorithm. Aiming at the strong mold-level coupling characteristics, Qiao et al. [5] proposed an auto-disturbance suppression algorithm based on neural network tuning. However, these prediction methods have not effectively overcome the effects of mold-level noise.

Precise mold level monitoring is regarded as the key to improving continuous casting production quality, as shown in Figure 1 [2–4]. It is an important source of reference data for casting speed control, segment roll gap control, mold-cooling water control, and stopper rod opening control. If the mold level fluctuates too much, the following will occur. First, it will cause impurities on the surface of the mold. Surface defects and internal defects of the slab are generated which affect the surface and internal quality of the slab. Second, it will affect the casting speed, affecting productivity and the production rhythm. Eventually, it will cause the slab and the continuous casting machine to stick together, damage the tundish slide, and even cause downtime. Accurate prediction of the mold level occupies an important position in the continuous casting production process. This paper proposes an advanced mold level signal denoising method to prepare accurate data input for future mold level prediction, realize the purpose of predictive control, and greatly reduce the occurrence of accidents affecting quality and safety in the continuous casting production process.

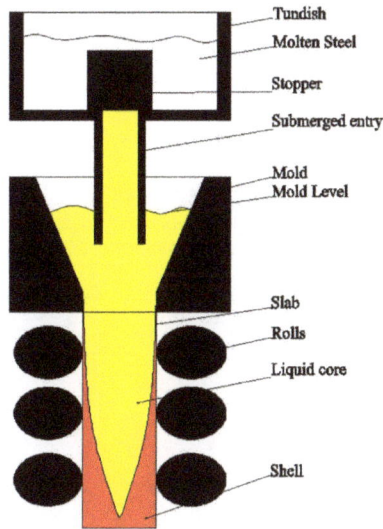

Figure 1. Mold-level model.

A data-driven method for mold-level prediction is proposed in this paper, which provides a new idea for mold-level control. The method takes variational mode decomposition (VMD) and support vector regression (SVR) as its core ideas, and creates mold-level predictions driven by data to overcome the influence of white noise caused by the casting speed and strong mold-level coupling.

Recent studies have shown that although there are many methods in the field of signal processing, none of them is applicable to all signal data. Wavelet transform (WT)-based signal processing methods are widely used, but wavelet denoising methods are limited by the selection of the wavelet basis function and affect the generalization ability of the wavelet. Although the method based on empirical mode decomposition (EMD) is widely used for the adaptability of its decomposition [6], the EMD method has serious pattern aliasing and boundary effects which seriously affect the signal decomposition. Especially in the process of signal noise processing, high-frequency components are often removed directly, resulting in loss of effective information. Signal processing techniques based on the VMD method have been widely used in recent years [7]. Compared with the EMD method, VMD effectively avoids mode aliasing and boundary effects and can realize the frequency domain splitting of signals and effective separation of components, which results in better noise and sample rate robustness.

For the prediction of time series, various prediction methods have appeared in the past several decades. Traditional time-series prediction methods, such as regression analysis and grey prediction [8], have some shortcomings, and the prediction accuracy of signals with large fluctuations needs to be improved [9]. The numerical weather prediction model for predicting future wind speed using mathematical models [10], multiple regression, exponential smoothing, the autoregressive moving average model (ARMA), and many others are used for wind-speed prediction, power prediction, stock-trend prediction, etc. Traditional time-series prediction methods have low precision and poor robustness to nonlinear disturbances. Mold level is non-linear and non-stationary in terms of the time scale and does not satisfy Gaussian normal distribution. Traditional time-series prediction methods are not suitable for mold-level prediction.

In recent years, with the rapid development of science and technology, artificial intelligence technology has been widely used and introduced into the prediction of time series, and good prediction results have been achieved [11]. Artificial neural networks (ANN) [12] and SVR [13] methods are the main tools for dealing with non-linear, non-stationary time series. SVR is a small-sample machine-learning method based on statistical learning theory, Vapnik–Chervonenkis (VC) dimension theory, and the minimum structural risk principle. Based on limited sample information, it seeks the best compromise between model complexity and learning ability to achieve the best promotion effect [14,15]. Liu and Gao [16] established a method for the online prediction of the silicon content in blast-furnace ironmaking processes. Compared with other soft sensors, the superiority of the proposed method is demonstrated in terms of the online prediction of the silicon content in an industrial blast furnace in China. Existing studies have shown that the ANN method takes a long time to calculate and is prone to localized minimization [17–20], leading to overfitting and poor prediction results. SVR is more robust to overfitting than ANN. The parameters of SVR can be improved by means of global optimization. It can be used to improve the prediction performance of SVR.

This paper focuses on the use of a hybrid algorithm for a time-series prediction model, and it is used for mold-level prediction. After comparing and discussing the hybrid algorithm for mold-level prediction, a new idea for continuous-casting process improvement is proposed. Firstly, the model uses EMD to decompose the original mold-level signal into several intrinsic mode functions (IMFs), and the key parameter of the VMD is obtained by the correlation analysis between the IMFs. VMD is performed based on the key parameter to obtain several IMFs, and the noise IMFs are denoised by wavelet threshold denoising (WTD). Then, SVR is used to predict each denoised component to obtain the predicted IMF. Finally, the predicted IMF reconstructs the predicted mold-level signal. The rest of this paper is organized as follows. The VMD algorithm is introduced in Section 2. VMD–SVR algorithms are introduced in Section 3. The performance of the three algorithms is compared through experiments in Section 4. Section 5 concludes this paper and makes recommendations.

2. Basic Algorithm Research

2.1. Variational Mode Decomposition

VMD is a new type of signal decomposition method. This method redefines an amplitude modulation-frequency modulation signal as an IMF, whose expression is

$$u_k(t) = A_k(t)\cos(\phi_k(t)) \tag{1}$$

where $\phi_k(t) \geq 0$ is the phase, $A_k(t)$ is the amplitude, $A_k(t) \geq 0$, $\omega_k(t) = \phi'_k(t)$, and $\omega_k(t)$ is the frequency.

In the interval range of $[t - \delta, t + \delta]$, $u_k(t)$ can be regarded as a harmonic signal with amplitude $A_k(t)$ and frequency $\omega_k(t)$, and $\delta = 2\pi/\phi'_k(t)$, where the prime denotes differentiation with respect to t.

The difference between VMD and EMD is that VMD is based on solving the variational problem and uses the variational model principle in the process of obtaining the IMFs, so that the sum of the estimated bandwidths of each IMF is minimized. The optimal solution of the constrained variational model is solved. The center frequency and bandwidth of the IMF are updated in the process of solving the variational model. The signal band is adaptively segmented based on the frequency domain of the signal itself. Further, a narrowband IMF is obtained.

The variational constraint model is as follows:

$$\min_{\{u_k\},\{\omega_k\}}\left\{\sum_k\left\|\partial_t\left[\left(\delta(t)+\frac{j}{\pi t}\right)\times u_k(t)\right]e^{-j\omega_k t}\right\|_2^2\right\}$$
$$\text{s.t.}\sum_k u_k = f \tag{2}$$

where $j=\sqrt{-1}$; $\{u_k\}:=\{u_1,u_2,\ldots u_K\}$ is the number of IMF; $\{\omega_k\}:=\{\omega_1,\omega_2,\ldots,\omega_K\}$ is the frequency center of each IMF; and $\sum_k:=\sum_{k=1}^K$ is the sum of all modes. $\|\|_2^2$ is the square of the 2-norm.

We introduce the Lagrange function as

$$L(\{u_k\},\{\omega_k\},\lambda)=\alpha\sum_k\left\|\partial_t\left[\left(\delta(t)+\frac{j}{\pi t}\right)\times u_k(t)\right]e^{-j\omega_k t}\right\|_2^2+\left\|f(t)-\sum_k u_k(t)\right\|^2+\left(\lambda(t),f(t)-\sum_k u_k(t)\right) \tag{3}$$

where α is the penalty factor and λ is the Lagrange multiplier. $\left\|f(t)-\sum_k u_k(t)\right\|_2^2$ is the second penalty.

$\langle\rangle$ is the integral mean of the variables.

The problem of solving the original minimum value can be transformed into the saddle point of the extended Lagrange expression by the alternating direction method, which is the optimal solution of the below formula:

$$u_k^{n+1}=\arg_{u_k}\min L(\{u_{i<k}^{n+1}\},\{u_{i\geq k}^{n+1}\},\{\omega_i^n\},\lambda^n) \tag{4}$$

$$\omega_k^{n+1}=\arg_{\omega_k}\min L(\{u_i^{n+1}\},\{\omega_{i<k}^{n+1}\},\{\omega_{i\geq k}^n\},\lambda^n) \tag{5}$$

$$\lambda^{n+1}=\lambda^n+\tau\left(f(t)-\sum_k u_k^{n+1}\right) \tag{6}$$

where $\sum_k\left\|u_k^{n+1}-u_k^n\right\|_2^2/\left\|u_k^n\right\|_2^2<\varepsilon$ is the convergence condition; n is the number of iterations; and τ is the update parameter.

Therefore, the original signal can be decomposed into K IMFs.

The calculation process of the VMD algorithm is as follows:

Step 1: Initialize $\{u_k^1\}$, $\{\omega_k^1\}$, λ^1 and n to zero;

Step 2: $n=n+1$, execute the entire loop;

Step 3: Execute the loop $k=k+1$ until $k=K$, update u_k: $u_k^{n+1}=\text{argmin}L(\{u_{i<k}^{n+1}\},\{u_{i\geq k}^n\},\{u_i^n\},\lambda^n)$;

Step 4: Execute the loop $k=k+1$, until $k=K$, update ω_k: $\omega_k^{n+1}=\text{argmin}L(\{\omega_{i<k}^{n+1}\},\{\omega_{i\geq k}^n\},\{\omega_i^n\},\lambda^n)$;

Step 5: Use $\lambda^{n+1}=\lambda_n+\tau\left(f(t)-\sum_k u_k(t)\right)$ to update λ;

Step 6: Given the discrimination condition $\varepsilon>0$, if the iteration stop condition is satisfied, all the cycles are stopped and the result is output, and K IMFs are obtained.

2.2. Support Vector Machine

SVM can not only solve the classification problem, but also solves the regression problem; the basic model is the largest linear classifier defined in the feature space. SVM aims to achieve a distinction between samples by constructing a hyperplane for classification so that the sorting interval between the samples is maximized and the sample to the hyperplane distance is minimized.

Set a training data set for a feature space $D=\{(x_1,y_1),(x_2,y_2),\ldots,(x_m,y_m)\}$, $x_i\in\chi=\mathfrak{R}^n$, $y_i\in y=\{+1,-1\}$, $i=1,2,\ldots,N$, where x_i is the i-th feature vector, y_i is the class tag of x_i.

The corresponding equation of the classification hyperplane is

$$h(x)=\omega\cdot x+b \tag{7}$$

where x is the input vector, w is the weight, and b is the offset.

The classification decision function is

$$\text{Sign}(h(x)) \tag{8}$$

$$\begin{cases} h(x) > 0, & y_i = 1 \\ h(x) < 0, & y_i = -1 \end{cases} \tag{9}$$

The support vector machine is implemented to find w and b when the interval between the separation hyperplane and the nearest sample point is maximized. When the training set is linearly separable, the sample points belonging to different classes can be separated by one or several straight lines with the largest interval. The maximum interval is solved by the following formula:

$$\max \gamma_i = y_i \left(\frac{w}{\|w\|} \cdot x_i + \frac{b}{\|w\|} \right) \tag{10}$$

$$\text{s.t. } y_i \left(\frac{w}{\|w\|} \cdot x_i + \frac{b}{\|w\|} \right) \geq \gamma, \ i = 1, 2, \dots, N \tag{11}$$

where γ is the geometric interval. Thus, we can obtain the linear separable support vector machine optimization problem.

$$\min_{w,b} \frac{1}{2} \|w\|^2 \tag{12}$$

$$\text{s.t. } y_i(w \cdot x_i + b) - 1 \geq 0, i = 1, 2, \dots, N \tag{13}$$

In the actual data set, there are many specific points, making the data set linearly inseparable; in order to solve this problem, we introduce a slack variable for each sample point $\xi_i \geq 0$, so that

$$y_i(w \cdot x_i + b) \geq 1 - \xi_i \tag{14}$$

For each slack variable ξ_i, pay a price ξ_i, and the optimization problem becomes

$$\min_{w,b,\varepsilon} \frac{1}{2} \|w\|^2 + C \sum_{i=1}^{N} \xi_i \tag{15}$$

where $C > 0$ is the penalty factor.

Most of the data are linearly inseparable; therefore, these data should be mapped to a high-dimensional feature space through non-linear mapping, letting the non-linear problem be transformed into a linear problem. The linear indivisible problem is transformed into a linearly separable problem.

Introduce kernel functions:

$$K(x_i, x_j) = \varphi(x_i) \cdot \varphi(x_j) \tag{16}$$

where the value of the kernel equals the inner product of two vectors, x_i and x_j.

At this point, we obtain

$$W(\alpha) = \frac{1}{2} \sum_{i=1}^{N} \sum_{j=1}^{N} \alpha_i \alpha_j y_i y_j K(x_i, x_j) - \sum_{i=1}^{N} \alpha_i \tag{17}$$

where α is the Lagrangian multiplier, $\alpha_i \geq 0$, $i = 1, 2, \dots, N$, and N is the number of samples.

In this paper, the radial basis function (RBF) is chosen as the SVR kernel function, and the expression is

$$K(x_i, x) = \exp\left(\frac{-\|x_i - x\|^2}{2g^2}\right) \tag{18}$$

where g is the kernel function coefficient.

At this point, the classification function becomes

$$f(x) = \text{sign}[\sum_{i=1}^{N} \alpha_i y_i \exp(\frac{-||x_i - x||^2}{2g^2}) + b]$$

(19)

2.3. Empirical Mode Decomposition

EMD is an adaptive signal processing technique suitable for non-linear and non-stationary processes [21]. In 1998, Huang et al. [6] proposed the empirical mode decomposition technology. Based on time scales, EMD local features such as local maxima, local minima, and zero-crossings, we decompose the signal into several IMFs and a residual; the IMFs are orthogonal to each other. Modal decomposition is determined by the signal itself.

EMD satisfies the following basic assumptions:

(1) In the entire data set, the number of extreme values and the number of zero crossings must be equal or at most have one point of difference.
(2) At any point, the average defined by the local maximum envelope and the minimum envelope is zero.

Finally, the original signal is decomposed into

$$x(t) = \sum_{i=1}^{N} c_i + r_N$$

(20)

where $x(t)$ is the original signal, c_i is the IMF, N is the number of IMFs, and r_N is the residual.

2.4. Wavelet Threshold Denoising

Suppose the model of denoising based on wavelet transform is

$$x = c + \sigma e$$

(21)

where x is the noise signal; c is the effective signal; e is the noise component in the noise signal; and σ is the noise intensity.

The wavelet transform and its denoising process are carried out in the following steps [22]:

(1) The noisy signal is transformed by wavelet transform. A wavelet basis is selected to determine the level N of the wavelet decomposition at the same time, and then the signal x is decomposed by the N-level wavelet.
(2) The wavelet coefficients are thresholder. In order to keep the overall shape of the signal unchanged and keep the effective signal, the hard threshold, soft threshold or other threshold methods are used to quantify the sparseness of each layer after decomposition.
(3) The inverse wavelet transform is performed, and the signal is reconstructed.

In this paper, a hard threshold denoising function is selected. Hard threshold processing compares the absolute value of wavelet transform coefficients with the threshold value. The coefficients smaller than or equal to the threshold value become zero, and the coefficients larger than the threshold value remain unchanged [23]. This method has better amplitude-preserving characteristics [24] and its expression is as follows:

$$S = \begin{cases} s, |s| \geqslant T \\ 0, |s| < T \end{cases}$$

(22)

where T is the threshold, and s is the wavelet decomposition coefficient.

3. Hybrid Algorithm Research

Mold-level prediction accuracy is influenced by many factors. In order to improve mold-level prediction accuracy, firstly, the noise in the original signal should be removed as much as possible. Then, we improve the prediction accuracy by using advanced prediction algorithms such as SVR. Thus, a prediction model based on the VMD–SVR algorithm for mold-level prediction is proposed in this paper. A hybrid algorithm flow chart is shown in Figure 2.

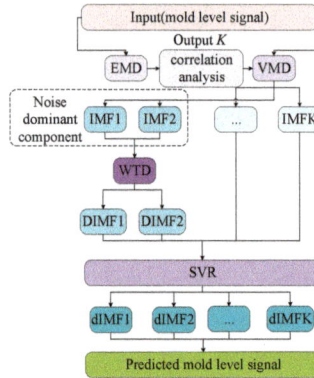

Figure 2. Hybrid algorithm flow chart. EMD—empirical mode decomposition; VMD—variational mode decomposition; IMF—intrinsic mode functions; WTD—wavelet threshold denoising; SVR—support vector regression.

Firstly, the original mold-level signal is subjected to data preprocessing to remove singular points. Then, all data are marked in the range of 0 to 1 to improve computational efficiency. Finally, the hybrid model is used for data prediction.

The hybrid algorithm flow is as follows:

Step 1: Adaptively decompose the mold-level data based on the EMD algorithm to obtain several IMFs;

Step 2: The K value of the key parameter of the VMD is obtained by the correlation analysis between the IMFs;

Step 3: Perform VMD decomposition on the original signal based on K to obtain K IMFs;

Step 4: Denoise the noise related component;

Step 5: Perform SVR on the denoised IMFs and other IMFs to obtain the predicted IMFs;

Step 6: Reconstruct the predicted component and obtain the predicted signal.

First, the mold-level signal is decomposed into several IMFs by the EMD, and the modal parameter K of the VMD is determined by correlation analysis between the IMFs. Then, the mold-level signal is decomposed into K IMFs by VMD, and the IMFs are analyzed to identify the noise dominant component, and the signal dominant component uses correlation analysis between the IMFs. Afterwards, in order to avoid the loss of effective information, the noise-related component is denoised by the WTD algorithm, and the effective information is effectively retained. SVR is performed on the denoised IMFs and other IMFs to obtain the predicted IMFs. Finally, the predicted IMFs are reconstructed to obtain the predicted signal.

The IMFs are obtained by adaptively decomposing the original mold-level data based on the novel VMD–SVR hybrid algorithm, the main purpose of which is to distinguish the noise-dominant IMFs and information-dominant IMFs. In order to preserve as much valid information as possible in the original mold-level data, denoising the noise-dominant IMFs can effectively remove the effects of white noise. Then, SVR is performed on all IMFs, the predicted IMFs are obtained for signal reconstruction, and the predicted mold-level data is obtained.

4. Experimental Studies

4.1. Problem Prescription

This paper presents a mold-level prediction model. This model is important for mold-level control and propose new ideas to improve continuous-casting automatic control. In order to clearly express the applicability, superiority, and generalization capability of the model application, the mold-level data of actual process parameters, collected from the continuous casting machine developed by the China National Heavy Machinery Research Institute Co., Ltd. (Xi'an, China), are used in this paper. We used an eddy current sensor to collect the mold-level signal at a steady cast speed. There are many uncertain disturbance factors in the mold-level control process, and the disturbance may change constantly at any time. Most of the disturbances are non-linear and non-stationary, and the long-term prediction model is difficult to establish.

A continuous casting production process data acquisition graph is presented in Figure 3. The time interval Δ*t* = 0.5 h, and the sampling frequency was 2.7 Hz.

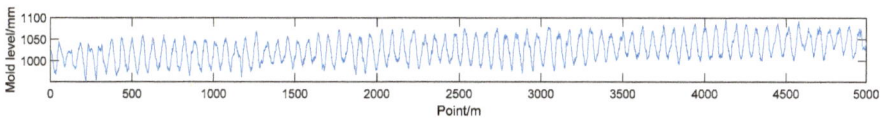

Figure 3. Mold level. The unit of the mold level is mm, while *m* is the number of points.

The main technical parameters of the continuous casting machine are shown in Table 1.

Table 1. Main technical parameters of the continuous casting machine.

Project	Specification
Continuous-casting machine model	Curved continuous caster
Secondary cooling category	Aerosol cooling, dynamic water distribution
Gap control	Remote adjustment, dynamic soft reduction
Basic arc radius/mm	9500
Mold length/mm	900
Metallurgical length/mm	39,200
Mold vibration frequency/time/min	25–400
Mold vibration amplitude/mm	2–10
Slab width/mm	900–2150
Slab thickness/mm	230/250
Working speed/m/min	0.8–2.03
Actual cast speed/m/min	1.3
Slab section size/mm × mm	230 × 1350
Mold oscillation frequency/Hz	1.36
Actual oscillation amplitude of mold/mm	60

4.2. Mold-Level Prediction Based on VMD–SVR Model

The VMD decomposition number is artificially determined, not adaptive. EMD is an adaptive decomposition method. Therefore, in order to minimize the interference of human factors, we decomposed the original data using EMD, and through the calculation of the correlation coefficient, a component having the largest correlation coefficient with the original signal was obtained as a boundary line between the high-frequency signal and the low-frequency signal, the high-frequency signal was integrated into one component, and the remaining components were retained to determine the number *K* of VMD decomposition.

First, the original data was subjected to EMD decomposition; the EMD decomposition results are shown in Figure 4.

(a)

(b)

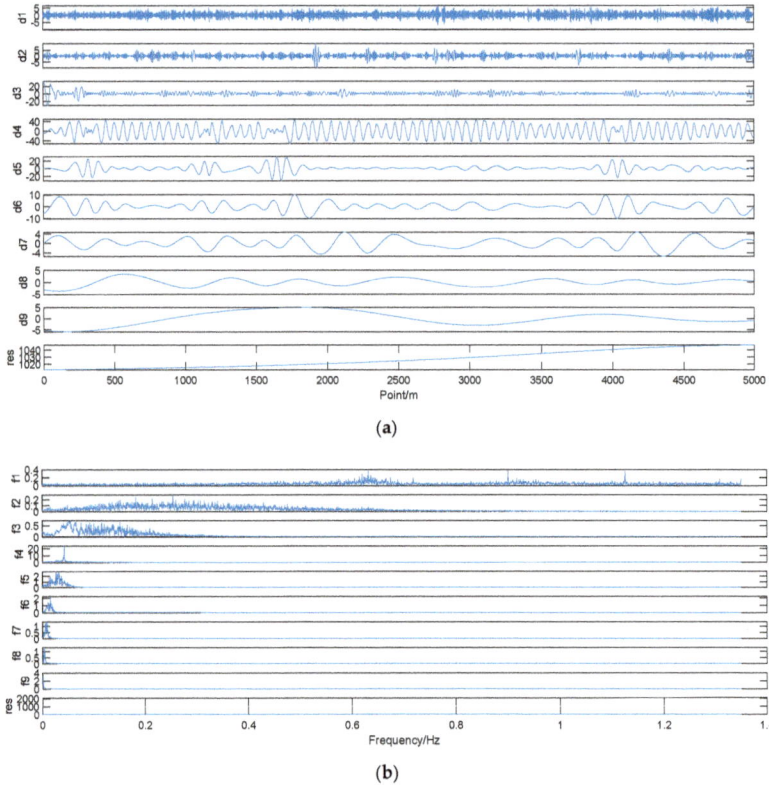

Figure 4. (**a**) Mold-level data EMD results; (**b**) spectrogram after EMD of the mold-level data; d_i is the *i*-th IMF, the unit of d_i is mm, *m* is the number of points, res is the residual, and f_i is the spectrum corresponding to the *i*-th IMF.

After the mold-level data is decomposed by the EMD as shown in Figure 3, the correlation coefficient between the original mold-level signal and the IMFs after EMD was determined, as shown in Table 2; IMFs 1–3 were seen to be weakly correlated with the original mold-level signal. There was a strong correlation between the original mold-level signal and the fourth IMF. We used IMFs 1–3 as a *K* value in the VMD decomposition, which is considered to be a high-frequency component of IMFs 1–3, and took the remaining IMF as 6 *K* values, thus obtaining *K* = 7, and performing VMD decomposition based on *K* = 7, which is not a simple direct merger of IMFs 1–3.

Table 2. The correlation coefficient between the original mold-level signal and the IMFs after EMD.

IMF	Correlation Coefficient
IMF 1	0.06
IMF 2	0.0906
IMF 3	0.1348
IMF 4	0.8474
IMF 5	0.1579
IMF 6	0.0196
IMF 7	0.0061
IMF 8	0.0598
IMF 9	0.0585

The VMD decomposition of the mold-level data was based on *K* = 7. The decomposition result is shown in Figure 5.

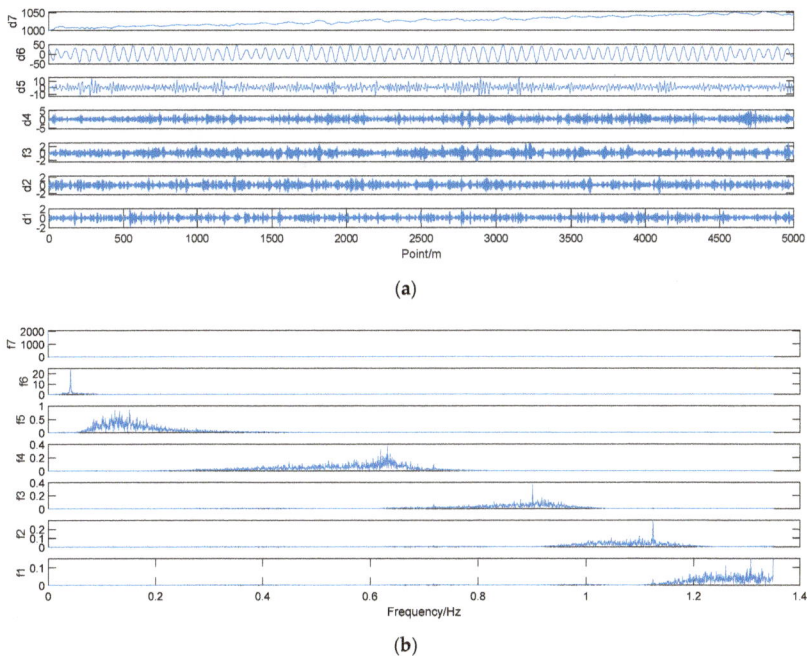

(a)

(b)

Figure 5. (a) Mold-level data VMD results; (b) spectrogram after VMD of the mold-level data; d_i is the *i*-th IMF, the unit of d_i is mm, *m* is the number of Point, and f_i is the spectrum corresponding to the *i*-th IMF.

It can be seen from Figure 4 that the mold-level data could clearly distinguish the center frequency of each IMF based on *K* = 7 decomposition, and no pattern aliasing occurred.

After the mold-level data was decomposed by the VMD, as shown in Figure 5, the correlation coefficient between the original mold-level signal and the IMFs after VMD was calculated, as shown in Table 3; IMFs 1–5 were weakly correlated with the original mold-level signal. There was a strong correlation between the original mold-level signal and the fourth IMF. Therefore, IMF 6 was a boundary line between the high-frequency signal and the low-frequency signal; high-frequency signals may also contain a small amount of effective information, and so, in order to minimize the loss of effective information, we performed wavelet threshold denoising on high-frequency signals (IMFs 1–5) instead of directly deleting them.

Table 3. The correlation coefficient between the original mold-level signal and the IMFs after VMD.

IMF	Correlation Coefficient
IMF 1	0.0279
IMF 2	0.0360
IMF 3	0.0429
IMF 4	0.0638
IMF 5	0.1769
IMF 6	0.8847
IMF 7	0.4560

It can be seen from Figure 6 that the noise reduction effect for IMFs 1–5 was very obvious. Both the main frequency and the amplitude had a large reduction.

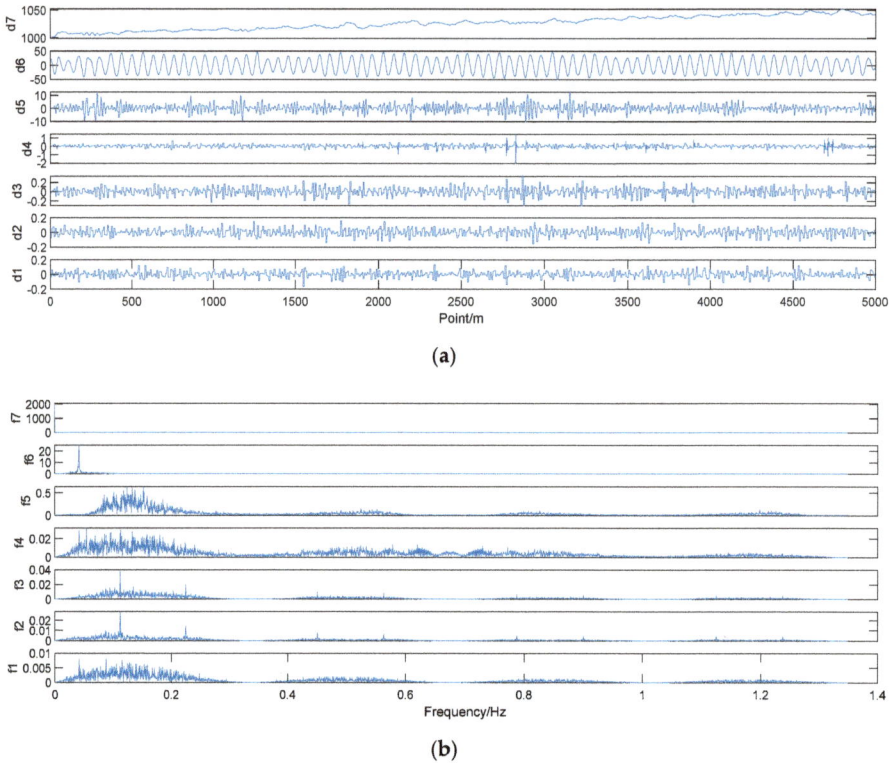

(a)

(b)

Figure 6. (**a**) Denoising result of IMFs 1–5; (**b**) spectrogram of the mold-level data after denoising; d_i is the i-th IMF, the unit of d_i is mm, m is the number of points, and f_i is the spectrum corresponding to the i-th IMF.

Then, SVR was performed on the all IMFs. In this section, the genetic algorithm was still used to globally optimize the model parameters C and g, so that the SVR model was determined. C was 15.2768 and g was 0.2018. The first 20 min of mold-level data was used as a training set, while the last 10 min of mold-level data was used as a test set in order to verify the prediction effect of the model. This method has high computational efficiency, high calculation accuracy, and can be run in real-time.

The optimization results of C and g are shown in Figure 7; fitness was the hit rate of the genetic algorithm. The predicted data of VMD–SVR are shown in Figure 8, and the VMD–SVR prediction error is shown in Figure 9.

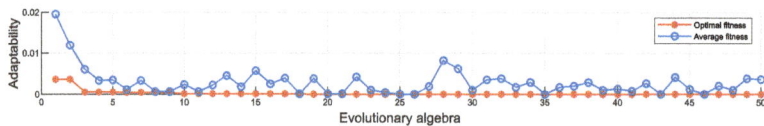

Figure 7. C and g optimization results. C is the penalty coefficient, g is the parameter of kernel function.

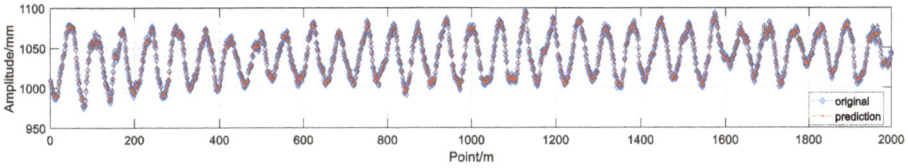

Figure 8. Comparison of VMD–SVR prediction results with original mold-level data. *m* is the number of points.

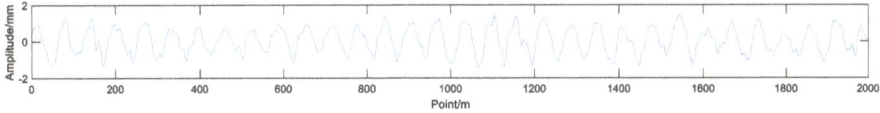

Figure 9. VMD–SVR prediction error. *m* is the number of points.

5. Prediction Results and Analysis

In this section, the performance of the three hybrid prediction algorithms is verified by the following four statistical indicators, which are the general purpose of the machine learning domain verification algorithm, and the optimal hybrid prediction model suitable for the mold steel level of the mold is selected.

Correlations between the original data and the predicted data, which is characterized by correlation coefficients (R):

$$R = \frac{\text{Cov}(P_i, A_i)}{\sqrt{\text{Var}(P_i) \cdot \text{Var}(A_i)}} \tag{23}$$

CC is defined as a statistical indicator and is used to reflect the close relationship between variables; the larger the CC, the better the algorithm performance.

Root mean square error (RMSE)

$$\text{RMSE} = \sqrt{\frac{\sum_{i=1}^{n} (P_i - A_i)^2}{n}} \tag{24}$$

RMSE is defined to reflect the degree of dispersion of a data set and to measure the deviation between the observed value and the true value; the smaller the RMSE, the better the algorithm performance.

Mean absolute error (MAE)

$$\text{MAE} = \frac{\sum_{i=1}^{n} |P_i - A_i|}{n} \tag{25}$$

MAE is defined as the average value of absolute error, better reflecting the actual situation of predicted error; the smaller the MAE, the better the algorithm performance.

Mean absolute percentage error (MAPE)

$$\text{MAPE} = \frac{\sum_{i=1}^{n} \left| \frac{P_i - A_i}{A_i} \right|}{n} \times 100 \tag{26}$$

MAPE can be used to measure the outcome of a model's predictions; the smaller the MAPE, the better the algorithm performance.

In Formulas (23)–(26), where P_i and A_i are the *i*-th predicted and actual values, respectively, and n is the total number of predictions.

From the test results in Table 4 and Figure 10, comparing the four indicators of the three algorithms, the test results of the average error in the algorithm described in this paper are inferior to the other two algorithms. However, in the test results of the other three indicators, the RMSE index is improved by 36.1%, the MAPE index is improved by 37.5%, the *R* is improved by 3%, and the MAE index is improved by 37.6%. Compared with WT and EMD, the VMD algorithm has shown great superiority, which not only rejects the dependence of the wavelet transform on basis function, but also avoids the boundary effect and pattern aliasing of empirical mode decomposition and improves the robustness of the algorithm and generalization ability.

Table 4. Test results comparison of prediction model. *R* is correlation coefficients; RMSE is root mean square error; MAE is mean absolute error; MAPE is mean absolute percentage error.

Algorithm	*R*	RMSE	MAE	MAPE
WT–SVR	0.9733	1.0824	0.9601	0.092316
EMD–SVR	0.9691	0.9480	0.7662	0.073558
VMD–SVR	0.9992	0.6910	0.5983	0.057686

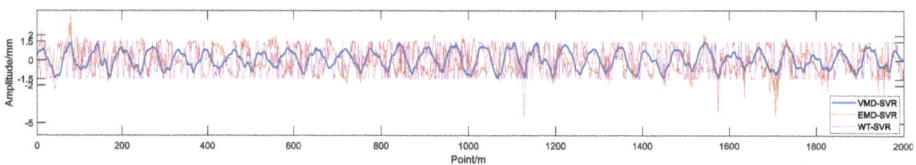

Figure 10. Prediction error between VMD–SVR and other methods. *m* is the number of points.

6. Conclusions

This paper proposes a prediction method based on VMD–SVR, which is suitable for mold-level prediction in continuous casting. In this method, the original mold-level data are adaptively decomposed by the EMD algorithm to obtain the effective IMF number *K*, via correlation coefficient analysis between the original mold-level signal and IMFs. The VMD decomposition of the original mold-level data is performed based on *K*, and the IMFs are obtained. Time-series prediction is performed for each IMF via SVR, and the VMD reconstruction is performed on the prediction result to obtain the final predicted mold-level signal. In order to verify the effectiveness of the proposed method, we compared the four statistical indicators of three algorithms; the conclusions are as follows.

(1) The VMD–SVR algorithm can be used to establish the prediction model, removing noise while retaining the effective information in the data, with good denoising performance and sampling rate robustness;

(2) In comparison with the results of the other two algorithms, the three indicators of the VMD–SVR algorithm are significantly better than those of the other two algorithms. The RMSE index is improved by 36.1%, the MAPE index are improved by 37.5%, the *R* is improved by 3%, and the MAE index is improved by 37.6%;

(3) The use of mold-level prediction methods in the research on mold prediction control represents a future research direction. Accurate mold-level prediction provides a new idea for mold-level prediction control, which has important practical significance;

(4) Using the accurately predicted mold-level data for mold-level control, the sliding nozzle and roller pressure disturbances can be well restrained. The anti-interference ability of the mold level control system is enhanced.

The potential feedback between the mold level controller and the mold level prediction will improve the accuracy and efficiency of the prediction model, which will be the focus of further research in a future paper.

Author Contributions: W.S. conceived and designed the experiments, Z.L. performed the experiments, L.Y. provided mold-level data, Q.H. analyzed the data, and Z.L. wrote the paper.

Metals **2019**, *9*, 458

Funding: This work was financially supported by the National Natural Science Foundation of China, grant number 51575429.

Acknowledgments: Q.G., X.L., H.Z., B.H., and Y.Z. are acknowledged for their valuable technical support.

Conflicts of Interest: The authors declare no conflicts of interest.

References

1. Ataka, M. Rolling technology and theory for the last 100 years: The contribution of theory to innovation in strip rolling technology. *ISIJ Int.* **2015**, *55*, 89–102. [CrossRef]
2. Jin, X.; Chen, D.F.; Zhang, D.J.; Xie, X. Water model study on fluid flow in slab continuous casting mould with solidified shell. *Ironmak. Steelmak.* **2011**, *38*, 155–159. [CrossRef]
3. Guo, G.; Wang, W.; Chai, T. Predictive mould level control in a continuous casting line. *Control Theory Appl.* **2011**, *18*, 714–717.
4. Tong, C.; Xiao, L.; Peng, K.; Li, J. Constrained generalized predictive control of mould level based on genetic algorithm. *Control Decis.* **2009**, *24*, 1735–1739.
5. Qiao, G.; Tong, C.; Sun, Y. Study on Mould level and casting speed coordination control based on ADRC with DRNN optimization. *Acta Autom. Sin.* **2007**, *33*, 641–648.
6. Huang, N.E.; Shen, Z.; Long, S.R.; Wu, M.C.; Shih, H.H.; Zheng, Q.; Yen, N.; Tung, C.C.; Liu, H.H. The empirical mode decomposition and the Hilbert spectrum for nonlinear and non-stationary time series analysis. *Proc. R. Soc. A-Math. Phys.* **1998**, *454*, 903–995. [CrossRef]
7. Konstantin, D.; Dominique, Z. Variational mode decomposition. *IEEE Trans. Signal Process.* **2014**, *62*, 531–544.
8. Lee, W.J.; Hong, J. A hybrid dynamic and fuzzy time series model for mid-term power load predicting. *Int. J. Electr. Power Energy Syst.* **2015**, *64*, 1057–1062. [CrossRef]
9. Dai, S.; Niu, D.; Li, Y. Daily peak load predicting based on complete ensemble empirical mode decomposition with adaptive noise and support vector machine optimized by modified grey wolf optimization algorithm. *Energies* **2018**, *11*, 163. [CrossRef]
10. Lynch, P. The origins of computer weather prediction and climate modeling. *J. Comput. Phys.* **2008**, *227*, 3431–3444. [CrossRef]
11. Gaudioso, M.; Gorgone, E.; Labbe, M.; Rodríguez-Chía, A.M. Lagrangian relaxation for SVM feature selection. *Comput. Oper. Res.* **2017**, *87*, 137–145. [CrossRef]
12. Wang, J.; Shi, P.; Jiang, P.; Hu, J.; Qu, S.; Chen, X.; Chen, Y.; Dai, Y.; Xiao, Z. Application of BP neural network algorithm in traditional hydrological model for flood predicting. *Water* **2017**, *9*, 48. [CrossRef]
13. He, F.; Zhang, L. Mold breakout prediction in slab continuous casting based on combined method of GA-BP neural network and logic rules. *Int. J. Adv. Manuf. Technol.* **2018**, *95*, 4081–4089. [CrossRef]
14. Fan, G.F.; Peng, L.L.; Hong, W.C.; Sun, F. Electric load predicting by the SVR model with differential empirical mode decomposition and auto regression. *Neurocomputing* **2016**, *173*, 958–970. [CrossRef]
15. Nie, H.; Liu, G.; Liu, X.; Wang, Y. Hybrid of ARIMA and SVMs for short-term load predicting, 2012 international conference on future energy, environment, and materials. *Energy Procedia* **2012**, *16*, 1455–1460. [CrossRef]
16. Liu, Y.; Gao, Z. Enhanced just-in-time modelling for online quality prediction in BF ironmaking. *Ironmak. Steelmak.* **2015**, *42*, 321–330. [CrossRef]
17. Shen, B.Z.; Shen, H.F.; Liu, B.C. Water modelling of level fluctuation in thin slab continuous casting mould. *Ironmak. Steelmak.* **2009**, *36*, 33–38. [CrossRef]
18. Hong, W.-C. Chaotic particle swarm optimization algorithm in a support vector regression electric load predicting model. *Energy Convers. Manag.* **2009**, *50*, 105–117. [CrossRef]
19. Ghosh, S.K.; Ganguly, S.; Chattopadhyay, P.P.; Datta, S. Effect of copper and microalloying (Ti, B) addition on tensile properties of HSLA steels predicted by ANN technique. *Ironmak. Steelmak.* **2009**, *36*, 125–132. [CrossRef]
20. Voyant, C.; Muselli, M.; Paoli, C.; Nivet, M.-L. Numerical weather prediction (NWP) and hybrid ARMA/ANN model to predict global radiation. *Energy* **2012**, *39*, 341–355. [CrossRef]
21. Lei, Y.G.; Lin, J.; He, Z.J.; Zuo, M.J. A review on empirical mode decomposition in fault diagnosis of rotating machinery. *Mech. Syst. Sig. Process.* **2013**, *35*, 108–126. [CrossRef]

22. Tomic, M. Wavelet transforms with application in signal denoising. *Ann. DAAAM Proc.* **2008**, 1401–1403.
23. El B'charri, O.; Latif, R.; Elmansouri, K.; Abenaou, A.; Jenkal, W. ECG signal performance de-noising assessment based on threshold tuning of dual-tree wavelet transform. *Biomed. Eng. Online* **2017**, *16*, 26. [CrossRef] [PubMed]
24. Varady, P. Wavelet-Based Adaptive Denoising of Phonocardiographic Records. In Proceedings of the 23rd Annual International Conference on IEEE-Engineering-in-Medicine-and-Biology-Society, Istanbul, Turkey, 25–28 October 2001; pp. 1846–1849.

metals

MDPI

Article

Modelling on Inclusion Motion and Entrapment during the Full Solidification in Curved Billet Caster

Yanbin Yin [1], Jiongming Zhang [1,*], Qipeng Dong [2] and Yuanyuan Li [1,3]

[1] State Key Laboratory of Advanced Metallurgy, University of Science and Technology Beijing, Beijing 100083, China; b20150490@xs.ustb.edu.cn (Y.Y.); lyy_job@163.com (Y.L.)
[2] School of Iron and Steel, Soochow University, Suzhou 215137, China; qpdong@outlook.com
[3] Liuzhou Iron & Steel Company, Ltd., Liuzhou 545002, China
* Correspondence: jmz2203@sina.com; Tel.: +86-010-8237-6597

Received: 12 April 2018; Accepted: 2 May 2018; Published: 6 May 2018

Abstract: Inclusions entrapped by the solidifying front during continuous casting would deteriorate the properties of the final steel products. In order to investigate the inclusion motion and the entrapment during the full solidification in curved billet caster, the present work has developed a three-dimensional numerical model coupling the flow, solidification, and inclusion motion. The predicted result indicates that the inclusion distribution inside the liquid pool of the mold is not perfectly symmetrical. Furthermore, the motion and the entrapment of micro inclusions in the mold are mainly affected by the molten steel flow pattern, however, those of macro inclusions depend both on the molten steel flow pattern and the buoyancy force of the inclusions. In the curved part of the strand, macro inclusions shift to the solidifying front of the inner radius as time goes on, while the solidifying front of the outer radius cannot entrap inclusions. The distributions of inclusions smaller than 5 μm in the solidified strand are even. However, for inclusions that are larger than 5 μm, their distributions become uneven. To validate the model, measurement of the strand surface temperature and the detection of inclusions in samples obtained from a plant have been performed. Good agreement is found between the predicted and experimental results.

Keywords: numerical simulation; molten steel flow; solidification; inclusion motion; inclusion entrapment; billet continuous casting

---

1. Introduction

Inclusions inside a steel matrix have a detrimental impact on the performance of steels, such as their strength, toughness, fatigability, surface appearance, etc. [1–4]. Hence, the removal of inclusions from the molten steel is a critical issue throughout the steelmaking process. In particular, continuous casting is the last opportunity for the removal of inclusions from the molten steel. Inclusions that were carried by the molten steel are injected into the continuous casting mold. It is believed that a portion of these inclusions would rise in the mold and finally be absorbed by the mold flux. However, the remainder would inevitably be entrapped by the solidifying front while they move inside the liquid pool of the strand. Therefore, an in-depth investigation on the inclusion motion, removal, and entrapment during the full solidification in a continuous casting strand is essential, and it can provide theoretical guidance for the improvement of the steel cleanliness and properties. It is very difficult to research the inclusion motion and entrapment during the continuous casting by plant measurements or physical experiments. Fortunately, numerical simulation is an appropriate method to study the inclusion motion and entrapment during the continuous casting.

In recent years, extensive research has been conducted on the particle (inclusion, bubble) behaviors during the continuous casting by method of numerical simulation [5–41]. Many previous studies have investigated particle behaviors in continuous casting strands using continuum [5,6] or

Eulerian–Eulerian multiphase flow models [7–9]. Lei H. et al. [5] studied the inclusion collision-growth in a slab continuous caster using a continuum model. The spatial distributions of the inclusion volume concentration and number density were revealed. Through coupling the electromagnetic force into the continuum model, Lei H. et al. [6] investigated the effect of in-mold electromagnetic stirring (M-EMS) on the inclusion collision-coalescence and the spatial distributions in a bloom caster. Due to the M-EMS, Archimedes force and Archimedes collision were considered for inclusions in the study. Liu Z. et al. [7–9] used the inhomogeneous Multiple Size Group (MUSIG) model or the average bubble number density (ABND) model under the Eulerian-Eulerian framework to describe the bubbly flow in slab continuous casting.

Many researchers [10–41] have studied the particle transport (motion, removal, and entrapment) in the continuous casting strand through Lagrangian descriptions for the inclusions or argon bubbles. Liu Z. et al. [14] studied the influence of electromagnetic brake (EMBr) on the transient fluid flow and inclusion transport (motion, removal) in a slab continuous casting mold through a three-dimensional (3D) mathematic model. In the study, the transport of inclusions inside the mold was calculated employing the Lagrangian approach. The results indicated that the inclusion transport inside the mold was asymmetric. When compared with no EMBr, the removal of inclusions for the EMBr arrangement was enhanced, nevertheless, that for the flow-control mold (FCM) arrangement was reduced.

Several previous works [24–41] have researched the entrapment of particles (inclusions, argon bubbles) by the solidifying front in continuous casting strand. Liu Z. et al. [37] developed a 3D numerical model coupling the fluid flow, solidification, and inclusion motion. In order to investigate the transport of inclusion clusters, an inclusion cluster model was developed based on the fractal theory and the conservation of mass. The flow and solidification of molten steel, the motion, and the entrapment of inclusion clusters in a vertical-bending continuous casting caster were studied through the coupling model. Thomas B. et al. [38] constructed a computational model coupling turbulent fluid flow and particle transports. In addition, a particle-capture model that was based on local force balances was also coupled into the computational model. Through the coupling model, the work simulated the entrapment of slag inclusions and bubbles during the thin-slab continuous casting. This paper adopted both Reynolds Average Navier-Stokes (RANS) and Large Eddy Simulation (LES) approaches to calculate the turbulence flow field. The results indicated that, particle capture depended on factors, such as particle size and density, molten steel transverse fluid, Primary Dendrite Arm Spacing (PDAS), solidification front orientation angle, and sulfur concentration gradient. Wang S. et al. [40] investigated the influence of electromagnetic parameters on the motion and the entrapment of inclusions in FCM continuous casting strands through a coupling mathematical model. The results suggested that the region about 5 mm beneath the slab surface became cleaner, and the aggregation of inclusions were eliminated while the upper and lower magnetic fields of the FCM increased. The above works researched the particle motion and the entrapment in the mold region or a part of the continuous casting strand.

Zhang L. et al. [41] conducted a study on the inclusion entrapment in the full length of a billet caster by means of numerical simulation. In the work, the geometry model was straight throughout the computational domain. However, most of the commercial continuous casters involve a curved part. The inclusion motion and entrapment in a curved continuous casting strand may have distinctive characteristics. At present, studies about the inclusion motion and entrapment during the full solidification in a curved continuous casting strand have not been reported.

The scope of the current work was to develop a three-dimensional numerical model to investigate the inclusion motion and entrapment during the full solidification in a curved billet caster. To validate the model, measurement of the strand surface temperature and the detection of inclusions in samples that were obtained from a plant have been performed.

2. Numerical Methodology

2.1. Assumptions

In the present work, the following assumptions are made in the cause of simplifying the numerical model:

1. treating the molten steel as an incompressible Newtonian fluid;
2. the influence of the mold taper and oscillation are not considered;
3. the mold flux and the level fluctuation of the molten steel are not neglected;
4. the free surface of the mold is assumed to be adiabatic;
5. the latent heat of the solid phase transformation is negligible, only the latent heat of solidification is considered;
6. the inclusion is treated as spherical alumina inclusion, and its density is constantly 3500 kg·m^{-3};
7. the aggregation and breakup of inclusions are not taken into account; and,
8. the influence of inclusion motion on the flow and the heat transfer of the molten steel is ignored.

2.2. The Model Details

The numerical model mainly involves two parts: the flow-solidification model and the inclusion motion model. The flow-solidification model can be found in the previous work [42], moreover, the inclusion motion model can be found in the previous work [31].

In the previous work [42], to calculate the molten steel heat transfer and solidification, the enthalpy-porosity technique was employed. In the enthalpy-porosity technique, the influence of solidification on the molten steel flow velocities is considered, through introducing a source term into the momentum conservation equation of the molten steel. The source term takes the following form:

$$S_U = \frac{(1-f)}{f^3 + e} A_{\text{mush}} \left(\vec{V} - \vec{V}_{\text{cast}} \right), \tag{1}$$

where f is liquid fraction of steel, e is a small number (0.001) to prevent division by zero, and A_{mush} is permeability coefficient. The calculation detail of A_{mush} can be found in our previous work [42].

The casting velocity, $\vec{V}_{\text{cast}}$, is a constant vector in the previous work. However, owing to the curved geometry model in the present work, the casting velocity should be treated as a variable varying with position.

As shown in Figure 1, in the curved part of the strand, the direction of the casting velocity at one point is parallel to the tangent of a circle at this point. Moreover, the center of the circle coincides with the center of the curved strand. Hence, the casting speed, $\vec{V}_{\text{cast}}$, can be expressed in Cartesian coordinate form, as follows:

$$(-\|\vec{V}_{\text{cast}}\| \cdot \sin\theta, 0, -\|\vec{V}_{\text{cast}}\| \cdot \cos\theta), \tag{2}$$

The magnitude of casting velocity at one point can be can be calculated by Equation (3):

$$\|\vec{V}_{\text{cast}}\| = V_{\text{center}} \cdot \frac{R}{R_{\text{caster}} - D/2}, \tag{3}$$

where R is the distance between the point and the caster center, R_{caster} is the radius of the caster, and D is the strand thickness. V_{center} is the magnitude of the casting velocity at the strand center, which is equal to the magnitude of the strand velocity at the caster exit. In the current work, V_{center} is set as 1.65 m·min^{-1}.

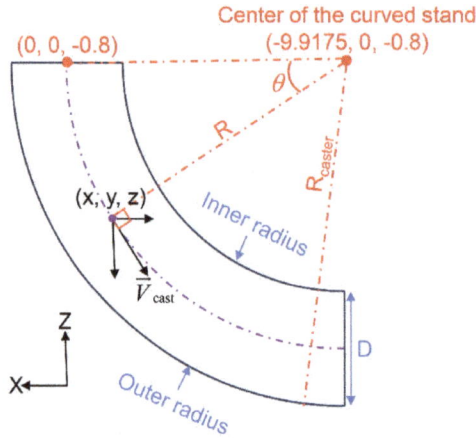

Figure 1. Schematic diagram of setting the casting velocity.

2.3. Geometry Model

As Figure 2 shows, in order to calculate the full solidification process of the billet caster, a full geometry model of the strand has been developed. The length of the computational domain is 10 m and the size of the strand cross section is 0.165 m × 0.165 m. The mold with a length of 0.8 m is vertical. The curved part of the caster initiates at the mold exit. The length of the foot roller zone (FRZ) is 0.3 m. Furthermore, the secondary cooling zone of the caster consists of three segments, which are SCZ1 (2.45 m), SCZ2 (2.4 m), and SCZ3 (1.5 m), respectively. Additionally, a part of the air cooling zone (ACZ) is involved in the computing domain, which is 2.55 m in length. The submerged entry nozzle (SEN) is a type of straight single port. The submerged depth of the SEN is 0.1 m. The inner and outer diameters of the SEN are 0.035 m and 0.075 m, respectively. To simulate the behavior of the solidified shell more accurately, the local grid refinement technology has been adopted.

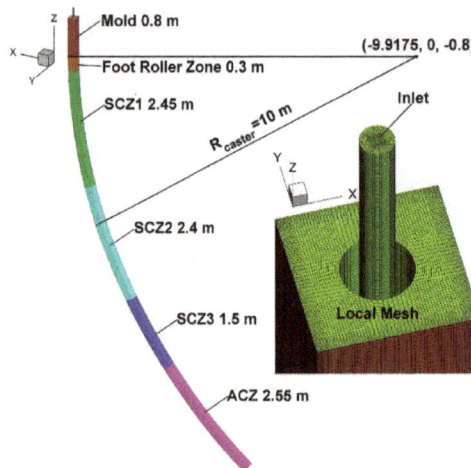

Figure 2. Geometry model and mesh in a billet caster.

2.4. Boundary Conditions

Boundary conditions for the flow-solidification simulation are set according to Dong's work [43]. To give full consideration to a statistically representative result of the inclusion entrapment and the computer capacity, 1000 inclusions per second with a consistent size are injected randomly into the SEN from the inlet. With regard to the inclusion motion, an escape boundary condition is defined for the top surface of the mold and the computational domain outlet. Moreover, a reflecting boundary is set for the walls inside and outside the SEN. In the current work, inclusions are assumed to be entrapped by the solidified shell if the local liquid fraction is below 0.6 [39,41], and the velocities of the captured inclusions are equal to the local casting velocity. Subsequently, the inclusions entrapped by the solidified shell would move together with the solidified shell.

The Material properties used in the present work are listed in Table 1.

Table 1. The Material properties and model parameters.

Parameters	Values	Dimensions
c_p, Specific heat	650	$J \cdot kg^{-1} \cdot K^{-1}$
k', Thermal conductivity	33.5	$W \cdot m^{-1} \cdot K^{-1}$
ρ, Steel density	7340	$kg \cdot m^{-3}$
L, Steel latent heat	231,637	$J \cdot kg^{-1}$
T_l, Liquid temperature	1827	K
T_s, Solid temperature	1636	K
μ, Molten steel molecular viscosity	0.00461	$kg \cdot s^{-1} \cdot m^{-1}$
T_{tun}, Tundish temperature	1758	K
d_p, Inclusion size	3.5, 5, 7, 10, 15, 20, 25, 50, 100, 200	μm

2.5. Numerical Procedure Details

The numerical model in the present work is solved using the CFD package OpenFOAM (Version 2.1.1). The solving process consists of two steps: first, the numerical simulation of the molten steel flow-solidification is executed for 600 s under transient mode, providing an initial condition for the second step; second, the calculation of inclusion motion coupling the molten steel flow and solidification is performed in the transient mode. With reference to the second step, the total calculation time was about 500 s, which could ensure that inclusions enter and exit the computational domain.

3. Results and Discussion

3.1. Solidification Model Validation

For the prediction of inclusion entrapment, an accurate solidification profile is of great importance. In order to validate the solidification model, four surface center temperatures of the strand were measured with infrared thermometers that were placed at the SCZ2, SCZ3, and at two positions in the ACZ. Figure 3 shows the variation of the predicted surface center temperature, which is in good agreement with the measured temperatures along the casting direction, and it therefore validates the solidification model. The solidified shell (liquid fraction: 0.6) thickness profile of the strand is also shown in Figure 3, where the predicted solidification end is located at 9.6 m below the meniscus. A re-melting zone of the solidified shell in the mold, resulting from the recirculation flow of the molten steel, will be discussed in the next section. After exiting from the mold, the thickness of the solidified shell increases with distance below the meniscus.

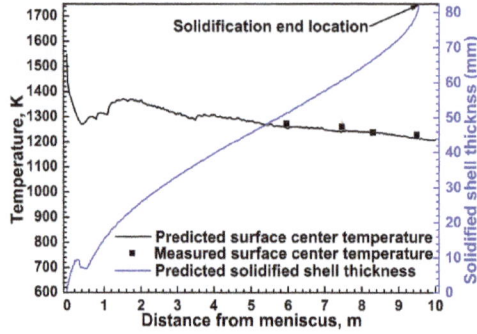

Figure 3. Solidification and temperature profiles along the casting direction.

3.2. Inclusion Motion and Entrapment in Mold

Figure 4 demonstrates the distribution of 5 μm inclusions inside the liquid pool of the mold at different moments. The solidifying front is represented by the gray iso-surface of the 0.6 liquid fraction of steel. The green spheres represent inclusions. It can be observed that inclusions that are injected into the mold are carried by the strong downward molten steel jet from the SEN port, at 0.5 s and 1.5 s after injection. At 5 s after injection, the inclusion motion that is carried by the molten steel jet flow continues, while some inclusions rise. Many inclusions concentrate in the lower part of the mold as a result of the weakening jet flow. At 15 s after injection, inclusions continue concentrating in the mold lower part, and some inclusions flow out of the mold, while the number of the rising inclusions increases. The rising inclusions move close to the solidifying front, which may be entrapped by the solidified shell, and then the inclusions move toward the mold top surface along the outer wall of the SEN. At 30 s after injection, some of the inclusions rise to the mold top surface and are removed, while many inclusions are transported into the liquid pool once again, following the molten steel. At 45 s after injection, the distribution of 5 μm inclusions in the liquid pool reaches a dynamic balance. It should be mentioned that the inclusion distribution inside the liquid pool of the mold is not perfectly symmetrical, which may be attributed to the random injection of inclusions from the inlet and the effect of the molten steel turbulence.

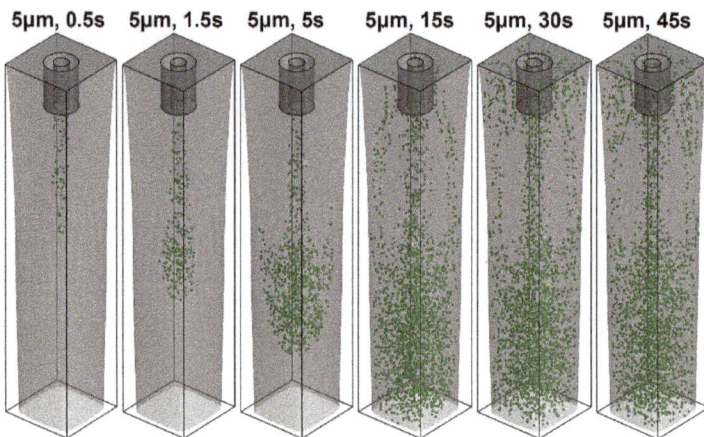

Figure 4. Transient distributions of 5 μm inclusions inside the liquid pool of the mold.

Figure 5 presents the inclusion distribution and the molten steel 3D streamline in the mold at 50 s after injection. As revealed by the predicted 3D streamline distribution, it can be seen that the molten steel is poured into the mold from the SEN port, leading to an impinging jet flow. A part of the poured molten steel flows straight downward and it exits the mold. However, the remainder flows upward along the solidified shell and creates an obvious recirculation flow zone (lower recirculation zone, LRZ) around the impinging jet in the upper zone of the mold. Subsequently, the molten steel flows toward the outer wall of the SEN, resulting in a relatively small recirculation zone (upper recirculation zone, URZ) around the SEN. It is the flow characteristic of the molten steel in the mold that leads to the transient motion and distribution of inclusions, as seen in Figure 4.

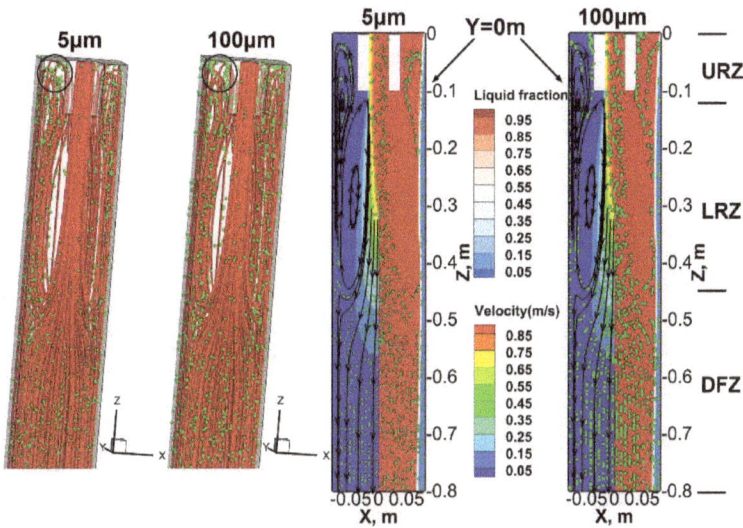

Figure 5. The inclusion (shown as green spheres) distribution, the molten steel streamline and contour plots of liquid fraction and velocity in the mold at 50 s after injection.

Figure 5 also shows the predicted two-dimensional (2D) molten steel streamline and contour plots of liquid fraction and velocity on the Y = 0 m plane in the mold at 50 s after injection. In addition, the inclusion distribution on the Y = 0 m plane, which is obtained by means of projecting the positions of inclusions between the Y = −0.005 m and the Y = 0.005 m planes onto the Y = 0 m plane, can be seen in Figure 5. The velocity magnitude of the molten steel in the impinging jet zone is obviously larger than that in other zones. Moreover, the velocity magnitude in the impinging jet zone decreases as the distance from the SEN port. The velocity magnitude of the molten steel becomes uniform at the mold lower region (downward flow zone, DFZ). Through the combination of the streamlines, liquid fraction contour plot, and the inclusion distribution on the Y = 0 m plane, it can be seen that inclusions that were carried by the downward molten steel would be entrapped by the solidifying front at the DFZ. Additionally, the lower recirculation flow of the molten steel leads to the solidified shell re-melting and inclusion entrapment. Similarly, at the URZ, the molten steel recirculation flow can also carry the inclusions to the solidifying front and make the inclusions be entrapped by the solidified shell.

At 50 s after injection, the 3D distribution of 5 μm inclusions has reached a dynamic balance. The 3D distribution of 100 μm inclusions in the mold has also achieved a dynamic balance. The inclusion distributions of the two size classes are similar, except at the black circle zone, where more inclusions in 100 μm rise to the top surface of the mold and are removed. The phenomenon is clearer on the Y = 0 m plane. Furthermore, the entrapping position of 100 μm inclusions is higher than that of 5 μm inclusions in the URZ. The entrapping positions of 100 μm inclusions at the DFZ and the LRZ are also relatively higher than those of the 5 μm inclusions. It is assumed that this is due to the buoyancy force of the inclusions.

The predicted initial entrapment positions of inclusions with different sizes in the mold reveal that the larger inclusions are closer to the billet surface (Figure 6). This is because the entrapping positions of larger inclusions are higher than those of the smaller inclusions in the URZ with the effect of the buoyancy force. A higher entrapping position of inclusions in the URZ indicates that these inclusions are entrapped by a thinner solidified shell. Figure 6 also shows the variation of the predicted removal ratio of the inclusions with different sizes from the mold top surface with time. It can be seen from the predicted result that the removal ratios of the micro inclusions (<= 25 μm) are apparently small, compared with those of the macro inclusions (>= 50 μm). It is concluded that the motion and the entrapment of micro inclusions in the mold are mainly affected by the molten steel flow pattern, since the buoyancy force of micro inclusions is negligible. However, the motion and the entrapment of macro inclusions in the mold depend both on the molten steel flow pattern and the buoyancy force.

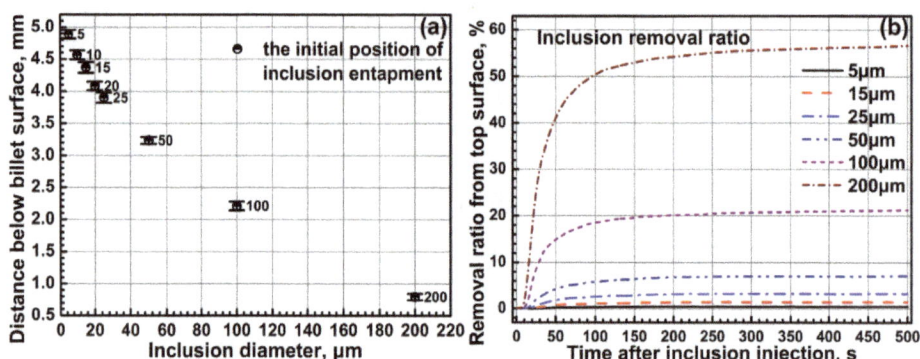

Figure 6. Predicted initial entrapment position (**a**) and removal ratio of inclusions with different sizes (**b**) in the mold.

3.3. Inclusion Motion and Entrapment in the Curved Part of the Strand

Figure 7 shows the distributions of 100 μm inclusions inside the liquid pool of the curved part of the strand at different times. The solidifying front is represented by the gray iso-surface of the 0.6 liquid fraction of steel. The green spheres represent inclusions. At 25 s after injection, the inclusions move into the FRZ, which is carried by the molten steel. At 50 s after injection, the inclusions move into the SCZ1. It is interesting to note that 100 μm inclusions shift to the solidifying front of the inner radius in the curved part of the strand as time goes on, while the solidifying front of the outer radius cannot entrap the inclusions. At 150 s after injection, the entrapment 100 μm inclusions terminates at the solidifying front of the inner radius.

In order to analyze the statistical entrapment positions of macro inclusions along the casting direction at 400 s after injection, the computational domains between the meniscus and the solidification end (9.6 m below the meniscus) are divided into 12 equal zones, respectively. Figure 8 shows the predicted entrapment ratios of the macro inclusions along the casting direction, and additionally, the terminal entrapment positions of 50 μm, 100 μm, and 200 μm inclusions at 400 s after injection. The entrapment ratio of the 100 μm inclusion decreases first, increases later, and then decreases to zero in the curved part of the strand. In addition, the entrapment ratio of the 100 μm inclusions reaches its peak value at the region of 2.4–3.2 m below the mold top surface. Moreover, the entrapment of the 100 μm inclusions terminates at the inner radius solidifying front at about 3.3 m below the meniscus. It is assumed that the flow pattern in the mold and the buoyancy force of inclusions lead to this phenomenon. A strong impinging jet flow of the molten steel is created by the straight port

SEN, resulting in many inclusions being transported into the curved part of the strand and then being entrapped by the local solidifying front. The buoyancy force leads to a deviation of the inclusion motion to the inner radius in the curved part of the strand. This phenomenon is more notable for 200 μm inclusions, as the buoyancy force of these inclusions is larger (Figure 8). This also explains why the entrapment of 50 μm inclusions terminates at a lower position under the meniscus.

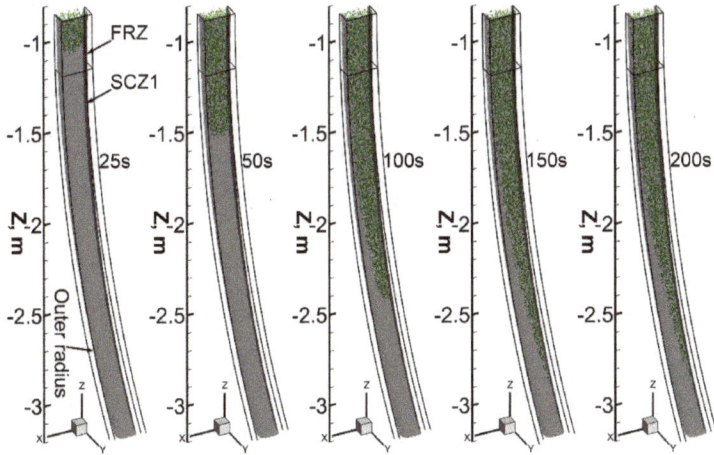

Figure 7. Transient distributions of 100 μm inclusions inside the liquid pool of the curved part of the strand.

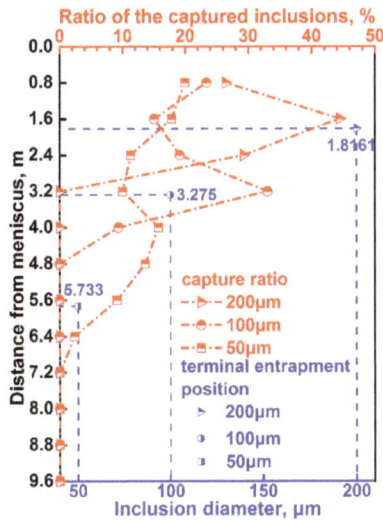

Figure 8. The predicted terminal entrapment positions and entrapment ratios of the macro inclusions along the casting direction.

3.4. Inclusion Distribution in the Solidified Stand

Due to the shift of the macro inclusions to the solidifying front of the inner radius while they move inside the liquid pool of the curved strand, an uneven distribution of the macro inclusions may exist in the solidified strand. Figure 9 presents the distribution of inclusions on the cross section of the solidified strand, through projecting the positions of inclusions of each size class in the region of 9.6–10 m below the meniscus at 500 s onto a plane. Rectangles were used in order to indicate the approximate concentrating zone of inclusions. Inclusions that are inside the solidified shell profile at the mold exit indicate that they are entrapped in the curved part of the strand. It can be observed that the distribution of 3.5 µm inclusions in the solidified strand is uniform. However, the distribution of 7 µm inclusions in the solidified strand becomes inhomogeneous. The aggregation of 7 µm inclusions is found between the billet center and the inner radius. Furthermore, with the increase of the inclusion diameter, the aggregation is intensified, and the inclusions are closer to the inner radius.

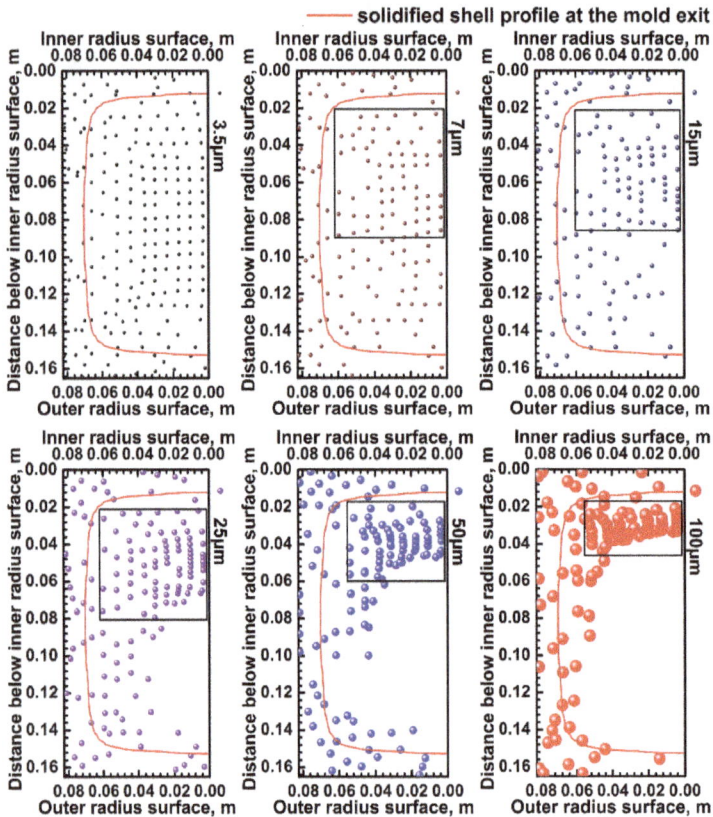

Figure 9. The predicted distributions of inclusions with different sizes on the cross section of the solidified strand.

In order to analyze the uneven distributions of inclusions in the solidified strand, a 15 mm thick zone in the center of the solidified strand has been selected (Figure 10). The positions of inclusions in each size class in the analysis zone have been gathered, and the median of the distances of these inclusions that are below the inner radius have been calculated. Moreover, a statistical parameter, coefficient of skewness, has been adopted to evaluate the distribution inhomogeneity of inclusions. To a certain extent, the distribution inhomogeneity of inclusions in the solidified strand can be reflected by the median and coefficient of skewness. The distributions of 3.5 µm and 5 µm inclusions in the solidified strand are even. However, for inclusions that are larger than 5 µm, the distributions become uneven. Furthermore, through the coefficient of skewness, it can be found that the inhomogeneity is enhanced with the increase of the inclusion diameter.

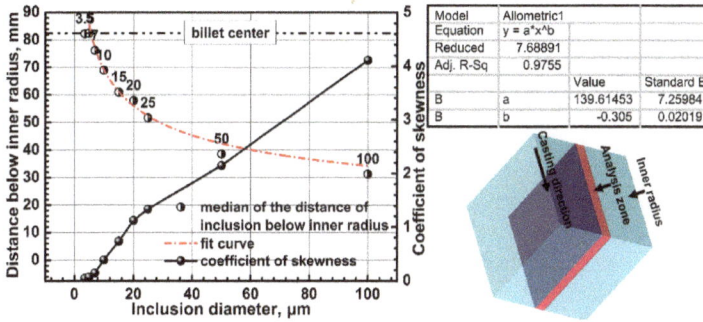

Figure 10. Variation of the distribution deviation of inclusions in the solidified strand.

3.5. Comparison Between the Predicted and the Experimental Results

Samples with the same process parameters and material properties adopted in the current work were obtained from a steel plant. The sampling schematic diagram can be seen in Figure 11. The steel samples were machined into 15 mm × 15 mm × 15 mm metallographic samples, and then these samples were polished and examined using EVO18, ZEISS scanning electron microscope (SEM). The numbers of inclusions in steel samples were automatically counted using the SEM control software INCAFeature (Oxford Instruments, Oxfordshire, England). The analyzed area of each steel sample was 11 × 11 mm². The inclusion size was determined by the equivalent circle diameter (ECD). Only inclusions that were above 2 μm ECD were counted. Manganese sulfide (MnS) inclusions that were precipitated during the solidification process were not counted. In total, 1263 inclusions were detected in the 11 steel samples (Figure 11). The number of inclusions in the samples decreases as the inclusion diameter increases. The largest of those detected inclusions is 38.78 μm. The measured inclusion number density reaches a maximum at the fourth steel sample from the inner radius, which is located between the inner radius and the center of the solidified strand.

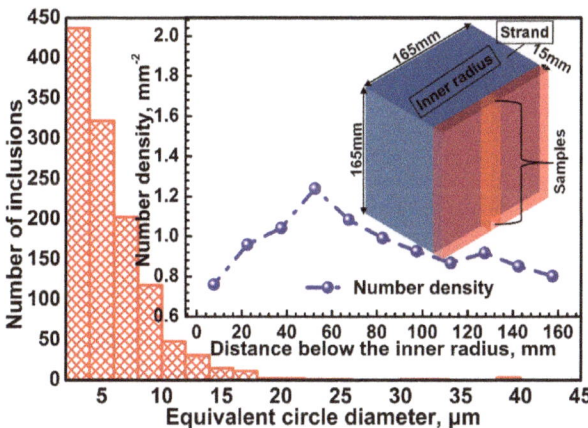

Figure 11. Sampling schematic diagram and the experimental result.

In order to compare the predicted result with the experimental result further, the detected inclusions in these steel samples are divided into four groups (2–5 μm, 5–10 μm, 10–20 μm, >20 μm), according to size. For the four groups, their size weighted averages are 3.30 μm, 6.90 μm, 12.67 μm, and 30.00 μm, respectively. The predicted results are collected in the analysis zone shown in Figure 11.

Figure 12 presents the comparison between the predicted and the experimental results. Good agreement is found between the predicted result of inclusions in 3.5 μm and the experimental result of inclusions in 2–5 μm. Likewise, the predicted distribution of inclusions in 7 μm is in good agreement with the distribution of inclusions in 5–10 μm, according to the experiment. For the predicted result of inclusions in 15 μm and the experimental result of inclusions in 10–20 μm, their variation trends are similar (Figure 12c). The relatively large deviation between the inclusion size in the predicted result and the weighted average size in the experimental result may result in the deviation between the experimental and predicted results. Inclusions that were larger than 20 μm were rarely detected in the steel samples. As Figure 12d shows, the variation trend in the experimental result of inclusions larger than 20 μm is similar to that in the predicted results of 25 μm, 50 μm, and 100 μm inclusions. Through the experimental result, it can also be found that the distribution of inclusions smaller than 5 μm in the solidified strand is even. However, for inclusions that were larger than 5 μm, their distributions become uneven. This is consistent with the predicted result discussed above. The comparison between the predicted and the experimental results indicates that the inclusion motion model is valid.

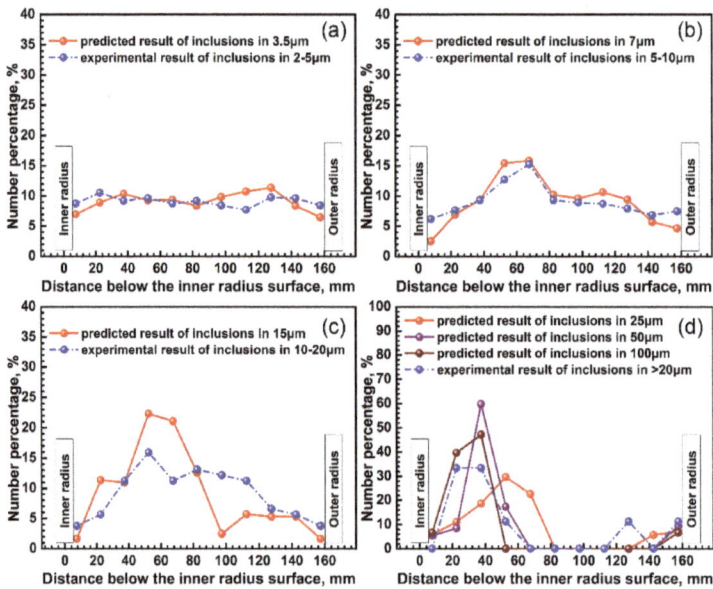

Figure 12. Comparison between the predicted and the experimental results: (**a**) 2–5 μm; (**b**) 5–10 μm; (**c**) 10–20 μm; (**d**) >20 μm.

4. Conclusions

The present work has developed a three-dimensional numerical model coupling the flow field, solidification, and inclusion motion for a curved strand. Through the coupling model, we have investigated the inclusion motion and entrapment during the full solidification in a curved billet caster. The characteristics of the inclusion motion and entrapment in the mold and the curved part of the strand have been revealed. Moreover, the inclusion distributions in the solidified strand have also been presented. The conclusions are as follows:

1. the inclusion distribution inside the liquid pool of the mold is not perfectly symmetrical, resulting from the random injection of inclusions from the inlet and the effect of the molten steel turbulence;

2. the entrapping positions of larger inclusions are higher than those of smaller inclusions in the URZ with the effect of the buoyancy force. As a result, the initial entrapping positions of larger inclusions are more close to the billet surface;

3. the motion and entrapment of micro inclusions in the mold are mainly affected by the molten steel flow pattern, since the buoyancy force of micro inclusions is negligible. However, the motion and entrapment of macro inclusions in the mold depend both on the molten steel flow pattern and the buoyancy force;

4. owing to the effect of the buoyancy force, macro inclusions shift to the solidifying front of the inner radius in the curved part of the strand as time goes on, while the solidifying front of the outer radius cannot entrap the inclusions;

5. the distributions of inclusions smaller than 5 μm in the solidified strand are even. However, for inclusions that are larger than 5 μm, their distributions become uneven. Furthermore, the inhomogeneity is enhanced with the increase of the inclusion diameter; and,

6. good agreement is found between the predicted and experimental results. The comparison between the predicted and the experimental results indicates that the inclusion motion model is valid.

Author Contributions: J.Z. and Y.Y. conceived and designed the study; Y.Y. and Q.D. performed the numerical calculation; Y.Y. and Y.L. conducted the experiment. Y.L. and Q.D. analyzed the experimental data; Y.Y. and J.Z. wrote the paper.

Acknowledgments: The authors gratefully express their appreciation to the National Natural Science Foundation of China (51474023) for sponsoring this work.

Conflicts of Interest: The authors declare no conflict of interest.

References

1. Zhang, L. Nucleation, growth, transport, and entrapment of inclusions during steel casting. *JOM* **2013**, *65*, 1138–1144. [CrossRef]

2. Hu, Y.; Chen, W.; Wan, C.; Wang, F.; Han, H. Effect of deoxidation process on inclusion and fatigue performance of spring steel for automobile suspension. *Metall. Mater. Trans. B* **2018**, *49*, 569–580. [CrossRef]

3. Deng, X.; Ji, C.; Cui, Y.; Tian, Z.; Yin, X.; Shao, X.; Yang, Y.; McLean, A. Formation and evolution of macro inclusions in IF steels during continuous casting. *Ironmak. Steelmak.* **2017**, *44*, 739–749. [CrossRef]

4. Wang, X.; Li, X.; Li, Q.; Huang, F.; Li, H.; Yang, J. Control of stringer shaped non-metallic inclusions of Cao-Al_2O_3 system in API X80 linepipe steel plates. *Steel Res. Int.* **2014**, *85*, 155–163. [CrossRef]

5. Lei, H.; Geng, D.; He, J. A continuum model of solidification and inclusion collision-growth in the slab continuous casting caster. *ISIJ Int.* **2009**, *49*, 1575–1582. [CrossRef]

6. Lei, H.; Jiang, J.; Yang, B.; Zhao, Y.; Zhang, H.; Wang, W.; Dong, G. Mathematical model for collision-coalescence among inclusions in the bloom continuous caster with M-EMS. *Metall. Mater. Trans. B* **2018**, *49*, 666–676. [CrossRef]

7. Liu, Z.; Li, B. Large-eddy simulation of transient horizontal gas–liquid flow in continuous casting using dynamic subgrid-scale model. *Metall. Mater. Trans. B* **2017**, *48*, 1833–1849. [CrossRef]

8. Liu, Z.; Qi, F.; Li, B.; Jiang, M. Multiple size group modeling of polydispersed bubbly flow in the mold: An analysis of turbulence and interfacial force models. *Metall. Mater. Trans. B* **2015**, *46*, 933–952. [CrossRef]

9. Liu, Z.; Li, B.; Qi, F.; Cheung, S.C.P. Population balance modeling of polydispersed bubbly flow in continuous casting using average bubble number density approach. *Powder Technol.* **2017**, *319*, 139–147. [CrossRef]

10. Yu, H.; Zhu, M. Three-dimensional magnetohydrodynamic calculation for coupling multiphase flow in round billet continuous casting mold with electromagnetic stirring. *IEEE Trans. Magn.* **2010**, *46*, 82–86. [CrossRef]

11. Ho, Y.; Hwang, W. Numerical simulation of inclusion removal in a billet continuous casting mold based on the partial-cell technique. *ISIJ Int.* **2003**, *43*, 1715–1723. [CrossRef]

12. Li, B.; Tsukihashi, F. Numerical estimation of the effect of the magnetic field application on the motion of inclusion in continuous casting of steel. *ISIJ Int.* **2003**, *43*, 923–931. [CrossRef]

13. Liu, Z.; Li, B.; Jiang, M.; Fumitaka, T. Euler-Euler-Lagrangian modeling for two-phase flow and particle transport in continuous casting mold. *ISIJ Int.* **2014**, *54*, 1314–1323. [CrossRef]

14. Liu, Z.; Li, L.; Li, B. Large eddy simulation of transient flow and inclusions transport in continuous casting mold under different electromagnetic brakes. *JOM* **2016**, *68*, 2180–2190. [CrossRef]

15. Liu, Z.; Sun, Z.; Li, B. Modeling of quasi-four-phase flow in continuous casting mold using hybrid Eulerian and Lagrangian approach. *Metall. Mater. Trans. B* **2017**, *48*, 1248–1267. [CrossRef]

16. Trindade, L.; Nadalon, J.; Vilela, A.; Vilhena, M.; Soares, R. Numerical modeling of inclusion removal in electromagnetic stirred steel billets. *Steel Res. Int.* **2007**, *78*, 708–713. [CrossRef]

17. Wang, S.; De Toledo, G.; Välimaa, K.; Louhenkilpi, S. Magnetohydrodynamic phenomena, fluid control and computational modeling in the continuous casting of billet and bloom. *ISIJ Int.* **2014**, *54*, 2273–2282. [CrossRef]

18. Wang, Y.; Zhang, L. Fluid flow-related transport phenomena in steel slab continuous casting strands under electromagnetic brake. *Metall. Mater. Trans. B* **2011**, *42*, 1319–1351. [CrossRef]

19. Yang, Y.; Jönsson, P.; Ersson, M.; Nakajima, K. Inclusion behavior under a swirl flow in a submerged entry nozzle and mold. *Steel Res. Int.* **2015**, *86*, 341–360. [CrossRef]

20. Yang, Y.; Jönsson, P.; Ersson, M.; Su, Z.; He, J.; Nakajima, K. The influence of swirl flow on the flow field, temperature field and inclusion behavior when using a half type electromagnetic swirl flow generator in a submerged entry and mold. *Steel Res. Int.* **2015**, *86*, 1312–1327. [CrossRef]

21. Yu, H.; Zhu, M. Influence of electromagnetic stirring on transport phenomena in round billet continuous casting mould and macrostructure of high carbon steel billet. *Ironmak. Steelmak.* **2012**, *39*, 574–584. [CrossRef]

22. Zhang, L.; Aoki, J.; Thomas, B. Inclusion removal by bubble flotation in a continuous casting mold. *Metall. Mater. Trans. B* **2006**, *37*, 361–379. [CrossRef]

23. Zhang, L.; Wang, Y.; Zuo, X. Flow transport and inclusion motion in steel continuous-casting mold under submerged entry nozzle clogging condition. *Metall. Mater. Trans. B* **2008**, *39*, 534–550. [CrossRef]

24. Wang, Q.; Zhang, L. Determination for the entrapment criterion of non-metallic inclusions by the solidification front during steel centrifugal continuous casting. *Metall. Mater. Trans. B* **2016**, *47*, 1933–1949. [CrossRef]

25. Liu, Z.; Li, B. Effect of vertical length on asymmetric flow and inclusion transport in vertical-bending continuous caster. *Powder Technol.* **2018**, *323*, 403–415. [CrossRef]

26. Pfeiler, C.; Wu, M.; Ludwig, A. Influence of argon gas bubbles and non-metallic inclusions on the flow behavior in steel continuous casting. *Mater. Sci. Eng. A* **2005**, *413–414*, 115–120. [CrossRef]

27. Liu, Z.; Li, L.; Li, B.; Jiang, M. Large eddy simulation of transient flow, solidification, and particle transport processes in continuous-casting mold. *JOM* **2014**, *66*, 1184–1196. [CrossRef]

28. Jin, K.; Vanka, S.; Thomas, B. Large eddy simulations of electromagnetic braking effects on argon bubble transport and capture in a steel continuous casting mold. *Metall. Mater. Trans. B* **2018**. [CrossRef]

29. Jin, K.; Thomas, B.; Ruan, X. Modeling and measurements of multiphase flow and bubble entrapment in steel continuous casting. *Metall. Mater. Trans. B* **2016**, *47*, 548–565. [CrossRef]

30. Lei, S.; Zhang, J.; Zhao, X.; He, K. Numerical simulation of molten steel flow and inclusions motion behavior in the solidification processes for continuous casting slab. *ISIJ Int.* **2014**, *54*, 94–102. [CrossRef]

31. Yin, Y.; Zhang, J.; Lei, S.; Dong, Q. Numerical study on the capture of large inclusion in slab continuous casting with the effect of in-mold electromagnetic stirring. *ISIJ Int.* **2017**, *57*, 2165–2174. [CrossRef]

32. Jin, K.; Thomas, B.; Liu, R.; Vanka, S.; Ruan, X. Simulation and validation of two-phase turbulent flow and particle transport in continuous casting of steel slabs. *IOP Conf. Ser. Mater. Sci. Eng.* **2015**, *84*, 012095. [CrossRef]

33. Pfeiler, C.; Thomas, B.; Wu, M.; Ludwig, A.; Kharicha, A. Solidification and particle entrapment during continuous casting of steel. *Steel Res. Int.* **2008**, *79*, 599–607. [CrossRef]

34. Lei, S.; Zhang, J.; Zhao, X.; Dong, Q. Study of molten steel flow and inclusions motion behavior in the solidification processes for high speed continuous casting slab by numerical simulation. *Trans. Indian Inst. Met.* **2016**, *69*, 1193–1207. [CrossRef]

35. Yuan, Q.; Thomas, B.; Vanka, S. Study of transient flow and particle transport in continuous steel caster molds: Part I. Fluid flow. *Metall. Mater. Trans. B* **2004**, *35*, 685–702. [CrossRef]

36. Yuan, Q.; Thomas, B.; Vanka, S. Study of transient flow and particle transport in continuous steel caster molds: Part II. Particle transport. *Metall. Mater. Trans. B* **2004**, *35*, 703–714. [CrossRef]

37. Liu, Z.; Li, B. Transient motion of inclusion cluster in vertical-bending continuous casting caster considering heat transfer and solidification. *Powder Technol.* **2016**, *287*, 315–329. [CrossRef]
38. Thomas, B.; Yuan, Q.; Mahmood, S.; Liu, R.; Chaudhary, R. Transport and entrapment of particles in steel continuous casting. *Metall. Mater. Trans. B* **2014**, *45*, 22–35. [CrossRef]
39. Liu, Z.; Li, B.; Zhang, L.; Xu, G. Analysis of transient transport and entrapment of particle in continuous casting mold. *ISIJ Int.* **2014**, *54*, 2324–2333. [CrossRef]
40. Wang, S.; Zhang, L.; Wang, Q.; Yang, W.; Wang, Y.; Ren, L.; Cheng, L. Effect of electromagnetic parameters on the motion and entrapment of inclusions in FC-mold continuous casting strands. *Metall. Res. Technol.* **2016**, *113*, 205. [CrossRef]
41. Zhang, L.; Wang, Y. Modeling the entrapment of nonmetallic inclusions in steel continuous-casting billets. *JOM* **2012**, *64*, 1063–1074. [CrossRef]
42. Dong, Q.; Zhang, J.; Liu, Q.; Yin, Y. Magnetohydrodynamic calculation for electromagnetic stirring coupling fluid flow and solidification in continuously cast billets. *Steel Res. Int.* **2017**, *88*, 1700067. [CrossRef]
43. Dong, Q.; Zhang, J.; Yin, Y.; Wang, B. Three-dimensional numerical modeling of macrosegregation in continuously cast billets. *Metals* **2017**, *7*, 209. [CrossRef]

metals

MDPI

Article

Laboratory Experimental Setup and Research on Heat Transfer Characteristics during Secondary Cooling in Continuous Casting

Yazhu Zhang [1,2], **Zhi Wen** [1,*], **Zengwu Zhao** [2,*], **Chunbao Bi** [2], **Yaxiang Guo** [2] and **Jun Huang** [2]

[1] School of Energy and Environment Engineering, University of Science and Technology Beijing, Beijing 100083, China; zhangyahzu212@imust.edu.cn
[2] Key Laboratory of Integrated Exploitation of Bayan Obo Multi-Metal Resources, Inner Mongolia University of Science and Technology, Baotou 014010, China; 15942914079@163.com (C.B.); 15847652053@163.com (Y.G.); hjun8420@imust.edu.cn (J.H.)
*   Correspondence: wenzhi@me.ustb.edu.cn (Z.W.); zhzengwu@imust.edu.cn (Z.Z.); Tel.: +86-137-0110-4376 (Z.W.); +86-186-0472-1886 (Z.Z.)

Received: 23 November 2018; Accepted: 4 January 2019; Published: 10 January 2019

Abstract: Spray cooling is a key technology in the continuous casting process and has a marked influence on the product quality. In order to obtain the heat transfer characteristics, which are closer to the actual continuous casting to serve the design, prediction and simulation, we created an experimental laboratory setup to investigate heat transfer characteristics of air mist spray cooling during the continuous casting secondary cooling process. A 200-mm thick sample of carbon steel was heated above 1000 °C, and then cooled in a water flux range of 0.84 to 3.0 L/(m^2·s). Determination of the boundary conditions involved experimental work comprising an evaluation of the thermal history and the heat flux and heat transfer coefficient (HTC) at the casting surface using inverse heat conduction numerical schemes. The results show that the heat fluxes were characterized via boiling curves that were functions of the slab surface temperatures. The heat flux was determined to be 2.9×10^5 W/m^2 in the range of 1100 to 800 °C with a water flux of 2.1 L/(m^2·s). The critical heat flux increased with the increase of water flux. The HTC was close to a linear function of water flux. We also obtained the relation between the HTC and the water flux in the transition boiling region for surface temperatures of 850 to 950 °C.

Keywords: air mist spray cooling; continuous casting; heat flux; HTC; secondary cooling

1. Introduction

In continuous casting, molten steel is poured from a tundish into a water-cooled mould and a partially solidified billet or slab is withdrawn from the bottom of the mould. The billet or slab is then cooled by a water spray (this is the secondary cooling process) so that the solidified billet or slab is produced constantly and continuously. Continuous casting is a bridge between steelmaking and rolling. Secondary cooling is an essential part of continuous casting and can strongly influence the quality of billets or slabs. The cooling must be controlled relative to the casting speed to avoid the formation of internal and surface cracks.

Steel solidification behavior is influenced by heat transportation under the specific cooling conditions. In the secondary cooling zone, the heat transfer behavior of the billet or slab surface is directly linked to the characteristics of the spray. These can be manipulated to control the solidification process, and in turn the billet or slab quality and the casting productivity. In an attempt to meet high billet or slab quality requirements, heat transfer in the mould and the secondary cooling have received much research attention. Due to requirements for high quality, the secondary cooling technology of continuous casting has been developed with specially designed nozzles, finer waterway control, and a more effective water injection strategy.

In recent years, numerical simulation has been widely applied to design and optimize the secondary cooling process. An accurate simulation, however, is strongly dependent on the level of understanding of the physical mechanism of the casting process, the high-temperature properties of the material involved, and thermal boundary conditions. Laboratory experiments can provide a database that can be used to specify the boundary conditions in mathematical models of secondary cooling. Thermal boundary conditions that characterize the boiling heat transfer in secondary cooling are applicable for industrial processes after validation with reliable experiments.

Important work on spray cooling can be found in the literature [1–6]. The main cooling effect derives from evaporation. Heat transfer is influenced by water flow, droplet size, and velocity. Most studies either focused on the macroscopic effect of the sprays on the heat transfer rate or attempted to characterize the heat transfer processes during the impact of the spray on a surface. Many efforts have been made to understand the effect of changing the magnitude of the heat flux or the heat transfer coefficient (HTC) in different boiling heat transfer regimes. Thomas et al. [7,8] carried out laboratory measurements of water flow and heat transfer during spray cooling. Their research focused on the conditions of the surface of the steel strand in the secondary spray cooling zones with water jet–air mist cooling. The steel surface temperature range was 1200 °C to 200 °C. Ramstorfer et al. [9] developed a dynamic spray cooling experimental platform where they measured the HTCs due to spray cooling using an experimental setup, allowing spray cooling up to a surface temperature of 1250 °C. Horský et al. [10] developed experimental methods and numerical models for spray-cooled surface heat transfer. Those papers discussed heat transfer during spray cooling and optimization of the cooling process. Ito et al. [11] also investigated the effects of the hydraulic pressure and water flow rate of a cooling water spray on cooling intensity and developed a more efficient secondary cooling system with a high-pressure water spray. Tsutsumi et al. [12] did laboratory experiments on the cooling capacities of hydraulic and mist spray cooling by several kinds of spray nozzles. They proposed an equation for HTC that considered the spray thickness and collision pressure. El-Bealy [13] studied the degree to which homogeneity of cooling conditions with air mist nozzles improved slab quality. The Brno University of Technology Heat Transfer and Fluid Flow Laboratory [14] established reliable techniques to measure the effects of individual nozzles and combinations of nozzles to determine the HTCs. They also discussed methods for determining the HTCs and using them in a solidification model.

Most industrial spray nozzles are polydispersed, which allows them to generate water droplets with a wide range of sizes and velocities. This, in addition to the complex interaction of spray droplets, makes it difficult to predict (by single or multiple sprays) the performance of an industrial spray nozzle. Extensive laboratory studies have been done on the heat transfer of drops and sprays at low mass flux or for a single spray nozzle. However, data on industrial spray nozzles with high mass flux or on multiple spray nozzles are very scarce. Meanwhile, thinner samples with short cooling times are used in laboratory studies, which have led to a measured data shortage in high-temperature ranges. Due to fewer measurement data, the measurement error has become larger. It has become essential to use thicker samples and closer to the actual billet or slab. Experiments closely emulating the actual process of secondary cooling in continuous casting are particularly important.

In the paper, we developed a multifunctional experimental setup, enabling a quantitative understanding of the secondary cooling of a continuous casting. The objective was to explore and obtain data that characterized heat transfer at high temperatures using multiple air mist spray nozzles. The calculated temperature profile, heat flux and HTCs provided insight and useful data for the development of cooling strategies for continuous casting.

2. Establishing the Experimental Setup

Compared with industrial secondary cooling of continuous casting, a laboratory-scale simulator is cost effective because it is not necessary to interrupt actual steel production. The laboratory simulator provides more flexibility to change the range of operational parameters required for optimizing the secondary cooling process and developing new continuous casting technologies. By separating the secondary cooling from the casting in the experimental setup, we avoid the operational risk associated with using liquid metal. This enables us to simulate air-mist spray cooling in the secondary cooling process.

An overview of the system configuration is shown in Figure 1a,b. The experimental setup consisted of an air–water spray, slab heating, slab cooling and data acquisition and analysis systems. To simulate the features of an industrial air mist spray, a laboratory air–water spray system was designed, as shown in Figure 2. The system consisted of water tanks, pumps, an air compressor, an air pressure tank, electromagnetic flowmeters, temperature transmitters, pressure transmitters, electric control valves, metal hoses and spray racks.

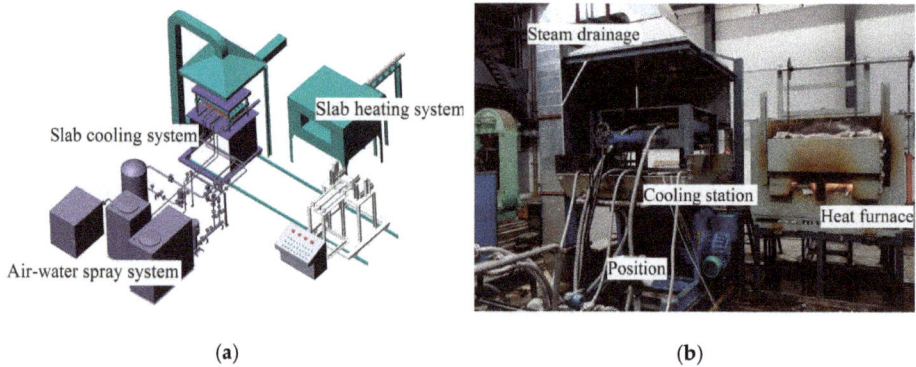

(a)   (b)

Figure 1. Schematic presentation (**a**) and equipment (**b**) of the experimental setup.

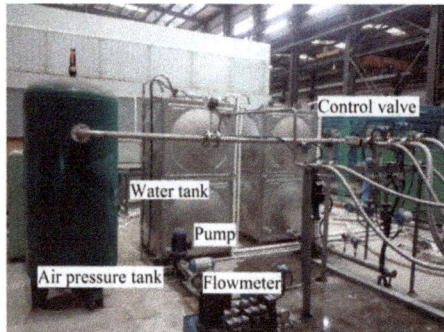

Figure 2. Air-water spray system.

Cooling water was supplied to the setup from the water tank. A valve was used to adjust the flow to the desired flowrate and pressure. The flow parameters were measured with the flowmeters and pressure gauges, which were controlled by the control software.

Nozzle alignments and cooling parameters are the main factors influencing the secondary cooling of continuous casting. The experimental setup was designed to meet the requirements of different nozzles with an extensive range of mass fluxes. The system could be adjusted to test the cooling effects of different spray configurations. The slab heating system comprised a feeding car, a heating furnace and a discharging electric actuator. After the desired temperature was reached, the slab was transported into the cooling station by a rack and pinion structure actuated by a motor. The slab cooling system included the cooling station, a slab depressing device, a reciprocating motion device, a drainage pipe, water pumps, electromagnetic flowmeters and the steam drainage system. It enabled spray rack replacement, slab movement simulation, adjustment of the slab cooling angle, and water and vapor discharge. Depending on the dynamic adjustment of the slab position by hydraulic pressure, it was possible to continuously simulate the position of the slab throughout the entire secondary cooling zone. The cooling station was placed in the positioner with 0° to 90° tilting to meet the slab arc change from initial mould position to final level position.

Figure 3 shows the top side cooling device that used three rows and four columns of nozzles with a 120° impingement angle. According to the nozzle angle, the overlap between the two nozzles was designed with nozzle manufacturer's suggestion. The jets coalesced into a mist curtain on the slab surface.

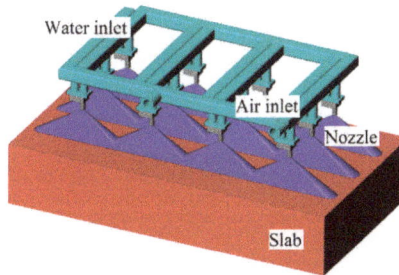

Figure 3. Schematic of the arrangement of the nozzles.

In this way, the slab contact conditions (e.g., surface structure and clamping roller) could be manipulated to be similar to those used in industrial practice. There were four clamping rollers with a 120-mm diameter and a 160-mm separation distance that could clamp a slab with a 100- to 350-mm thickness.

Figure 4 shows the clamping rollers with the top spray configuration. During the tests, the slab surface was exposed to the air mist nozzles through the gap between the rollers. The heat was conducted to the slab surface and removed by the cooling air and water. The cooling process consumed much water and air, and heat was released during the process. Vapor and cooling water were collected and directed into a drain while the apparatus operated.

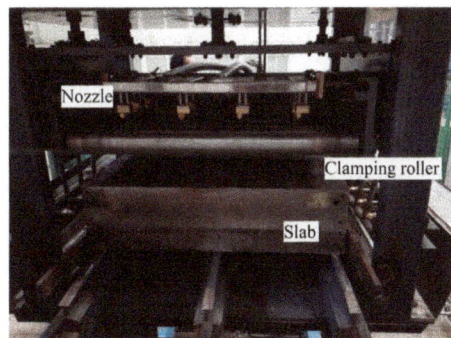

Figure 4. The clamping rollers with no contact slab.

A data acquisition system was used to monitor the temperature, pressure, flow rate, and so forth during heating and cooling. It comprised a computer, a data acquisition box, K-type thermocouples and a software package, which were used for process parameter acquisition, calculation and data display. The thermocouples were inserted into the subsurfaces of the slabs. The slabs were then preheated to a desired initial temperature and transported to the cooling system. The measured temperature was used as an input for calculating the inverse heat conduction. By using the calculated results, we could evaluate the local heat flux at the boundaries. The heat flux was expressed via a boiling curve with different variables. The experimental setup provided the flexibility to adjust the operational parameters (mass flux, heating temperature, casting speed, etc.), enabling study of the heat transfer in secondary cooling. The technical parameters of the setup are shown in Table 1.

Table 1. Technical parameters of the pilot experimental set-up.

Condition	Parameter
Max. heating slab size	$1100 \times 600 \times 350$ mm(thickness)
Max. heating temperature	1250 °C
Max. water flow	18.0 m³/h
Max. air flow	6.0 m³/min
Max. water pressure	1.0 MPa
Max. air pressure	0.4 MPa
Max. nozzle array	3 rows and 4 columns
Max. temperature synchronous acquisition frequency	5 Hz
Cooling condition	Single or multi-side cooling

3. Inverse Heat Conduction Problem

Normally, it is difficult to measure surface heat flux and temperature of a slab undergoing air mist spray cooling. When thermocouples are set at a certain distance from the surface of the slab to measure temperatures at different positions, a mathematical model can be used to calculate the surface heat flux and surface temperature [15,16]. The heat boundary condition can be calculated by recording the temperature as a function of time, which is in the form of an inverse heat transfer problem [17].

We used thermocouples embedded in the slab at certain depths to measure cooling curves at different positions. The surface temperature and heat flux of the slab were calculated using the Beck's sequential function specification method [18,19]. As a result of the calculation, the relation between the HTC and the slab surface temperature could be obtained. In this study, we regarded the internal heat transfer in the slab as one-dimensional unsteady heat conduction along the direction of the spray cooling, ignoring the heat loss from the side of the slab. Figure 5 shows the physical model of the inverse heat conduction problem during the cooling.

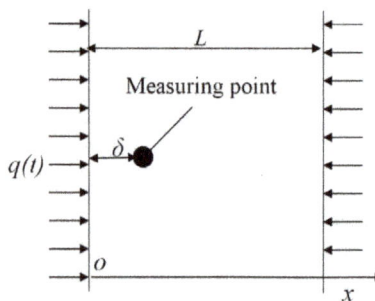

Figure 5. Schematic of the physical model of the inverse heat conduction problem.

The slab was heated to T_0 and cooled by the heat flux $q(t)$. A thermocouple was embedded at $x = \delta$ to record the temperature history at a time interval of Δt. The aim was to calculate $q(t)$ using the measured temperature data with known initial conditions and the thermophysical properties of the materials involved in the system. The governing equation and boundary conditions were

$$\rho C_p(T)\frac{\partial T}{\partial t} = \frac{\partial}{\partial x}\left(k(T)\frac{\partial T}{\partial x}\right)$$
$$T(x,0) = T_0$$
$$T|_{x=\delta} = Y(t)$$
$$-k\partial T/\partial x|_{x=L} = 0$$
$$q(t) = -k\partial T/\partial x|_{x=0}$$

(1)

where ρ is the slab density, δ is the distance between the outer surface of the slab and the thermocouple, k is the conductivity, T is the slab temperature, T_0 is the initial slab temperature, C_p is heat capacity, x is the thickness direction coordinate, and $Y(t)$ is the measured temperature. In the model, T_0 and $Y(t)$ were known; only the surface heat flux $q(t)$ had to be solved. Since air mist cooling is a transient process, the temperature field and the heat flux q_M at a certain time t_M can be solved for if the temperature field $T_{M-1}(x)$ and the heat flux q_{M-1} at t_{M-1} have been determined. Assuming $q(t) = q_M$, which is constant when $t_{M-1} < t < t_M$, Equation (1) can be modified as Equation (2):

$$\rho C_p(T)\frac{\partial T}{\partial t} = \frac{\partial}{\partial x}\left(k(T)\frac{\partial T}{\partial x}\right)$$
$$-k\partial T/\partial x\Big|_{x=0} = \begin{cases} q_M = Const & t_{M-1} < t < t_M \\ q(t) & t > t_M \end{cases}$$
$$-k\partial T/\partial x\big|_{x=L} = 0$$
$$T(x, t_{M-1}) = T_{M-1}(x)$$

(2)

In this work, a sensitivity coefficient $Z(x,t) = \partial T(x,t)/\partial q_M$ was introduced to evaluate the sensitivity of the measured temperature point error. In Beck's method, the heat flux $q(t)$, is discretized into a series of q_i over a measurement interval Δt. The heat flux guess is kept constant within a period of time, $r\Delta t$, where r is the number of future time steps. Solved as a forward heat conduction problem, the predicted temperature by using the applied heat flux guess q_i can be obtained at $t = i + r\Delta t$. In this way, the value of q_i in each time interval can be found to minimize the difference between the measured and calculated temperatures. The surface heat flux at different times is therefore calculated by defining the least squares error function. A program was written in Fortran to solve the one-dimensional transient heat conduction problem. In the process of calculation, the measurement interval Δt was 0.25 s, and the number of future time steps r was 10. Thermocouples were embedded at $x = 20$ mm to record the temperature history. The thermal properties of the materials used in the calculation were functions of temperature and are shown in Table A1. In the course of calculation, it was considered that the steel density 7850 Kg/m^3 was constant. After programming, one of the parameters about the number of mesh points needed to be tested to investigate the grid independence of the results. When the number of mesh points was greater than 400, the results do not change. The number of mesh points used in this paper was 1000.

4. Experimental Results and Discussion

One of the goals of the setup was to obtain experimental data under industrial heat flux boundary conditions for describing the boiling heat transfer in the secondary cooling of continuous casting. Precise temperature measurement was a key for a reliable heat transfer measurement [20]. Several 8-mm diameter holes were drilled 180 mm deep from the side of the 200-mm-thick slab perpendicular to the casting direction. The holes were drilled with flat bottoms to ensure that the tips of the thermocouples could touch the bottoms of the holes. A short distance from the cooling surface to the holes was required to reduce the time delay as the thermocouples responded to the changes in surface heat flux. In this experiment, the thermocouples were inserted 20 mm from the cooling surface. The thermocouples' fastening devices were welded to the surface of the slab to push the thermocouples' tips into contact with the bottoms of the holes. The gaps in the holes were filled with refractory cotton.

Figure 6 shows the armored K-type thermocouples that were inserted into the holes and "locked" inside by fastening devices. The locations of thermocouples numbered 1 to 5 are also shown in the figure. The thermocouples were far away from the edge of the slab to ensure a reduced heat loss effect. The thermocouple NO.1 and NO.3 were arranged inside the slab under the clamping roller respectively. The NO.2 was in the middle of the two clamping rollers. The slab was tilted 15° from the horizontal position to simulate the slab position.

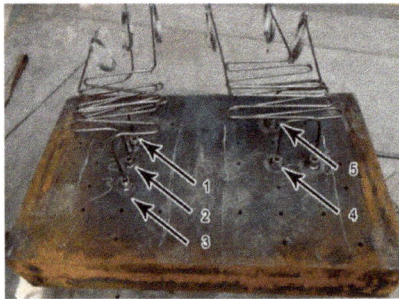

Figure 6. Sketch of the locations of thermocouples.

The air mist spray characteristics, such as water distribution profiles, droplet velocity and the Sauer mean diameter, had an effect on heat transfer. These parameters depended on atomization and gas-water parameters of the nozzle. The main experiment parameters are shown in Table 2. Due to the swing of the spray cooling racks and the overlapping between the nozzles, the uniformity of the water distribution in the vertical and horizontal casting direction was guaranteed. It could be considered that the slab surface cooling was homogeneous. The paper focused on the average convection heat transfer coefficient in secondary cooling. The average water flux was defined as a ratio of water flow to the area of the cooling surface of the slab. The water flux was 0.84 to 3.0 L/(m²·s) in the experiment. The nozzle arrangements with three rows and four columns were derived from the industrial continuous casting in this paper. The experimental nozzle model used was HPZ5.0-120B2.

Table 2. Parameters.

Experimental Condition	Parameter Value
Slab size	$1100 \times 600 \times 200$ mm
Material	ASTM A572 Gr.50
Air pressure	0.2 MPa
Water flux	2.1 L/(m²·s)
Nozzle distance from slab surface	180 mm
Heating target temperature	1150 °C
Temperature sampling frequency	4 Hz
Cooling condition	Top cooling
Water temperature	15.2 °C
Experimental nozzle model	HPZ5.0-120B2

Figure 7. shows a typical slab cooling in the experiment setup for a top spray configuration at a water temperature of 15.2 °C. Clearly, much vapor was generated during cooling, and the slab was cooled from the surface.

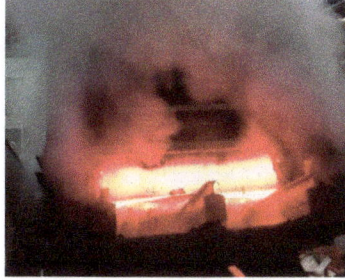

Figure 7. Slab cooling process.

Typical temperature readings relative to time curves during the spray cooling are shown in Figure 8. Regardless of the thermocouple location, the time–temperature curves were similar. A high cooling rate was observed, and the slab temperature dropped drastically. Due to the strong relation between the heat flux and the surface temperature, heat extraction rates changed rapidly with time.

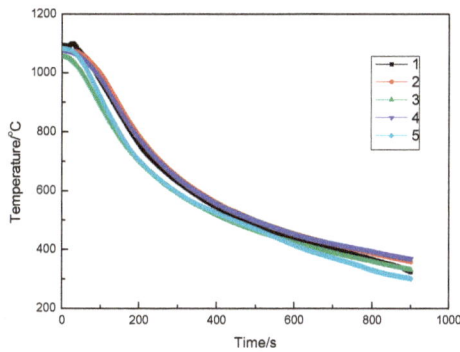

Figure 8. Temperature evolution during cooling at positions 1 to 5.

The inverse heat transfer algorithm was applied to calculate heat fluxes by using the measured temperature profile data. The calculated heat fluxes were plotted as functions of surface temperatures at each thermocouple location to determine the boiling curve.

Figure 9 shows the relation between the surface temperature and the heat flux at different measuring positions. It can be seen that the heat flux was not linearly related to the surface temperature. The five thermocouples' average heat flux of the slab first increased to a maximum value when the slab was cooled to approximately 574.2 °C, and then decreased from the peak value with further cooling. The maximum average heat flux was 5.0×10^5 W/m^2 at 572.4 °C. At 1000 °C, the average heat flux was 2.2×10^5 W/m^2. In the temperature range from 1100 to 800 °C, the average heat flux was 2.9×10^5 W/m^2. Comparing the heat flux at the different thermocouple positions, better vapor discharge conditions resulted in faster cooling and a higher heat flux.

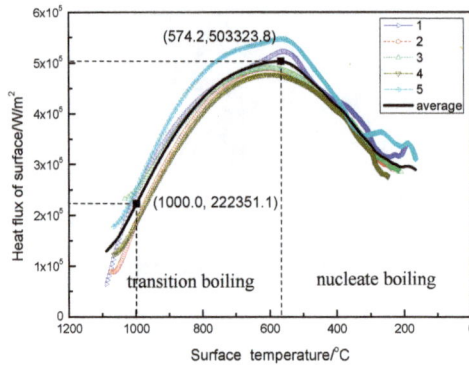

Figure 9. Relationship between surface temperature and heat flux.

As the simulated casting speed was 1.8 m/min, all thermocouples areas were cooled by sprays and support rolls. Each thermocouple reflected the average heat transfer on the surface. The NO.5 thermocouple was located in the lower part of the sample. Good steam exhaust conditions resulted in the greater heat flux.

The surface temperature history experienced two distinct regimes. The boiling curve allowed a better understanding of the physical state of the surface [21]. The first turn in the curve coincided with the starting time of the spray cooling, from radiation to transition boiling. The second turn coincided with the critical heat flux (CHF), from transition boiling to nucleate boiling. At high surface temperatures, where film boiling was dominant, a vapor film near the solid hot surface minimized direct droplet contact time with the surface, resulting in a low heat-transfer rate. As the surface temperature decreased, the droplets began to penetrate the vapor film, and a sharp increase in the heat transfer rate was observed. After reaching the CHF, the heat flux decreased, going through the nucleate boiling regime and finally to convection or one-phase cooling.

The air mist spray heat transfer curve was similar to that of pool boiling in all the boiling regimes. However, we found that the cooling curve did not experience the Leidenfrost point due to the absence of stable film boiling. This can be explained by the following four circumstances. First, the distance between the thermocouple and the cooled surface was 20 mm, the thermocouple's measurement delay resulted in the experiment not showing any Leidenfrost temperature. Second, the air and water flow rate are high in industrial conditions, but the residence time was short in the high-temperature region under experiment conditions. Therefore, the temperature point of Leidenfrost was not obvious. Third, the liquid droplets on the slab did not form a layer under the air mist spray cooling because the spray carried substantial momentum, which pushed the residual droplets away [22]. Fourth, the cooling steel surface oxidized, forming an oxide layer that increased the slab roughness. The oxidation of the surface had the effect of raising the gasification nucleation number.

Figure 10 shows the relation between surface temperature and HTC. It could be seen that the HTC linearly varies with the temperature from 1100 to 500 °C. As temperatures decreased from 1100 to 800 °C, the average HTC linearly increased from 120.0 $W/m^2 \cdot K$ to 542.6 $W/m^2 \cdot K$. At a CHF temperature of 574.2 °C, the average HTC was 888.9 $W/m^2 \cdot K$.

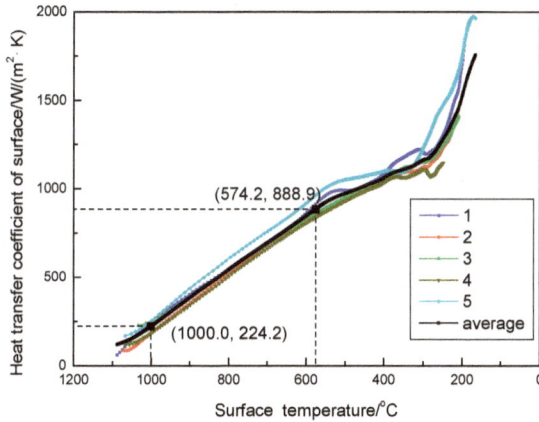

Figure 10. Relationship between surface temperature and HTC.

Using the method developed by this paper, we measured the heat transfer boiling curve of another condition with water fluxes of 0.84 L/(m^2·s) and 3.0 L/(m^2·s). The effect of the cooling water flux was assessed by comparing the heat transfer characteristics of the tests.

Similar slopes for the transition boiling and nucleate boiling regime were observed for different water fluxes during cooling and shown in Figure 11a. The HTC curve showed a typical linear relation, as shown in Figure 11b. The heat flux with a water flux of 3.0 L/(m^2·s) was higher, especially the CHF. There was a clear trend of increased CHF with water flux. Higher water flux led to more droplets breaking the vapor film and accelerating the steam discharge, strengthening the heat transfer on the slab surface.

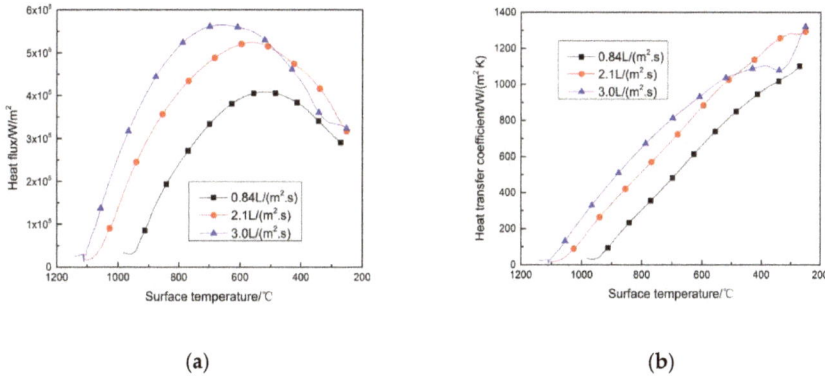

(a)

(b)

Figure 11. Heat characteristics with different water flux for (**a**) heat flux; (**b**) HTC.

The HTC as a function of the temperature difference between the steel surface and the water flux was found to be particularly useful to apply to the solidification model, as shown in Figure 12. We concluded that the HTC was a close linear function of water mass flux for surface temperatures of 850 to 950 °C. The integrated HTC for the section can be described by the generic equation HTC = aG^b, where a and b are constants determined experimentally. The following relation for heat transfer in the transition film boiling region for surface temperatures of 850 to 950 °C is suggested:

HTC = 152 × G$^{1.06}$ (W/m^2·K), where 0.84 ≤ G ≤ 3.0 (L/m^2·s)

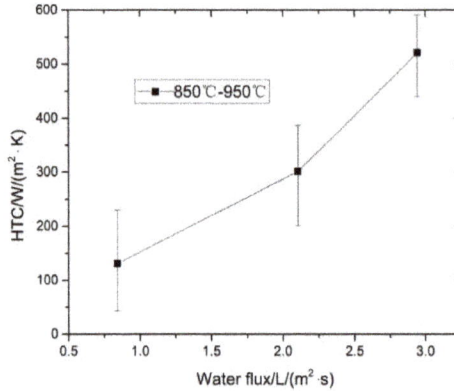

Figure 12. The HTC as a function of water flux.

HTC indicated the heat transfer efficiency between the surface of slab and cooling water, and the effect of heat transfer was high when HTC was large. In general, it needed to be measured by experiment and statistics, and expressed by empirical formula. Significant work regarding spray cooling could be found in the literature. Most researchers studied the macroscopic heat transfer rate. Different researchers had given different empirical formulas according to the test conditions and these formulas had different forms. Gan [23] summarized some of the formula as follows:

$HTC = 360 \times G^{0.556}$, where $0.8 < G < 2.5$ (L/m^2·s), 727 °C $< T_s < 1027$ °C

$HTC = 423 \times G^{0.556}$, where $1 < G < 7$ (L/m^2·s), 627 °C $< T_s < 927$ °C

$HTC = 581G^{0.451}(1 - 0.0075T_w)$, where T_w is the water temperature

$HTC = \alpha(708G^{0.75}T_s^{-1.2} + 0.116)$, where α is a calibration factor, T_s is the slab temperature

$HTC = 157G^{0.55}(1 - 0.0075T_w)$, where T_w is the water temperature

$HTC = 130 + 350G$

Compared with the formula in the paper, there were some differences between the researches. In some cases, the difference was striking. HTC was related to the factors of water flux, spray pressure, spray distance, nozzle structure, surface temperature of slab, water temperature etc. All of these factors had an impact on the relation.

An optimized transition boiling can help to improve heat transfer for specific spray characteristics. The test model was for carbon steel, but the results equally apply to other materials [24]. However, deviations should be expected due to the surface characteristics of the spray-cooled surface [25].

In summary, heat flux and HTC are critical for the theoretical heat design of an industrial process. The experimental setup provides a powerful tool for a quantitative understanding of the heat transportation phenomena of not only continuous casting but also other industries.

5. Conclusions

In this work, an experimental setup to simulate the secondary cooling of continuous casting was developed. This provided a powerful tool for optimizing the continuous casting operation and for better continuous casting machine designs. An inverse heat conduction algorithm was used to calculate the surface heat fluxes during the slab cooling. The main conclusions derived from this study are:

(1) Transition boiling was the primary heat transfer characteristic in the range from 1100 to 800 °C during secondary cooling for the continuous casting of steel.

(2) In the experimental spray cooling, the average heat flux was measured to be 2.9×10^5 W/m^2 in the range of 1100 to 800 °C with water injection of 2.1 L/(m^2·s). The surface HTC increased linearly from 120.0 W/m^2·K to 542.6 W/m^2·K as the temperature decreased from 1100 to 800 °C.

(3) The relation between HTC and water flux in the transition boiling region for surface temperatures of 850 to 950 °C was suggested to be HTC = $152 \times G^{1.06}$.

Author Contributions: Conceptualization, Y.Z. and Z.W.; methodology, Y.Z., J.H. and Z.W.; formal analysis, Y.Z.; investigation, Y.Z. and J.H.; resources, Z.Z.; data curation, C.B. and Y.G.; writing—original draft preparation, Y.Z.; writing—review and editing, Z.Z.; visualization, C.B. and Y.G.; supervision, Y.Z.; project administration, J.H.; funding acquisition, Z.Z. and Z.W.

Funding: The work was supported by National Natural Science Foundation of China under Grant 51264030; National Key R & D Program of China under Grant 2016YFC0401201; and Natural Science Foundation of Inner Mongolia under Grant 2017MSLH0534.

Conflicts of Interest: The authors declare no conflicts of interest.

Appendix A

Table A1. Thermophysical property.

$T/°C$	$C_P/J(kg·K)^{-1}$	$k/W(m·K)^{-1}$
20	462	44.55
100	481	42.95
200	508	41.02
300	530	38.23
400	560	35.74
500	605	33.20
600	680	30.81
700	824	29.39
765	1360	38.38
800	718	25.39
900	615	26.13
1000	604	25.57
1100	685	23.71
1200	858	20.55

References

1. Wendelstorf, J.; Spitzer, K.H.; Wendelstorf, R. Spray water cooling heat transfer at high temperatures and liquid mass fluxes. *Int. J. Heat Mass Transf.* **2008**, *51*, 4902–4910. [CrossRef]

2. Petrus, B.; Zheng, K.; Zhou, X.; Thomas, B.G.; Bentsman, J. Real-Time, Model-Based Spray-Cooling Control System for Steel Continuous Casting. *Metall. Mater. Trans. B* **2011**, *42*, 87–103. [CrossRef]

3. Ramstorfer, F.; Roland, J.; Chimani, C.; Mörwald, K. Investigation of Spray Cooling Heat Transfer for Continuous Slab Casting. *Mater. Manuf. Process* **2011**, *26*, 165–168. [CrossRef]

4. Hauksson, A.T.; Fraser, D.; Prodanovic, V.; Samarasekera, I. Experimental study of boiling heat transfer during subcooled water jet impingement on flat steel surface. *Ironmak Steelmak* **2004**, *31*, 51–56. [CrossRef]

5. Minchaca, J.I.; Castillejos, A.H.; Acosta, F.A. Size and Velocity Characteristics of Droplets Generated by Thin Steel Slab Continuous Casting Secondary Cooling Air-Mist Nozzles. *Metall. Mater. Trans. B* **2011**, *42*, 500–515. [CrossRef]

6. Zhang, J.; Chen, D.F.; Zhang, C.Q.; Wang, S.G.; Hwang, W.S.; Han, M.R. Effects of an even secondary cooling mode on the temperature and stress fields of round billet continuous casting steel. *J. Mater. Process. Technol.* **2015**, *222*, 315–326. [CrossRef]

7. Hernandez, C.A.; Minchaca, J.I.; Humberto, C.E.; Acosta, F.A.; Zhou, X.; Thomas, B.G. Measurement of heat flux in dense air-mist cooling: Part II—The influence of mist characteristics on steady-state heat transfer. *Exp. Therm. Fluid Sci.* **2013**, *44*, 161–173. [CrossRef]

8. Hernandez, C.A.; Castillejos, A.H.; Acosta, F.A.; Zhou, X.; Thomas, B.G. Measurement of heat flux in dense air-mist cooling: Part I—A novel steady-state technique. *Exp. Therm. Fluid Sci.* **2013**, *44*, 147–160. [CrossRef]

9. Ramstorfer, F.; Roland, J.; Chimani, C.; Mörwald, K. Modelling of air-mist spray cooling heat transfer for continuous slab casting. *Int. J. Cast Met. Res.* **2013**, *22*, 39–42. [CrossRef]

10. Horský, J.; Raudenský, M.; Pohanka, M. Experimental Study of Heat Transfer in Hot Rolling and Continuous Casting. *Mater. Sci. Forum* **2005**, *473–474*, 347–354. [CrossRef]

11. Ito, Y.; Murai, T.; Miki, Y.; Mitsuzono, M.; Goto, T. Development of Hard Secondary Cooling by High-pressure Water Spray in Continuous Casting. *ISIJ Int.* **2011**, *51*, 1454–1460. [CrossRef]

12. Tsutsumi, K.; Kubota, J.; Hosokawa, A.; Ueoka, S.; Nakano, H.; Kuramoto, A.; Sumi, I. Effect of Spray Thickness and Collision Pressure on Spray Cooling Capacity in a Continuous Casting Process. *Steel Res. Int.* **2018**, *89*, 9. [CrossRef]

13. El-Bealy, M.O. Air-Water Mist and Homogeneity Degree of Spray Cooling Zones for Improving Quality in Continuous Casting of Steel. *Steel Res. Int.* **2011**, *82*, 1187–1206. [CrossRef]

14. Moravec, R.; Blazek, K.; Horsky, J.; Graham, C.; Fiegle, S.; Dombovic, T.; Kaurich, T. Coupling of Solidification model And Heat Transfer Coefficients to Have Valuable Tool for Slab Surface Temperatures Prediction. In Proceedings of the METEC 7th InSteelCon, Düsseldorf, Germany, 27 June–1 July 2011; pp. 1–9.

15. Chen, L.; Askarian, S.; Mohammadzaheri, M.; Samadi, F. Simulation and Experimental study of Inverse Heat Conduction Problem. *Fund. Chem. Eng.* **2011**, *233–235*, 2820–2823. [CrossRef]

16. Buczek, A.; Telejko, T. Inverse determination of boundary conditions during boiling water heat transfer in quenching operation. *J. Mater. Process. Technol.* **2004**, *155–156*, 1324–1329. [CrossRef]

17. Tapaswini, S.; Chakraverty, S.; Behera, D. Numerical solution of the imprecisely defined inverse heat conduction problem. *Chin. Phys. B* **2015**, *24*, 050203. [CrossRef]

18. Beck, J.V.; Blackwell, B.; St Clair, C.R. *Inverse Heat Conduction, Ill-Posed Problems*; John Wiley & Sons: Hoboken, NJ, USA, 1985.

19. Woodbury, K.A.; Beck, J.V.; Najafi, H. Filter solution of inverse heat conduction problem using measured temperature history as remote boundary condition. *Int. J. Heat Mass Trans.* **2014**, *72*, 139–147. [CrossRef]

20. Stetina, J.; Mauder, T.; Klimes, L.; Masarik, M.; Kavicka, F. Operational Experiences with the Secondary Cooling Modification of Continuous Slab Casting. In Proceedings of the Metal 2013: 22nd International Conference on Metallurgy And Materials, Brno, Czech Republic, 15–17 May 2013; pp. 62–67.

21. Raudensky, M.; Horsky, J. Secondary cooling in continuous casting and Leidenfrost temperature effects. *Ironmak Steelmak* **2005**, *32*, 159–164. [CrossRef]

22. Timm, W.; Weinzierl, K.; Leipertz, A. Heat transfer in subcooled jet impingement boiling at high wall temperatures. *Int. J. Heat Mass Trans.* **2003**, *46*, 1385–1393. [CrossRef]

23. Gan, Y. *Practical Manual for Continuous Casting*; Metallurgical Industry Press: Beijing, China, 2010; pp. 68–69, ISBN 978-7-5024-5044-1.

24. Fang, Q.; Ni, H.; Zhang, H.; Wang, B.; Liu, C. Numerical Study on Solidification Behavior and Structure of Continuously Cast U71Mn Steel. *Metals* **2017**, *7*, 483. [CrossRef]

25. Teodori, E.; Pontes, P.; Moita, A.; Georgoulas, A.; Marengo, M.; Moreira, A. Sensible Heat Transfer during Droplet Cooling: Experimental and Numerical Analysis. *Energies* **2017**, *10*, 790. [CrossRef]

metals

MDPI

Article

Control of Upstream Austenite Grain Coarsening during the Thin-Slab Cast Direct-Rolling (TSCDR) Process

Tihe Zhou [1,*], Ronald J. O'Malley [2], Hatem S. Zurob [1], Mani Subramanian [1], Sang-Hyun Cho [3] and Peng Zhang [3]

[1] Department of Materials Science and Engineering, McMaster University, 1280 Main Street West, Hamilton, ON L8S 4L7, Canada; zurobh@mcmaster.ca (H.S.Z.); subraman@mcmaster.ca (M.S.)

[2] Department of Materials Science and Engineering, Missouri University of Science & Technology, 1400 N. Bishop Ave., Rolla, MO 65409-0330, USA; omalleyr@mst.edu

[3] Algoma Inc. 105 West Street, Sault Ste. Marie, ON P6A 7B4, Canada; Sang-Hyun.Cho@algoma.com (S.-H.C.); peng.zhang@algoma.com (P.Z.)

* Correspondence: tom.zhou@stelco.com; Tel.: +1-519-587-4541 (ext. 5398)

Received: 27 December 2018; Accepted: 29 January 2019; Published: 1 February 2019

Abstract: Thin-slab cast direct-rolling (TSCDR) has become a major process for flat-rolled production. However, the elimination of slab reheating and limited number of thermomechanical deformation passes leave fewer opportunities for austenite grain refinement, resulting in some large grains persisting in the final microstructure. In order to achieve excellent ductile to brittle transition temperature (DBTT) and drop weight tear test (DWTT) properties in thicker gauge high-strength low-alloy products, it is necessary to control austenite grain coarsening prior to the onset of thermomechanical processing. This contribution proposes a suite of methods to refine the austenite grain from both theoretical and practical perspectives, including: increasing cooling rate during casting, liquid core reduction, increasing austenite nucleation sites during the delta-ferrite to austenite phase transformation, controlling holding furnace temperature and time to avoid austenite coarsening, and producing a new alloy with two-phase pinning to arrest grain coarsening. These methodologies can not only refine austenite grain size in the slab center, but also improve the slab homogeneity.

Keywords: thin-slab cast direct-rolling; austenite grain coarsening; grain growth control; liquid core reduction; secondary cooling; two-phase pinning

1. Introduction

Owing to low capital and operating cost, thin-slab cast direct-rolling (TSCDR) has become a major process for hot flat-rolled production since Nucor started the first thin slab caster, directly linked to a hot rolling mill, back in 1989. This process is based on a novel funnel mold caster, which can produce a thin slab of thickness from 50 to 90 mm, instead of the conventional continuous casting slab thicknesses of 200 to 250 mm [1,2]. Figure 1 is an example of the TSCDR process at Algoma Inc. The liquid steel is fed via the ladle and tundish into a funnel-shaped copper mold with primary cooling control. Solidification initiates on the mold wall and the external solidified shell increases in thickness as the steel strand passes through the mold. Based on the steel grades and slab thickness, the casting speed can be from 2.5 to 6.5 m/min. Once leaving the mold, the thin slab passes through the secondary-cooling zone and solidification continues. The secondary-cooling zone has eight segments with multi-point bending and unbending, with 12.7-m containment length, by using air mist cooling. The liquid core reduction system can refine the as-cast microstructure and reduce the centerline segregation and solidification-related defects. The continuous slab is cut to length and then sent to the roller hearth soaking furnace. To maximize the use of the rolling capacity, the thin slab caster has two strands along with two shuttle furnaces, which can transfer the slab sideways to allow the two strands to feed a single rolling mill. The slab is rolled in a single pass roughing mill after descaling, then goes through the heated transfer table and is rolled in the six-stand finishing mill; the resulting hot strip then passes through the run-out table with a laminar cooling system and is coiled at the down coiler [3].

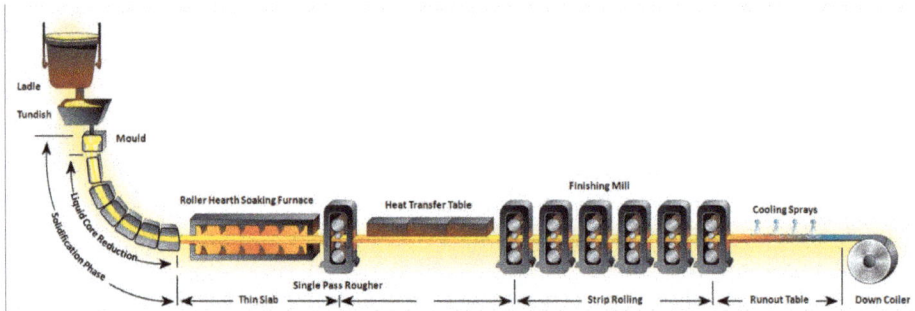

Figure 1. Thin slab casting direct strip production complex at Algoma Inc. [3].

TSCDR mills currently produce a variety of steel grades, including interstitial free steel, low carbon to medium carbon steel, high-strength low-alloy, and advanced high strength multiphase steel grades [1]. Recently, great effort has been placed on using this process to produce high grade micro alloyed steels that can be utilized in bridge guard rails, wind turbine towers, rail cars, and oil and gas pipelines, with stringent low-temperature ductile to brittle transition (DBTT) and drop weight tear test (DWTT) requirements to maintain structural integrity and safety over several decades of service [4,5]. It well established that refining austenite grain size before pancaking can improve DBTT and DWTT properties [6]. In most cases, these applications require a hot band thickness of 10 mm or more. This requirement challenges the TSCDR process, because the ratio of the thickness of the initial as-cast slab to that of the final product is only of the order of 5–7 to 1. It has been proven that production of higher high strength low alloyed (HSLA) grades is very difficult, owing to the presence of extremely large austenite grains in the center of the slab prior to thermomechanical processing [7]. The limited number of thermomechanical deformation passes available in the TSCDR process cannot refine these larger grains [8,9]. In order to achieve uniform and finer microstructure, it is very important to control the upstream austenite grain coarsening before the slab enters the roughing mill. This contribution will focus on refining austenite grain by increasing the cooling rate during solidification, increasing nucleation sites for delta-ferrite by liquid core reduction, and increasing austenite nucleation sites during the delta-ferrite to austenite phase transformation, as well as controlling austenite coarsening inside the holding furnace. In addition, the possibility of rolling new alloy with two-phase pinning is also discussed.

2. Materials, Experimental Procedure, and Model Setup

The experimental materials in this study consisted of an HSLA based American Petroleum Institute (API) X70 cast slab sample from industry, as well as an Fe-Al model alloy with 1.5% Al addition to generate delta/austenite two-phase microstructure at different temperatures. The two chemistries are compared in Table 1. The addition of Al in the Fe-Al model alloy can stabilize delta-ferrite down to room temperature. A two-phase mixture of delta-ferrite and austenite will exist at temperatures between 1310 °C and the eutectoid temperature. The Fe-Al model alloy was prepared by induction melting at CANMET Materials Technology Lab (Hamilton, ON, Canada); the as-received microstructure was delta-ferrite with grain size of approximately 85 μm, after pilot mill hot rolling to the thickness of 10 mm.

Table 1. Chemical composition of the new alloy used in this investigation (wt %).

wt %	C	Mn	Si	Al	Ti	Nb	N
API X70	0.05	1.60	0.30	0.0037	0.0012	0.07	0.0060
Fe-Al model alloy	0.051	1.00	0.36	1.5	0	0	0

In order to quantify the grain coarsening at high temperature, a simple non-isothermal grain growth model [9] was utilized to capture the evolution of grain growth at different stages of the TSCDR process. Starting with the simple equation:

$$\frac{d\overline{R}}{dt} = \alpha M(t) \frac{2\gamma_{gb}}{\overline{R}} \tag{1}$$

then integrating with respect to time, which led to:

$$\overline{R}^2 = \overline{R}_0^2 + 4\gamma_{gb} \int_0^t \alpha M(t) dt \tag{2}$$

where $\overline{R}$ is the mean radius of an individual grain, $\overline{R}_0$ is the initial grain radius, and γ_{gb} denotes the grain boundary energy per unit of area. A reasonable value of 0.8 J·m^{-2} [10] was used for the calculation. α is a shape factor with value of ~1.5 [11], and $M(t)$ is the mobility of the grain boundaries [9,12]. Delta-ferrite grain boundary mobility is shown as [9]:

$$M_\delta(t) = \frac{0.7075}{T(t)} \times \exp\left(\frac{-20,995.43}{T(t)}\right) \tag{3}$$

While austenite grain boundary mobility is listed as [9]:

$$M_\gamma(t) = \frac{0.3072}{T(t)} \times \exp\left(\frac{-20,837.14}{T(t)}\right) \tag{4}$$

In this equation, $T(t)$ is an expression for the temperature as a function of time, which was obtained either experimentally from the data recorded, using a thermocouple, or the temperature profiles during the TSCDR process predicted by the heat transfer model. The details are in the Appendix A.

During a typical TSCDR practice, for instance, Table 1 chemistry, Ti concentration is very low and large TiN particles are formed in the liquid during the late stages of solidification. These particles coarsen during the subsequent solid-state process at high temperature [13,14]. These large TiN particles exert a very small particle pinning effect [15]. Strong particle pinning conditions are not encountered until fine Nb(C,N) precipitates are formed during thermomechanical processing [16–18]. In addition, according to Zurob et al [19], the solute-drag effect of all alloying elements was shown to be negligible at temperatures above 1200 °C. Therefore, Equation (2) can be used to model the grain-size evolution during the TSCDR process up to the point where the slabs exit the homogenization furnace prior to thermomechanical processing at the roughing mill.

To validate the grain growth model, a 70 mm and an 85 mm thickness slab of API X70 were sampled after solidification prior to entering the twin roller hearth tunnel furnaces. The slab crops were sectioned to measure the austenite grain size at various distances from the slab surface. All the samples were prepared using standard metallographic techniques. The prior austenite grain boundaries were revealed using an aqueous solution of picric acid with sodium dodecylbenzene sulfonate, with additions of hydrochloric acid for the different chemistry. Microstructure was investigated using optical and scanning electron microscopy. The image analysis was performed using Clemex PE5.0 software (Clemex Technologies Inc., Longueuil, QC, Canada). The grain size was measured using the area intercept method and the true three-dimensional grain diameter was calculated as 1.382 times the linear intercept diameter [20].

3. Results and Discussions

3.1. Microstructure and Model Validation

The microstructure of the industrially-supplied TSCDR 85- and 70-mm thick slabs of API X70 steel are shown in Figures 2 and 3, respectively. At the surface of the 85 mm API X70 slab, the prior austenite grain size was about 50 μm. At the center of the slab, the prior austenite grain size was as large as 1151 μm. While, the prior austenite grain size at the surface of the 70 mm API X70 slab was about 14 μm, and at the center of the slab was about 858 μm.

Figures 2 and 3 indicated that the industrial TSCDR as-cast microstructure was non-uniform with extremely large grains at the slab center, and that the 70 mm thick slab had a finer austenite grain size compared to the 80 mm slab.

The grain growth model Equation (2) can be used to calculate the delta-ferrite and austenite grain size evolution at different positions in the slab. The important points that are required to be considered is that of the initial grain size $\overline{R}_0$ and the cooling path ($T(t)$, in Equations (3) and (4)), which varies from the surface to the center of the slab, leading to grain-size variations. To understand the HSLA microstructure evolution during the TSCDR process, THERMO-CALC (Thermo-Calc Software AB, Solna, Sweden) was used to predict the relevant transformation temperatures for API X70 and the Fe-Al model alloy using the TCFE6 database [21]. The results are given in Table 2.

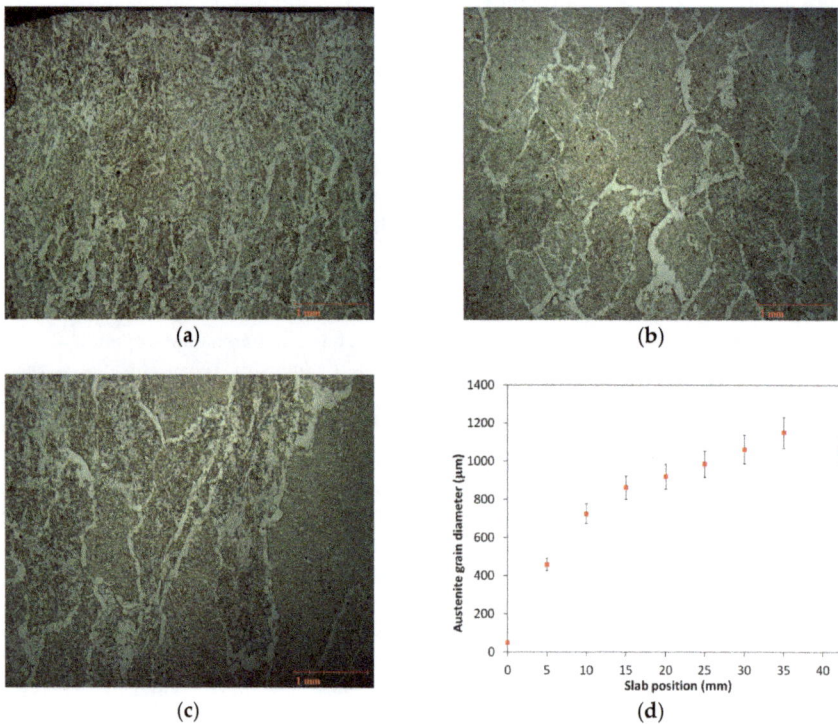

Figure 2. 85-mm slab austenite grain size evolution of American Petroleum Institute (API) X70 from the industrial thin slab casting process: (**a**) close to slab surface; (**b**) 20 mm from the slab surface; (**c**) close to the slab center; and (**d**) summary of the measured austenite grain size with distance from slab surface to center.

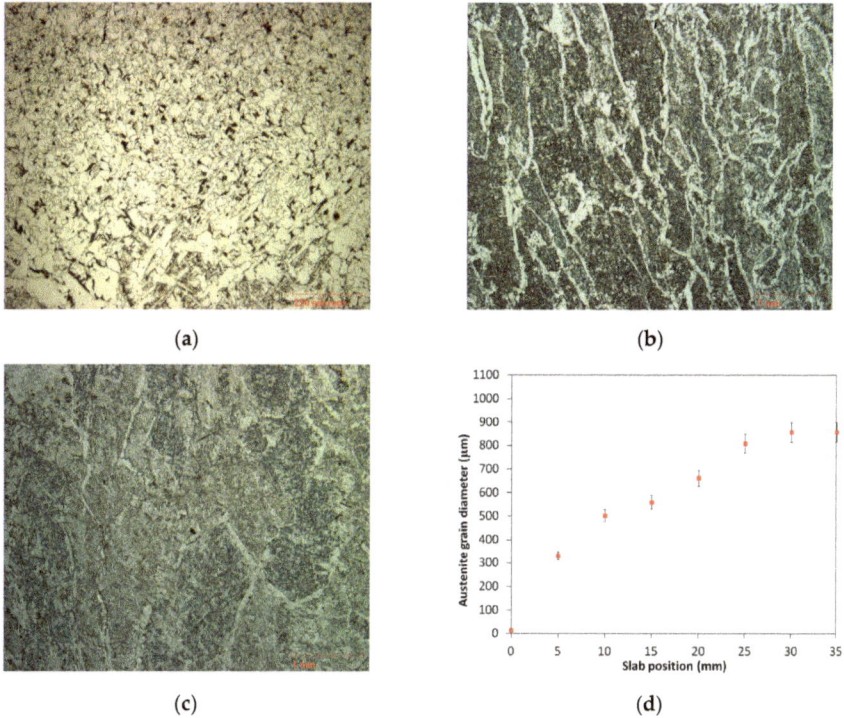

Figure 3. 70-mm slab austenite grain size evolution of API X70 from the industrial thin slab casting process: (**a**) close to slab surface; (**b**) 20 mm from the slab surface; (**c**) close to the slab center; and (**d**) summary of the measured austenite grain size with distance from slab surface to center.

Table 2. Phase transformation temperature of API X 70.

Phase	Liquid	Liquid + Delta	Delta	Delta + Austenite	Austenite
API X 70 (°C)	>1524	1524–1496	1496–1477	1477–1448	1448–852
Fe-Al alloy (°C)	>1530	1530–1500	1500–1412	1412–734	-

In this study, we have not attempted to model the delta-ferrite to austenite transformation. Instead we simply assumed that the delta grain growth occurred down to 1477 °C and austenite grain growth occurred after the delta-ferrite to austenite phase transformation. The secondary dendrite arm spacing (SDAS) was used as the initial delta-ferrite grain size, $\overline{R}_0$, which was calculated using the CON1D V7.0 slab casting heat transfer model, assuming the casting speed of 3.4 m/min [22], as shown in Figure 4a. The initial austenite grain size was presumed to be smaller than the final delta grain size, and was divided by a factor of 3 [23,24] to account for the effect of grain refinement due to the delta-ferrite to austenite transformation in the grain size calculation. Finally, the cooling path $T(t)$ at each point of the slab was also estimated by the CON1D V7.0 model [22]. For example, Figure 4b shows the temperature paths at the surface, 5, 10, and 20 mm below the surface, and the center of the API X70 85 mm slab that was cast with 3.4 m/min casting speed. Due to the spray jet cooling and the high local heat extraction, when the segment rolls directly contacted the slab, the temperature curve on the slab surface showed an irregular trend. Nonetheless, this irregular trend should not interfere with the interpretation of the grain growth with the slab position during the TSCDR process.

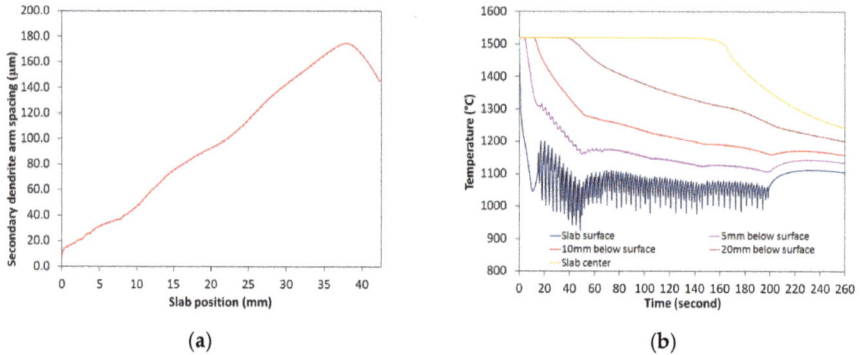

Figure 4. CON1D V7.0 slab casting heat transfer model predicted: (**a**) secondary dendrite arm spacing as a function of position within the API X70 slab; and (**b**) temperature paths at the surface, 5, 10, and 20 mm below the surface, and the center of the slab.

Using Equations (2) and (3), an example of delta grain size evolution with time at 5 mm below the slab surface is shown in Figure 5a. The austenite grain size evolution at the same position (5 mm below the surface) with time just before leaving the holding furnace, by using Equations (2) and (4), is shown in Figure 5b. The solid line represents the model predicted austenite grain diameter and the dashed line represents the temperature profile in the TSCDR process at the corresponding slab position.

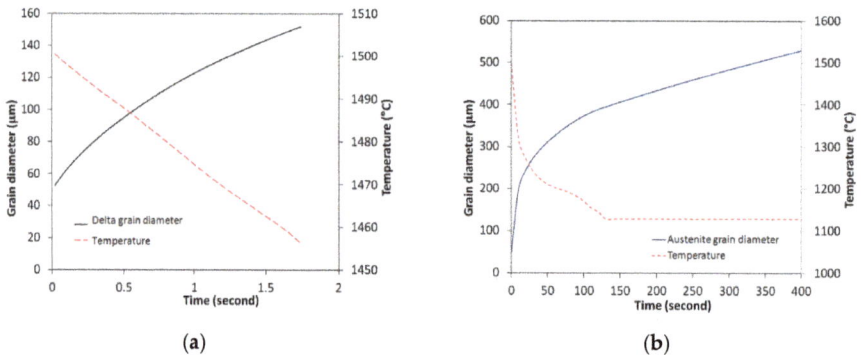

Figure 5. (**a**) The predicted delta grain size evolution with time 5 mm below the slab surface; and (**b**) the predicted austenite grain size evolution at the same position before entering the roller hearth holding furnace.

Similar calculations were conducted as a function of the slab thickness. Figure 6a shows the predicted delta grain size as a function of slab position, just before the onset of the delta-ferrite to austenite transformation. The solid diamonds are the calculated delta grain sizes; the solid line is used to highlight the trends of grain size change with distance from the surface to center of the slab. In addition, the austenite grain size could be calculated when the slab was about to enter the holding furnace, and upon leaving the holding furnace prior to entering the roughing mill.

Figure 6b shows the calculated austenite grain size with slab thickness when the slab is about to enter the holding furnace. The solid diamonds and solid lines follow the same notation for delta grains as noted previously. The experimentally measured austenite grain size from the 85 mm slab from Figure 2d (solid squares) is superimposed for comparison. It indicates that very good agreement was obtained between the model prediction and the measured austenite grain size as a function of the slab position. This provides strong support that the normal grain growth model developed here can be used to predict austenite grain growth at high temperatures prior to the thermomechanical processing. The calculated austenite grain size for the 70 mm slab and the corresponding experimental measurements (Figure 3d) with slab position is shown in Figure 6c. Once again, the predicted austenite grain size was in good agreement with the experimental data as a function of slab position. This agreement further validates the grain growth treatment employed here and supports the applications of the model in the following sections.

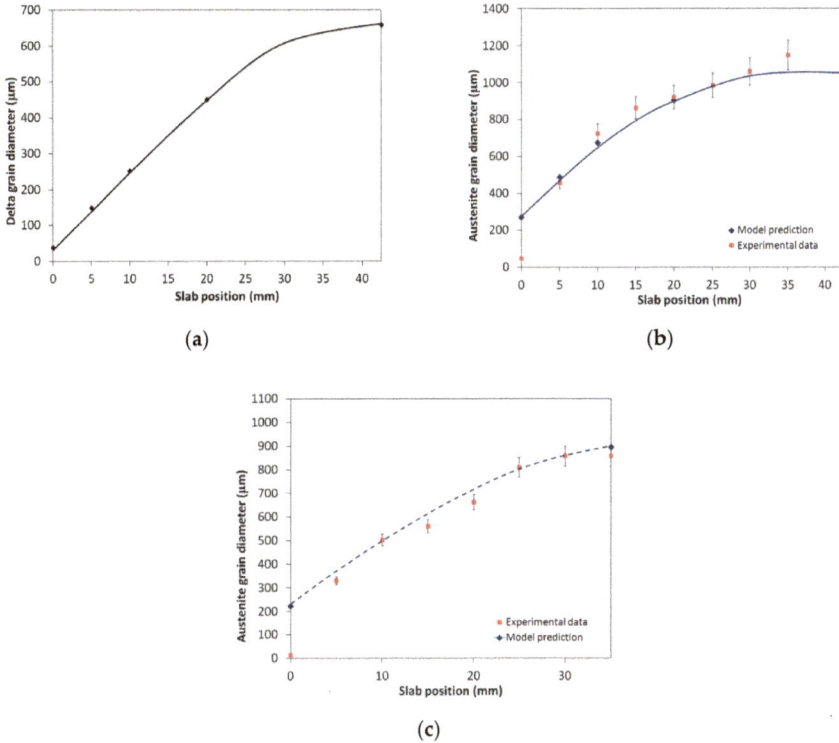

(a)

(b)

(c)

Figure 6. (**a**) The predicted delta grain size as a function of slab position; (**b**) comparison of model prediction and experimental measurement of austenite grain size with slab position before the 85 mm slab enters the holding furnace; and (**c**) comparison of model prediction and experimental measurement of austenite grain size with slab position before the 70 mm slab entering the holding furnace.

3.2. Increasing Cooling Rate to Refine As-Cast Microstructure

Reducing the slab thickness can increase the cooling rate at the slab center, which can refine the austenite grain at the slab center and reduce the non-uniformity of the as-cast microstructure. In what follows, the consequences of reducing the slab thickness from 85 mm to 50 and 30 mm are examined assuming that the only change is the enhanced cooling rate of the slab [25–27]. In order to determine the cooling rate and the initial secondary dendrite arm spacing, the CON1D V7.0 slab casting heat transfer model [22] was used for slabs of 85, 50, and 30 mm slab thicknesses, as shown in Figure 7.

The initial delta-ferrite grain size was, once again, taken to be SDAS; therefore, the model prediction of delta grain size as a function of position for the 30 and 50 mm thin slabs just before the onset of the delta to gamma transformation are shown in Figure 8a, which also includes, for comparison, the results shown earlier for the 70 and 85 mm slabs. The calculated austenite grain size before entering the homogenization furnace is shown in Figure 8b. The symbols in these Figures have the same meaning as discussed previously.

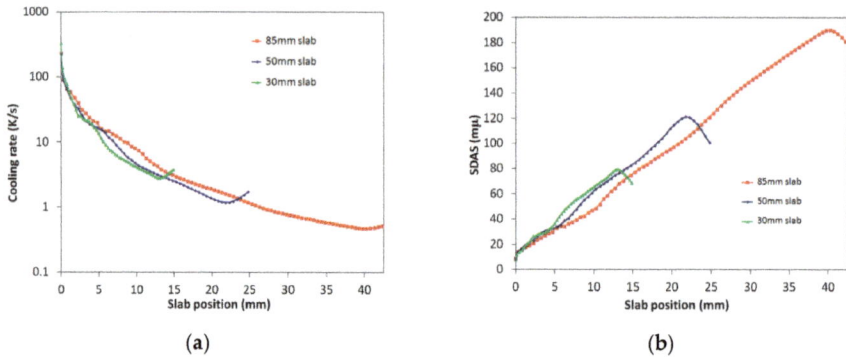

Figure 7. CON1D V7.0 slab casting heat transfer model predicted: (**a**) cooling curves; and (**b**) secondary dendrite arm spacing (SDAS) at different positions of 30, 50, and 85 mm slabs.

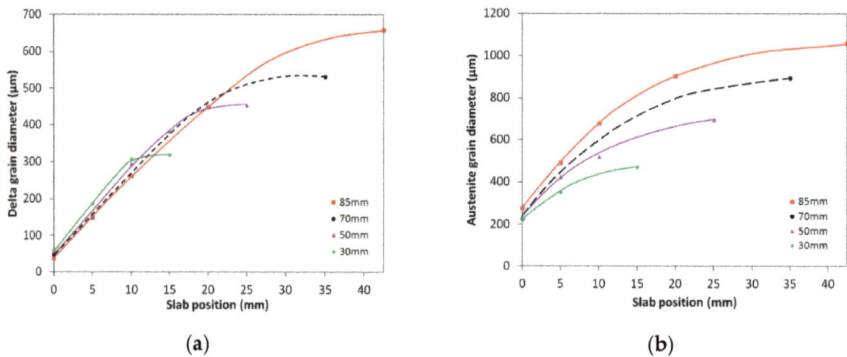

Figure 8. (**a**) The predicted delta grain size of 30, 50, 70, and 85 mm slabs as a function of slab position just before the onset of the delta to gamma transformation; (**b**) the predicted austenite grain size of 85, 70, 50, and 30 mm slabs as a function of slab position when the slab is about to enter the homogenization furnace.

It is clear from these calculations that, due to the enhanced cooling rate, austenite grains at the center of the thinner slabs had greatly reduced in size. When the slabs were about to enter the homogenization furnace, the austenite grain size at the center of the 85 mm thick slab was about 1058 μm. However, the grain size was 896 μm at the center of the 70 mm thick slab, 693 μm at the center of the 50 mm thick slab, and 470 μm at the center of the 30 mm thick slab. In addition, the homogeneity of the microstructure had improved by increasing the cooling rate; the ratio of largest grain size to smallest grain size was 3.8 to 1 for the 85 mm thick slab, 2.8 to 1 for the 50 mm thick slab, and 2.1 to 1 for the 30 mm slab. Therefore, one can conclude that reducing the slab thickness can refine and homogenize the as-cast microstructure due to the enhanced cooling rate at the center of the slab. Experimental measurements of 85 mm (Figure 2) and 70 mm (Figure 3) industrial slab austenite grain sizes, using the distance from the slab surface to center, demonstrated the validity and technological merit of the increased cooling on reducing austenite grains at the center of the slab. The austenite grain size could be reduced from 1151 to 858 μm if the casting slab thickness was reduced from 85 to 70 mm. The main difficulty in applying this method is that it requires changing the layout of the TSCDR process for casting thinner slabs, such as 50 and 30 mm thick slabs. In addition, the smaller slab thickness will further reduce the amount of thermomechanical processing that can be performed downstream, resulting in a larger average grain size, and possibly more grain size non-uniformity despite the improved initial microstructure. Thus, an optimum thickness could be determined by considering both the solidification and grain growth (as described above) as well as the subsequent thermomechanical processing.

It is well established that as-cast microstructure is a function of the solidification rate (V) and temperature gradient (G) ahead of the solid–liquid front. The effect of the temperature gradient and velocity on the primary dendrite arm spacing can be summarized in the following equation [28,29]:

$$\lambda_1 = A_1 G^{-m} V^{-n} \tag{5}$$

where λ_1 is the primary arm spacing, G is the average temperature gradient in front of tip of dendrite in the liquid side, and V is average solidification velocity. A_1, m, and n are constants. For the secondary dendrite arm spacing λ_2, the most widely accepted expression for the relationship between λ_2 and cooling rate (GV) [30,31] is:

$$\lambda_2 = B_1 (GV)^{-n} \tag{6}$$

where B_1 and n are constants. Increasing secondary cooling with restricted casting speed will increase the cooling rate (GV). As a result, the primary and secondary dendrite arm spacing will decrease based on Equations (5) and (6). Figure 9a shows the Algoma DSPC model-predicted slab surface temperatures using different secondary cooling set-ups (the water loops configuration of DSPC is illustrated in Figure 9b). The solid diamonds are the slab surface temperatures for the standard spray cooling set-up used for low carbon steels, 0.6 L/kg of hot steel, and the increased secondary cooling, 1.4 L/kg of hot steel, is used for HSLA steels (solid squares in Figure 9a). When the liquid superheat was above 15°C, the casting speed was restricted to 3.0 m/min; once the liquid superheat was below 10 °C, the casting speed could be increased to 3.4 m/min. Based on the above setup, the predicted 85-mm slab surface and center temperature profiles using increased secondary cooling, with a casting speed of 3.0 m/min, using the CON1D V7.0 model are shown in Figure 9c. Figure 9d shows austenite grain size with slab position using standard cooling and increased secondary cooling. The solid line highlights the trends of grain refinement with distance from the slab surface to the center. The increased secondary cooling practice was predicted to have a much finer austenite grain size than the standard spray practice at both the surface and center of the slab; austenite grain size was reduced from 276 to 253 μm at the surface and from 1058 to 808 μm at the slab center.

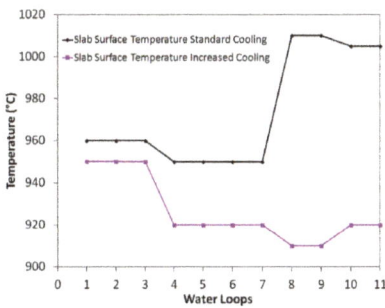

(a)

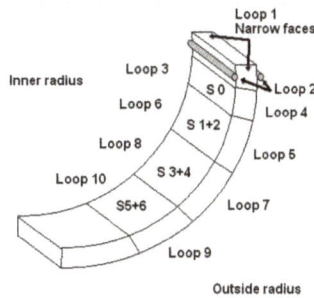

(b)

Figure 9. *Cont.*

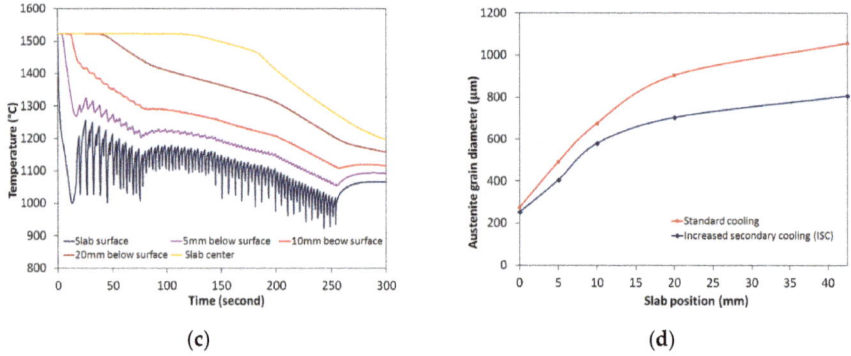

(c)

(d)

Figure 9. (**a**) the Algoma direct strip production complex (DSPC) model-predicted slab surface temperature using different secondary cooling set-ups; (**b**) the water loops configuration at Algoma DSPC; (**c**) the calculated 85 mm slab temperature profile using increased secondary cooling; and (**d**) comparison of 85 mm slab austenite grain sizes using standard cooling and increased secondary cooling.

3.3. Liquid Core Reduction

The TSCDR process at Algoma Inc. can adjust the strand gap dynamically during the casting process. The strand thickness can be reduced just below the mold by a tapered roll guide configuration of the "0" segment. Approximately a 10–30 mm strand reduction can be achieved with liquid core by means of many hydraulically-adjustable roll support segments. In this way, the slab thickness can be reduced from 98 mm to either 85 or 70 mm (Figure 10a). The liquid core reduction during casting produces a convective movement, which mixes the solidified dendrite structure and the liquid steel. A melt flow introduced by convection will generate strong shear stresses, which will shed away the newly formed dendrite arms near the solidification front. The newly formed dendrite crystals are then transported into the hot liquid pool by convective movement. Some of the dendrites are re-melted, while others survive and are transported back to the solidifying region. These surviving broken dendrite tips then form additional nucleation sites for delta-ferrite [32]. The finer delta-ferrite, in turn, will provide more nucleation sites for austenite during the delta-ferrite to austenite phase transformation resulting austenite grain refinement. This basic grain multiplication mechanism induced by liquid core reduction is shown schematically in Figure 10b.

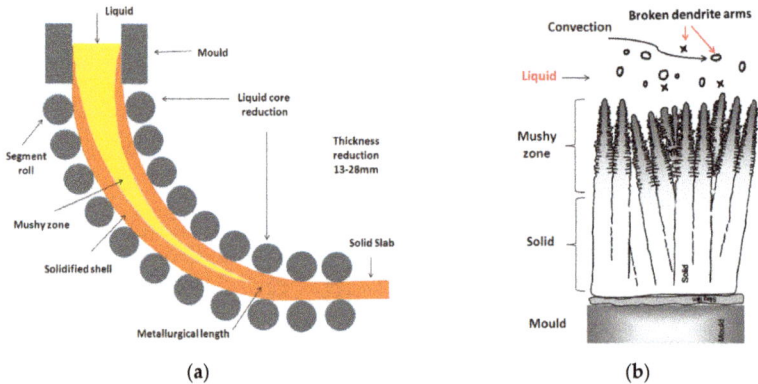

(a)

(b)

Figure 10. (**a**) Sketch diagram of liquid core reduction during thin-slab cast direct-rolling (TSCDR) process; (**b**) the basic mechanism of grain multiplication or grain refinement from liquid core reduction in the solidification region. The convective movement generates shear, to break the dendrite tips, and circulates the debris in the liquid pool.

The liquid core reduction can not only refine as cast microstructure, but also can reduce center line segregation, as well as other solidification-related defects, such as shrinkages and porosity. The metallurgical length for HSLA steels is about 10.0 m which is at the end of segment "6". The liquid core reduction system together with a dynamic control of the liquid pool length can predict the best squeezing point during casting for these HSLA grades. For API X70, the liquid core reduction was set to occur at the fraction solid between 0.4–0.6. Figure 11a,b show API chemistry slab macro etch from strand 3 and strand 4 at the slab center position, respectively. The strand 3 slab sample shows centerline segregation and solidification shrinkages, which indicates that the squeezing was done late with higher solid fraction; however, the strand 4 sample had the diffused centerline, which confirms that the liquid core reduction was carried out at the optimum set point in the solid fraction.

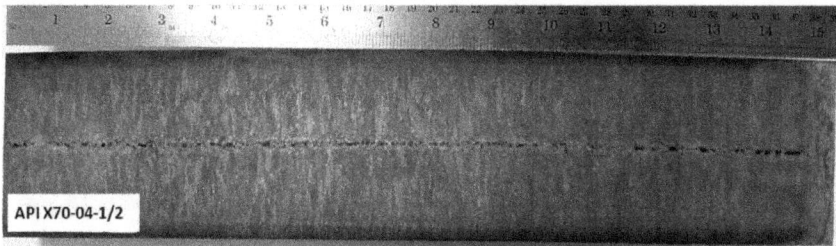

(a)

(b)

Figure 11. API X70 slab macro etch from: (**a**) strand 3; and (**b**) strand 4, with different set points of liquid core reductions.

3.4. Increasing the Number of Austenite Nucleation Sites during Delta-Ferrite to Austenite Phase Transformation

The HSLA steels are low carbon steels (<0.08 wt %) which solidify as delta-ferrite. The ThermoCalc predicted the delta to austenite phase transformation was about 1477 °C (Table 2) and occurred during the liquid core reduction stage. Very little information is available concerning the kinetics of this transformation and its effect on the grain size. To demonstrate the refinement of as-cast microstructure using deformation, the Fe-Al model alloy (Table 1) was studied using a quenching dilatometer at the CANMET Materials Technology Lab [33]. The specimen was reheated into the delta region for 60 s; a compressive strain of 0.2 was applied, and then cooled to 1125 °C at a cooling rate of 50 °C/s. The deformed sample was quenched to room temperature as soon as it reached 1125 °C. Figure 12a shows fine sub-grains that were present within the original delta-ferrite grains. Higher magnification SEM image (Figure 12b) confirmed that the sub-grain boundaries and the delta-ferrite grain boundaries were decorated by fine austenite precipitates. An electron backscatter diffraction (EBSD) map (Figure 12c) showed that austenite grains nucleated along the delta grain and sub-grain boundaries (red line is large angle grain boundary ($\theta > 12°$) and the white is low angle grain boundary

$(12° > θ > 2°))$. This demonstration suggested that sub-grains formed as a result of deformation prior to the delta-ferrite to austenite transformation. During this phase transformation, the sub-grain boundaries and the original delta grain boundaries provided nucleation sites for austenite grains, which would lead to the refinement of the austenite grain structure.

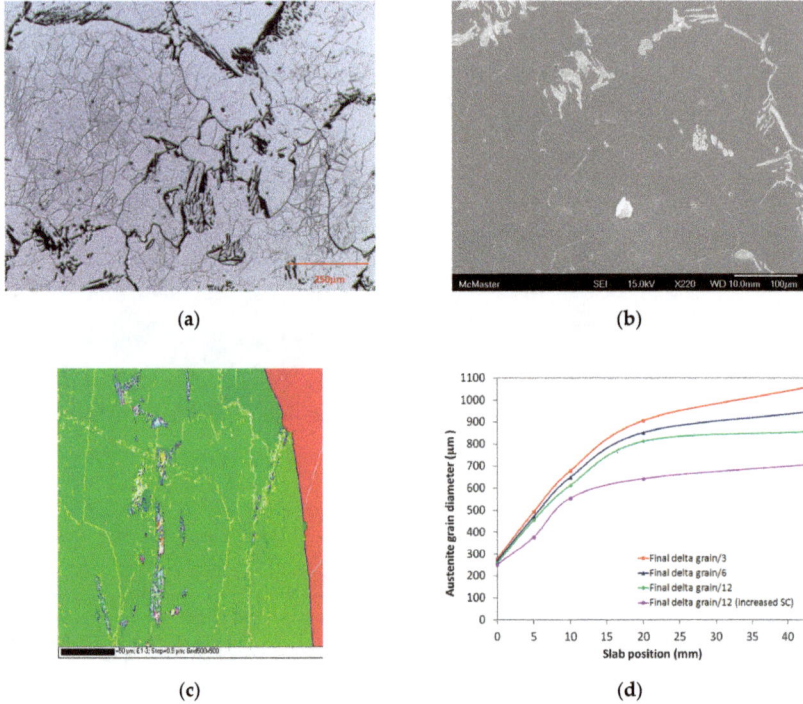

(a)

(b)

(c)

(d)

Figure 12. Microstructure of model Fe-Al alloy reheated into the delta region for 60 s, a compressive strain of 0.2 was applied followed by cooling at 50 °C/s to 1125 °C; the sample was quenched to room temperature as soon as it reached 1125 °C: (**a**) Fine sub-grains present within the original delta-ferrite grains; (**b**) austenite nucleates along the original delta grain boundaries; (**c**) electron backscatter diffraction (EBSD) shows austenite grains nucleated along the delta grain and sub-grain boundaries, and (**d**) predicted austenite grain size when the slab is about to enter the homogenization furnace, with different austenite nucleation sites.

To capture the effect of austenite nucleation sites on the austenite grain coarsening kinetics, the validated grain growth model can be used to calculate austenite grain evolution at different stages of the TSCDR process. Figure 12d summarizes the predicted austenite grain sizes when the slab is about to enter the holding furnace for the different densities of austenite nucleation sites. The various lines represent the grain size achieved when 12, 6, and 3 austenite grains nucleate within each delta grain. The calculation reveals that these extra nucleation sites had little effect on the final grain size at the surface of the slab; however, the austenite grain did decrease with increasing the austenite nucleation sites; the austenite grain in the 85 mm slab center could be reduced from 1058 μm to 945 μm and 856 μm, respectively. The purple line in Figure 12d indicates the austenite grain size trend by using the austenite nucleation density 12 and increased secondary cooling. The austenite grain size in the slab center could be refined from 1058 to 705 μm. The calculated results predicted the potential of increasing the number of austenite nucleation sites during the delta-ferrite to austenite phase transformation to refine the austenite grain size. Much more work is still needed; however, to design new alloys and the deformation schedule during casting that could take advantage of this novel approach.

3.5. Control of Holding Furnace Temperature and Holding Time

Once the continuous slab leaves the secondary cooling zone it is cut to length and then sent to the roller hearth holding furnace, waiting for thermomechanical processing. The holding furnace standard set-up for HSLA at Algoma DSPC has a holding temperature of 1150 °C for 18 min. Austenite grains continue to coarsen inside the holding furnace. The experimental measurement of austenite grain size from the 85 and 70 mm slabs (Figures 2d and 3d) were used as the initial grain size, and the effects of holding temperature and time on austenite grain coarsening by using Equations (2) and (4) are summarized in Figure 13.

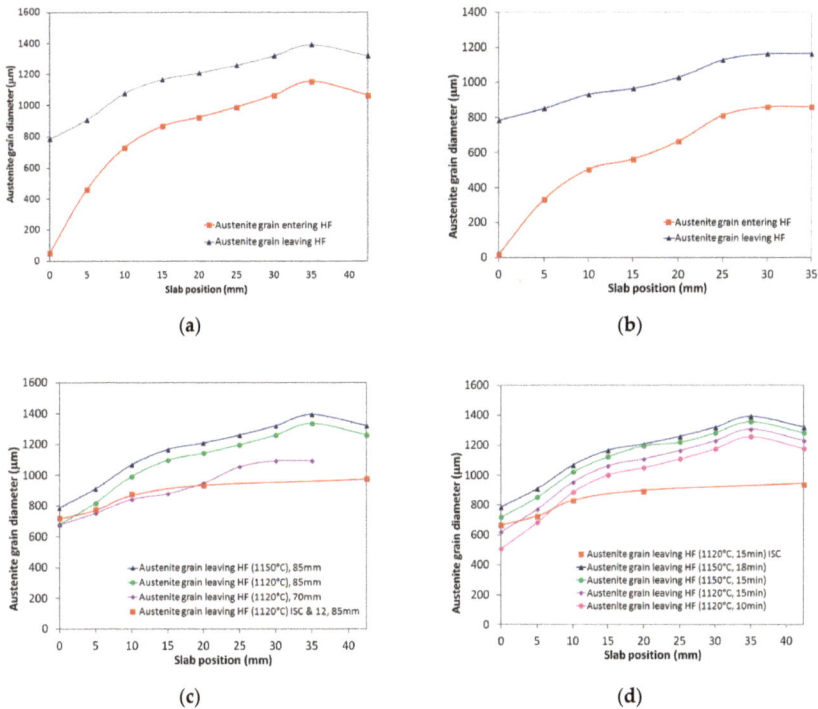

Figure 13. (a) Comparison of DSPC 85-mm thick slab austenite grain size before entering and after leaving the holding furnace (1150 °C, 18 min); (b) comparison of DSPC 70-mm thick slab austenite grain size before entering and after leaving the holding furnace (1150 °C, 18 min); (c) the effect of holding temperature 1120 and 1150 °C on austenite size after leaving the holding furnace; and (d) the effect of holding time and temperature on austenite grain size after leaving the holding furnace.

Figure 13a shows that, for a DSPC 85 mm slab, austenite grain diameter increased from 50 to 784 μm at the slab surface and from 1161 to 1391 μm in the slab center when the slabs were held at 1150 °C for 18 minutes. Extremely large austenite grains existed at the slab center before thermomechanical processing. Austenite grains coarsen much faster on the slab surface than in the slab center due to the larger driving force on the surface. Austenite grain size close to the center of the slab could be reduced from 1391 to 1161 μm by casting a 70 mm slab instead of 85 mm slab (Figure 13b). The effect of holding temperatures on austenite coarsening kinetics is shown in Figure 13c. Once again, the experimental measurement data were used as initial grain size, and the holding time,

18 min, was used for the calculation. The austenite grain size could be reduced from 1391 to 975 μm if increased secondary cooling was used, as well as using the austenite nucleation density 12 during the delta-ferrite to austenite phase transformation when casting an 85 mm slab. The effect of holding time at 1150 and 1120 °C is summarized in Figure 13d. The best combination to control austenite coarsening was to use increased secondary cooling during casting and to set up the holding furnace at 1120 °C for 15 min; thus, the austenite grain size could be reduced from 1391 to 935 μm.

3.6. The Possibility of Producing a New Alloy with Two-Phase Pinning

To refine HSLA steels' austenite grain size during the TSCDR process, carbides/nitrides of Ti, Nb, and V are extensively used to retard grain growth at high temperatures [34,35]. However, these precipitates are ineffective at pinning grain growth when the steel is held at a high temperature for a long time, due to the dissolution of fine particles and rapid particle coarsening. Zhou et al [36,37] proposed a new steel system that can automatically pin the delta grain growth by using a small volume fraction of austenite phase at high temperature. The grain growth is controlled by the austenite phase coarsening rate, which is determined by the bulk diffusion.

Figure 14a shows the CON1D calculated temperature profile at the slab surface as well as those at 5 and 10 mm below the surface of an 85 mm API X70 slab, at Algoma Inc., using the DSPC process cast at a speed of 3.4 m/min. The recorded thermal profile obtained from the laboratory solidification experiment using the new Fe-Al model alloy (Table 1) was superimposed for comparison. The reordered cooling rates, using water quenching, forced air cooling, air cooling, and in mold cooling used in the lab processing were similar to the cooling rates calculated at the slab surface, 5 and 10 mm below the surface, and at the slab center of the TSDCR cast 85 mm slab. In this way, one can compare the average grain sizes obtained from the solidification simulation tests to the grain-sizes measured at 0, 5, and 10 mm below the surface and the center of the industrial slab. These comparisons are shown in Figure 14b, in which the grain-size prior to entry into the soaking furnace was obtained by directly measuring the prior austenite grain size as a function of distance from the surface of the slab (Figure 2d). The data points for the new steel were positioned by matching the cooling rates in the solidification simulation test to the position at which these cooling rates would be observed within the slab. It can be seen that the expected grain size at the center would be 280 μm, compared to 1475 μm for API X70, if the new steel was cast in the form of an 85 mm slab. This clearly demonstrates the potential advantage of this new alloy. One could also compare the grain-sizes within the API X70 slab after exiting the soaking furnace, to those expected in the new alloy, and observe that the grain size of the new alloy was essentially unchanged as a result of soaking, which can prevent excessive grain growth prior to the onset of thermomechanical processing.

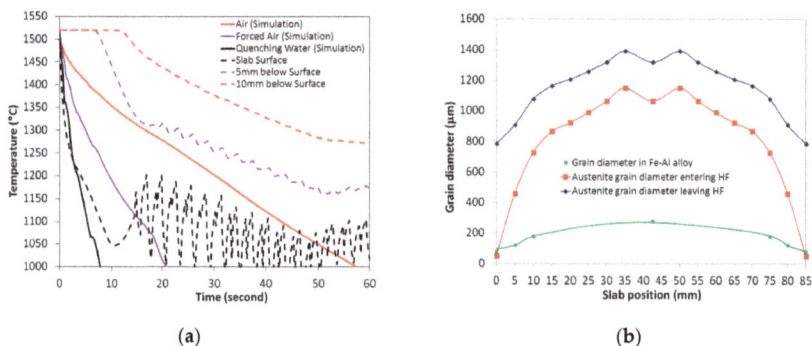

(a)

(b)

Figure 14. (a) Comparison of CON1D predicted temperature profiles on the surface and 5 and 10 mm below the API X70 85 mm slab, and the recorded thermal profile during simulation process [37]; (b) comparison of grain size evolution with slab distance using the TSCDR process to produce API X70 and Fe-Al new alloy.

4. Conclusions

The developed grain growth model successfully reproduced grain growth as a function of position within the API X70 slab in the TSCDR process. The results suggest that it is essential to control grain coarsening in each step, from solidification to the holding furnace, in order to maintain a required fine and uniform austenite grain size prior to the onset of thermomechanical processing.

Reducing the slab thickness can increase the cooling rate at the slab center during the TSCDR process. Predictions of the grain growth model suggest that austenite grain diameter can be reduced from 1345 to 500 µm if a 30 mm slab high cooling rate is produced. In addition, this would lead to less non-uniformity in the as-cast microstructure by refining the grains at the center of the slab. Increasing secondary cooling with restricted casting speed will increase cooling rate, resulting in primary and secondary dendrite arm spacing refinement. Increased secondary cooling, from 0.6 L/kg of hot steel to 1.4 L/kg of hot steel, can reduce the grain size at the center of an 85-mm thick slab, from 1345 to 942 µm.

Liquid core reduction together with dynamic control of the liquid pool length can not only reduce center line segregation and solidification-related defects, but also provides the potential for generating more nucleation sites for delta-ferrite resulting in austenite grain refinement.

Increasing the number of austenite nucleation sites during the delta-ferrite to austenite phase transformation is an effective method of refining and homogenizing the as-cast microstructure of the TSCDR micro alloyed steels. When the Fe-Al model alloy was deformed prior to the onset of the delta to gamma transformation austenite, nucleated prolifically along the original delta grain boundaries and the newly recrystallized delta grain boundaries. The application of 20% deformation generated more than 30 recrystallized grains in each original delta grain. The calculations confirmed that the austenite grain in the 85 mm slab center can be reduced from 1345 to 1001 µm by doubling the nucleation sites.

Austenite grains continue growing inside the holding furnace. Optimizing the holding temperature and time can control austenite coarsening. The austenite grain size can be reduced from 1475 to 1072 µm if casting a 70 mm slab with a soaking temperature of 1120 °C for 10 min, instead of 1150 °C for 18 min.

The use of a delta-ferrite/austenite duplex microstructure is an effective method to retard grain growth at high temperatures. In the delta-ferrite/austenite duplex microstructure, the delta grain growth rate is very slow and controlled by the rate of coarsening of second phase particles. The developed grain growth model predicts that the delta grain size is 10 times smaller in a duplex microstructure than that in materials without pinning. Laboratory validation shows that the delta grains are pinned throughout the TSCDR process, starting from the final stages of solidification. The concept of dual phase to retard grain coarsening, as demonstrated by the delta-ferrite/austenite duplex microstructure, has great potential for producing more uniform as-cast microstructure for the TSCDR process.

Author Contributions: Conceptualization, T.Z., H.S.Z., and M.S.; methodology, T.Z., H.S.Z., M.S., and R.J.O.; software, R.J.O.; validation, T.Z., H.S.Z., and R.J.O.; formal analysis, T.Z., H.S.Z., and R.J.O.; investigation, T.Z., H.S.Z., and R.J.O.; resources, S.-H.C. and P.Z.; data curation, T.Z. and R.J.O.; writing—original draft preparation, T.Z.; writing—review and editing, R.J.O. and M.S.; visualization, H.S.Z.; supervision, H.S.Z. and M.S.; project administration, M.S.; funding acquisition, M.S.

Funding: This research received no external funding.

Acknowledgments: The authors wish to acknowledge with thanks (i) technical support in making model alloys by CANMET (Hamilton, ON, Canada), (ii) assistance in material characterization from Canadian Centre for Electron Microscopy (CCEM) at McMaster University (Hamilton, ON, Canada), and (iii) the technological support from the DSPC operation team and New Product Development Department at Algoma Inc. (Sault Ste. Marie, ON, Canada).

Conflicts of Interest: The authors declare no conflict of interest.

Appendix A

To calculate the grain boundary mobility $M(t)$ in Equation (A1), the Turnbull mobility was used as an initial estimation:

$$M_{\text{pure}} = \frac{wD_{\text{GB}}V_{\text{m}}}{b^2RT} \tag{A1}$$

In the above equation, w is the grain boundary thickness, D_{GB} is the grain boundary self-diffusion coefficient, V_{m} is the molar volume, b is the magnitude of the Burgers vector, R is the gas constant and T is the absolute temperature. The delta-ferrite has body-centered-cubic (BCC) crystal structure, the Burgers vector is $b = 1/2<111>$ and $b = \sqrt{3}a/2$, where a is the lattice parameter of delta-ferrite, 0.286 nm. The molar volume, $V_m = 7.11$ cm^3. The activation energy for diffusion along the grain boundary was taken to be $Q_{\text{GB}} = 0.68Q$, where $Q = 256$ kJ/mole is the activation energy for bulk diffusion in BCC. Finally, $w = 1$ nm and $Q_{\text{GB0}} = 1.67 \times 10^{-4}$ m^3/s. Given that Turnbull mobility does not take into account attachment kinetics, the grain boundary mobility in this way overestimates the experimental grain growth kinetics. The best fit of the experimental data was obtained using a mobility, which is 1/3 of the Turnbull estimate [9]. Therefore, the delta grain mobility used in this work was:

$$M_\delta(t) = \frac{0.7075}{T(t)} \times \exp\left(\frac{-20,995.43}{T(t)}\right) \tag{A2}$$

To estimate the mobility of the austenite grain boundaries, the austenite with face-centered cubic (FCC) crystal structure has $b = 1/2<110>$; therefore, $b = \sqrt{2}a/2$, where a is 0.357 nm. The molar volume, V_m, is 6.85 cm^3, the bulk diffusion activation energy in FCC $Q = 284$ kJ/mole, and that of grain boundary diffusion is: $Q_{\text{GB}} = 0.61Q$. $w = 1$ nm and $Q_{\text{GB0}} = 0.49 \times 10^{-4}$ m^3/s. Once again, the Turnbull mobility leads to an overestimation of the austenite grain growth kinetics. The best fit of the experimental data was obtained with a mobility, which is 0.96 times the Turnbull estimate [9]. Thus, the austenite grain boundary mobility used in this calculation was:

$$M_\gamma(t) = \frac{0.3072}{T(t)} \times \exp\left(\frac{-20,837.14}{T(t)}\right) \tag{A3}$$

The cooling path $T(t)$ at each point of the slab and secondary dendrite arm spacing, thermophysical properties, and spray heat transfer coefficients were calculated using the CON1D V7.0 slab casting heat transfer model. The casting speeds for different slab thicknesses used in the simulations are listed as the following:

(1) 85 mm slab, casting speed 3.0–3.4 m/min;
(2) 70 mm slab, casting speed 3.4–4.0 m/min;
(3) 50 mm slab, casting speed 4.5–5.5 m/min;
(4) 30 mm slab, casting speed 4.5–6.5 m/min.

References

1. Klinkenberg, C.; Kintscher, B.; Hoen, K.; Reifferscheid, M. More than 25 years of experience in thin slab casting and rolling current state of the art and future developments. *Steel Res. Int.* **2017**, *88*, 1700272. [CrossRef]
2. Arvedi, G.; Mazzolari, F.; Siegl, J.; Hohenbichler, G.; Holleis, G. Arvedi ESP first thin slab endless casting and rolling results. *Ironmak. Steelmak.* **2010**, *37*, 271–275. [CrossRef]
3. Zhou, T.; Zhang, P.; Kuuskman, K.; Cerilli, E.; Cho, S.H.; Burella, D.; Zurob, H.S. Development of medium-high carbon hot rolled steel strip on a thin slab casting direct strip production complex. *Ironmak. Steelmak.* **2018**, *45*, 603–610. [CrossRef]
4. Bhattacharya, D.; Misra, S. Development of microalloyed steels through thin slab casting and rolling (TSCR) route. *Trans. Indian Inst. Met.* **2017**, *70*, 1647–1659. [CrossRef]

5. Challa, V.S.A.; Misra, R.D.K.; O'Malley, R.; Jansto, S.G. The Effect of Coiling Temperature on the Mechanical Properties of Ultrahigh-Strength 700 MPa Grade Processed via Thin-Slab Casting. In Proceedings of the AISTech 2014 Proceedings, Indianapolis, IN, USA, 5–8 May 2014; pp. 2987–2997.

6. Nie, W.J.; Xin, W.F.; Xu, T.M.; Shi, P.J.; Zhang, X.B. Enhancing the toughness of heavy thick X80 pipeline steel plates by microstructure control. *Adv. Mater. Res.* **2011**, *194–196*, 1183–1191. [CrossRef]

7. Reip, C.P.; Hennig, W.; Kempken, J.; Hagmann, R. Development of CSP processed high strength pipe steels. *Mater. Sci. Forum.* **2005**, *500–501*, 287–294. [CrossRef]

8. Wang, R.; Garcia, C.I.; Hua, M.; Zhang, H.; DeArdo, A.J. The Microstructure Evolution of Nb,Ti Complex Microalloyed Steel During the CSP Process. *Mater. Sci. Forum.* **2005**, *500–501*, 229–236. [CrossRef]

9. Zhou, T.; O'Malley, R.J.; Zurob, H.S. Study of grain-growth kinetics in delta-ferrite and austenite with application to thin-slab cast direct-rolling microalloyed steels. *Metall. Mater. Trans. A* **2010**, *41*, 2112–2120. [CrossRef]

10. Martin, J.W.; Doherty, R.D.; Cantor, B. *Stability of Microstructure in Metallic Systems*; Cambridge University Press: Cambridge, UK, 1997; pp. 219–231.

11. Humphreys, F.J.; Hatherly, M. *Recrystallization and Related Annealing Phenomena*, 2nd ed.; Elsevier Ltd.: Oxford, UK, 2004; pp. 11–25.

12. Turnbull, D. Theory of grain boundary motion. *Trans. AIME* **1951**, *191*, 661–665.

13. Köthe, A.; Kunze, J.; Backmann, G.; Mickel, C. Precipitation of TiN and (Ti,Nb)(C,N) during solidification, cooling and hot direct deformation. *Mater. Sci. Forum.* **1998**, *284–286*, 493–500. [CrossRef]

14. Nagata, M.T.; Speer, J.G.; Matlock, D.K. Titanium nitride precipitation behavior in thin-slab cast high-strength low-alloy steels. *Metall. Mater. Trans. A* **2002**, *33*, 3099–3109. [CrossRef]

15. Smith, C.S. Grains, Phases, and interfaces: An interpretation of microstructure. *Trans. Metall. Soc. AIME* **1948**, *175*, 15–51.

16. Kwon, O.; DeArdo, A.J. Interactions between recrystallization and precipitation in hot-deformed microalloyed steels. *Acta Metall.* **1991**, *39*, 529–538. [CrossRef]

17. Palmiere, E.J.; Garcia, C.I.; DeArdo, A.J. Compositional and microstructural changes which attend reheating and grain coarsening in steels containing niobium. *Metall. Mater. Trans. A* **1994**, *25*, 277–286. [CrossRef]

18. Poths, R.M.; Rainforth, W.M.; Palmiere, E.J. Strain Induced precipitation in model and conventional microalloyed steels during thermomechanical processing. *Mater. Sci. Forum.* **2005**, *500–501*, 139–145. [CrossRef]

19. Zurob, H.S.; Hutchinson, C.R.; Brechet, Y.; Purdy, G. Modeling recrystallization of microalloyed austenite: Effect of coupling recovery, precipitation and recrystallization. *Acta Mater.* **2002**, *50*, 3075–3092. [CrossRef]

20. Gladman, T. *The Physical Metallurgy of Microalloyed Steel*; Institute of Metals: London, UK, 1997.

21. Thermo-Calc Software. Available online: www.thermocalc.com (accessed on 2 November 2018).

22. Meng, Y.; Thomas, B.G. Heat-transfer and solidification model of continuous slab casting: CON1D. *Metall. Mater. Trans. B* **2003**, *34*, 685–705. [CrossRef]

23. Yin, H.; Emi, T.; Shibara, H. Morphological Instability of δ-ferrite/γ-austenite interphase boundary in low carbon steel. *Acta Mater.* **1999**, *47*, 1523–1535. [CrossRef]

24. Kim, H.S.; Kobayashi, Y.; Nagai, K. Prediction of prior austenite grain size of high-phosphorous steels through phase transformation simulation. *ISIJ Int.* **2006**, *46*, 854–858. [CrossRef]

25. Holzhauser, J.F.; Spitzer, K.H.; Schwerdtfeger, K. Study of heat transfer through layers of casting flux: experiments with a laboratory set-up simulating the conditions in continuous casting. *Steel Res.* **1999**, *70*, 252–257. [CrossRef]

26. Gonzalez, M.; Goldschmit, M.B.; Assanelli, A.P.; Berdaguer, E.F.; Dvorkin, E. Modeling of the solidification process in a continuous casting installation for steel slabs. *Metall. Mater. Trans. B* **2003**, *34*, 455–473. [CrossRef]

27. Louhenkilpi, S.; Makinen, M.; Vapalahti, S.; Raisanen, T.; Laine, J. 3D state and transient simulation tools for heat transfer and solidification in continuous casting. *Mater. Sci. Eng. A* **2005**, *413–414*, 135–138. [CrossRef]

28. McCarrney, D.G.; Hunt, J.D. Measurements of cell and primary dendrite arm spacing in directionally solidified aluminum alloys. *Acta Metall.* **1981**, *29*, 1851–1863. [CrossRef]

29. Bouchard, D.; Kirkaldy, J.S. Prediction of dendrite arm spacings in unsteady and steady-state heat flow of unidirectionally solidified binary alloys. *Metall. Mater. Trans. B.* **1997**, *28*, 651–663. [CrossRef]

30. Taha, M.A. Influence of solidification parameters on dendrite arm spacings in low carbon steels. *J. Mater. Sci. Lett.* **1986**, *5*, 307–310. [CrossRef]
31. Cahn, R.W.; Haasen, P. *Physical Metallurgy*, 4th ed.; North-Holland Physics Publishing: Amsterdam, The Netherlands, 1996.
32. Sobral, M.D.C.; Mei, P.R.; Santos, R.G.; Gentile, F.C.; Bellon, J.C. Laboratory simulation of thin slab casting. *Ironmak. Steelmak.* **2003**, *30*, 412–416. [CrossRef]
33. Zhou, T.H.; Gheribi, A.E.; Zurob, H.S. Austenite particle coarsening and delta-ferrite grain growth in model Fe-Al alloy. *Can. Metall. Q.* **2013**, *52*, 90–97. [CrossRef]
34. Zhou, T.; Overby, D.; Badgley, P.; Martin-Root, C.; Wang, X.; Liang, S.L.; Zurob, S.H. Study of processing, microstructure and mechanical properties of hot rolled ultra high strength steel. *Ironmak. Steelmak.* **2018**. [CrossRef]
35. Hillert, M. Inhibition of grain growth by second-phase particles. *Acta Metall.* **1988**, *36*, 3177–3181. [CrossRef]
36. Zhou, T.; Zurob, H.S.; O'Malley, R.J.; Rehman, K. Model Fe-Al steel with exceptional resistance to high temperature coarsening. Part I: Coarsening mechanism and particle pinning effects. *Metall. Mater. Trans. A* **2015**, *41*, 178–189. [CrossRef]
37. Zhou, T.; Zhang, P.; O'Malley, R.J.; Zurob, H.S.; Subramanian, M. Model Fe-Al steel with exceptional resistance to high temperature coarsening. Part II: Experimental validation and applications. *Metall. Mater. Trans. A* **2015**, *41*, 190–198. [CrossRef]

![metals logo] *metals*

MDPI

Article

Simulation of Crack Initiation and Propagation in the Crystals of a Beam Blank

Gaiyan Yang [1,2,3], Liguang Zhu [2,3], Wei Chen [2,3,*], Gaoxiang Guo [2,3] and Baomin He [2,3]

[1] School of Metallurgical and Ecological Engineering, University of Science and Technology Beijing, Beijing 100083, China; gaiyanyang-1@163.com
[2] North China University of Science and Technology, Tangshan 063210, Hebei, China; zhulg@ncst.edu.cn (L.Z.); 17330544382@163.com (G.G.); 18232527312@163.com (B.H.)
[3] Hebei Engineering Research Center of High Quality Steel Continuous Casting, Tangshan 063009, Hebei, China
* Correspondence: chenwei@ncst.edu.cn or hblgdxzzb@163.com; Tel.: +86-315-880-5053

Received: 5 October 2018; Accepted: 1 November 2018; Published: 5 November 2018

Abstract: Surface cracking seriously affects the quality of beam blanks in continuous casting. To study the mechanism of surface crack initiation and propagation under beam blank mesoscopic condition, this study established a polycrystalline model using MATLAB. Based on mesoscopic damage mechanics, a full implicit stress iterative algorithm was used to simulate the crack propagation and the stress and strain of pores and inclusions of the polycrystalline model using ABAQUS software. The results show that the stress at the crystal boundary is much higher than that in the crystal, cracks occur earlier in the former than in the latter, and cracks extend along the direction perpendicular to the force. When a polycrystalline model with pores is subjected to tensile stress, a stress concentration occurs when the end of the pores is perpendicular to the stress direction, and the propagation and aggregation direction of the pores is basically perpendicular to the direction of the tensile stress. When a polycrystalline model with impurities is subjected to force, the stress concentrates around the impurity but the strain here is minimal, which leads to the crack propagating along the impurity direction. This study can provide theoretical guidance for controlling the generation of macroscopic cracks in beam blanks.

Keywords: polycrystalline model; pores; inclusions; mechanism; beam blank; crystal; propagation

---

1. Introduction

As the raw material of H-beam production, the beam blank has the advantages of fewer rolling passes, high production capacity, low maintenance cost, and high product quality [1,2]. However, because of the complexity of its section, there are often more quality problems than with other blank types [3], particularly surface cracks [4] as shown in Figures 1 and 2 [5]. To meet the new requirements of a new era in terms of the quality of the beam blank and reduce the surface cracks of the billet, it is urgent to conduct in-depth theoretical research on the surface crack propagation mechanism of the billet at different scales, and thus to play a good guiding role in the production of the beam blanks.

At present, the research on cracks of beam blanks has mainly focused on optimizing the process parameters and improving the equipment structure [6–10]. Numerical simulation of crack initiation and propagation of beam blanks remains less studied and the study of the stress concentration effect on the grain interior and boundary is lacking; study of the effects of inclusions and pores on crack initiation and propagation under mesoscopic conditions does not exist. Lasko et al. simulated the crack generation and expansion process of $Al_2O_3/6061$ Al composites under different mechanical properties and damage parameters using ABAQUS [11]. Jiang et al. introduced the nodal enrichment functions in the extended finite element method and used the level set method to track the crack propagation path of the compact tension specimen [12]. Wang and Schwalbe studied the transition from intercrystalline to transcrystalline fatigue crack propagation under different ageing conditions of the alloy Cu-35%Ni-3.5%Cr [13]. Shen simulated the crack propagation of the slab under a mesoscopic condition [14]. Xue et al. used the Abaqus program and its user subroutine UVARM to simulate and analyze the single micro-crack mode in a microcosmic view field, and found that the crack propagation form is different with different presetting angles under different stress states; the higher the stress triaxiality value, the more easily the crack propagates, while the crack tends to close when the value is negative. The deformation energy ratio can precisely reflect the influence degree of stress states on crack propagation [15]. The aforementioned literature addresses crack initiation and propagation. Although some of these studies were completed under microscopic observation, the studies of the Q235B beam blank, particularly under the polycrystalline model, were nearly unable to study the effect of inclusions and pores on crack initiation and propagation. The polycrystalline model combined with metallographic experiments is closer to reality, and can do what other model simulations could not do prior. In this study, through the establishment of a polycrystalline model, the crack propagation process in the crystal of a beam blank is analyzed, and the stress and strain conditions with pores or inclusions are also analyzed, providing theoretical guidance for the generation of macroscopic cracks and process optimization.

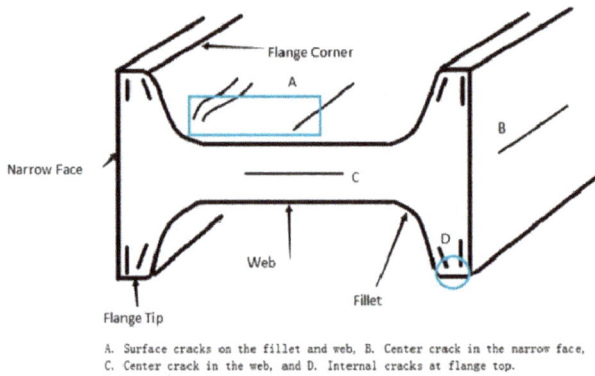

A. Surface cracks on the fillet and web, B. Center crack in the narrow face,
C. Center crack in the web, and D. Internal cracks at flange top.

Figure 1. Cross-section schematic of a beam blank [5].

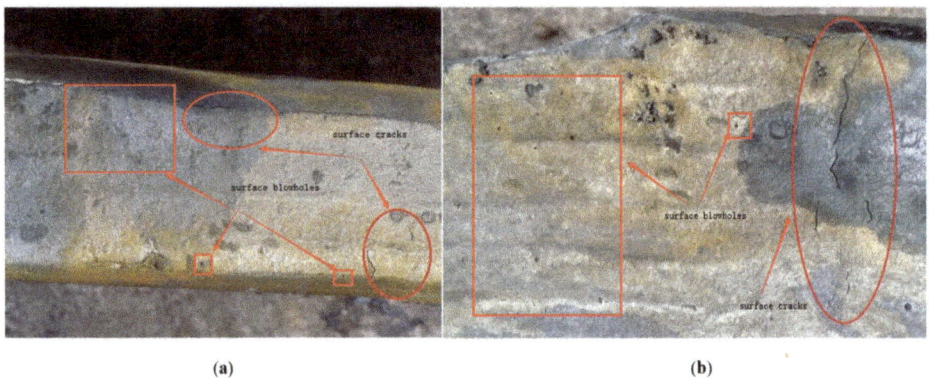

(a)   (b)

Figure 2. Surface cracks and blowholes on (**a**) web and (**b**) fillet of actual beam blank [5].

2. Establishment of a Polycrystalline Model

The polycrystalline model in this study was mainly programmed using MATLAB software (USA). A Voronoi diagram was drawn using the program; its process is shown in Figure 3. The main principle of Voronoi diagram construction is to first randomly generate N points within a certain coordinate range, and then form a triangle with the three points closest to each other, which will generate an infinite number of triangles in the range of coordinates, forming a network. Finally, the midline of the three sides of each triangle in the triangle net is created, and these midlines are connected to form the final polycrystalline model. We termed this method the dual generation method. As can be seen from Figure 4, the polycrystalline model drawn using this method can not only show the irregularity in the size and distribution of grains, but also be consistent in terms of actual grain morphology [16]. Therefore, it is reasonable to draw a polycrystalline model using this method.

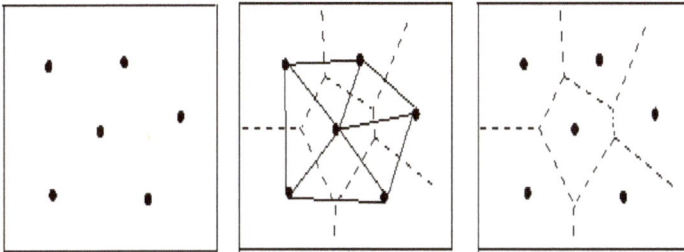

Figure 3. Implementation process of the Voronoi diagram.

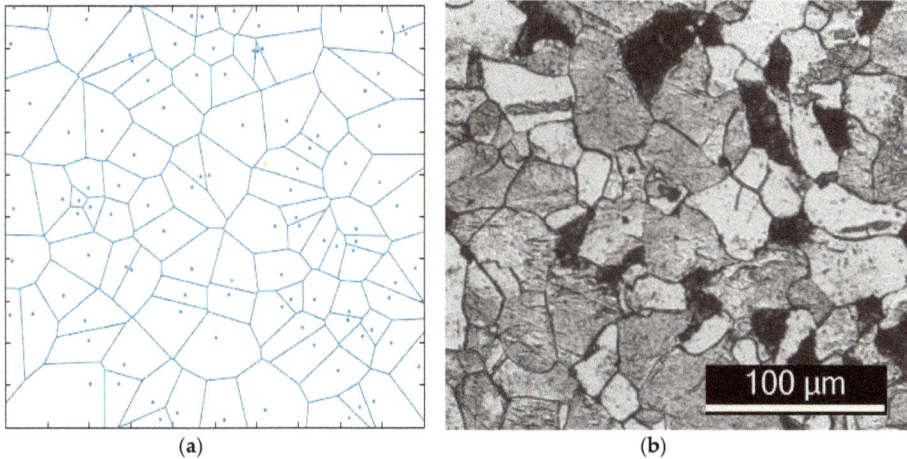

(a)   (b)

Figure 4. Grain size diagram of (**a**) polycrystalline Voronoi diagram in MATLAB and (**b**) actual grain size of the beam blank.

The coordinates of each point obtained from MATLAB are written into the input file and imported into ABAQUS software (6.12 Abaqus, France) to obtain the polycrystalline model. The size of the model area was 0.1 mm × 0.1 mm, as shown in Figure 5; the related parameters of the Q235B materials in the simulation are shown in Table 1.

Table 1. Main parameters used in the simulation.

Parameter	Data
Steel grade	Q235B
Steel compositions	C: 0.19%, Mn: 0.43%, Si: 0.20%, P: 0.025%, S: 0.007%
Elastic Modulus of matrix (E_m)	206 GPa [16]
Elastic Modulus of inclusion (E_i)	20,600 GPa [16]
Poisson's ratio (ν)	0.28 [16]
Specific heat (C_p)	687 J/kg·°C [14]
Thermal conductivity (λ)	32 W/m·°C [14]
Thermal expansion Coefficient (α)	1.30×10^{-5} m/K [14]
Ultimate tensile strength (σ_b)	390 MPa [14]
Yield stress (σ_s)	235 MPa [14]
The density of strand (ϱ)	7400 kg/m^3 [14]

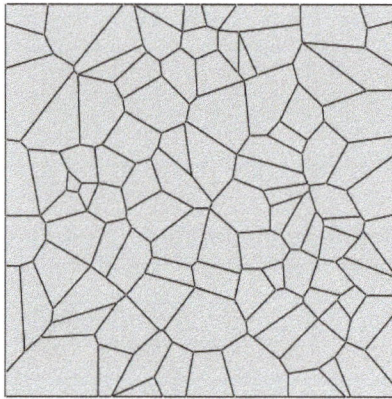

Figure 5. Two-dimensional polycrystalline model in ABAQUS.

2.1. Polycrystalline Model with Cracks

Four types of cracks of different angles and different positions were created on the polycrystalline model of the beam blank. The location of the crack was in the grain interior and the grain boundary; the angles between the crack and tensile stress were 15°, 45°, 75°, and 90°. Figure 6 shows the two typical positions of cracks and loading direction into a 45° angle.

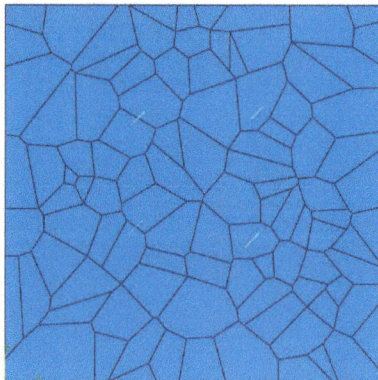

Figure 6. Crack position distribution diagram of the crack with the tensile stress at an angle of 45°.

2.2. Polycrystalline Model with Pores

During the process of casting the beam blank, the generation of pores is mostly caused by improper protective casting, wet raw materials, poor baking in tundish, poor deoxidization of the liquid steel, and poor degassing in refining [17,18]. The existence of pores will have a great impact on the tensile properties of the materials; therefore, the simulation of the stress and strain of a polycrystalline model with pores is of special significance to prevent the generation of cracks.

Models with pores are mainly created in component modules. When creating the model, material properties such as the elastic modulus and Poisson ratio of the model are assigned to the attribute module, while those with holes are not assigned to properties. Based on the original polycrystalline model, a hole was obtained by cutting into the assembly module, and a model with pores was established to simulate the stress and strain. Figure 7 shows the pores distribution on the polycrystalline model components.

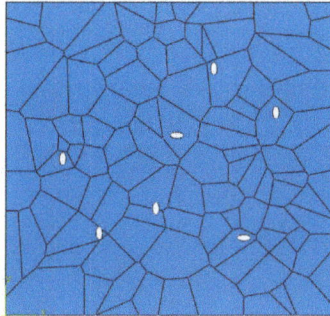

Figure 7. The distribution of pores on the polycrystalline model components.

2.3. Polycrystalline Model with Inclusions

Inclusions in the billet are typically harder than the matrix, which may cause stress concentration around the impurities, leading to metal damage or fractures. Therefore, the study of a polycrystalline model with impurities has a great effect in reducing cracks in the beam blank.

Models with inclusions were also created in the component module. In the attribute module, the matrix was endowed with corresponding elasticity and plasticity. However, because of the high hardness of the inclusions, it was difficult to produce deformation; thus, only the elastic modulus is provided, which is 100 times the size of the matrix [16]. Based on the original polycrystalline model, inclusions were obtained using an incorporating method in the assembly module. Figure 8 shows the distribution of the inclusions on the polycrystalline model components.

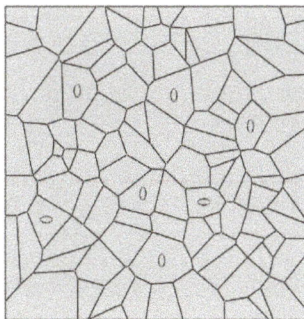

Figure 8. Inclusions distribution on the polycrystalline model components.

2.4. Meshing and Initial Boundary Conditions

By referring to the relevant data, it was found that during the process of elastoplastic analysis, when the object of the study is an incompressible material, non-coordinated units and linear reduction integral units are mainly used. A linear reduction integral can make the calculation easier and save time. The element type selected using this calculation method in ABAQUS was CPS4R, that is, a four-node bilinear plane stress quadrilateral reduced integration element. The control property of the grid was quadrilateral. In this case, the total number of units in the model was 6562; the meshing of the polycrystalline model is shown in Figure 9.

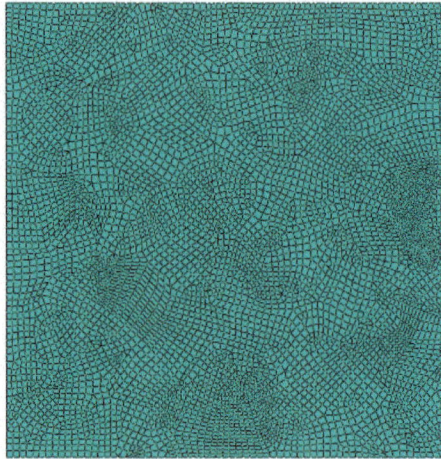

Figure 9. Meshing of the polycrystalline model.

In this study, the boundary condition set in ABAQUS was to select the displacement/rotation angle, the freedom of the boundary UR3 in the X direction was set to 0, velocity loads were separately applied on both sides, and speed/acceleration was selected with a magnitude of 0.05 m/s and a velocity load in the Y direction of 0 [16]. The boundary conditions can better reflect the actual loading conditions. The specific load is shown in Figure 10.

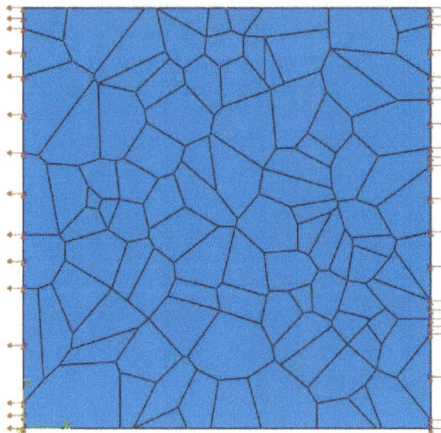

Figure 10. Load distribution.

3. Analysis of Crack Initiation and Propagation in a Crystal

3.1. Stress and Strain Analysis of Defect-Free Crystals

As can be seen in Figures 11 and 12, in the polycrystalline model without any defect, the stress and strain distribution at the boundary of the polycrystalline model is obviously different from that in the crystal, and the stress and strain at the boundary are obviously greater than those in the crystal. This is because of the interaction of external factors and the grain boundary. For example, under a high temperature condition, the grain boundary will undergo a structural change, which will weaken the grain boundary. Another reason is that the properties of the grain boundary and crystal matrix are quite different. As the tensile stress gradually increases, stress corrosion will first occur at the grain boundary, making the grain boundary more prone to fracturing.

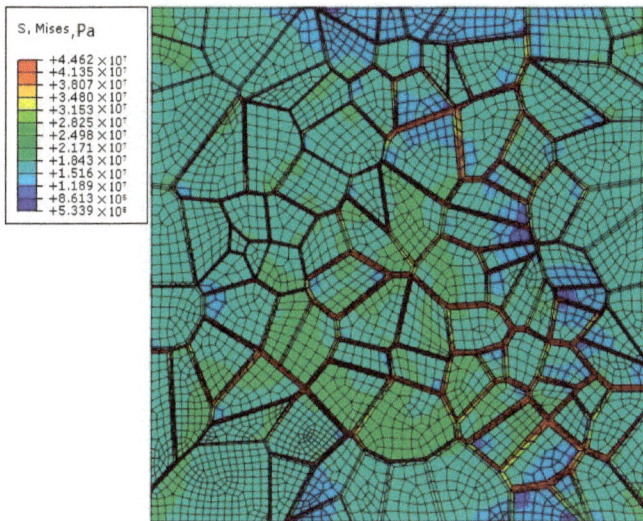

Figure 11. Stress distribution in the polycrystalline model without defects.

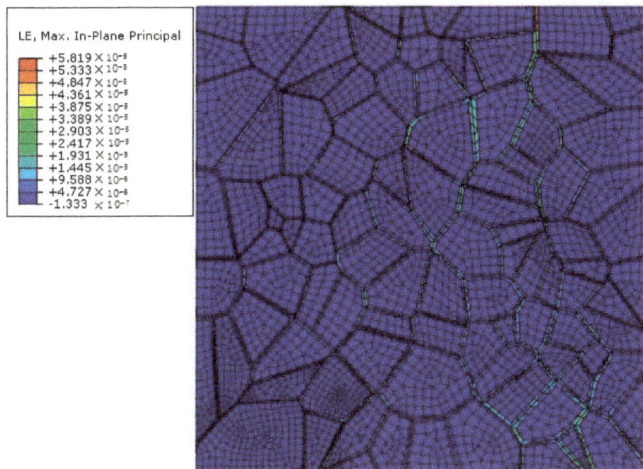

Figure 12. Strain diagram of the polycrystalline model without defects.

3.2. Analysis of a Polycrystalline Model with Multiple Cracks

Cracks are created at two typical positions with four angles, different from the tensile stress. The two typical positions are the crystal interior and boundary, respectively, at four angles between the crack and tensile stress, namely, 15°, 45°, 75°, and 90°. Figures 13–16 show the crack propagation in the two typical positions at different angles from the load direction.

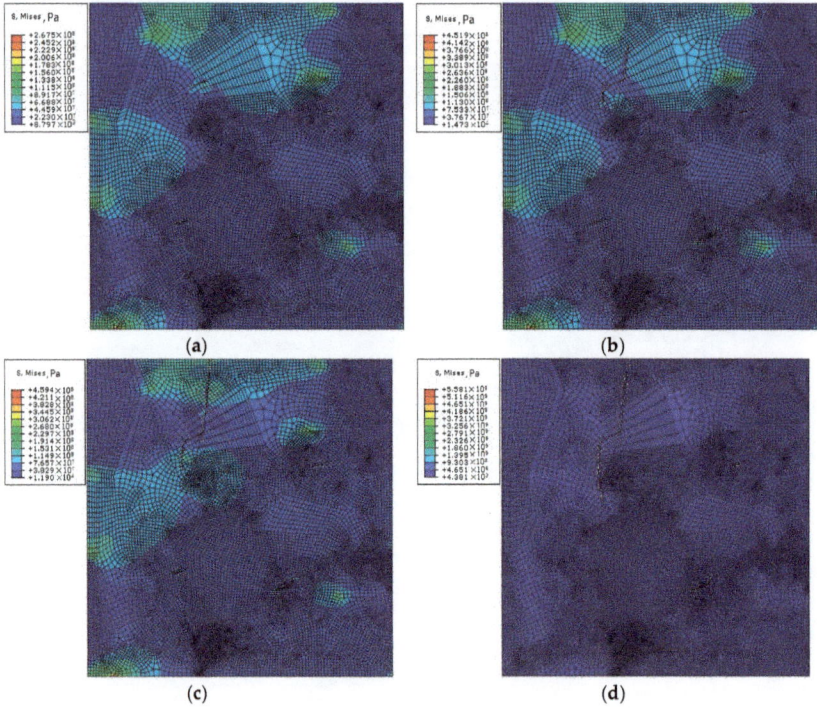

Figure 13. The (**a**) first stage, (**b**) second stage, (**c**) third stage, and (**d**) fourth stage of crack propagation with a load direction of 15°.

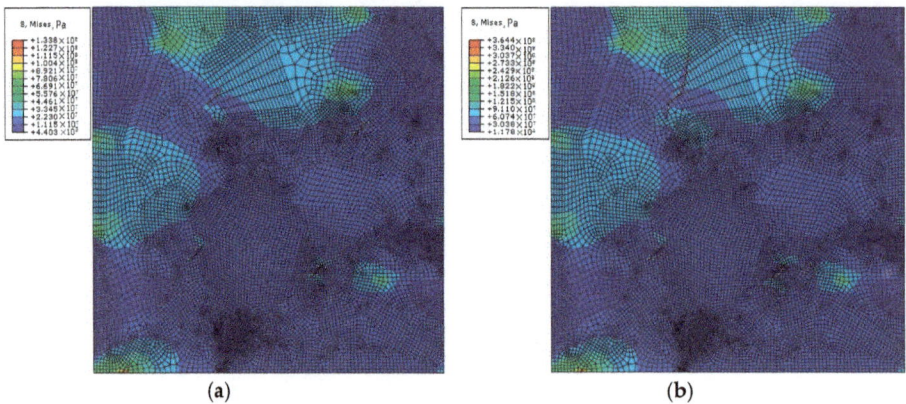

Figure 14. *Cont.*

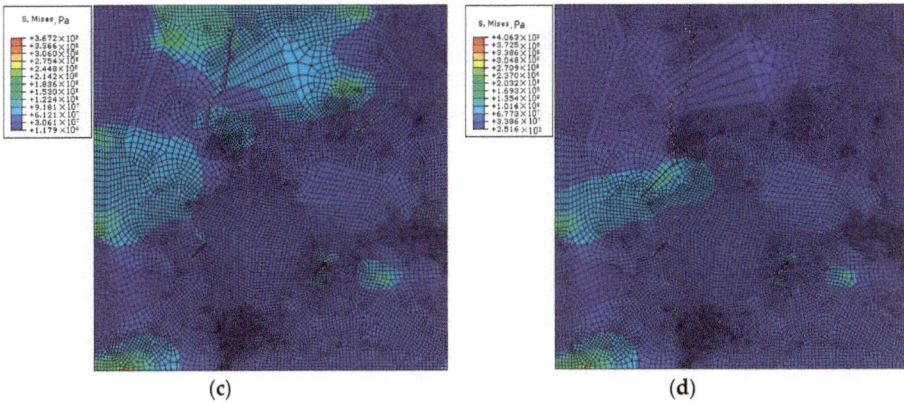

Figure 14. The (**a**) first stage, (**b**) second stage, (**c**) third stage, and (**d**) fourth stage of crack propagation with a load direction of 45°.

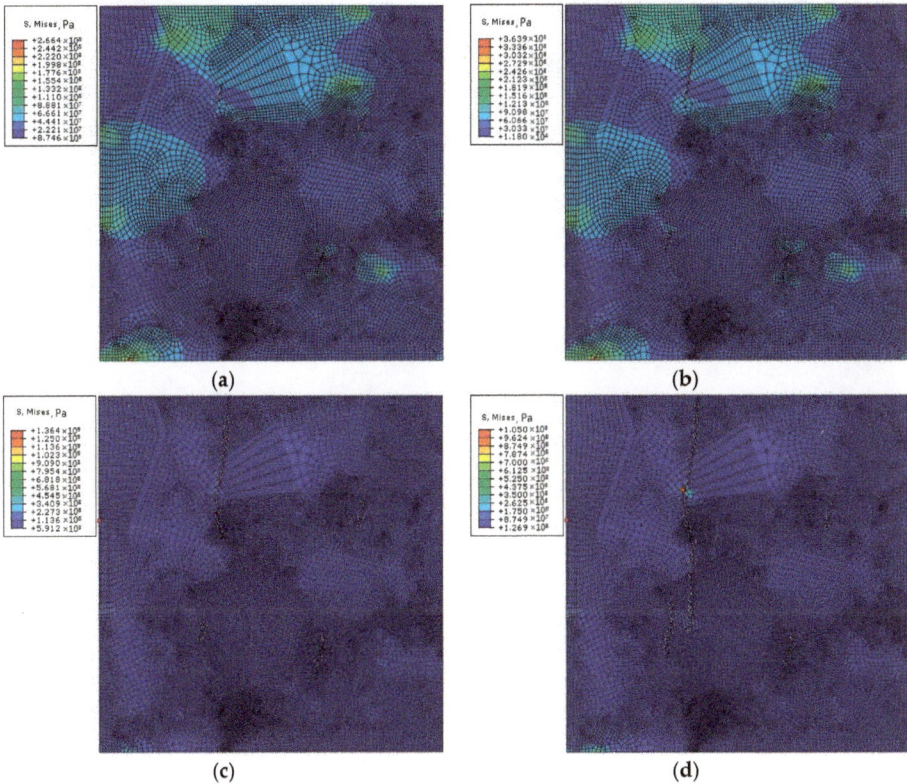

Figure 15. The (**a**) first stage, (**b**) second stage, (**c**) third stage, and (**d**) fourth stage of crack propagation with a load direction of 75°.

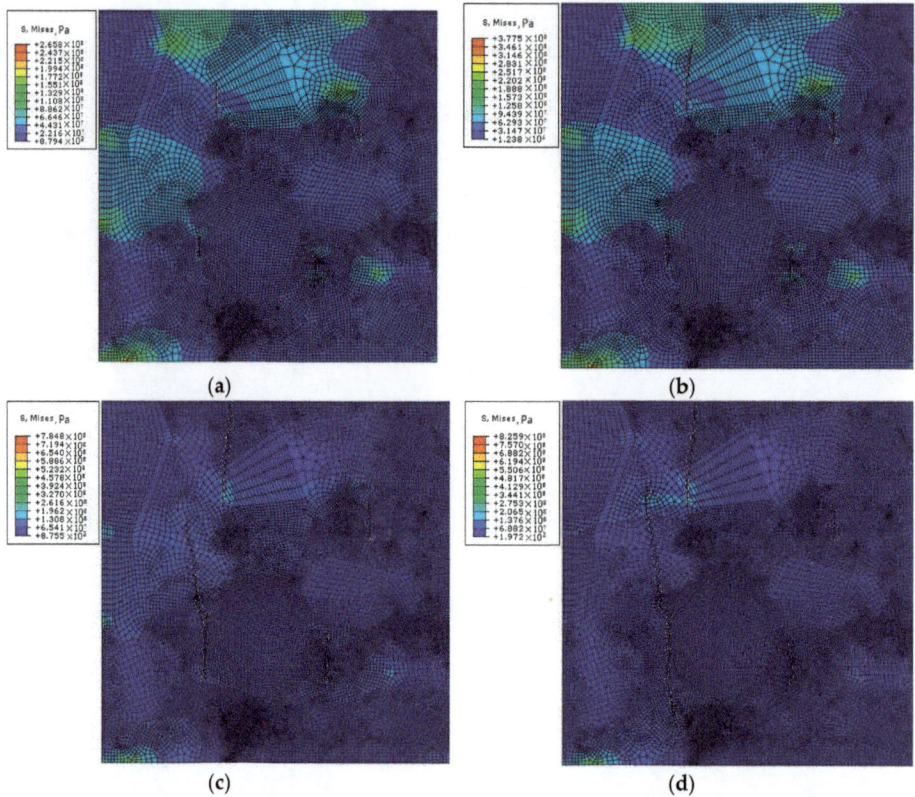

Figure 16. The (**a**) first stage, (**b**) second stage, (**c**) third stage, and (**d**) fourth stage of crack propagation with a load direction of 90°.

According to the aforementioned Figures 13–16, crack propagation occurs at the crystal boundary prior to that in the crystal, which is a result of crystal interaction that makes the stress at the crystal boundary more concentrated than that in the crystal. Thus, crack propagation is accelerated. Through the analysis of crack propagation under the aforementioned conditions, the following can be found:

- The direction of the crack propagation is basically along with the direction perpendicular to the tensile stress, and the amount of crack propagation along this direction is also the greatest, consistent with the study of Yang et al. [19], Wang and Schwalbe [13].
- The amount of crack propagation at the grain boundary is higher than that in the crystal.
- The angle and tensile stress of the crack are unrelated to the direction of crack propagation, and their influence on crack propagation is mainly reflected in the time and sequence of the crack propagation and the final crack propagation displacement.

3.3. Analysis of a Polycrystalline Model of Different Porosities

Damage to metallic materials when they contain pores can be approximately divided into three processes: first, nucleation occurs around the grain boundary or the second phase particle, then the holes begin to grow, and finally they connect together to form cracks. Therefore, it is of great significance to study the stress and strain process of a polycrystalline model with holes to show the mechanism of crack formation.

Figure 17 shows the stress distribution of the polycrystalline model with different porosities, from which certain regularity can be found in pores. In the direction perpendicular to tensile stress, the maximum stress can be observed at the end of pores; while in the same direction of the force, the minimum stress and strain can be calculated around other pores with insignificant deformation. It indicates that when the polycrystalline model deforms, stress concentration at the end of the pores is mostly perpendicular to the direction of tensile stress.

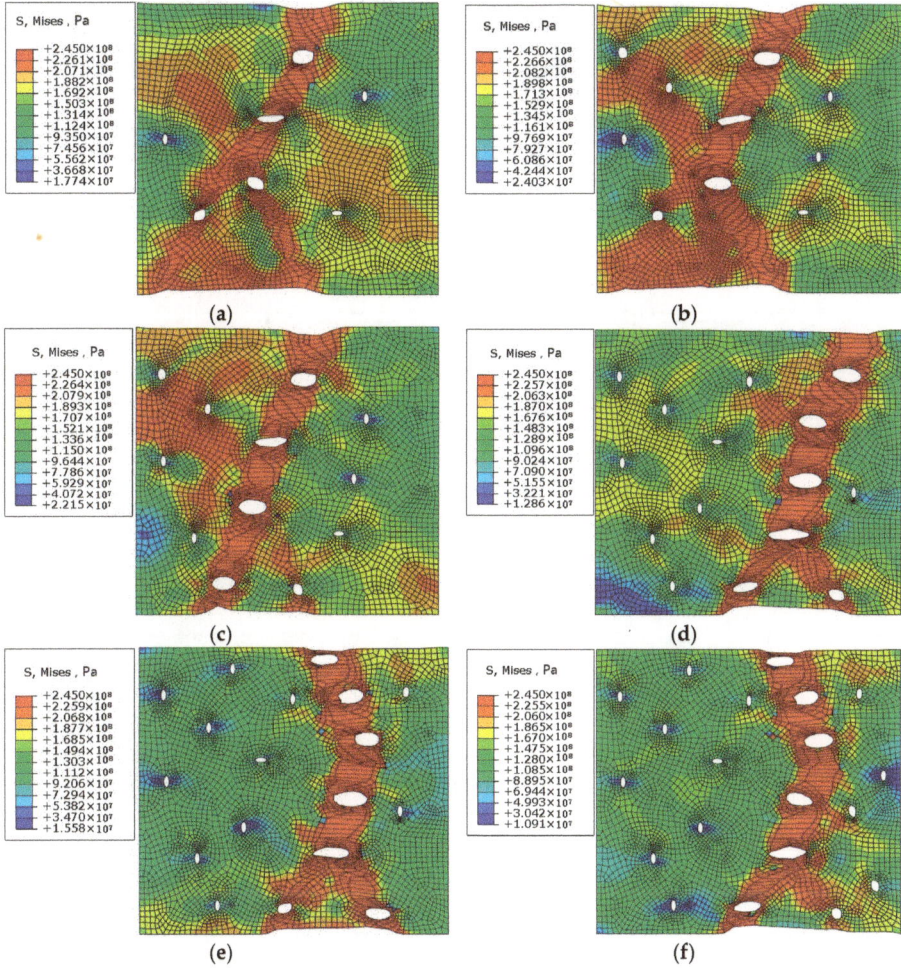

Figure 17. Stress distribution diagram of a polycrystalline model with different porosity of (**a**) 0.40%, (**b**) 0.60%, (**c**) 0.75%, (**d**) 0.90%, (**e**) 1.10%, and (**f**) 1.30%.

Figure 18 shows the strain results of polycrystalline models when the porosity is 0.4% and 1.3%, respectively. It can be found from the variation diagram of strain that the plastic strain is more concentrated around the pores than the matrix, and the polycrystalline model with a porosity of 1.3% is significantly larger than the polycrystalline model with a porosity of 0.4%, which indicates that the effects of different porosity on metal damage are different. At the same time, under different porosity, the pores begin to propagate and aggregate in the direction perpendicular to the tensile stress. The main reason for this phenomenon is that in the direction perpendicular to the tensile stress, the stress concentration occurs at the end of the pores, which resulting in larger equivalent plastic strain around the pores, and the localization of the strain results in continuous aggregation and propagation of the pores, so the crack always starts from this direction.

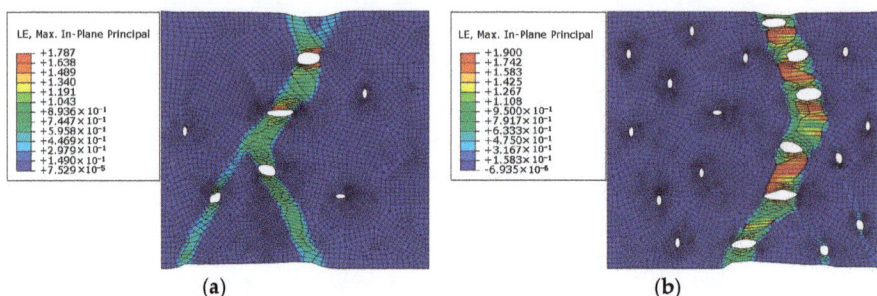

(a)   (b)

Figure 18. Tensile strain diagram with porosity of (**a**) 0.4% and (**b**) 1.3%.

3.4. Analysis of Polycrystal Model with Different Proportion Inclusions

The size and number of inclusions in steel is not only an important condition to evaluate steel quality, but also one of the main factors causing defects, and is the unavoidable existence in steel materials. Because the properties of the inclusions are different from the properties of the steel matrix materials, it has a great influence on the deformation and fracture process of the steel materials. In this paper, the stress and strain behavior of polycrystalline model with different percentage of inclusions is simulated to analyze its effect on material failure.

On the polycrystalline model parts, different proportion areas are divided as inclusions, and then inclusions material properties are given in the property module. Because it is a hard phase relative to the matrix of the polycrystalline model, only elastic properties are assigned to it. In this method, models with 0.44%, 0.57%, 0.70%, and 0.82% percentage of impurities were created respectively, and the same boundary conditions were set for simulation. Figure 18 shows the stress distribution in matrix with inclusions. As can be seen from the figure, the stress mainly concentrates at the inclusions, and the stress at the inclusions is much larger than that at the matrix. The reason for this phenomenon is that the elasticity and plasticity differences between impurities and the matrix are relatively large in the polycrystalline model. When the polycrystalline model is under stress, impurities can hardly meet the deformation requirements when the matrix deforms. Therefore, the impurities cannot deform simultaneously with the matrix, as a result, the stress accumulates in and around impurities, making the stress distribution at the impurities more obvious. Combined with Figures 17 and 19, it can be seen that the influence of pores on the crack is greater than that of inclusions.

As can be seen from the strain of a polycrystalline model of different impurity percentages in Figure 20, the deformation of the matrix of the polycrystalline model is much different from that of the impurities; the deformation of the impurities is much less than that of the matrix. This is because of the large differences in the elasticity and plasticity of the impurities and polycrystalline matrix. After the loading is applied, the impurity cannot deform because of its properties; thus the stress concentration will occur around the impurity and the stress value will be increasingly greater, resulting in the separation of the impurity and the matrix, and thus a hole. Therefore, when a metal material is subjected to a large external force, the crack will often expand along the direction of the impurities, resulting in the material damage and fracture.

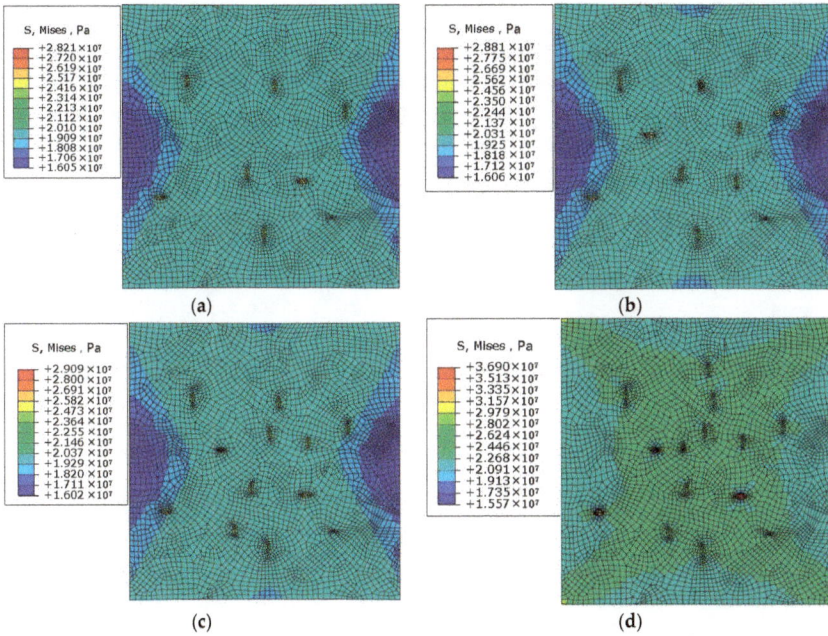

Figure 19. Stress distribution diagram of a polycrystalline model with different impurity percentages of (**a**) 0.44%, (**b**) 0.57%, (**c**) 0.70%, and (**d**) 0.82%.

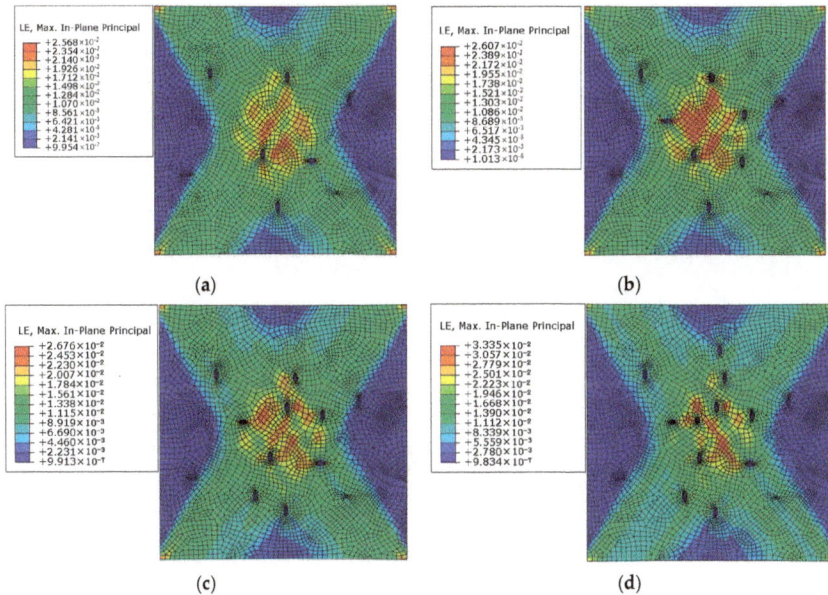

Figure 20. Strain distribution diagram of a polycrystalline model with different impurity percentages of (**a**) 0.44%, (**b**) 0.57%, (**c**) 0.70%, and (**d**) 0.82%.

4. Conclusions

In this study, a polycrystalline model of a beam blank was established using MATLAB. Based on mesoscopic damage mechanics, the full implicit stress iterative algorithm was used to simulate the crack propagation of the polycrystalline model using ABAQUS software, as well as a stress and strain simulation of a polycrystalline model with pores and impurities. The conclusions are as follows:

1. The stress at the crystal boundary is much greater than that in the crystal after loading on the defect-free polycrystalline model.
2. Crack propagation occurs at the crystal boundary prior to that in the crystal, and the propagation of the former is greater than that of the latter.
3. The direction of crack propagation is basically along the direction perpendicular to the tensile stress, and the amount of crack propagation along this direction is also the greatest.
4. When a polycrystalline model with pores is subjected to tensile stress, stress concentration occurs at the position where the end of pores is perpendicular to the stress direction, and the propagation and aggregation direction of the pores are basically perpendicular to the direction of the tensile stress. When the pores ends are parallel to the stress direction, the stress values at these locations are relatively small.
5. When a polycrystalline model with impurities is subjected to a force, a large stress concentration will occur at the impurities, while the strain generated by the impurities is the smallest. This often causes the crack to propagate along the direction of the impurities, thus causing damage to and fracturing of the material.
6. Both inclusions and pores affect crack initiation and propagation. Through mesoscopic simulation, it was found that the influence of pores on cracks is greater than that of inclusions.
7. The simulation of mesoscopic crack initiation and propagation of a beam blank can provide theoretical guidance for the generation of macrocracks and process optimization.

Author Contributions: W.C. and L.Z. conceived and designed the experiments and research ideas; G.Y., G.G., and B.H. performed the experiments, the simulations, the early stage of the investigation, and the data collection; W.C. and G.Y. analyzed the data and graphics; W.C. and L.Z. supervised the whole work; G.Y. wrote and revised the paper.

Funding: This research was funded by the National Natural Science Foundation of China (51574103 and 51574106).

Acknowledgments: The authors would like to thank the technical support of Hebei Engineering Research Center of High Quality Steel Continuous Casting.

Conflicts of Interest: The authors declare no conflict of interest.

References

1. Xu, M.G.; Zhu, M.Y. Transport Phenomena in a Beam-Blank Continuous Casting Mold and a New Design of Submerged Entry Nozzle. *ISIJ Int.* **2015**, *55*, 791–798. [CrossRef]
2. Xu, H.L.; Wen, G.H.; Sun, W.; Wang, K.Z.; Yan, B. Analysis of thermal behavior for beam blank continuous casting mold. *J. Iron Steel Res. Int.* **2010**, *17*, 17–22. [CrossRef]
3. Lee, J.E.; Yeo, T.J.; Kyu Hwan, O.H.; Yoon, J.K.; Yoon, U.S. Prediction of cracks in continuously cast steel beam blank through fully coupled analysis of fluid flow, heat transfer, and deformation behavior of a solidifying shell. *Metall. Mater. Trans. A* **2000**, *31*, 225–237. [CrossRef]
4. Xu, H.L. Simulation and Optimization of Cooling Process for Continuous Casting Beam Blank. Ph.D. Thesis, Chongqing University, Chongqing, China, 2010.
5. Yang, G.Y.; Zhu, L.G.; Chen, W.; Yu, X.W.; He, B.M. Initiation of Surface Cracks on Beam Blank in the Mold during Continuous Casting. *Metals* **2018**, *8*, 712. [CrossRef]
6. Lv, M.; Lu, B.; Wang, X.X.; Shan, Z.G.; Zhang, Q.; Chen, Y.S. Research and control on web plate crack of the large type beam blank. *Iron Steel* **2010**, *45*, 99–102. [CrossRef]
7. Wu, J.; Xu, W.; Yang, Y.D. Technics study of reduing longitudinal surface crack of beam blank. *Iron Steel* **2009**, *45*, 95–97. [CrossRef]

8. Xu, H.L.; Wen, G.H.; Sun, W.; Wang, K.Z.; Yan, B.; Luo, W. Thermal behaviour of moulds with different water channels and their influence on quality in continuous casting of beam blanks. *Ironmak. Steelmak.* **2013**, *37*, 380–386. [CrossRef]

9. Du, Y.P.; Yang, J.W.; Shi, R.; Cui, X.C. Effect of submerged entry nozzle (sen) parameters and shape on 3-d fluid flow in mould for beam blank continuous casting. *Acta Metall. Sin.* **2004**, *17*, 705–712.

10. De Santis, M.; Cristallini, A.; Rinaldi, M.; Sgrò, A. Modelling-based innovative feeding strategy for beam blanks mould casting aimed at as-cast surface quality improvement. *ISIJ Int.* **2014**, *54*, 496–503. [CrossRef]

11. Lasko, G.; Weber, U.; Schmauder, S. Finite Element Simulations of Crack Propagation in Al_2O_3/6061 Al Composites. *Acta Metall. Sin.* **2014**, *27*, 853–861. [CrossRef]

12. Jiang, Z.W.; Wan, S.; Cheng, C. Analysis of the Crack Propagation Based on Extended Finite Method. *Appl. Mech. Mater.* **2013**, *275–277*, 169–173. [CrossRef]

13. Wang, G.X.; Schwalbe, K.H. A study of the transition from intercrystalline to transcrystalline fatigue crack propagation in different ageing conditions of the alloy Cu-35%Ni-3.5%Cr. *Int. J. Fatigue* **1993**, *15*, 3–8. [CrossRef]

14. Shen, J.L. Study on Simulation of Surface Cracks Growth for Continuous Casting Q235 Slab. Mater's Thesis, Chongqing University, Chongqing, China, 2013.

15. Xue, F.M.; Li, F.G.; Li, J.; He, M.; Yuan, Z.W.; Wang, R.T. Numerical modeling crack propagation of sheet metal forming based on stress state parameters using xfem method. *Comput. Mater. Sci.* **2013**, *69*, 311–326. [CrossRef]

16. Wang, F.G. Crack Initiatiation and Propagation of Duplex Stainless Steel Finite Element Simulation. Master's Thesis, Xi'an Technological University, Xi'an, China, 2014.

17. Chen, L.Y.; Zhu, Y.C. Cause for bubbles occurring in top CC slab & countermeasures. *Contin. Cast.* **1999**, *2*, 19–20. [CrossRef]

18. Geng, M.S.; Wang, X.H.; Zhang, J.M.; Wang, W.J.; Liu., Z.M. Study of Surface Blow Hole Defects of Continuous Casting Slab. *Iron Steel* **2010**, *45*, 45–50. [CrossRef]

19. Yang, G.Y.; Zhu, L.G.; Chen, W.; Yu, X.W.; Guo, G.X. Propagation of Surface Cracks on Beam Blank in the Mould during Continuous Casting. *Ironmak. Steelmak.* **2018**. [CrossRef]

![metals logo] *metals*

MDPI

Article

Applied Mathematical Modelling of Continuous Casting Processes: A Review

Michael Vynnycky [†]

Division of Processes, Department of Materials Science and Engineering, Brinellvägen 23, KTH Royal Institute of Technology, 100 44 Stockholm, Sweden; michaelv@kth.se; Tel.: +46-73-765-2037

† Current address: Department of Applied Mathematics and Statistics, Institute of Mathematical and Computer Sciences, University of São Paulo at São Carlos, P.O. Box 668, São Carlos 13560-970, São Paulo, Brazil.

Received: 21 October 2018; Accepted: 2 November 2018; Published: 9 November 2018

Abstract: With readily available and ever-increasing computational resources, the modelling of continuous casting processes—mainly for steel, but also for copper and aluminium alloys—has predominantly focused on large-scale numerical simulation. Whilst there is certainly a need for this type of modelling, this paper highlights an alternative approach more grounded in applied mathematics, which lies between overly simplified analytical models and multi-dimensional simulations. In this approach, the governing equations are nondimensionalized and systematically simplified to obtain a formulation which is numerically much cheaper to compute, yet does not sacrifice any of the physics that was present in the original problem; in addition, the results should agree also quantitatively with those of the original model. This approach is well-suited to the modelling of continuous casting processes, which often involve the interaction of complex multiphysics. Recent examples involving mould taper, oscillation-mark formation, solidification shrinkage-induced macrosegregation and electromagnetic stirring are considered, as are the possibilities for the modelling of exudation, columnar-to-equiaxed transition, V-segregation, centreline porosity and mechanical soft reduction.

Keywords: asymptotic analysis; numerical simulation; continuous casting

---

1. Introduction

Continuous casting is a process whereby molten metal is solidified into semi-finished billets, blooms, slabs or strips for subsequent rolling in finishing mills; it is the most frequently used process to cast not only steel, but also aluminum and copper alloys, with the main configurations being as shown in Figure 1. Although originally a process that was developed on an industrial scale in the early decades of the 20th century for the casting of non-ferrous alloys, its use has been widespread also for steel since the 1950s. Compared to casting in moulds, continuous casting is more economical, as it consumes less energy and produces less scrap. Furthermore, the properties of the products can be easily modified by changing the casting parameters. As all operations can be automated and controlled, continuous casting offers numerous possibilities to adapt production flexibly and rapidly to changing market requirements and to combine it with digitization technologies. Nevertheless, new challenges continuously arise, as ways are sought to minimize casting defects and to cast new alloys.

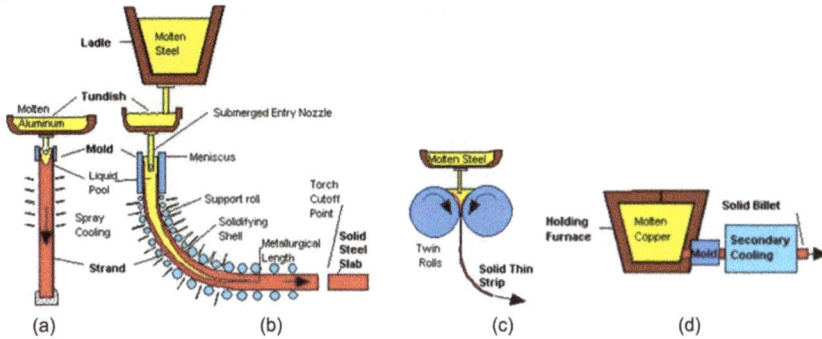

Figure 1. Continuous casting: (**a**) vertical; (**b**) curved; (**c**) strip; and (**d**) horizontal. Reproduced with permission from Brian G. Thomas, 2018 [1].

An important aspect in meeting these challenges is the use of modelling and simulation, as reviewed recently for the case of steel in [2]. With readily available and ever-increasing computational resources, the majority of modelling and simulation activity for continuous casting processes focuses on large-scale numerical simulation. Whilst there is certainly a need for this type of modelling, the focus in this paper is on an alternative approach more grounded in applied mathematics, asymptotic methods in particular, which lies between overly simplified analytical models and multi-dimensional simulations. In this approach, the governing equations are nondimensionalized and systematically simplified to obtain a formulation which is numerically much cheaper to compute, yet does not sacrifice any of the physics that was present in the original problem; in addition, the results should agree also quantitatively with those of the original model. This approach is well-suited to the modelling of continuous casting processes, which often involve the interaction of complex multiphysics.

There are several motivations for this hybrid asymptotic/numerical approach. The first is that it has been applied successfully in other areas of science and technology, as witnessed by the growth of an activity known as practical asymptotics [3–8]; note, however, that this is not identical to the model order-reduction approach [9], which is also prevalent in industrial mathematics. The second is the Moore's law effect in modelling [10]: namely, that, although computational power doubles every 18 months, it will still be many years before all the length scales in a casting process will be numerically resolved [11]. Moreover, although the use of asymptotic methods for the modelling of continuous casting is not unknown, it remains somewhat limited to fluid flow and heat transfer aspects [12–20]. Thus, the goal here is to review recent activity in this area, based primarily on the activities of the author and co-workers [21–33].

The layout of the paper is as follows. Section 2 focuses on models for the determination of metallurgical length. Section 3 is on the role of the air gap and implementation of mould taper. Section 4 focuses on oscillation-mark formation in the continuous casting of steel, whereas Section 5 is on solidification shrinkage-induced macrosegregation. Section 6 is on electromagnetic stirring, with conclusions being drawn in Section 7. Finally, Appendix A shows in more detail how the practical asymptotics approach is applied to one of the sub-problems considered in the paper.

2. Metallurgical Length

2.1. Pure Metals or Eutectic Alloys

To be able to correctly dimension a continuous casting process, it is of primary importance to have an estimate of where complete solidification will occur. This requires a model that takes into account fluid flow, heat transfer and phase change; a generalized two-dimensional (2D) schematic for this is shown in Figure 2. The inlet drawn here does not correctly reflect the situation in all continuous casting processes. For example, for the continuous casting of steel blooms or billets, a submerged entry nozzle is used, whereas, for the strip casting of copper and its alloys, multiple jets are located above the molten metal surface. For the latter, a 2D model was considered in [34,35] which employed the k-ε model to describe the turbulent flow of molten metal; in addition, experimental measurements were carried out to obtain data necessary for modelling the heat transfer between the solidified copper shell and the surrounding cooling mould. The resulting equations were solved using the commercial software CFX [36]. In much later work [21], the original model equations were reconsidered in the light of a much simpler model which simply took account of the streaming of the melt at the casting speed, V_{cast}; through asymptotic reduction, the original 2D time-independent model becomes effectively a 1D time-dependent, model, with the usual understanding that the coordinate in the casting direction corresponds to the product of V_{cast} and time, if the casting geometry is slender. Moreover, since the original model was for an almost pure copper melt, the simpler model was formulated in terms of a sharp interface between liquid and solid phases [21]; this approach would also be suitable for eutectic alloys.

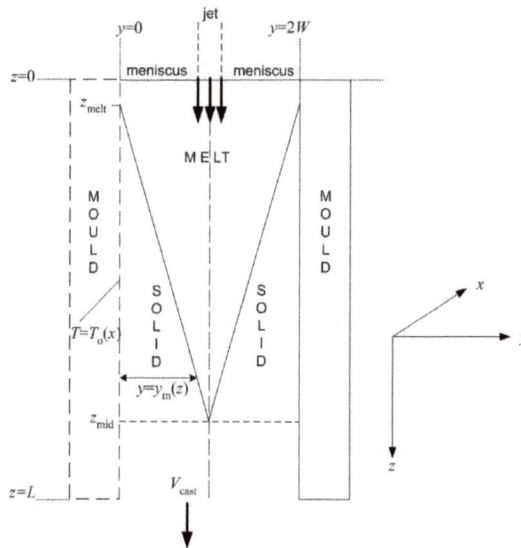

Figure 2. Schematic of a vertical continuous casting process.

Figure 3 shows a comparison of the key results from [21,34,35]. Thus, although significantly more computational effort was required in generating the results in [34,35], there is scant difference between them. Whilst this may be due to the limited effects of turbulence lower down in the caster, it is notable that all quantities agree well even near the top, where the effects of turbulence should still be quite strong.

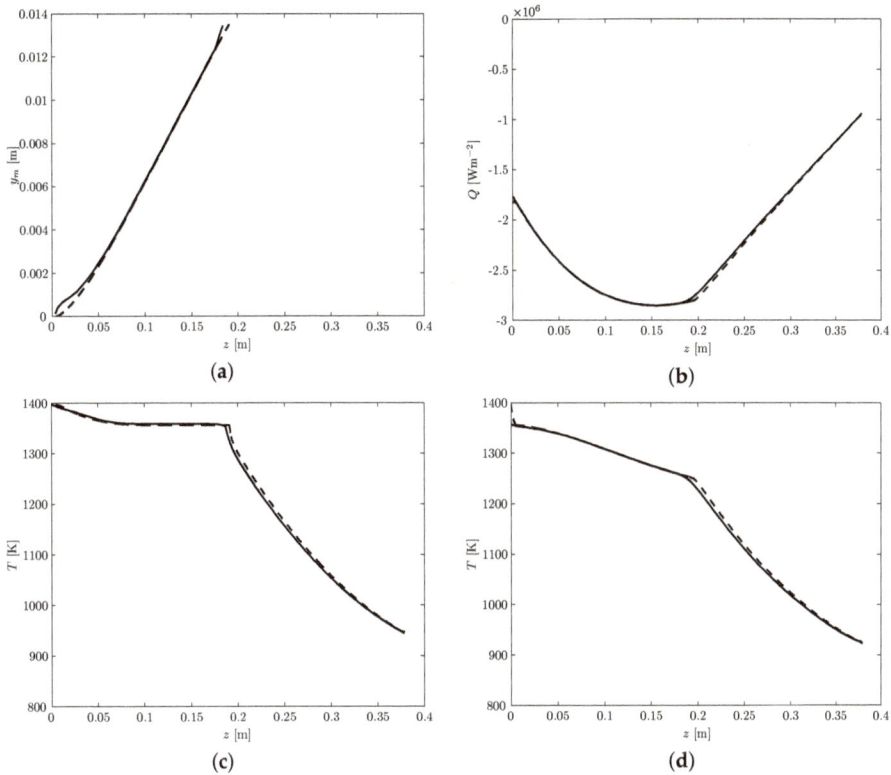

Figure 3. Comparison of: (**a**) the location of the solidification front, y_m; (**b**) the heat flux, Q, at the outer edge of the copper strip; (**c**) the temperature at the centreline ($y = W$); and (**d**) the temperature at the outer edge of the copper strip. In all cases, the solid line is for the full model, and the dashed line is for the reduced model; z denotes the distance along the casting direction from the meniscus. Reproduced from [21], with copyright permission from Elsevier, 2017.

2.2. Alloys

In general of course, the alloys that are cast have a substantial solidification interval, leading to the formation of a mushy zone between the melt and the solid, in which both coexist; thus, the situation is now more akin to Figure 4 than Figure 2. Typically, the numerical solution of the governing equations is handled by introducing an auxiliary variable, commonly the local liquid fraction, and using an enthalpy formulation on a fixed grid [37–41]. This generally functions well enough if a basic description of fluid flow and heat transfer is required—all the more so if the solidification interval of the alloy is large. However, it is often required to determine quantities that affect the quality of the final solidified alloy, such as the degree of macrosegregation of an alloy's solute elements, the ratio of equiaxed-to-columnar crystals or the thermomechanical stress; all of these are associated with processes in the mush. Thus, to compute all of these quantities accurately in the region where they really matter, it would be convenient to be able to resolve the locations of the solid-mush and mush-liquid interfaces explicitly.

The methodology to address this situation for the case of continuous casting processes was developed in [22]. This was done by applying a boundary immobilization method for the solidus and liquidus isotherms, transforming from (y, z) to (η, z) variables, as shown in Figure 5, with η given by

$$
\eta = \begin{cases} y/y_s\,(z)\,, & 0 \le y \le y_s\,(z) \\ 1 + \frac{y - y_s(z)}{y_l(z) - y_s(z)}, & y_s\,(z) \le y \le y_l\,(z) \\ 2 + \frac{y - y_l(z)}{W - y_l(z)}, & y_l\,(z) \le y \le W \end{cases} ,
$$

where y_l and y_s denote the locations of the liquidus and solidus isotherms, respectively. A benchmark problem was solved using three different formulations:

(**A**)   an enthalpy-like formulation just mentioned using the full 2D time-independent equations;
(**B**)   an enthalpy-like formulation using a reduction of the full 2D time-independent equations to a 1D transient-like formulation of the type used in [21];
(**C**)   the proposed new formulation, which is also 1D transient-like but resolves the locations of the solidus and liquidus isotherms explicitly.

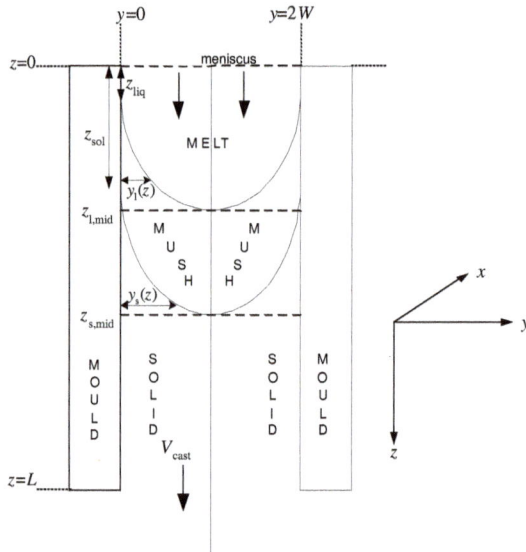

Figure 4. 2D schematic of the vertical continuous casting of an alloy.

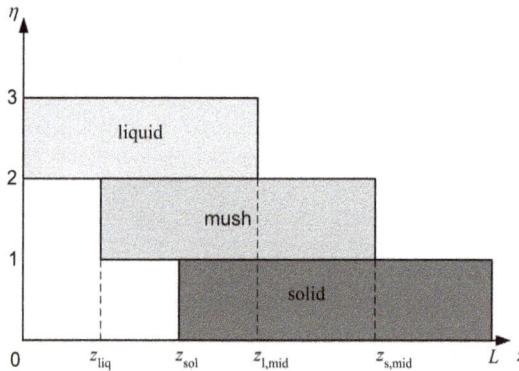

Figure 5. Computational domain in (η, z) variables.

The key results are summarized in Figure 6. Figure 6a–f shows, respectively, the location of the liquidus isotherm (y_l), the location of the solidus isotherm (y_s), the mould wall temperature, the heat flux at the mould wall, the centreline temperature and the temperature gradient in the casting direction at the centreline, and indicates that the novel formulation *C* gives results that agree very well with those of the more conventional formulations *A* and *B*. In future, it is therefore hoped to extend this formulation for modelling the multiphysical phenomena that are present in the mushy zone, as mentioned above.

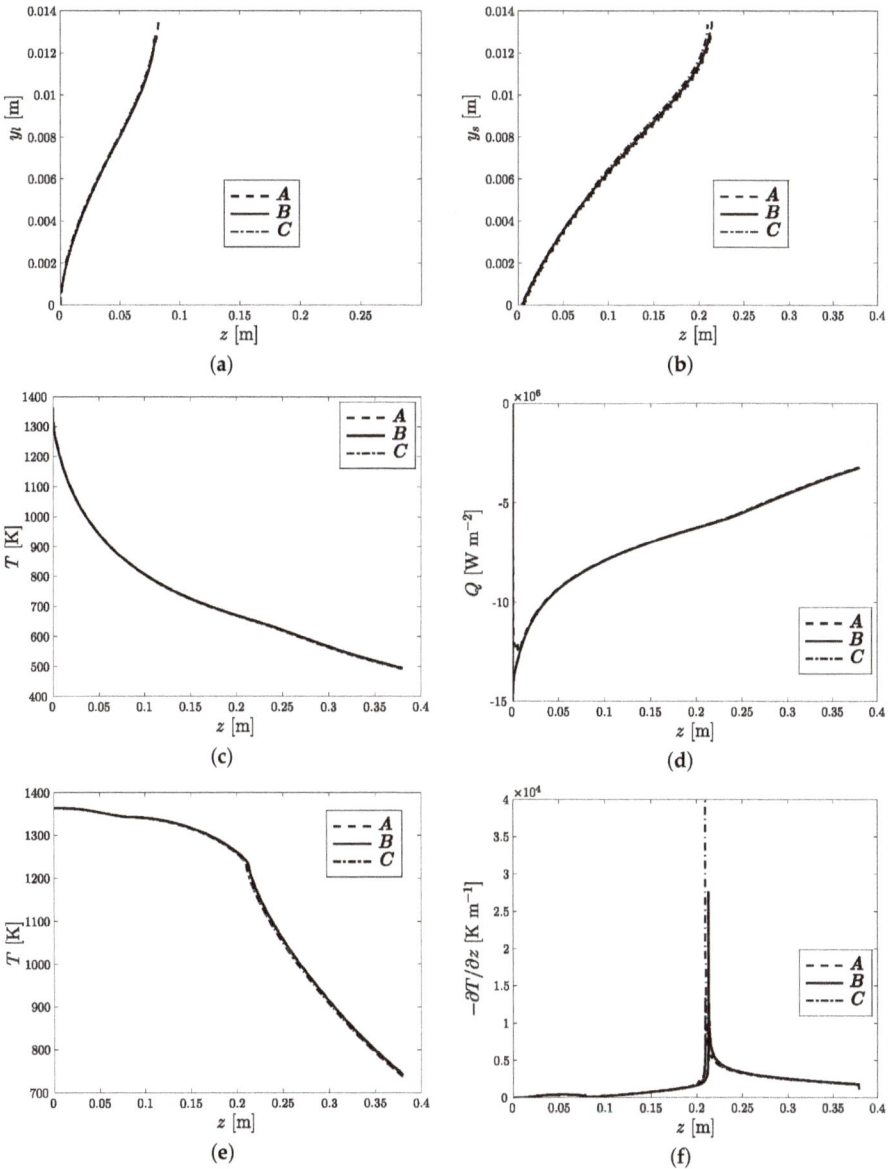

Figure 6. (a) The location of the liquidus isotherm, y_l; (b) the location of the solidus isotherm, y_s; (c) mould wall temperature; (d) mould wall heat flux, Q; (e) centreline temperature; and (f) $-\partial T/\partial z$ at the centreline. All quantities are given as functions of z. Reproduced from [22], with copyright permission from Elsevier, 2017.

3. Air Gap and Mould Taper

In the models presented thus far, heat transfer between the solidified shell and the mould wall is characterized by an experimentally measured heat transfer coefficient and the surface of the mould wall is assumed to be parallel to the casting direction. In practice, however, it is thought that an air gap forms between the mould wall and the solidified shell when the latter, in going from a plastic to elastic state, is strong enough to withstand the metallostatic pressure of the adjacent molten metal, thereby receding from the mould as a result of contraction; a contributing factor is also thought to be the expansion of the mould itself. Air-gap formation prohibits effective heat transfer between the mould and shell, leading to longer solidification lengths and requiring supplementary process design considerations, such as mould tapering. Consequently, the actual situation is more similar to that shown in Figure 7, which shows the idea for the case of the casting of pure metal or eutectic alloy. Now, in addition to determining the location of the solid/melt interface, it is also necessary to determine the location where the air gap begins to form, denoted by z_{gap}, as well as the width of the air gap as a function of z, given in the figure as $r_w(z) - r_a(z)$.

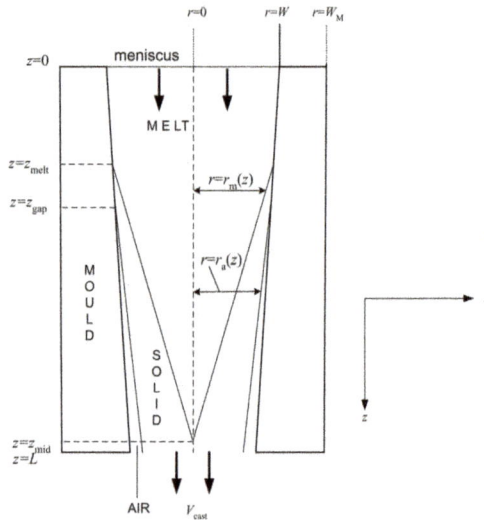

Figure 7. A 2D schematic of the continuous casting process with tapered mould walls and superheat.

Early analytical models made preliminary headway on the air gap problem [42–45]. Later papers have been solely numerical studies [46–52]. In the latter, the common approach is to employ multidimensional finite element models to describe the coupled thermomechanical interaction that occurs between the solidified shell and the mould wall. However, in contrast to all of these, in [24–27], the fact that the air gap is slender was used, together with the generalized plane strain approximation, to derive an asymptotic model for the aforementioned interaction in the case of an untapered mould; in particular, closed-form expressions can be found for the stress and strain components. Ultimately, the only numerical burden is the computation of a moving boundary problem for the temperature, although with back-coupling to the structure mechanical problem via the boundary conditions; the case of a tapered mould fits easily into this framework, since the taper is always small enough to be amenable to asymptotic methods: for example, in the case of steel casters, the taper is usually of the order of 2%/m [47].

In [28,29], the interplay of melt superheat and mould taper was considered for an axisymmetric continuous casting geometry, although we just focus on the effect of mould taper itself for one particular value of superheat here. Figure 8a shows the different taper profiles considered; here, M is a dimensionless parameter related to the actual mould taper, in per cent/m, by

$$M = \frac{LX}{100\alpha\Delta T}$$

where

$$X = \frac{100}{L}\left(\frac{r_w(0) - r_w(L)}{r_w(0)}\right),$$

where α is the thermal expansion coefficient of the metal and ΔT is an appropriate temperature scale for the problem (see [28,29] for exact details). Figure 8b–d shows, respectively, how the position of the solid/air interface (r_m), the position of the solid/air interface (r_a) and the width of the air gap, $a_g := r_w(z) - r_a(z)$, vary with z for different values of M. Lastly, Figure 9 gives an idealized result: the minimum taper necessary to avoid any air gap at all, with Figure 8b also showing the profile for r_m, denoted there by $\bar{\rho}_w$, that this would correspond to. However, in Figure 9, a taper profile that is much more extreme than those currently in use would be required.

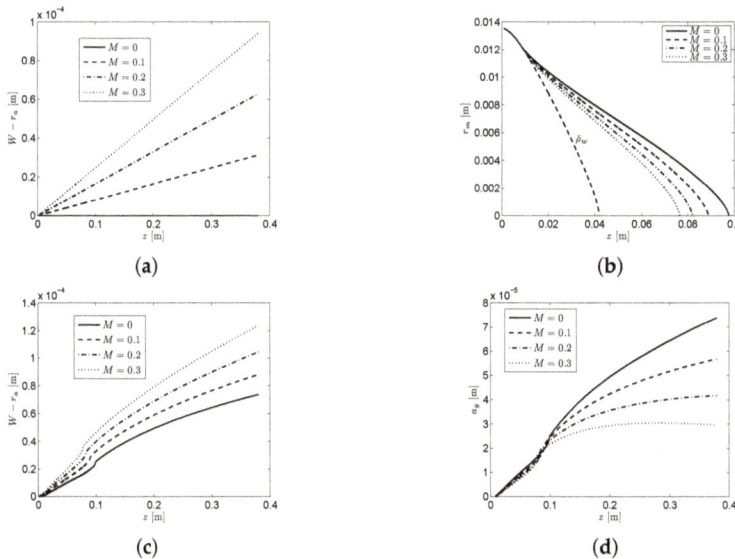

Figure 8. Dependence on M of: (**a**) the taper profile, $r_w(z)$, relative to W; (**b**) the position of the solid/air interface, $r_m(z)$; (**c**) the position of the solid/air interface, $r_a(z)$; and (**d**) the width of the air gap, $a_g(z)$.

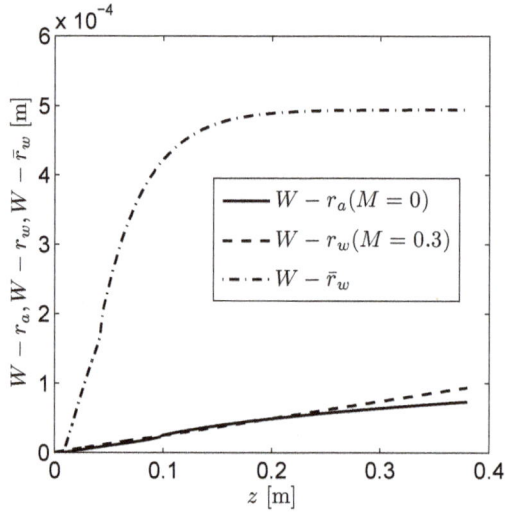

Figure 9. Comparison of the size of the air gap for the untapered system (solid line) with the $M = 0.3$ linear taper (dotted line), and the ideal taper required to completely eliminate the air gap (dashed line).

Finally, we note that, although the results in Figures 8 and 9 are for a somewhat fictitious situation, work is nevertheless underway to extend the results to the more realistic case of the continuous casting of round steel billets; a good source for validating the approach will be the experimental and theoretical results in [47]. A further future extension would then be for mould taper for blooms, billets and slabs, in which the generalized plane strain approximation can also be used, as well as the fact that the air gap is slender and the taper is small enough for asymptotic methods to be applicable.

4. Oscillation-Mark Formation

Although the continuous casting configurations shown in Figures 2, 4 and 7 indicate a mould that is stationary, in reality, it is made to oscillate in the casting direction; this is particularly so for the continuous casting of steel. Moreover, a mould powder, often termed flux, is introduced over the steel melt; the powder becomes molten and the flows between the steel and the oscillating mould, with the resulting configuration being as in Figure 10, which depicts the initial stages of solidification in continuous casting. In this way, solidified steel is prevented from contacting the mould directly and resolidified flux is shaken from the surface of the mould. However, as is well-known and hinted at in Figure 10, a detrimental effect of this construct is the formation of depressions on the surface of the solidified steel, commonly referred to as oscillation marks; an experimental example is shown in Figure 11. Moreover, although the mould oscillation is normally periodic in time, this does not in general mean that the marks which form are identical and periodically spaced. For example, Figure 12 shows the spacing between thirteen marks, also termed as the pitch, taken from a continuously cast steel sample, from which it is clear that, more often than not, the pitch obtained was close to the average value. This may in itself not be noteworthy, other than that this average value is also close to the theoretical value, as we show below. Further samples of experimental oscillation-mark data, arguably the largest ever collected from a single continuous casting process, can be found in [53,54].

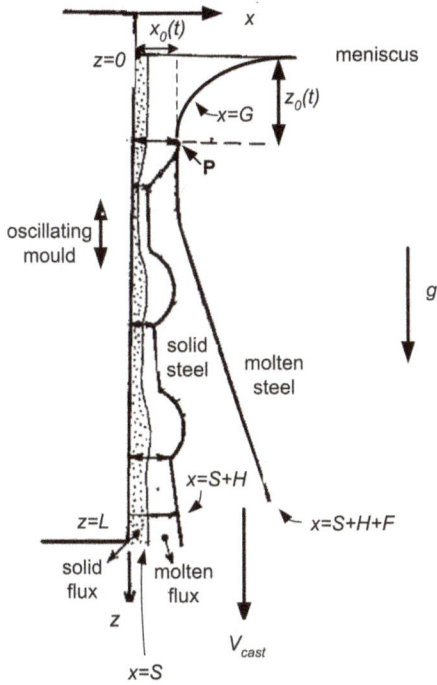

Figure 10. A sketch showing how an oscillation mark forms.

Figure 11. Two adjacent oscillation marks and typically observed microstructure underneath them.

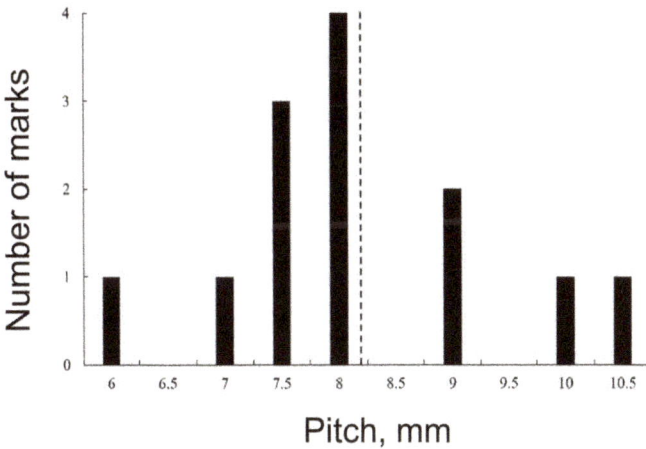

Figure 12. Experimental results showing the number of marks associated with a given pitch, taken from [30]. The vertical dashed line shows the average value.

Oscillation marks have been subject of modelling in the continuous casting literature since the early 1980s [55], with the most recent review being given in [56]. At the centre of this modelling is the need to relate it to the early experimental work of Tomono [57], which classified oscillation marks as being either of overflow-type, in which case the solidified shell is strong enough to avoid deformation, causing the steel meniscus to overflow the solid tip, or fold-type, in which case the solid shell is too thin to prevent its tip from bending back under the rim pressure. Recent attempts at modelling this problem have tended towards the use of computational fluid dynamics [56,58–61], although results which give either fold or overflow marks remain elusive, as is a criterion, in terms of process parameters, for when one or the other should occur.

In this context, Vynnycky et al. [30] recently revisited an earlier model for oscillation-mark formation by Hill et al. [18], with a view to providing a detailed and systematic asymptotic analysis; the model used lubrication theory coupled to heat conduction in the solid flux, molten flux and solid steel regions, and was able to predict the oscillation-mark shape. This resulted in a model involving fifteen dimensionless parameters which was able to give good agreement with the experimental oscillation-mark profiles shown in Figure 11; the comparison is shown in Figure 13. Note that the pitch in this figure is given by V_{cast}/f, where f is the mould oscillation frequency, and that this is close to the dashed line in Figure 11 for the values $V_{cast} = 0.013$ ms^{-1} and $f = 19/12$ s^{-1} used in [30,53]. Furthermore, the profiles and the character of the model, which neglects the presence of the meniscus completely, indicated that the marks were of fold-type.

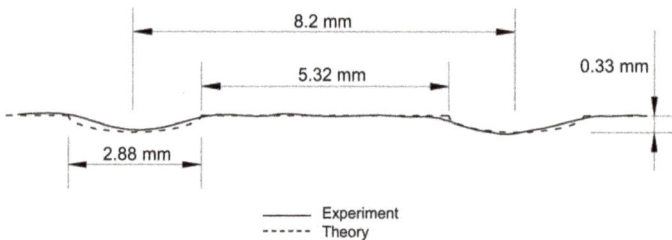

Figure 13. Comparison of the theoretically calculated oscillation mark profiles and those measured experimentally by Saleem [53].

In further work [32], an attempt was made to extend the model to include the effect of the meniscus, thereby allowing the possibility of model parameters to drive whether the model predicts fold-type or overflow-type marks. For this purpose, a novel "moving-point" formulation was derived, which unlike the volume-of-fluid (VOF) formulation adopted in [56,58–61], attempts to track explicitly the location of the point at which solidification first begins. This approach suggests that if solidification has not occurred within a capillary length from the top of the meniscus, then the meniscus profile predicted by Bikerman [62], and which emerges naturally in the asymptotic analysis in [32], will collapse, leading to overflow; however, further work is required to extend the isothermal model derived in [32] to determine where the temperature first becomes low enough for the steel to start to solidify.

5. Macrosegregation

Macrosegregation is the term used to denote to variations in composition that occur in alloy castings or ingots and range in scale from several millimetres to even metres; it is a central problem, since it strongly influences the further workability of the cast products and their mechanical properties. As was already well-established as early as the 1960s [63,64], it arises as a consequence of the nature of the solidification process for alloys, which involves the formation of a mushy network of solid dendrites through which there is the slow flow of interdendritic melt, and the transport of alloying elements. In particular, as solidification occurs, if the solute—for example, Cu in an Al-4.5 wt% Cu alloy [65] or Sn in a Cu-8 wt% Sn alloy [66]—is more soluble in the liquid phase than it is in the solid phase, then it is rejected into the melt, resulting in a non-uniform distribution of solute in the final solidified casting.

Over the years, there has developed a substantial body of modelling work on macrosegregation in one-dimensional transient solidification [67–72], some of which is of relevance to steady-state continuous or direct chill casting processes. In higher dimensions, computational fluid dynamics (CFD) is often used, although, as indicated in [73], numerical dispersion and diffusion are present in the simulated macrosegregation profiles reported in the literature, hindering the interpretation of CFD results. In particular, it has been found that unstructured computational meshes can eliminate the numerical dispersion that is present when structured meshes are used; however, undesirable numerical diffusion is introduced instead. On the other hand, refining a structured mesh alleviates problems with numerical oscillations, but instead results in a dramatic increase in computation time [74]. More details on recent efforts to come to grips with these difficulties can be found in [73,75–79].

One particular cause of macrosegregation in continuous casting is due to solidification shrinkage, which tends to dominate that due to natural convection induced by thermal and solutal gradients. The modelling of this phenomenon is particularly amenable to asymptotic methods since the macrosegregation is driven by a dimensionless parameter, $\varepsilon = \rho_s/\rho_l - 1$, that is often no greater than around 0.1; here, ρ_s and ρ_l are the solid and liquid densities, respectively. Consequently, the resulting non-uniform solute profile can be obtained as the second term in a regular perturbation series in ε. As a result, and as shown in [23], there is therefore no need to resort to CFD to compute the Darcy-damped Navier–Stokes equations for modelling the melt and the mushy zone. Although the analysis is algebraically somewhat lengthy, the computational load is not; moreover, because of the hybrid analytical-numerical method used, there is no possibility for any numerical diffusion or dispersion. The situation considered in [23] is similar to that in Figure 4, although with no superheat, so that there is only a mushy zone and a fully solidified region. Moreover, the lever rule is assumed for microsegregation at the microscale, and it is demonstrated analytically that the leading-order solute at the macroscale is also given by the lever rule.

The results of the approach were compared with those obtained in [80] using CFD for the continuous casting of an Al-4.5 wt% Cu alloy, and are shown in Figure 14. Several features are noteworthy in this figure. First, we see so-called inverse segregation at $y = 0$: this is the well-established phenomenon wherein the solute concentration is higher at the outer surface of the casting than it is elsewhere [65,74,81]. Moreover, near $y = W$, we would expect to see negative segregation, i.e., the concentration would be less than the initial composition. Note that two curves have been included from [80]: one using the linearized phase diagram for the Al-Cu system, and one using the non-linearized phase diagram. As can be seen, the agreement with the results of the asymptotic model is very reasonable for $y \leq 0.02$ m, although rather less so thereafter. This could be due to a variety of factors:

- the fact that the model in [23] does not include superheat, whereas the model in [80] had a superheat of 27 K;
- in [80], numerical issues associated with the use of CFD, of the type mentioned earlier; and
- the fact that the geometry in question, which has an aspect ratio of six, is not slender enough for the asymptotic approach to be valid.

It is clear that the method based on the asymptotic approach is far from complete, with the most urgent extensions being the inclusion of superheat and the possibility to assume Scheil equation- or back diffusion-based microsegregation at the microscale.

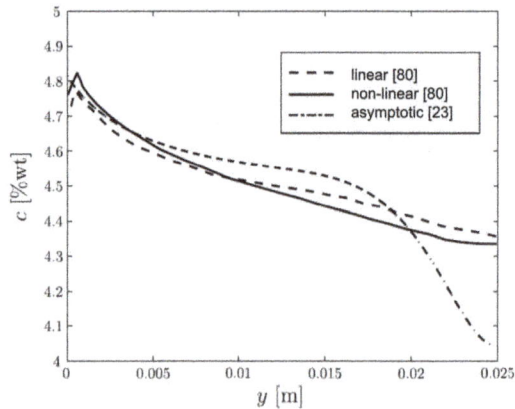

Figure 14. Comparison of the macrosegregation profiles obtained in [80], using the linearized and non-linearized phase diagrams for the Al-Cu system, and using the asymptotic model in [23].

6. Electromagnetic Stirring

Electromagnetic stirring (EMS) has been used in the continuous casting of steel [82] since the 1970s as a way to control solidification structures. Since roughly the same time, mathematical modelling has been used to elucidate the actual role of EMS in affecting the motion of the steel melt; most noteworthy is a sequence of papers by Schwerdtfeger and co-workers [83–88] which explore, both experimentally and theoretically, the effect of stirring in the round-billet, rectangular-bloom and slab geometries that are characteristic for the continuous casting of steel. In particular, the models considered consist of the turbulent Navier–Stokes equations for the velocity field of the molten metal and Maxwell's equations for the induced magnetic flux density. In principle, there is two-way coupling between the models, since the alternating magnetic field gives rise to a Lorentz force which drives the velocity field, which can in turn affect the magnetic field. Typically, the magnetic Reynolds number is rather low, and rarely greater than unity, meaning that the velocity-free Maxwell's equations can be solved; the output is then used to constitute the Lorentz force which drives the velocity field. Moreover, the frequency of the magnetic field is typically great enough to allow the use of the time average of the Lorentz force as input to the Navier–Stokes equations.

In calculating the induced magnetic field, an assumption is necessary as regards the applied oscillating field surrounding the domain of interest, typically the steel strand. In the models mentioned above [83–88], the assumption comes in the form of a boundary condition for the normal component of the magnetic flux density at the surface of the strand. However, revisiting the problem for the continuous casting of round billets [83,88], the configuration for which is shown in Figure 15a, Vynnycky [31] showed that prescribing the normal component of the magnetic flux density at the surface of the strand leads to a non-unique solution for the components of the magnetic flux density, and hence for the components of the time-averaged Lorentz force. On the other hand, it was found that prescribing the tangential component of the magnetic flux density would lead to a unique solution. Moreover, for the circular configuration, it was found that it was possible to choose the tangential component so that the original expressions for the components of the time-averaged Lorentz force, first given in [83,88], would be recovered. In addition, although the analysis was carried out for when the magnetic Reynolds number is small enough that there is only one-way coupling between the fluid flow to the magnetic field, the result concerning non-uniqueness would hold even when there is genuine two-way coupling.

At present, work is under way to re-evaluate the corresponding situation for rectangular strands [84], which is algebraically more complex. Nevertheless, a resolution of the issue is timely, since the expressions derived in [83,88] for the components of the time-averaged Lorentz force have been cited and used on numerous occasions since, even up to the present day [89–94]. Moreover, the results are even more significant in the case of modulated EMS [95–97], where magnetic fields of different frequencies are applied and it is the intention that the resulting Lorentz force should have a constant time-averaged and a time-varying component; in this case, posing the correct boundary conditions for the magnetic field is vital for obtaining meaningful results from modelling.

A further activity concerns obtaining a better idea of how the solid and mushy zones that form during continuous casting, giving the situation shown in Figure 15b, affect the ability of the applied magnetic field to stir the remaining melt [33]. Some preliminary results are given in Figure 16 for the case of round billets. For these results, the model parameters from [83,88] have been used. The axisymmetric Navier–Stokes equations have been solved, although a Darcy-like damping term has been included so as to take into account the effect of the mushy zone. In Figure 16a, the outer radius of the mushy zone, r_m, has been taken to be 60% of the radius of the mould; note that this value of r_m corresponds to the value used for the radius of the melt in [88]. Figure 16a compares solutions for the azimuthal velocity, v_θ, for three different values of the inner radius of the mushy zone, r_0— 0, $r_m/3$, $2r_m/3$—with the solution when the Darcy term is neglected completely, so that there is no mush. Here, the curve for the no-mush case corresponds to that computed in [88]. It turns out that the profile given here has a maximum value that is around six times lower than that in [88], which was around 1.5 m s^{-1}, as well as having its maximum displaced further away from $r = r_m$; it is not surprising that there is a difference, since the turbulence model used, which adopts the Prandtl mixing length hypothesis [98], was one of simplest possible, whereas a more sophisticated two-equation k-W model was used in [88]. Nevertheless, even the value obtained here is not unreasonable in the context of electromagnetically-stirred melts. As regards the other curves in this plot, it is clear that the presence of the mushy layer reduces v_θ significantly: for example, for $r_0 = 0$, corresponding to mush occupying the entire region from $r = 0$ to r_m, the maximum value of v_θ is around 0.02 m s^{-1}. For Figure 16b, $r_m = r_b$; here, the profile for the no-mush curve resembles much more closely that computed in [88], with even the maximum value for v_θ comparing favourably. There is also a proportionately greater drop in v_θ when going from the no-mush case to when $r_0 = 2r_m/3$.

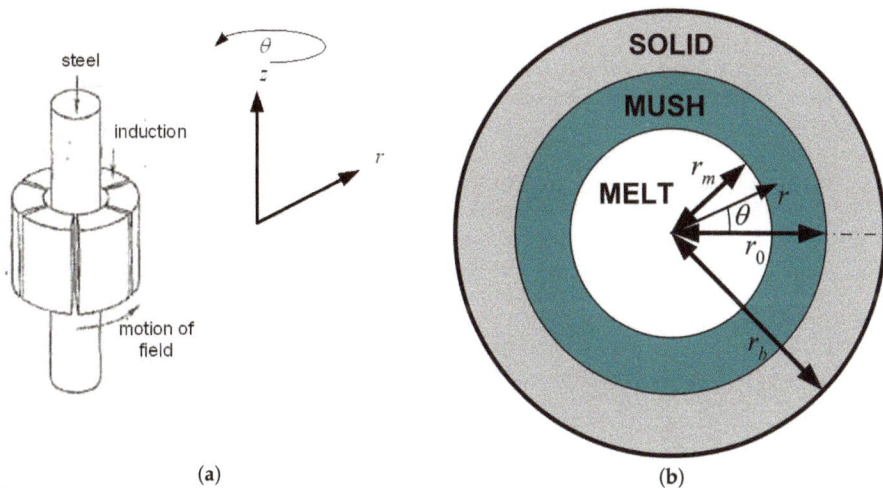

(a) (b)

Figure 15. (**a**) Schematic of an arrangement of an inductor around a circular steel strand for inducing rotating fields; and (**b**) schematic of circular solid, mush and melt regions in the cross-section of the strand.

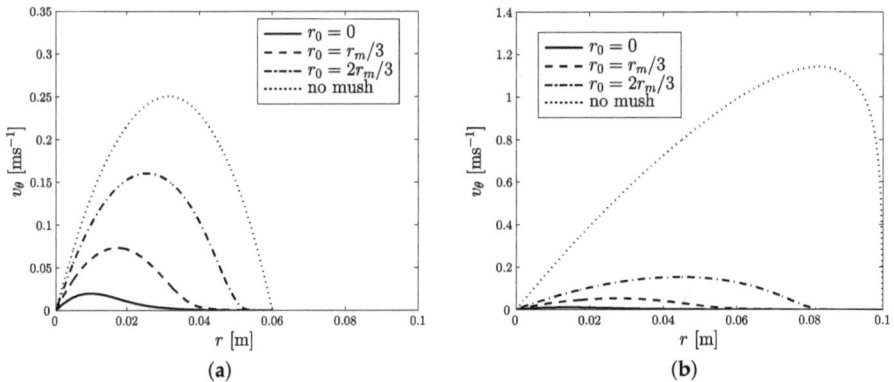

Figure 16. Azimuthal velocity, v_θ, vs. radial distance, r, for $r_0 = 0$, $r_m/3$, $2r_m/3$ and no mush, with: (a) $r_m = 0.06$ m; and (b) $r_m = 0.1$ m.

The model in [33] can be extended in a number of ways. First, although the casting velocity was not included in this analysis, it can be incorporated without affecting the analysis. After that, it would be of interest to include an equation for the conservation of heat, so that the temperature and hence the liquid fraction are computed as part of the model, rather than the liquid fraction simply being prescribed, as was the case in [33]. A nondimensional analysis would make it clear whether the effect of the magnetic field affects the heat transfer in the problem, through either Joule heating or convection. Once these models are in place, it should be possible to assess the role of stirring in macrosegregation and crystal structure formation: with respect to white-band formation in the case of the former [87,99,100], and columnar-to-equiaxed crystal transition in the case of the latter. A parallel line of activity would be to adapt these ideas to linear travelling magnetic fields for rectangular slabs, billets and blooms.

7. Conclusions

This paper has reviewed recent efforts in employing applied mathematical techniques, predominantly analysis and asymptotic methods, for the modelling of continuous casting processes. The focus has been on using these techniques in order not only to infer the qualitative behaviour of models, but also with a view to obtaining reduced-model formulations that require considerably less computational time than the original models, whilst still retaining their salient physical features. The work presented was divided into five topics: metallurgical length; air-gap formation and mould taper; oscillation-mark formation; macrosegregation; and electromagnetic stirring. Whilst work on all of these is still ongoing, a certain degree of validation has already been achieved:

- For the determination of metallurgical length, the proposed approach already yielded good agreement with a more computationally intensive approach using CFD and turbulence modelling [21], for the case of a pure metal, and with a more conventional enthalpy-formulation approach for alloys [22].
- For solidification shrinkage-induced macrosegregation in a binary alloy, reasonably good quantitative agreement was achieved for the cross-sectional macrosegregation profile, and even better agreement is to expected when the model has been developed to include superheat [23].
- For oscillation-mark formation, very good agreement was obtained with experimental results for fold-type marks [30].

For air-gap formation and mould taper, validation will soon be attempted against the results of Kelly et al. [47] for the continuous casting of round billets. For electromagnetic stirring, validation will be possible against the original experimental results of Dubke et al. [84], although this work is perhaps more embryonic, in view of an anomaly in the modelling of EMS in continuous casting that was recently found in [31].

In addition to the possible extensions of the work already carried out in each of these topics, which are detailed towards the end of Sections 2–6, these topics also form the starting points of investigations into a variety of as yet under-researched areas in continuous casting. Amongst these are the following:

Exudation For the case of alloys that have a long solidification interval—for example, alloys of copper or aluminium [41,66,70,101,102]—interdendritic melt seeps out through the air gap that forms between the solidifying shell and the chilled mould surface; this is known as exudation. Mathematically, this means that the region $z_{liq} \leq z \leq z_{sol}$ at $y = 0$ in Figure 4 is effectively an outlet for the interdendritic melt. The situation is an example of macrosegregation that requires the methods developed in Sections 2 and 5.

Columnar-equiaxed transition In the continuous casting of steel, it is well-established that the solidified structure consists of equiaxed crystals in the centre of the casting, surrounded by a columnar crystals, and that EMS is often to used to increase the extent of the equiaxed zone, by means, it is believed, of dendrite fragmentation. However, models which can predict the columnar-to-equiaxed transition (CET) and how it can be affected by EMS are non-existent, although there is some work on CET using the cellular automaton method [103–105].

V-segregation V-segregation is the name given to the eponymously-shaped channels that are formed in the centre of the equiaxed zone of continuously cast steel slabs, blooms and billets, in which the macrosegregation level of interstitial elements such as carbon and sulphur, and of substitutional elements such as molybdenum and chromium, can be quite extensive. Even today, the mechanism behind the development of V-shaped segregate channels or lines is still not well understood [106]. The appearance of the V-segregates differs between different cast sizes, width of the equiaxed zone and casting speed. V-segregates can be found in low alloyed steels, as well as in stainless and high-alloyed steels.

Centreline porosity Continuously cast products develop centreline porosity along the strand direction, as a result of an extended mushy zone in the centre of the solidifying material. The length of the mushy zone depends mainly on solidification mode, degree of solidification shrinkage, extent of the equiaxed zone, cross section dimensions and casting speed. In particular, the centre pore develops due to a lack of feeding to compensate for the solidification shrinkage. When the pressure in the centre decreases below the equilibrium pressure for gas phase, porosity will develop, with the pore growing along the centre line. Although internal porosity is well-studied for ingot and component casting [107–109], this is far from the case for continuous casting [110].

Mechanical soft reduction A method used to counteract V-segregations, centreline porosity and centreline segregation is mechanical soft reduction [111–113], whereby the cast steel cross section is reduced by pinch rollers to compensate for the downward liquid flow in the mushy zone due to solidification shrinkage.

Lastly, we point out the relevance of the above to the thermo-fluid dynamics topics that are relevant to as-cast quality, particularly in the continuous casting of steel. Although such topics are numerous, the work on oscillation-mark formation begins to address surface quality; moreover, it provides a relatively cheap computational framework for considering the formation and propagation of surface cracks, as are known to occur, for example, in peritectic steels [114]. Furthermore, although the work on macrosegregation considered only solidification shrinkage, it already contains the full array of momentum, heat and solute conservation equations that would be required to model macrosegregation due to other causes: for example, thermosolutal convection [115], or electromagnetic stirring, which is known to produce so-called white bands in the casting of steel [87,99,100,116,117], but which has never been properly modelled. Moreover, the work on solidification shrinkage-induced macrosegregation may help to inform on centreline porosity. Although the analysis in [23] did not dwell on the second term in the regular perturbation series for pressure, this quantity becomes negative and unbounded as full solidification is reached at the centreline, if the lever rule is used to describe segregation at the microscale; this would indicate that the pressure as a whole tends to zero, meaning the appearance of a pore. In addition, it is apparent that the pressure does not behave in this way for other microsegregation rules. It would be impossible to deduce this qualitative behaviour from numerical simulations alone, which further highlights the benefit of thinking asymptotically.

Funding: The author would like to thank FAPESP (Fundação de Amparo a Pesquisa do Estado de São Paulo) for the award of a visiting researcher grant [Grant Number 2018/07643-8].

Conflicts of Interest: The author declares no conflict of interest.

Appendix A. Notes on Practical Asymptotics

As indicated in Section 1, this article has been underpinned by the use of practical asymptotics; however, there has not been space to show details of how it works, and this issue is now addressed here.

To begin with, one would need to demonstrate the following four steps in action in the topics considered in Sections 2–6:

(a) Nondimensionalization of the original governing equations;
(b) Analysis of the nondimensionalized governing equations, and identification of the key dimensionless parameters and asymptotic reduction;
(c) Evidence that the computation of the reduced model is cheaper than the computation of the original model would have been (if the reduced model does not have an analytical solution); and
(d) Evidence of agreement between the results of the original model and the asymptotically reduced model.

Moreover, point (c) presupposes that one has indeed solved the problem in both the computationally intensive way and the cheap way; however, in practice, the visibility of "computational cheapness" comes from the observation that it ought to be cheaper to compute the numerical solutions for:

1. Fewer partial differential equations (PDEs) rather than more;
2. One-dimensional models rather than two-dimensional models, and two-dimensional models rather than three-dimensional models;
3. Ordinary differential equations (ODEs) rather than PDEs; and
4. Problems having fewer model parameters rather than more, with regard to the need for parameter studies to obtain a complete understanding of model behaviour.

Even so, there are examples in other areas of science and technology where a given problem has been solved both ways and the computational cost, in terms of CPU (central processing unit) time and RAM (random access memory), has been compared [118–123], although this was not done for the problems considered here.

In this context, we return to the problem of determining the metallurgical length in Section 2.1:

• The original model was two-dimensional steady-state, consisted of six PDEs and contained 32 model parameters [34,35].
• After nondimensionalization and asymptotic reduction, the model was one-dimensional and transient-like, consisted of two PDEs (or effectively different representations of the same PDE) and contained six model parameters [21].
• A direct comparison of CPU time and RAM was not carried out, but we note that one computation in [35] required 30 h of CPU time on a Cray J932 Supercomputer—with current computational architectures, this would no doubt take much less, but still would not be as short as the few seconds required for the formulation in [21].
• The agreement between the model results was very good, as shown in Figure 3.

Thus, this example demonstrates quite concretely how the asymptotic approach has been successfully applied.

References

1. Return to Introduction to Continuous Casting. Available online: http://ccc.illinois.edu/introduction/overview.html#fig1 (accessed on 6 November 2018).
2. Thomas, B.G. Review on modeling and simulation of continuous casting. *Steel Res. Int.* **2018**, *89*, 1700312. [CrossRef]
3. Kuiken, H.K. *Practical Asymptotics*; Kluwer Academic Publishers: Dordrecht, The Netherlands, 2001.
4. Holmes, M.H.; King, J.R. Practical Asymptotics II. *J. Eng. Math.* **2003**, *45*, 155–404. [CrossRef]
5. Witelski, T.P.; Rienstra, S.W. Introduction to Practical Asymptotics III. *J. Eng. Math.* **2005**, *53*, 199. [CrossRef]
6. McCue, S.W. Preface to the fourth special issue on practical asymptotics. *J. Eng. Math.* **2009**, *63*, 153–154. [CrossRef]
7. Korobkin, A. Preface to the fifth special issue on practical asymptotics. *J. Eng. Math.* **2011**, *69*, 111–112. [CrossRef]
8. Smith, W.R. Preface to the sixth special issue on "Practical Asymptotics". *J. Eng. Math.* **2017**, *102*, 1–2. [CrossRef]
9. Schilders, W.H.A.; van der Vorst, H.A.; Rommes, J. (Eds.) *Model Order Reduction: Theory, Research Aspects and Applications*, 1st ed.; Mathematics in Industry 13; Springer: Berlin/Heidelberg, Germany, 2008.
10. Voller, V.R.; Porte-Agel, F. Moore's law and numerical modeling. *J. Comput. Phys.* **2002**, *179*, 698–703. [CrossRef]
11. Pickering, E.J.; Chesman, C.; Al-Bermani, S.; Holland, M.; Davies, P.; Talamantes-Silva, J. A comprehensive case study of macrosegregation in a steel ingot. *Metall. Mater. Trans. B* **2015**, *46*, 1860–1874. [CrossRef]
12. DiLellio, J.A.; Young, G.W. An asymptotic model of the mold region in a continuous steel caster. *Metall. Mater. Trans. B* **1995**, *26B*, 1225–1241. [CrossRef]
13. Smelser, R.E.; Johnson, R.E. An asymptotic model of slab casting. *Int. J. Mech. Sci.* **1995**, *37*, 793–814. [CrossRef]
14. Johnson, R.E.; Cherukuri, H.P. Vertical continuous casting of bars. *Proc. R. Soc. A* **1999**, *455*, 227–244. [CrossRef]
15. Cherukuri, H.P.; Johnson, R.E. Modelling vertical continuous casting with temperature-dependent material properties. *Int. J. Mech. Sci.* **2001**, *43*, 1243–1257. [CrossRef]
16. Bland, D.R. Flux and the continuous casting of steel. *IMA J. Appl. Maths* **1984**, *32*, 89–112. [CrossRef]
17. Hill, J.M.; Wu, Y.H. On a nonlinear Stefan problem in the continuous casting of steel. *Acta Mech.* **1994**, *107*, 183–198. [CrossRef]
18. Hill, J.M.; Wu, Y.H.; Wiwatanapataphee, B. Analysis of flux flow and the formation of oscillation marks in the continuous caster. *J. Eng. Math.* **1999**, *36*, 311–326. [CrossRef]
19. King, J.R.; Lacey, A.A.; Please, C.P.; Wilmott, P.; Zoryk, A. The formation of oscillation marks on continuously cast steel. *Math. Eng. Ind.* **1993**, *4*, 91–106.
20. Howison, S.D. *Practical Applied Mathematics: Modelling, Analysis, Approximation*; Cambridge University Press: Cambridge, UK, 2005.
21. Mitchell, S.L.; Vynnycky, M. Verified reduction of a model for a continuous casting process. *Appl. Math. Mod.* **2017**, *48*, 476–490. [CrossRef]
22. Vynnycky, M.; Saleem, S. On the explicit resolution of the mushy zone in the modelling of the continuous casting of alloys. *Appl. Math. Mod.* **2017**, *50*, 544–568. [CrossRef]
23. Vynnycky, M.; Saleem, S.; Fredriksson, H. An asymptotic approach to solidification shrinkage-induced macrosegregation in the continuous casting of binary alloys. *Appl. Math. Mod.* **2018**, *54*, 605–626. [CrossRef]
24. Vynnycky, M. An asymptotic model for the formation and evolution of air gaps in vertical continuous casting. *Proc. R. Soc. A* **2009**, *465*, 1617–1644. [CrossRef]
25. Vynnycky, M. Air gaps in vertical continuous casting in round moulds. *J. Eng. Math.* **2010**, *68*, 129–152. [CrossRef]
26. Vynnycky, M. On the role of radiative heat transfer in air gaps in vertical continuous casting. *Appl. Math. Mod.* **2013**, *37*, 2178–2188. [CrossRef]
27. Vynnycky, M. On the onset of air-gap formation in vertical continuous casting with superheat. *Int. J. Mech. Sci.* **2013**, *73*, 69–76. [CrossRef]

28. Florio, B.J.; Vynnycky, M.; Mitchell, S.L.; O'Brien, S.B.G. Mould-taper asymptotics and air gap formation in continuous casting. *Appl. Math. Comput.* **2015**, *268*, 1122–1139. [CrossRef]

29. Florio, B.J.; Vynnycky, M.; Mitchell, S.L.; O'Brien, S.B.G. On the interactive effects of mould taper and superheat on air gaps in continuous casting. *Acta Mech.* **2017**, *228*, 233–254. [CrossRef]

30. Vynnycky, M.; Saleem, S.; Devine, K.M.; Florio, B.J.; Mitchell, S.L.; O'Brien, S.B.G. On the formation of fold-type oscillation marks in the continuous casting of steel. *R. Soc. Open Sci.* **2017**, *4*, 176002. [CrossRef] [PubMed]

31. Vynnycky, M. On an anomaly in the modeling of electromagnetic stirring in continuous casting. *Metall. Mater. Trans. B* **2018**, *49B*, 399–410. [CrossRef]

32. Vynnycky, M.; Zambrano, M. Towards a "moving-point" formulation for the modelling of oscillation-mark formation in the continuous casting of steel. *Appl. Math. Mod.* **2018**, *63*, 243–265. [CrossRef]

33. Vynnycky, M. Porous-media braking of electromagnetic stirring in the continuous casting of steel. In Proceedings of the 24th ABCM International Congress of Mechanical Engineering, Curitiba, Brazil, 3–8 December 2017.

34. Mahmoudi, J.; Vynnycky, M.; Fredriksson, H. Modelling of fluid flow, heat transfer and solidification in the strip casting of a copper base alloy: (III). Solidification—A theoretical study. *Scand. J. Metall.* **2001**, *30*, 136–145. [CrossRef]

35. Mahmoudi, J.; Vynnycky, M.; Sivesson, P.; Fredriksson, H. An experimental and numerical study on the modelling of fluid flow, heat transfer and solidification in a copper continuous strip casting process. *Mater. Trans.* **2003**, *44*, 1741–1751. [CrossRef]

36. AEA Technology. *CFX 4.2 Flow Solver User Guide*; AEA Technology: Harwell, UK, 1995.

37. Swaminathan, C.R.; Voller, V.R. A general enthalpy method for modeling solidification processes. *Met. Trans. B* **1992**, *23B*, 651–664. [CrossRef]

38. Voller, V.R.; Peng, S. An enthalpy formulation based on an arbitrarily deforming mesh for solution of the Stefan problem. *Comput. Mech.* **1994**, *14*, 492–502. [CrossRef]

39. Aboutalebi, M.R.; Hasan, M.; Guthrie, R.I.L. Numerical study of coupled turbulent flow and solidification for steel slab casters. *Numer. Heat Transf.* **1995**, *28*, 279–297. [CrossRef]

40. Aboutalebi, M.R.; Hasan, M.; Guthrie, R.I.L. Coupled turbulent flow, heat and solute transport in continuous casting processes. *Metall. Mater. Trans. B* **1995**, *26*, 731–744. [CrossRef]

41. Thevik, H.J.; Mo, A.; Rusten, T. A mathematical model for surface segregation in aluminum direct chill casting. *Metall. Mater. Trans. B* **1999**, *39*, 135–142. [CrossRef]

42. Savage, J. A theory of heat transfer and air gap formation in continuous casting molds. *J. Iron Steel Inst.* **1962**, *198*, 41–47.

43. Richmond, O.; Tien, R.H. Theory of thermal stresses and air-gap formation during the early stages of solidification in a rectangular mold. *J. Mech. Phys. Solids* **1971**, *19*, 273–284. [CrossRef]

44. Kristiansson, J.O. Thermal stresses in the early stage of the solidification of steel. *J. Therm. Stresses* **1982**, *5*, 315–330. [CrossRef]

45. Tien, R.H.; Richmond, O. Theory of maximum tensile stresses in the solidifying shell of a constrained regular casting. *J. Appl. Mech.* **1982**, *49*, 481–486. [CrossRef]

46. Kim, K.Y. Analysis of gap formation at mold-shell interface during solidification of aluminium alloy plate. *ISIJ Int.* **2003**, *43*, 647–652. [CrossRef]

47. Kelly, J.E.; Michalek, K.P.; O'Connor, T.G.; Thomas, B.G.; Dantzig, J.A. Initial development of thermal and stress fields in continuously cast steel billets. *Metall. Mater. Trans. A* **1988**, *19A*, 2589–2602. [CrossRef]

48. Grill, A.; Sorimachi, K.; Brimacombe, J.K. Heat flow, gap formation and break-outs in the continuous casting of steel slabs. *Metall. Mater. Trans. B* **1976**, *7B*, 177–189. [CrossRef]

49. Bellet, M.; Decultieux, F.; Menai, M.; Bay, F.; Levaillant, C.; Chenot, J.L.; Schmidt, P.; Svensson, I.L. Thermomechanics of the cooling stage in casting processes: Three-dimensional finite element analysis and experimental validation. *Metall. Mater. Trans. B* **1996**, *27*, 81–99. [CrossRef]

50. Huespe, A.E.; Cardona, A.; Fachinotti, V. Thermomechanical model of a continuous casting process. *Comput. Methods Appl. Mech. Eng.* **2000**, *182*, 439–455. [CrossRef]

51. Li, C.; Thomas, B.G. Thermomechanical finite-element model of shell behavior in continuous casting of steel. *Metall. Mater. Trans. B* **2004**, *35B*, 1151–1172. [CrossRef]

52. Sun, D.; Annapragada, S.R.; Garimella, S.V.; Singh, S.K. Analysis of gap formation in the casting of energetic materials. *Numer. Heat Transf.* **2007**, *51*, 415–444. [CrossRef]

53. Saleem, S. On the Surface Quality of Continuously Cast Steels and Phosphor Bronzes. Ph.D. Thesis, KTH Royal Institute of Technology, Stockholm, Sweden, 2016.

54. Saleem, S.; Vynnycky, M.; Fredriksson, H. A study of the oscillation marks' characteristics of continuously cast Incoloy alloy 825 blooms. *Metall. Mater. Trans. A* **2016**, *47*, 4068–4079. [CrossRef]

55. Takeuchi, E.; Brimacombe, J.K. The formation of oscillation marks in the continuous casting of steel slabs. *Metall. Mater. Trans. B* **1984**, *15*, 493–509. [CrossRef]

56. Jonayat, A.S.M.; Thomas, B.G. Transient thermo-fluid model of meniscus behavior and slag consumption in steel continuous casting. *Metall. Mater. Trans. A* **2014**, *45*, 1842–1864. [CrossRef]

57. Tomono, H. Elements of Oscillation Mark Formation and Their Effect on Transverse Fine Cracks in Continuous Casting of Steel. Ph.D. Thesis, École Polytechnique Fédérale de Lausanne, Lausanne, Switzerland, 1979.

58. Ramirez-Lopez, P.E.; Lee, P.D.; Mills, K.C. Explicit modelling of slag infiltration and shell formation during mould oscillation in continuous casting. *ISIJ Int.* **2010**, *50*, 425–434. [CrossRef]

59. Ramirez-Lopez, P.E.; Lee, P.D.; Mills, K.C.; Santillana, B. A new approach for modelling slag infiltration and solidification in a continuous casting mould. *ISIJ Int.* **2010**, *50*, 1797–1804. [CrossRef]

60. Lee, P.D.; Ramirez-Lopez, P.E.; Mills, K.C.; Santillana, B. Review: The "butterfly effect" in continuous casting. *Ironmak. Steelmak.* **2012**, *39*, 244–253. [CrossRef]

61. Ramirez-Lopez, P.E.; Mills, K.C.; Lee, P.D.; Santillana, B. A unified mechanism for the formation of oscillation marks. *Metall. Mater. Trans. B* **2012**, *43B*, 109–122. [CrossRef]

62. Bikerman, J.J. *Physical Surfaces*; Academic Press: New York, NY, USA, 1970.

63. Flemings, M.C.; Nereo, G.E. Macrosegregation. I. *AIME Met. Soc. Trans.* **1967**, *239*, 1449–1461.

64. Flemings, M.C.; Mehrabian, R.; Nereo, G.E. Macrosegregation. PT. 2. *AIME Met. Soc. Trans.* **1968**, *242*, 41–49.

65. Reddy, A.V.; Beckermann, C. Modeling of macrosegregation due to thermosolutal convection and contraction-driven flow in direct chill continuous casting of an Al-Cu round ingot. *Metall. Mater. Trans. B* **1997**, *28*, 479–489. [CrossRef]

66. Saleem, S.; Vynnycky, M.; Fredriksson, H. Formation of the tin rich layer and inverse segregation in phosphor bronzes during continuous casting. In Proceedings of the Minerals, Metals and Materials Society (TMS) 2015: 144th Annual Meeting and Exhibition, Orlando, FL, USA, 15–19 March 2015; pp. 15–22.

67. Diao, Q.Z.; Tsai, H.L. Modelling of solute redistribution in the mushy zone during solidification of aluminium-copper alloys. *Metall. Trans.* **1993**, *24A*, 963–973. [CrossRef]

68. Chen, J.H.; Tsai, H.L. Inverse segregation for a unidirectional solidification of aluminium-copper alloys. *Int. J. Heat Mass Transf.* **1993**, *36*, 3069–3075. [CrossRef]

69. Diao, Q.Z.; Tsai, H.L. The formation of negative- and positive-segregated bands during solidification of aluminum-copper alloys. *Int. J. Heat Mass Transf.* **1993**, *36*, 4299–4305. [CrossRef]

70. Mo, A. Mathematical modelling of surface segregation in aluminum DC casting caused by exudation. *Int. J. Heat Mass Transf.* **1993**, *36*, 4335–4340. [CrossRef]

71. Voller, V.R.; Sundarraj, S. A model of inverse segregation: The role of microporosity. *Int. J. Heat Mass Transf.* **1995**, *38*, 1009–1018. [CrossRef]

72. Minakawa, S.; Samarasekera, I.V.; Weinberg, F. Inverse segregation. *Metall. Trans.* **1985**, *16*, 595–604. [CrossRef]

73. Du, Q.; Eskin, D.G.; Katgerman, L. Numerical issues in modelling macrosegregation during DC casting of a multi-component aluminium alloy. *Int. J. Numer. Methods Heat Fluid Flow* **2009**, *19*, 917–930. [CrossRef]

74. Jalanti, T.; Swierkosz, M.; Gremaud, M.; Rappaz, M. Modelling of macrosegregation in continuous casting of aluminium. In *Continuous Casting*; Ehrke, K., Schneider, W., Eds.; WILEY-VCH Verlag GmbH: Weinheim, Germany, 2006; pp. 191–198.

75. Založnik, M.; Xin, S.; Šarler, B. Verification of a numerical model of macrosegregation in direct chill casting. *Int. J. Numer. Methods Heat Fluid Flow* **2008**, *18*, 308–324. [CrossRef]

76. Venneker, B.C.H.; Katgerman, L. Modelling issues in macrosegregation predictions in direct chill castings. *J. Light Met.* **2002**, *2*, 149–159. [CrossRef]

77. Eskin, D.G.; Zuidema, J.; Savran, V.I.; Katgerman, L. Structure formation and macrosegregation under different process conditions during DC casting. *Mater. Sci. Eng. A* **2004**, *384*, 232–244. [CrossRef]

78. Eskin, D.G.; Du, Q.; Katgerman, L. Relationship between shrinkage-induced macrosegregation and the sump profile upon direct-chill casting. *Scr. Mater.* **2006**, *55*, 715–718. [CrossRef]

79. Du, Q.; Eskin, D.G.; Katgerman, L. Modeling macrosegregation during direct-chill casting of multicomponent aluminum alloys. *Metall. Mater. Trans. A* **2007**, *38A*, 180–189. [CrossRef]

80. Jalanti, T. Etude et Modélisation de la Macroségrégation dans la Coulée Semi-Continue des Alliages d'Aluminium. Ph.D. Thesis, École Polytechnique Fédérale de Lausanne, Lausanne, Switzerland, 2000.

81. Fredriksson, H.; Åkerlind, U. *Materials Processing during Casting*; Wiley: Chichester, UK, 2006.

82. Tzavaras, A.A.; Brody, H.D. Electromagnetic stirring and continuous-casting—Achievements, problems, and goals. *J. Met.* **1984**, *36*, 31–37. [CrossRef]

83. Spitzer, K.H.; Dubke, M.; Schwerdtfeger, K. Rotational electromagnetic stirring in continuous-casting of round strands. *Metall. Mater. Trans. B* **1986**, *17*, 119–131. [CrossRef]

84. Dubke, M.; Tacke, K.H.; Spitzer, K.H.; Schwerdtfeger, K. Flow fields in electromagnetic stirring of rectangular strands with linear inductors: Part I. Theory and experiments with cold models. *Metall. Mater. Trans. B* **1988**, *19B*, 581–593. [CrossRef]

85. Dubke, M.; Tacke, K.H.; Spitzer, K.H.; Schwerdtfeger, K. Flow fields in electromagnetic stirring of rectangular strands with linear inductors: Part II. Computation of flow fields in billets, blooms, and slabs of steel. *Metall. Mater. Trans. B* **1988**, *19*, 595–602. [CrossRef]

86. Dubke, M.; Spitzer, K.H.; Schwerdtfeger, K. Spatial-distribution of magnetic-field of linear inductors used for electromagnetic stirring in continuous-casting of steel. *Ironmak. Steelmak.* **1991**, *18*, 347–353.

87. Tacke, K.H.; Grill, A.; Miyazawa, K.; Schwerdtfeger, K. Macrosegregation in strand cast steel—Computation of concentration profiles with a diffusion-model. *Arch. Eisenhüttenw.* **1981**, *52*, 15–20. [CrossRef]

88. Tacke, K.H.; Schwerdtfeger, K. Stirring velocities in continuously cast round billets as induced with rotating electromagnetic-fields. *Stahl und Eisen* **1979**, *99*, 7–12.

89. Zhang, C.; Shatrov, V.; Priede, J.; Eckert, S.; Gerbeth, G. Intermittent behavior caused by surface oxidation in a liquid metal flow driven by a rotating magnetic field. *Metall. Mater. Trans. B* **2011**, *42*, 1188–1200. [CrossRef]

90. Liu, H.; Xu, M.; Qiu, S.; Zhang, H. Numerical simulation of fluid flow in a round bloom mold with in-mold rotary electromagnetic stirring. *Metall. Mater. Trans. B* **2012**, *43*, 1657–1675. [CrossRef]

91. Yang, J.; Xie, Z.; Ning, J.; Liu, W.; Ji, Z. A framework for soft sensing of liquid pool length of continuous casting round blooms. *Metall. Mater. Trans. B* **2014**, *45*, 1545–1556. [CrossRef]

92. Poole, G.M.; Heyen, M.; Nastac, L.; El-Kaddah, N. Numerical modeling of macrosegregation in binary alloys solidifying in the presence of electromagnetic stirring. *Metall. Mater. Trans. B* **2014**, *45*, 1834–1841. [CrossRef]

93. Ren, B.Z.; Chen, D.F.; Wang, H.D.; Long, M.J.; Han, Z.W. Numerical simulation of fluid flow and solidification in bloom continuous casting mould with electromagnetic stirring. *Ironmak. Steelmak.* **2015**, *42*, 401–408. [CrossRef]

94. Fang, Q.; Ni, H.; Zhang, H.; Wang, B.; Lv, Z. The effects of a submerged entry nozzle on flow and initial solidification in a continuous casting bloom mold with electromagnetic stirring. *Metals* **2017**, *7*, 146. [CrossRef]

95. Wang, X.; Fautrelle, Y.; Etay, J.; Moreau, R. A periodically reversed flow driven by a modulated traveling magnetic field: Part I. Experiments with GaInSn. *Metall. Mater. Trans. B* **2009**, *40*, 82–90. [CrossRef]

96. Eckert, S.; Nikrityuk, P.A.; Raebiger, D.; Eckert, K.; Gerbeth, G. Efficient melt stirring using pulse sequences of a rotating magnetic field: Part I. Flow field in a liquid metal column. *Metall. Mater. Trans. B* **2007**, *38*, 977–988. [CrossRef]

97. Beitelman, L.S.; Curran, C.P.; Lavers, J.D.; Tallback, G. Modulated Electromagnetic Stirring of Metals at Advanced Stage of Solidification. EP Patent EP080783247, 22 August 2011.

98. Versteeg, H.; Malalasekera, W. *An Introduction to Computational Fluid Dynamics: The Finite Volume Method*, 2nd ed.; Pearson: Harlow, UK, 2007.

99. Bridge, M.R.; Rogers, G.D. Structural effects and band segregate formation during the electromagnetic stirring of strand-cast steel. *Met. Trans. B* **1984**, *15*, 581–589. [CrossRef]

100. Kor, G.J.W. Influence of circumferential electromagnetic stirring on macrosegregation in steel. *Ironmak. Steelmak.* **1982**, *9*, 244–251.

101. M'Hamdi, M.; Håkonsen, A. Experimental and numerical study of surface macrosegregation in DC casting of aluminium sheet ingots. In *Modeling of Casting, Welding and Advanced Solidification Processes-X, Proceedings of the 10th International Conference on Modeling of Casting, Welding and Advanced Solidification Processes, Destin, FL, USA, 25–30 May 2003*; Stefanescu, D.M., Warren, J.A., Jolly, M.R., Krane, M.J.M., Eds.; Minerals, Metals & Materials Soc.: Warrendale, PA, USA, 2003; pp. 505–512.

102. Haug, E.; Mo, A.; Thevik, H.J. Macrosegregation near a cast surface caused by exudation and solidification shrinkage. *Int. J. Heat Mass Transf.* **1995**, *38*, 1553–1563. [CrossRef]

103. Luo, Y.Z.; Zhang, J.M.; Wei, X.D.; Xiao, C.; Hu, Z.F.; Yuan, Y.Y.; Chen, S.D. Numerical simulation of solidification structure of high carbon SWRH77B billet based on the CAFE method. *Ironmak. Steelmak.* **2012**, *39*, 26–30. [CrossRef]

104. Luo, S.; Zhu, M.; Louhenkilpi, S. Numerical simulation of solidification structure of high carbon steel in continuous casting using cellular automaton method. *ISIJ Int.* **2012**, *52*, 823–830. [CrossRef]

105. Wang, W.; Ji, C.; Luo, S.; Zhu, M. Modeling of dendritic evolution of continuously cast steel billet with cellular automaton. *Metall. Mater. Trans. B* **2018**, *49*, 200–212. [CrossRef]

106. Guan, R.; Ji, C.; Zhu, M.; Deng, S. Numerical simulation of V-shaped segregation in continuous casting blooms based on a microsegregation model. *Metall. Mater. Trans. B* **2018**, *49*, 2571–2583. [CrossRef]

107. Lee, P.; Chirazi, A.; See, D. Modeling microporosity in aluminum-silicon alloys: A review. *J. Light Met.* **2001**, *1*, 15–30. [CrossRef]

108. Dantzig, J.A.; Rappaz, M. *Solidification*; EPFL Press: Lausanne, Switzerland, 2009.

109. Stefanescu, D.M. Computer simulation of shrinkage related defects in metal castings—A review. *Int. J. Cast Met. Res.* **2005**, *18*, 129–143. [CrossRef]

110. Du, P. Numerical Modeling of Porosity and Macrosegregation in Continuous Casting of Steel. Ph.D. Thesis, University of Iowa, Iowa City, IA, USA, 2013.

111. Rogberg, B.; Ek, L. Influence of soft reduction on the fluid flow, porosity and center segregation in CC high carbon- and stainless steel blooms. *ISIJ Int.* **2018**, *58*, 478–487. [CrossRef]

112. Domitner, J.; Wu, M.; Kharicha, A.; Ludwig, A.; Kaufmann, B.; Reiter, J.; Schaden, T. Modeling the effects of strand surface bulging and mechanical softreduction on the macrosegregation formation in steel continuous casting. *Metall. Mater. Trans. A* **2014**, *45*, 1415–1434. [CrossRef]

113. Mayer, F.; Wu, M.; Ludwig, A. On the formation of centreline segregation in continuous slab casting of steel due to bulging and/or feeding. *Steel Res. Int.* **2010**, *81*, 660–667. [CrossRef]

114. Saleem, S.; Vynnycky, M.; Fredriksson, H. The influence of peritectic reaction/transformation on crack susceptibility in the continuous casting of steels. *Metall. Mater. Trans. B* **2017**, *48*, 1625–1635. [CrossRef]

115. Sun, H.; Zhang, J. Study on the macrosegregation behavior for the bloom continuous casting: Model development and validation. *Metall. Mater. Trans. B* **2014**, *45B*, 1133–1149. [CrossRef]

116. Hurtuk, D.J.; Tzavaras, A.A. Some effects of electromagnetically induced fluid-flow on macrosegregation in continuously cast steel. *Metall. Trans. B Proc. Met.* **1977**, *8*, 243–251. [CrossRef]

117. Sasaki, K.; Sugitani, Y.; Kobayashi, S.; Ishimura, S. The effect of fluid flow on the formation of the negative segregation zone in steel ingots. *Tetsu Hagane* **1979**, *65*, 60–69. [CrossRef]

118. Vynnycky, M.; Shugai, G.; Yakubenko, P.; Mellgren, N. Asymptotic reduction for numerical modeling of polymer electrolyte fuel cells. *SIAM J. Appl. Math.* **2009**, *70*, 455–487. [CrossRef]

119. Ly, H.; Birgersson, E.; Vynnycky, M.; Sasmito, A.P. Validated reduction and accelerated numerical computation of a model for the proton exchange membrane fuel cell. *J. Electrochem. Soc.* **2009**, *156*, B1156–B1168. [CrossRef]

120. Ly, H.; Birgersson, E.; Vynnycky, M. Asymptotically reduced model for a proton exchange membrane fuel cell stack: Automated model generation and verification. *J. Electrochem. Soc.* **2010**, *157*, B982–B992. [CrossRef]

121. Ly, H.; Birgersson, E.; Vynnycky, M. Computationally efficient multi-phase models for a proton exchange membrane fuel cell: Asymptotic reduction and thermal decoupling. *Int. J. Hydrog. Energy* **2011**, *36*, 14573–14589. [CrossRef]

122. Vynnycky, M.; Sharma, A.K.; Birgersson, E. A finite-element method for the weakly compressible parabolized steady 3D Navier-Stokes equations in a channel with a permeable wall. *Comput. Fluids* **2013**, *81*, 152–161. [CrossRef]
123. Sharma, A.K.; Birgersson, E.; Vynnycky, M. Towards computationally-efficient modeling of transport phenomena in three-dimensional monolithic channels. *Appl. Math. Comput.* **2015**, *254*, 392–407. [CrossRef]

MDPI

St. Alban-Anlage 66

4052 Basel

Switzerland

Tel. +41 61 683 77 34

Fax +41 61 302 89 18

www.mdpi.com

Metals Editorial Office

E-mail: metals@mdpi.com

www.mdpi.com/journal/metals

Structural Control of Mineral Deposits

Structural Control of Mineral Deposits

Theory and Reality

Special Issue Editor

Alain Chauvet

MDPI • Basel • Beijing • Wuhan • Barcelona • Belgrade

MDPI

Special Issue Editor
Alain Chauvet
University of Montpellier
France

Editorial Office
MDPI
St. Alban-Anlage 66
4052 Basel, Switzerland

This is a reprint of articles from the Special Issue published online in the open access journal *Minerals* (ISSN 2075-163X) from 2018 to 2019 (available at: https://www.mdpi.com/journal/minerals/special_issues/structural_control_deposits).

For citation purposes, cite each article independently as indicated on the article page online and as indicated below:

> LastName, A.A.; LastName, B.B.; LastName, C.C. Article Title. *Journal Name* **Year**, *Article Number, Page Range*.

ISBN 978-3-03897-784-1 (Pbk)
ISBN 978-3-03897-785-8 (PDF)

Cover image courtesy of Alain Chauvet.

Contents

About the Special Issue Editor

Alain Chauvet (Dr. HDR) is a senior CNRS Researcher at the Géosciences Montpellier laboratory, France. After a Ph.D. in Structural Geology devoted to the late-orogenic extension in Norway, he specialized in Tectonic Control of Ore Deposits with a focus on the perigranitic mineralizations of South America, China, Europe, ..., and the characterization of the relationships between Large Igneous Provinces and Mineralizations in North Africa. He is involved in several collaborative projects and expertises with industrial mining companies.

Preface to "Structural Control of Mineral Deposits"

This compilation of publication results from more than 20 years of questioning and of applying structural geology in mining geology by the guest editor. If it is common to place the various deposits of the earth into large classes that allow recognizing and identifying some characters useful to detect, explore, and find other similar deposits, experience demonstrates that each deposit is unique and cannot answer perfectly to a generic model. This is why we suspect that there exists a gap between theory (i.e., the classical model) and reality that needs to be estimated and taken into account in any type expertise or study of an unknown mineral deposit. The following publications try to be concerned by this way of working.

My knowledge and interest in the structural control of mineral deposits benefited from several discussions, suggestions, and field trip shared with a lot of persons that are greatly acknowledged here. An exhaustive list is impossible, but I want to particularly acknowledge the CVRD (Vale), Buenaventura, Cedimin, Managem, Kasbah Resource, CMS, SMI, and CTT mining companies and all geologists that took the time to discuss with me of structural problems in mining geology, with a special mention of A.S. André, L. Badra, L. Bailly, L. Barbanson, Y. Branquet, X. Charonnat, A. Ennaciri, C. Ennaciri, M. Faure, L. Fontboté,. A. Gaouzi, S. Gialli, E. Gloaguen, M. Iseppi; K. Kouzmanov, J. Onezime, P. Piantone, S. Sizaret, E. Tourneur, J. Tuduri, N. Volland. All the reviewers that significantly improved the quality of this book are also warmly acknowledged.

Alain Chauvet

Special Issue Editor

minerals

MDPI

Editorial

Editorial for Special Issue "Structural Control of Mineral Deposits: Theory and Reality"

Alain Chauvet

UMR 5243, Géosciences Montpellier, University of Montpellier, cc 60, CEDEX 5, 34095 Montpellier, France; chauvet@gm.univ-montp2.fr; Tel.: +33-(0)4-67-14-48-57

Received: 7 March 2019; Accepted: 8 March 2019; Published: 11 March 2019

"Structural Control" remains a crucial point that is frequently absent in scientific and/or economic analyses of ore deposits, whatever their type and class, although a selection of references illustrates its importance [1–5]. The case of lode deposits is particularly adapted, but other types, like breccia pipes, stockwork, massive sulphides, skarn, etc., also concern Structural Control. Works on the Structural Control of ore deposits are not abundant in the recent literature, and, as frequently suggested, structural geology often is not sufficiently developed in the exploration programs of many mining camp's strategies. A few compilations have been devoted to this theme in the last two decades, such as (i) the special publication of the Geological Society of London, concerned with the link between fracturing, flow, and mineralization [6], (ii) the review of the Society of Economic Geology, devoted to Structural Control [7], (iii) a special publication of the Geological Society of London, looking to study the genetic link that can exist between mineralization and orogenic domains [8], and finally, (iv) a special issue of the Journal of Structural Geology, devoted to the application of Structural Geology in mineral exploration and mining [9]. In addition to these four compilations, only a few publications have been concerned with this theme, and most of them are dated before the year 2000. These publications mostly concerned vein internal infilling textures [10,11], the vein formation model, with the contribution and controversy of the crack seal, dissolution-precipitation, diffusion, and seismic-valves mechanisms (e.g., [12–16]). In his review, Chauvet [17] discussed of some of these concepts, in order to highlight the role and the significance of pre-existing structures in the formation of vein-style deposits.

Three publications of this volume explore the development of mineralization in the specific context of orogenic domains. Cugerone et al. [18] offer a detailed study of a rather complex Pb–Zn mineralisation developed within the orogenic Hercynian Pyrenees during two mineralization stages, each of them linked with a deformational event. The syntectonic primary mineralization is remobilized and helps the formation of the second one. The same approach is used within the two following contributions on the same theme [19,20]. Funedda et al. [19] and Fridovsky et al. [20] also used a detailed description of the relationships between mineralization and deformation in deformed domains, such as the Variscan domain of Sardinia and the Verkhovansk-Kolyma folded region of NE Russia. Funedda et al. [19] pay close attention to the opening process of structures that will serve as traps for mineralised fluid catching, a fact that is fundamental in any tectonic understanding of a mineralised vein system [17]. Fridovsky et al. [20] also proposed a pluri-deformational model associated with multiple stages of mineralisation formation.

The relationship between magmatism, regional tectonic context, and mineralization remain a question that has still been debated in several recent publications [21,22], thus demonstrating that this question is still relevant and may help in the distinction between intrusion-related, orogenic deposits and the Cu–Au-rich porphyry types. Two contributions explore new methods of investigation that provide an innovative vision of the relationship between magmatism and mineralization. Song et al. [23] examine the consequences of the telescoping of two mineralized systems (a subsequent epithermal system affects a primary porphyric one within the Tiegelongnan Porphyry and the epithermal overprinting Cu (Au) deposit, Central Tibet, China) with a focus on the role of the dislocation effects on ore reserve calculations and future deposits discoveries. Tuduri et al. [24] suggest an original way to demonstrate the genetic link between mineralization and magmatism by establishing that both are developed in the same regional tectonic context, in the highly mineralised Moroccan Anti-Atlas. This contribution represents an indirect but efficient way to relatively date the emplacement of magmatism and mineralization formations, and their relationships.

In the past, the concept of a gold-bearing shear zone has not given satisfying results in terms of our understanding of gold deposits, and has been more or less totally abandoned, except within few specific sectors of the Canadian shield in which the role of major crustal faults is still at the centre of the accepted models [25]. In the domain of economic geology, faults are fundamental structures that can have two contrasting behaviours: (i) Hydrogeological barriers that help the concentration of ore, as demonstrated by the contribution of Grare et al. [26] in the case of the Kiggavik uranium example (Canada), and (ii) a zone of permeability that can favour fluid circulation and can serve as a guide for the mineralisation trapping. The work of Maciel et al. [27] proposes a surprising example in which fault occurrences have a negative role for clay authigenesis efficiency; this work also discusses the consequence on reservoir characteristics. Sun et al. [28] end the section on relations with brittle tectonics by presenting an innovative GIS-based spatial analysis of mineral deposit patterns in correlation with detailed structural features, in order to propose some implications on Structural Control. The chosen example was provided from the Copper deposit of the Tongling Ore district of Eastern China.

Concerning other orebodies than vein-type ones, volcanic-hosted massive sulphide deposits (VHMS) have been recently the subject of much debate, specifically with the suggestion of a significant contribution of "replacement processes" in their modes of formation [29,30]. In addition, it has been demonstrated that stockwork within VHMS environments can result in subsequent syntectonic veining instead of earlier veins related to feeder zones [31]. Indeed, the observation of stockwork within a VHMS context needs to be considered with particular attention because of the possible coexistence of the two types of stockwork: the one related to the feeder zone and the other the result of subsequent deformation [17]. It has been suggested that the second event and associated metal may contribute significantly to a relative enrichment in VHMS environment. Without any reference to some replacement process, the contribution of Admou et al. [32] ends the special issue with a very attractive formation model of the Moroccan Guemassa VHMS deposit, strongly involving the active role of normal faults and Structural Control, since the beginning of the volcanic activity. In fact, it appears that most of the VHMS deposits certainly do not present the classical geometrical model exhibited within all teaching books, but instead form by wall-rock replacement (metasomatism) strongly helped by the re-using of pre-existing structures, such as folds, unconformities, and/or fault and deformation features. Such a contribution is frequently underestimated.

Conflicts of Interest: The author declares no conflicts of interest.

References

1. Forde, A.; Bell, T.H. Late structural control of mesothermal vein-hosted gold deposits in Central Victoria, Australia: Mineralization and exploration potential. *Ore Geol. Rev.* **1994**, *9*, 33–59. [CrossRef]
2. Davis, B.K.; Hippertt, J.F.M. Relationships between gold concentration and structure in quartz veins from the Hodgkinson Province, northeastern Australia. *Mineral. Depos.* **1998**, *33*, 391–405. [CrossRef]
3. Chauvet, A.; Piantone, P.; Barbanson, L.; Nehlig, P.; Pedroletti, I. Gold deposit formation during collapse tectonics: Structural, mineralogical, geochronological, and fluid inclusion constraints in the Ouro Preto gold mines, Quadrilátero Ferrífero, Brazil. *Econ. Geol.* **2001**, *96*, 25–48. [CrossRef]
4. Chauvet, A.; Bailly, L.; André, A.S.; Monié, P.; Cassard, D.; Llosa Tajada, F.; Vargas, J.R.; Tuduri, J. Internal vein texture and vein evolution of the epithermal Shila-Paula district, southern Peru. *Mineral. Dep.* **2006**, *41*, 387–410. [CrossRef]
5. Tunks, A.J.; Cooke, D.R. Geological and structural controls on gold mineralization in the Tanami District, Northern Territory. *Mineral. Dep.* **2007**, *42*, 107–126. [CrossRef]

6.   McCaffrey, K.; Lonergan, L.; Wilkinson, J. *Fractures, Fluid Flow and Mineralization*; Geological Society of London Special Publication: London, UK, 1999; Volume 155, 328p.

7.   Richards, J.P.; Tosdal, R.M. Structural Controls on Ore Genesis. In *Reviews in Economic Geology*; Society of Economic Geologists, Inc.: Littleton, CO, USA, 2001; 181p.

8.   Blundell, D.J.; Neubauer, F.; Von Quadt, A. *The Timing and Location of Major Ore Deposits in an Evolving Orogen*; Geological Society of London Special Publication: London, UK, 2002; Volume 204, 358p.

9.   Vearncombe, J.R.; Blenkinsop, T.G.; Reddy, S.M. Applied Structural Geology for Mineral Exploration and Mining. *J. Struct. Geol.* **2004**, *26*, 989–994. [CrossRef]

10. Dowling, K.; Morrison, G. Application of quartz textures to the classification of gold deposits using North Queensland examples. *Econ. Geol. Mon.* **1990**, *6*, 342–355.

11. Dong, G.; Morrison, G.; Jaireth, S. Quartz textures in epithermal veins, Queensland - classification, origin, and implication. *Econ. Geol.* **1995**, *90*, 1841–1856. [CrossRef]

12. Ramsay, J.G. The crack-seal mechanism of rock deformation. *Nature* **1980**, *284*, 135–139. [CrossRef]

13. Cox, S.F.; Etheridge, M.A. Crack-seal fibre growth mechanism and their significance in the development of oriented layer silicate microstructures. *J. Struct. Geol.* **1983**, *92*, 147–170. [CrossRef]

14. Bons, P.D.; Jessell, M.W. Experimental simulation of the formation of fibrous veins by localised dissolution-precipitation creep. *Mineral. Mag.* **1997**, *61*, 53–63. [CrossRef]

15. Boullier, A.M.; Robert, F. Paleoseismic events recorded in Archean gold quartz vein networks, Val-Dor, Abitibi, Quebec, Canada. *J. Struct. Geol.* **1992**, *14*, 161–177. [CrossRef]

16. Bons, P.D.; Elburg, M.A.; Gomez-Rivas, E. A review of the formation of tectonic veins and their microstructures. *J. Struct. Geol.* **2012**, *43*, 33–62. [CrossRef]

17. Chauvet, A. Structural Control of Ore Deposits: The Role of Pre-existing Structures on the Formation of Mineralised Vein Systems. *Minerals* **2019**, *9*, 56. [CrossRef]

18. Cugerone, A.; Oliot, E.; Chauvet, A.; Gavaldà Bordes, J.; Laurent, A.; Le Goff, E.; Cenki-Tok, B. Structural Control on the Formation of Pb-Zn Deposits: An Example from the Pyrenean Axial Zone. *Minerals* **2018**, *8*, 489. [CrossRef]

19. Funedda, A.; Naitza, S.; Buttau, C.; Cocco, F.; Dini, A. Structural Controls of Ore Mineralization in a Polydeformed Basement: Field Examples from the Variscan Baccu Locci Shear Zone (SE Sardinia, Italy). *Minerals* **2018**, *8*, 456. [CrossRef]

20. Fridovsky, V.Y.; Kudrin, M.V.; Polufuntikova, L.I. Multi-Stage Deformation of the Khangalas Ore Cluster (Verkhoyansk-Kolyma Folded Region, Northeast Russia): Ore-Controlling Reverse Thrust Faults and Post-Mineral Strike-Slip Faults. *Minerals* **2018**, *8*, 270. [CrossRef]

21. Dressel, B.C.; Chauvet, A.; Trzaskos, B.; Biondi, J.C.; Bruguier, O.; Monié, P.; Villanova, S.N.; Newton, J.B. The Passa Tres lode gold deposit (Parana State, Brazil): An example of structurally-controlled mineralisation formed during magmatic-hydrothermal transition and hosted within granite. *Ore Geol. Rev.* **2018**, *102*, 701–727. [CrossRef]

22. Tuduri, J.; Chauvet, A.; Barbanson, L.; Labriki, M.; Dubois, M.; Trapy, P.H.; Lahfid, A.; Poujol, M.; Melleton, J.; Badra, L.; et al. Structural control, magmatic-hydrothermal evolution and formation of hornfels-hosted, intrusion-related gold deposits: Insight from the Thaghassa deposit in Eastern Anti-Atlas, Morocco. *Ore Geol. Rev.* **2018**, *97*, 171–198. [CrossRef]

23. Song, Y.; Yang, C.; Wei, S.; Yang, H.; Fang, X.; Lu, H. Tectonic Control, Reconstruction and Preservation of the Tiegelongnan Porphyry and Epithermal Overprinting Cu (Au) Deposit, Central Tibet, China. *Minerals* **2018**, *8*, 398. [CrossRef]

24. Tuduri, J.; Chauvet, A.; Barbanson, L.; Bourdier, J.L.; Labriki, M.; Ennaciri, A.; Badra, L.; Dubois, M.; Ennaciri-Leloix, C.; Sizaret, S.; Maacha, L. The Jbel Saghro Au(–Ag, Cu) and Ag–Hg Metallogenetic Province: Product of a Long-Lived Ediacaran Tectono-Magmatic Evolution in the Moroccan Anti-Atlas. *Minerals* **2018**, *8*, 592. [CrossRef]

25. Poulsen, K.H.; Robert, F. Shear zones and gold: Practical examples from the southern Canadian Shield. *Geol. Assoc. Can. Short Course Notes* **1989**, *6*, 239–266.

26. Grare, A.; Lacombe, O.; Mercadier, J.; Benedicto, A.; Guilcher, M.; Trave, A.; Ledru, P.; Robbins, J. Fault Zone Evolution and Development of a Structural and Hydrological Barrier: The Quartz Breccia in the Kiggavik Area (Nunavut, Canada) and Its Control on Uranium Mineralization. *Minerals* **2018**, *8*, 319. [CrossRef]

27. Maciel, I.B.; Dettori, A.; Balsamo, F.; Bezerra, F.H.R.; Vieira, M.M.; Nogueira, F.C.C.; Salvioli-Mariani, E.; Sousa, J.A.B. Structural Control on Clay Mineral Authigenesis in Faulted Arkosic Sandstone of the Rio do Peixe Basin, Brazil. *Minerals* **2018**, *8*, 408. [CrossRef]

28. Sun, T.; Xu, Y.; Yu, X.; Liu, W.; Li, R.; Hu, Z.; Wang, Y. Structural Controls on Copper Mineralization in the Tongling Ore District, Eastern China: Evidence from Spatial Analysis. *Minerals* **2018**, *8*, 254. [CrossRef]

29. Aerden, D.G.A.M. Formation of Massive Sulfide Lenses by Replacement of folds: The Hercules Pb–Zn Mine, Tasmania. *Econ. Geol.* **1993**, *88*, 377–396. [CrossRef]

30. Perkins, W.G. Mount Isa lead-zinc orebodies: Replacement lodes in a zoned syndeformational copper–lead–zinc system? *Ore Geol. Rev.* **1997**, *12*, 61–110. [CrossRef]

31. Chauvet, A.; Onézime, J.; Charvet, J.; Barbanson, L.; Faure, M. Syn- to late-tectonic stockwork emplacement within the Spanish section of the Iberian Pyrite Belt: Structural, textural and mineralogical constraints in the Tharsis-La Zarza areas. *Econ. Geol.* **2004**, *99*, 1781–1792. [CrossRef]

32. Admou, S.; Branquet, Y.; Badra, L.; Barbanson, L.; Outhounjite, M.; Khalifa, A.; Zouhair, M.; Maacha, L. The Hajjar Regional Transpressive Shear Zone (Guemassa Massif, Morocco): Consequences on the Deformation of the Base-Metal Massive Sulfide Ore. *Minerals* **2018**, *8*, 435. [CrossRef]

minerals

Review

Structural Control of Ore Deposits: The Role of Pre-Existing Structures on the Formation of Mineralised Vein Systems

Alain CHAUVET

CNRS-UMR 5243, Géosciences Montpellier, University of Montpellier, cc 60,
34095 Montpellier CEDEX 5, France; alain.chauvet@univ-montp2.fr; Tel.: +3-34-6714-4857

Received: 29 October 2018; Accepted: 15 January 2019; Published: 17 January 2019

Abstract: The major role played by pre-existing structures in the formation of vein-style mineral deposits is demonstrated with several examples. The control of a pre-existing decollement level on the formation of a crustal extension-related (collapse) gold deposit is first illustrated in the Quadrilátero Ferrífero from Brazil. Shear zone and decollement structures were also examined and shown to control veins formation by three distinct processes: (i) re-aperture and re-using of wrench shear zones in the case of Shila gold mines (south Peru); (ii) remobilisation of metal in volcanic-hosted massive sulphide (VHMS) deposit by subsequent tectonic events and formation of a secondary stockwork controlled by structures created during this event (Iberian Pyrite Belt, Spain); (iii) formation of economic stockwork by contrasting deformation behaviours between ductile black schist versus brittle more competent dolomite (Cu-Ifri deposit, Morocco). Two examples involve changing of rheological competence within zones affected by deformation and/or alteration in order to receive the mineralisation (case studies of Achmmach, Morocco, and Mina Soriana, Spain). The last case underscores the significance of the magmatic–hydrothermal transition in the formation of mesothermal gold deposits (Bruès mine, Spain). All these examples clearly demonstrate the crucial role played by previously formed structures and/or texture in the development and formation of ore deposits.

Keywords: vein; structure; textures; infilling; breccia; comb quartz; pull-apart; exploration; pre-existing structures; decollement

1. Introduction

Numerous studies have been devoted to the process of vein formation mainly because of their significance in term of tectonics and deformation (stress and strain determination), e.g., [1–5], but also because of their significant economic interest in the case of metal-bearing veins, e.g., [6–8]. Several works have concentrated on the external geometry of veins and their relationships with the mode of opening and, consequently, the local or regional stress field during vein formation [9–11]. Complementary studies have also integrated information that can be deduced from vein infilling textures, such as the classical tripartite division in syntaxial (inward growth), antitaxial (outward growth) and stretching veins (complex pattern with no consistent growth direction) [12–14]. In economic geology, and particularly in vein-type deposits, the study of the nature and texture of vein infilling is particularly important because it lies at the base of the ore-forming process itself. No recent works have been concerned by this kind of analysis. The latest contributions [15–17] only deal with the internal texture of ore deposits without considering the (external) geometry of the veins themselves.

This paper, as an introduction to the Minerals Special Issue "Structural Control of Ore Deposits, Theory and Reality" focuses on the relationships between the shape and internal texture of ore-bearing veins with the objectives to better understanding vein formation processes and, consequently, to improve mining exploration strategies. I will present and discuss seven case studies of metal-bearing veins with different modes of formation that highlight the role of pre-existing features on their development. This aspect seems to be frequently underestimated, at least in the case of vein deposits, and this work aims to demonstrate its significance in the development of exploration and exploitation programs. The re-using of some previously formed structures has, in that case, a significant but passive role with respect to the formation of the economic feature. This concept is already exemplified in another contribution of this volume [18]. Each of the seven cases presented herein include a brief overview of the regional geology and deformation history, followed by a detailed geometrical and textural analysis of ore-bearing veins and a regional-scale genetic model that integrates these data. The relationships between neo-formed structures versus pre-existing ones will be highlighted in each case as well as its implications for regional vein distribution and, consequently, exploration programs.

2. Methodology: Structural Analysis Applied to Metallogeny

The literature on structurally controlled vein-style mineral deposits, e.g., [7,8], has provided a number of theoretical concepts that link vein orientation and geometry with respect to the tectonic stresses. However, only a few studies have integrated regional tectonic context, vein shape and internal vein texture [19–21].

The present work emphasises the importance of multidisciplinary work on vein-style mineralisation combining (micro) structural, textural, and mineralogical analyses, in order to answer to the following questions:

- What tectonic context is responsible for trap formation? (the geometrical analysis)
- What is the mode and condition of filling? (the internal analysis)

2.1. Trap Formation

Figure 1 illustrates three classic tectonic traps found frequently in the literature [22] and described in this work. We can distinguish (i) vein opening during flexural folding of multilayers with contrasted lithology with a void being created within competent levels by (re-)opening of the coeval axial plane cleavage (Figure 1a); (ii) gaps formed by differential shearing due to fracturing and boudinage of more competent layers leading to stockwork development (Figure 1b); (iii) aperture controlled by extensional relay (pull-apart) associated with either fibrous/comb veins or breccia veins (Figure 1c). The factors determining these alternative types will be treated further. These three mechanisms are the ones most commonly invoked to explain vein formation in economic interest with probably more cases resembling Figure 1c. Figure 2 illustrates the mechanism of re-using a previously formed structure left lateral shear zone showing a main shearing vein with breccia and secondary cleavage parallel veins that opened during a subsequent tectonic event with different stress directions as when the shear zone originally formed. Internal filling for each stage is drastically different: brecciation without infilling in the case of the main shearing plane and comb or fibrous crystallisation in the secondary cleavage parallel veins (Figure 2).

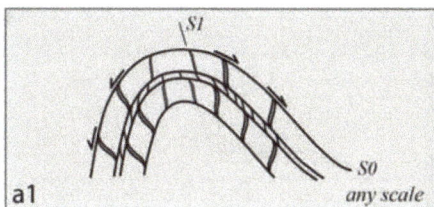

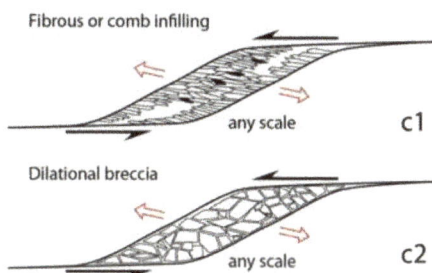

Figure 1. Three examples of traps and voids mode of formation. (**a**) Model of formation of sigmoidal veins by antithetic bedding-controlled gliding during fold development (**a1**). Example from the Bourg-d'Oisans area (**a2**). (**b**) Stockwork formation in accommodation of shearing affecting some multi-layers rocks with contrast of competence (**b1**, **b2** and **b3**). Conceptual sketch (**b4**) in which yellow veins represent the stockwork formed in more competent layers. (**c**) Trap opening within left lateral pull-apart (**c3**). The two sketches **c1** and **c2** illustrate the process and show the two types of filling encountered within these structures. See text for explanation.

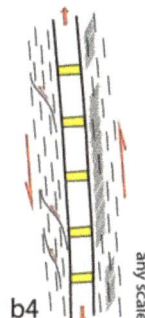

It may be surprising to see that Figures 1a and 2, propose that a cleavage plane, reputed to be a plane of maximum flattening, is used as a syntectonic trap for mineralisation during its formation.

However, this solution has been adopted and demonstrated within two works on this topic [23,24]. We will see, in the following examples, what are our arguments to defend this unconventional concept. The case of Figure 2 should not raise the same concern because, here, cleavage serves as a receptacle for ore concentration in a late deformation event that causes dilation parallel to a pre-existing cleavage.

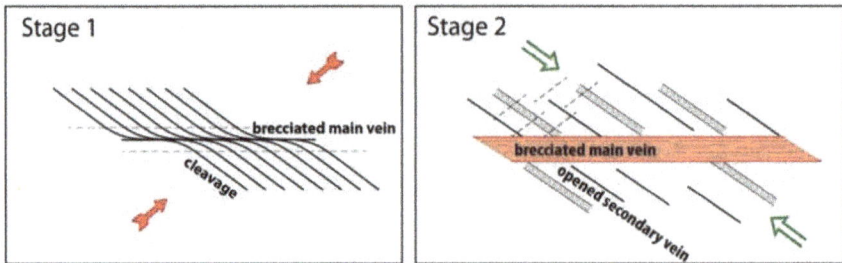

Figure 2. Example of re-using and re-opening of a previously formed left-lateral shear zone during two different states of stress. Red and green arrows correspond to the shortening direction of each stage. Note that main infilling is realised during stage 2.

2.2. Internal Texture

The following internal texture of ore-bearing veins can be distinguished.

- *Massive or buck texture:* rare examples of this texture have been interpreted to result where voids are filled after, not during their formation. This texture is frequently characterised by euhedral or anhedral grain of variable size throughout the vein [15] due to uniform growth rates. Grain orientation can also be highly variable. In fact, such a texture provides limited information about the tectonic conditions prevailing during vein formation.
- A *fibrous or comb crystal shape* corresponds to crystallisation coeval with vein opening and represents the more interesting texture for the topics of this study (Figure 3). Whether comb or fibrous textures develop depends on the rate of trap opening versus crystal growth (see below).

Comb quartz is commonly related to (Figure 3a) (i) a supersaturated fluid invading an open space (the initial fracture) with competitive crystal growth normal to the walls [16,25,26]; and (ii) a slow opening rate of the fracture keeping pace with the rate of crystal growth [15,27]. Veins formed by this process only differ from crack-seal veins [28–30] by the lack of fibrous crystallisation and evidence for incremental cracking, such as successive and parallel inclusion trails. Indeed, fibrous textures result from the same process as comb infilling, except that crystal growth is incremental instead of continuous. In this case, the crack is caused by fluid overpressure and crystallisation occurs immediately after the aperture with a unique free direction for crystal growth—the vein centre. The succession of cracking event and, consequently, of immediate filling, explains the continuous crystallisation and, therefore, the fibres (Figure 3b) [31]. By contrast, comb texture is supposed to form where the rate of crystallisation is lower than the opening rate. In this case, the crystallisation only covers the vein wall, and crystals are larger as in the case of fibrous veins and can develop during multiple growth stages (Figure 3a), sometimes associated with a change of fluid composition and chemistry [15]. It is still uncertain, though, if all fibrous veins form by the crack-seal mechanism, or whether they can also form by continuous fibre growth, where diffusion keeps pace with the rate of dilation [26,32].

a. Comb texture

b. Fibrous texture

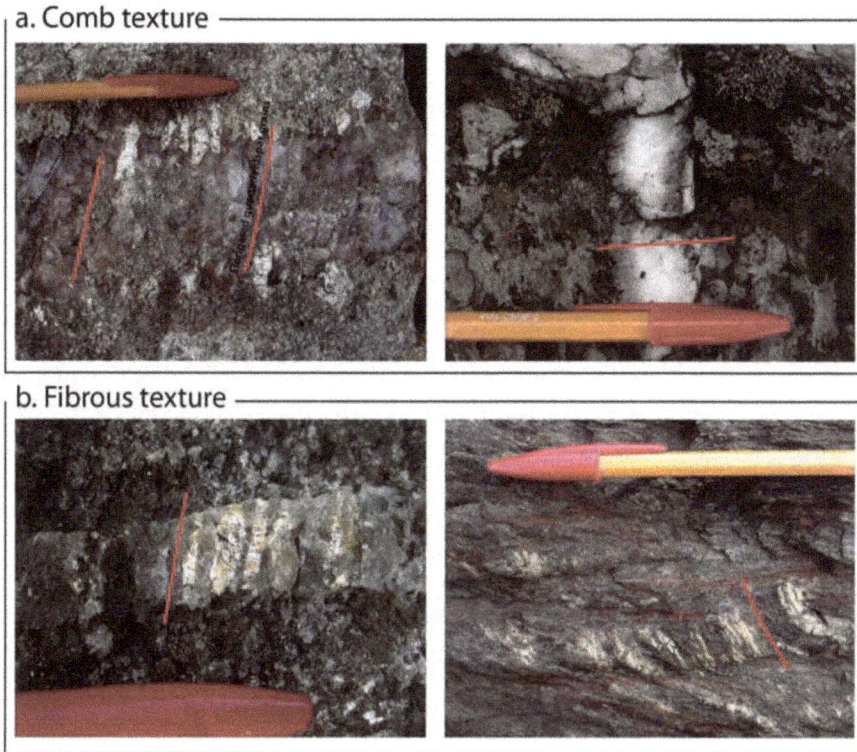

Figure 3. Examples of comb and fibrous textures from the Hercynian mining district of Tras-os-Montes, Galicia, Spain. Red lines illustrate the elongated quartz and feldspar grains indicate the opening direction. Note the difference between fibres that cross-cut the veins and comb grains that do not traverse the vein.

- *Breccia textures* are witnesses of complex processes for which we have to take into account three parameters based on the recognition between fragments and matrix in order to understand their process of formation:

  ○ The nature of the matrix or cement (rock flour, sediment, volcanic, magmatic, hydrothermal, ...);

  ○ The nature and shape of the fragments (circularity, size, distribution, fabrics, monogenic or polygenic, lithological nature);

  ○ The relationships between fragments and matrix/cement (matrix-supported or grain-supported).

As a function of these three parameters, a genetic classification of breccia has been proposed by [33], which is frequently used as an indicator of the conditions of vein formation [34–39]. The most used in lode-related economic geology are the tectonic, hydrothermal, magmatic, collapse-related, crackle, hydraulic and dilational breccia. Their recognition is based on the following features (Figure 4):

- *Tectonic breccia* is easily recognisable because of grain reduction and oriented fragments (Figure 4a). Depending of its maturity (function of the strain intensity), fragments can be in contact (grain-supported breccia, beginning of fragmentation and subsequent comminution) or finally flooded in a largely developed matrix (matrix-supported breccia). Tectonic breccia is more frequently monogenic. With respect to the intensity of the deformation and the presence or lack of clay minerals, they can be called cataclasite, ultracataclasite or gouge.

- *Hydrothermal breccia* is characterised by more-or-less rounded fragments of the same nature (not always) in place within a hydrothermal matrix. Frequently, the final voids are filled by cement that can frequently contain some metals in economic contexts (Figure 4b).

- *Magmatic breccia* is more or less similar to the hydrothermal ones, except that the matrix is only magmatic and there is no cement (Figure 4c). In this case, the fragments are rounded and never in contact (matrix supported breccia). Due to the explosive processes, magmatic breccias are frequently polygenic. The differentiation between matrix and fragments, both magmatic, is sometimes difficult, especially in thin sections.

- *Collapse breccia* is easy to recognise because they show a large variation of fragment size, the presence of cement, and grain-supported texture (Figure 4d). Their geometry is clearly consistent with their mode of formation: (i) collapse of the fragments in response to an underlying explosion or void formation by dissolution and (ii) posterior cementation.

- *Crackle breccia* is an early stage of what is going becoming a hydrothermal, tectonic or hydraulic breccia. Due to their mode of formation, they are monogenic, with a low matrix and they can be assimilated to early fragmentation in response to either tectonic stress or fluid-related fracturing. Some parts frequently exhibit the host rocks being not totally disrupted whether other parts can be more mature with well-expressed breccia texture (Figure 4e).

- *Hydraulic breccia* is the result of hydraulic fracturing. It exhibits typical jigsaw geometry with a monogenic character and a very regular pattern (Figure 4f). The matrix is well represented, and fragments are never in contact. The mode of formation is only due to cracking due to fluid overpressure. *Dilational breccia* forms within extensional relay or pull-apart (Figures 1c, 2 and 5). In this case, breccia formation is explained by the fact that void creation causes the fragmentation of the hosted rock affected by the pull-apart formation. Fragments are weakly transported and sometimes rotated and the occurrence of cement is common. Why some pull-aparts are filled by fibrous/comb crystals or dilational breccia remains an open question (Figure 2c). The outcrop in Figure 5 can help because the two types of infilling have been observed within the same structure. Since dilational breccia has been observed on the wall of the secondary formed comb infilling (Figure 5b), we suspect that both types of texture can be developed in the same structural context. Field relationships demonstrate that dilational breccia texture can form at the beginning of the process, when rates of aperture are weak and late and rapid opening can explain the superposition of fibrous/comb infilling. Indeed, the alternative formation of dilational breccia or comb texture in the core of pull-apart can appear as a function of opening velocity, crystal growth rate, and fluid saturation. We guess that dilational breccia in the core of pull-apart can be created during all main tectonic contexts (i.e., compression, extension, transtension, etc.) and not restricted to the only case of wrench tectonics, as this has been established for the large-scale pull-apart-related basin formation along crustal-scale faults [40].

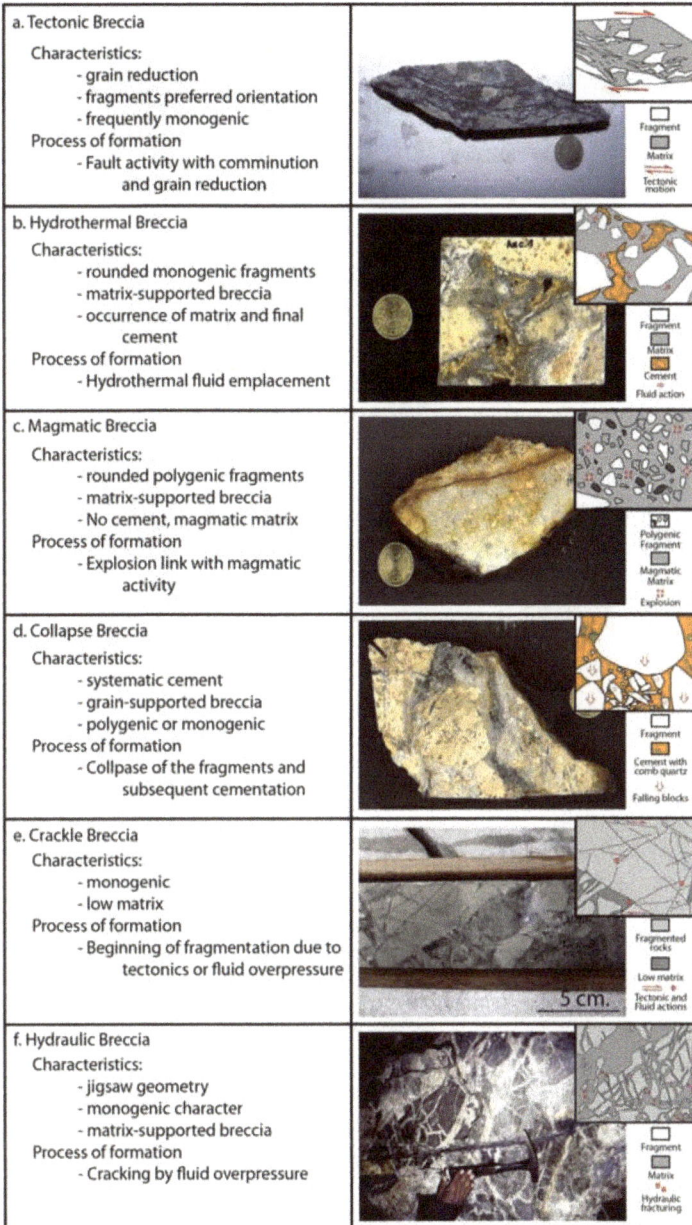

Figure 4. Breccia texture classification frequently used in ore geology. Each case is described in detail in the text. A conceptual sketch is indicated for each photograph in order to correctly interpret the image. The scale of these sketches is the same as the corresponding photograph. In red, some indications about the process responsible for breccia formation are provided.

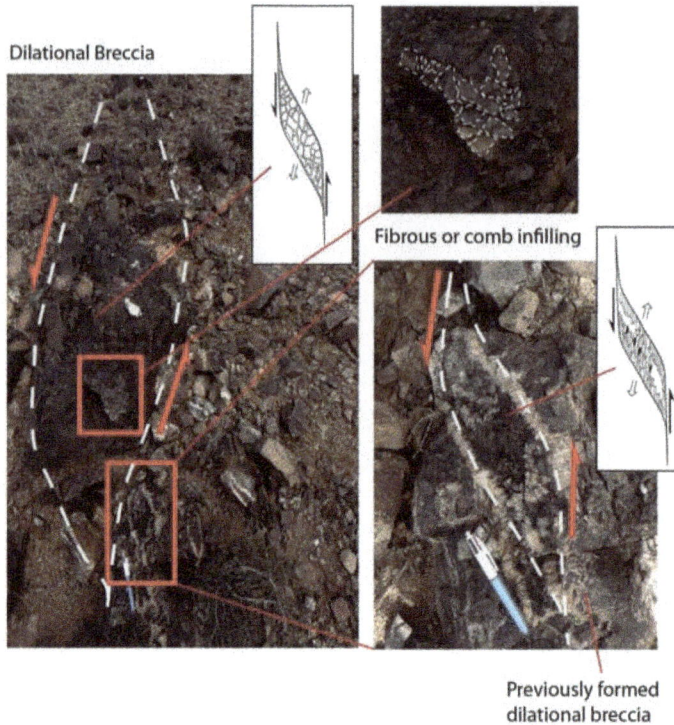

Figure 5. Superposition of dilational breccia texture and fibre/comb ones within similar left-lateral pull-apart structures (see text for explanation of the cause of occurrence of fibrous or breccia texture).

3. Vein Formation Process and Tectonics: Examples from Ore Deposit Study

3.1. Gold Concentration during Collapse Tectonics

A pluridisciplinary approach has been undertaken with the study of gold-bearing quartz veins of the Quadrilátero Ferrífero (Minas Gerais, Brazil) [41]. This deposit has been classically interpreted as a strongly deformed pre-tectonic one on the basis of the sigmoid shape (Figure 6) exhibited by ore-bearing veins and, also, as a typical Archean orogenic gold deposit [42]. The internal vein texture study combined with regional tectonic constraints suggests an alternative model in which veins are formed in response to the late collapse tectonics, later with respect to the nappes emplacement [41]. During these movements, sigmoid voids were created and filled by elongate quartz grains associated with sulphides, tourmaline and carbonates (Figure 6a). Such a normal motion has been facilitated by the existence of mica-rich levels (Figure 6b) that underline the foliation of the hosted meta-sediments. This result has been obtained essentially because the internal texture has revealed that quartz grains are un-deformed and that their formation was the result of only one opening episode, within free tectonic stress context. The superposition of sigmoidal pull-apart formation on earlier shearing deformational event has been clearly observed (Figure 6c). The process of formation is similar to the conceptual sketch of the Figure 2 with re-using and re-opening of pre-existing shear zone. Here, the sigmoidal shape of the mica-rich level due to the earlier thrust event is re-used. Inferred hydrothermal effects were responsible for the formation of illite, sulphides, carbonate, garnet and green biotite within the matrix.

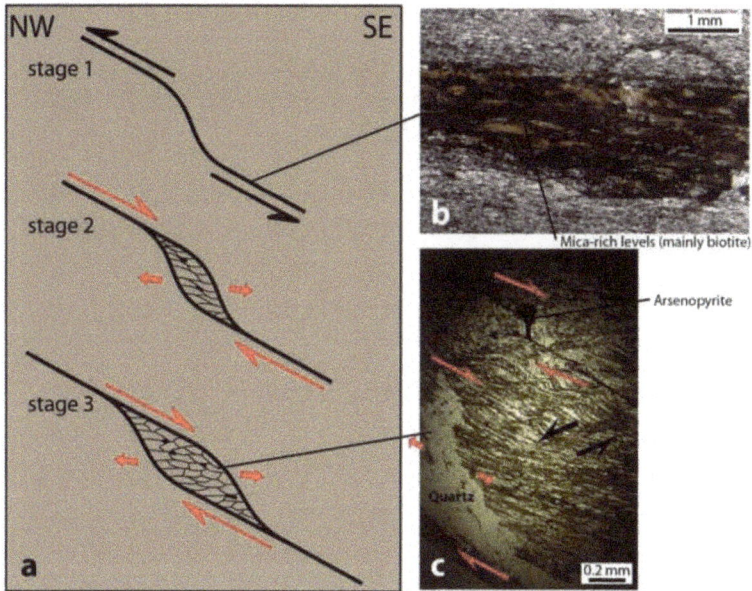

Figure 6. (**a**) Formation of sigmoid vein during late-orogenic collapse tectonics (red arrows) within the Quadrilátero Ferrífero (Minas Gerais, Brazil). Note that vein re-used an early level formed by mica alignment related to the nappe emplacement (**b**). The sigmoidal shape reflects thrust-related emplacement (black arrows). Extensional pull-apart post-dated thrust-related structures and controlled gold-bearing quartz-sulphide veins (**c**).

This study demonstrates that the Ouro Preto mesothermal gold deposit was formed in context of late-orogenic collapse, drastically different from the conventional auriferous shear-zone model, model that has been intensively used during the 1990s [43,44]. A recent study confirms this hypothesis by the demonstration of resetting of older zircons by ca. 496 Ma old hot fluid rock interactions in the area of Passagem [45].

This example clearly illustrates the importance of the systematic observation of the internal vein texture before concluding on the mechanism of vein formation based only on the geometrical analysis.

3.2. Vein Opening and Filling Controlled by Regional-Scale Structures within Volcanic Domains

The Shila-Paula district is one of the numerous Au/Ag low sulfidation epithermal one of Southern Peru. It is characterised by numerous veins hosted by the tertiary subaerial volcanics of the Western Cordillera. Field studies shown that most of the mineralised bodies consist of the systematic association of main E–W veins and secondary N120–135°E veins (Figure 7) [46]. Two main stages of ore deposition are identified [47]. Stage 1 consists of a quartz–adularia–pyrite–galena –sphalerite–chalcopyrite–electrum–Mn silicates and carbonates assemblage that fills the main E–W veins (Figure 7a,b,e). Stage 2, also called the bonanza stage, carries most of the precious mineralisation and consists of quartz, Fe-poor sphalerite, chalcopyrite, pyrite, adularia, galena, tennantite–tetrahedrite, polybasite–pearceite, and electrum. This stage is mainly observed within secondary veins, in final geodic filling (Figure 7c,f) and in veinlets that cut stage 1 assemblage (Figure 7d,g). In main veins, the ore is systematically brecciated, whereas tectonic-free environment characterised the filling of secondary veins. The age of veins was estimated to be around 10.8 Ma using 40Ar/39Ar ages on adularia crystals from different veins [47].

A two-stage model is proposed to explain vein formation. The first stage was assumed to correspond to the development of E–W sinistral shear zones and associated N120°E cleavages under the effects of a NE–SW trending shortening direction, which has been previously recognised at the Andean scale (Quechua II phase) (Figure 7a). These structures serve as a receptacle for the emplacement of stage 1 ore assemblage that was brecciated during ongoing deformation (Figure 7b). The second event operates a re-opening of the previously formed structures under a NW–SE trending shortening direction that allowed the re-opening of pre-existing cleavage and the formation of scarce N50°E trending S2 cleavages (Figure 7c), such as in the model in Figure 2. This stage was followed by the bonanza ore emplacement both within geodes in core of the main E–W veins and in secondary N120–135°E veins (Figure 7d). The two directions of shortening, NE–SW for the first event and NW–SE for the second one, are also recorded by the orientation of fluid inclusion planes within quartz crystals from the host rocks.

This study represents a unique example, constrained by combined tectonic, textural, mineralogical, geochronological, and fluid inclusion data, of the establishment of a complete model of deposit formation in which the re-using of previously formed tectonic features as a factor of gold concentration in epithermal environment is evidenced.

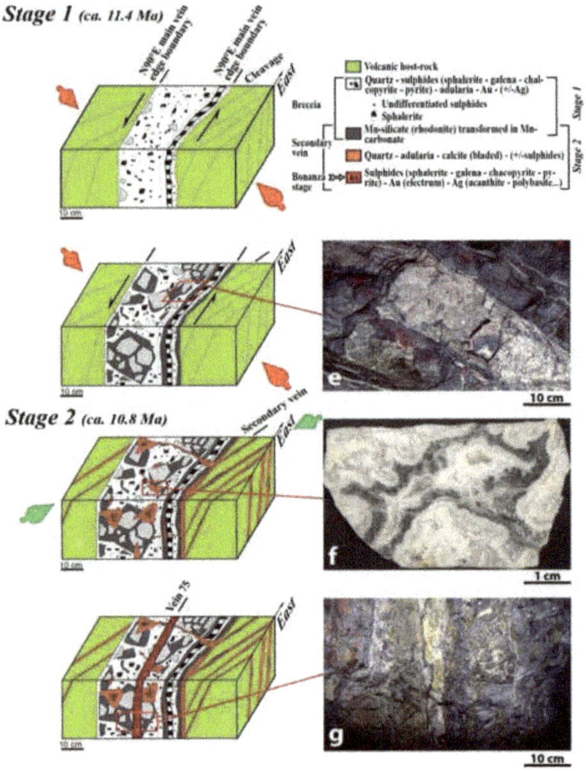

Figure 7. Example of vein formation by the re-using of pre-existing structures within the Volcanic domain of South Peru (see text for explanation). (**a**) Formation and filling of sinistral shear zone and creation of associated cleavage under the effect of NE–SW shortening direction (red arrows). (**b**) Formation of Mn-rich breccia under the same shortening direction. (**c**) Formation of secondary veins by re-opening of the cleavage by N120°E trending shortening direction (green arrows). Formation of geodic structures filled by stage 2-related paragenesis and subsequent "bonanza" stage and richer veins (Veta 75) (**d**).

3.3. Tectonic Stockwork Development in Fold and Thrust Belt Environment

3.3.1. The Iberian Pyrite Belt Example

The Variscan Iberian Pyrite Belt is affected by a continuous and progressive deformation that integrates the formation of south-directed folds and thrust vs. north-verging features [48]. In the light of this structural model, a microtectonic, textural, and mineralogical study of the stockworks associated with the volcanic-hosted massive sulphide (VHMS) deposits has been performed [24].

In addition to the first and primary stockwork assimilated to represent the feeder veins of the VHMS, a second stockwork occurs under the form of undeformed veins frequently emplaced within cleavage and shear planes characteristic of the south-verging deformation (Figure 8a). The characteristic mineral of both stages is pyrite. Within the primary stockwork, sulphides have Co–As-rich rims that are interpreted as overgrowths formed during the second event. Instead of that, the base metal assemblage that characterised the second stockwork is similar to the one seen within the primary stockwork, especially when developed at its contact. Conversely, second stockwork is observed far away from the primary one and VHMS is only filled by quartz and pyrite. The internal texture indicates that pyrite is systematically fractured by ongoing deformation and quartz fibres develop within asymmetric pressure consistent with the south-verging tectonics. Geochemical analysis of pyrites emphasises the discrimination between both stockworks and pyrites from early stockwork are S-depleted/Fe-enriched with respect to those of the second stockwork. Since sulphides are located within the cleavage and shearing plane (Figure 8b,c) and quartz fibres systematically develop around pyrite grains, the second stockwork is interpreted as developing during the south-directed tectonics [24]. Moreover, the second stockwork has been encountered within axial plane cleavage related to the south-directed tectonics (Figure 8d), similar to the conceptual trap formation within cleavage of Figure 1a. This is strongly confirmed by sulphide textural analysis that demonstrates the co-existence of deformational effects (pyrite subgrain boundary, blow-up pull-apart, Figure 8e) and post-tectonic pyrite overgrowth (Figure 8f). If the mineralised fluids responsible for this second stockwork result from the VHMS and early stockwork remobilisation or from external metamorphic source remain uncertain.

3.3.2. The Moroccan Palaeozoic High-Atlas Example

The mineralised district of the High Seksaoua (Western High Atlas, Morocco) is characterised by a lithological succession marked by an alternation of schists and limestones attributed to the Cambrian and affected by the Hercynian orogeny. The existence of stratiform masses of pyrite first suggests that this deposit can be a VHMS (Figure 9a) [49]. However, copper mineralisation, the first economic goal, is absent from the stratiform pyrite levels but systematically localised close to the dolomite/black schist contact in which a top-to-the-N–NW décollement-type tectonics [50,51] has been identified (Figure 9b). The economic mineralisation is a syntectonic stockwork (Figure 9c,d) formed in response to this top-to-the-N–NW shearing event that only affects the black schist layers (Figure 9). We suggest that vein formation is a brittle response, within competent dolomite levels, of the ductile deformation that affects the black schist (Figure 9a) [51], as described in the conceptual model of the Figure 1b.

This tectonic has been correlated to the late-Hercynian tectonics on the base of the Permo-Triassic age (ca. 270 Ma) given by $^{40}Ar/^{39}Ar$ dating realised on white micas related to the stockwork [51]. This important result questions the syngenetic interpretation accepted until now for this mineralisation, and allows us to propose a new model of formation in which the "décollement"-type tectonics represent the main factor of ore concentration.

The Iberian and Moroccan examples illustrated the role of decollement structures in the emplacement of stockworks associated with metal-rich minerals. If no economic concentration can be deduced from the remobilisation process during Hercynian orogeny in the case of the South Iberian Pyrite Belt and, moreover, compared to the huge metal mass represented by the VHMS, a similar process applied to the Ifri deposit led to the formation of the Cu-rich economic orebodies.

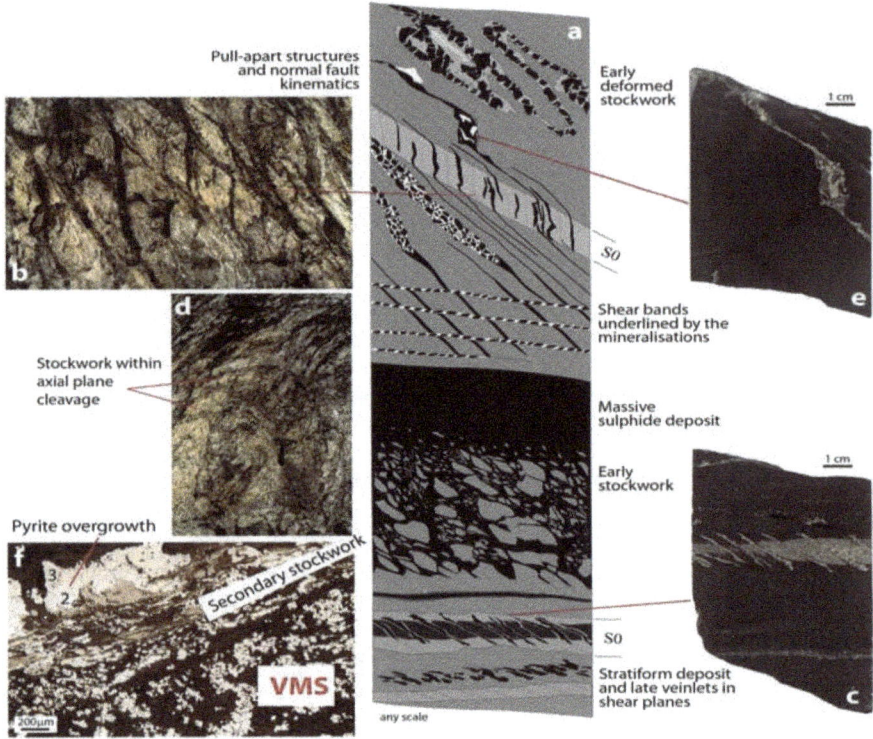

Figure 8. Schematic illustration showing the relationships between first stockwork and secondary one around the VHMS of the Iberian Pyrite Belt (south Spain). (**a**) Schematic distribution of the different stockwork and mineralised features close to a VHMS body. Occurrence of second stockwork within meter-scale shear bands (**b**) and within axial planar cleavage (**d**). (**c**) Stratiform pyrite-rich level cut by secondary pyrite-rich veins. (**e**) Small pull-apart filled by syntectonic quartz and pyrite. (**f**) Pyrite metablasts and overgrowths (2 and 3) formed close to a synkinematic second stockwork.

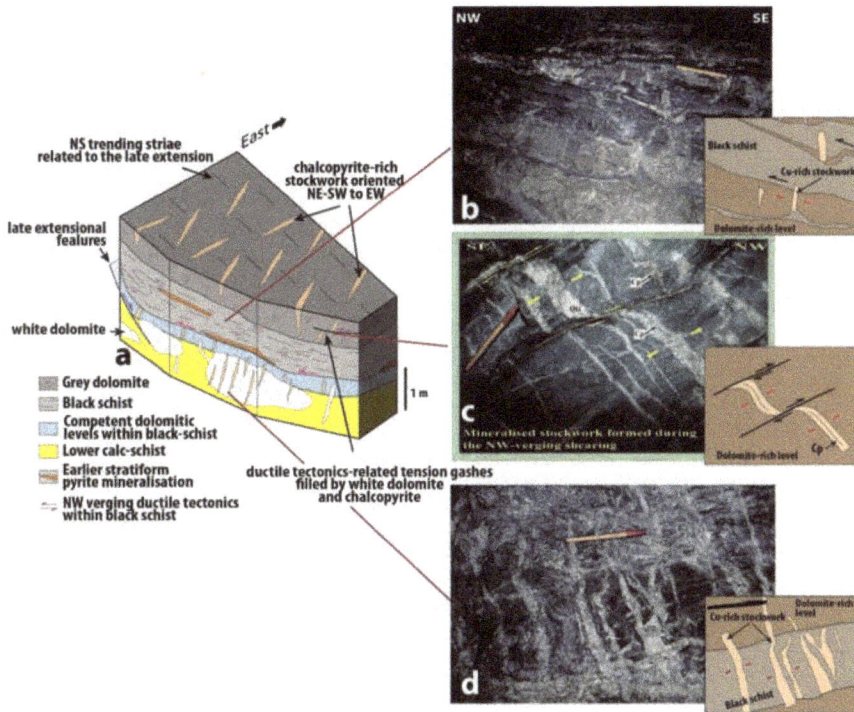

Figure 9. (**a**) Illustrations of the Ifri Cu mine model showing some images of the Cu-rich stockwork formed in response of ductile decollement within black schist. Each photograph is associated to a schematic cartoon in order to explain the geological process. (**b**) Example of vein formation within competent dolomite rich level in response of the ductilely deformed black schist. (**c**) Opening of chalcopyrite/quartz-rich veins due to NW-directed shearing. (**d**) Stockwork formation by contrasted behaviour of dolomite and black schist.

3.4. Rheological Control on Ore Concentration

In this chapter, we illustrate the significant role of rheological conditions on the development of ore deposits. Here, we will highlight two cases in which changing rheological conditions due to intensive tourmaline-rich alteration process favoured the formation of lode-related deposits.

3.4.1. Sn-rich Breccia Formation of the Achmmach Prospect (Moroccan Central Massif)

The Achmmach tin mineralisation occurs within the NE part of the Massif Central domain of Morocco, hosted by Ordovician, Silurian, and Devonian low-grade meta-sediments affected by the Hercynian deformational events, weakly represented in this area. The region concerned by the deposit exhibits a regular, N030–045°E trending cleavage mainly within the Silurian calc-schists. The mineralisation is the result of a long-lived process that includes four events that occurred in a transtensional tectonic regime associated with the late magmatic-hydrothermal evolution of the Hercynian orogeny (see details in [52]).

- The first event is the formation of tourmaline-rich halos in core of the calc-schist. These halos have ellipsoidal shape resembling tension gashes and are supposed to have formed during E–W trending shortening. Since they follow the N070°E trend of the cleavage, most halos are "en echelon" and indicators of a right-lateral potential shearing. Conjugate left-lateral "en echelon" tourmaline halos also exist but are less common. The rock shown in Figure 10 was collected in the core of one of the alteration halos and is entirely affected by the tourmalinisation.

- The second event is link to the development of right-lateral shearing only in levels affected by the tourmalinisation (Figure 10a,c). It is noteworthy that this deformation is consistent with the same tectonic context and therefore probably result from ongoing transtension controlled by E–W shortening. Main shear bands are oriented N070°E.
- Third, we have evidence of transformation of the previously formed shear band in tourmaline-rich breccia levels (Figure 10a,c). Such levels can reach thickness of 2 or 3 meters. The breccia is matrix-supported with a well-developed tourmaline-rich matrix, and can exhibit some domains with fragment-preferred orientation thus translating to tectonic- and hydrothermal-type breccia.
- The fourth texture is the most important because it is associated with cassiterite and thus representative of the economic stage. Transtension is transformed in extension and normal faults developed with the formation of a clast-supported breccia with numerous voids formation and cassiterite crystallisation (Figure 10b). These mineralised structures are systematically formed at the core of the first breccia levels and always in association with the tourmaline halos.

To conclude, mineralisation in the case of Achmmach prospect is clearly the result of polyphase deformation during the late orogenic evolution of this Hercynian domain and certainly associated with some granite emplacement. Granite remains hidden except for the occurrence of some rare outcrops. Magmatic affinity is demonstrated by the ubiquitous presence of tourmaline at each stage of the process. It is suggested in this case that ore concentration benefitted from the change of rheology due to tourmaline invasion (tourmaline-rich halos). The process was achieved by ongoing successive structures until the final formation of mineralised orebodies (shearing, brecciation and cassiterite crystallisation).

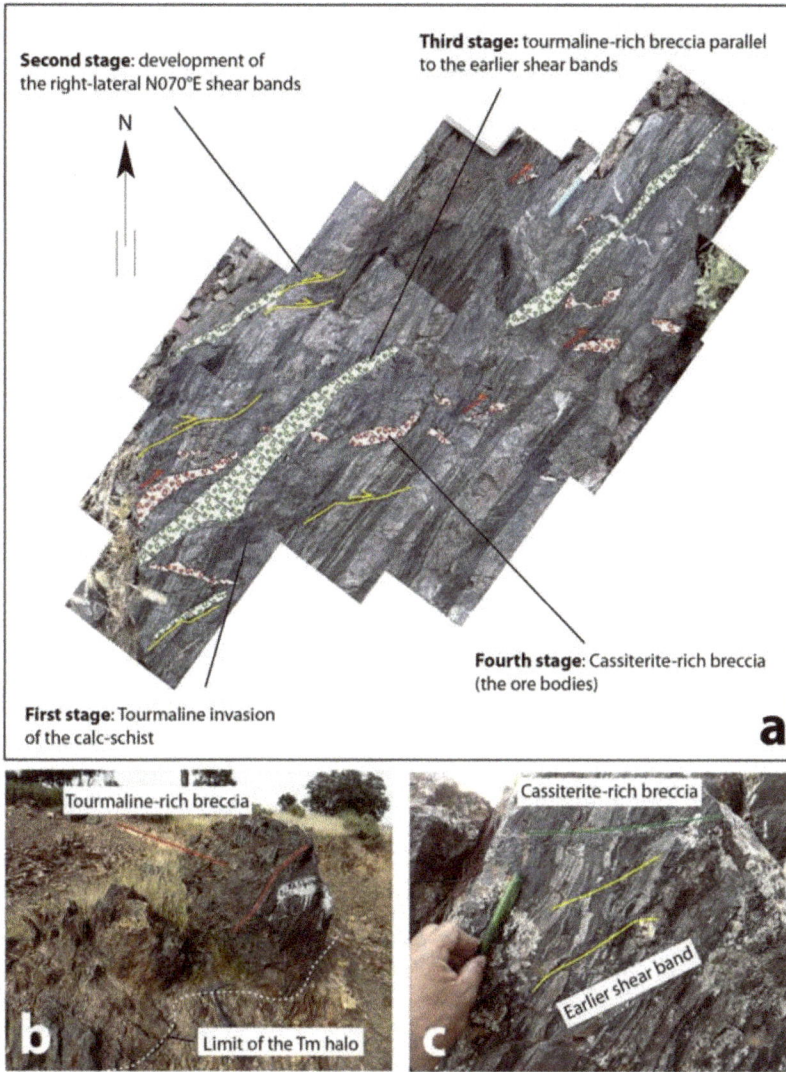

Figure 10. Deposit history and evolution of the four stages that explain the formation of the Achmmach tin deposit (Morocco) recognised in a unique block (**a**). (**b**) Image of the "North Zone" area in which the succession of structure can be observed. Note the limit of the tourmaline alteration halo that delimitate the zone where mineralisation developed. (**c**) Parallelism between ductile shear bands and mineralised breccia showing that breccia re-used the earlier plane of deformation to develop.

3.4.2. The Sn–W-rich Perigranitic Mineralisations of Beariz (Galicia, Spain)

The Hercynian orogenic cycle is reputed for abundant calc-alkaline granites that were emplaced during late orogenic Carboniferous extension (collapse of the thickened crust) [53]. This event also related to an intensive ore formation frequently starting with Sn–W mineralisation close to the granitic bodies grading to As–Au towards more distal parts [54]. This event has been called "Or 300" by the French school, and recognised throughout the European Hercynides from the Bohemia Massif to Maghreb [55,56]. In Galicia (NW Spain), the Tras-os-Montes area is a segment of the Hercynian orogen and characterised by voluminous magmatic complexes emplaced between 325 and 300 Ma (from G1 to G4 granitic events, [57]). The area is the site of abundant ore deposits (Au, W, Sn, REE). The example shown herein concerns the Sn–W deposit of the Mina Soriana in which mineralised veins were formed during a tectonic context dominated by NS extension linked to EW shortening [58]. The veins formed with E–W strike normal to a NS stretching lineation. This phenomenon is clearly exposed within the Mina Soriana outcrop. The Mina Soriana outcrop (Figure 11a) exposes a horizontal sill of leucogranite that was injected into mica schist during the N–S lineation-related tectonic event. Tourmaline halos (50 cm of thickness) are developed along the upper and lower host rocks (Figure 11b). Steeply dipping veins mainly filled by quartz and tourmaline occur normal to the magmatic sill and only within the tourmaline-rich halo (Figure 11c). This indicates that micaschist affected by the alteration had changed its competence and reacted differently than the surrounding unaltered micaschist. Veins are limited to the alteration halo although some of them cross cut the magmatic sill, as in boudinage-related structures (Figure 11a). Since tourmaline grains are aligned N–S within the alteration halo but also at the margin of quartz vein (Figure 11d), all features, i.e., magmatic sill emplacement, tourmaline-rich halo, and quartz vein development, are coeval and controlled by the same tectonic event.

In this case, the different behaviour, as explained in Figure 1b, lies at the origin of vein formation and, hence, of the formation of the Mina Soriana main vein, which outcrops further north, with the same orientation as the small-scale quartz veins and is mined as the main orebody. The difference with Figure 1b is that, in this case, the variation of rheology is not a pre-existing lithological feature, but was created during the same process that produced the ore concentration.

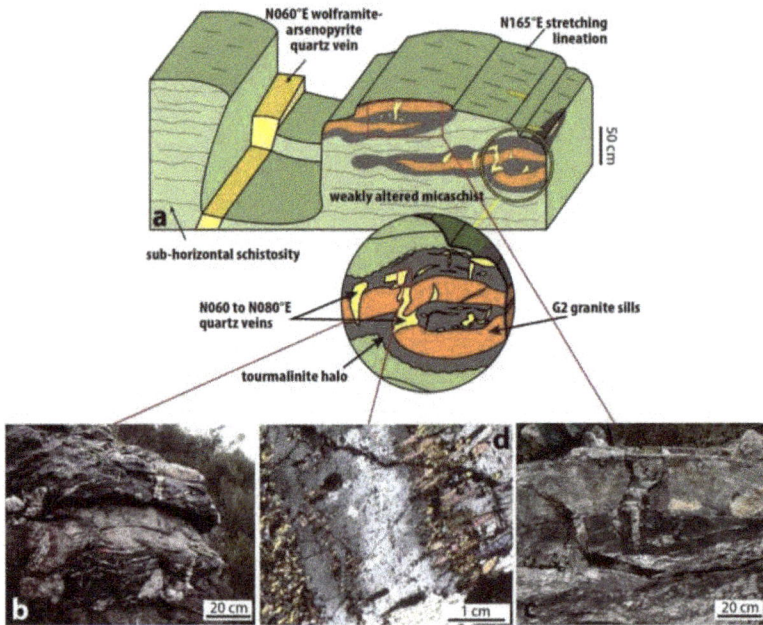

Figure 11. (**a**) Mina Soriana outcrop, Galicia (Spain), showing the development of mineralised quartz veins thanks to tourmaline-rich alteration halos developed in response to granitic sill emplacement (modified from [59]). (**b**) Close view of the granitic sill, tourmaline halo and vertical quartz veins. (**c**) Microscopic view of quartz vein rim showing tourmaline syntectonic growing. (**d**) Development of vertical quartz mineralised veins limited to the tourmaline-rich alteration halo.

3.5. Re-Use of Magmatic Structures: The Magmatic-Hydrothermal Transition

In the same geological context as the Mina Soriana outcrop, gold mineralisation exists, but close to the G3 granites (the Bruès one, in this case). Gold is associated with quartz-bearing veins that form a regular network emplaced at the Northern cupola of the Bruès G3 granite [58]. This network is also E–W oriented, and veins dip steeply to the north. A link with the G3 granite emplacement has been demonstrated by field and microstructure analysis [57,58]. Few direct arguments have been advanced, but the fact that mineralised quartz veins are systematically emplaced at the contact with the aplite–pegmatite dikes (Figure 12a), which are both controlled by the same structural context, is probably the strongest evidence. In thin sections, the transition from the magmatic stage to the hydrothermal is illustrated (Figure 12b) by an intermediate stage (magmatic-hydrothermal) marked by the crystallisation of K-feldspars that are larger than the ones in the surrounding granite (Figure 12b). In addition, comb shapes confirm that feldspars are syntectonic, neo-formed, and indicate the vein opening direction. Internal fractures show that opening continued after feldspar crystallisation, as evidence for a continuous process. In the centre of the vein and after comb quartz crystallisation (the hydrothermal stage), we find a thin fracture in which small white micas, recrystallised quartz, and sulphides (essentially arsenopyrite and pyrite) occur. White micas and oblique fabrics of recrystallised quartz (Figure 12c) also indicate a normal shearing consistent with the direction of vein opening. This demonstrates that a similar tectonic control prevailed from the late magmatic stages (border of the vein) until the final ore formation (sulphide and white mica at the centre of the vein).

This example confirms the indisputable spatial relationship between early magmatism, late magmatism and hydrothermalism. We highlight, here, the fact that we can follow, mineralogically and texturally, the continuity between late magmatic features (the border of the vein with K-feldspar formation), hydrothermal quartz and late sulphide growth. Once more, the role of the pre-existing network (the magmatic dike on top of the Bruès granite) seems essential for the development of the mineralised system.

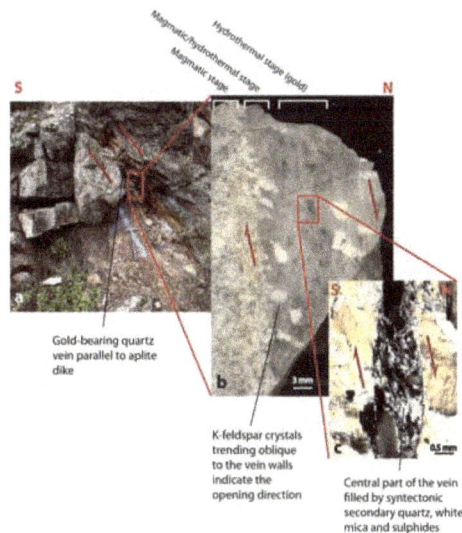

Figure 12. Relationships between gold-bearing quartz vein and granitic dike within the Bruès granite cupola, Galicia. (**a**) Outcrop view showing the close relationship between quartz vein and granitic dike. (**b**) Sample view showing the transitional contact between granite and quartz hydrothermal vein. (**c**) Thin section of the central part of the hydrothermal vein representative of the mineralised stage. Note the occurrence of dynamic recrystallised quartz and white mica indicating a normal sense of shearing. Red arrows indicate the sense of motion.

4. Discussion and Conclusions

(i)   Detailed study of geometry and composition of vein associated with ore deposits, combined with mineralogical and textural constraints, is indispensable in order to understand the mode of formation of mineralised systems. In the case of the epithermal veins of the Shila deposit, the model of formation suggests that the formation of the economic deposit is strongly dependent on the pre-existing structuring of the area.

(ii)   Without studying internal textures, the interpretation of (external) vein shape can be ambiguous and is not enough to constrain the vein formation process. The example shown of epithermal veins in southern Peru is highly illustrative in this sense. Vein geometry (main vein and associated cleavage) indicates left-lateral shearing, but the opposite conclusion is deduced when taking into account the fact that the veins and, particularly, the secondary ones, are characterised by aperture and stress-free textures that are not consistent with the classical status of what we call a cleavage. This highlights the importance of examining internal vein texture in addition to tectonic and geometrical analyses of any type of ore deposit. A similar conclusion can be drawn from the Passagem gold-bearing veins that were originally interpreted as pre-tectonic but later recognised to have formed during late-orogenic collapse affecting the area. This has significant implications for exploration and exploitation strategies because of the different age and predicted local geometries (angle, elongation) of the potential orebodies.

(iii) Two examples demonstrate the existence of syntectonic stockwork, i.e., metal remobilisation within the huge VHMS of the Iberian Pyrite Belt and the copper mineralisation of the Moroccan High Atlas. Few studies have really demonstrated this hypothesis, but our results provide strong evidence for the synchronism between stockwork formation/emplacement and deformation. Even if secondary stockwork formation does not represent an economic goal within the Iberian Pyrite Belt, such a process led to the formation of the economic orebodies of Ifri (Moroccan High Atlas). This highlights the importance of detailed study of any type of mineralised veins, even if at first inspection they do not seem to be of direct economic interest.

(iv)   Although the term "magmatic-hydrothermal transition" may sound old fashioned [60], we demonstrate with the example of Bruès (the last one) that, even though we cannot prove that the mineralising fluid were magmatic, ore formation is intimately associated with the late evolution of magmatic systems in many orogenic and/or mesothermal gold deposits [61]. The Bruès outcrop is a wonderful demonstration of continuity between late magmatic process and hydrothermal mineralisation. It is remarkable that, although detailed absolute geochronology is lacking, the evidence for the same tectonic control from the earliest magmatic stages to the latest hydrothermal stage strongly favours a continuous process. This cannot be enough for affirming the link between mineralisation and granite activity but strongly argued for this and re-addressed the discussion concerning the characteristics of orogenic and intrusion-related gold deposit (IRGD) [62,63].

(v)   Competency contrasts in a volume of rock also appear to be a favourable factor for ore concentration and vein formation [64], as shown herein for the Achmmach tin deposit and Mina Soriana W. In these cases, rheological variation was not due to original lithological differences, but induced during early stages of the mineralisation event itself, by heterogeneous alteration. It has been argued that tin mineralisation could not have formed in the Achmmach domain without earlier development of a tourmaline halo within the monotonous calc-schist. These alteration halos, formed during an early stage of transtension tectonics, create a drastic contrast in competency contrast, which controlled the partitioning of ongoing deformation and, eventually, the mineralisation. The case of Mina Soriana is similar, but the link with magmatism is, here, highlighted by the occurrence of a granitic sill responsible for the tourmaline-rich alteration. Nonetheless, in both cases, a link with late magmatic activity can be inferred in view of the above discussion about the role of the magmatic-hydrothermal transition.

In the light of these results, the importance of tectonic and microtectonic analysis at different scales in modern metallogenic studies is underlined. This work should be realised at the regional scale down to the scale of internal vein textures. The complementarity nature of pluridisciplinary works, even though already adopted in many previous studies, has been again demonstrated by the examples proposed and discussed in this paper. Change of scale and integration within the regional, geological and tectonic context are two additional conditions for a comprehensive analysis. The benefits of such an approach are both fundamental, leading to a better understanding of the mechanism of vein formation process, and economics, leading to better knowledge of specific orebody geometry and distribution and hence highly recommended in any type of exploration program.

Funding: This research was partly funded by Projects CAPES-COFECUB and CNRS GDR Transmet.

Acknowledgments: The mining companies SEIMSA (Iberian Pyrite Belt, Spain), CVRD (Vale group, Brazil), CEDIMIN and BUENAVENTURA (Peru), MANAGEM (Morocco), SMS (Seksaoua, Morocco) and KASBAH RESSOURCES (Achmmach, Morocco) are gratefully acknowledged for their constant help, support and fruitful discussions. L. Badra, L. Bailly, L. Barbanson, Y. Branquet, P. Chaponnière, P. Couderc, M. Dardennes, A. Gaouzi, J.M. Georgel, E. Gloaguen, M. Majhoubi, M. Menezes, J. Onezime, J. Rosas, and J. Tuduri, are tanked for their contribution. Two anonymous reviewers and D. Aerden are kindly acknowledged for their fruitful and constructive review.

Conflicts of Interest: The authors declare no conflict of interest. The funders had no role in the design of the study; in the collection, analyses, or interpretation of data; in the writing of the manuscript, or in the decision to publish the results.

References

1. Durney, D.W.; Ramsay, J.G. Incremental strains measured by syntectonic crystal growths. In *Gravity and Tectonics*; De Jong, K.A., Scholten, K., Eds.; Wiley: New York, NY, USA, 1973; pp. 67–96.
2. Durney, D.W. Pressure solution and crystallization deformation. *Philos. Trans. R. Soc. Lond.* **1976**, *A283*, 229–240. [CrossRef]
3. Beach, A. The geometry of en-echelon vein arrays. *Tectonophysics* **1975**, *28*, 245–263. [CrossRef]
4. Bons, P.D. The formation of large quartz veins by rapid ascent of fluids in mobile hydrofractures. *Tectonophysics* **2001**, *336*, 1–17. [CrossRef]
5. Bons, P.D.; Elburg, M.A.; Gomez-Rivas, E. A review of the formation of tectonic veins and their microstructures. *J. Struct. Geol.* **2012**, *43*, 33–62. [CrossRef]
6. Fisher, D.; Byrne, T. The character and distribution of mineralized fractures in the Kodiak Formation, Alaska: Implications for fluid flow in an underthrust sequence. *J. Geophys. Res.* **1990**, *95*, 9069–9080. [CrossRef]
7. McCaffrey, K.; Lonergan, L.; Wilkinson, J. *Fractures, Fluid Flow and Mineralization*; Geological Society of London Special Publication: London, UK, 1999; 155p.
8. Richards, J.P.; Tosdal, R.M. Structural controls on ore genesis. *Rev. Econ. Geol.* **2001**, *14*, 181.
9. Smith, J.V. En echelon sigmoidal vein arrays hosted by faults. *J. Struct. Geol.* **1996**, *18*, 1173–1179. [CrossRef]
10. Smith, J.V. Geometry and kinematics of convergent conjugate vein array systems. *J. Struct. Geol.* **1996**, *18*, 1291–1300. [CrossRef]
11. Cox, S.F. Deformational controls on the dynamics of fluid flow in mesothermal gold systems. In *Fractures, Fluid Flow and Mineralization*; McCaffrey, K., Lonergan, L., Wilkinson, J., Eds.; Geological Society of London Special Publication: London, UK, 1999; Volume 155, pp. 123–140.
12. Spencer, S. The use of syntectonic fibres to determine strain estimates and deformation paths: An appraisal. *Tectonophysics* **1991**, *194*, 13–34. [CrossRef]
13. Hilgers, C.; Urai, J.L. Microstructural observations on natural syntectonic fibrous veins: Implications for the growth process. *Tectonophysics* **2002**, *352*, 257–274. [CrossRef]
14. Barker, S.L.L.; Cox, S.F.; Eggins, S.M.; Gagan, M.K. Microchemical evidence for episodic growth of antitaxial veins during fracture-controlled fluid flow. *Earth Planet. Sci. Lett.* **2006**, *250*, 331–344. [CrossRef]
15. Dowling, K.; Morrison, G. Application of quartz textures to the classification of gold deposits using North Queensland examples. *Econ. Geol. Monogr.* **1990**, *6*, 342–255.
16. Dong, G.; Morrison, G.; Jaireth, S. Quartz textures in epithermal veins, Queensland—Classification, origin, and implication. *Econ. Geol.* **1995**, *90*, 1841–1856. [CrossRef]

17. Taylor, R. *Ores Textures, Recognition and Interpretation*; Economic Geology Research Unit and Springer: Berlin, Germany, 2009; 288p.

18. Funedda, A.; Naitza, S.; Buttau, C.; Cocco, F.; Dini, A. Structural Controls of Ore Mineralization in a Polydeformed Basement: Field Examples from the Variscan Baccu Locci Shear Zone (SE Sardinia, Italy). *Minerals* **2018**, *8*, 456. [CrossRef]

19. De Roo, J.A. Mass transfer and preferred orientation development during extensional microcracking in slate-belt folds, Elura Mine, Australia. *J. Metamorph. Geol.* **1989**, *7*, 311–322. [CrossRef]

20. Davis, B.K.; Hippertt, J.F.M. Relationships between gold concentration and structure in quartz veins fr'om the Hodgkinson Province, northeastern Australia. *Miner. Depos.* **1998**, *33*, 391–405. [CrossRef]

21. Forde, A. The late orogenic timing of mineralisation in some slate belt gold deposits, Victoria, Australia. *Miner. Depos.* **1991**, *26*, 257–266. [CrossRef]

22. Cox, S.F.; Knackstedt, M.A.; Braun, J. Principles of structural control on permeability and fluid flow in hydrothermal systems. In *Structural Controls on Ore Genesis*; Richards, J.P., Tosdal, R.M., Eds.; Reviews in Economic Geology; Society of Economic Geologists: Littleton, CO, USA, 2001; Volume 14, pp. 1–24.

23. Gratier, J.P.; Vialon, P. Deformation pattern in a heterogeneous material: Folded and cleaved sedimentary cover immediately overlying a crystalline basement (Oisans, French Alps). *Tectonophysics* **1980**, *65*, 151–180. [CrossRef]

24. Chauvet, A.; Onézime, J.; Charvet, J.; Barbanson, L.; Faure, M. Syn- to late-tectonic stockwork emplacement within the Spanish section of the Iberian Pyrite Belt: Structural, textural and mineralogical constraints in the Tharsis—La Zarza areas. *Econ. Geol.* **2004**, *99*, 1781–1792. [CrossRef]

25. Cox, S.F.; Etheridge, M.A. Crack-seal fibre growth mechanism and their significance in the development of oriented layer silicate microstructures. *J. Struct. Geol.* **1983**, *92*, 147–170. [CrossRef]

26. Fisher, D.M.; Brantley, S.L. Models of quartz overgrowth and vein formation: Deformation and episodic fluid flow in an ancient subduction zone. *J. Geoph. Res.* **1992**, *97*, 20043–20061. [CrossRef]

27. Hilgers, C.; Köhn, D.; Bons, P.D.; Urai, J.L. Development of crystal morphology during unitaxial growth in a progressively widening vein: II. Numerical simulations of the evolution of antitaxial fibrous veins. *J. Struct. Geol.* **2001**, *23*, 873–885. [CrossRef]

28. Ramsay, J.G. The crack-seal mechanism of rock deformation. *Nature* **1980**, *284*, 135–139. [CrossRef]

29. Cox, S.F. Antitaxial crack-seal vein microstructures and their relationship to displacement paths. *J. Struct. Geol.* **1987**, *9*, 779–787. [CrossRef]

30. Hilgers, C.; Urai, J.L. On the arrangement of solid inclusions in fibrous veins and the role of the crack-seal mechanism. *J. Struct. Geol.* **2005**, *27*, 481–494. [CrossRef]

31. Boullier, A.M.; Robert, F. Paleoseismic events recorded in Archean gold quartz vein networks, Val-Dor, Abitibi, Quebec, Canada. *J. Struct. Geol.* **1992**, *14*, 161–177. [CrossRef]

32. Bons, P.D.; Jessell, M.W. Experimental simulation of the formation of fibrous veins by localised dissolution-precipitation creep. *Mineral. Mag.* **1997**, *61*, 53–63. [CrossRef]

33. Jébrak, M. Hydrothermal breccias in vein-type ore deposits: A review of mechanisms, morphology and size distribution. *Ore Geol. Rev.* **1997**, *12*, 111–134. [CrossRef]

34. Sibson, R.H. Brecciation Processes in Fault Zones: Inferences from Earthquake Rupturing. *Pure Appl. Geophys.* **1986**, *124*, 159–175. [CrossRef]

35. Taylor, R.G.; Pollard, P.J. *Mineralized Breccia Systems—Methods of Recognition and Interpretation, Economic Geology Research Unit Contribution 46*; James Cook University of North Queensland: Douglas, Australia, 1993; 36p.

36. Clark, C.; James, P. Hydrothermal brecciation due to fluid pressure fluctuations: Examples from the Olary Domain, South Australia. *Tectonophysics* **2003**, *366*, 187–206. [CrossRef]

37. Tarasewicz, J.P.T.; Woodcok, N.H.; Disckson, J.A.D. Carbonate dilation breccias: Examples from the damage zone to the Dent Fault, northwest England. *Geol. Soc. Am. Bull.* **2005**, *117*, 736–745. [CrossRef]

38. Davies, A.G.S.; Cooke, D.R.; Gemmell, J.B.; Van Leeuwen, T.; Cesare, P.; Hartshorn, G. Hydrothermal Breccias and Veins at the Kelian Gold Mine, Kalimantan, Indonesia: Genesis of a Large Epithermal Gold Deposit. *Econ. Geol.* **2008**, *103*, 717–757. [CrossRef]

39. Woodcok, N.H.; Mort, K. Classification of fault breccias and related fault rocks. *Geol. Mag.* **2008**, *145*, 435–440. [CrossRef]

40.	Burchfiel, B.C.; Stewart, J.H. "Pull-apart" origin of the central segment of Death Valley, California. *Geol. Soc. Am. Bull.* **1966**, *77*, 439–442. [CrossRef]

41.	Chauvet, A.; Piantone, P.; Barbanson, L.; Nehlig, P.; Pedroletti, I. Gold deposit formation during collapse tectonics: Structural, mineralogical, geochronological, and fluid inclusion constraints in the Ouro Preto gold mines, Quadrilátero Ferrífero, Brazil. *Econ. Geol.* **2001**, *96*, 25–48. [CrossRef]

42.	Souza Martins, B.; Lobato, L.M.; Rosière, C.A.; Hagemann, S.G.; Schneider Santos, J.O.; dos Santos Peixoto Villanova, F.L.; Figueiredo e Silva, R.C.; de Ávila Lemos, L.H. The Archean BIF-hosted Lamego gold deposit, Rio das Velhas greenstone belt, Quadrilátero Ferrífero: Evidence for Cambrian structural modification of an Archean orogenic gold deposit. *Ore Geol. Rev.* **2016**, *72*, 963–988. [CrossRef]

43.	Kerrich, R. Geodynamic setting and hydraulic regimes: Shear zone hosted mesothermal gold deposits. In *Mineralisation and Shear Zones*; Bursnall, J.T., Ed.; Geological Association of Canada Short Course Notes; Mineralogical Association of Canada: Quebec City, QC, Canada, 1989; Volume 6, pp. 89–128.

44.	Poulsen, K.H.; Robert, F. Shear Zones and Gold: Practical Examples from the Southern Canadian Shield. In *Mineralisation and Shear Zones*; Bursnall, J.T., Ed.; Geological Association of Canada Short Course Notes; Mineralogical Association of Canada: Quebec City, QC, Canada, 1989; Volume 6, pp. 239–266.

45.	Cabral, A.R.; Zeh, A. Detrital zircon without detritus: A result of 496-Ma-old fluid–rock interaction during the gold-lode formation of Passagem, Minas Gerais, Brazil. *Lithos* **2015**, *212–215*, 415–427. [CrossRef]

46.	Cassard, D.; Chauvet, A.; Bailly, L.; Llosa, F.; Rosas, J.; Marcoux, E.; Lerouge, C. Structural control and K/Ar dating of the Au-Ag epithermal veins in the Shila Cordillera, southern Peru. *C. R. Acad. Sci. Paris* **2000**, *330*, 23–30. [CrossRef]

47.	Chauvet, A.; Bailly, L.; André, A.S.; Moni#xE9;, P.; Cassard, D.; Llosa Tajada, F.; Rosas Vargas, J.; Tuduri, J. Internal vein texture and vein evolution of the epithermal Shila-Paula district, southern Peru. *Miner. Depos.* **2006**, *41*, 387–410. [CrossRef]

48.	Onézime, J.; Charvet, J.; Faure, M.; Bourdier, J.L.; Chauvet, A. A new geodynamic interpretation for the South Portuguese Zone (SW Iberia) and the Iberian Pyrite Belt genesis. *Tectonics* **2003**, *22*, 1027. [CrossRef]

49.	Barbanson, L.; Chauvet, A.; Gaouzi, A.; Badra, L.; Mechiche, M.; Touray, J.C.; Oukarou, S. Les minéralisations Cu–(Ni–Bi–U–Au–Ag) d'Ifri (district du Haut Seksaoua, Maroc): Apport de l'étude texturale au débat syngenèse versus épigenèse. *C. R. Géosci.* **2003**, *335*, 1021–1029. [CrossRef]

50.	Gaouzi, A.; Chauvet, A.; Barbanson, L.; Badra, L.; Touray, J.C.; Oukarou, S.; El Wartiti, M. Mise en place syntectonique des minéralisations cuprifères du gîte d'Ifri (District du Haut Seksaoua, Haut-Atlas occidental, Maroc). *C. R. Acad. Sci. Paris* **2001**, *333*, 277–284.

51.	Chauvet, A.; Barbanson, L.; Gaouzi, A.; Badra, L.; Touray, J.C.; Oukarou, S. Example of a structurally controlled copper deposit from ther Hercynian Western High-Atlas (Morocco): The High Seksaoua mining district. In *The Timing and Location of Major Ore Deposits in an Evolving Orogen*; Blundell, D.J., Neuber, F., Von Quadt, A., Eds.; Geological Society of London Special Publication: London, UK, 2002; Volume 204, pp. 247–271.

52.	Mahjoubi, E.M.; Chauvet, A.; Badra, L.; Sizaret, S.; Barbanson, L.; El Maz, A.; Chen, Y.; Amman, M. Structural, mineralogical, and paleoflow velocity constraints on Hercynian tin mineralization: The Achmmach prospect of the Moroccan Central Massif. *Miner. Depos.* **2015**, *51*, 431–451. [CrossRef]

53.	Chauvet, A.; Volland-Tuduri, N.; Lerouge, C.; Bouchot, V.; Monié, P.; Charonnat, X.; Faure, M. Geochronological and geochemical characterization of magmatic-hydrothermal events within the southern Variscan external domain. *Intern. J. Earth Sci.* **2012**, *101*, 69–86. [CrossRef]

54.	Poulsen, K.H.; Robert, F.; Dubé, B. *Geological Classification of Canadian Gold Deposits*; Geological Survey of Canada, Bulletin: Ottawa, ON, Canada, 2000; Volume 540, 113p.

55.	Bouchot, V.; Milési, J.P.; Lescuyer, J.L.; Ledru, P. Les minéralisations aurifères de la France dans leur cadre géologique autour de 300 Ma. *Chron. Rech. Min.* **1997**, *528*, 13–62.

56.	Bouchot, V.; Ledru, P.; Lerouge, C.; Lescuyer, J.L.; Milési, J.P. Late Variscan mineralizing systems related to orogenic processes: The French Massif. *Ore Geol. Rev.* **2005**, *27*, 169–197. [CrossRef]

57.	Gloaguen, E. Apport D'une Étude Intégrée sur les Relations Entre Granites et Minéralisations Filoniennes (Au et Sn-W) en Contexte Tardi Orogénique (Chaîne Hercynienne, Galice Centrale, Espagne). Ph.D. Thesis, University of Orléans, Orléans, France, 2006.

Minerals **2019**, *9*, 56

58. Gloaguen, E.; Branquet, Y.; Chauvet, A.; Bouchot, V.; Barbanson, L.; Vigneresse, J.L. Tracing the magmatic/hydrothermal transition in regional low-strain zones: The role of magma dynamics in strain localization at pluton roof, implications for intrusion-related gold deposits. *J. Struct. Geol.* **2014**, *58*, 108–121. [CrossRef]

59. Sizaret, S.; Branquet, Y.; Gloaguen, E.; Chauvet, A.; Barbanson, L.; Arbaret, L.; Chen, Y. Estimating the local paleo-fluid flow velocity: New textural method and application to metasomatism. *Earth Planet. Sci. Lett.* **2009**, *280*, 71–82. [CrossRef]

60. Audétat, A.; Günther, D.; Heinrich, C.A. Formation of magmatic-hydrothermal ore deposits: Insights from LA-ICP-MS analysis of fluid inclusions. *Science* **2008**, *279*, 2091–2094.

61. Pe-Piper, G.; Piper, D.J.W.; McFarlane, C.R.M.; Sangster, C.; Zhang, Y.; Boucher, B. Petrology, chronology and sequence of vein systems: Systematic magmatic and hydrothermal history of a major intracontinental shear zone, Canadian Appalachians. *Lithos* **2018**, *304–307*, 299–310. [CrossRef]

62. Groves, D.I.; Goldfarb, R.J.; Gebre-Mariam, M.; Hagemann, S.; Robert, F. Orogenic gold deposits: A proposed classification in the context of their crustal distribution and relationship to other gold deposit types. *Ore Geol. Rev.* **1998**, *13*, 7–27. [CrossRef]

63. Lang, J.R.; Baker, T. Intrusion-related gold systems: The present level of understanding. *Miner. Depos.* **2001**, *36*, 477–489. [CrossRef]

64. Everall, T.J.; Sanislav, I.V. The Influence of Pre-Existing Deformation and Alteration Textures on Rock Strength, Failure Modes and Shear Strength Parameters. *Geosciences* **2018**, *8*, 124. [CrossRef]

minerals

MDPI

Article

Structural Control on the Formation of Pb-Zn Deposits: An Example from the Pyrenean Axial Zone

Alexandre Cugerone [1,*], Emilien Oliot [1], Alain Chauvet [1], Jordi Gavaldà Bordes [2], Angèle Laurent [1], Elisabeth Le Goff [3] and Bénédicte Cenki-Tok [1]

[1] Géosciences Montpellier, UMR CNRS 5243, Université de Montpellier, Place E. Bataillon, CC 60, 34095 Montpellier, France; emilien.oliot@umontpellier.fr (E.O.); alain.chauvet@umontpellier.fr (A.C.); angele.laurent@etu.umontpellier.fr (A.L.); benedicte.cenki-tok@umontpellier.fr (B.C.-T.)

[2] Conselh Generau d'Aran, Vielha, 25530 Lleida, Spain; j.gavalda@aran.org

[3] Bureau de Recherches Géologiques et Minières (BRGM), Territorial Direction Languedoc-Roussillon, 1039 Rue de Pinville, 34000 Montpellier, France; e.legoff@brgm.fr

* Correspondence: alexandre.cugerone@umontpellier.fr; Tel.: +33-643-983-585

Received: 21 September 2018; Accepted: 23 October 2018; Published: 26 October 2018

Abstract: Pb-Zn deposits and specifically Sedimentary-Exhalative (SEDEX) deposits are frequently found in deformed and/or metamorphosed geological terranes. Ore bodies structure is generally difficult to observe and its relationships to the regional structural framework is often lacking. In the Pyrenean Axial Zone (PAZ), the main Pb-Zn mineralizations are commonly considered as Ordovician SEDEX deposits in the literature. New structural field analyzes focusing on the relations between mineralization and regional structures allowed us to classify these Pb-Zn mineralizations into three types: (I) Type 1 corresponds to minor disseminated mineralization, probably syngenetic and from an exhalative source. (II) Type 2a is a stratabound mineralization, epigenetic and synchronous to the Variscan D_1 regional deformation event and (III) Type 2b is a vein mineralization, epigenetic and synchronous to the late Variscan D_2 regional deformation event. Structural control appears to be a key parameter in concentrating Pb-Zn in the PAZ, as mineralizations occur associated to fold hinges, cleavage, and/or faults. Here we show that the main exploited type 2a and type 2b Pb-Zn mineralizations are intimately controlled by Variscan tectonics. This study demonstrates the predominant role of structural study for unraveling the formation of Pb-Zn deposits especially in deformed/metamorphosed terranes.

Keywords: Pb-Zn deposits; Pyrenean Axial Zone; SEDEX; remobilization; structural control; sphalerite

1. Introduction

The world's most important Pb-Zn resources consist in Sedimentary-Exhalative (SEDEX) mineralizations [1]. These types of ore deposits are syngenetic sedimentary to diagenetic. Occurrence of laminated sulfides parallel to bedding associated to sedimentary features (graded beds, etc.) are the key geological argument [2]. These important deposits occur often in ancient metamorphosed and highly deformed terranes for example in Red Dog, Alaska [3,4]; Rampura, India [5]; or Broken Hill, Australia [6]. In these cases, the processes of ore formation are still largely debated. In consequence, unraveling the relationships between mineralization and orogenic remobilization(s) is essential in order to understand the genesis of Pb-Zn deposits in deformed and metamorphosed environments. For example, in Broken Hill [6–8] and Cannington [9] deposits in Australia some authors argued for a metamorphogenic and epigenetic mineralization as large metasomatic zones may have refined pre-existing Pb-Zn rich rocks. Other authors consider a pre-metamorphic and syngenetic origin with only limited remobilization linked to tectonic events [10–12]. In the world-class Jinding Pb-Zn deposit, the host rock has undergone a complex tectonic deformation [13]. Some authors proposed a syngenetic origin of the deposit [14,15] whereas others argued for an epigenetic genesis of the deposit based on field study, textural evidences [16–19], fluid inclusion [19,20], and paleomagnetic age [13]. Nowadays, these high-tonnage Pb-Zn deposits are the preferential target of numerous academic and industrial studies also for the presence of rare metals like Ge, Ga, In, or Cd associated with sulfides.

The Pb-Zn deposits hosted in the Pyrenean Axial Zone (PAZ) area that has suffered Variscan tectonics [21–23] are usually considered to be SEDEX. As an example, due to their geometry and the presence of distal volcanic rocks, Bois et al. [24] and Pouit et al. [25] considered as SEDEX the Pb-Zn mineralizations located in the Pierrefitte anticlinorium. In Bentaillou area, Fert [26] and Pouit [27,28] demonstrated that the stratigraphic and sedimentary controls were dominant processes during the genesis of these mineralizations. In the Aran Valley, deposits (Liat, Victoria-Solitaria, and Margalida) have been studied by Pujals [29] and Cardellach et al. [30,31]. These authors concluded on a stratiform and possibly exhalative formation of Pb-Zn mineralizations associated with a poor remobilization during Variscan deformation. Only few authors have documented the impact of Variscan tectonics on the genesis of these mineralizations. These are Alonso [32] in Liat, Urets, and Horcalh deposits or Nicol [33] for Pierrefitte anticlinorium deposits. In the Benasque Pass area, south of the Bossòst anticlinorium, Garcia Sansegundo et al. [34] indicated probable Ordovician stratiform or stratabound Pb-Zn mineralizations intensely reworked during Variscan tectonics. The Pb isotopes study realized by Marcoux [35] showed a unique major event of Pb-Zn mineralization interpreted as sedimentary-controlled and Ordovician or Devonian in age. Remobilization processes of Pb isotopes seem however poorly constrained and a complete structural study related to these analyzes is lacking.

Pyrenean sulfide mineralizations are an excellent target for investigating the links between orogenic deformation(s) and the genesis of associated mineralization(s), as well as finding key arguments to make the distinction between strictly syngenetic or rather epigenetic mineralizations and structurally remobilized mineralizations. In this work we will demonstrate that Pb-Zn deposits from five districts in the PAZ, previously largely considered SEDEX, were actually formed through processes involving a strong structural control.

2. Geological Setting

The Pyrenean Axial Zone (PAZ, Figure 1) is the result of the collision between the Iberian and Eurasian plates since the Lower Cretaceous. Deep parts of the crust were exhumed during this orogeny. The PAZ is composed of Paleozoic metasedimentary rocks locally intruded by Ordovician granites deformed and metamorphosed during the Variscan orogeny, like the Aston or Canigou gneiss domes [23,36].

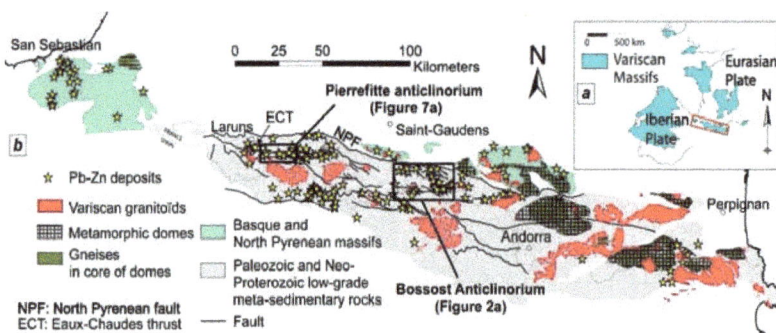

Figure 1. (**a**) Location of the Pyrenean Axial Zone (PAZ) within the Variscan belt of Western Europe. (**b**) Schematic map of the Pyrenean Axial Zone (PAZ) and location of all recognized Pb-Zn deposits (based on BRGM (French geological survey) and IGME (Spanish geological survey) databases). Note the abundance of these deposits especially in the central and western domains of the PAZ.

The PAZ is generally divided in two domains [21,36–39]: (i) a deep-seated domain called Infrastructure, which contains medium to high-grade metamorphic rocks and (ii) a shallow-seated domain called the Superstructure, which is composed of low-grade metamorphic rocks. The Infrastructure presents flat-lying foliations but highly deformed domains appear locally with steep and penetrative crenulation foliations. Alternatively, the Superstructure presents moderate deformation associated to a slaty cleavage [40,41] These two domains are intruded by Late-Carboniferous granites, like the Bossòst and the Lys-Caillaouas granites [37,42,43].

In the PAZ several deformation phases essentially Variscan in age (325–290 Ma) are recognized. The first deformation event (D_1) is marked by a cleavage (S_1) that is often parallel to the stratification (S_0). Regional M_1 metamorphism is of Medium-Pressure and Low-Temperature (MP/LT) and synchronous of this first D_1 deformation [22]. The second deformation event (D_2) is expressed by a moderate to steep axial planar (S_2) cleavage. M_2 is a Low-Pressure and High-Temperature (LP/HT) metamorphism linked to the Late-Variscan granitic intrusions, and it is superposed to the M_1 metamorphism [44,45]. Late-Variscan and/or Pyrenean-Alpine D_3 deformations are locally expressed as fold and shear zones like the Merens and/or probably the Bossòst faults [41,46,47].

The Pyrenean Pb-Zn regional district is the second largest in France with ~400,000 t Zn and ~180,000 t Pb extracted [48,49]. These sulfides deposits are localized in the PAZ in the Pierrefitte and Bossòst anticlinoriums (Figure 1b). Sphalerite (ZnS) and galena (PbS) are essentially present in Ordovician and Devonian metasediments. Few Pb-Zn deposits are hosted in granitic rocks [50].

This study focuses on Pb-Zn deposits located in the Bossòst anticlinorium (Figure 1) [42,44,51] and includes a comparison with Pb-Zn deposits occurring in the Pierrefitte anticlinorium. The southern part of the Bossòst anticlinorium forms the Aran Valley synclinorium. The northern part is limited by the North Pyrenean fault (Figure 2a). It is mostly composed of Cambrian to Devonian rocks and an intruding Late-Variscan leucocratic granite named the Bossòst granite.

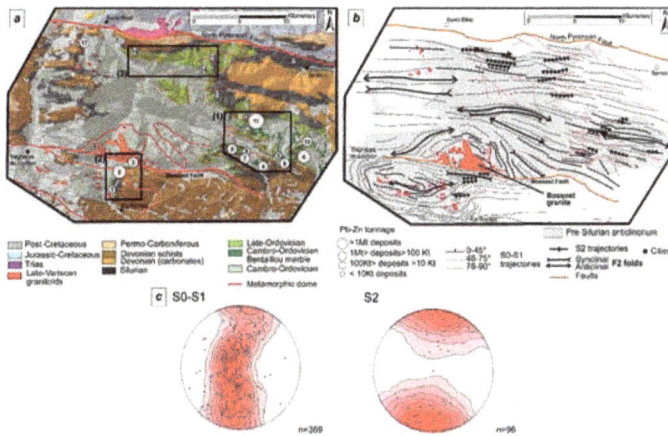

Figure 2. The Bossòst Anticlinorium. (**a**) Geological map with positions of the three districts: (1) Bentaillou-Liat-Urets district, see Figures 3 and 4; (2) Margalida-Victoria-Solitaria district, see Figure 5; and (3) Pale Bidau-Argut-Pale de Rase district, see Figure 6. Pb-Zn deposits are numbered as follows: 1: Solitaria; 2: Victoria; 3: Margalida; 4: Plan del Tor; 5: Urets; 6: Horcall; 7: Mauricio-Reparadora; 8: Estrella; 9: Liat; 10: Malh de Bolard; 11: Bentaillou; 12: Crabere; 13: Uls; 14: Pale Bidau; 15: Pale de Rase; 16: Argut. Lithologies are based on geological map of BRGM (France [52–54]) and IGME (Spain, Aran Valley; Garcia-Sansegundo et al. [55]). Metamorphic dome boundaries are related to andalousite isograd presented by Zwart; (**b**) Structural map with foliation trajectories of S_0-S_1, subvertical S_2, and related F_2 folds. Note preferential apparition of Pb-Zn deposits when S_2 cleavage is well-expressed. (**c**) Schmidt stereographic projections (lower hemisphere) of poles to S_0-S_1 and S_2 subvertical foliation planes.

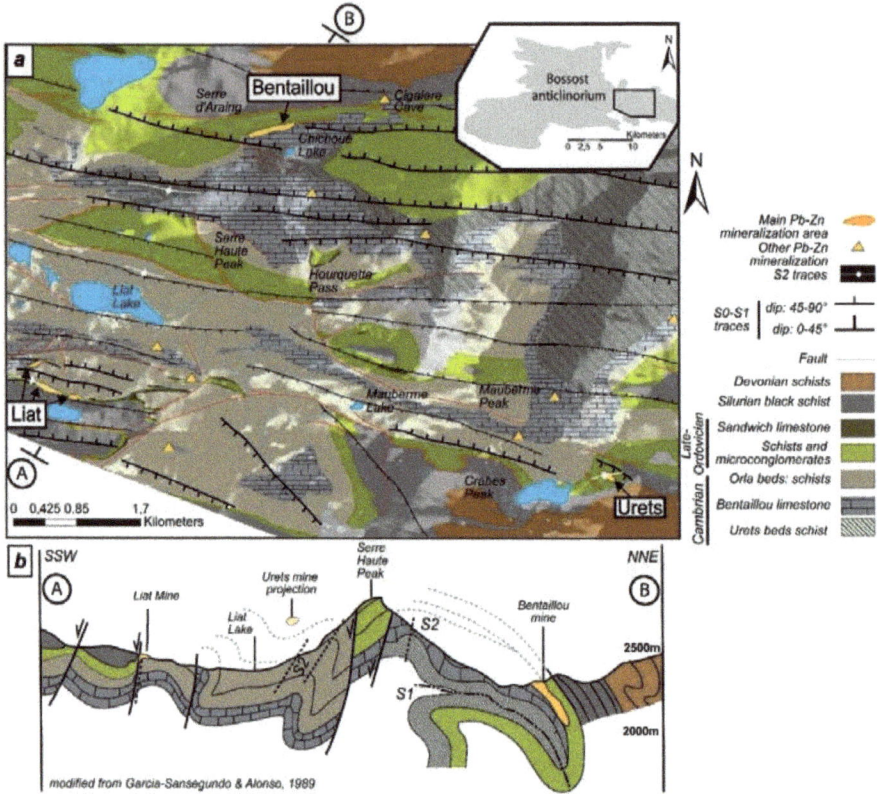

Figure 3. (**a**) Structural map of the Bentaillou-Liat-Urets district based on field study and BRGM/IGME geological maps. Location in Bossòst anticlinorium is indicated in the small sketch map (see also location on Figure 2a); (**b**) Structural NNE-SSW cross-section of the Liat-Bentaillou area. Location of the cross-section is indicated in the Figure 3a (modified from Garcia-Sansegundo and Alonso [56]). Note presence of Pb-Zn mineralization at rock competence interface and close to F1 fold hinge in Bentaillou mine.

Three main Pb-Zn districts are recognized in the Bossòst anticlinorium (Figure 2): (I) The Bentaillou-Liat-Urets district is located in the eastern part of the anticlinorium and was the most productive in the Bossòst anticlinorium, ~1.4 Mt at 9% of Zn and 2% of Pb metals [32,33]. (II) The Margalida-Victoria-Solitaria district is located in the southern part of the anticlinorium, close to the Bossòst granite. Production reached ~555,000 t with 11% Zn and 0.1% Pb [49]. (III) The Pale Bidau-Argut-Pale de Rase district is located in the northern part of the anticlinorium. Pb-Zn production did not exceed ~7000 t of Zn and ~3000 t of Pb [57].

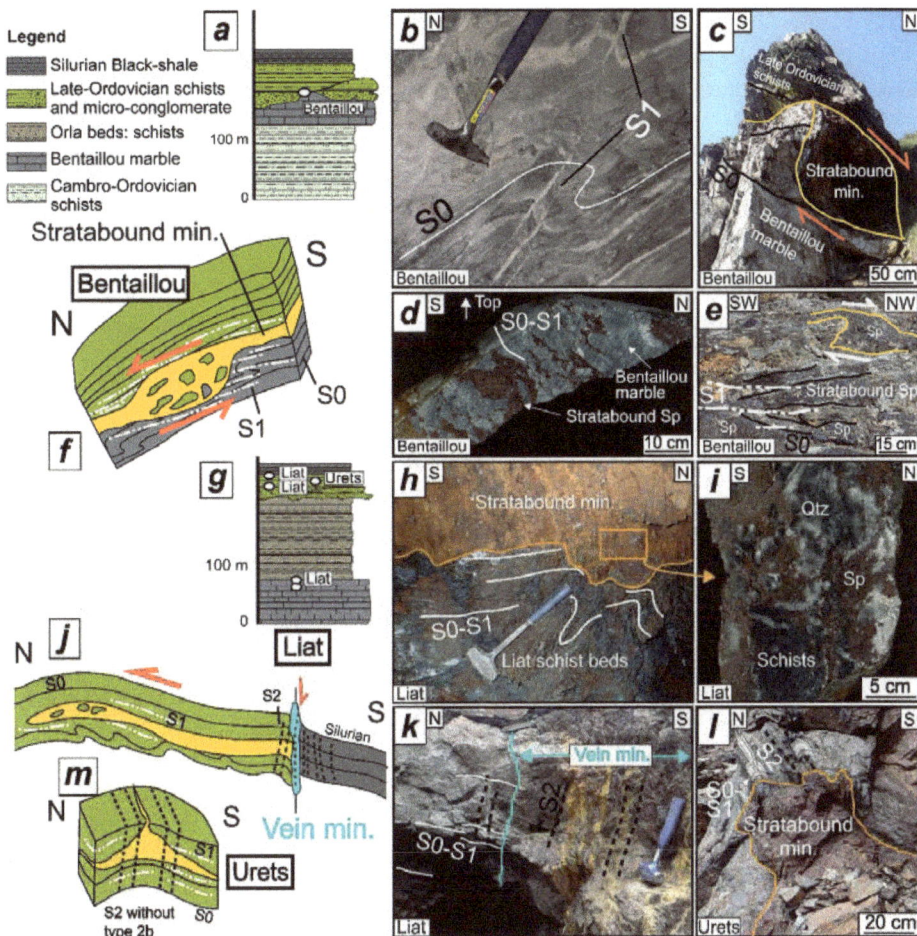

Figure 4. Field observation and structural models in the Bentaillou-Liat-Urets deposits (see location in Figure 3a). (**a**) Stratigraphic log of Bentaillou areas with position of the Pb-Zn deposits; (**b**) F_1 fold in Bentaillou marble with S_1 cleavage marked by calcite recrystallization; (**c**) pull-apart geometry of Pb-Zn mineralization in Bentaillou area; (**d**) oriented sample of typical mineralization in Bentaillou marble; and (**e**) relationship between sphalerite mineralization and host rock structure. Note that sphalerite is not folded by F_1 folds and intersect S0 stratification; (**f**) 3D structural model of Bentaillou deposits with Pb-Zn mineralization in cm to pluri-m pull apart geometry; (**g**) stratigraphic log of Liat-Urets area with position of the Pb-Zn deposits; (**h**) stratabound mineralization in top of folded schist beds in Liat area; (**i**) brecciated Pb-Zn mineralization in Liat area with clast of schist and quartz in sphalerite matrix; (**j**) 3D structural model of Liat deposit with dm to m stratabound and vein mineralizations; (**k**) vein Pb-Zn mineralization in Liat deposit; (**l**) stratabound mineralization in F_2 fold hinge in Urets deposit; (**m**) 3D structural model of Urets deposit with pluri-dm to m Pb-Zn mineralization in F_2 fold hinge. Mineral abbreviations: Qtz-quartz; Sp-sphalerite.

Pierrefitte anticlinorium is located north of the Cauteret granite and intersected by the Eaux-Chaudes thrust (ECT; Figure 1). It is essentially composed of Ordovician rocks in the West and Devonian terranes in the East. Two districts are studied: (I) Pierrefitte mines is the largest district in the PAZ which produced ~180,000 t of Zn, ~100,000 t of Pb and ~150 t of Ag [48]. (II) Arre and Anglas mines are located west to Pierrefitte mines. Pb-Zn production did not exceed ~6500 t of Zn [48].

3. Structural Analysis of Three Pb-Zn Districts in the Bossòst Anticlinorium

The Bossòst anticlinorium is a 30 × 20 km E-W-trending asymmetric antiform hosting a metamorphic dome (Figure 2a). Pre-Silurian lithologies are dominated by Cambro-Ordovician schists. Locally, other lithologies are present like the Cambro-Ordovician Bentaillou marble or the late-Ordovician microconglomerate and limestone (Figure 2b).

Two distinct cleavages can be observed in the Bossòst anticlinorium. S_1 transposes the S_0 stratification and is roughly oriented N090–N120°E with varied dip angles both to the north and to the south (Figure 2b,c). S_0-S_1 dip angles are low in the metamorphic dome (Figure 2b) but this pattern is not restricted to the core of the anticlinorium. In the eastern part of the anticlinorium, foliation is generally low to moderately dipping (0–45° N or S, Figure 2b) and Garcia-Sansegundo and Alonso [54] supposed the presence of large recumbent F_1 folds in the Bentaillou and Horcalh-Malh de Bolard areas. The presence of a Late Ordovician microconglomerate at the base of Bentaillou limestone is described by Garcia-Sansegundo and Alonso [56] and confirms this hypothesis. Furthermore, the presence of these folds is inferred by the observation of dm- to pluri-m north-verging recumbent F_1 folds in Bentaillou marble in the underground levels of the mine and also by their presence in the Devonian schists.

Close to the southern boundary of the Bossòst granite, S_0-S_1 foliation in high-grade schists is steeply dipping (Figure 2b). The S_2 cleavage trends N080–120°E and is generally sub-vertical (Figure 2c) as axial plane of F_2 south-verging folds. S_2 cleavage and related F_2 folds are particularly well developed in the southern part of the Bossòst anticlinorium (Figure 2b).

In the PAZ districts, three Pb-Zn mineralization types are commonly observed and two of these will be described below: Stratabound mineralization is subparallel to S_0-S_1 and Vein mineralization is parallel to S_2. Disseminated mineralization is not a key mineralization type and is spread in the host rocks.

3.1. District of Bentaillou-Liat-Urets

This district is located in the southeastern part of the Bossòst anticlinorium. Three main extraction areas are present in this district: (i) Bentaillou mine is located in the north of the district (Figure 3a). Exploitation finished in 1953 and produced ~110,000 t of Zn and ~40,000 t of Pb. At that time, it was the second largest mine in the Pyrenees [58], (ii) the Liat mine lays southwest of the district and (iii) Urets is located southeast of the district (Figure 3a). Both produced ~60,000 t of Zn [49]. Mineralization occurrences will be described in the following parts.

3.1.1. Bentaillou Area

Mineralization lays close to the hinge of a N090–110°E kilometer-size F_1 recumbent fold (Figure 3b) and is essentially located at the top of the Cambro-Ordovician marble, below the Late-Ordovician schists (Figure 4a). Mineralized stratabound bodies are broadly parallel to S_0-S_1 which is sub horizontal with a progressive increase of the dip from 45°N to 80°N to the lowest underground mine levels in the north (Figure 3b). Relicts axial planar S_1 of F_1 recumbent isoclinal folds are locally underlined by recrystallized calcite in N090–100°E axial planes (Figure 4b).

Pb-Zn stratabound mineralizations are present in cm- to pluri-m N-S open-filling structures which can be assimilated to pull-apart features (Figure 4c) that were formed in association with a dextral top to the north kinematic. These mineralized bodies show typical impregnation textures (Figure 4d) and sphalerite presents mm to cm grain sizes. Pb-Zn mineralization is absent in weakly D_1 deformed areas whereas it occurs in highly deformed domains associated to the appearance of S_1 cleavage in F_1 fold hinges (Figure 4e,f).

3.1.2. Liat Area

Pb-Zn mineralization is located at the rock interface (Figure 4g) and can be hosted in Bentaillou marble, especially on top of the marble, between the microconglomerate and Liat beds or between

Liat beds and the Silurian black-shale. The large hm-size open F_2 fold is bordered to the south by a Silurian synclinorium (Figure 3a,b). S_1 cleavage is strictly parallel to S_0 in the area. D_2 deformation is well expressed in the south at the contact between Silurian black-shale and Late Ordovician schists.

Mineralized stratabound bodies with pluri-dm to m-thickness appear parallel to the shallow dipping S_0-S_1. Folds in Liat schists are present locally at the base of the mineralization (Figure 4h). It presents a brecciated texture (Figure 4i) with clasts of quartz and schists. Sphalerite presents cm grain sizes. At the contact with the Silurian black-shale the dip of Late Ordovician schist increases and a normal fault is inferred. Vertical Pb-Zn vein mineralization parallel to S_2 is present in this fault. It intersects S_0 stratification, S_1 cleavage as well as stratabound mineralizations (Figure 4j,k). Vein mineralization also presents a brecciated texture and sulfide grains are oriented parallel to S_2. Sphalerite presents an infra-mm grain size.

3.1.3. Urets Area

This Pb-Zn mineralization is hosted in Liat schist. D_2 deformation is intensively present in this area, with numerous N100–130°E F_2 open to isoclinal folds associated to a subvertical N90–120°E S_2 cleavage. Stratabound pluri-dm to m Pb-Zn mineralization is mainly located in F_2 fold hinges (Figure 4l) and can locally intersect S_0 stratification (Figure 4m). Pb-Zn mineralization has a brecciated texture with mm sphalerite grains and mm to cm quartz clasts.

3.2. District of Margalida-Victoria-Solitaria

This district is located south of the Bossòst anticlinorium (Figure 2a). Three main extraction areas are present in this district from north to the south (Figure 5a): (i) Margalida mine is located close to the Bossòst granite next to the Bossòst fault, (ii) Victoria mine, and (iii) Solitaria mine lays south of the granite and north and west to Arres village. Margalida and Solitaria mine produced less than 50,000 t of ore with ~10% of Zn and 1% of Pb [49]. Victoria produced ~504,000 t with 11% of Zn and 1% of Pb [49].

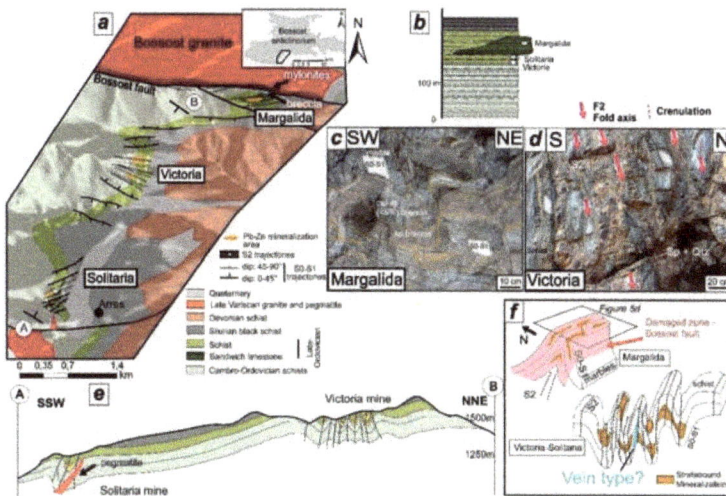

Figure 5. Margalida-Victoria-Solitaria district (see location on Figure 2a). (**a**) Structural map (lithologies are based on IGME geological map (Spain, Aran Valley; Garcia-Sansegundo et al. [1]) and location on the Bossòst anticlinorium (pre-Silurian rocks); (**b**) stratigraphic log; (**c**) stratabound Pb-Zn mineralization in Margalida mine hosted in Sandwich limestone level; (**d**) typical stratabound folded mineralization (F_2 isoclinal folds) in Victoria; (**e**) structural NNE-SSW cross-section of Victoria-Solitaria area; and (**f**) structural model of Margalida and Victoria-Solitaria mines. Mineral abbreviations: Cal—Calcite; Qtz—Quartz; Sp—Sphalerite.

3.2.1. Margalida Area

Pb-Zn mineralization is located in Late-Ordovician sandwich limestone (Figure 5a,b) which forms the core of an anticlinal presenting a vertical N100°E-trending axial plane (supposed F_2 fold). Mineralization is located in the damaged zone of the Bossòst N090°E-trending fault. Mineralization appears as pluri-dm lenses generally parallel to S_0-S_1. Still mineralization is not always concordant to S_0-S_1 (Figure 5c). The texture of sulfide mineralization in Margalida area is different to this in Victoria-Solitaria area as sulfide grain size is infra-mm.

3.2.2. Victoria-Solitaria Areas

Pb-Zn mineralization is hosted by Late Ordovician schists (Figure 5a,b,d) and generally parallel to S_0-S_1. Locally S_0-S_1 is intensively folded by F_2 asymmetrical isoclinal N090–N120°E folds and a vertical S_2 N070–110°E axial planar cleavage can be observed. Stratabound mineralization appears only in domains where F_2 folds imprint is intense (Figure 5e). Furthermore, in Victoria and Solitaria mines exploitation was preferentially undertaken in F_2 fold hinges. Pb-Zn mineralization is thicker in fold hinge (dm to m in thickness) and probably reworked during this D_2 deformation phase (Figure 5e,f). Sphalerite grains are often sub-millimetric. The presence of vein mineralization cannot be completely excluded as vertical galleries are present.

3.3. District of Pale Bidau-Argut-Pale de Rase

The general structural description of the district is given in [53]. In this section more details are given on the structural features of the Pale Bidau area (see location on Figure 2a).

Two different Pb-Zn mineralization geometries appear: a first stratabound mineralization is hosted only in F_2 fold pelitic level and concordant to S_0-S_1, marked by cm to pluri-m box-work texture. The second mineralization consists of veins oriented N090–120°E and consists of dm to m veins largely developed when D_2 deformation is important. Various dips are present for this mineralization but is mainly subvertical. Geometry of this mineralization can be interpreted as a pull-apart (Figure 6a) opened in a dextral top to the north movement and controlled by S_2 cleavage. Where S_2 cleavage is less pronounced, mineralization is thinner and seems to present in the sub-horizontal to 45°N S_0-S_1 cleavage (Figure 6b,c). Sphalerite crystals did not reach mm grain size.

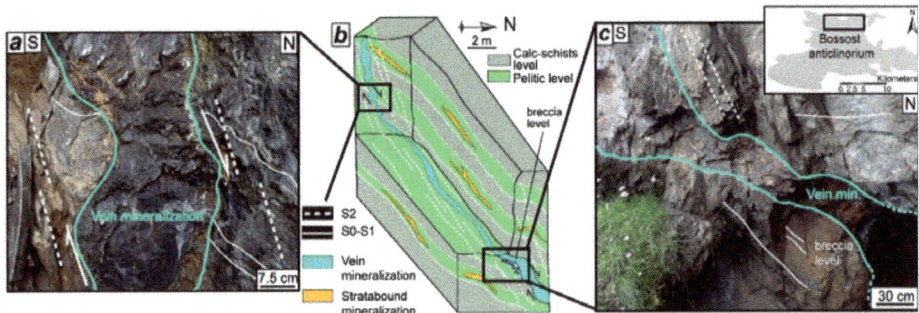

Figure 6. Field observations and 3D structural model of Pale Bidau deposit (see location in Figure 2a). (a) Vein Pb-Zn mineralization that occurs in pull-apart geometry; (b) 3D model presenting the relations between stratabound and vein Pb-Zn mineralizations; and (c) vein mineralization with presence of breccia at the base of a pull-apart mineralized structure.

4. Comparison with the Pierrefitte Anticlinorium: Pierrefitte and Arre-Anglas-Uzious Districts

The Pierrefitte anticlinorium is a 25 × 10 km NNW-SSE anticlinorium located in the western part of the PAZ (Figures 1b and 7a). Its core is composed of Ordovician schists and Late-Ordovician carbonated breccias. Upper stratigraphic levels are made of Silurian black-shales and Devonian rocks.

In western parts km-scale Valentin NNW-SSE anticlinal is included in the Pierrefitte anticlinorium. Compared to the Bossòst anticlinorium, the volume of late-Variscan granite or pegmatitic rocks outcropping is smaller and there is no metamorphic dome in the core (Figure 7a).

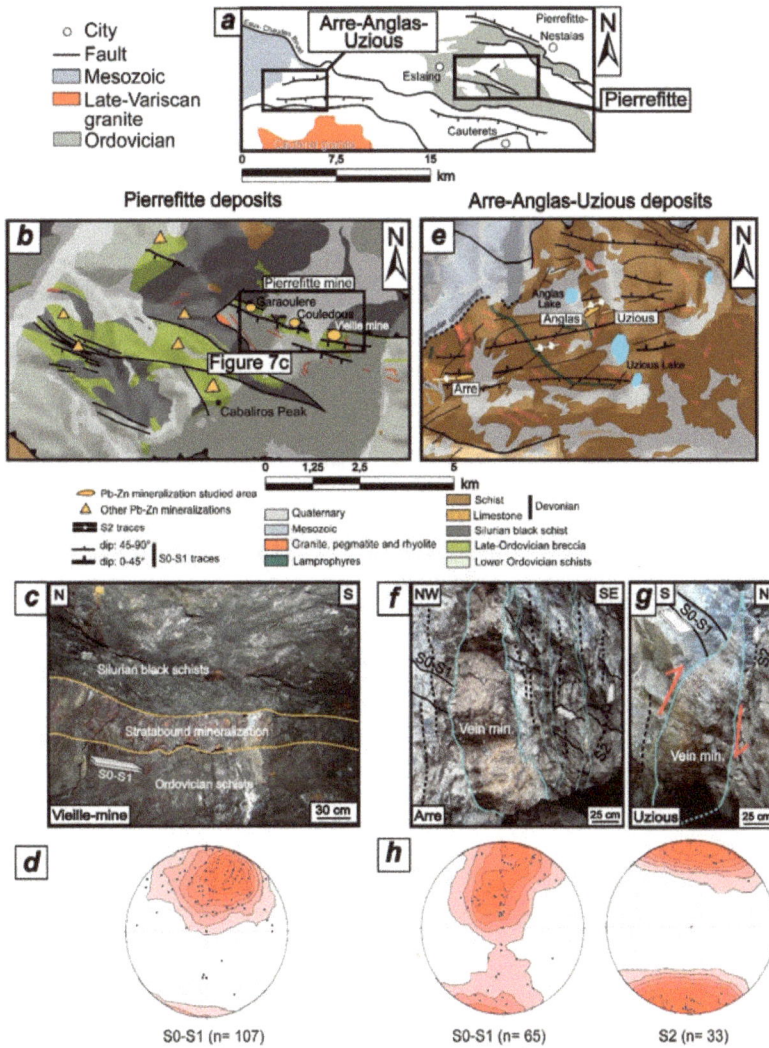

Figure 7. The Pierrefitte anticlinorium. (**a**) Simplified structural map showing the location of (**b**,**e**); (**b**) structural map zoomed on Pierrefitte mine (Lithologies are based on BRGM geological maps [59]; (**c**) photograph of typical stratabound Pb-Zn mineralization in Pierrefitte mine; (**d**) Schmidt stereographic projections (lower hemisphere) of poles to S_0-S_1 foliations measured in the Pierrefitte anticlinorium; (**e**) structural map of the Pierrefitte-Valentin anticlinorium zoomed on Anglas-Uzious and Arre mines (Lithologies are based on BRGM geological maps); (**f**) photograph of Arre vein mineralization parallel to S_2 cleavage; (**g**) photograph of Anglas-Uzious vein mineralization; and (**h**) Schmidt stereographic projections (lower hemisphere) of poles to S_0-S_1 and S_2 foliations measured in Anglas-Uzious and Arre areas.

The Pierrefitte anticlinorium is structured by several thrusts within Silurian levels (Figure 7b,c) associated to D_1 deformation. S_2 vertical N090–100°E cleavage is well expressed in Devonian levels at the rim of the anticlinorium but is less visible in the Ordovician core.

Numerous Pb-Zn mines are present in Late-Ordovician and Devonian terranes. These have produced ~3 Mt (average 9% of Zn and 5% of Pb).

4.1. Pierrefitte District

The Pierrefitte mines (Garaoulere, Couledous, Vieille-Mine) are located at the contact with Late Ordovician rocks mainly carbonate breccia. N100–110°E S_0-S_1 foliation moderately dips (20° to 60°) to the south (Figure 7d).

Stratabound mineralization lays at the top of the Late Ordovician series at the contact or within the Silurian black-shales (Figure 7c), which follows a regional thrust parallel to S_0-S_1. The presence of a thrust in Pierrefitte area is reported in [21,60,61] and this observation is supported in galleries by the occurrence of dm-scale dextral shear bands with a top-to-the-north-east kinematic. The mine galleries and the main exploited ore follow this regional thrust zone. S_1 cleavage often transposed S_0 stratification and corresponds to axial planes of isoclinal recumbent F_1 N090–120°E folds.

4.2. Arre-Anglas-Uzious District

Arre and Anglas-Uzious mines are hosted by Devonian schists and Lower Devonian limestone respectively (Figure 7e). S_2 cleavage is well-expressed even in Devonian limestone in the area and subvertical with a N090–100°E trend.

Arre mine is located in the western hinge of the Pierrefitte anticlinorium close to the contact of limestone and schistose rocks. The mineralization is composed of two ore bodies showing a trend of N040–090°E and a dip of 70°N to 90°N. Mineralization appears parallel to S_2 cleavage and discordant to S_0-S_1 (Figure 7f) which is typical of a vein mineralization. Anglas and Uzious mines are located in the northern part of the Pierrefitte anticlinorium. Mineralization consists in multiple pluri-centimeters to m vein orebodies, with several orientations from N060° to N100°E and subvertical dips. Uzious veins intersect magmatic aplite with a N050°E trend and have a pull-apart geometry (Figure 7g) linked to the presence of N090–100°E S_2 weak structures (Figure 7h). Many conjugate fractures N030–50°E with various dips are filled with mineralization close to the veins but their extension is limited to few dm.

5. Ore Petrology and Microstructures

A synthetic paragenetic sequence of the three Pb-Zn mineralization geometries investigated in this study is presented in the Figure 8. Disseminated mineralization represents the primary layered ore that is essentially composed of sparsely disseminated pluri-μm to mm grains of sphalerite, pyrite, magnetite, and galena. In all the studied deposits this mineralization is minor and does not constitute the exploited ore. Sulfides may appear in graded-beds or have a typical framboidal appearance (Figure 9a).

Stratabound and vein mineralizations constitute the main sulfides mineralizations. Sphalerite is the more widespread sulfide in these two mineralizations. Pyrite, galena pyrrhotite, chalcopyrite, and arsenopyrite are present in minor amounts. Metamorphic muscovite, chlorite, or biotite are intimately associated to sulfide mineralization. In Victoria-Solitaria, metamorphic Zn-spinel or gahnite is present in the host rocks and in breccia clasts in stratabound sulfide mineralization. The presence of gahnite in Victoria was previously reported [62]. In the host rock gahnite is elongated parallel to S_1 and is intersected by stratabound mineralization (Figure 9b).

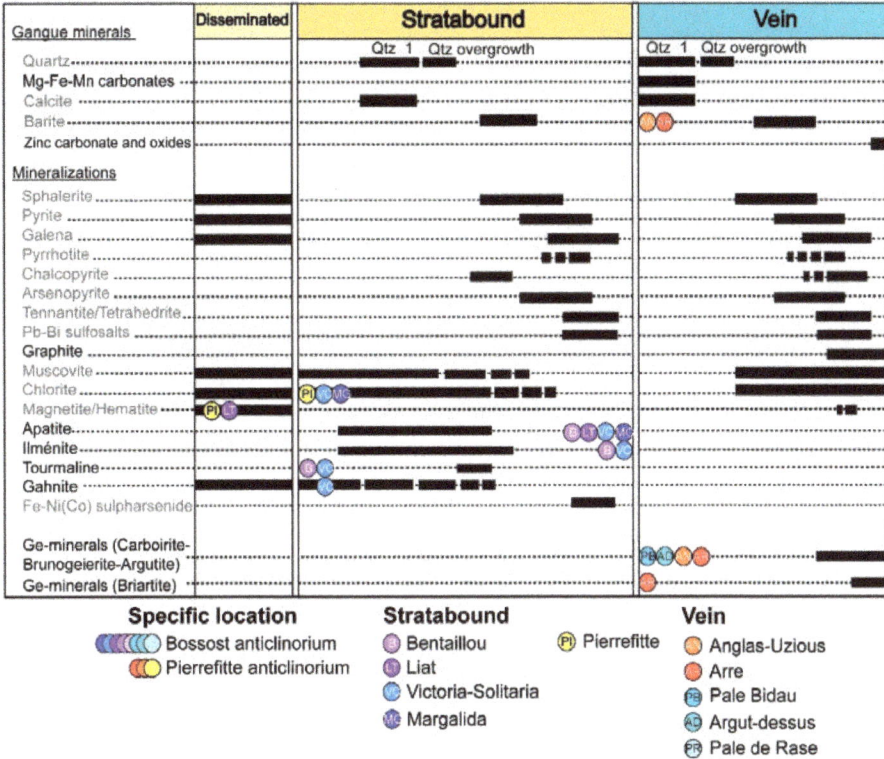

Figure 8. Paragenetic succession of ore and gangue minerals for all the eleven Pyrenean-studied Pb-Zn deposits. Minerals in grey are common to both stratabound and vein mineralizations and minerals in black are only present in stratabound or vein. Minerals only reported in a deposit are noted with the deposit circle. Several minerals like apatite, ilmenite, or tourmaline are only present in stratabound mineralization. Ge-minerals, graphite zinc carbonates or oxides, and Mg-Fe-Mn carbonates are only observed in vein mineralization (*n* = 110).

Stratabound Pb-Zn mineralization is a post-disseminated mineralization. SEM images show a primary framboidal galena intersected by a secondary stratabound pull-apart mineralization in the Bentaillou mine (Figure 9a).

In the Pierrefitte anticlinorium stratabound magnetite is abundant, especially in the Pierrefitte mine. It has crystallized prior to sphalerite. In the Pierrefitte mine syn-cinematic sphalerite crystallizes in asymmetric pressure shadows around a clast of magnetite (Figure 9c). Sphalerite appears parallel to S_1 cleavage and intersects S_0 stratification in an isoclinal F_1 fold hinge (Figure 9d). In the Bossòst anticlinorium and especially in Liat mine, sphalerite and quartz mineralization intersect F_2 folded pelitic rocks (Figure 9e). The same quartz associated to sphalerite is present in crack and seal veins (Figure 9e). In Margalida a typical durchbewegung texture with quartz spheroids in a sphalerite matrix shows a deformational imprint on this mineralization.

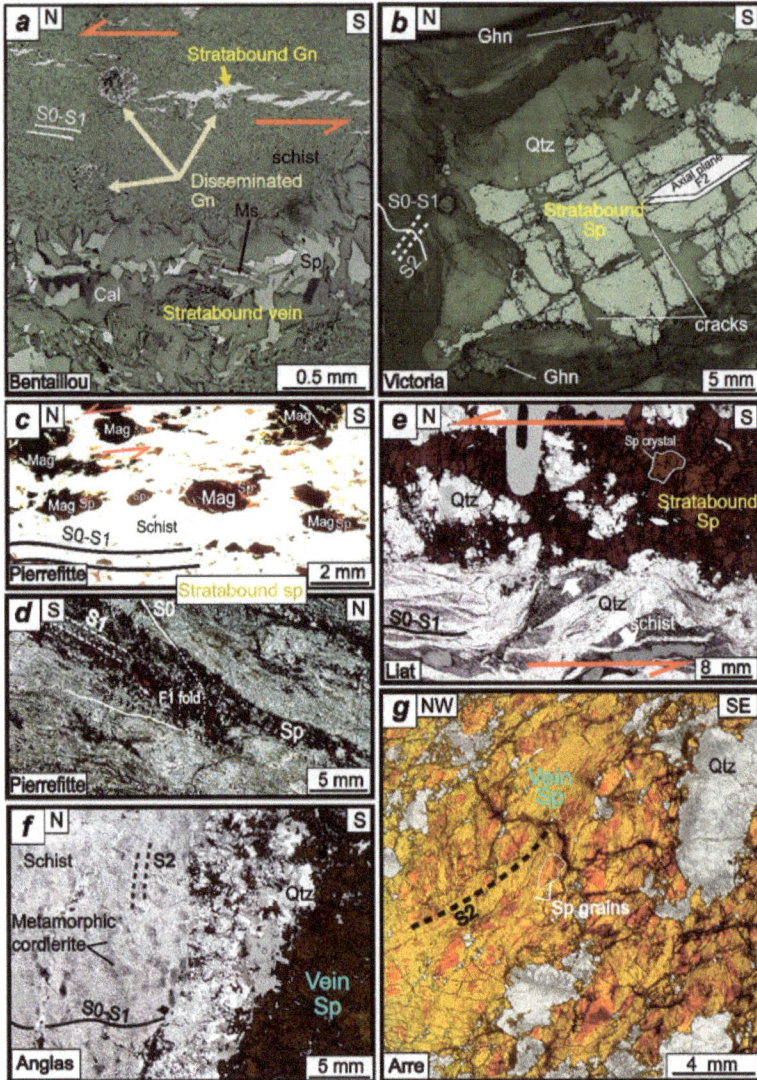

Figure 9. Microphotographs showing characteristics textures in the three mineralization types. (**a**) Bentaillou disseminated mineralization truncated by stratabound mineralizations (reflected light); (**b**) stratabound Victoria folded mineralization which intersects gahnite D_1 metamorphic mineral (reflected light); (**c**) syn-kinematic sphalerite which crystallizes in asymmetric pressure shadows around clasts of magnetite in Pierrefitte mine (transmitted light); (**d**) sphalerite mineralization from Pierrefitte mine parallel to S_1 and in F_1 isoclinal fold hinge. Mineralization intersects S0 stratification (transmitted light); and (**e**) stratabound sphalerite and quartz mineralizations which intersect pelitic host rock in Liat mine. Sphalerite is interpreted syn-kinematic D_1 (transmitted light); (**f**) vein mineralization in Anglas mine which intersect S_0-S_1 and parallel to S_2 cleavage marked by metamorphic cordierite (transmitted light); and (**g**) deformed and recrystallized vein sphalerite in Arre deposit. The two textures are identified with white line Recrystallization area in yellow marked S_2 cleavage (transmitted light). Mineral abbreviations: Cal—Calcite; Ghn—Gahnite; Gn—Galena; Mag—Magnetite; Ms—Muscovite; Qz—Quartz; Sp—Sphalerite.

Stratabound mineralization contain apatite, ilmenite, and tourmaline minerals that are only observed in this mineralization. In the Pierrefitte mineralization, the abundance of chlorite and muscovite associated to the mineralization is remarkable compared to other Pyrenean deposits.

Vein mineralization intersects S_0 at the micron scale (Figure 9f). In the Anglas deposit, vein mineralization is essentially composed of sphalerite, galena, quartz and calcite. The hanging wall of the vein is parallel to S_2 foliation and is marked by cordierite crystallization. Sphalerite in vein mineralization appears highly deformed and recrystallized with mm relictual grains and recrystallized μm-size crystals (like in Arre deposit, see Figure 9g). In Pale Bidau deposit, vein mineralization is only present in domains where the S_2 cleavage is well-marked. Note that Ge-minerals are exclusively present in the vein mineralization (Figure 8).

6. Discussion

6.1. Types of Pb-Zn Mineralizations in the PAZ

The presence of three major types of Pb-Zn mineralizations is demonstrated in this study: Disseminated but layered mineralization, which is now defined as Type 1, appears with graded-beds and framboidal appearance (Figure 9a). Stratabound mineralization (now defined as Type 2a) is a syn-D_1 mineralization concordant to the S_1 foliation. Vein mineralization (now defined as Type 2b) is a syn- to post-D_2 vein-type mineralization, parallel to the subvertical S_2 foliation. Type 2a and Type 2b are undoubtedly epigenetic and were formed as a consequence of Variscan tectonics.

The first and earlier Type 1 mineralization (Figure 10) is recognized in all the studied deposits in the Bossòst and Pierrefitte anticlinoriums, but it is not the main exploited resources. Its formation may be linked to the early volcanic Ordovician or Devonian events as proposed by Pouit [27] and Reyx [63]. In Pierrefitte anticlinorium, Nicol [55] proposed a unique Devonian source for the Pb-Zn mineralizations. Syngenetic formation is preferred for the Type 1 mineralization as sulfides appear layered and with sedimentary affinities. Nevertheless, framboidal texture may as well form in a post-sedimentation environment like in hydrothermal veins [64].

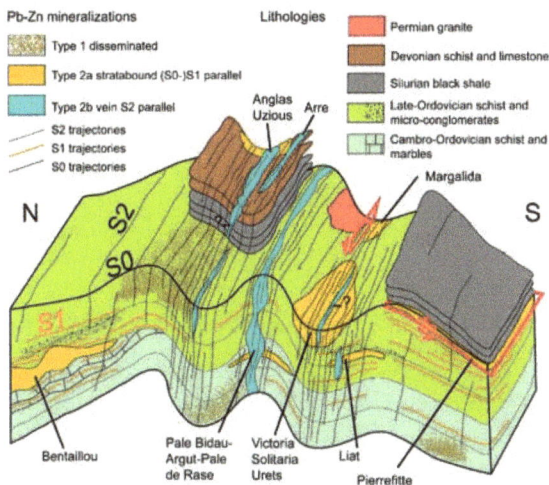

Figure 10. Schematic 3D sketch displaying the three main mineralization types which are typically observed in the studied area and related to each studied deposit. Note cm to pluri-m pull apart geometries in Bentaillou Type 2a mineralizations and in Type 2b mineralizations. Type 2b vein mineralizations are located in intensely S2 deformed domains. Other structural traps like saddle-reef formation in fold hinge or rock competence interfaces are represented for Victoria-Solitaria, Urets, Pierrefitte, and Liat deposits respectively.

The second stratabound Type 2a mineralization (Figure 10) is deposited parallel to S_0-S_1. It corresponds to the main Pb-Zn mineralization episode in the PAZ (~95% of the total exploited ore volume). In the Bentaillou area, Type 2a mineralization intersects S_0 stratification and is hosted by S_1 cleavage (Figure 4e), which is axial planar to isoclinal F_1 folds. Fert [26] and Pouit [27,28] proposed a syngenetic model for the Bentaillou deposit and described a normal stratigraphic succession that has been later folded by F_2 folds. F_1 isoclinal recumbent N090°E folds are absent in their model. Here we observe that Bentaillou Pb-Zn mineralization is localized essentially close to F_1 fold hinges at the interface between marble and schist or microconglomerate (Figure 4c). The source for Type 2a sulfides may be related to layered and supposed syngenetic Type 1 sulfides that are disseminated in the Ordovician and Devonian neighboring metasediments, or to Late-Variscan granitic intrusions, probably at least temporally close to the Type 2a mineralizations. Opening of top to the north cm to pluri-m pull-apart-type structures (Figure 4c) enables the formation of the large amount of mineralizations in Bentaillou. Pb-Zn ore is not observed at the base of Bentaillou marbles due to important karstification (Cigalere cave, Figure 3a), however it is deposed at Bularic [65] both above and below this marble level. In the Liat area, Pujals [29] described a syngenetic or diagenetic mineralization with apparent limited reworking. Our model shows that Type 2a stratabound mineralization is linked to the Variscan D_1 deformation. In the Victoria-Solitaria area, Type 2a stratabound mineralization occurs where D_2-related structures are present and can be locally remobilized in fold hinges. These thicker mineralizations in fold hinge may be linked to the saddle-reef process [66–68] associated with formation of the dilatation zone during folding. These deposits have been studied by Pujals [29], Cardellach et al. [30,69], Alvarez-Perez et al. [70], and Ovejero-Zappino [49,71]. These authors argued for a SEDEX origin based on syngenetic mineralization associated to the presence of syn-sedimentary faults. These models differ from our hypothesis: here we report that S_1 cleavage is parallel to axial plane of recumbent km-size isoclinal folds and transposes the S_0 stratification. F_2 folded Type 2a stratabound mineralization is thicker in fold hinge and intersects metamorphic minerals as gahnite. Presence of this Zn-spinel may be linked to a primary minor sulfide mineralization (Type 1, Figure 10) or to a D_1 metamorphic fluid rich in Zn. Chemistry of gahnite was analyzed by Pujals [29] and its composition is typical to metamorphosed zinc deposits. This testifies that Type 2a Pb-Zn mineralization is syn- to post-M_1 metamorphism. Alonso [32] demonstrated a predominant role of mechanical remobilization associated to deformation in the Bossòst anticlinorium and, especially, F_2 folds in Horcalh and fault in Liat. Our model is similar as we consider that the Variscan D_2 deformation locally remobilized Type 2a mineralization. The Margalida deposit records an additional deformational event compared to Victoria and Solitaria. Hosted in a ductile deformed marble and close to the Bossòst ductile fault, the Type 2a mineralization appears largely deformed with a typical durchbewegung texture. No sedimentary structure is recognized in the marble [70]. This attests for a Late Hercynian and/or Pyrenean deformation associated to the fault on the mineralization. Comparison with the Pierrefitte anticlinorium shows the same syn-D_1 Type 2a mineralization associated to regional thrust tectonics. The main exploited Pb-Zn mineralization in Pierrefitte mine was pluri-m scale levels parallel to S_0-S_1 and the regional thrust (Figure 10). Our work comforts the study of Nicol [60] which has shown an important remobilization of the mineralization in Ordovician and Devonian metasediments linked to D_1 deformation. On the contrary, Bois et al. [24] proposed a syngenetic deposition related to the activity of Late-Ordovician syn-sedimentary faults and volcanism that may have induced these mineralizations. In this case, remobilization is weak and sulfides crystalize prior to Variscan metamorphism [24]. But the presence of sphalerite parallel to S_1 cleavage and in pressure shadows around magnetite clast concordant to S_1 rather attests for a syn-D_1 mineralization event.

The third Type 2b vein mineralization (Figure 10) is parallel to S_2 cleavage. It intersects S_0-S_1 cleavage and former Type 2a stratabound mineralization. It has been recognized in the Pale Bidau-Argut-Pale de Rase districts [57] and Arre-Uzious-Anglas districts. It appears in a limited number of deposits in the PAZ. Type 2b mineralization is present in pluri-dm veins with restricted extension and highly differs structurally and mineralogically. The presence of Ge-minerals and absence of apatite, tourmaline, or ilmenite are remarkable here. Nonetheless, possible Type 2a remobilization with external contribution is not excluded in the Type 2b vein formation. In the Uzious mine mineralization intersects magmatic aplite. Therefore it has probably emplaced syn- or post-Cauteret granite and is certainly late-Variscan in age (aplite from late-Variscan Cauteret granite) as supposed by Reyx [63]. Deformation of sphalerite, which is supposed to be syn-D_2 and/or syn-D_3, and the unusual sulfide paragenesis are inconsistent with a Mesozoic age as described in Aulus-Les Argentieres undeformed sphalerite [72]. Other Pb-Zn deposits, like the La Gela deposit [73] or Carboire deposit, could be attached to this third type as they are characterized by vertical Pb-Zn veins and presence of Ge-minerals. These late-Variscan Pb-Zn deposits have been recognized in Saint-Salvy (cf. M_2 mineralization) even if the main Pb-Zn mineralization event is Mesozoic [74].

6.2. Genetic Model of PAZ Pb-Zn Deposits Formation Linked to Regional Geology

The genetic model comprises four stages (Figure 11) based on the regional tectonic event model of Mezger and Passchier [22] and Garcia-Sansegundo and Alonso [56].

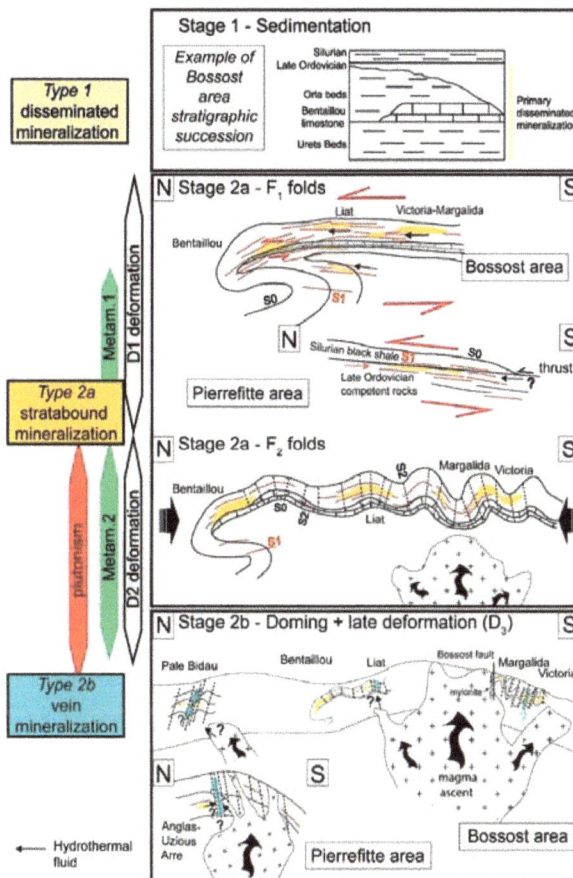

Figure 11. Genetic model for the formation of the three main Pb-Zn mineralization types. **Stage 1** is the disseminated Type 1 mineralization that is supposed to be syn-sedimentary. **Stage 2a** is the syn-D_1 Type 2a stratabound mineralization which is followed by the formation of F_2 folds and local remobilization of Pb-Zn mineralizations (saddle reef). **Stage 2b** represents the Type 2b late-Variscan vein mineralizations.

Stage 1 represents the syn-sedimentary layered mineralization (SEDEX deposit, Pb-Zn Type 1 disseminated mineralization). Primary sulfides were recognized in all pre-Silurian stratigraphic succession in the Bossòst area (Figure 11) and in Devonian rocks in Anglas-Uzious-Arre district. In the Pierrefitte area primary sphalerite is absent, which is probably linked to important hydrothermal low-grade alteration and D_1 overprint.

Stage 2a starts during the D_1 Variscan deformation and induces Type 2a stratabound mineralization. This mineralization occurs preferentially where a rheological contrast exists between two lithologies (e.g., marble-schist; schist-microconglomerates) and in highly D_1 deformed area (Figure 11). Stage 2a continues with D_2 Variscan deformation and the formation of N090–110°E F_2 upright folds. Granitic intrusions occur at that stage (Figure 11). This D_2 deformation locally reworked mineralization like in Victoria mines where the mineralization is remobilized in fold hinges. Horcalh mineralized fault [32] is interpreted as synchronous to D_2 deformation.

Stage 2b occurs during the doming phase and the late-Variscan Type 2b vein mineralizations (Figure 11). This mineralization type preferentially occurs parallel to the vertical S_2 cleavage and is mostly observed in the Pierrefitte and Bossòst anticlinoriums. Pull-apart-type structures are observed in Pale Bidau and Uzious mines. A late deformation D_3 corresponds to faults like the Bossòst mylonitic fault close to Margalida district.

We have shown that the Pb-Zn deposits in the PAZ were polyphased and closely linked to Variscan tectonics. There are at least three Pb-Zn mineralization-forming events, and two of them are evidently structurally controlled. Type 1 may be syngenetic, but little ore is present. The main exploited ores are Type 2a and Type 2b which have emplaced under a marked structurally control, either associated to S_1 and trapped in F_1 fold hinge, at lithology interface or in highly D_1 or D_2 deformed areas.

6.3. Is Pb-Zn Deposits Emplacement Sedimentary- or Structurally-Controlled?

SEDEX deposits are sedimentary controlled and syn- to diagenetic, and sulfides in them are laminated and included into the bedding [2]. In our study area, Pyrenean Pb-Zn mineralizations have been previously described as SEDEX by many authors [24,28–30,75,76]. The origin of several world-class Pb-Zn deposits is debated as well. For example, the geneses of Broken Hill-type deposit [6–12] or Jinding deposit [14–20] are still not understood and the authors have not yet decided between syngenetic or epigenetic models. In the Pyrenees, authors interpreted stratiform and lenses ore body shapes. The stratiform argument is not relevant because frequently S_0 stratification is parallel to the S_1 axial plane of isoclinal recumbent folds, typical of intensively deformed areas. Crystallization of sphalerite secant to isoclinal recumbent fold hinges attests that the main mineralization is parallel to S_1 and not to S_0. Structural observations are supported by the mineralogical study. The three PAZ Pb-Zn mineralization types contain the same constitutive minerals, like sphalerite, galena, and pyrite, but various trace minerals are present according to the type. These mineralogical differences are key parameters to distinguish between different Pb-Zn mineralization events in a single deposit.

In intensely deformed and metamorphosed terranes, the simple geometric link between mineralization and stratification is not relevant enough to distinguish between sedimentary or structural control. Structures are often parallelized due to pervasive tectonic events which makes the structural analysis difficult. Reworking of the ore-body during deformation can have obliterated geochemical tracers like isotopic data, especially Pb isotopes [77–79]. Consequently, a detailed structural study from regional to micro-scale focusing on the relationships between mineralization and cleavages is crucial. Pinpointing locations where structures like cleavage are secant (fold hinge), as well as deciphering textural relations between metamorphic minerals and mineralization, will lead to a better understanding of the ore-body genesis.

7. Conclusions

Three main types of Pb-Zn mineralizations have been distinguished in the Pyrenean Axial Zone. A minor type (Type 1) is a stratiform disseminated mineralization that presents syngenetic

Minerals **2018**, *8*, 489

characteristics. The two other mineralization types, previously described as SEDEX, are in reality post-sedimentation and formed as a result of Variscan polyphased tectonics: Type 2a is a syn-D_1 stratabound mineralization that is parallel to S_1 foliation. Type 2b is a syn to post-D_2 vein-type mineralization that is parallel to subvertical S_2 cleavage. Structural control is thus a key parameter for the remobilization of Pb-Zn mineralizations in this area like in (D_1 and D_2) fold hinges (saddle reef), high (D_1) deformed zones, rock contrast interfaces, and S_2 cleavages. A multiscale detailed structural study is essential for unraveling the formation of Pb-Zn deposits, especially in deformed and/or metamorphosed terranes.

Author Contributions: A.C. (Alexandre Cugerone) and B.C.-T. conceived the research within the framework of the A.C. (Alexandre Cugerone)'s PhD project; A.C. (Alexandre Cugerone), E.O., A.C. (Alain Chauvet), B.C.-T., J.G.B., and A.L. participated in field work; A.C. (Alexandre Cugerone) acquired the samples and performed all the analytical work under the guidance of A.C. (Alain Chauvet) and B.C.-T.; A.C. (Alexandre Cugerone) wrote the paper with contributions from E.O., B.C.-T., E.L.G., and J.G.B.

Funding: This research was funded by the French Geological Survey (Bureau de Recherches Géologiques et Minières; BRGM) through the national program "Référentiel Géologique de France" (RGF-Pyrénées).

Acknowledgments: The authors gratefully acknowledge Kalin Kouzmanov and Stefano Salvi for their involvement in the project. We thank the ARSHAL association for Bentaillou mine access and Jean-Marc Poudevigne, Louis de Pazzis, and Bernard Lafage for their precious knowledge of the Pyrenean Pb-Zn mines. We acknowledge Christophe Nevado and Doriane Delmas for thin section preparation. The authors are thankful for the editorial handling of Jax Jiang and for the constructive comments of the three anonymous reviewers.

Conflicts of Interest: The authors declare no conflicts of interest.

References

1. Wilkinson, J.J. Sediment-Hosted Zinc-Lead Mineralization: Processes and perspectives. *Treatise Geochem.* **2013**, *13*, 219–249.
2. Leach, D.L. Sediment-hosted lead-zinc deposits: A global perspective. *Econ. Geol.* **2005**, *100*, 561–607.
3. Moore, D.W.; Young, L.E.; Modene, J.S.; Plahuta, J.T. Geologic setting and genesis of the Red Dog zinc-lead-silver deposit, western Brooks Range, Alaska. *Econ. Geol.* **1986**, *81*, 1696–1727. [CrossRef]
4. Kelley, K.D.; Jennings, S. A special issue devoted to barite and Zn-Pb-Ag deposits in the Red Dog district, Western Brooks Range, northern Alaska. *Econ. Geol.* **2004**, *99*, 1267–1280. [CrossRef]
5. Hazarika, P.; Upadhyay, D.; Mishra, B. Contrasting geochronological evolution of the Rajpura-Dariba and Rampura-Agucha metamorphosed Zn-Pb deposit, Aravalli-Delhi Belt, India. *J. Asian Earth Sci.* **2013**, *73*, 429–439. [CrossRef]
6. Lawrence, L.J. Polymetamorphism of the sulphide ores of Broken Hill, NSW, Australia. *Miner. Depos.* **1973**, *8*, 211–236. [CrossRef]
7. Gibson, G.M.; Nutman, A.P. Detachment faulting and bimodal magmatism in the Palaeoproterozoic Willyama Supergroup, south-central Australia; keys to recognition of a multiply deformed Precambrian metamorphic core complex. *J. Geol. Soc.* **2004**, *161*, 55–66. [CrossRef]
8. Hobbs, B.E.; Walshe, J.L.; Ord, A.; Zhang, Y.; Carr, G.C. The Broken Hill orebody: A high temperature, high pressure scenario. *AGSO Rec.* **1998**, *2*, 98–103.
9. Walters, S.; Bailey, A. Geology and mineralization of the Cannington Ag-Pb-Zn deposit: An example of Broken Hill-Type mineralization in the eastern succession, Mount Isa Inlier, Australia. *Econ. Geol.* **1998**, *93*, 1307–1329. [CrossRef]
10. Webster, A.E. The Structural Evolution of the Broken Hill Pb-Zn-Ag Deposit, New South Wales, Australia. Ph.D. Thesis, University of Tasmania, Hobart, Australia, 2004.
11. Bodon, S.B. Paragenetic relationships and their implications for ore genesis at the Cannington Ag-Pb-Zn deposit, Mount Isa inlier, Queensland, Australia. *Econ. Geol.* **1998**, *93*, 1463–1488. [CrossRef]
12. Haydon, R.C.; Mcconachy, G.W. The stratigraphic setting of Pb-Zn-Ag mineralization at Broken Hill. *Econ. Geol.* **1987**, *82*, 826–856. [CrossRef]
13. Yalikun, Y.; Xue, C.; Symons, D.T.A. Paleomagnetic age and tectonic constraints on the genesis of the giant Jinding Zn-Pb deposit, Yunnan, China. *Miner. Depos.* **2018**, *53*, 245–259. [CrossRef]

14. Shi, J.X.; Yi, F.H.; Wen, Q.D. The rock-ore characteristics and mineralisation of Jinding lead-zinc deposit, Lanping. *J. Yunnan Geol.* **1983**, *2*, 179–195.

15. Wang, J.B.; Li, C.Y.; Chen, X. A new genetic model for the Jinding lead-zinc deposit. *Geol. Explor. Non Ferr. Met.* **1992**, *1*, 200–206. (In Chinese)

16. Leach, D.L.; Song, Y.C.; Hou, Z.Q. The world-class Jinding Zn–Pb deposit: Ore formation in an evaporite dome, Lanping Basin, Yunnan, China. *Miner. Depos.* **2017**, *52*, 281–296. [CrossRef]

17. Kyle, J.R.; Li, N. Jinding: A giant tertiary sandstone-hosted Zn–Pb deposit, Yunnan, China. *SEG Newsl.* **2002**, *50*, 1–9.

18. Chi, G.; Xue, C.; Qing, H.; Xue, W.; Zhang, J.; Sun, Y. Hydrodynamic analysis of clastic injection and hydraulic fracturing structures in the Jinding Zn-Pb deposit, Yunnan, China. *Geosci. Front.* **2012**, *3*, 73–84. [CrossRef]

19. Xue, C.; Zeng, R.; Liu, S.; Chi, G.; Qing, H.; Chen, Y.; Yang, J.; Wang, D. Geologic, fluid inclusion and isotopic characteristics of the Jinding Zn-Pb deposit, western Yunnan, South China: A review. *Ore Geol. Rev.* **2007**, *31*, 337–359. [CrossRef]

20. Chi, G.; Qing, H.; Xue, C. An overpressured fluid system associated with the giant sandstone-hosted Jinding Zn-Pb deposit, western Yunnan, China Chapter. In *Mineral Deposit Research: Meeting the Global Challenge*; Mao, J., Bierlein, F.P., Eds.; Springer: Berlin, German, 2005; pp. 93–96.

21. Zwart, H.J. The Geology of the Central Pyrenees. *Leidse Geol. Meded.* **1979**, *50*, 1–74.

22. Mezger, J.E.; Passchier, C.W. Polymetamorphism and ductile deformation of staurolite-cordierite schist of the Bossòst dome: Indication for Variscan extension in the Axial Zone of the central Pyrenees. *Geol. Mag.* **2003**, *140*, 595–612. [CrossRef]

23. Denèle, Y.; Laumonier, B.; Paquette, J.-L.; Olivier, P.; Gleizes, G.; Barbey, P. Timing of granite emplacement, crustal flow and gneiss dome formation in the Variscan segment of the Pyrenees. *Geol. Soc. Lond. Spec. Publ.* **2014**, *405*, 265–287. [CrossRef]

24. Bois, J.P.; Pouit, G. Les minéralisations de Zn (Pb) de l'anticlinorium de Pierrefitte: Un exemple de gisements hydrothermaux et sédimentaires associés au volcanisme dans le Paléozoïque des Pyrénées centrales. *Bureau Rech. Geol. Min.* **1976**, *6*, 543–567. (In French)

25. Pouit, G.; Fortuné, J.-P. Métallogénie comparée des Pyrénées et du Sud du Massif-central. In Proceedings of the 26ème Congrès Géologique International, Paris, France, 7–17 July 1980; p. 61. (In French)

26. Fert, D. Un Aspect de la Métallogénie du Zinc et du Plomb Dans l'Ordovicien des Pyrénées Centrales: Le District de Sentein (Ariège, Haute-Garonne). Ph.D. Thesis, University Pierre Marie Curie, Paris, France, 1976.

27. Pouit, G. Différents Modèles de Minéralisations «Hydrothermale Sédimentaire», à Zn (Pb) du Paléozoïque des Pyrénées Centrales. *Miner. Depos.* **1978**, *13*, 411–421. (In French) [CrossRef]

28. Pouit, G. Les minéralisations Zn-Pb exhalatives sédimentaires de Bentaillou et de l'anticlinorium paléozoïque de Bosost (Pyrénées ariégeoises, France). *Chron. Rech. Min.* **1986**, *485*, 3–16. (In French)

29. Pujals, I. Las Mineralizaciones de Sulfuros en el Cambro-Ordovicico de la Val d'Aran (Pirineo Central, Lérida). Ph.D. Thesis, University Autónoma Barcelona, Barcelona, Spain, 1992.

30. Cardellach, E.; Phillips, R.; Ayora, C. Metamorphosed stratiform sulphides of the Liat area, Central Pyrenees, Spain. *Inst. Min. Metall. Trans.* **1982**, *91*, 90–95. (In Spanish)

31. Cardellach, E. Estudio microscópico de las mineralizaciones de Pb-Zn de Liat, Baguergue y Montoliu. *Acta Geol. Hisp.* **1977**, *12*, 120–122. (In Spanish)

32. Alonso, J.L. Deformaciones Sucesivas en el Area Comprendida Entre Liat y el Puerto de Orla—Control Estructural de los Depositos de Sulfuros (Valle de Aran, Pirineos Centrales). Master's Thesis, University Oviedo, Oviedo, Spain, 1979. (In Spanish)

33. Nicol, N. Etude Structurale des Minéralisations Zn-Pb du Paléozoïque du Dôme de Pierrefitte (Hautes-Pyrénées). Goniométrie de Texture Appliquée aux Minéraux Transparents et Opaques. Ph.D. Thesis, University Orléans, Orléans, France, 1997. (In French)

34. García-Sansegundo, J.; Martin-Izard, A.; Gavaldà, J. Structural control and geological significance of the Zn-Pb ores formed in the Benasque Pass area (Central Pyrenees) during the post-late Ordovician extensional event of the Gondwana margin. *Ore Geol. Rev.* **2014**, *56*, 516–527. [CrossRef]

35. Marcoux, E. Isotope du plomb et paragénèses métalliques, traceurs de l'histoire des gites minéraux. *Bur. Rech. Geol. Min.* **1986**, *117*, 1–289. (In French)

36. Zwart, H.J. Metamorphic history of the Central Pyrenees, Part II, Valle de Aran. *Leidse Geol. Meded.* **1963**, *28*, 321–376.

37. Kleinsmiede, W.F.J. Geology of the Valle de Aran (Central Pyrenees). *Leidse Geol. Meded.* **1960**, *25*, 129–245.

38. Cochelin, B.; Lemirre, B.; Denèle, Y.; De Saint Blanquat, M.; Lahfid, A.; Duchêne, S. Structural inheritance in the Central Pyrenees: The Variscan to Alpine tectonometamorphic evolution of the Axial Zone. *J. Geol. Soc. Lond.* **2017**, *175*, 16. [CrossRef]

39. De Sitter, L.U.; Zwart, H.J. Tectonic development in supra and infra-structures of a mountain chain. In *Structure of the Earth's Crust and Deformation of Rocks*; Det Berlingske Bogtrykkeri: Copenhagen, Denmark, 1960; Volume 18, pp. 248–256.

40. Carreras, J.; Capellà, I. Tectonic levels in the Palaeozoic basement of the Pyrenees: A review and a new interpretation. *J. Struct. Geol.* **1994**, *16*, 1509–1524. [CrossRef]

41. Carreras, J.; Druguet, E. Framing the tectonic regime of the NE Iberian Variscan segment. *Geol. Soc. Lond. Spec. Publ.* **2014**, *405*, 249–264. [CrossRef]

42. Cochelin, B. Champ de déformation du socle Paléozoïque des Pyrénées. Ph.D. Thesis, Université Toulouse 3 Paul Sabatier, Toulouse, France, 2016. (In French)

43. Mezger, J.E.; Gerdes, A. Early Variscan (Visean) granites in the core of central Pyrenean gneiss domes: Implications from laser ablation U-Pb and Th-Pb studies. *Gondwana Res.* **2016**, *29*, 181–198. [CrossRef]

44. Pouget, P. Hercynian tectonometamorphic evolution of the Bosost dome (French Spanish Central Pyrenees). *J. Geol. Soc. Lond.* **1991**, *148*, 299–314. [CrossRef]

45. Mezger, J.E. Comparison of the western Aston-Hospitalet and the Bossòst domes: Evidence for polymetamorphism and its implications for the Variscan tectonic evolution of the Axial Zone of the Pyrenees. *J. Virtual Explor.* **2005**, *19*, 1–19. [CrossRef]

46. Mezger, J.E.; Schnapperelle, S.; Rölke, C. Evolution of the Central Pyrenean Mérens fault controlled by near collision of two gneiss domes. *Hallesches Jahrb.* **2012**, *34*, 11–29.

47. Carreras, J. Zooming on Northern Cap de Creus shear zones. *J. Struct. Geol.* **2001**, *23*, 1457–1486. [CrossRef]

48. *BRGM International Report: Les Gisements de Pb-Zn Français (Situation en 1977)*; BRGM: Orléans, France, 1984; pp. 1–278. (In French)

49. Ovejero Zappino, G. Mineralizaciones Zn-Pb ordovícicas del anticlinorio de Bossost. Yacimientos de Liat y Victoria. Valle de Arán. Pirineo (España). *Bol. Geol. Min.* **1991**, *102*, 356–377. (In Spanish)

50. Castroviejo Bolibar, R.; Serrano, F.M. Estructura y metalogenia del campo filoniano de Cierco (Pb-Zn-Ag), en el Pirineo de Lérida. *Bol. Geol. Min.* **1983**, *1983*, 291–320. (In Spanish)

51. Aerden, D.G.A. Kinematics of orogenic collapse in the Variscan Pyrenees deduced from microstructures in porphyroblastic rocks from the Lys-Caillaouas massif. *Tectonophysics* **1994**, *238*, 139–160. [CrossRef]

52. Barrère, P.; Bouquet, C.; Debroas, E.J.; Pelissonnier, H.; Peybernes, B.; Soulé, J.C.; Souquet, P.; Ternet, Y. Arreau. In *BRGM Geological Map 1/50,000 with Note*; BRGM: Orléans, France, 1984; p. 60. (In French)

53. Clin, M.; Taillefer, F.; Pouchan, P.; Muller, A. Bagnères de Luchon. In *BRGM Geological Map 1/50,000 with Note*; BRGM: Orléans, France, 1989; p. 78. (In French)

54. Lavigne, J. Pic de Mauberme. In *BRGM Geological Map 1/50,000 with Note*; BRGM: Orléans, France, 1972; p. 24. (In French)

55. García-Sansegundo, J.; Merino, J.R.; Santisteban, R.R.; Leyva, F. Canejan-Vielha Mapa geologico 1:50,000. *Inst. Geol. Min. Espana* **2013**, *1*, 66. (In French)

56. Garcia-Sansegundo, J.; Alonso, J.L. Stratigraphy and structure of the southeastern Garona Dome. *Geodin. Acta* **1989**, *3*, 127–134. [CrossRef]

57. Cugerone, A.; Cenki-Tok, B.; Chauvet, A.; Le Goff, E.; Bailly, L.; Alard, O.; Allard, M. Relationships between the occurrence of accessory Ge-minerals and sphalerite in Variscan Pb-Zn deposits of the Bossost anticlinorium, French Pyrenean Axial Zone: Chemistry, microstructures and ore-deposit setting. *Ore Geol. Rev.* **2018**, *95*, 1–19. [CrossRef]

58. Dubois, C. *Mangeuses d'homme: L'épopée des Mines de Bentaillou et de Bulard en Ariège*; Privat Edition: Toulouse, France, 2015; p. 320. (In French)

59. Barrère, P.; Bois, J.-P.; Soulé, J.-C.; Ternet, Y. Argelès-Gazost. In *BRGM Geological Map 1/50,000 with Note*; BRGM: Orléans, France, 1980; pp. 1–48. (In French)

60. Nicol, N.; Legendre, O.; Charvet, J. Les minéralisations Zn-Pb de la série paléozoïque de Pierrefitte (Hautes-Pyrénées) dans la succession des évènements tectoniques hercyniens. *C. R. Acad. Sci.* **1997**, *324*, 453–460. (In French)

61. Calvet, P. Etude Structurale et Métallogénique de L'anticlinorium de Pierrefitte: Influence de la Déformation sur les Minéralisations Stratiformes. Ph.D. Thesis, University d'Orléans, Orléans, France, 1988; p. 283. (In French)

62. Alvarez-Perez, A.; Campa-Vineta, J.A.; Montoriol-Pous, J. Sobre la presencia de gahnita ferrífera en Bossost (Vall D'Aran, Lérida). *Acta Geol. Hisp.* **1974**, *9*, 111–113. (In Spanish)

63. Reyx, J. Relations entre Tectonique, Métamorphisme de Contact et Concentrations Metalliques dans le Secteur des Anciennes Mines d'Arre et Anglas (Hautes-Pyrénées-Pyrénées atlantiques). Ph.D. Thesis, University Paris VI, Paris, France, 1973; p. 83. (In French)

64. Scott, R.J.; Meffre, S.; Woodhead, J.; Gilbert, S.E.; Berry, R.F.; Emsbo, P. Development of framboidal pyrite during diagenesis, low-grade regional metamorphism, and hydrothermal alteration. *Econ. Geol.* **2009**, *104*, 1143–1168. [CrossRef]

65. Vernhet, Y. Les Minéralisations Zincifères de l'Ordovicien et du Dévonien du Val d'Orle (District de Sentein, Ariège) et de la région de Fourcaye (Val d'Aran, Espagne). Ph.D. Thesis, University Pierre Marie Curie, Paris, France, 1981; p. 279. (In French)

66. Windh, J. Saddle Reef and Related Gold Mineralization, Hill End Gold Field, Australia: Evolution of an Auriferous Vein System during Progressive Deformation. *Econ. Geol.* **1995**, *90*, 1764–1775. [CrossRef]

67. Bull, S.W.; Large, R.R. Setting the stage for the genesis of the giant Bendigo ore system. *Geol. Soc. Lond. Spec. Publ.* **2015**, *393*, 161–187. [CrossRef]

68. Zeng, M.; Zhang, D.; Zhang, Z.; Li, T.; Li, C.; Wei, C. Structural controls on the Lala iron-copper deposit of the Kangdian metallogenic province, southwestern China: Tectonic and metallogenic implications. *Ore Geol. Rev.* **2018**, *97*, 35–54. [CrossRef]

69. Cardellach, E.; Alvarez-Perez, A. Interpretación genética de las mineralizaciones de Pb-Zn del Ordovícico Sup. de la Vall de Aran. *Acta Geol. Hisp.* **1979**, *14*, 117–120. (In Spanish)

70. Alvarez-Perez, A.; Campa-Vineta, J.A.; Montoriol-Pous, J. Mineralogénesis de los yacimientos del área de Bossost (Vall d'Aran, Lérida). *Acta Geol. Hisp.* **1977**, *4–6*, 123–126. (In Spanish)

71. Ovejero Zappino, G. Mineralizaciones Zn-Pb del Ordovicico Superior del Valle de Aran (Anticlinorio de Bossost). Pireneo de Lerida (Espana). *Bol. Soc. Esp. Mineral.* **1987**, *10*, 35–37.

72. Munoz, M.; Baron, S.; Boucher, A.; Béziat, D.; Salvi, S. Mesozoic vein-type Pb-Zn mineralization in the Pyrenees: Lead isotopic and fluid inclusion evidence from the Les Argentières and Lacore deposits. *C. R. Geosci.* **2015**, *348*, 322–332. [CrossRef]

73. Militon, C. Métallogénie polyphasée à Zn, Pb, Ba, F et Mg, Fe de ma région de Gèdre-Gavarnie-Barroude (Hautes-Pyrénées). Ph.D. Thesis, University d'Orléans, Orléans, France, 1987. (In French)

74. Munoz, M.; Boyce, A.J.; Courjault-Rade, P.; Fallick, A.E.; Tollon, F. Multi-stage fluid incursion in the Palaeozoic basement-hosted Saint-Salvy ore deposit (NW Montagne Noire, southern France). *Appl. Geochem.* **1994**, *9*, 609–626. [CrossRef]

75. Pouit, G. Les minéralisations Zn-Pb dans l'Ordovicien des Pyrénées centrales-Etude préliminaire. *Rapp. BRGM* **1974**, *74*, 50. (In French)

76. Pouit, G.; Bois, J.P. Arrens Zn (Pb), Ba Devonian deposit, Pyrénées, France: An exhalative-sedimentary-type deposit similar to Meggen. *Miner. Depos.* **1986**, *21*, 181–189. [CrossRef]

77. Wagner, T.; Schneider, J. Lead isotope systematics of vein-type antimony mineralization, Rheinisches Schiefergebirge, Germany: A case history of complex reaction and remobilization processes. *Miner. Depos.* **2002**, *37*, 185–197. [CrossRef]

78. Marcoux, E.; Moelo, Y. Lead isotope geochemistry and paragenetic study of inheritance phenomena in metallogenesis: Examples from base metal sulfide deposits in France. *Econ. Geol.* **1991**, *86*, 106–120. [CrossRef]

79. Kamona, A.F.; Lévêque, J.; Friedrich, G.; Haack, U. Lead isotopes of the carbonate-hosted Kabwe, Tsumeb, and Kipushi Pb-Zn-Cu sulphide deposits in relation to Pan African orogenesis in the Damaran-Lufilian Fold Belt of Central Africa. *Miner. Depos.* **1999**, *34*, 273–283. [CrossRef]

minerals

MDPI

Article

Structural Controls of Ore Mineralization in a Polydeformed Basement: Field Examples from the Variscan Baccu Locci Shear Zone (SE Sardinia, Italy)

Antonio Funedda [1,*], Stefano Naitza [1], Cristina Buttau [1], Fabrizio Cocco [1] and Andrea Dini [2]

[1] Dipartimento di Scienze Chimiche e Geologiche, Università degli Studi di Cagliari, Cittadella Universitaria (Blocco A), S.S. 554 bivio per Sestu, 09042 Monserrato (CA), Italy; snaitza@unica.it (S.N.); cbuttau@gmail.com (C.B.); fabrcocco@gmail.com (F.C.)
[2] Consiglio Nazionale delle Ricerche (CNR)-Istituto di Geoscienze e Georisorse, 56100 Pisa, Italy; andrea.dini@igg.cnr.it
* Correspondence: afunedda@unica.it

Received: 2 July 2018; Accepted: 11 October 2018; Published: 16 October 2018

Abstract: The Baccu Locci mine area is located in a sector of the Variscan Nappe zone of Sardinia (the Baccu Locci shear zone) that hosts several type of ore deposits mined until the first half of the last century. The orebodies consist of lenses of Zn–Cu sulphides, once interpreted as stratabound, and Qtz–As–Pb sulphide ± gold veins; the implication of structural controls in their origin were previously misinterpreted or not considered. Detailed field mapping, structural analyses, and ore mineralogy allowed for unraveling how different ore parageneses are superimposed each other and to recognize different relationships with the Variscan structures. The sulphide lenses are parallel to the mylonitic foliation, hosted in the hinges of minor order upright antiforms that acted as traps for hydrothermal fluids. The Qtz–As–Pb sulphide veins crosscut the sulphide lenses and are hosted in large dilatational jogs developed in the hanging wall of dextral-reverse faults, whose geometry is influenced by the attitude of reverse limbs of late Variscan folds. The ores in the Baccu Locci shear zone are best interpreted as Variscan orogenic gold-type; veins display mutual crosscutting relationships with mafic dikes dated in the same district at 302 ± 0.2 Ma, a reliable age for the mineralizing events in the area.

Keywords: sulphide lenses; hinge trap; dilational jogs; orogenic gold; mafic dikes; mineralization chronology; arsenopyrite; late Variscan strike-slip faults

1. Introduction

Structural controls of ore deposits hosted in poly-deformed low-grade metamorphic basements are often not easy to recognize, even more so when there are multiple generations of mineralization, mostly because the relationships between structures and ore bodies are often not clear. The correct understanding of the structural controls is, in fact, relevant for defining the characteristics of the deposit, its emplacement style, its age, and therefore, its origin. In some cases, the difficulties in unravelling the tectonic structures prevent the understanding of the ore bodies' geometry, leading to mistakes in mineral exploration, evaluation of ore deposits, mine planning, and even mineral exploitation. Recognizing whether the structural controls on ore deposits are "passive", and therefore attributable to tectonic structures developed before mineralization, or, conversely, "active," i.e., related to tectonic structures progressively evolving with mineralization, can provide valuable indications in this sense. Indeed, passive structural controls imply that there are no direct correlations between tectonic and mineralizing events; the physical parameters that characterized the deformational events, e.g., thermo-baric conditions, presence of fluids, state of stress, etc., were independent from mineralization processes and they had no influence on ore formation. Conversely, in the case of active controls, these features were critical for mineralization and shaped both the deformational context and the genesis of the ores.

In the nappe zone of the Variscan basement of SE Sardinia, several Sb–W, As-Au, and As–Pb–(Cu, Zn, Ag, Au) ore deposits are located along a lower green-schist facies mylonitic belt, folded by a plurikilometric antiform (the Flumendosa Antiform) during the Variscan collisional phases and then affected by extensional tectonics. During the postcollisional extension, early phases were characterized by ductile-type structures that, with progressive exhumation of deeper tectonic units, further evolved in brittle–ductile to brittle regimes [1], supporting large-scale hydrothermal fluid circulation in the crust. In this frame, mineralization processes and ore deposition were structurally controlled, both passively and actively, at all scales.

In this work, we present a case study of the mineralized systems occurring in the Baccu Locci mine area, located in the core of the Flumendosa Antiform. In this area, several stacked tectonic units are separated by thick mylonitic zones and folded together [2]. Exhumation of the antiformal core is evidenced by the superposition of brittle–ductile and brittle structures over ductile ones; accordingly, different kinds of structurally controlled mineral deposits exposed in the area allow a distinction between passive and active emplacement styles, also leading to a relative chronology of tectonic and mineralizing events. Recent studies in the Baccu Locci mine area [1,3–5] highlighted that: (1) ore formation was structurally controlled and related to the tectonic evolution, during late Variscan extensional phases, of the Baccu Locci regional shear zone, and (2) the relationships between different ores materialize a superposition of tectonic events, further complicated by mafic magmatism and diking. Apart from this general picture, however, several unsolved issues still persist: (1) type and timing of different mineralization, (2) their relationships with tectonic structures and structural controls, (3) ore mineral associations and mafic dikes, and (4) relationships with regional and local stress fields.

In this paper, we intend to contribute to the first two issues and provide some new data for the third one.

2. Geological and Ore Deposits Outline of Baccu Locci Mine District

2.1. Regional Geology

The section of Variscan chain exposed in Sardinia consists of three tectono-metamorphic zones: (1) an inner zone in the north, with medium- to high-grade regional metamorphism, thrust on (2) a nappe zone in central-south, with low-grade regional metamorphism, in turn thrust on (3) a foreland zone in the southwest, non-metamorphosed or with very low-grade regional metamorphism (Figure 1a,b). The emplacement of the tectonic units is generally from the top to the south, with some exceptions [6]. At the map scale, in the nappe zone a distinction can be made between external nappes in the south, displaying a well recognizable lithostratigraphic succession, and internal nappes in the central-north, where it is not easy to recognize the litho-stratigraphic succession. The study area is part of the external nappe zone (Figure 2) and it consists of three tectonic units (from the bottom: Riu Gruppa, Gerrei, and Meana Sardo units) stacked one above the other during the Variscan shortening phases and then affected by late orogenic extension (Figure 3).

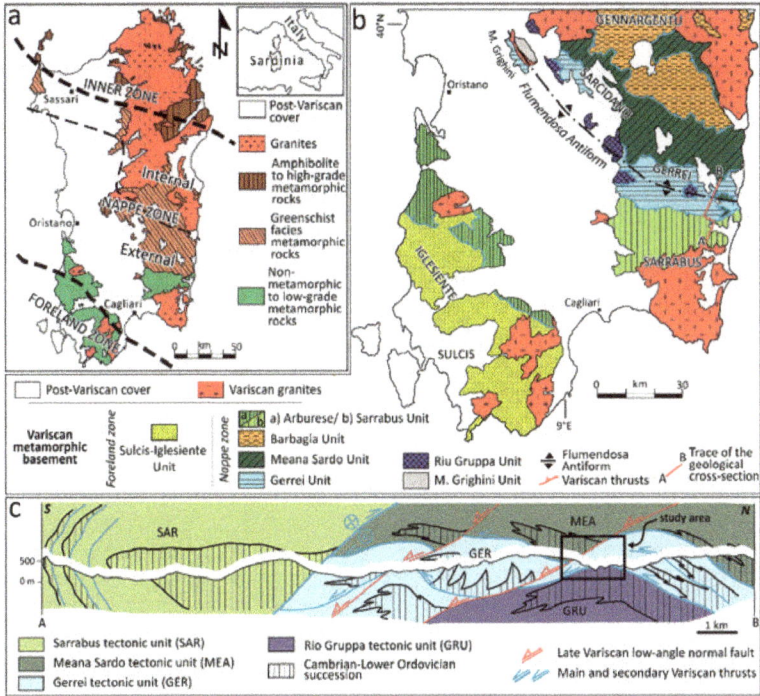

Figure 1. (**a**) Geological sketch map of the Variscan basement of Sardinia. (**b**) Tectonic sketch of the SE Sardinia Variscan basement. (**c**) Geological cross-section of the Gerrei-Sarrabus region in SE Sardinia; modified after [7], reprinted with permission.

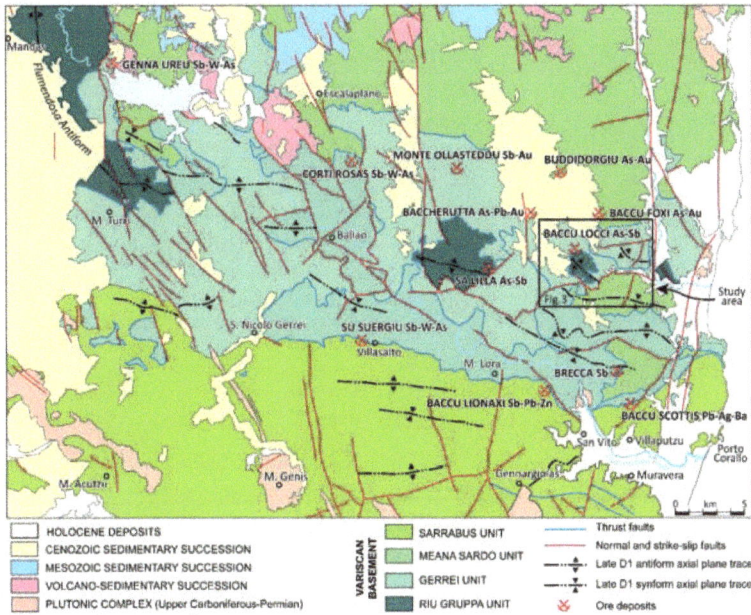

Figure 2. Tectonic sketch map and mineral deposits of the Gerrei district.

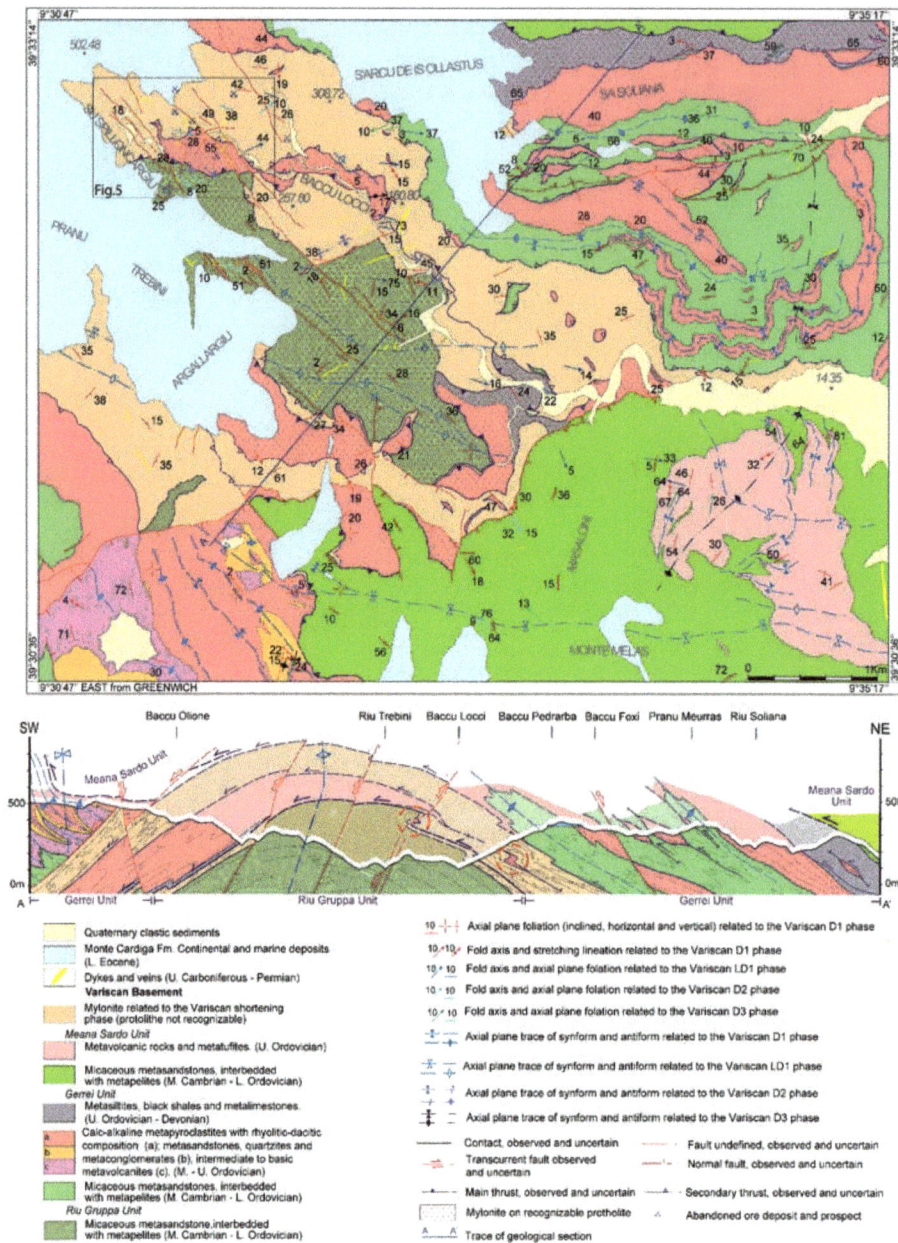

Figure 3. Structural schematic map of the Baccu Locci shear zone and geological cross-section, after [2]. In the geological cross-section, red circles point out D2 folds that are schematically represented in Figure 15.

2.1.1. Stratigraphic Outline

The tectonic units share a similar lithostratigraphic outline (with some differences, especially in the Middle Ordovician volcanic sequences), which consists of four middle Cambrian to lower Carboniferous successions, separated by three large regional unconformities [8]. They are:

- a mainly siliciclastic succession, with rare interlayered volcanic rocks, of middle Cambrian to Early Ordovician age (*Arenarie di San Vito* Fm.);
- a Middle Ordovician volcano-sedimentary succession with tuffites, metavolcanoclastites, and interlayered epiclastites, andesitic in the lower part (M. S. Vittoria Fm.) and rhyolithic in the upper part (*"Porfiroidi" Auct.* Fm.); the basal contact of this succession is marked by some discontinuous conglomerates (*Metaconglomerato di Muravera* Fm.);
- a siliciclastic to carbonatic succession of Late Ordovician to early Carboniferous age, with lithic sandstones and arkosic arenites (*Metarcose di Genna Mesa* Fm., Late Ordovician), siltstones and marls (*Scisti di Rio Canoni* Fm., Late Ordovician.), black shales and limestones (*"Scisti neri a graptoliti" Auct.* Fm., Silurian to Early Devonian; *"Calcari di Villasalto"* Auct. Fm., Early Devonian to early Carboniferous);
- a lower Carboniferous siliciclastic sequence with conglomerates, sandstones, and olistoliths of the older formations that unconformable rests on the Devonian formations, which does not crop out in the study area (*Pala Manna Fm.*, early Carboniferous); it is the youngest formation involved by the Variscan orogeny in Sardinia.

The metamorphic basement of SE Sardinia is then intruded by an upper Palaeozoic (upper Carboniferous to lower Permian) intrusive complex. Lower Permian leucogranitic rocks outcrop close to the study area in the Quirra sector [9]; they belong to a calc-alkaline, ferroan, F-bearing, ilmenite-series intrusion, part of a magmatic suite dated at 286 ± 2 Ma [10]. The entire period of granitoid magmatism is associated with calc-alkaline volcanism [11] and with widespread mafic and felsic diking [12]. Early diking phases are represented by swarms of calc-alkaline (spessartitic) mafic dikes; one of them, crosscutting the same tectonic unit in a nearby area, is dated at 302 Ma [13].

The early Eocene *Monte Cardiga* Fm., made up of conglomerates, sandstones, and marls that are deposited in littoral environments lies unconformably on the metamorphic basement and the intrusive complex. In the surrounding areas, the oldest deposits that unconformably cover the Variscan basement are Middle Pennsylvanian continental deposits [14].

2.1.2. Structural Outline

The oldest tectonic features that are recognizable in the Baccu Locci mine district involve the lower Carboniferous rocks and are related to the Variscan orogeny. Indeed, there is no evidence of the pre-Middle Ordovician tectonic structures recognized in adjacent areas [7], as they were probably obliterated by Variscan deformation. The overlap of several early to late Carboniferous deformation events is evidence that a D1 collisional phase with crustal thickening and subhorizontal shortening occurred under ductile conditions and a D2 postcollisional extension with the reactivation of some of D1 structures occurred in the ductile–brittle transition (Figure 4) [6,8,15]. The D1 phase is characterized by a general SSW-directed nappe emplacement, regional folding and thrusting, and syntectonic regional lower green-schist facies metamorphism. An exception is represented by the Sarrabus unit, the shallowest nappe of the stack, which crops out south of the study area and it is emplaced from the top to the west [16]. The D1 early shortening structures are large kilometer-scale recumbent isoclinal folds facing southward, with well-developed penetrative axial plane foliation—generally a slaty cleavage produced in lower green-schist facies metamorphism—followed by almost contemporary south-southwest thrusts that produced thick mylonitic belts. Development of such shear zones between the different tectonic units, thick up to several hundred meters, is common in the Variscan basement of Sardinia [17–24]. Among them, the Baccu Locci shear zone is one of the most noteworthy [3]. It can be followed in the field for more than 30 km; the study area is located on its eastern side. The Baccu

Locci shear zone is characterized by widespread and penetrative mylonitic foliation with a mineral assemblage typical of lower green-schist facies metamorphism that is parallel to the large thrust and generally cuts at a low angle the early D1 axial plane foliation [3]. The deformation is highly partitioned; in the core of the shear zone, it is not possible to recognize the mylonite protolith, although several slices of less deformed rocks have been recognized and mapped [2,3,19]. (Note that, following the choice by [2,3], in Figures 3 and 5 we distinguished the mylonite whose protolith is not recognizable from the mylonite whose protolith is still recognizable). At a later stage during the collisional phase, a late-D1 (LD1) N-S shortening event led to development at a regional scale of large, weakly east-plunging upright folds, with axes up to 50 km long that refolded both isoclinal folds and ductile shear zones. Of these late folds, the main fold is the Flumendosa Antiform [1] (Figures 1 and 2), which runs roughly from WNW to ESE for more than 50 km, folding the D1 nappe stack. In detail, the Flumendosa Antiform consists of some minor order antiforms and synforms with km-size wavelength. One such northern minor order folds crops out in the study area. The LD1 axis is generally east-plunging, and at the hinge zone, a subvertical spaced crenulation cleavage discontinuously developed. Then, the LD1 folds were in turn deformed during the D2 postcollisional extensional phase. The limbs of the LD1 antiforms are deformed by several asymmetric recumbent folds with subhorizontal axial planes and axes parallel to the LD1 limb attitude [1] (Figure 4). The D2 folds are overturned away from the antiformal hinge zone [1]. They are often associated with low-angle normal shear zones that allowed for the exhumation of deeper units and enhanced the antiformal structure. They have opposite structural-facing direction in the fold flanks: north-facing in the northern limb and south-facing in the southern limb. Their major order wavelength exposed in the field is about 30 to 40 m. Folds with "outer" structural facing and low-angle normal faults are interpreted as produced by vertical shortening of steeply inclined bedding and earlier foliations. During exhumation, rocks were progressively carried to shallower structural levels, where brittle behavior became prevalent. Thus, the deformation style changed, and the final stage of postcollisional extension was accommodated by high-angle normal faults [6]. A D3 folding event, with vertical axial planes and axis trend changing from N-S to N 40°, is also recognized, but it is still not clear whether it could be related to a final stage of postcollisional extension or to the following D4 strike slip faulting [8]. Finally, D4 strike-slip faults affected the exhumed basement but did not involve the Permian to Eocene successions that crop out in the study area. At the Variscan Realm scale, a late strike-slip tectonics is from far recognized. Moreover, is clearly observable in the field that the lower Permian granitoids postdate the D1, D2, and D3 structures, whereas there is less evidence about their relationship with strike-slip tectonics. As we will describe below, LD1 and D2 folds, as well as late faults, played a significant role in controlling the geometry of orebodies.

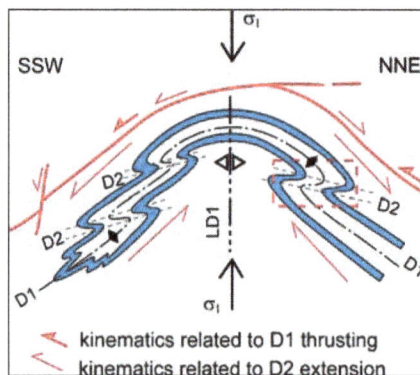

Figure 4. Schematic relationships between D1, LD1, and D2 structure. Note the development of recumbent D2 folds in the limb of upright LD1 antiform (modified after [1]). The red dashed box indicates the location of the scheme in Figure 15.

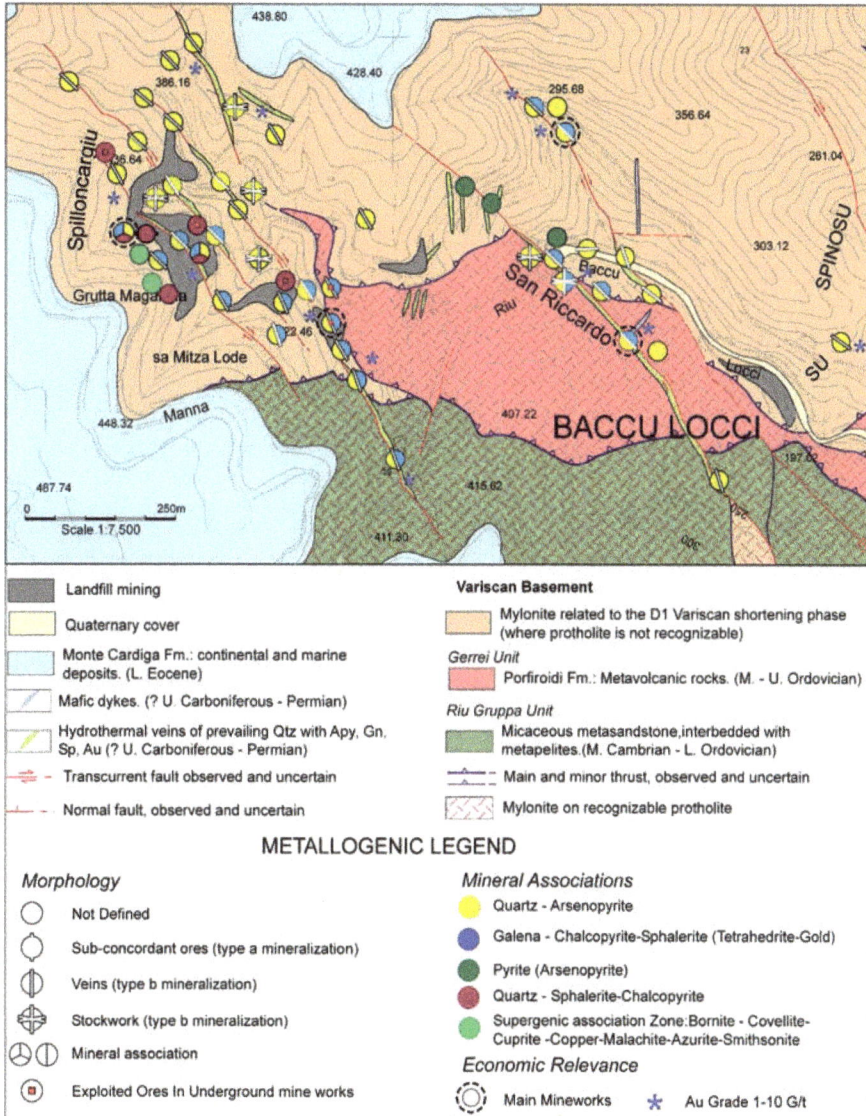

Figure 5. Mineralized outcrops in the Baccu Locci mine area, after [2].

2.2. The Gerrei-Sarrabus Metallogenic District

The Gerrei-Sarrabus region has been historically the second most important mining area in Sardinia and the most important antimony district in Italy (Villasalto–Ballao district). Several reviews of the district's mineral deposits have been done in the past [25–27], including a recent attempt to interpret some of the main mineral associations in the Variscan metallogenic epoch of Sardinia [28]. The main mineral deposits of the region include the following (Figure 2):

(a) Zn–Cu–Pb sulphide lenses, disseminations, and ores;

(b) Sb–W, As–Au, and As–Pb–(Cu, Zn, Ag, Au) mesothermal systems with quartz–sulphide veins, stockwork, and disseminated ores;

(c)   Mo–W–F greisen and vein-type ores;

(d)   F–Ba ± Pb–Ag–Cu–Zn vein systems.

Ores of (a) and (b)-types are hosted in the Palaeozoic metamorphic basement, and mostly occur in the northern part of the district (Gerrei); (c) type ores are hosted in and/or related to the suite of F-bearing Permian granites of San Vito and Quirra intrusions [9,29]; and, (d) type ores occur prevalently in the southern part of the district (Sarrabus), are broadly typified by the "Sarrabus Silver Lode" [30] and Silius [31] deposits, and they are possibly related to another suite of F-rich Permian granites (Sette Fratelli and Monte Genis intrusions [9]). Although still very lacking and non-systematic, some isotope and fluid inclusion data on different Gerrei-Sarrabus ore deposits are available from several recent studies [10,25,30–32].

Only (a) and (b) type ores are present in the study area, and thus will be considered in detail in this work. Both types are structurally controlled and are located at different structural levels.

3. Materials and Methods

To reconstruct the orebodies and their relationships with the tectonic structures, detailed field work was performed, mapping at 1:5000 and 1:10,000 scale both mineral deposits and tectonic features (foliations, folds, faults, kinematic indicators). The surface data have been integrated with the subsurface data, in rare cases directly sampled in the mine when still accessible, but generally taken from 1:500 scale exploitation maps related to the last period of mining activities in 1961. The grade of detail reported in those maps allows us, in some cases, to recognize not only the mine works, but also the geometries of the ores. Importing and processing these data in the three-dimensional (3D) workspace of Rhinoceros© and Move (©Midland Valley, Edinburgh, Scotland, UK), allow minimization of the unbalanced area and validation of the geological-structural map (see description in [2]).

The kinematic analysis of fault slip data was performed using FaultKin 5.2 by Allmendinger [33]. The slip direction and the sense of slip were inferred from several types of kinematic indicators—slickenlines, tension gashes, drag folds, and S-C structures—all collected along the D4 transpressive faults and related orebodies. They all indicate similar kinematics, suggesting that the paleostress remained the same during vein formation. The occurrence of different types of kinematics indicators during the same tectonic event (i.e., D4 faulting), sometimes exactly along the same fault, can depend on the mechanic behavior of the involved rocks. In general, when the damage zone is large enough and there is a relevant grain-size reduction in cataclased rocks, S-C-like fabric easily developed, although physical condition and deformation mechanisms are those typical of shallow structural levels [34,35]. The orientation of the principal strain axes that was achieved from fault slip data analysis gives an idea about the paleostress field accounting for the fault kinematics.

In this study, ore mineralogy and microtextural studies were performed on polished sections by optical microscopy in reflected light. Several transmitted light studies on thin sections of ores and their host rocks have been performed in past works [3,5,7,19].

4. Results

4.1. Structures and Mineralization in the Baccu Locci Mine Area

The mineralized deposits of the Baccu Locci mine area (NE Gerrei, Figures 2 and 5) occur in a key area for studying the structural and metallogenic evolution of the region. In Baccu Locci, the ores are exposed at different structural levels along the hinge zone of the Flumendosa Antiform, hosted by the strongly mylonitized siliciclastic metasediments and felsic metavolcanics of the Gerrei tectonic unit [3]. Both Zn–Cu–Pb sulphide lenses (*type a* mineralization) and Qtz–As–Pb sulphide hydrothermal vein deposits (*type b* mineralization) are represented in the area, where it is also possible to recognize spatial relationships of ore deposits with coeval mafic dikes [4].

4.1.1. The Zn–Cu–Pb Sulphide Lenses (*Type a* Mineralization)

Zn–Cu–Pb sulphide lenses occur in the Su Spilloncargiu mine sector (Figure 5), hosted in phyllitic quartz–mylonite rocks (Figure 6a–e), whose protolith was made mainly by siliciclastic sediments, with a high occurrence of recrystallized quartz and muscovite, metamorphosed in lower green-schist facies during the Variscan Orogeny. Due to supergene alteration (Figure 7b), the Spilloncargiu primary ore is not well exposed on the surface; access to underground works is not easy, but some samples can be found in the dumps of the lowermost mineworks.

Ore Mineralogy and Textures

At the outcrop scale, the mineralization shows a laminated texture, with the sulphides being disposed along the mylonitic foliation. At the microscopic scale, it is possible to recognize that the deposition of ore minerals is polyphasic. The first event is characterized by dark ferroan sphalerite with abundant and finely dispersed chalcopyrite and pyrrhotite inclusions (*chalcopyrite disease* [36]). This early sphalerite is the prevailing sulphide in the ore (Figure 7a) and it displays an evident cataclastic texture. Voids and fractures in cataclased sphalerite aggregates are filled by abundant galena, chalcopyrite, and pyrite, while arsenopyrite and tetrahedrite-group minerals are usually subordinate (Figure 7b). Galena crystals show only slight evidence of deformation and have not suffered the cataclastic deformation that involved the sphalerite. Gangue minerals are subordinate to sulphides and consist of two recognized generations of gray quartz: early quartz, macrocrystalline, and cataclased, is related to the sphalerite stage of mineralization; late quartz is microcrystalline, fills the microfractures, and is associated with galena and other sulphides (Figure 7c). Wall rock hydrothermal alteration is difficult to recognize, due to late superposed supergene phenomena, but it certainly included silicification and pervasive dispersion of carbonaceous matter, possibly precipitated during mineralization, which gave the sulphide-bearing mylonite a distinctive black color. No free gold was ever identified in this ore, whose Au geochemical contents are very low [2,26,37].

Structures

The old mine exploited several lenticular sulphide orebodies that were located in the hinge zone of a km-size open, upright antiform (a minor order structure of the largest Flumendosa Antiform) related to the LD1 collisional phase (see cross-section in Figure 3 and the scheme in Figure 4). The antiform axis weakly plunges toward N120 (Figures 3 and 8); the structure deformed the foliated isoclinal folds related to the early stage of shortening (D1 Variscan phase) and the mylonitic shear zone that separates the Riu Gruppa and Gerrei tectonic units. The exploited sulphide lens-shaped orebodies developed parallel to the D1 mylonitic foliation, which in this area is at a low angle with the D1 axial plane foliation (Figure 9). There is no evidence of the primary bedding, completely transposed during the D1 tectonic phase. The lenses attain a maximum thickness of 6–7 m for maximum extension in a strike of 80–100 m. Although located in the hinge zone of an antiform, they cannot simply be classified as typical saddle reefs, as they do not display the classical triangular shape related to hinge collapse. The main dip of the mineralized lenses is up to 20° toward N 140°, so more or less in the axis plunge direction. The shape of the orebodies is well detectable by studying the old room-and-pillar mineworks, still being partially accessible.

In detail, it is possible to recognize crosscut relationships between different events of mineralization. The large lens-shaped orebodies are mainly characterized by abundant Zn–Cu sulphides forming veinlets 1 to 10 cm in size hosted in the mylonite (Figure 6a), being generally parallel to the mylonitic foliation of the hinge zone (Figure 9). Galena, chalcopyrite, pyrite, and other sulphides are arranged in cm-sized veinlets that involve the hosting rocks for about 1 m of thickness; they cut at low angle the Zn–Cu lenses (Figure 6e), thus postdating them. In the field, these veins became progressively steeper and more arsenopyrite-rich (Figure 7c).

Figure 6. Outcrop pictures of the Baccu Locci mineral deposits: (**a**) Su Spilloncargiu mineworks: Zn–Cu–Pb sulphide lens ore (*type a* mineralization). The thin sulphide "beds" (dashed line) are parallel to the Variscan mylonitic foliation (solid line). (**b**) Baccu Trebini outcrop (*type b* mineralization): Qtz–As–Pb sulphide sheeted veins along a SW-dipping brittle shear zone. (**c**) Along Rio Baccu Locci, close to San Riccardo mineworks (*type b* mineralization): Qtz–As–Pb dm-size vein with subhorizontal slickenlines. (**d**) Baccu Trebini outcrop (*type b* mineralization): Fault breccia with Qtz–As–Pb sulphides wrapping wall rock clasts along SW-dipping fault. (**e**) Su Spilloncargiu mineworks (*type a* mineralization): in a pillar of the exploited mine a *type b* Qtz–As–Pb sulphide vein (white dotted line) crosscuts a sulphide lens that is parallel to Variscan foliation (white lines).

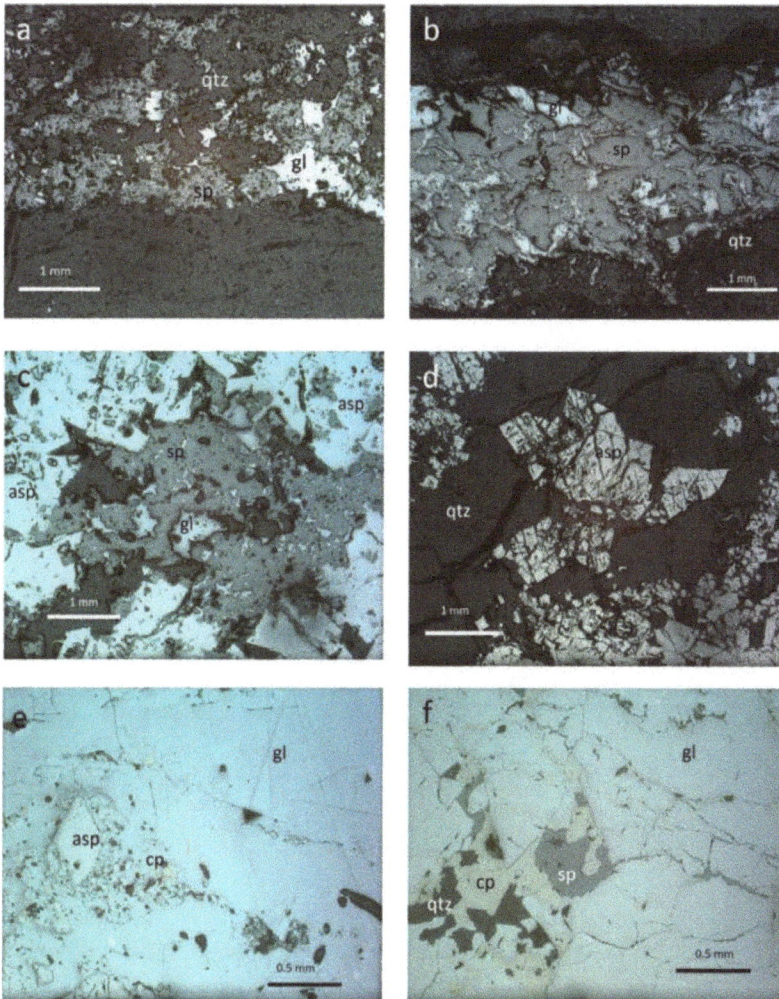

Figure 7. Microtextural features of ores in the study area (polished sections, reflected light): (a) Su Spilloncargiu mineworks, early mixed sulphide ore (*type a* mineralization): Brecciated quartz–sphalerite layer infilled by late galena; the mineralized layer follows the S_m foliation and exhibits sharp contact with a phyllosilicate layer in the mylonitic matrix; high-Fe sphalerite shows a distinct *chalcopyrite disease* with large chalcopyrite and pyrrhotite exsolutions. (b) Su Spilloncargiu mineworks, early mixed sulphide ore (*type a* mineralization cementation zone): Brecciation of sphalerite, infilled by primary galena, is particularly highlighted by cementing fine veinlets of secondary galena. (c) Su Spilloncargiu mineworks, quartz–arsenopyrite veins (*type b* mineralization crosscutting *type a* mineralization): Arsenopyrite aggregates enveloping early sphalerite aggregates (with some galena) from mixed sulphide ore. (d) Su Spilloncargiu mineworks, quartz–arsenopyrite veins (*type b* mineralization): Typical brecciated arsenopyrite texture, with large aggregates of fractured sub-idiomorphic crystals. (e) San Riccardo mineworks (type b mineralization): large galena (chalcopyrite) enveloping arsenopyrite crystals. (f) San Riccardo mineworks (type b mineralization): Detail of the late ore, with abundant fractured galena infilled by late chalcopyrite and inclusion poor sphalerite; sphalerite only displays very fine chalcopyrite exsolutions along crystallographic planes. Qtz, quartz, gl, galena, sp, sphalerite, asp, arsenopyrite.

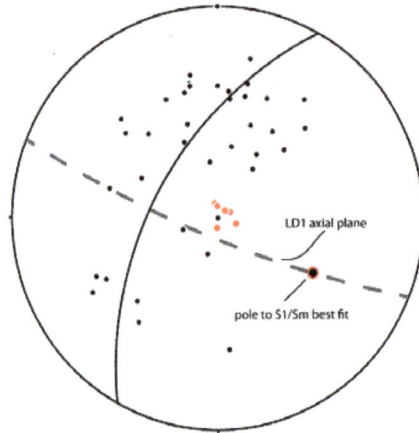

Figure 8. Stereographic projection (equal area, lower hemisphere) of D1 tectonic foliation (black dot), LD1 axial surface (dashed line), and calculated axis (red circle), attitude of sulphide orebodies (red dot).

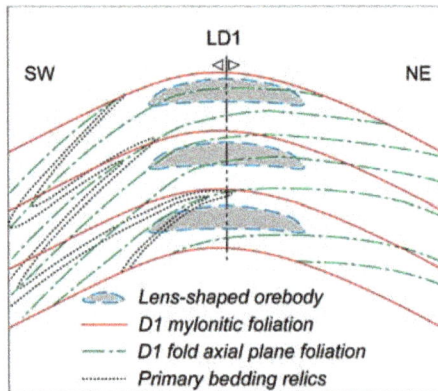

Figure 9. Schematic relationships between lens-shaped sulphide orebodies, D1 fold axial plane foliation, and D1 mylonitic foliation in the hinge zone of LD1 antiform at Spilloncargiu mine works (not to scale).

4.1.2. Qtz–As–Pb Sulphide Vein Systems (*Type b* Mineralization)

Swarms of quartz–As–Pb (Zn, Cu, Ag, Au) sulphide hydrothermal veins occur frequently in the Baccu Locci mine area and in several nearby localities along the highly mineralized Baccu Locci shear zone. The best exposition occurs in the San Riccardo/Su Spinosu mine (Figure 5), where the veins reach the maximum mapped thickness.

Ore Mineralogy and Textures

Textural and mineralogical observations of the Qtz–As–Pb sulphide vein ores are the basis for a general paragenetic sequence that includes: (1) a *pre-ore stage*, with large precipitation of coarse-grained white quartz and rare scheelite [37]; (2) an *As–Fe sulphide stage*, following the diffuse cataclasis of white quartz, with abundant arsenopyrite (I) and pyrite in quartz macro- and microfractures, after an initial arsenopyrite–pyrite dissemination in the wall rock; (3) a *sulphide/sulfosalts stage*, after new cataclasis (Figure 7), with infilling of gray microcrystalline quartz, Pb–Zn–Cu–Ag sulphides (galena, sphalerite, chalcopyrite, argentite), euhedral arsenopyrite (II), phyrrothite, Cu–Ag–Sb–As sulfosalts (tetrahedrite–freibergite, bournonite, stephanite), rare stibnite, and gold/electrum [37,38];

and, (4) a late stage, with cryptocrystalline quartz and pyrite. In the sulphide/sulfosalts stage, textural evidence indicates an initial deposition of galena, which is the most abundant mineral (Figure 7); sphalerite, chalcopyrite, and other sulphides followed. Fine disseminations of euhedral (sometimes needle-shaped) arsenopyrite (II) in cataclased wall rocks are probably related to this stage. Sphalerite in *type b* ores is distinctly different from that in *type a* sulphide lenses, being much less abundant, less deformed, less dark (it shows more evident internal reflections), and much less affected by *chalcopyrite disease*. Wall rock alteration occurred from the earliest stages, but it is substantially limited to narrow zones close to the veins, commonly marked by intense sericitization, sulphidation (fine pyrite dispersion), and silicification; in many outcrops, the footwall of the veins is strongly cataclastic and displays a characteristic black color (*black cataclasite*), further indication of diffuse precipitation of carbonaceous matter during some of the mineralizing phases. A system of E-W quartz–feldspar cm-thick veinlets has been locally observed; it distinctly crosscuts the quartz–sulphide veins and may be related to a different (and very late) phase of fluid circulation in the area. Gold grades in sulphide veins are 1–12 g/t [27,38], with good persistence in the whole mineralized vein system; silver grades are 1000–1200 g/t [27]. The Au/Ag ratio in gold grains is <1, in opposition to regional trends [37]. According to Bakos et al. [37], 10 μm sized gold grains are particularly associated with chalcopyrite and galena/bournonite myrmekitic intergrowths that infill microfractures in cataclased arsenopyrite aggregates.

Structures

Qtz–As–Pb sulphide veins are generally hosted in narrow brittle shear zones, usually not thicker than 10 m, of high-angle faults dipping about 70° toward N 230°, confirming the structural trend recognizable at the scale of the entire district for *type b* mineralization [26,27]. The faults generally involve both the quartz mylonite, whose protolith is not possible to ascribe to one of the mapped formations, and the Ordovician rhyolitic volcanites with augen-textures ("Porfiroidi" Fm.) that constitute some hectometer-sized tectonic slices inside the shear zone (Figure 3).

The faults clearly cut all the ductile D1, LD1, D2, and D3 structures (shear zone, folds, and foliations) and are sealed by the lower Eocene sediments. From the structural map (Figure 3), it is evident that the mineralized faults are parallel to the LD1 antiform axis and are mainly located in the hinge zone. We can interpret this occurrence considering two likely types of reactivation of previous structures, both occurring at a shallower structural level than the LD1 phase. They could be hinge-parallel fractures developed in the fold outer arc, parallel to the *bc* plane according the fold-related joints classification by Hancock [39]; or, more probably, the faults reactivated the noncontinuous, spaced crenulation cleavage that discontinuously developed just in the hinge zone of LD1 antiforms.

The mineralized bodies are lenticular, elongated to laminated veins that are typical of a fault-fill vein system. They can vary from isolated veinlets no more than 1 cm thick to sheeted veinlets and laminated veins in which the hydrothermal mineral component prevails over the host rock component (Figure 6b). Along the fault zones hosting the veins, several kinematic indicators are found, frequently at the contact between veins and wall rocks. In some damage zones there are fault breccia with wall rock clasts wrapped by dominant hydrothermal quartz (Figure 6d); furthermore, some shear zone is characterized by a foliated black cataclasite showing S-C type fabric. Slickensides and striated surfaces occur also on the contact between the fault-fill veins (Figure 6c). Slickenlines, tension gashes and S-C type complex foliations collected along the SW-dipping faults hosting the main quartz–sulphide veins all indicate a dextral strike-slip kinematic with a small reverse component (a tectonic transport direction from the top to the NW, some data are plotted in Figure 10b). A kinematic analysis was performed also considering faults hosting Qtz–As–Pb sulphide veins but with a different orientation allow for us to reconstruct a strain ellipsoid with a subvertical intermediate axis (λ_2, and subhorizontal shortest (λ_3) and longest (λ_1) axes, respectively, oriented roughly N-S and E-W (Figure 10). Although parallelism between strain and stress ellipsoid cannot be demonstrated and we did not perform a paleostress inversion, the kinematics suggests a paleostress field with a subhorizontal σ_1 in agreement with strike-slip tectonics.

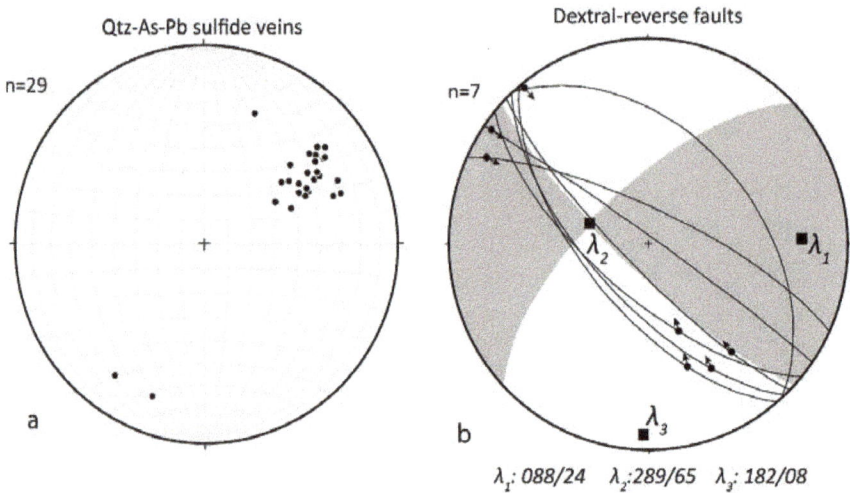

Figure 10. (a) Stereographic projection (equal area, lower hemisphere) of poles to Qtz–As–sulphide main orebodies in the Baccu Locci zone; (b) Kinematic analysis of the SW dipping faults hosting quartz Qtz–As–sulphide veins. λ_1, λ_2, and λ_3 are the directions of maximum, intermediate, and minimum strain ellipsoid axes, respectively.

Typical economic orebodies exploited in the past mine were sulphide-rich ore shoots up to 8–9 m thick, extending along the SW-dipping faults 100–300 m along strike, and over 100 m down dip. By using the available detailed maps of mineworks, it is possible to construct a 3D model of these orebodies, particularly for the San Riccardo/Su Spinosu mine (Figure 11b). Along the stretched mineralized zones, the orebody thickness increases where the fault surface is less inclined, almost subhorizontal, and decreases where the fault is steeper, generally becoming no thicker than 2 m. At the underground level 214.68 in the San Riccardo mineworks, which is unfortunately hardly accessible today, this geometry has been observed exactly along section C-C' in Figure 11a, where the fault plane and the sulphide veins are subhorizontal (Figure 12). Although no longer accessible, from the 3D model and from the old mine reports, the room-and-pillar exploitation between levels 201.68 and 179.68 can be considered as a less inclined sector of the fault, marked by an increase in thickness of the orebody (section A-A' in Figure 11a). From these observations, the more relevant economic orebodies of *type b* mineralization may be associated with very large dilational jogs developed on the hanging wall of the transpressive faults (Figure 11a), where the occurrence of less inclined segments connecting the subvertical ones (Figure 11c) produced room for the emplacement of the orebodies during the fault activity.

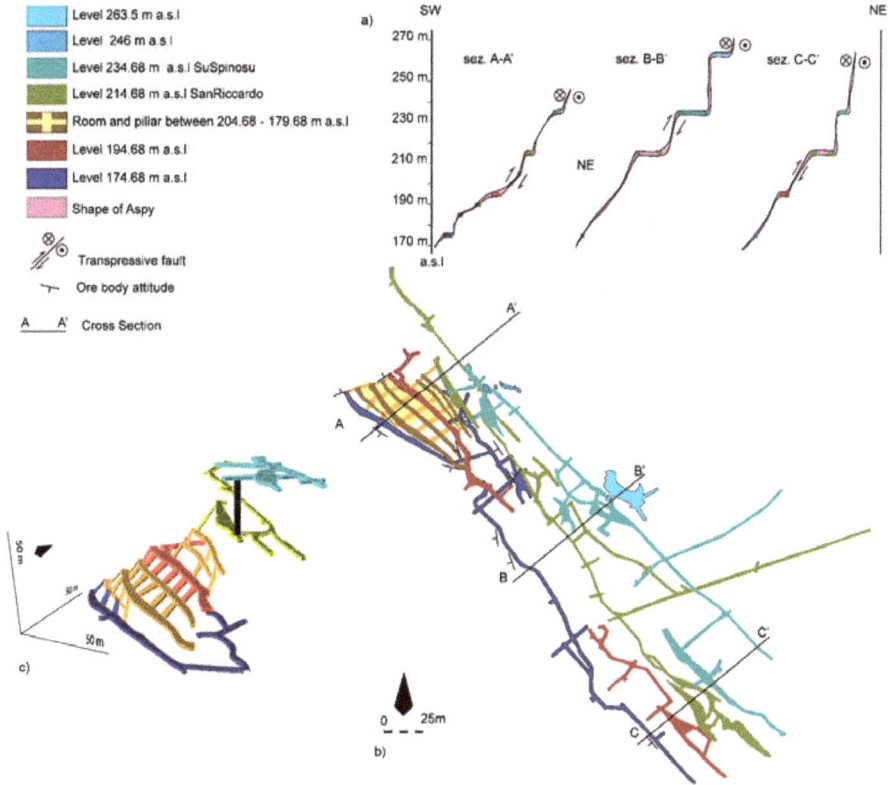

Figure 11. The Baccu Locci/San Riccardo mine, *type b* mineralization (numbers are elevation above sea level, a.s.l.): (**a**) schematic vertical sections that point out the sigmoid shape of the sulphide veins of the San Riccardo orebodies (see trace in b), interpreted as large dilational jogs; (**b**) minework plans based on the original mine maps (different levels are identified by colors; and, (**c**) three-dimensional (3D) model of the northern part of mineworks in b, in the area of section A-A'; note that the orientation is different from the map in b.

4.1.3. Mafic Dikes

Detailed mapping of the Baccu Locci mine area [2] revealed a wide occurrence of mafic dikes, verifying complex mutual geometrical relationships among them and the ores (see also [26,38]). Dikes are variable in size (0.1–10 m thick) and orientation (from subvertical to subhorizontal), with a prevalence of N-S direction. They distinctly crosscut the foliation of the hosting mylonites, extending along for tens of meters. In different outcrops there is clear evidence that the dikes cut across and partly follow the structural pattern of the quartz–sulphide veins. Moreover, they are also locally crosscut by *type b* quartz–sulphide veins. Of particular interest is the relationships along the large dilational jogs that were observed at the San Riccardo mineworks; there, a mafic dike abruptly changes its attitude from subvertical to subhorizontal, following the tectonic foliation in the reverse limb of a D2 recumbent fold, and becoming parallel to the fault plane hosting the *type b* veins (Figure 12a).

Zucchetti [38] first classified the mafic rocks as spessartitic lamprophyres: under the microscope they show an aphanitic to porphyritic (doleritic) texture, with small labradoritic–bytownitic plagioclase, idiomorphic hornblende phenocrysts, and corroded quartz xenocrysts in a strongly altered groundmass; accessory phases are apatite, titanite, magnetite, ilmenite. These features allow for us to frame them among the calc-alkaline mafic dikes widely intruded in the SE Sardinia basement during the late Variscan extension [12] and recently dated, in the Gerrei district, at 302 ± 0.2 Ma (U–Pb dating on zircon [13]).

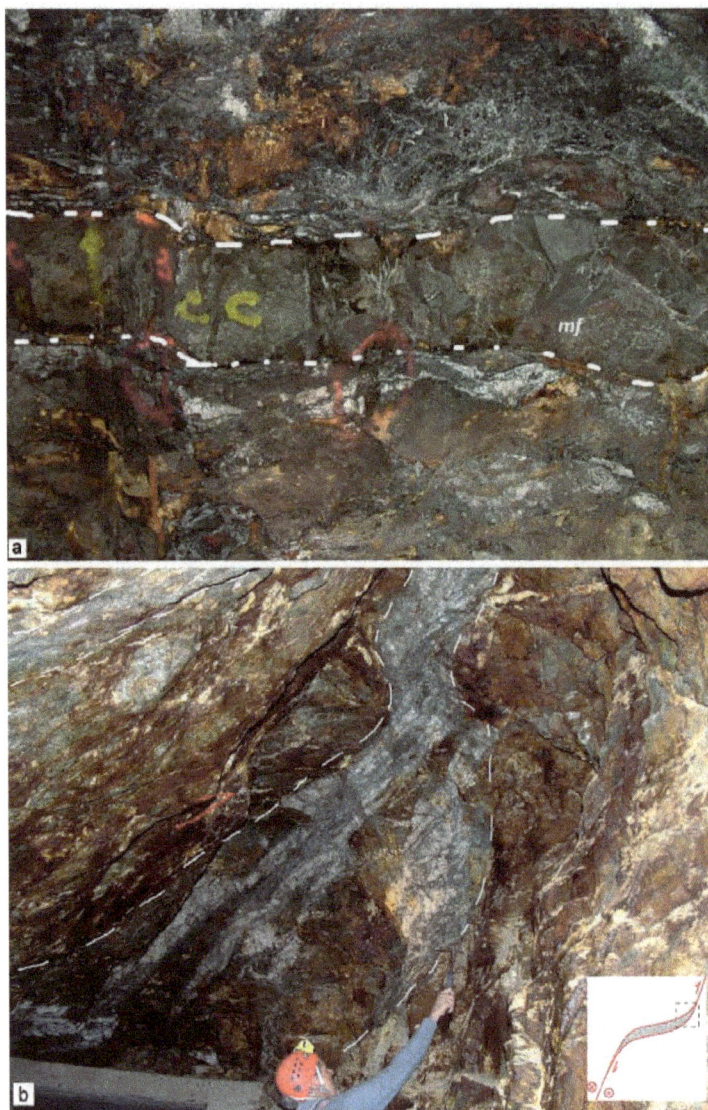

Figure 12. Outcrop relationships between Late Variscan mafic dikes and *type b* ores in the southern branch of San Riccardo underground mineworks: (**a**) sub-horizontal Qtz–As–Pb sulphides veins parallel to a mafic dike (mf) about 1.0 m thick (contact is highlighted by dashed white line); and, (**b**) Qtz–As–Pb sulphide ore (underlined by dashed white line) that progressively became less steep toward the left (ESE). The dashed box in the small picture indicates the location of the outcrop respect to the whole lens-shaped orebody. Photo in a is about 2 m to the left (i.e., west) of photo in b.

5. Discussion

As previously discussed, the basic approach of this study was mainly focused on finding field and ore microscopy elements that are able to: (1) unravel the reciprocal relationships between ore deposit types and (2) provide indications on the controls operated by the Variscan tectonic structures during mineralization events in the Baccu Locci mine area.

Previous works have usually considered the occurrence of *type a* sulphide lenses in the Baccu Locci mine area as a single ore. This was interpreted in different ways, essentially trying to establish genetic links with different phases of mineralization and with the *type b* ores. Thus, according to a proposed "synsedimentary" model, the Zn–Cu–Pb sulphides would represent an initial (predeformation) concentration of metals (*proto-ore*) that, remobilized during Variscan deformation, provided the sulphide component to *type b* veins [37,40]. An interesting issue was first raised by the study of Zucchetti [38], which evidenced a partial superposition of mineral assemblages in both ores. This was considered as indicative of an origin of Zn–Cu–Pb sulphides by lateral infilling from Qtz–As–Pb sulphide veins, but without clearly distinguishing the two mineralization in space and time.

Structural data and ore mineralogy observations that were carried out with this study document a complex mineralization history, including two distinct kinds of polyphasic ores (here, *type a* and *type b* ores in this text) that show different minerals assemblage and different spatial relationships with tectonic structures (Figure 13).

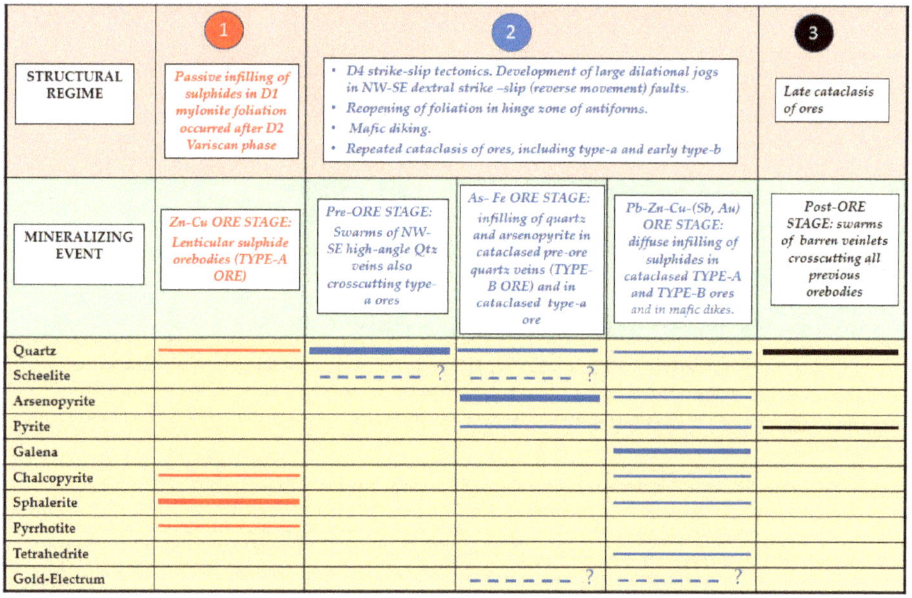

	①	**②**			**③**
STRUCTURAL REGIME	*Passive infilling of sulphides in D1 mylonite foliation occurred after D2 Variscan phase*	• *D4 strike-slip tectonics. Development of large dilational jogs in NW-SE dextral strike–slip (reverse movement) faults.* • *Reopening of foliation in hinge zone of antiforms.* • *Mafic diking.* • *Repeated cataclasis of ores, including type-a and early type-b*			*Late cataclasis of ores*
MINERALIZING EVENT	*Zn-Cu ORE STAGE: Lenticular sulphide orebodies (TYPE-A ORE)*	*Pre-ORE STAGE: Swarms of NW-SE high-angle Qtz veins also crosscutting type-a ores*	*As-Fe ORE STAGE: infilling of quartz and arsenopyrite in cataclased pre-ore quartz veins (TYPE-B ORE) and in cataclased type-a ore*	*Pb-Zn-Cu-(Sb, Au) ORE STAGE: diffuse infilling of sulphides in cataclased TYPE-A and TYPE-B ores and in mafic dikes.*	*Post-ORE STAGE: swarms of barren veinlets crosscutting all previous orebodies*
Quartz		———————————			———————
Scheelite		– – – – ?	– – – – ?		
Arsenopyrite			———————		
Pyrite			———————		
Galena				———————	
Chalcopyrite	———————			———————	
Sphalerite	———————			———————	
Pyrrhotite	———————				
Tetrahedrite				———————	
Gold-Electrum			– – – – ?	– – – – ?	

Figure 13. Evolution of mineralizing events in the Baccu Locci mine area and their relationships with tectonic events.

5.1. Relationships between Ore Type Deposits

The *type a* ore crops out in the Spilloncargiu mine, where the Zn–Cu sulphide lenses are cut by a swarm of *type b* veins (Figure 14). Ore microscopy evidence corroborates the relationships at the outcrop scale; in fact, *type a* minerals (mainly sphalerite, calcopyrite, and pyrrhotite) are strongly cataclased and *type b* minerals (mainly galena and chalcopyrite with subordinate arsenopyrite and tetrahedrite) infill the voids that were created by cataclasis. These textures confirm that *type b* postdates *type a* mineralization and that a brittle tectonic event occurred in between. In turn, the *type b* minerals show mutual crosscutting relationships that demonstrate the progressive and polyphased deposition of white quartz (pre-ore stage in Figure 13), arsenopyrite (As–Fe stage), and galena (Pb–Zn–Cu–Sb–Au stage) as the most abundant mineral phases. This *type b* mineral assemblage is well represented in the San Riccardo/Su Spinosu ore, where the ore paragenesis is not associated with a *type a* ore and linkage between the ore and transpressive dextral faults is manifest. Furthermore, ore microscopy shows that between the several *type b* mineralizing stages, there are progressive mutual relationships of cataclasis and successive mineral infilling, showing that *type b* was synkinematic with a brittle deformation. The post-ore stage of mineralization (cryptocrystalline quartz, pyrite) is widespread and crosscuts all of the previous mineral assemblages, so it postdates the main mineralizing events. Only in *type b* veins does gold occurrence assume economic relevance. On the contrary, a lack of significant gold grades in *type a* mineralization is noteworthy; it could be related to lower content in arsenopyrite, which has been recorded in other areas of the Gerrei–Sarrabus district as a probable first carrier for gold (Brecca mine: [5,41]).

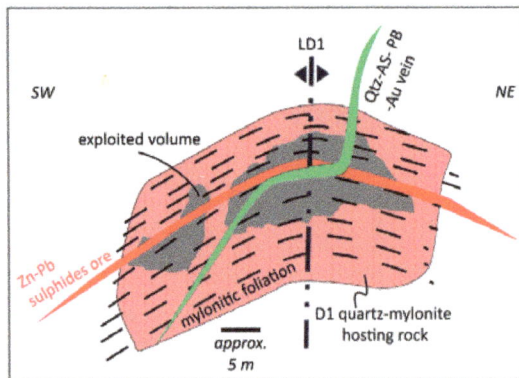

Figure 14. Sketch of the mutual relationships between *type a* and *type b* ores located at the top of LD1 antiform hinge zone.

5.2. Orebody/Tectonic Structure Relationships and Structural Control

The *type a* ore is hosted in the mylonites that characterize the Baccu Locci shear zone and it is located at the top of the hinge of a large LD1 antiform that folded together bedding, D1 folding axial-plane foliation, and mylonitic foliation. Interpreting the mylonite as Silurian black shales, some previous studies considered the tectonic foliation as primary bedding and classified the ores as stratiform and synsedimentary of probable volcano-exhalative origin [37,40]. From the definition of the Baccu Locci mylonitic shear zone [3], it is well established that the Zn–Cu sulphide lenses developed parallel to D1 Variscan foliations, in both the axial plane and mylonitic (Figure 6a). Furthermore, *type a* Zn–Cu sulphides show no physical relationship with the Ordovician metavolcanics ("Porfiroidi" Fm., Figures 3 and 5) that crop out as less deformed tectonic slices in the Baccu Locci shear zone. On the whole, these geometric relationships clearly rule out the possible syngenetic ("stratiform" or "stratabound") options for mineralization, excluding any connection between the primary bedding (now completely obliterated by the Variscan tectonic imprint) and orebodies. Similarly, field relationships and ore textures exclude possible syn-D1 ("syn-metamorphic") deposition; in fact, the sulphides infill the D1 foliations but they are not in turn affected by them. In this way, the LD1 hinge zone operated as a typical structural trap, but it essentially exerted "passive" structural control, because there is no evidence of synkinematic formation of mineral deposits. At the Gerrei–Sarrabus district scale, *type a* ores have not been recognized in places other than hinge zones of LD1 folds, also by the extensive surveys that were made during mine exploitation. Further confirmation may be found in an adjacent LD1 antiform, a few kilometers west of the study area (Riu Gruppa antiform, Sa Lilla mine; Figure 2), where previously reputed "stratoid" orebodies of sulphides, similar by mineral association and texture [42] to those of Baccu Locci/Spilloncargiu, are located at the top of a comparable hinge zone, in which development of sulphide "beds" was once again parallel to the D1 tectonic foliations.

Some speculation can be made about the modality that allowed the brittle reactivation of horizontal tectonic foliations to create space for *type a* mineral deposition. Structural data point out that *type a* ores postdate D1 and LD1 structures. Further, the creation of horizontal lens-shaped voids now filled with mineral deposits during the D2 extensional phase sounds unrealistic, because the vertical stress σ_1 (that we can image roughly higher than the lithostatic stress) operating during the rocks' exhumation would have prevented the opening of subhorizontal discontinuities. So, we can argue that *type a* ores could have been developed after the Gerrei and Rio Gruppa tectonic units got to shallower crustal levels, where the fluid pressure could overcome the lithostatic stress, possibly when the vertical stress σ_1 related to the extensional dynamics ceased. The role of decreased lithostatic stress in allowing for the development of open spaces that are suitable for mineralization should be confirmed by the absence of similar orebodies in deeper parts of the hinge zone. Anyway, this interpretation can be demonstrated when data about the baric environment are available. Up to now, the only available data from fluid inclusions [10] highlight that a metamorphic environment can be excluded, thus mineralization might postdate D2 deformation.

The *type b* Qtz–As–Pb veins are hosted in narrow brittle–ductile shear zones, generally developed inside the quartz–mylonitic rocks. We described their occurrence in the Spilloncargiu mine at the top of an LD1 antiform, where they cut at low angle *type a* ore, but they reach their main thickness in the San Riccardo/Su Spinosu mine. There, the availability of detailed mineworks plans permitted the recognition of the tectonic relationship and highlighted the occurrence of large dilational jogs (Figure 13). Interestingly, there are close relationships between jog geometry and older D1–LD1–D2 structures, although jogs clearly postdate them. In fact, the San Riccardo main fault is a generally WSW-steepening dipping transpressive fault that mostly cuts at high angle the NE-dipping D1 foliation in the northeast limb of LD1 antiform (Figure 15). The fault abruptly changes the dip direction to WSW, gently dipping just where it crosses gently SW-dipping D1 foliation in the reverse limb of a D2 recumbent fold (Figure 15). So, dilational jogs developed when the D4 fault reactivated preexisting anisotropies (in this case, D1 tectonic foliations) if they had the right attitude. Ore mineralogy shows repeated cataclasis and mineral infilling during the several mineralizing stages of *type b* veins, suggesting a progressive brittle deformation that in some cases produced large dilational jogs that were suitable to host orebodies with economic relevance [43,44]. Actually, the most important mineral assemblage—the galena- (and gold-) rich ore related to the Pb–Zn–Cu–Sb–Au stage of mineralization (Figure 13)—is associated with the largest jogs. The jog structures are not perfectly cylindrical, so their geometries can change slightly along strikes (Figure 11). This is probably due to a change in the D1 foliation attitude in the reverse limbs of D2 recumbent folds or to a change of the local stress field.

The recognition of such large dilational jogs is not very common [44] and it has been possible by the availability of observations at different scales. Moreover, the case of Baccu Locci/San Riccardo reveals that the occurrence of previous reactivable foliation, possibly with different attitude due to a polyphase deformation, can be one of the situations suitable for the development of large jogs. In the study area, the understanding of such active structural control would allow a different way to find new, possibly more fruitful, orebodies. In the case study, the change in dip direction of D1 tectonic foliation on the limbs of LD1 folds could be ignored, because it is not directly linkable with the mineralized veins, but it could be a clue to identify the occurrence of large dilational jogs.

Finally, some considerations might be given to the relationships between mafic dikes and the Qtz–As–Pb sulphide vein system, which is problematic. The definition of possible genetic links between the mafic magmatism and the mineralization processes falls well outside the scope of this work, requiring a wider geochemical study at the district scale. However, field mapping and explorations into the old mineworks showed several mutual relationships between dikes and ore veins (Figure 12).

These spatial relationships suggest a coeval emplacement. Mafic dikes may have intruded in an interval between the main mineralizing events of *type b* ore, in particular, between the first and second mineralization stages (pre-ore/As–Fe sulphide stages) and the third stage (sulphide/sulfosalts stage) of the paragenetic sequence, producing apparently contrasting timing relationships in the field. Under this hypothesis, considering the age available for these rocks in the neighboring areas [13], mafic dikes assume a chronological constraint, suggesting an age of mineralizing events around 302 Ma; that is, an age in which the Variscan basement of southern Europe suffered a widespread tectonic extension. This age is, in fact, consistent with: (a) ^{40}Ar-^{39}Ar dating of hydrothermal white mica in quartz–arsenopyrite–gold veins in the nearby Monte Ollasteddu area (307 ± 3 Ma) [10], and (b) several geological constraints occurring in the whole Gerrei district [26], indicating that the quartz–arsenopyrite–gold vein systems clearly predate the previously described mineralized vein systems related to F-bearing granites, dated at 286 Ma [9,12], and they are unconformably covered by lower Permian sediments dated at 295 Ma [13]. However, it cannot be excluded that the latest stages of mineralization in Baccu Locci (including the *type b* Pb–Zn–Cu–Sb–Au sulphide/sulfosalts stage that affects the mafic dikes) could be related to events referable to the younger part of the outlined chronological interval. Overall, these constraints allow for us to consider the ore deposits of the Baccu Locci mine area as part of a much wider metallogenic event that affected various massifs of the Variscan orogen between 310 and 300 Ma [45,46]. From their geological, mineralogical, and geochemical characteristics, they can be best classified as Variscan orogenic gold type [47]. As in other massifs of European Variscides, in Sardinia this metallogenic event involved a crustal-scale flow of fluids during the late Variscan extension, resulting in widespread mineralizing processes in major regional structures, such as the Flumendosa Antiform. Large shear zones, such as the Baccu Locci shear zone, are part of the main plumbing system through which deep fluids were focused toward the shallower parts of the crust.

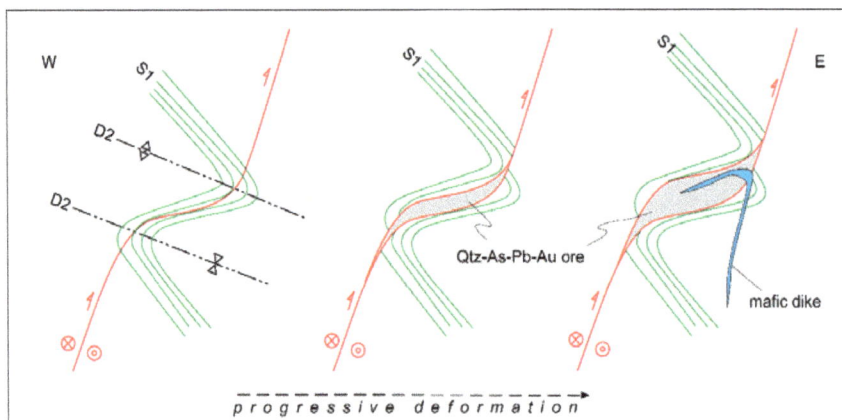

Figure 15. Baccu Locci/San Riccardo mine: scheme of the geometric relationships between foliation, faults, mafic dikes, and Qtz–As–Pb sulphide veins. See the location of this structure in the larger LD1 antiform in the cross-section in Figures 3 and 4.

6. Conclusions

The ores in Baccu Locci are a good example of structurally controlled mineralization in a basement characterized by the overprinting of several tectonic phases, from ductile to brittle, during both compressive and extensional regimes. The control exerted was either a passive reactivation of older foliations to create space for mineral deposition, or an active syn-kinematic deposition of minerals during the progressive evolution of the hosting structure. In particular, the emplacement of the older types of ores exposed in Baccu Locci emerges as the result of the opening of previous discontinuities (foliations) in the hinge zone of large antiforms, where they are subhorizontal, after their exhumation to shallow structural levels when postcollisional extension ceased. Afterward, a different stress field produced transpressive faults that reactivated anisotropic surfaces parallel to the axial plane in the hinge zone. Along the transpressive dextral faults, large dilational jogs developed, whose geometry was influenced by sudden changes of the attitude of Variscan foliation in the reverse limbs of recumbent folds. The jogs formed together with mineral deposits, exerting in this way an "active" control of mineralization and hosting the more economically relevant ores. As a general statement, the occurrence of older tectonic foliations and folds might be taken into account not only because they can be directly presumed to be hosting mineralized veins, a common concept in the study of structure–orebody relationships, but also when considering the influence that they could have in modifying hosting structures and favoring the formation of more significant orebodies, being in this way a good tool for prospecting new relevant ores.

Although the overprinting relationships between the different ores and the mafic dikes are now clearer, more data are needed to better constrain the thermobaric environment in which the minerals were deposited, the time interval between them, the source of the ore fluids, and finally the role of mafic dikes in the large frame of the Late Variscan metallogenic epoch in Sardinia.

Author Contributions: A.F., A.D., and S.N. conceptualized the study. A.F. performed the geological mapping. A.F., C.B, and F.C. performed the structural analysis and 3D structural modelling. A.D. and S.N. performed the orebodies' survey. S.N. performed the macro to micro-scale ore mineralogy. A.F. and S.N. wrote the original draft and with C.B. and F.C. reviewed and edited the draft. Funding acquisition and project administration were performed by A.F. and S.N.

Funding: This research was funded by FdS-RAS Fondazione di Sardegna and Regione Autonoma della Sardegna grant number F72F16003080002 and by Italian Government, project PRIN-2005 grant number 2005047008.

Acknowledgments: The authors are grateful to two anonymous reviewers for their comments that improved the quality of the paper and thank the editors for the careful editorial management.

Conflicts of Interest: The authors declare no conflict of interest.

References

1. Conti, P.; Carmignani, L.; Oggiano, G.; Funedda, A.; Eltrudis, A. From thickening to extension in the Variscan belt—Kinematic evidence from Sardinia (Italy). *Terra Nova* **1999**, *11*, 93–99. [CrossRef]
2. Funedda, A.; Naitza, S.; Conti, P.; Dini, A.; Buttau, C.; Tocco, S.; Carmignani, L. The geological and metallogenic map of the Baccu Locci mine area (Sardinia, Italy). *J. Maps* **2011**, *2011*, 103–114. [CrossRef]
3. Conti, P.; Funedda, A.; Cerbai, N. Mylonite development in the Hercynian basement of Sardinia (Italy). *J. Struct. Geol.* **1998**, *20*, 121–133. [CrossRef]
4. Funedda, A.; Naitza, S.; Conti, P.; Dini, A.; Buttau, C.; Tocco, S.; Carmignani, L. Structural control of ore deposits: The Baccu Locci shear zone (SE Sardinia). *Rend. Online SGI* **2011**, *15*, 66–68.
5. Lerouge, C.; Bouchot, V.; Douguet, M.; Naitza, S.; Tocco, S.; Funedda, A. *Variscan Gold Mineralisation of Baccu Locci and Brecca, Southeastern Sardinia: Petrographic and Geochemical Studies*; BRGM Report N RP-54431-FR; BRGM: Orleans, France, 2007; p. 47. Available online: http://infoterre.brgm.fr/rapports/RP-54431-FR.pdf (accessed on 13 October 2018).
6. Conti, P.; Carmignani, L.; Funedda, A. Change of nappe transport direction during the Variscan collisional evolution of central-southern Sardinia (Italy). *Tectonophysics* **2001**, *332*, 255–273. [CrossRef]
7. Cocco, F.; Funedda, A. The Sardic Phase: Field evidence of Ordovician tectonics in SE Sardinia, Italy. *Geol. Mag.* **2017**, 1–14. [CrossRef]
8. Carmignani, L.; Conti, P.; Barca, S.; Cerbai, N.; Eltrudis, A.; Funedda, A.; Oggiano, G.; Patta, E.D.; Ulzega, A.; Orrù, P.; et al. *Foglio 549-Muravera. Note Illustrative*; Servizio Geologico d'Italia: Roma, Italy, 2001; p. 140.
9. Conte, A.M.; Cuccuru, S.; D'Antonio, M.; Naitza, S.; Oggiano, G.; Secchi, F.; Casini, L.; Cifelli, F. The post-collisional late Variscan ferroan granites of southern Sardinia (Italy): Inferences for inhomogeneity of lower crust. *Lithos* **2017**, *294–295*, 263–282. [CrossRef]

10. Dini, A.; Di Vincenzo, G.; Ruggieri, G.; Rayner, J.; Lattanzi, P. Monte Ollasteddu, a new late orogenic gold discovery in the Variscan basement of Sardinia (Italy)—Preliminary isotopic (40Ar-39Ar, Pb) and fluid inclusion data. *Miner. Depos.* **2005**, *40*, 337–346. [CrossRef]

11. Cortesogno, L.; Cassinis, G.; Dallagiovanna, G.; Gaggero, L.; Oggiano, G.; Ronchi, A.; Seno, S.; Vanossi, M. The Variscan post-collisional volcanism in Late Carboniferous-Permian sequences of Ligurian Alps, Southern Alps and Sardinia (Italy): A synthesis. *Lithos* **1998**, *45*, 305–328. [CrossRef]

12. Ronca, S.; Del Moro, A.; Traversa, G. Geochronology, Sr-Nd isotope geochemistry and petrology of Late Hercynian dike magmatism from Sarrabus (SE Sardinia). *Period. Mineral.* **1999**, *68*, 231–260.

13. Dack, A. Internal Structure And geochronology of the Gerrei Unit in the Flumendosa Area, Variscan External Nappe Zone, Sardinia, Italy. Master's Thesis, Boise State University, Boise, ID, USA, 2009.

14. Cassinis, G.; Ronchi, A. Upper Carboniferous to Lower Permian continental deposits in Sardinia (Italy). *Geodiversitas* **1997**, *19*, 217–220.

15. Carmignani, L.; Carosi, R.; Di Pisa, A.; Gattiglio, M.; Musumeci, G.; Oggiano, G.; Pertusati, P.C. The Hercynian chain in Sardinia (Italy). *Geodin. Acta* **1994**, *7*, 31–47. [CrossRef]

16. Conti, P.; Patta, E.D. Large scale W-directed tectonics in southeastern Sardinia. *Geodin. Acta* **1998**, *11*, 217–231. [CrossRef]

17. Carosi, R.; Musumeci, G.; Pertusati, P.C. Senso di trasporto delle unità tettoniche erciniche della Sardegna dedotto dagli indicatori cinematici nei livelli cataclastico-milonitici. *Rend. Soc. Geol. Ital.* **1990**, *13*, 103–106.

18. Casini, L.; Funedda, A. Potential of pressure solution for strain localization in the Baccu Locci Shear Zone (Sardinia, Italy). *J. Struct. Geol.* **2014**, *66*, 188–204. [CrossRef]

19. Casini, L.; Funedda, A.; Oggiano, G. A balanced foreland–hinterland deformation model for the Southern Variscan belt of Sardinia, Italy. *Geol. J.* **2010**, *45*, 634–649. [CrossRef]

20. Funedda, A. Foreland- and hinterland-verging structures in fold-and-thrust belt: An example from the Variscan foreland of Sardinia. *Int. J. Earth Sci.* **2009**, *98*, 1625–1642. [CrossRef]

21. Funedda, A.; Meloni, M.A.; Loi, A. Geology of the Variscan basement of the Laconi-Asuni area (central Sardinia, Italy): The core of a regional antiform refolding a tectonic nappe stack. *J. Maps* **2015**, *11*, 146–156. [CrossRef]

22. Montomoli, C.; Iaccarino, S.; Simonetti, M.; Lezzerini, M.; Carosi, R. Structural setting, kinematics and metamorphism in a km-scale shear zone in the Inner Nappes of Sardinia (Italy). *Ital. J. Geosci.* **2018**, *137*, 294–310. [CrossRef]

23. Carmignani, L.; Pertusati, P.C. Analisi strutturale di un segmento della catena ercinica: Il Gerrei (Sardegna sud-orientale). *Boll. Soc. Geol. Ital.* **1977**, *96*, 339–364.

24. Carmignani, L.; Oggiano, G.; Barca, S.; Conti, P.; Salvadori, I.; Eltrudis, A.; Funedda, A.; Pasci, S. *Geologia della Sardegna. Note Illustrative della Carta Geologica in Scala 1:200.000*; Servizio Geologico d'Italia: Roma, Italy, 2001.

25. Carmignani, L.; Cortecci, G.; Dessau, G.; Duchi, G.; Oggiano, G.; Pertusati, P.; Saitta, M. The antimony and tungsten deposit of Villasalto in South-Eastern Sardinia and its relationship with Hercynian tectonics. *Schweiz. Mineral. Petrogr. Mitt.* **1978**, *58*, 163–188.

26. Funedda, A.; Naitza, S.; Tocco, S. Caratteri giacimentologici e controlli strutturali nelle mineralizzazioni idrotermali tardo-erciniche ad As-Sb-W-Au del basamento metamorfico paleozoico della Sardegna Sud-orientale. *Resoconti dell'Associazione Mineraria Sarda* **2005**, *110*, 25–46.

27. Garbarino, C.; Naitza, S.; Tocco, S.; Farci, A.; Rayner, J. Orogenic Gold in the Paleozoic Basement of SE Sardinia. In *Mineral Exploration an Sustainable Development*; Eliopoulos, D.G., Ed.; Mill Press: Rotterdam, The Netherlands, 2003; pp. 767–770.

28. Naitza, S.; Oggiano, G.; Cuccuru, S.; Casini, L.; Puccini, A.; Secchi, F.; Funedda, A.; Tocco, S. Structural and magmatic controls on Late Variscan Metallogenesis: Evidences from Southern Sardinia (Italy). In *Mineral Resources in a Sustainable World, Proceedings of the 13th Biennial SGA Meeting, Nancy, France, 24–27 August 2015*; André-Mayer, A.S., Cathelineau, M., Muchez, P.H., Pirard, E., Sindern, S., Eds.; The Society for Geology Applied to Mineral Deposits (SGA): Nancy, France, 2015; pp. 161–164.

29. Naitza, S.; Conte, A.M.; Cuccuru, S.; Oggiano, G.; Secchi, F.; Tecce, F. A Late Variscan tin province associated to the ilmenite-series granites of the Sardinian Batholith (Italy): The Sn and Mo mineralisation around the Monte Linas ferroan granite. *Ore Geol. Rev.* **2017**, *80*, 1259–1278. [CrossRef]

30. Belkin, H.E.; De Vivo, B.; Valera, R. Fluid inclusion study of some Sarrabus fluorite deposits, Sardinia, Italy. *Econ. Geol.* **1984**, *79*, 409–414. [CrossRef]
31. Boni, M.; Balassone, G.; Fedele, L.; Mondillo, N. Post-Variscan hydrothermal activity and ore deposits in southern Sardinia (Italy): Selected examples from Gerrei (Silius Vein System) and the Iglesiente district. *Period. Mineral.* **2009**, *78*, 19–35.
32. Giamello, M.; Protano, G.; Riccobono, F.; Sabatini, G. The W-Mo deposit of Perda Majori (SE Sardinia, Italy): A fluid inclusion study of ore and gangue minerals. *Eur. J. Mineral.* **1992**, *4*, 1079–1084. [CrossRef]
33. Allmendinger, R.W.; Cardozo, N.; Fisher, D. *Structural Geology Algorithms: Vectors and Tensors*; Cambridge University Press: Cambridge, UK, 2012; p. 302.
34. Lena, G.; Barchi, M.R.; Alvarez, W.; Felici, F.; Minelli, G. Mesostructural analysis of S-C fabrics in a shallow shear zone of the Umbria–Marche Apennines (Central Italy). *Geol. Soc. Lond. Spec. Publ.* **2018**, *409*, 149–166. [CrossRef]
35. Rutter, E.H.; Maddock, R.H.; Hall, S.H.; White, S.H. Comparative microstructures of natural and experimentally produced clay-bearing fault gouges. *Pure Appl. Geophys.* **1986**, *124*, 3–30. [CrossRef]
36. Barton, P.B.; Bethke, P.M. Chalcopyrite disease in sphalerite: Pathology and epidemiology. *Am. Mineral.* **1987**, *72*, 451–467.
37. Bakos, F.; Carcangiu, G.; Fadda, S.; Mazzella, A.; Valera, R. The gold mineralization of Baccu Locci (Sardinia, Italy): Origin, evolution and concentration processes. *Terra Nova* **1990**, *2*, 232–237. [CrossRef]
38. Zucchetti, S. The lead-arsenic-sulfide ore deposit of Bacu Locci (Sardinia-Italy). *Econ. Geol.* **1958**, *53*, 867–876. [CrossRef]
39. Hancock, P.L. Brittle microtectonics: Principles and practice. *J. Struct. Geol.* **1985**, *7*, 437–457. [CrossRef]
40. Schneider, H.-J. Schichtgebundene NE-Metall- und F-Ba-Lagerstätten im Sarrabus-Gerrei-Gebiet, SE-Sardinien. I. Bericht: Zur Lagerstättenkunde und Geologie. *Neues Jahrb. Mineral. Monatshefte* **1972**, *1972*, 529–541.
41. Lerouge, C.; Naitza, S.; Bouchot, V.; Funedda, A.; Tocco, S. Invisible gold in arsenopyrite of the Variscan Au-Sb Brecca mineralization (Gerrei district, Southeastern Sardinia). *Geol. Fr.* **2007**, *2*, 124.
42. Violo, M. Contributo alla conoscenza dei giacimenti stratoidi polimetallici, in area metamorfica. Il giacimento di Sa Lilla (San Vito, Cagliari-Sardegna). *Resoconti dell'Associazione Mineraria Sarda* **1966**, *71*, 5–110.
43. Cox, S.F.; Braun, J.; Knackstedt, M.A. Principles of structural control on permeability and fluid flow in hydrothermal systems. *Rev. Econ. Geol.* **2001**, *14*, 1–24.
44. Robert, F.; Poulsen, K.H. Vein formation and deformation in greenstone gold deposits. *Rev. Econ. Geol.* **2001**, *14*, 111–155.
45. Bouchot, V.; Ledru, P.; Lerouge, C.; Lescuyer, J.L.; Milesi, J.P. Late Variscan mineralizing systems related to orogenic processes: The French Massif Central. *Ore Geol. Rev.* **2005**, *27*, 169–197. [CrossRef]
46. De Boorder, H. Spatial and temporal distribution of the orogenic gold deposits in the Late Palaeozoic Variscides and Southern Tianshan: How orogenic are they? *Ore Geol. Rev.* **2012**, *46*, 1–31. [CrossRef]
47. Bouchot, V.; Milesi, J.P.; Ledru, P. Crustal Scale Hydrothermal Palaeofield and Related Au, Sb, W Orogenic Deposits at 310–305 Ma (French Massif Central, Variscan Belt). *SGA News* **2000**, *10*, 6–12.

minerals

MDPI

Article

Multi-Stage Deformation of the Khangalas Ore Cluster (Verkhoyansk-Kolyma Folded Region, Northeast Russia): Ore-Controlling Reverse Thrust Faults and Post-Mineral Strike-Slip Faults

Valery Y. Fridovsky [1,*], Maxim V. Kudrin [1] and Lena I. Polufuntikova [2]

[1]   Diamond and Precious Metal Geology Institute, SB RAS, Yakutsk 677000, Russia; kudrinmv@mail.ru
[2]   M.K. Ammosov North-Eastern Federal University, Yakutsk 677000, Russia; pli07@list.ru
*   Correspondence: 710933@list.ru; Tel.: +7-4112-33-58-72

Received: 4 May 2018; Accepted: 22 June 2018; Published: 26 June 2018

Abstract: This study reports the results of the analysis of multi-stage deformation structures of the Khangalas gold ore cluster, northeast Russia. Four Late Mesozoic-Early Eocene deformation stages were identified. The first deformation event (D1) was characterized by the development of NW-striking tight to isoclinal folds of the first generation (F1) and interstratal detachment thrusts. Major folds, extensive thrusts, boudinage, cleavage, auriferous mineralized fault zones and quartz-vein gold mineralization were formed in the reverse and thrust fault stress field during the progressive deformation stage (D1), with NE-SW-oriented σ1. Post-ore deformation is widely manifested in the region. Structures D2 and D3 are coaxial. Sinistral strike-slip motions (D2 and D3) occurred along NW-trending faults under prevailing W-E compression. They were accompanied by the formation of NS- and NE-striking F2–3 folds with steep hinges and by bending of the earlier formed structures, among them ore-controlling ones. The last deformation event (D4) was represented by normal-dextral strike-slip faulting, refolding of rocks, pre-existing structures and ore bodies and by the development of folds with steep hinges. Key structural elements of varying age are described, the chronology of deformation events and mineralization reconstructed and their relation to geodynamic events in northeast Asia established.

Keywords: Verkhoyansk-Kolyma folded region; Khangalas ore cluster; orogenic gold mineralization; deformation structure; thrust fault; strike-slip fault

---

1. Introduction

The Khangalas ore cluster (KOC) is located in the southeastern part of the Kular-Nera slate belt of the Verkhoyansk-Kolyma folded region in northeast Russia. It was discovered in the 1940s. Rich placer deposits with large gold nuggets are known there. Commercial exploitation of the KOC commenced in the latter half of the 20th Century and continues to present day. At an early stage of geologic investigation (1940–1980), considerable attention was given to concordant ore bodies confined to the limbs of brachyanticlinal folds [1]. Then, in the late 1980s to the early 2000s, mineralized fault zones of complex structure with diverse mineralization were identified and investigated, which considerably enlarged the mineral resource potential of the KOC [2,3]. Southeasterly, in the Upper Kolyma gold district with a similar geological-structural setting, several small- and medium-sized gold deposits are found, such as Vetrenskoye, Chay-Yuruye, Svetloye, etc. [4,5]. Gold mineralization of the KOC is of the orogenic type, which is characterized by a close relationship with Late Jurassic-Neocomian tectonomagmatic events in the Verkhoyansk-Kolyma folded region [6–10]. The paper presents new data on the KOC geology obtained by the authors in the last few years, which provide a better understanding of the relations between folds, faults and mineralization in the region.

2. Materials and Methods

Structural-tectonic factors are among the most important in controlling localization of orogenic Au-quartz deposits [11,12]. Structural-kinematic studies in the Khangalas ore cluster were conducted using up-to-date methods [13–17]. The morphology of ore veins in natural exposures and mine workings was studied and their relations to geological structures established. Interactions between the veins and faults were studied using the belt method [18]. This is based on the regular position of vein poles relative to faults in stereographic projection. The method enables determination of the sense of displacement during vein formation [18]. Measurements of planar and linear structures (bedding, cleavage, vein-veinlet bodies, faults, ore zones, jointing, fold hinges, boudinage, slickenlines, etc.) were made. The kinematics of main deformation stages and the paleo-orientation of stress were reconstructed relative to major deformation structures of NW strike. Structural data were statistically analyzed and plotted on the upper hemisphere of the Wulff stereographic net.

In 2005, detailed field studies and structural mapping were conducted at the Nagornoye deposit and the Dvoinoye, Ampir and Klich-Kontrolnoye occurrences, at Mudeken and other localities and in 2014 and 2017 at the Khangalas deposit and Ozhidaniye occurrence. The nomenclature of structural elements is taken from [19]. Planar structures (S) are given as dip azimuth/dip angle (e.g., 90/60 denotes eastward dip at 60°). For linear deformation elements (*l*), denotation plunge azimuth/plunge angle is used (e.g., 215/45 means plunge azimuth of 215° and plunge angle of 45°). Signs S1 and *l*1 denote the relation of a structural element to a particular deformation stage (D1) event. The studies enabled refining the general architecture of the region, revealing the particularities of the ore-controlling structures, identifying key structural elements, reconstructing the chronology of deformational events and mineralization and establishing their relationship with regional geodynamic events in northeast Asia.

3. Geology of the Southeastern Part of the Kular-Nera Slate Belt and the Khangalas Ore Cluster

The Kular-Nera slate belt (KNSB) is situated in the central part of the Verkhoyansk-Kolyma folded region [6]. It is mainly composed of Upper Permian, Triassic and Lower Jurassic terrigenous rocks. Extensive faults separate the belt from adjacent tectonic structures. In the northeast, it is separated from the In'yali-Debin synclinorium by the Charky-Indigirka and Chai-Yureye faults, and in the southwest, the Adycha-Taryn fault separates it from structures of the passive margin of the North Asian craton. The structural pattern of KNSB is defined by linear folds and faults of NW strike that developed over several deformation stages. Within the KOC, NW-striking faults represent branches of the Nera Fault, which manifests itself as 4 km-wide zones of intensive deformations and subvertical foliation of the rocks. Dextral strike-slip motions have been reported along the Nera Fault [20].

Magmatism is poorly manifested in KNSB. It is mainly represented by granitoid massifs, subvolcanic magma of dacite composition and dikes belonging to the NNW-striking Tas-Kystabyt magmatic belt. They were formed in Late Jurassic-Albian times [21]. Various tectonomagmatic events characteristic of the Late Jurassic-Late Cretaceous history of the eastern margin of the North Asian craton are manifested within KNSB [6,22,23]. The Late Jurassic-Early Cretaceous was marked by accretion and collision of the Kolyma-Omolon microcontinent against the craton margin and by subduction processes in the Uda-Murgal island arc. These events produced different-age fold and thrust structures, S- and I-type granitoids and orogenic gold deposits. In the Late Neocomian, the direction of the Kolyma-Omolon microcontinent motion and of subduction in the Uda-Murgal arc changed [6]. At that time, left-lateral strike slip motions first occurred in KNSB along NW-trending faults. Post-accretionary tectonic events and Au-Sb and Ag mineralization events were related to Late Cretaceous subduction within the Okhotsk-Chukotka arc [24].

The Khangalas ore cluster is located in the arch of the Nera anticlinorium that is represented in the study area by the NW-striking Dvoinaya anticline composed of dislocated Upper Permian and Lower-Middle Triassic terrigenous rocks (Figure 1). The Upper Permian (P_2) deposits make up the core of the Dvoinaya anticline. The lower part of the section consists of massive brownish-grey and grey greywacke sandstones with thin siltstone interbeds. The upper part is dominated by an 800 m thick sequence of dark-grey and black siltstones with inclusions of pebbles of sedimentary, magmatic and metamorphic rocks. The limbs of the Dvoinaya anticline are made of 680–750 m thick Lower-Middle Triassic deposits (T_1), mainly dark-grey shales, mudstones and siltstones with rare interbeds of light-grey sandstones. The Middle Triassic deposits of the Anisian stage (T_2a) are represented by a 700–800 m thick sequence of alternating sandy siltstones and siltstones with rare fine-grained sandstone interbeds. The Ladinian strata (T_2l) are chiefly made of interbedded siltstones and sandstones with a total thickness of 850–950 m.

The main ore-controlling rupture dislocations are the Khangalas, Dvoinoy and Granitny faults represented by zones of breccia and fracture, low sulfidation of rocks and quartz-carbonate vein mineralization (Figure 1). The Khangalas Fault crosscuts the Khangalas ore cluster in a northwest direction. It controls localization of the Khangalas deposit and Ampir and Klich-Kontrolnoye occurrences. Within the study area, the exposed fault changes its strike from NW-SE to E-W and has a dip direction to S-W and S. The bedding of rocks exposed in the S-W wall strikes N-W, and rocks of the N-E wall strike NE-SW and E-W. The Dvoinoy Fault strikes E-W, and its fault plane is subvertical. In the central part of the KOC, northward of the Klich-Kontrolnoye occurrence, the Dvoinoy Fault adjoins the Khangalas Fault. The northeastern branch of the Dvoinoy Fault controls mineralization at the Nagornoye deposit. The rocks of the S wall of the fault have a N-E strike, while those of the N one strike E-W. The Granitny Fault is located in the southwestern part of the KOC. Outside the Khangalas ore cluster, the Ala-Chubuk massif of biotite granites is confined to it.

Magmatic activity is manifested by rare mafic and intermediate dikes of the normal and subalkaline series of Late Jurassic (Nera Complex (J_3n)) and Late Cretaceous (Khulamrinsk Complex (K_2ch)) age (Figure 1). The Nera magmatic complex includes basalt, gabbro and diorite porphyry dikes that extend for a distance from a few tens of meters to 2 km and have NE strike and a thickness of 1–20 m. The dikes underwent alteration. They contain quartz-carbonate veinlets. The Khulamrinsk magmatic complex consists of rare trachybasalt dikes extending for 200–500 m. They have a NW strike and are 1–10 m thick.

At 7 km to the northwest of the Khangalas ore cluster is the exposed Ala-Chubuk massif of biotite granites. The K-Ar age of the massif determined on orthoclase from porphyry phenocrysts is 145.0 ± 3.0 Ma and on biotite from the groundmass 149.0 ± 3.0 Ma [25]. The available geophysical data imply the presence of unexposed intrusions of similar composition at the Nagornoye and Khangalas deposits [25]. The Khangalas, Dvoinoye and Duk ore fields are identified within the KOC. The first field occurs in the southeast of the ore cluster and includes the Khangalas deposit and the Ozhidaniye occurrence. To the northwest of them are the Klich-Kontrolnoye, Dvoinoye and Ampir occurrences belonging to the Dvoinoye ore field. The Duk ore field includes the Nagornoye deposit. The ore bodies consist of extensive mineralized fault zones and concordant and cross-cutting gold-quartz veins and veinlets with simple mineral composition. The amount of ore minerals does not exceed 1–3%. These are arsenopyrite, pyrite, galena, sphalerite, chalcopyrite and native gold with 820–830‰ fineness and rare antimonite and Pb-sulfosalts. Quartz is the main gangue mineral, with less abundant carbonates (calcite and siderite) and chlorite. A series of successive mineral assemblages are identified in the ores of the deposits. These are pyrite-arsenopyrite-quartz metasomatic, quartz-pyrite-arsenopyrite vein, chalcopyrite-sphalerite-galena and sulfosalt-carbonate assemblages. The early pyrite-arsenopyrite-quartz mineralization is developed in wall-rock metasomatites. It is represented by irregular disseminations of pyrite and arsenopyrite metacrysts and by thin quartz streaks. Pyrite prevails over arsenopyrite. Minerals of the metasomatic assemblage are characterized by euhedral and subhedral crystals and a streaky-disseminated structure. Pyrite and arsenopyrite grains show evidence of deformation and corrosion. Pyrite and arsenopyrite of the early vein assemblage occur as disseminated euhedral grains and intergrowths. Pyrites of the vein assemblage contain Co, Sb, As, Ni, Cu and Zn trace contaminants. Minerals of the productive chalcopyrite-sphalerite-galena assemblage sporadically occur as disseminations and small aggregates in milk-white quartz and as microinclusions in early sulfides. The principal mineral of the sulfosalt-carbonate assemblage is siderite. Sulfosalts are represented by boulangerite, tetrahedrite and bournonite.

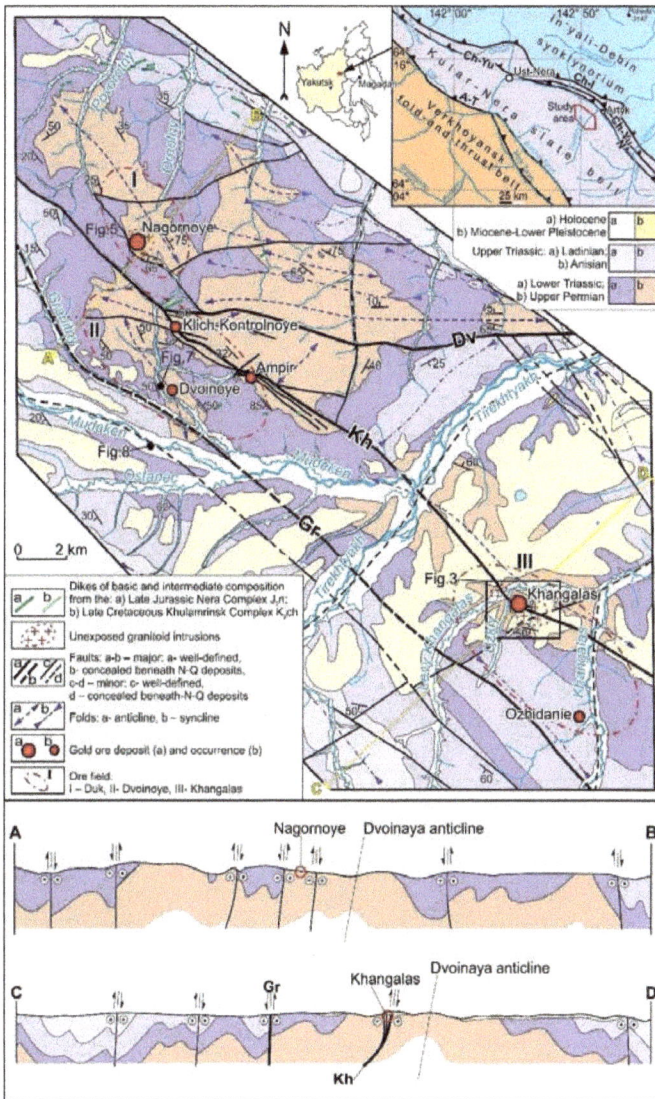

Figure 1. Geological sketch map, sections and location of gold deposits and occurrences of the Khangalas ore cluster, modified and supplemented from [2]. The inset map shows the position of the Khangalas ore cluster. Faults: Ch-I, Charky-Indigirka; Ch-Yu, Chai-Yureye; N, Nera; A-T, Adycha-Taryn.

Microthermometric studies of the fluid were conducted at the laboratory of the M.K. Ammosov North-Eastern Federal University on an optical microscope AxioScope.Al fitted with a motorized heating stage (up to 600 °C) and a liquid nitrogen sample cooling system (down to −196 °C) (LNP95). Analyses were made on milk-white quartz samples from the ore veins. The results of thermo-

and cryo-metric studies showed that ore-forming fluids of the KOC originated at temperatures of 310–330 °C and a pressure of 0.9 kbar. The data obtained suggest that gold mineralization was formed at a depth of about 3.5 km.

Age determinations are scarce for the KOC mineralization. The K-Ar sericite age of the Nagornoye deposit is 130.0 ± 4.0 Ma [23]. The authors of the given paper conducted Re-Os dating of gold (Sample Number X-45-14) from a quartz vein from the Yuzhnaya ore zone of the Khangalas deposit at the Center of Isotope Research of the Karpinsky All-Russian Scientific-Research Geological Institute (St. Petersburg, Russia). The isochron Re-Os age of gold was 137.0 ± 7.6 Ma. This indicates that productive orogenic gold-quartz mineralization of the region was formed in Valanginian-Hauterivian times.

The mineralized brittle fault zones consist of breccias and blocks of quartzose sandstones and siltstones and are often accompanied by concordant (a few cm to 1–2 m thick, in swells up to 5 m) and cross-cutting quartz veins and veinlets and disseminated sulfide mineralization (Figure 2). They are mainly localized in sandstones and at their contacts with siltstones and have conformable and crossing relations with the host rocks. The ore zones underwent strong supergene alteration as indicated by the presence of Fe oxides, sulfates, clay minerals, etc. Nesterov N.V. [26] has reported on a secondary gold enrichment at the Khangalas deposit. The quartz veins are often deformed in the mineralized fault zones, which is indicative of post-ore deformation. The host siltstones and sandstones contain disseminated sulfide mineralization (Figure 2D) represented by fine to coarse crystalline arsenopyrite and pyrite occurring as crystals, nests and veinlets.

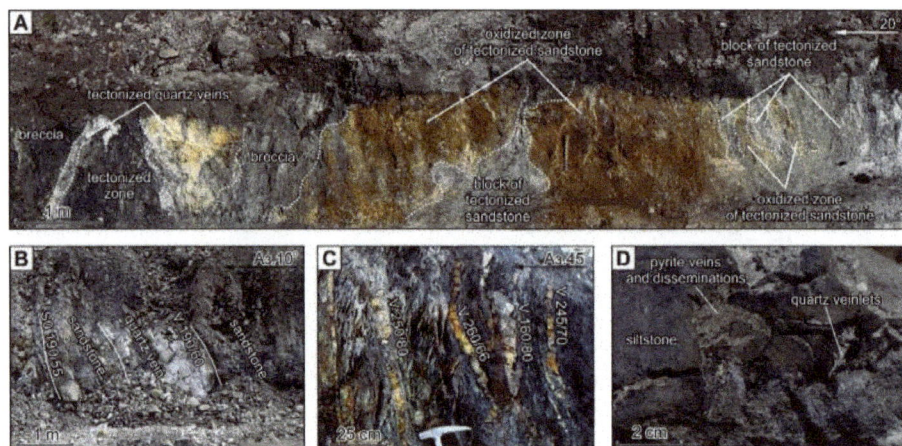

Figure 2. Khangalas Fault (**A**) and types of mineralization in the Khangalas ore cluster (KOC): (**B**) concordant veins, Centralnaya zone of the Khangalas deposit; (**C**) vein-veinlet mineralization; (**D**) veinlet-disseminated mineralization.

4. Deformation Structures of Key Deposits and Localities of the Khangalas Ore Cluster

This section presents the results of the analysis of the deformation structures of accretionary-collisional and post-accretionary stages in the formation history of the Khangalas and Nagornoye deposits, Dvoinoye occurrence and Mudeken locality.

4.1. Khangalas Deposit

The deposit occurs in the southeastern part of the KOC, on the right bank of Levy Khangals Creek, in the area between its Uzkiy and Zimniy tributaries (Figure 1). The host rocks are represented by Upper Permian sandstones and, more rarely, siltstones. Mineralization is localized in five fault zones

(Severnaya, Promezhutochnaya, Centralnaya, Yuzhnaya and Zimnyaya, length up to 1400 m) with concordant and cross quartz veins 0.1–5 m thick in the crest of the Dvoinaya anticline (Figures 3 and 4). Ore zones with a thickness of up to 32 m dip S-W, S and S-E at 30–50° to 70–80°. The reserves exceed 11 tons of gold with an average grade of 11.2 g/t Au [27].

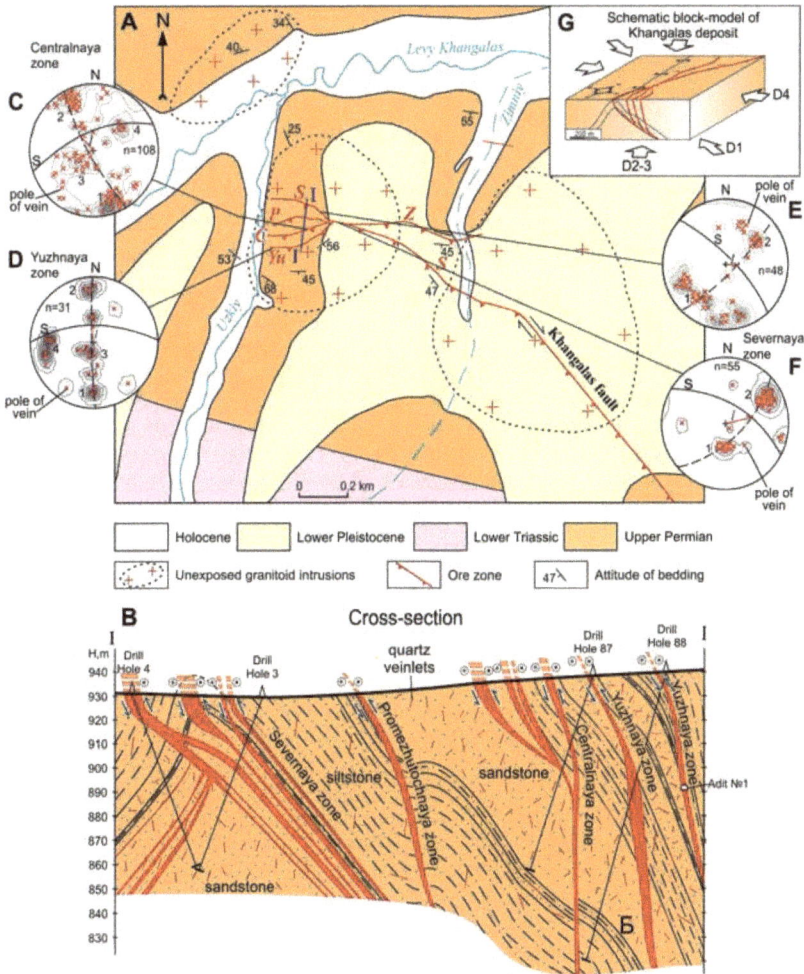

Figure 3. Geological sketch map (**A**), cross-section (**B**) and stereograms of the quartz veins poles (**C–F**), Khangalas deposit, (**G**) Schematic block-model of Khangalas deposit. In (**A**), the I-I line shows the position of the cross-section (**B**). Mineralized fault zones: S, Severnaya; P, Promezhutochnaya; C, Centralnaya; Yu, Yuzhnaya; Z, Zimnyaya. Symbols in stereograms and figures hereafter are: S, position of fault or ore zone; *l*, calculated direction of rock motion; n, number of measurements; dashed line, belt of vein poles. Contours of poles to planes (per 2% area).

Various deformation structures are manifested at the Khangalas deposit (Figures 3 and 4). Early isoclinal and tight concentric folds (F) with N-W strike and subhorizontal hinges (b) occur in narrow (up to a few tens of meters) zones (Figure 4D). In the study area, such folds were first mapped on the northeast limb of the Nera anticlinorium, in the zone influenced by the Chay-Yureye

Fault [8]. Early folds are draped during progressive deformation into late folds, so that crests of late folds can often be seen on the limbs of early folds (Figure 4C). F1 folds are the most widespread in the KOC area; they have NW-SE strike and gentle hinges (b1) (Figure 4A). These are for the most part open folds that pass into tight ones nearby the fault zones. On the right side of Uzkiy Creek, folds and ore zones have E-W and, less frequently, N-E strike due to superimposed strike-slip deformation. F1 folds are accompanied by *Cl* cleavage (Figure 4). It is platy, rarely shelly-platy, and its intensity depends on the rock composition. The most intense cleavage is observed in siltstones, whereas in sandstones, it becomes coarse-platy. Its regional NW strike changes to NE-SW and E-W in areas of superposed strike-slip deformation.

The Severnaya mineralized zone is the most extensive one. On the western side of the Khangalas deposit, the Promezhutochnaya, Centralnaya and Yuzhnaya zones branch off from the Severnaya zone forming a horse tail termination structure, and on the eastern side, the Zimnyaya zone diverges from it in the E-W direction. The ore-controlling structures are confined to the core of the Dvoinaya anticlinal fold. The strike of the ore zones varies from NW-SE to S-W and, locally, to NE-SW.

Analysis of the attitude of quartz veins and veinlets revealed five variously-oriented systems (Figures 2 and 3). Veins of the first system have persistent parameters; they are conformable with the host rocks (Figures 2B,C and 3). Quartz veins of the second system follow the orientation of the bedding plane and ore zones, but they dip in the opposite direction. Low-angle veins of the third system localized in tension fractures in sandstones are rather common. The orientation of the fourth vein system is normal to the strike of mineralized faults. In some areas, all five systems of veins and veinlets are present, which form stockworks. Such systems of quartz vein mineralization, which are related to the reverse and thrust fault stress field, are also found at other gold deposits in the central part of the Kular-Nera slate belt (Bazovskoye, Malo-Tarynskoye, Levoberezhnoye and Sana) [7,28–32].

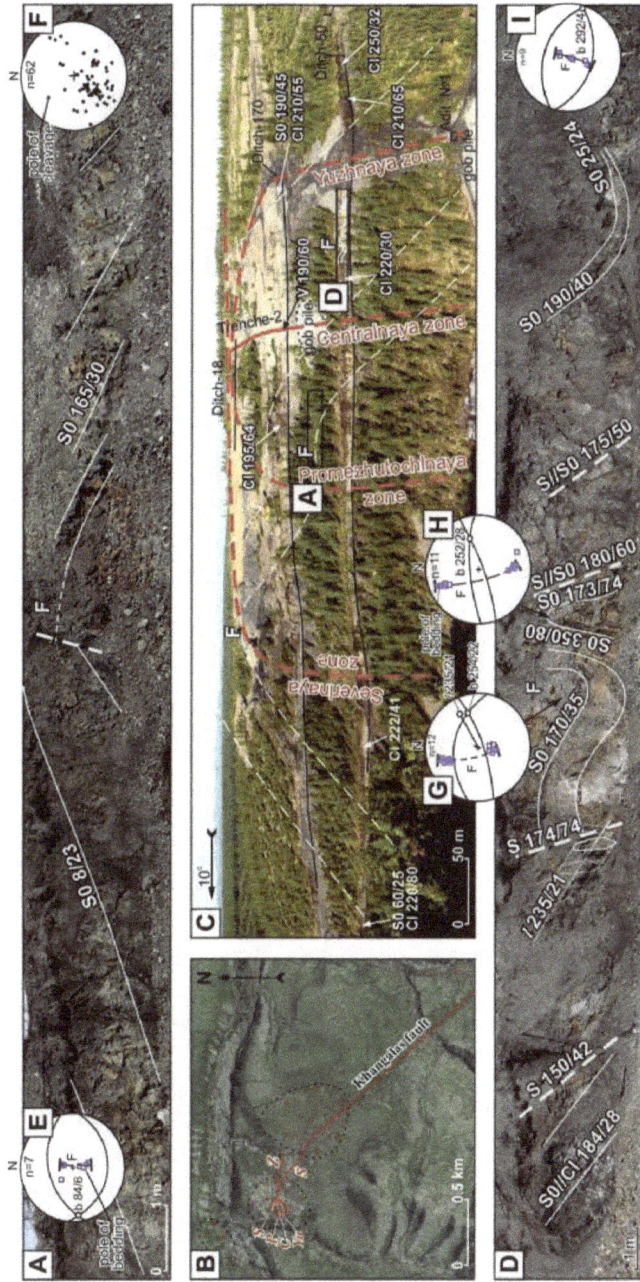

Figure 4. Bing Maps-satellite image (**B**), panoramic photo (**C**) and folds (**A,D**) of the Khangalas deposit. (**E–I**) Stereograms of bedding poles; Cl, cleavage. Mineralized fault zones: S, Severnaya; P, Promezhutochnaya; C, Centralnaya; Yu, Yuzhnaya; Z, Zimnyaya.

4.2. Nagornoye Deposit

The deposit is located in the northwestern part of the KOC within the Duk ore field (Figure 5). Ore bodies are between bedding planes within faults of an ESE-WNW trend. In some places, the fault zones are accompanied by concordant (up to 100 m in extent) and feathering quartz veins and veinlets (Figure 5E,F). The faults have reverse fault and strike-slip kinematics. The thickness of ore zones on the Nagornoye deposit varies from 0.6 to 3–4 m (average 1.0 m). Ore minerals include native gold, pyrite, arsenopyrite, Fe-gersdorffite, galena, sphalerite, chalcopyrite, tetrahedrite, bournonite and the rutile-group minerals [33]. They constitute up to 3% of the veins' volume. The host rocks are mostly represented by Upper Permian sandstones with siltstone interbeds. They have an ESE-WNW strike and a steep (70–75°) to vertical, sometimes overturned bedding. They are deformed into ESE-WNW F folds with horizontal b hinges in which limbs with NNE-striking open folds with steep hinges have developed (Figure 5C,D).

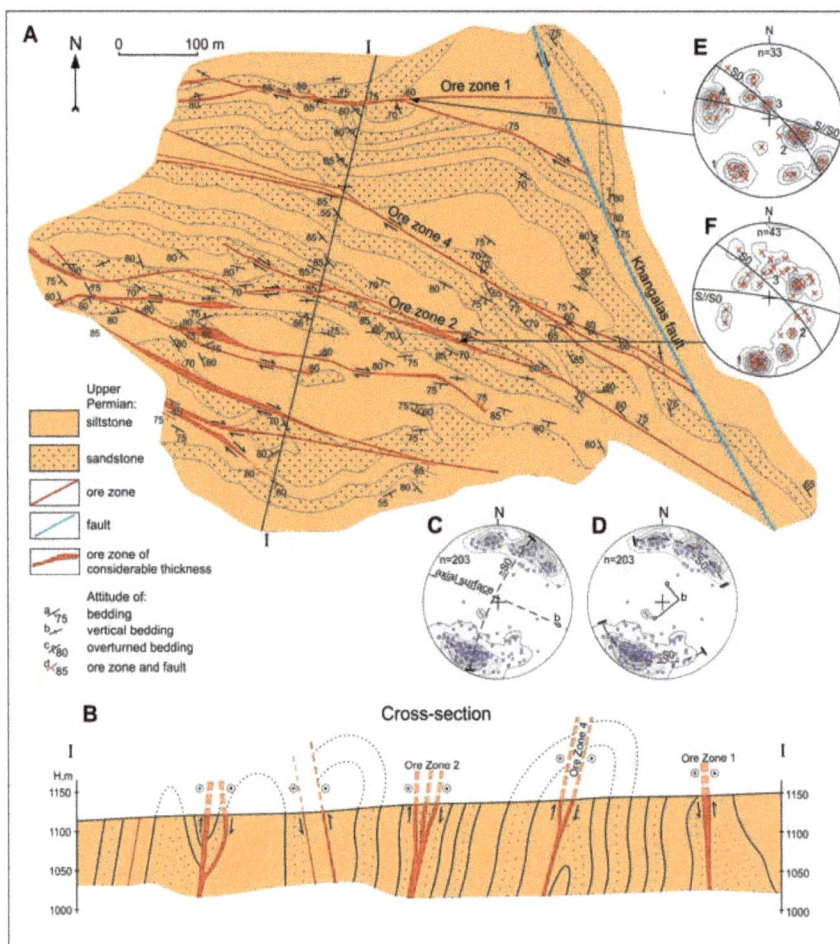

Figure 5. Geological sketch map and section of the Nagornoye occurrence. In (**A**), the I-I line shows the position of the cross-section (**B**). The diagrams show: (**C,D**) bedding poles; (**E**) quartz vein poles in Ore Zone 1; (**F**) quartz vein poles in Ore Zone 2. Contours of poles to planes (per 2% area).

The most extensively studied is Ore Zone 2 (Figures 5 and 6A). It is exposed over much of its length in a mining trench, where it parallels a subvertical sequence of quartzy sandstones interlayered with siltstones (Figure 6A). Statistical analysis of the attitude of quartz veins showed that on the diagrams, fields of poles of veins are grouped, in spite of their significant scatter, along the great circle arcs corresponding to the projection of mineralized ore zones (S) (190/85) conformable with the rock bedding (S0) (Figure 5E,F). This indicates that feathering veins are mainly localized in extension fractures oriented at an obtuse angle to ore-bearing structures. Concordant veins of the first system are common (Figure 5E,F).

Variously-oriented slickenlines are established (Figure 6B). One can observe subvertical slickenlines (*l*-178/81) of the early thrust-faulting stage of deformation on E-W fault planes (Figure 6B). Strike slip accretionary slickenlines are manifested at the contacts of ore bodies (Figure 6C). These structural elements are associated with low-amplitude zones of warping observed on the northern wall of a trench that exposed Ore Zone 1. The axes of the warping zones plunge to SE (120/59) and are orthogonal to (*l*-290/30).

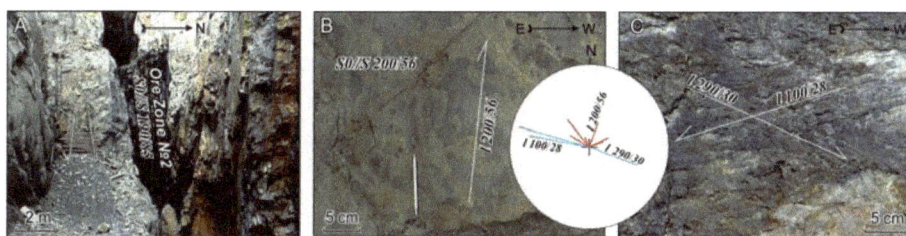

Figure 6. Ore Zone 2 (**A**), reverse-faulting (**B**) (red lines on the diagram) and strike-slip faulting (**C**) (blue lines on the diagram) slickenlines, Nagornoye deposit. Arrows show the direction of displacement of the faults' hanging walls.

4.3. Dvoinoye Occurrence

The occurrence is located on the left side of the valley of Dvoinoy Creek (Figure 1). The host rocks are dominantly Late Permian sandstones with lesser siltstones. The ore bodies are represented by mineralized fault zones with quartz veins and veinlets ranging from 0.3–1.8 m (average 1.0 m) in thickness. The gold grade varies from 0.5–30.7 g/t Au.

At Dvoinoye, one can observe two systems of boudinage-structures on the limbs of folds (Figure 7). The first system is characterized by a subhorizontal long axis (Figure 7B), and the second system plunges to SE (120/54) (Figure 7C).

Figure 7. Dextral strike-slip fault (**A**) and boudinage-structures (**B,C**), Dvoinoy Creek.

4.4. Mudeken Locality

Mudeken is located on the right side of the same name creek, ~6 km upstream from the Dvoinoy Creek mouth (Figure 1). Two faults and several folds are seen in an exposure of interlayered sandstones and siltstones along the creek bank (Figure 8). Fault S (77/60) is made of brecciated siltstones and sandstones with quartz veinlets. Two shelly-platy cleavages are manifested in its walls. The first *Cl* cleavage dips to W (265/80) and is subconformable with S0 (278/50). The second *Cl* cleavage dips gently to SE (130/35–49) following the strike of fault S (160/40), which is traced by an 8 cm thick quartz-carbonate vein of banded structure. In the lying and hanging walls of the fault are observed several en echelon systems of quartz-carbonate veins: V-SE (220–260/55), V-NE (75/80) and V-SE (136/42). Kinematic reconstructions of the vein systems revealed their relation to dextral strike-slip motions (Figure 8H).

To the east, in the footwall of S fault, one can observe widely developed folds with the axial planes conformable with the fault plane. The folds are asymmetric, with gentle and extensive SE limbs and steeper SW limbs; commonly, they are overturned due to dextral strike-slip motions along the bedding plane (Z-shaped folds) (Figure 8D). Bedding poles measured on the fold limbs form a belt along the great circle arc, which is characteristic of cylindrical folds (Figure 8F). Fold hinges (b) mostly dip S-SE at 35–60°. Early cleavage (*Cl*) is deformed in folds (Figure 8G). The cleavage poles presented in Figure 8G show that early cleavage was deformed during dextral strike-slip motions.

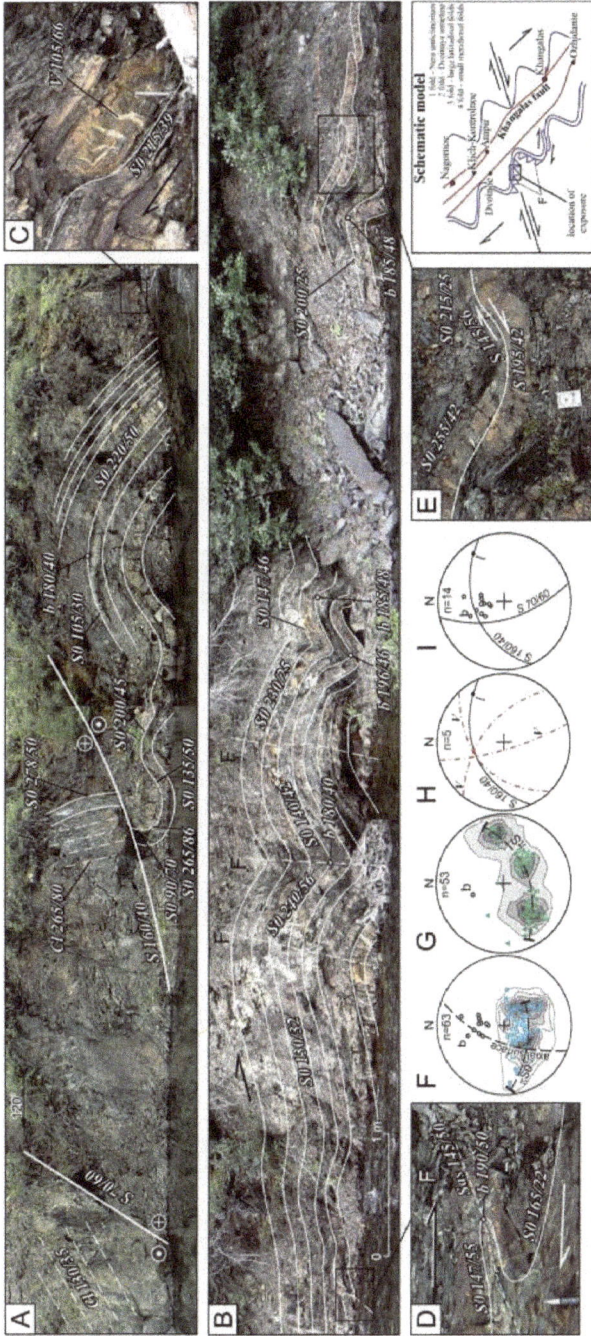

Figure 8. Dextral strike-slip deformations in Upper Triassic sandy siltstones, Mudeken Creek. (**A**,**B**) Fold and fault structures; (**C**) ladder veins in sandstone beds (plan view); (**D**) shaped overturned fold (plan view); (**E**) dextral strike-slip fault (plan view); (**F**–**I**) diagrams: (**F**) bedding poles, (**G**) cleavage poles, (**H**) projection of quartz-carbonate veins shown in (**C**,**I**) projections of faults and hinges of folds; *l*, direction of tectonic transportation. Contours of poles to planes (per 2% area).

5. Discussion

Structural-kinematic analysis of deformation elements within the KOC revealed specific structures of deposits and occurrences. Figure 9 shows stereograms of major structural elements of the KOC (bedding, cleavage, quartz veins and veinlets, as well as mineralized fault zones). Fold structures of the Khangalas deposit have NW-SE to E-W strike. Fold hinges (b1) dip to WSW at angles varying from 4–28°. Steep dip angles are due to superposed strike-slip deformations. Hinges of the third order F1 folds smoothly undulate in the direction of NW regional folding within the Nera anticlinorium. *Cl1* cleavage at the Khangalas deposit has a NW-SE strike. In the fault zones, cleavage is deformed by late strike-slip faulting, as well as bedding. In stereographic projections, poles of quartz veins and veinlets are arranged along subvertical belts. From the aforesaid, it appears that the formation of auriferous quartz veins and veinlets is related to major fold and thrust deformations of D1 stage (J3-Knc). Faults and ore zones at the Khangalas deposit have sublatitudinal and, more rarely, northeast and northwest strike. The majority of them dip S at 30–60°.

F1 compressed folds of sublatitudinal strike are deformed, like *Cl1* cleavage at the Nagornoye deposit, by late strike-slip faults. This led to the formation on the limbs of F1 folds of F4 open folds with steep (b4) hinges plunging to NNE and SSW. On the stereogram, vein bodies form a steeply-dipping belt of poles. Faults and ore zones are, for the most part, interstratal and have latitudinal to NE-SW, rarely NW-SE orientation.

Bedding of rocks in the Dvoinoy ore field is characterized by NW-SE and NE-SW strike related to two different deformation stages: D1 and D4, respectively. Cleavages of NW-SE and E-W orientation are recognized. The first cleavage *Cl1* is associated with the D1 stage. It was formed in relation to early fold-and-thrust dislocations. Cleavage *Cl4* is related to the right-lateral strike-slip stage (D4). Fault zones within the Dvoinoye ore field have mostly WNW-ESE strike.

Object	Bedding	Cleavage	Vein-veinlets	Fault
Khangalas deposit	n=61	n=62	n=242	n=50
Nagornoye deposit	n=191	n=48	n=60	n=12
Dvoinoye ore field	n=104	n=73	n=60	n=15

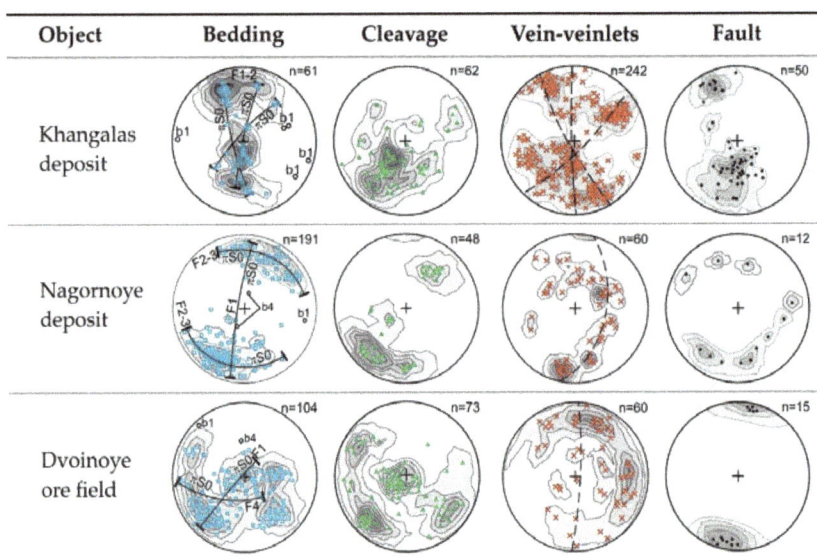

Figure 9. Stereograms of poles of bedding, cleavage, veins and faults, Khangalas ore cluster. Dashed line, belt of vein poles. Contours of poles to planes (per 2% area).

The results of studying other deposits within the Kular-Nera slate belt [22,31–36] in combination with the available information on the general tectonic and metallogenic evolution of the Verkhoyansk-Kolyma folded region [6,7,9,37] and the data on the relationships between the mapped structural elements obtained in this study indicate that deformation occurred in four stages: D1, D2, D3 and D4 (Table 1).

Table 1. Evolution of tectonic events and associated mineralization in the southwestern part of the Kular-Nera slate belt.

Characteristic	Deformation Event			
	D1	D2	D3	D4
Geodynamic setting	Frontal accretion	Oblique accretion	Post-accretionary	Post-accretionary
Kinematics of NW faults	Thrust	Sinistral strike-slip	Sinistral strike-slip	Dextral strike-slip
Structural paragenesis	Interstratal detachment thrust, interstratal ramps, thrusts, cross and oblique ramps, NW tight and isoclinal folds, NW-SE open and tight folds (F1) with horizontal hinges (b1), boudinage, fault cleavage, downdip slickenlines	Sinistral strike-slip, NE-SW and NW-SE folds (F2), horizontal slickenlines	Sinistral strike-slip, NE-SW and NW-SE folds (F3), horizontal slickenlines	Dextral strike-slip, W-E and NW-SE fold (F4), sublatitudinal cleavage horizontal slickenlines
Veins	V1	-	V3	V4
Mineralization	Au	-	Sb	Ag
Attitude of σ1	Subhorizontal, NE-SW	Subhorizontal, E-W	Subhorizontal, E-W	Subhorizontal, N-S
Graphic model				

D1 and D2 deformations occurred in the Late Jurassic-Early Cretaceous in response to frontal (D1) and then oblique (D2) accretion of the Kolyma-Omolon microcontinent to the North Asian craton margin, as well as subsynchronous subduction-related processes within the Uda-Murgal active margin. The first Late Jurassic deformation event (D1) is characterized by development of NW-trending tight and isoclinal folds (F1), interstratal detachment thrusts at the contacts of rocks with contrasting physical-mechanical properties and ramps. The sense of tectonic transportation is SW. Mineralization consists of rare non-auriferous quartz-chlorite or quartz-chlorite-calcite veins occurring throughout the area. The thickness of the veins does not exceed a few tens of centimeters. Often, the veins are concordant with the host rock bedding.

During subsequent progressive deformation, the early interstratal detachment thrusts are deformed into reverse faults. Thrust deformations are associated with linear open and tight folds (F1) of the concentric type with prevailing NW-SE strike and platy cleavage. Within the ore zones, F1 folds form bands of intense deformation up to a few hundreds of meters wide. On the limbs of F2 folds, one can observe rootless intrafolio fold hinges of the early F1 folds [8] and slickenlines oriented in the direction of dip. This stage was marked by the origination of orogenic ore-magmatic systems and intrusion of granitoids and mafic and intermediate dikes of the Late Jurassic Nera complex, which produced vein, vein-veinlet and veinlet-disseminated gold-quartz (gold-sulfide-quartz) mineralization. According to the mineralogical data, mineralization was formed at a depth of about 3.5 km. The ore bodies are localized in mineralized fault zones, in the crests of folds (saddle veins) and sandstone strata (vein-veinlet bodies).

D2: The Late Neocomian is the period when the direction of motion of the Kolyma-Omolon microcontinent and of subduction in the Uda-Murgal arc began changing [6]. Within the Kular-Nera slate belt, the second stage of accretion (Aptian, Lower-Cretaceous) is characterized by the first left-lateral strike-slip motions (D2) on NW-striking faults, which occurred under prevailing W-E compression. At that time, intrusions of subvolcanic granite-porphyry were formed [24]. Strike-slip deformations reworked hydrothermal-metamorphogenic and gold-quartz (gold-sulfide-quartz) mineralizations causing their corrosion and remobilization, as well as the dynamic metamorphism of ore bodies. In association with strike-slip faults, F2 open folds were formed. The width of the mapped folds ranges up to a few hundred meters, but wider folds are likely, as well. Fold hinges plunge to N-E and N. At the contacts of mineralized fault zones, one can see variously-oriented subhorizontal slickenlines superposed on vertical ones.

D3: Post-accretionary tectonic events are associated with Late Cretaceous subduction in the Okhotsk-Chukotka arc [6,24]. This stage is marked by the formation of N-S and NE-SW folds (F3) with steep hinges and by activation of earlier structures, including ore-controlling ones. Note the similar kinematics of D2 and D3 deformations. F3 folds have various morphologies: from open symmetric to tight overturned ones. Characteristic are sinistral motions most widely manifested along the axial part of the Adycha-Taryn Fault [36].

At that stage, gold mineralization was superposed by Late Mesozoic antimony mineralization [24]. As shown in [36], a strong influence of Sb-bearing fluid caused significant reworking of the mineral complex of gold mineralization in the Adycha-Taryn zone. Formation of Sb mineralization was related to leaching and replacement processes that led to the formation of new mineral parageneses. Quartz-sericite metasomatites of the gold ore stage are characterized by the superposed late carbonate-paragonite-pyrophyllite-dickite paragenesis. Pyrite and arsenopyrite metacrysts are replaced by a mixture of antimonite and pyrophyllite. In the ore zones, late ore deposition is manifested by berthierite and antimonite. They form numerous streaks in milk-white quartz cementing its fragments and forming brecciated zones. The broken down milk-white quartz is freed from impurities along the fluid conductors. Solution occurs along the quartz grain boundaries. Transparent regenerated quartz forms streaks and aggregates of small (up to 1–2 mm) prismatic crystals, with pyramidal terminations. The early sulfides and sulfosalts also underwent intensive corrosion, leaching and redeposition, which led to the formation of an association of regenerated minerals. Microcrystals of the regenerated high-fineness (900–1000%) gold are often surrounded by reaction rims of aurostibite and antimony gold (Sb up to 8%) [36].

D4: The latest and fourth deformation event (D4) is characterized by dextral strike-slip faulting, refolding of rocks, reactivation of the earlier ore-controlling structures, as well as the formation of E-W folds and cleavage. Maastrichtian-Early Eocene strike-slip deformation is inferred to be related to oblique subduction of the Pacific Ocean plates beneath the eastern margin of north Asia [6] and/or to the formation of a transform margin in northeast Asia [38].

6. Conclusions

Studies of deformation structures of the Khangalas gold-ore cluster showed that they were forming over a long time during the course of the Late Jurassic-Neocomian accretionary and Late Cretaceous-Early Paleocene post-accretionary events in the Verkhoyansk-Kolyma folded region.

The first deformation event (D1) was characterized by the development of NW-striking tight to isoclinal folds of the first generation (F1) and interstratal detachment thrusts. Major folds, extensive thrusts, boudinage, cleavage, Au-bearing mineralized fault zones and quartz-vein mineralization were formed in the conditions of the tectonic stress field characteristic of reverse and thrust faulting, with the horizontal $\sigma 1$ and vertical $\sigma 3$. The D1 stage was progressive deformation under a contractional regime. In the zones of regional faults, where deformations are most intensely manifested, inter- and intra-stratal reverse and thrust faults developed in sandstones and at their contacts with siltstones, which were accompanied by intense, small-scale folding. These were favorable structural conditions for localization of mineralized fault zones with concordant and cross Au-quartz veins.

Post-ore deformations are widely manifested within the KOC. The D2 and D3 structures are co-axial. Sinistral strike-slip motions (D2–3) occurred along NW-striking faults. Associated with them were submeridional and NE-trending folds (F2–3) with steep hinges, as well as warping of the earlier, including ore-controlling, structures. The sinistral strike-slip stage is poorly manifested within the

KOC, but its presence is established from the analysis of fractures, slickenlines and fault-line folds. Faults of NE strike can also be assigned to this structural paragenesis. It is likely that sinistral strike-slip deformations changed to dextral ones that are strongly manifested within the KOC.

The fourth event (D4) is represented by normal-dextral strike-slip motions, refolding of rocks, earlier structural elements and ore bodies. At this stage, latitudinal structures (F4 folds, *Cl4* cleavage and S4 faults) were formed, which are more widespread here than in other metallogenic zones of the Upper Indigirka district (Adycha-Taryn, Mugurdakh-Selerikan). It is assumed that large-scale dextral strike-slip faults modified the structure of deposits and occurrences in the KOC. In the most strongly-deformed areas, the strike of the structures changed to sublatitudinal and, more rarely, to NE (Khangalas deposit). Ore zones of the Khangalas deposit were previously considered as "horse tail" structures [2], but detailed analysis of the relationships between auriferous quartz veins, ore zones and rock bedding permitted assigning them to the first-stage paragenesis (D1). The large scale and long duration of post-ore strike-slip motions can be inferred from the observation that early Au-bearing quartz veins are ground to "quartz flour" in the zones of later strike-slip faults. Also observed are quartz breccias in which early milk-white quartz is cemented by later chalcedony-like grey quartz typical for Ag-Sb mineralization known from the Verkhoyansk-Kolyma folded region [39,40].

Thus, it can be concluded that tectogenesis within the Verkhoyansk-Kolyma folded region followed a regular change from the Late Jurassic-Neocomian frontal accretionary regime to the Aptian-Early Eocene strike-slip regime and that gold mineralization was related to orogenic processes, as is exemplified by the Khangalas ore cluster described in this article.

Author Contributions: Idea of the study conceived by V.Y.F. Collection of field materials by V.Y.F., M.V.K. and L.I.P. Treatment of data and writing the text of the paper by V.Y.F. and M.V.K. Figure drawing by M.V.K.

Funding: This research was funded by Diamond and Precious Metal Geology Institute, Siberian Branch of the Russian Academy of Sciences, project number [No. 381-2016-004] and by Russian Foundation for Basic Research, grant number [No. 18-35-00336].

Conflicts of Interest: The authors declare no conflicts of interest.

References

1. Rozhkov, I.S.; Grinberg, G.A.; Gamyanin, G.N.; Kukhtinskiy, Y.G.; Solovyev, V.I. *Late Mesozoic Magmatism and Gold Mineralization of the Upper Indigirka District*; Nauka: Moscow, Russia, 1971; p. 238. (In Russian)

2. Oxman, V.S.; Suzdalova, N.I.; Kraev, A.A. *Deformation Structures and Dynamic Conditions for the Formation of Rocks in Upper Indigirka District*; Yakut Scientific Center Siberian Branch of the Russian Academy of Sciences: Yakutsk, Russia, 2005; p. 200, ISBN 5-463-00128-6. (In Russian)

3. Fridovsky, V.Y.; Kudrin, M.V. Deformation structures of the Khangalas ore cluster. In Proceedings of the All-Russian Scientific-Practical Conference Geology and Mineral Resources of Northeast Russia, Yakutsk, Russia, 31 March–2 April 2015; pp. 537–540. (In Russian)

4. Voroshin, S.V.; Tyukova, E.E.; Newberry, R.J.; Layer, P.W. Orogenic gold and rare metal deposits of the Upper Kolyma District, Northeastern Russia: Relation to igneous rocks, timing, and metal assemblages. *Ore Geol. Rev.* **2014**, *62*, 1–24. [CrossRef]

5. Petrov, O.V.; Morozov, A.F.; Mikhailov, B.K.; Orlov, V.P.; Militenko, N.V.; Mezhelovsky, N.V.; Feoktistov, V.P.; Shatov, V.V.; Molchanov, A.V.; Migachev, I.F.; et al. *Mineral Potential of the Russian Federation*; Petrov, O.V., Ed.; FSBI A.P. Karpinsky Russian Geological Research Institute: St. Petersburg, Russia, 2009; p. 223, ISBN 978-5-93761-156-7. (In Russian)

6. *Tectonics, Geodynamics, and Metallogeny of the Sakha Republic (Yakutia) Territory*; Parfenov, L.M.; Kuzmin, M.I. (Eds.) MAIK Nauka/Interperiodika: Moscow, Russia, 2001; p. 571, ISBN 5-7846-0046-X. (In Russian)

7. Fridovsky, V.Y.; Prokopiev, A.B. Tectonics, geodynamics and gold mineralization of the eastern margin of the North Asia Craton. In *The Timing and Location of Major Ore Deposits in an Evolving Orogen*; Blundel, D.J., Neuber, F., von Quadt, A., Eds.; Geological Society: London, UK, 2002; Volume 204, pp. 299–317.

8. Fridovsky, V.Y.; Polufuntikova, L.I.; Solovyev, E.E. Dynamics of the formation and structures of the southeastern sector of the Adycha-Nera metallogenic zone (northeast Yakutia). *Russ. J. Domest. Geol.* **2003**, *3*, 16–21. (In Russian)

9. Fridovsky, V.Y. Structural control of orogenic gold deposits of the Verkhoyansk-Kolyma folded region, northeast Russia. *Ore Geol. Rev.* **2017**, in press. [CrossRef]

10. Goryachev, N.A.; Pirajno, F. Gold deposits and gold metallogeny of Far East Russia. *Ore Geol. Rev.* **2014**, *59*, 123–151. [CrossRef]

11. Groves, D.I.; Goldfarb, R.J.; Gebre-Mariam, M.; Hagemann, S.G.; Robert, F. Orogenic gold deposits: A proposed classification in the context of their crustal distribution and relationship to other gold deposit types. *Ore Geol. Rev.* **1998**, *13*, 7–27. [CrossRef]

12. Groves, D.I.; Condie, K.C.; Goldfarb, R.J.; Hronsky, J.M.A.; Vielreicher, R.M. Secular changes in global tectonic processes and their influence on the temporal distribution of gold-bearing mineral deposits. *Econ. Geol.* **2005**, *100*, 203–224. [CrossRef]

13. Ramsay, J.G.; Huber, M.I. *The Techniques of Modern Structural Geology*; Academic press: London, UK, 1987; Volume 2, p. 704, ISBN 0-12-576922-9.

14. Fossen, H. *Structural Geology*; Cambridge University Press: Cambridge, UK, 2010; p. 463, ISBN 978-0-521-51664-8.

15. Prokopiev, A.V.; Fridovsky, V.Y.; Gaiduk, V.V. *Faults (Morphology, Geometry, Kinematics)*; Parfenov, L.M., Ed.; Yakut Scientific Center Siberian Branch of the Russian Academy of Sciences: Yakutsk, Russia, 2004; p. 148, ISBN 5-463-00016-6. (In Russian)

16. Price, N.J.; Cosgrove, J.W. *Analysis of Geological Structures*; Cambridge University Press: Cambridge, UK, 2005; p. 502, ISBN 0521319587.

17. Fossen, H.; Cavalcante, G.C.G.; Pinheiro, R.V.L.; Archanjo, C.J. Deformation—Progressive or multiphase. *J. Struct. Geol.* **2018**, in press. [CrossRef]

18. Sherman, S.I.; Dneprovsky, Y.I. *Stress Fields of the Earth's Crust and Geological Structural Methods of Their Study*; Nauka: Novosibirsk, Russia, 1989; p. 158. (In Russian)

19. Spencer, E.W. *Introduction to the Structure of the Earth*; McGraw-Hill Book Company: New York, NY, USA, 1977.

20. Gusev, G.S. *Folded Structures and Faults of the Verkhoyansk-Kolyma System of the Mesozoic*; Nauka: Moscow, Russia, 1979; p. 208. (In Russian)

21. Bakharev, A.G.; Zaitsev, A.I. The Tas-Kystabyt magmatic belt. In *Tectonics, Geodynamics and Metallogeny of the Rupublic of Sakha (Yakutia)*; Parfenov, L.M., Kuzmin, M.I., Eds.; MAIK Nauka/Interperiodika: Moscow, Russia, 2001; pp. 263–269, ISBN 5-7846-0046-X. (In Russian)

22. Fridovsky, V.Y. Structures of gold ore fields and deposits of the Yana-Kolyma ore belt. In *Metallogeny of Collisional Geodynamic Settings*; Mezhelovsky, N.V., Gusev, G.S., Eds.; GEOS: Moscow, Russia, 2002; Volume 1, pp. 6–241. (In Russian)

23. Sokolov, S.D. Tectonics of northeast Asia: An overview. *Geotectonics* **2010**, *44*, 493–509. (In Russian) [CrossRef]

24. Bortnikov, N.S.; Gamynin, G.N.; Vikent'eva, O.V.; Prokof'ev, V.Y.; Prokop'ev, A.V. The Sarylakh and Sentachan gold-antimony deposits, Sakha-Yakutia: A case of combined mesothermal gold-quartz and epithermal stibnite ores. *Geol. Ore Depos.* **2010**, *52*, 339–372. (In Russian) [CrossRef]

25. Akimov, G.Y. New data on the age of Au-quartz mineralization in the Upper-Indigirka district. *Dokl. Acad. Nauk.* **2004**, *398*, 80–83. (In Russian)

26. Nesterov, N.V. Secondary zonation of gold ore deposits in Yakutia. *Izvestiya Tomsk Polytech. Univ.* **1970**, *239*, 242–247. (In Russian)

27. GeoInfoComLLC. Available online: http://mestor.geoinfocom.ru/publ/1-1-0-55 (accessed on 5 April 2017).

28. Fridovsky, V.Y. Collisional metallogeny of gold deposits of the Verkhoyansk-Kolyma orogenic region. *Izvestiya VUZOV Geol. Explor.* **2000**, *4*, 53–67. (In Russian)

29. Fridovsky, V.Y.; Gamyanin, G.N.; Polufuntikova, L.I. Dora-Pil ore field: Structure, mineralogy, and geochemistry of the ore-formation environment. *Ores Met.* **2012**, *5*, 7–21. (In Russian)

30. Fridovsky, V.Y.; Gamyanin, G.N.; Polufuntikova, L.I. The Sana Au-quartz deposit within the Taryn ore cluster. *Raz. I Okhrana Nedr.* **2013**, *2*, 3–7. (In Russian)

31. Fridovsky, V.Y.; Gamyanin, G.N.; Polufuntikova, L.I. The structure, mineralogy, and fluid regime of ore formation in the polygenic Malo-Taryn gold field, northeast Russia. *Russ. J. Pac. Geol.* **2015**, *9*, 274–286. [CrossRef]

32. Fridovsky, V.Y.; Polufuntikova, L.I.; Goryachev, N.A.; Kudrin, M.V. Ore-controlling thrusts of the Bazovskoe gold deposit (East Yakutia). *Dokl. Earth Sci.* **2017**, *474*, 617–619. [CrossRef]

33. Akimov, G.Y. Lithological-structural control of Au-quartz ores of the Nagornoye deposit, East Yakutia. *Ores Met.* **2000**, *4*, 42–46. (In Russian)

34. Fridovsky, V.Y. Strike-slip duplexes on the Badran deposit. *Izvestiya VUZOV Geol. Explor.* **1999**, *1*, 60–65. (In Russian)

35. Fridovsky, V.Y. Analysis of deformational structures of the El'gi ore cluster (East Yakutia). *Russ. J. Domest. Geol.* **2010**, *4*, 39–45. (In Russian)

36. Fridovsky, V.Y.; Gamyanin, G.N.; Polufuntikova, L.I. Gold quartz and antimony mineralization in the Maltan deposit in northeast Russia. *Russ. J. Pac. Geol.* **2014**, *8*, 276–287. [CrossRef]

37. Prokopiev, A.V.; Tronin, A.V. Structural and sedimentation characteristics of the zone of junction of the Kular-Nera slate belt and Inyali-Debin synclinorium. *Russ. J. Domest. Geol.* **2004**, *5*, 44–48. (In Russian)

38. Khanchuk, A.I.; Ivanov, V.V. Meso-Cenozoic geodynamic settings and gold mineralization of the Russian Far East. *Russ. Geol. Geophys C/C Geol. Geofiz.* **1999**, *40*, 1607–1617.

39. Gamyanin, G.N.; Goryachev, N.A. Subsurface mineralization of eastern Yakutia. *Tikhook. Geol.* **1988**, *2*, 82–89. (In Russian)

40. Goryachev, N.A.; Gamyanin, G.N.; Prokofiev, V.Y.; Velivetskaya, A.V.; Ignatiev, A.V.; Leskova, N.V. Silver-antimony mineralization of the Yana-Kolyma belt (Northeast Russian). *Tikhook. Geol.* **2011**, *30*, 12–26. (In Russian)

minerals

MDPI

Article

Tectonic Control, Reconstruction and Preservation of the Tiegelongnan Porphyry and Epithermal Overprinting Cu (Au) Deposit, Central Tibet, China

Yang Song [1,*], Chao Yang [2], Shaogang Wei [3], Huanhuan Yang [1,4], Xiang Fang [1,2] and Hongtao Lu [1]

1   Key Laboratory of Metallogeny and Mineral Assessment, Institute of Mineral Resources, Chinese Academy of Geological Sciences, Beijing 100037, China; zggyhh@163.com (H.Y.); francisfx@126.com (X.F.); hai5987@163.com (H.L.)
2   Department of Geology and Geological Engineering, University Laval, QC G1V 0A6, Canada; yangchao2012cags@126.com
3   First Crust Monitoring and Application Center, China Earthquake Administration, Tianjin 300180, China; shaogang_wei@yahoo.com
4   College of Earth, Ocean, and Atmospheric Sciences, Oregon State University, Corvallis, OR 97331, USA
*   Correspondence: songyang@mail.cgs.gov.cn; Tel.: +86-010-6899-9087

Received: 9 May 2018; Accepted: 6 September 2018; Published: 10 September 2018

Abstract: The newly discovered Tiegelongnan Cu (Au) deposit is a giant porphyry deposit overprinted by a high-sulfidation epithermal deposit in the western part of the Bangong–Nujiang metallogenic belt, Duolong district, central Tibet. It is mainly controlled by the tectonic movement of the Bangong–Nujiang Oceanic Plate (post-subduction extension). After the closure of the Bangong–Nujiang Ocean, porphyry intrusions emplaced at around 121 Ma in the Tiegelongnan area, which might be the result of continental crust thickening and the collision of Qiangtang and Lhasa terranes, based on the crustal radiogenic isotopic signature. Epithermal overprinting on porphyry alteration and mineralization is characterized by veins and fracture filling, and replacement textures between two episodes of alteration and sulfide minerals. Alunite and kaolinite replaced sericite, accompanied with covellite, digenite, enargite, and tennantite replacing chalcopyrite and bornite. This may result from extension after the Qiangtang–Lhasa collision from 116 to 112 Ma, according to the reopened quartz veins filled with later epithermal alteration minerals and sulfides. The Tiegelongnan deposit was preserved by the volcanism at ~110 Ma with volcanic rocks covering on the top before the orebody being fully weathered and eroded. The Tiegelongnan deposit was then probably partly dislocated to further west and deeper level by later structures. The widespread post-mineral volcanic rocks may conceal and preserve some unexposed deposits in this area. Thus, there is a great potential to explore porphyry and epithermal deposit in the Duolong district, and also in the entire Bangong–Nujiang metallogenic belt.

Keywords: tectonic control; overprinting; preservation; vein-filling; replacement; porphyry; epithermal; Tiegelongnan; Tibet

1. Introduction

In the past two decades, some large porphyry deposits have been found in Tibet, China, such as Yulong deposit (6.22 Mt at 0.99% Cu) [1], Qulong deposit (7.1 Mt at 0.5% Cu) [2], Jiama deposit (7.4 Mt at 0.5% Cu) [3], Duobuza deposit (2.9 Mt at 0.46% Cu) [4], and Bolong deposit (3.8 Mt at 0.5% Cu) [5]. This indicates that Tibet can be considered one of the most significant potential porphyry Cu systems in the world. Recently, epithermal deposits have also been discovered and reported in Tibet. Epithermal deposits are genetically associated with porphyry Cu deposits, especially high and intermediate sulfidation epithermal ones, which could be discovered at upper or lateral locations of porphyry deposits in some cases [6]. However, the epithermal deposits do not always occur close to porphyry deposits, because epithermal deposits are normally at shallow crustal levels (surface to 1–2 km depth), therefore they could be easily eroded by later orogenesis [6].

The Duolong porphyry Cu-Au district is located in the Bangong–Nujiang metallogenic belt (BNMB), central Tibetan Plateau, which was discovered in 2007 and hosts several large porphyry and epithermal deposits and ore prospects (Figure 1). The Tiegelongnan deposit was discovered in 2013, containing the largest scale Cu resource within this district, and it was documented as a porphyry Cu (Au) deposit overprinted by high-sulfidation mineralization [7]. The total Cu content persevered in the Tiegelongnan deposit is around 1600 Mt at 0.51% Cu. The Au content is small at about 280 Mt with a low grade of 0.13g/t Au on average.

Despite numerous studies on the metallogeny of the Tiegelongnan deposit, the tectonic control of this type deposit has rarely been demonstrated. Formation of the porphyry and epithermal deposits in the Duolong district was indicated to be associated with the magma arising from the closure of the Bangong–Nujiang Ocean (BNO) in the Early Cretaceous [8,9]. However, how the tectonic activities control the formation of the porphyry Cu system is controversial. The Tiegelongnan deposit is the first high-sulfidation epithermal deposit being discovered in the Tibetan Plateau. Previous studies suggested that the limited number of epithermal deposits found in the Tibetan Plateau is due to the dramatic uplift and deep level erosion. The Tiegelongnan deposit is an example to study the tectonic control, reconstruction and preservation process of porphyry Cu systems in the Tibetan Plateau. In this study, we reviewed history of the tectonic setting, magma emplacement, multiple episodes' mineralization, exhumation, and preservation of the Tiegelongan deposit, based on published literatures and the detailed drill core logging and deposit 1:500 scale mapping. Besides, we discussed the implications of this study on exploration of porphyry and epithermal deposits in the Duolong district and other places in Tibet.

2. The Duolong District

The Duolong ore district is located approximately at 100 km northwest of Gerze county, on the western BNMB (Figure 1). This belt is supposed to be a suture zone as the remnants of the Bangong–Nujiang Ocean (BNO) which records the evolution of the BNO during the period of Permian to Cretaceous. This belt is over 2000 km-long striking to the east, and it is dominated by Jurassic–Cretaceous flysch, mélange, and ophiolitic fragments [10,11]. The Bangong–Nujiang suture zone extends across the central Tibetan Plateau, which separates the Qiangtang and Lhasa terranes (Figure 1a) [12,13].

There are several porphyry deposits, epithermal deposits, and porphyry and epithermal ore prospects in the Duolong district (Figure 1). The Duobuza [4], the Bolong [5], and the Naruo [14] are porphyry Cu (Au) deposits. Whereas, the Tiegelongnan deposit is a porphyry deposit overprinted by epithermal deposit [7,15]. In the Dibao, Nadun, Sena, Saijiao, and the Ga'erqin areas, there are porphyry or epithermal ore prospects (Figure 1b). Li et al., (2011) and Lin et al., (2017) suggested these deposits are related to granodiorite and quartz diorite porphyry intrusions being emplaced during 123–116 Ma, and their mineralization timing is between 120 Ma and 118 Ma [16,17].

The intrusions in the Duolong district are dominated by intermediate to felsic rocks with minor gabbro. The granodiorite porphyry and quartz-diorite porphyry are more widespread than other rock types—including diorite, granodiorite, and gabbro. The porphyry and epithermal deposits in the Duolong district are hosted by the granodiorite and quartz diorite porphyries, and also by the contact zone between these intrusions and the Jurassic quartz-feldspar sandstones. Volcanic rocks like basalt, andesite, and basaltic andesite are also widespread in the Duolong district. These volcanic rocks in this district all belong to the Meiriqiecuo Formation (K_1m) unconformably overlying on the sedimentary rocks.

The sedimentary sequences in the Duolong ore district are dominated by Mesozoic pelagic sediments and Cenozoic continental sediments including conglomerates and sandstones.

These sequences are composed of the Upper Triassic Riganpeicuo Formation (T_3r), the Lower Jurassic Quse Formation (J_1q), the Lower to Middle Jurassic Sewa Formation ($J_{1-2}s$), the Upper Cretaceous Abushan Formation (K_2a), and the Upper Oligocene Kangtuo Formation (E_3k). The Riganpeicuo Formation dominated by limestone is unconformably overlain by the Quse and Sewa formations. The Quse Formation mainly occurred in the center and southwestern part of the ore district as the main host formation of the Duobuza and Bolong porphyry deposits. It conformably contacts with the overlying Sewa Formation, which is the predominant host formation of the Tiegelongnan deposit. These two Jurassic formations are thought to be part of the metamorphosed accretionary complex formed by north-dipping subduction of the BNO plate under the Qiangtang terrane [18]. They were also interpreted to be bathyal to abyssal flysch succession, implying a stable shallow-marine continental-shelf sedimentary environment along the southern continental margin of the South Qiangtang terrane [10]. Furthermore, Wei et al. (2017) [9] proposed that a continental margin arc setting in the southern Qiangtang terrane during the Early Cretaceous.

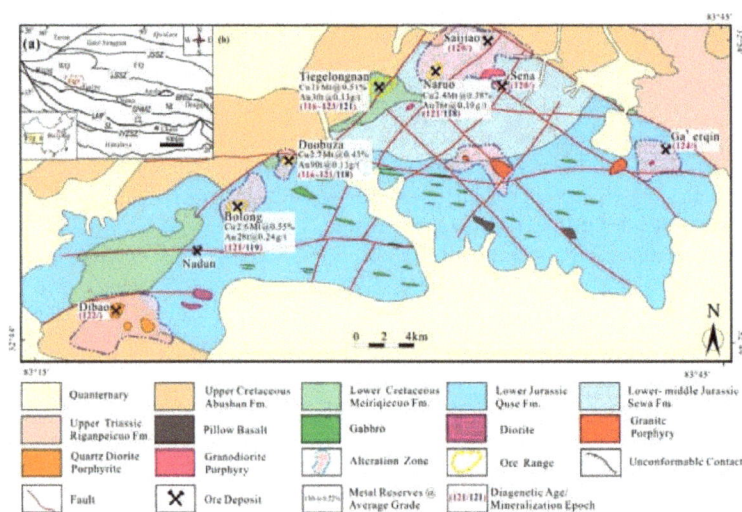

Figure 1. Regional geological map of the Duolong ore cluster, modified after [8], ages are from [17].

There are three main faults in the Duolong district striking at NE–SW, E–W, and NW–SE, respectively (Figure 1). The NE–SW fault is a major ore-controlling structure. A number of ore-bearing granodioritic porphyry intrusions emplaced along this fault, and therefore, many large porphyry copper deposits such as the Bolong, Duobuza, Tiegelongnan, and Naruo deposits occurred.

Most E–W thrust faults are large scale and traverse across the entire Duolong district, dipping to the south with an angle between 49° and 16° [19]. There are some granodiorite porphyries (125–120 Ma, unpublished data) beaded along this NE–SW fault. A mylonite sample obtained from the fault zone was well constrained with a $^{40}Ar/^{39}Ar$ plateau age at 127.8 ± 1.1 Ma [19], which represents an early period of thrusting. The NW–SE faults are normal slip faults dipping to the south, which might be related to the neo-tectonic movements. These faults are characterized by sunken landform and valleys with fault breccia exposed.

3. Tiegelongnan Alteration and Mineralization

The Tiegelongnan deposit is hosted by the Sewa Formation quartz-feldspar sandstones and some intermediate and felsic porphyry intrusions. These rocks were mostly concealed by the andesite of the Meiriqiecuo Formation which is currently well exposed (Figure 2a). The earliest diorite porphyry within this deposit was intruded at 123 Ma before the mineralization, which is mainly distributed at the eastern and southern margins of this area [8]. Several phases of granodiorite porphyries are syn-mineral intrusions with ages ranging from 121 Ma to 116 Ma [8,20,21]. They are indistinguishable from petrology and crosscutting relationships, because subsequent strong alteration weakened their differences and boundaries. Therefore, the geochemistry data, especially the mobile elements, could not be used to discriminate their geochemical features.

3.1. Alteration

Drill core logging reveals concealed features of the Tiegelongnan deposit (Figure 2b). Five phases of hydrothermal alteration were identified in the Tiegelongnan deposit, according to the dominant alteration mineral assemblage, including: biotite alteration, sericite-pyrite-quartz (phyllic) alteration, chlorite alteration, alunite alteration, and kaolinite-dickite alteration [22]. Alunite-kaolinite-dickite assemblages are also named as advanced argillic alteration in high-sulfidation epithermal deposits [23].

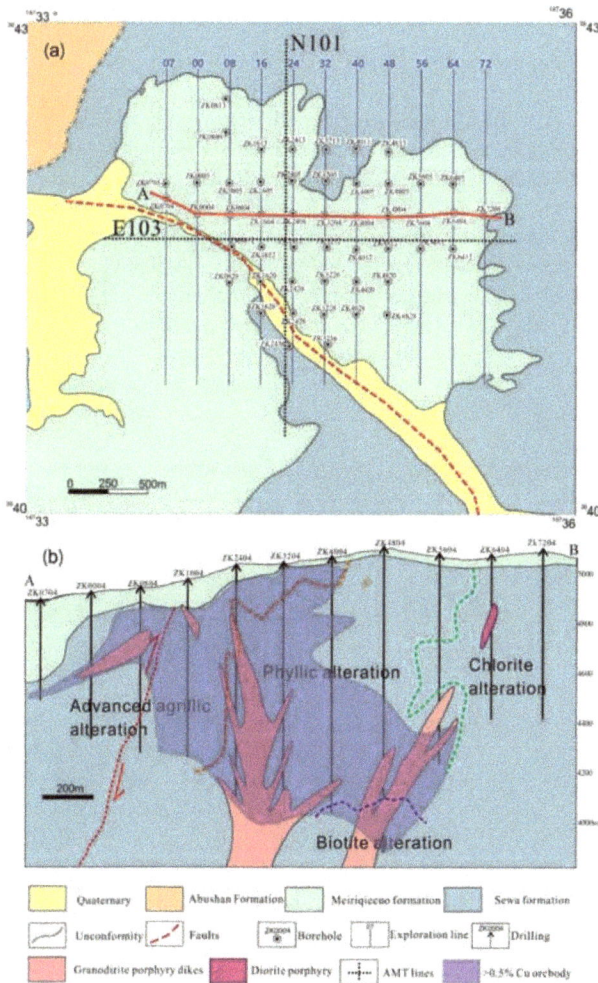

Figure 2. Ground surface and cross-section map of the Tiegelongnan porphyry Cu (Au) deposit.

Hydrothermal biotite occurs in wall-rock sandstones beneath an elevation of ~4100 m as disseminated fine grains and less commonly as vein biotite-quartz. It is not typical potassic alteration with no K-feldspar, and rare magnetite, rutile, and anhydrite. Sericite-pyrite-quartz (phyllic) alteration is widespread in the Tiegelongnan deposit, hosting most of the ore minerals. These granodiorite porphyries and their surrounding sandstones are mostly altered to sericite and quartz. Some fluorite, rutile, anhydrite, and magnetite can be identified under the microscope. There are some quartz-chalcopyite-pyrite and quartz-molybdenite veins occurring in the phyllic alteration zone, not intensely. Chlorite alteration is in the southern and eastern part of the deposit, which scattered in the sandstone and diorite and pyrite-chlorite veins. Alunite mostly occurs in the alunite-sulfides veins, locally with kaolinite and dickite, distributed in a narrow and shallow place at a height ranging from 4500 m to 4950 m. Kaolinite-dickite alteration occurs as more widespread kao-dic veins than the alunite veins, which is featured with kaolinite replacing sericite (fine grained muscovite) grains. Except for those minerals, some pyrophyllite occurs in the sericite-pyrite-quartz alteration zone, and the rutile is widespread from the biotite alteration to alunite alteration zones. Biotite, sericite, and chlorite are typical porphyry stage alteration minerals [24]. It is typical that these porphyry alteration stage minerals are overprinted by epithermal alteration minerals in the Tiegelongnan deposit, and these minerals formed in different occurrences, such as breccia, vein-filling, or as replacement, which corresponds with the multiple dating results on the altered minerals. Biotite and sericite display a ^{40}Ar-^{39}Ar age at ~121 Ma, whereas alunite indicates the age ranging from 117 to 100 Ma [21]. Alunite-kaolinite-dickite breccia is late epithermal stage products, breaking earlier phyllic altered rocks (Figure 3a). Kaolinite-dickite veins cut the early phyllic and biotite alteration stages' quartz-sulfide veins. However, they are most commonly shown as kaolinite filling in biotite veins (Figure 3b), and kaolinite fills the barren quartz-pyrite veins in phyllic alteration zone (Figure 3c). Kaolinite also replaces fine muscovite grains in the phyllic alteration zone (Figure 3d). In some cases, a mineral sequence is shown as a single vein, firstly with quartz crystalizing, followed by alunite alteration, and ending up with kaolinite crystalized in the center of the vein. This alunite, kaolinite, and dickite assemblage was documented as acidic minerals and epithermal products in condition of low temperature and low pH value [25].

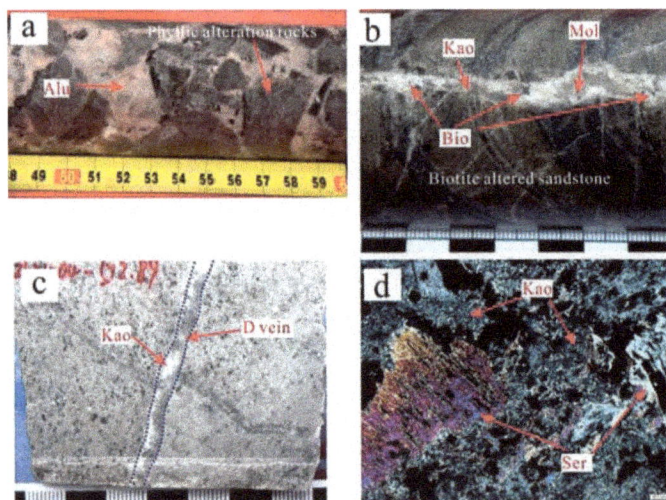

Figure 3. The epithermal alteration overprinted on the porphyry alteration. (**a**) Alunite breccia break phyllic altered and mineralized host rocks, (**b**) kaolinite vein crosscutting biotite altered host rocks and filling inbiotite-molybdenite vein, (**c**) kaolinite filling in cavity of quartz-pyrite vein, (**d**) sericite replaced by kaolinite grains. Alu: alunite, Bio: biotite, Ser: sercite, Kao: kaolinite, Mol: molybdenite.

3.2. Mineralization

Chalcopyrite, bornite, and pyrite are sulfide assemblage precipitating in biotite and phyllic alteration zone with minor molybdenite, whereas the Cu (Fe)-As-S minerals enargite, tennantite, and Cu-S covellite, digenite are the dominant sulfides in advanced argillic alteration zone [26,27]. Chalcopyrite and bornite in the biotite and phyllic alteration zone are the main Cu mineralization of the porphyry stage, including quartz-chalcopyrite ± bornite veins and disseminated chalcopyrite and bornite. This is the main porphyry Cu oreody, hosted as quartz-sulfides veins, and disseminated sulfides in wall rocks. The enargite-tennantite-covellite-digenite assemblage mostly occurs in the alunite and kaolinite alteration zone, which are the products of the high-sulfidation epithermal Cu orebody [8]. Epithermal stage Cu sulfides are mainly presented as alunite-kaolinite-sulfides veins.

These two stages of sulfide mineral assemblages display a complicated overprinting and cross-cutting relationship. Kaolinite-sulfide veins crosscut quartz veins (Figure 4a), those sulfides are mostly Cu (Fe)-(As)-S minerals, also some chalcopyrite and bornite were reported as result of solid solution from those minerals [25]. Under the microscope, we find some enargite filling in the fractures of quartz veins along with kaolinite. The pyrite occurs as early phyllic alteration product, because it is the most easily being replaced by the enargite. Replacement textures of chalcopyrite, bornite, and pyrite affected by Cu (Fe)-As-S and Cu-S minerals are common in the Tiegelongnan deposit. The pyrite is replaced from the edge firstly by bornite, and then the bornite is replaced by digenite and covellite (Figure 4b). Enargite and tennantite replace chalcopyrite (Figure 4c). There are some arguments that replacement relationship between sulfides is supergene replacement textures, because covellite and digenite are typical supergene sulfides also. However, δ^{65}Cu of covellite and digentie in the Tiegelongnan are averaging at 0.25‰ [28], which is similar to the hypogene copper sulfides δ^{65}Cu value [29]. In some cases, the Cu (Fe)-As-S and Cu-S minerals are filled in the fractures instead of replacing Fe-bearing minerals (Figure 4d), which might indicate a brittle force condition before the epithermal mineralization. This corresponds with the alunite and kaolinite breccia in Figure 3a. The Cu (Fe)-As-S and Cu-S sulfides even cut through post-mineral porphyry and breccia rocks. Generally, overprinting of the Cu (Fe)-As-S and Cu-S on chalcopyrite-bornite-pyrite assemblage is common in the Tiegelongnan deposit, and it was demonstrated in different occurrences, including the former replacing the latter minerals, the former filling in fractures of the latter sulfides, and the former cutting the chalcopyrite-bornite mineralized rocks or veins.

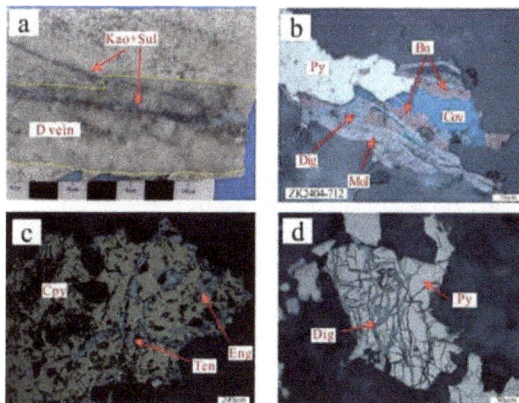

Figure 4. Textures of epithermal sulfides overprinting porphyry sulfides. (**a**) Kaolinite-sulfide veinlets cross-cut quartz-pyrite veins, (**b**) later covellite and digenite replacing borinite and pyrite, (**c**) tennantite and engargite replacing chalcopyrite, (**d**) pyrite fractures filled with digenite. Bn: bornite; Cpy: chalcopyrite, Cov: covellite, Dig: digenite, Eng: enargite, Mol: molybdenite; Kao: kaolinite, Py: pyrite, Sul: sulfides, Ten: tennantite.

4. Structures

The Tiegelongnan deposit and most of other deposits in the Duolong district are along the NE–SW faults. It is widely accepted that these faults mainly controlled the emplacement of magma and hydrothermal fluids in the Duolong district [30]. However, there were few convincing studies clarifying the overlying volcanic rocks which conceal the whole porphyry and epithermal ore bodies. The NW–SE faults, named as Rongna Fault, are characterized by geomorphologically linear sunken terrain, valleys, with fault springs seen at the ground level. In the deposits area, the fault occurred as a river valley, which is called the Rongna Valley (Figure 5). The fault divides the Meiriqiecuo Formation andesite into two parts, suggesting the structural movement took place after andesite eruption, which is dated at ~110 Ma [8].

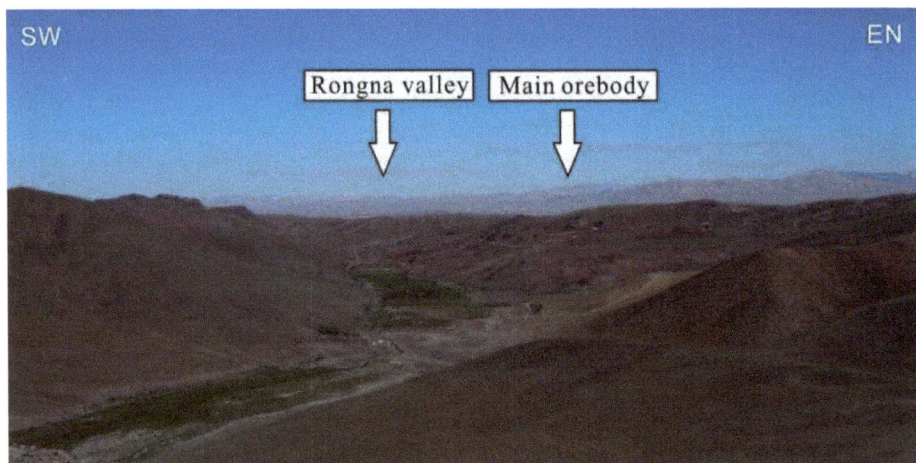

Figure 5. Rongna Valley, topography of the Rongna fault in the Tiegelongnan deposit.

The audio-frequency magneto-telluric method (AMT) was applied to understand the fault features, and further to predict the occurrences of the orebody in a deep level on the south side. From the ATM tests, electrical properties of different rocks obviously vary from each other in the Tiegelongnan deposit. The Cretaceous volcanic cap-rocks have low polarizability, whereas extremely high resistivity is shown in the paleo-weathering crust. The Jurassic sandstone showed low resistivity and high polarization, while the advanced argillic altered sandstone has high resistivity. We found the >0.5% grade Cu whole porphyry and epithermal orebodies correspondent with the low resistivity zone (Figure 6). There are two low-resistivity anomalies (C1 and C2) in the E103 AMT cross-section and the C1 anomaly coincides with the explored orebody. Therefore, the C2 low-resistivity anomaly could be another part of the whole orebody. The fault plane shown in the AMT cross-section is dipping to the south with an angle of 70° to 80° (Figure 6).

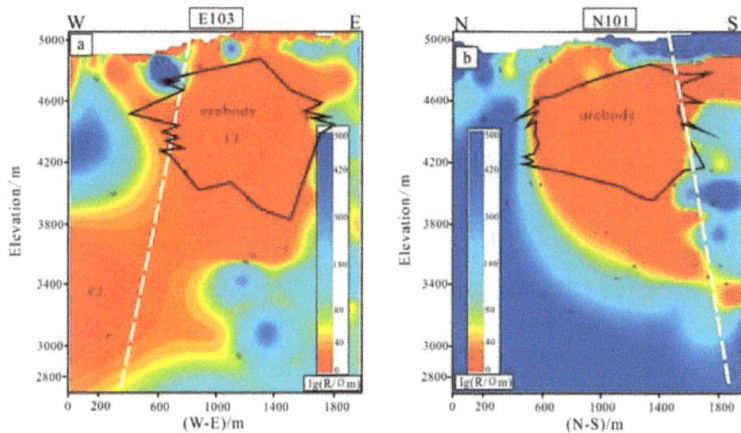

Figure 6. 2D inversion resistivity section of the audio-frequency magneto-telluric method (AMT) test (**a**) north–south cross-section; (**b**) east–west cross-section.

5. Post-Mineral Weathering and Erosion

The epithermal Cu (Au) mineralization orebody found by diamond drilling is covered by the Meiriqiecuo Formation volcanic rocks (Figure 7a). The dome-shaped and gently dipping andesite layer was discovered by drilling at an elevation of 5110–4930 m above sea level at the ground surface and 5080–4620 m at the bottom. The average thickness of the andesite is 90 m, thinning from southwest to northeast (Figure 7b).

A layer of weathered paleosoil is observed between the andesite and the underlying porphyry and epithermal Cu (Au) orebodies, which suggests that prominent weathering occurred after mineralization before the overlying andesite. Three types of erosional surfaces are recognized (Figure 8). The first type is a weakly weathered eluvium without movement, containing detrital sandstone, with some malachite and azurite. The second type of erosional surface is residual ancient soil, defined as a complex of clay soil and illuvial soil with a small amount of debris, which is generally formed in the watershed or on slope landforms. The third type is slope washes, which is weakly weathered eluvial material transported by water, accumulated on a slope, and incorporated rounded fragments of the basement. Slope washes form in the transitional area between erosional and depositional zones in this area.

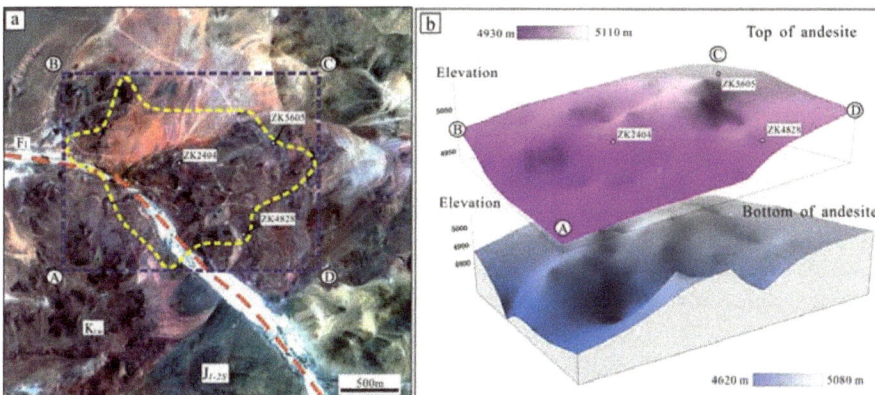

Figure 7. (**a**) High-resolution remote sensing image, and (**b**) a 3-D map of the Meiriqiecuo Formation andesite [31]. Red dotted lines: faults; the yellow dotted line: the boundary of the mineralized body.

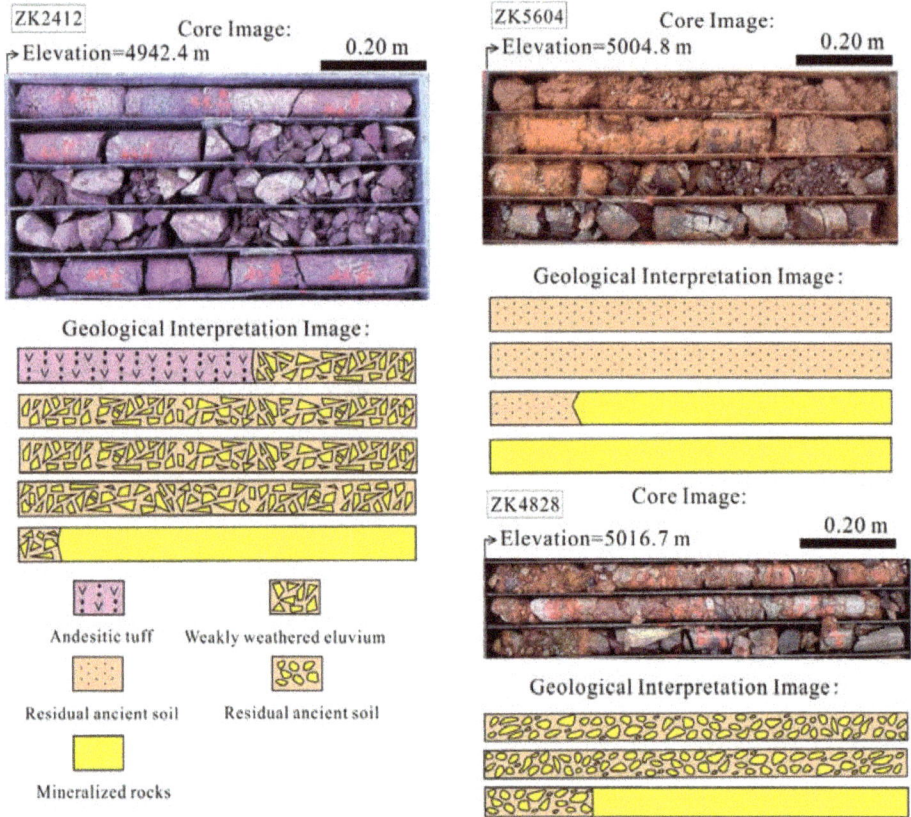

Figure 8. Photographs of typical weathering paleo-crusts in the Tiegelongnan deposit. ZK2412—The first type weathering weakly weathered eluvium. ZK5604—The second type weathering residual ancient soil. ZK4828—The third type weathering slope washes.

6. Discussion

6.1. Magmatism Indication of Tectonic Setting in the Duolong District

The Bangong–Nujiang ocean crust has started subducting northwards beneath the southern Qiangtang terrane since the Middle Jurassic. This leads to the formation of the large intermediate to felsic magmatic arc that emplaced inboard of the southern continental margin of the south Qiangtang terrane in the Middle to Late Jurassic (170–145 Ma) [32–36]. Some authors argued that the Bangong–Nujiang oceanic crust may subduct in two directions, both northward beneath southern Qiangtang Terrane and southward beneath northern Lhasa Terrane, respectively [36,37]. From 145 to 130 Ma, however, there is a noticeable magmatic gap in the southern Qiangtang terrane. Similar magmatic gaps occur in the Andes [38] and southern Gangdese areas [39] in response to the low-angle or flat-slab subduction of the oceanic crust. After that, the Bangong–Nujiang oceanic basin was closed. Although the closure time is controversial, it is generally as accepted as the period from Middle Jurassic to Late Cretaceous [12,40–43]. Recent research narrowed the closure time within 10 Ma from 140 Ma to 130Ma [37]. Based on the time constraints, it suggests that the collision between the Qiangtang and Lhasa terranes occurred through an arc–arc 'soft' collision from the east to the west after the BNO closure [32,40,41]. The Jurassic to Lower Cretaceous (<125 Ma) marine sedimentary rocks were transposed, intruded by granitoids, and were uplifted above sea level before around 118 Ma [40]. The extensive magmatism in the Duolong district is associated with the Qiangtang–Lhasa collision event [9,37].

Numerous igneous rocks such as gabbro, basalt, basaltic andesite, andesites, rhyolite, and intermediate to felsic porphyries are widely distributed in vicinity of the Tiegelongnan deposit. Zircon U-Pb ages of the porphyry intrusions in the Tiegelongnan deposit range from 115.9 Ma to 123.1 Ma, which is consistent with the mineralization (molybdenite Re-Os) ages (119.0 ± 1.4 Ma [20]; 121.2 ± 1.2 Ma [8]. This is also consistent with porphyry intrusions and mineralization ages of other deposits in the Duolong district, such as Bolong and Duobuza deposits [17]. They are temporally associated with this younger generation of magmatic emplacement which is related with the Qiangtang–Lhasa collision. Intrusion rocks geochemistry and isotope studies have been conducted to understand the genetic association between those porphyry deposits and tectonic settings. Geochemistry data mostly obtained from the ore-bearing porphyritic intrusions in the Duolong district indicate they are magmatic rocks and adakite-like rocks [11,14,44,45]. They have relatively high oxygen fugacity (f_{O2}) and high H_2O contents that are critical to the formation of porphyry and epithermal deposits [46]. During the process of magma upwelling, adakite-like melts might get mixed with large amount of copper and other metals and sulfur from either interaction with hot peridotite in the mantle wedge region [47] or mixing with mantle-derived melts [48]. It eventually resulted in mantle-derived juvenile materials, which are thought to bring heat and materials to generate juvenile mafic lower crust. The magmas experience various degrees of fractional crystallization and crustal contamination during its emplacement, when it is derived from the remelting of the juvenile mafic lower crust as a result of previous arc magmatism. Some of these hybrid magmas formed calc-alkaline ore bearing porphyries via the shallow magma emplacement, leading to the formation of giant porphyry and epithermal Cu (Au) deposits [49–52].

The Jurassic (170–145 Ma) intermediate–felsic intrusive rocks of the southern Qiangtang terrane primarily exhibit negative whole-rock $\varepsilon_{Nd}(t)$ and zircon $\varepsilon_{Hf}(t)$ and old Hf isotope crustal model ages, indicating that those Jurassic rocks were largely derived from mature or recycled continental crust materials [35,36,53]. This is compatible with what been observed in the Early Cretaceous Fuye pluton, Caima pluton and Qingcaoshan pluton in the Qiangtang Terrane [53]. In contrast, $\varepsilon_{Nd}(t)$ and $\varepsilon_{Hf}(t)$ value and Hf isotope model ages of the Early Cretaceous (~126–116 Ma) magmatic rocks from the Duolong district indicate they were probably derived from magma as a mixture of the juvenile lower crust and mature crustal materials [9]. In addition, previous studies on Pb isotopic compositions of the porphyry intrusions, sulfides, and sulfate in the Tiegelongnan deposit suggest that the Pb of this deposit is mainly derived from a crust–mantle mixed subduction zone [26,54,55].

Although there is no specific research on the geochemistry of the porphyry intrusions in the Tiegelongnan deposit, owing to their strong alteration and leaching erasing its geochemical signature, the features of the intrusions in the Duolong district could represent that in the Tiegelongnan deposit. Thus, it suggests that plenty of juvenile crust materials are involved in the intrusions in the Tiegelongnan porphyry Cu (Au) deposit. The juvenile crust materials have been becoming gradually dominated during the Late Mesozoic since the vertical growth and thickening of the continental crust of the southern Qiangtang terrane during the Early Cretaceous [9,56]. It is commonly accepted that variable sources conjunctly contributed to the formation of magma in this district [53]. The dominant crust signature from radiogenic isotopes is reported as features of the post-subduction products, which is well explained by Richards (2009) [57]. All of these suggest that magmatism and mineralization of the Tiegelongnan porphyry Cu (Au) deposit probably occurred in an active continental margin environment after the subduction of BNO plate, as result of continental crust thickening and terranes collision (Figure 9).

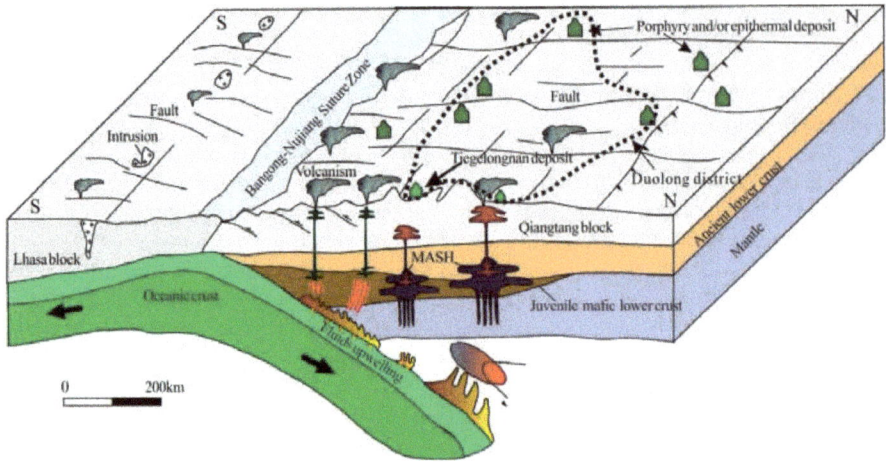

Figure 9. Tectonic setting model for the formation of the Duolong deposit, Tibet.

6.2. Tectonic Reconstruction of Epithermal Overprinting Porphyry

The continental crust thickening resulted from the Qiangtang–Lhasa collision might be the reason for the emplacement of magma and formation of porphyry deposits in the Duolong district. An extensional condition is common during the post-subduction stage of two plates or terranes [57], thus Sillitoe and Hedenquist (2003) explained that some arcs are subjected to neutral stress condition or mild extension where high-sulfidation epithermal deposits formed [23]. This might be one of the reasons that porphyry mineralization is overprinted by epithermal mineralization in the Tiegelongnan deposit. Besides, water table decline is another possible reason for epithermal mineralization formed at deep level where porphyry mineralization formed earlier at depth [58]. The paleosoil occurred between the whole mineralization Cu orebody and the overlying andesite in the Tiegelongan deposit indicates weathering and erosion took place before andesite covering at 110 Ma, which is possible to decrease the water table. Furthermore, fluid inclusion pressure evidence suggests that rapid mountains lifting and erosion took place between two stages of porphyry intrusions at 121 and at 116 Ma respectively in the Tiegelongnan deposit [22]. The occurrences of overprinting of epithermal on porphyry alteration and mineralization in the Tiegelongnan deposit are mainly characterized by veins and fracture filling, and the replacement between the typical minerals of two episodes. From the absolute geochronology data, alunite was formed between 117 and 100 Ma, which postdates the ages of biotite and sercite at 121 Ma [21]. Early formed veins were reopened under an extensional situation, giving channels for later epithermal fluids. Reopening of early formed textures can improve rock permeability. It enables magmatic-hydrothermal fluid to cool with low temperature isotherm dropping. When the isotherm line decline to deep level of earlier formed porphyry deposit, epithermal minerals formed and replaced porphyry minerals. Specifically, the low temperature minerals alunite and kaolinite and high sulfidation state sulfides (tennantite, enargite, digenite, covellite, etc.) precipitated and replaced porphyries stage alteration and sulfide minerals at a higher temperature and in a deeper level. Although most porphyry deposits are formed in convergent and compressive settings [49], high sulfidation epithermal can be generated in calc-alkaline andesitic-dacitic arcs under neutral stress or mild extension conditions [23]. During the formation of this porphyry and epithermal deposit, the extensional strain in the Duolong district might be related with the Bangong–Nujiang subduction zone retreating after the BNO closure [9,10,44].

During the epithermal stage, some breccia formed, and straight veins break phyllic altered rocks, indicate that epithermal events overprinting on the porphyry events. Plenty of breccia implies that rocks are brittle in the epithermal stage. Hydrothermal processes in ductile and brittle rocks in a magmatic–epithermal environment study suggest that the ductile-brittle transition commonly occurs about 370–400 °C [59]. This is higher than epithermal hydrothermal fluid temperature 160–270 °C [6]. Therefore, before epithermal events taking place, the wall rocks of the Tiegelongnan deposit is brittle and could be easily broken by the accumulation of hydrothermal fluid or by fault events in the Duolong district. This gives access to epithermal fluid arriving at shallow sites, and overprints porphyry system along those faults, fractures, and other opened space.

6.3. Post-Mineral Erosion and Preservation

The crust thickening and terranes collision in the Bangong–Nujiang suture zone, combining with later coming India–Asia plate collisions, resulted in rapid uplifting and erosion of the deposits in the Duolong district. It also affects the uplifting and erosion of most parts of the Tibetan Plateau [33,41]. The large-scale uplift was activated by the Lhasa–Qiangtang collision [60,61], that resulted in the angular unconformity contact between the continental Abushan Formation and the underlying marine sediments [33]. The erosion of the Tiegelongnan deposit happened right after its formation, which is observed from the weathered rocks under the andesite. The fluid inclusion study suggested the eroded layers may reach a thickness of 600–1200 m before cover of the andesite at 110 Ma [27]. In the following exhumation rate calculations, the youngest formation age of the Tiegelongnan deposit was used at ~118 Ma dated from the molybdenite. The exhumation rate of ground surface increased with the elevation [62] and it has functional relationship with elevation. The smallest exhumation thickness and interval of the protection can be used to calculate the smallest exhumation rate. Thus the Tiegelongnan deposit has experienced exhumation for 6 to 7 m.y. and the exhumation rate is 0.1–0.2 mm/y.a, averagely at 0.15 mm/y.a. This is consistent with the common exhumation rate of the epithermal deposit at ~0.167 mm/y.a, and the porphyry deposit at 0.158 mm/y.a [63].

Northern Tibet has been elevated more than 5000 m, and it is continuously affected by the India–Asia collision system, thus leading to a more intensive exhumation than other parts of Tibet, while the southern Tibetan Plateau attained a 3–4 km elevation at ~99 Ma [64]. The Tiegelognnan deposit was overlain by the andesite after small interval of exhumation which protected the Tiegelongnan deposit from totally erosion. The erosion after andesite settle is still non-negligible. The apatite HeFTy program [65] was used to model the prolonged thermal history of the Tiegelongnan deposit [66]. It suggested that the Tiegelongnan deposit has experienced four cooling stages: (i) relatively slow cooling from Early Cretaceous to Late Cretaceous (120–75 Ma); (ii) fast cooling in Late Cretaceous (75–60 Ma); (iii) moderately fast cooling from Eocene to Oligocene (45–30 Ma); (iv) very fast cooling since Late Miocene (<7.8 Ma). The exhumation thickness is at least 3600 m since Late Cretaceous in the Duolong district. Yin et al., (2000) and Kapp et al., (2007) reported the subsequent India–Asia collision led to 1400 km of shortening within recent 70 m.y. [40,41]. The andesite has a thickness of ~500 m [67], which is not thick enough to withstand that intensive exhumation. Therefore, the andesite is not the solely protective cover for the Tiegelongnan deposits. Post-mineral sediments, such as the Upper Cretaceous Abushan Formation (K$_2$a) and the Upper Oligocene Kangtuo Formation (E$_3$k), might also act as significant caps for protecting orebodies.

Along with the thickening of the continental arc, collapse of the crust may destruct the whole porphyry and epithermal Cu (Au) orebodies. The Rongna Fault might be the result of the collapse of the accumulated crust rocks. The inversion resistivity cross-section in Figure 6a showed that a normal high angle fault breaks the high resistivity zone into two separated C1 and C2 zones. It also cuts through the andesite from the Rongna Valley, and may dislocate and conceal the western part of the Cu (Au) orebody to deeper places.

A simplified model (Figure 10) was applied to describe the porphyry Cu system in the Tiegelongnan deposit. Epithermal Cu (Au) mineralization veins are mostly retained in the advanced argillic alteration zone, which overprints phyllic alteration zone on top of porphyry Cu orebody (Figure 10a). This is a transitional zone between the disseminated high-sulfidation epithermal precious metal deposit and the porphyry Cu (Au) orebody (Figure 10b). There might have been disseminated epithermal precious metal deposit existing in theory. The reason of not being detected on the deposit is that it either was not well preserved because of severe erosion or has not been found yet. Deep in the porphyry Cu (Au) system, the bottom of the porphyry Cu orebody has not been detected yet. We suggest there could be economic potential at depth, because biotite alteration is shown at depth, if it could represent the typical potassic alteration, which usually is the core of mineralization orebody in the porphyry Cu system [68]. Obviously, we did not reveal the whole biotite alteration, which might be concealed at a deeper level.

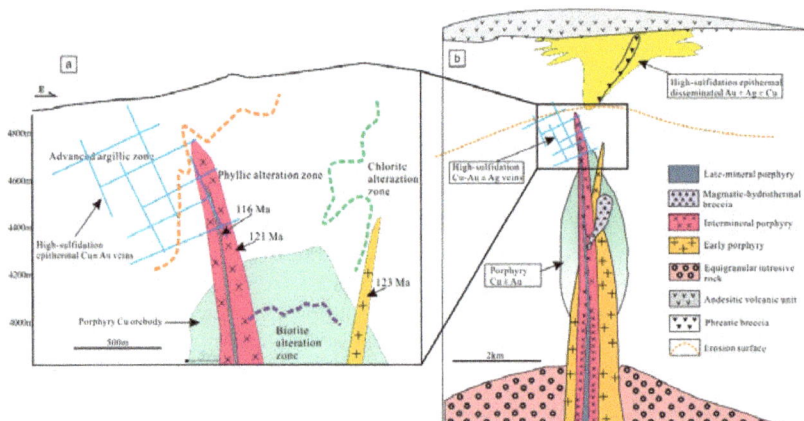

Figure 10. (**a**) Simplified geology map of the A-B sections in Tiegelongnan [31]. (**b**) Anatomy of a condensed porphyry Cu system showing the spatial interrelationships of porphyry Cu orebody, high-sulfidation epithermal Cu ± Au orebody, and late-mineralization andesitic volcanic rocks [31].

7. Implications and Conclusions

The geological history of the Tigelongnan deposit is as follows: (i) Early Jurassic active continental margin sandstone (pre-mineral stage); (ii) emplacement of multiple porphyry intrusions and formation of the porphyry ore system (first mineral stage); (iii) overprinting of epithermal alteration and mineralization (second mineral stage); (iv) weathering and erosion (≥600 m) and volcanic extrusion (preservation); (v) continuous lifting and erosion/movement of the Rongna fault (dislocation). This process is associated with the movement of the Bangong–Nujiang oceanic plate, following Qiangtang-Lhasa terranes collision and the effect of the India–Lhasa terranes subduction and collision. Marine sandstone was deposited before or during subduction of the oceanic plate in Jurrassic. Ignous rocks emplacement and eruption are the results of the subduction of oceanic plate during 170–145 Ma, and the Qiangtang–Lhasa terrane collision between 126 and 116 Ma. Hydrothermal fluid is induced by those porphyry intrusions, and it contributes to porphyry mineralization. Shortly after that, younger stage of epithermal fluid overprinted the porphyry mineralization because of erosion of the system due to the extensional structures at around 116 Ma. With continuous collision of Qiangtang and Lhasa terranes, rapid lifting and strong exhumation partly eroded the ore deposit. Volcanic andesite (~110 Ma) covered on the top of the orebody protects it from entire erosion. Subsequent India and Lhasa plates subduction and collision uplifted the plateau and caused the erosion of those deposits in the Duolong district again. Post-mineral structure such as the Rongna Fault possibly dislocated the main orebody of the Tiegelongnan deposit (Figure 6a).

This review of the Tiegelongnan deposit is significant for the future exploration programs, even for exploration of porphyry and epithermal deposits in the Duolong district or even the Bangong–Nujiang suture zone. The Rongna Fault dislocated the Tiegelongnan deposit in its western part. Another part of the orebody on the hanging wall of the fault might slip southwestward and be concealed to a deeper domain. The nearly 11 Mt of Cu resource currently being explored might be part of the entire Tiegelongnan porphyry and epithermal Cu (Au) orebody, which is similar to the San Manuel-Kalamazoo porphyry copper deposit in South America [69]. Comparably, we are confident on the great potential of the Tiegelongnan deposit. Further understanding of the dislocation of the Rongna Fault should be conducted later, which would contribute to increasing the ore reserves of the deposit at depth.

The volcanic rocks unconformably overlie on the whole porphyry and epithermal Cu (Au) orebody, which prevents the orebody from being subject to further erosion. This might be the reason for only epithermal copper orebody being found at the top the Tiegelongnan deposit so far, but not anywhere else. There might be epithermal mineralization on the Duobuza and Bolong deposits, but they might be fully eroded away due to the lack of overlying protection. Furthermore, due to the large range of volcanic rocks in the Duolong district, more porphyry and epithermal copper and gold deposits are of great potential to be preserved, and that could be the future exploration direction in the Duolong district.

Author Contributions: Y.S. carried out the project and coordinated this study. All authors took part in the field work, analyzed the results and wrote the manuscript; Y.S. and C.Y mainly revised and edited the paper.

Funding: This research was partly funded by the National Key R & D Program of China (No. 2018YFC0604106), the National Natural Science Foundation of China (No. 41402178), and the Chinese Geological Survey Program (DD20160026).

Acknowledgments: We thank three anonymous reviewers, the Editor Shi, and the academic editor of *Minerals* Alain Chauvet for their positive and constructive comments that are helpful to improve the manuscript significantly. Qing Zhang from the University of Wollongong provided major contributions to the English editing of the paper.

Conflicts of Interest: The authors declare no conflict of interest.

References and Notes

1. Hou, Z.; Xie, Y.; Xu, W.; Li, Y.; Zhu, X.; Khin, Z.; Beaudoin, G.; Rui, Z.; Huang, W.; Luo, C. Yulong deposit, eastern Tibet: A high-sulfidation Cu-Au porphyry copper deposit in the eastern Indo-Asian collision zone. *Int. Geol. Rev.* **2007**, *49*, 235–258.

2. Yang, A.; Hou, Z.; White, C.N.; Chang, Z.; Li, Z.; Song, Y. Geology of the post-collisional porphyry copper–molybdenum deposit at Qulong, Tibet. *Ore Geol. Rev.* **2009**, *36*, 133–159. [CrossRef]

3. Zheng, W.; Tang, J.; Zhong, K.; Ying, L.; Leng, Q.; Ding, S.; Lin, B. Geology of the Jiama copper–polymetallic system, Lhasa Region, China. *Ore Geol. Rev.* **2016**, *74*, 151–169. [CrossRef]

4. Zhu, X.; Li, G.; Chen, H.; Ma, D.; Zhang, H.; Zhang, H.; Liu, C.; Wei, L. Petrogenesis and metallogenic setting of porphyries of the Duobuza porphyry Cu–Au deposit, central Tibet, China. *Ore Geol. Rev.* **2017**, *89*, 858–875. [CrossRef]

5. Zhu, X.; Li, G.; Chen, H.; Ma, D.; Huang, H. Zircon U–Pb, Molybdenite Re–Os and K-feldspar ^{40}Ar/^{39}Ar Dating of the Bolong Porphyry Cu–Au Deposit, Tibet, China. *Resour. Geol.* **2015**, *65*, 122–135. [CrossRef]

6. Hedenquist, J.W.; Arribas, A.; Gonzalez-Urien, E. Exploration for epithermal gold deposits. *Rev. Econ. Geol.* **2000**, *13*, 245–277.

7. Tang, J.; Sun, X.; Ding, S.; Wang, Q.; Wang, Y.; Yang, C.; Chen, H.; Li, Y.; Li, Y.; Wei, L.; et al. Discovery of the epithermal deposit of Cu (Au-Ag) in the Duolong ore concentrating area, Tibet. *Acta Geosci. Sin.* **2014**, *35*, 6–10. (In Chinese) [CrossRef]

8. Lin, B.; Tang, J.-X.; Chen, Y.-C.; Song, Y.; Hall, G.; Wang, Q.; Yang, C.; Fang, X.; Duan, J.-L.; Yang, H.-H. Geochronology and Genesis of the Tiegelongnan Porphyry Cu (Au) Deposit in Tibet: Evidence from U–Pb, Re–Os Dating and Hf, S, and H–O Isotopes. *Resour. Geol.* **2017**, *67*, 1–21. [CrossRef]

9. Wei, S.-G.; Tang, J.-X.; Song, Y.; Liu, Z.-B.; Feng, J.; Li, Y.-B. Early Cretaceous bimodal volcanism in the Duolong Cu mining district, western Tibet: Record of slab breakoff that triggered ca. 108–113 Ma magmatism in the western Qiangtang terrane. *J. Asian Earth Sci.* **2017**, *138*, 588–607. [CrossRef]

10. Geng, Q.; Zhang, Z.; Peng, Z.; Guan, J.; Zhu, X.; Mao, X. Jurassic–Cretaceous granitoids and related tectono-metallogenesis in the Zapug–Duobuza arc, western Tibet. *Ore Geol. Rev.* **2016**, *77*, 163–175. [CrossRef]

11. Li, J.X.; Qin, K.; Li, G.; Xiao, B.; Zhao, J.; Chen, L. Petrogenesis of Cretaceous igneous rocks from the Duolong porphyry Cu–Au deposit, central Tibet: Evidence from zircon U–Pb geochronology, petrochemistry and Sr–Nd–Pb–Hf isotope characteristics. *Geol. J.* **2016**, *51*, 285–307. [CrossRef]

12. Pan, G.; Wang, L.; Li, R.; Yuan, S.; Ji, W.; Yin, F.; Zhang, W.; Wang, B. Tectonic evolution of the Qinghai-Tibet plateau. *J. Asian Earth Sci.* **2012**, *53*, 3–14. [CrossRef]

13. Zhu, D.; Zhao, Z.; Niu, Y.; Dilek, Y.; Hou, Z.; Mo, X. The origin and pre-Cenozoic evolution of the Tibetan Plateau. *Gondwana Res.* **2013**, *23*, 1429–1454. [CrossRef]

14. Ding, S.; Chen, Y.; Tang, J.; Zheng, W.; Lin, B.; Yang, C. Petrogenesis and Tectonics of the Naruo Porphyry Cu (Au) Deposit Related Intrusion in the Duolong Area, Central Tibet. *Acta Geol. Sin.* **2017**, *91*, 581–601. [CrossRef]

15. Li, J.; Qin, K.; Li, G.; Noreen, J.; Zhao, J.; Cao, M.; Huang, F. The Nadun Cu–Au mineralization, central Tibet: Root of a high sulfidation epithermal deposit. *Ore Geol. Rev.* **2016**, *78*, 371–387. [CrossRef]

16. Li, G.; Li, J.; Qin, K.; Duo, J.; Zhang, T.; Xiao, B.; Zhao, J. Geology and Hydrothermal Alteration of the Duobuza Gold-Rich Porphyry Copper District in the Bangongco Metallogenetic Belt, Northwestern Tibet. *Resour. Geol.* **2012**, *62*, 99–118. [CrossRef]

17. Lin, B.; Chen, Y.; Tang, J.; Wang, Q.; Song, Y.; Yang, C.; Wang, L.; He, W.; Zhang, L. ^{40}Ar/^{39}Ar and Rb-Sr Ages of the Tiegelongnan Porphyry Cu-(Au) Deposit in the Bangong Co-Nujiang Metallogenic Belt of Tibet, China: Implication for Generation of Super-Large Deposit. *Acta Geol. Sin.* **2017**, *91*, 602–616. [CrossRef]

18. Li, G.; Duan, Z.; Liu, B.; Zhang, H.; Dong, S.; Zhang, L. The discovery of Jurassic accretionary complexes in Duolong area, northern Bangong Co-Nujiang suture zone, Tibet, and its geologic significance. *Geol. Bull. China* **2011**, *30*, 1256–1260. (In Chinese)

19. Liu, Y.; Wang, M.; Li, C.; Xie, C.; Chen, H.; Li, Y.; Fan, J.; Li, X.; Xu, W.; Sun, Z. Cretaceous structures in the Duolong region of central Tibet: Evidence for an accretionary wedge and closure of the Bangong–Nujiang Neo-Tethys Ocean. *Gondwana Res.* **2017**, *48*, 110–123. [CrossRef]

20. Fang, X.; Tang, J.; Song, Y.; Yang, C.; Ding, S.; Wang, Y.; Wang, Q.; Sun, X.; Li, Y.; Wei, L.; et al. Formation epoch of the South Tiegelong superlarge epithermal Cu (Au-Ag) deposit in Tibet and its geological implications. *Acta Geosci. Sin.* **2015**, *36*, 168–176. (In Chinese)

21. Yang, C.; Beaudoin, G.; Tang, J.; Song, Y. An extreme long life span of porphyry and epithermal Cu deposit: The Tiegelongnan deposit, Tibet, China. 2018; in preparation.

22. Yang, C.; Beaudoin, G.; Tang, J.; Song, Y. Geology and genesis of Tiegelongnan porphyry and epithermal base metal deposit in Duolong district, Tibet, China: From stable isotope and fluid inclusions constrains. 2018; in preparation.

23. Sillitoe, R.H.; Hedenquist, J.W. Linkages between volcanotectonic settings, ore-fluid compositions, and epithermal precious metal deposits. *Spec. Publ. Soc. Econ. Geol.* **2003**, *10*, 315–343.

24. Seedorff, E. Porphyry deposits: Characteristics and origin of hypogene features. *Econ. Geol.* **2005**, *29*, 251–298.

25. Stoffregen, R. Genesis of Acid-Sulfate Alteration and Au-Cu-Ag Mineralization at Summitville, Colorado. *Econ. Geol.* **1987**, *82*, 1575–1591. [CrossRef]

26. Wang, Y.; Tang, J. The First Discovery of Colusite in the Tiegelongnan Supper-large Cu (Au, Ag) Deposit and Significance for the Genesis of the Deposit. *Acta Geol. Sin.* **2018**, *92*, 400–401. [CrossRef]

27. Yang, C.; Tang, J.; Wang, Y.; Yang, H.; Wang, Q.; Sun, X.; Feng, J.; Yin, X.; Ding, S.; Fang, X.; et al. Fluid and geological characteristics researches of Southern Tiegelong epithermal porphyry Cu-Au deposit in Tibet. *Miner. Depos.* **2014**, *33*, 1287–1305. (In Chinese)

28. Duan, J.; Tang, J.; Li, Y.; Liu, S.; Wang, Q.; Yang, C.; Wang, Y. Copper isotopic signature of the Tiegelongnan high-sulfidation copper deposit, Tibet: Implications for its origin and mineral exploration. *Miner. Depos.* **2016**, *51*, 591–602. [CrossRef]

29. Mathur, R.; Munk, L.; Nguyen, M.; Gregory, M.; Annell, H.; Lang, J. Modern and paleofluid pathways revealed by Cu isotope compositions in surface waters and ores of the Pebble porphyry Cu-Au-Mo deposit, Alaska. *Econ. Geol.* **2013**, *108*, 529–541. [CrossRef]

30. Chen, H.Q.; Qu, X.M.; Fan, S.F. Geological characteristics and metallogenic prospecting model of Duolong porphyry copper gold ore concentration area in Gerze County, Tibet. *Miner. Depos.* **2015**, *34*, 321–332.

31. Song, Y.; Yang, H.H.; Lin, B.; Liu, Z.B.; Qin, W.; Ke, G.; Chao, Y.; Xiang, F. The Preservation System of Epithermal Deposits in South Qiangtang Terrane of Central Tibetan Plateau and Its Significance: A Case Study of the Tiegelongnan Superlarge Deposit. *Acta Geosci. Sin.* **2017**, *38*, 659–669. (In Chinese)

32. Kapp, P.; Yin, A.; Manning, C.; Harrison, T.; Taylor, M.; Ding, L. Tectonic evolution of the early Mesozoic blueschist-bearing Qiangtang metamorphic belt, central Tibet. *Tectonics* **2003**, *22*. [CrossRef]

33. Kapp, P.; Yin, A.; Harrison, T.; Ding, L. Cretaceous-Tertiary shortening, basin development, and volcanism in central Tibet. *Geol. Soc. Am. Bull.* **2005**, *117*, 865–878. [CrossRef]

34. Pullen, A.; Kapp, P.; Gehrels, G.; Ding, L.; Zhang, Q. Metamorphic rocks in central Tibet: Lateral variations and implications for crustal structure. *Geol. Soc. Am. Bull.* **2011**, *123*, 585–600. [CrossRef]

35. Liu, D.; Huang, Q.; Fan, S.; Zhang, L.; Shi, R.; Ding, L. Subduction of the Bangong–Nujiang Ocean: Constraints from granites in the Bangong Co area, Tibet. *Geol. J.* **2014**, *49*, 188–206. [CrossRef]

36. Hao, L.; Wang, Q.; Wyman, D.A.; Ou, Q.; Dan, W.; Jiang, Z.; Wu, F.; Yang, J.; Long, X.; Li, J. Underplating of basaltic magmas and crustal growth in a continental arc: Evidence from Late Mesozoic intermediate–felsic intrusive rocks in southern Qiangtang, central Tibet. *Lithos* **2016**, *245*, 223–242. [CrossRef]

37. Zhu, D.; Li, S.; Cawood, P.; Wang, Q.; Zhao, Z.; Liu, S.; Wang, L. Assembly of the Lhasa and Qiangtang terranes in central Tibet by divergent double subduction. *Lithos* **2016**, *245*, 7–17. [CrossRef]

38. Allmendinger, R.; Jordan, T.; And, S.; Isacks, B. The evolution of the Altiplano-Puna plateau of the Central Andes. *Annu. Rev. Earth Planet. Sci.* **1997**, *25*, 139–174. [CrossRef]

39. Zhu, D.; Pan, G.; Wang, L.; Mo, X.; Zhao, Z.; Zhou, C.; Liao, Z.; Dong, G.; Yuan, S. Tempo-spatial variations of Mesozoic magmatic rocks in the Gangdese belt, Tibet, China, with a discussion of geodynamic setting-related issues. *Geol. Bull. China* **2008**, *27*, 1535–1550. (In Chinese)

40. Kapp, P.; Decelles, P.; Gehrels, G.; Heizler, M.; Lin, D. Geological records of the Lhasa-Qiangtang and Indo-Asian collisions in the Nima area of central Tibet. *Geol. Soc. Am. Bull.* **2007**, *119*, 917–933. [CrossRef]

41. Yin, A.; Harrison, T.M. Geologic evolution of the Himalayan-Tibetan orogen. *Annu. Rev. Earth Planet. Sci.* **2000**, *28*, 211–280. [CrossRef]

42. Qu, X.-M.; Wang, R.; Xin, H.; Jiang, J.; Chen, H. Age and petrogenesis of A-type granites in the middle segment of the Bangonghu–Nujiang suture, Tibetan plateau. *Lithos* **2012**, *146*, 264–275. [CrossRef]

43. Li, J.-X.; Qin, K.; Li, G.; Xiao, B.; Zhao, J.; Cao, M.; Chen, L. Petrogenesis of ore-bearing porphyries from the Duolong porphyry Cu–Au deposit, central Tibet: Evidence from U–Pb geochronology, petrochemistry and Sr-Nd–Hf–O isotope characteristics. *Lithos* **2013**, *160*, 216–227. [CrossRef]

44. Li, X.; Li, C.; Sun, Z.; Wang, M. Origin and tectonic setting of the giant Duolong Cu–Au deposit, South Qiangtang Terrane, Tibet: Evidence from geochronology and geochemistry of Early Cretaceous intrusive rocks. *Ore Geol. Rev.* **2017**, *80*, 61–78. [CrossRef]

45. Li, G.; Qin, K.; Li, J.; Evans, N.; Zhao, J.; Cao, M.; Zhang, X. Cretaceous magmatism and metallogeny in the Bangong–Nujiang metallogenic belt, central Tibet: Evidence from petrogeochemistry, zircon U–Pb ages, and Hf–O isotopic compositions. *Gondwana Res.* **2017**, *41*, 110–127. [CrossRef]

46. Hou, Z.; Mo, X.; Gao, Y.; Qu, X.; Meng, X. Adakite, a possible host rock for porphyry copper deposits: Case studies of porphyry copper belts in Tibetan Plateau and in Northern Chile. *Miner. Depos.* **2003**, *22*, 1–12. (In Chinese)

47. Defant, M.J.; Drummond, M.S. Derivation of some modern arc magmas by melting of young subducted lithosphere. *Nature* **1990**, *347*, 662. [CrossRef]

48. Sillitoe, R. Epochs of intrusion-related copper mineralization in the Andes. *J. S. Am. Earth Sci.* **1988**, *1*, 89–108. [CrossRef]

49. Sillitoe, R.H. A plate tectonic model for the origin of porphyry copper deposits. *Econ. Geol.* **1972**, *67*, 184–197. [CrossRef]

50. Hou, Z.; Gao, Y.; Qu, X.; Rui, Z.; Mo, X. Origin of adakitic intrusives generated during mid-Miocene east–west extension in southern Tibet. *Earth Planet. Sci. Lett.* **2004**, *220*, 139–155. [CrossRef]

51. Hou, Z.; Yang, Z.; Lu, Y.; Kemp, A.; Zheng, Y.; Li, Q.; Tang, J.; Yang, Z.; Duan, L. A genetic linkage between subduction-and collision-related porphyry Cu deposits in continental collision zones. *Geology* **2015**, *43*, 247–250. [CrossRef]

52. Oyarzun, R.; Márquez, A.; Lillo, J.; López, I.; Rivera, S. Reply to Discussion on "Giant versus small porphyry copper deposits of Cenozoic age in northern Chile: Adakitic versus normal calc-alkaline magmatism" by Oyarzun R, Márquez A, Lillo J, López I, Rivera S (Mineralium Deposita 36: 794–798, 2001). *Miner. Depos.* **2002**, *37*, 795–799. [CrossRef]

53. Li, J.; Qin, K.; Li, G.; Richards, J.; Zhao, J.; Cao, M. Geochronology, geochemistry, and zircon Hf isotopic compositions of Mesozoic intermediate–felsic intrusions in central Tibet: Petrogenetic and tectonic implications. *Lithos* **2014**, *198–199*, 77–91. [CrossRef]

54. Lv, L.; Zhao, Y.; Song, L.; Tian, Y.; Xin, H. Characteristics of C, Si, O, S and Pb isotopes of the Fe-rich and Cu (Au) deposits in the western Bangong–Nujiang metallogenic belt, Tibet, and their geological significance. *Acta Geol. Sin.* **2011**, *85*, 1291–1304. (In Chinese)

55. Xin, H.; Qu, X.; Wang, R.; Liu, H.; Zhao, Y.; Wei, H. Geochemistry and Pb, Sr, Nd isotopic features of ore-bearing porphyries in Bangong Lake porphyry copper belt, western Tibet. *Miner. Depos.* **2009**, *28*, 785–792. (In Chinese)

56. Hawkesworth, C. The generation and evolution of the continental crust. *J. Geol. Soc.* **2010**, *167*, 229–248. [CrossRef]

57. Richards, J.P. Postsubduction porphyry Cu-Au and epithermal Au deposits: Products of remelting of subduction-modified lithosphere. *Geology* **2009**, *37*, 247–250. [CrossRef]

58. Sillitoe, R. Styles of high-sulphidation gold, silver and copper mineralisation in porphyry and epithermal environments. In Proceedings of the Australasian Institute of Mining and Metallurgy, Melbourne, Australia, 11–13 September 2000; Volume 305, pp. 19–34.

59. Fournier, R.O. Hydrothermal processes related to movement of fluid from plastic into brittle rock in the magmatic-epithermal environment. *Econ. Geol.* **1999**, *94*, 1193–1211. [CrossRef]

60. Zhang, K.; Zhang, Y.; Tang, X.; Xia, B. Late Mesozoic tectonic evolution and growth of the Tibetan plateau prior to the Indo-Asian collision. *Earth Sci. Rev.* **2012**, *114*, 236–249. [CrossRef]

61. Li, Y.; Wang, C.; Li, Y.; Ma, C.; Wang, L.; Peng, S. The Cretaceous tectonic event in the Qiangtang Basin and its implications for hydrocarbon accumulation. *Pet. Sci.* **2010**, *7*, 466–471. [CrossRef]

62. Yang, K.; Ma, C. Some advances in the rates of continental erosion and mountain uplift. *Geol. Sci. Technol. Inf.* **1996**, *15*, 89–96.

63. Kesler, S.E.; Wilkinson, B.H. The role of exhumation in the temporal distribution of ore deposits. *Econ. Geol.* **2006**, *101*, 919–922. [CrossRef]

64. Murphy, M.; Yin, A.; Harrison, T.; Dürr, S.; Chen, Z.; Ryerson, J.; Kidd, F.; Wang, X.; Zhou, X. Did the Indo-Asian collision alone create the Tibetan plateau? *Geology* **1997**, *25*, 719–722. [CrossRef]

65. Ketcham, R.A. Forward and inverse modeling of low-temperature thermochronometry data. *Rev. Mineral. Geochem.* **2005**, *58*, 275–314. [CrossRef]

66. Yang, H.H.; Tang, J.; Dilles, J.; Song, Y. Temperature Study of the Duolong Porphyry Cu-Au District and its implications for the Evolution of the Qiangtang Terrane in Tibet, China. *Int. Geol. Rev.* **2018**. submitted.

67. Li, G. High temperature, salinity and strong oxidation ore-forming fluid at Duobuza gold-rich porphyry copper in the Bangonghu tectonic belt, Tibet: Evidence from fluid inclusions study. *Acta Petrol. Sin.* **2007**, *23*, 935–952.

68. Sillitoe, R.H. Porphyry Copper Systems. *Econ. Geol.* **2010**, *105*, 3–41. [CrossRef]

69. Lowell, J.D.; Guilbert, J.M. Lateral and vertical alteration-mineralization zoning in porphyry ore deposits. *Econ. Geol.* **1970**, *65*, 373–408. [CrossRef]

minerals

MDPI

Review

The Jbel Saghro Au(–Ag, Cu) and Ag–Hg Metallogenetic Province: Product of a Long-Lived Ediacaran Tectono-Magmatic Evolution in the Moroccan Anti-Atlas

Johann Tuduri [1,2,]*, Alain Chauvet [3], Luc Barbanson [2], Jean-Louis Bourdier [2], Mohamed Labriki [4], Aomar Ennaciri [4], Lakhlifi Badra [5], Michel Dubois [6], Christelle Ennaciri-Leloix [4], Stanislas Sizaret [2] and Lhou Maacha [4]

[1] BRGM, F-45060 Orléans, France
[2] ISTO, UMR7327, Université d'Orléans, CNRS, BRGM, F-45071 Orléans, France; luc.barbanson@univ-orleans.fr (L.B.); jean-louis.bourdier@univ-orleans.fr (J.-L.B.); stanislas.sizaret@univ-orleans.fr (S.S.)
[3] Géosciences Montpellier, cc. 060, Université de Montpellier 2, CEDEX 5, 34095 Montpellier, France; alain.chauvet@univ-montp2.fr
[4] MANAGEM, Twin Center, BP 5199, Casablanca 20100, Morocco; m.labriki@managemgroup.com (M.L.); a.ennaciri@managemgroup.com (A.E.); cleloix@yahoo.fr (C.E.-L.); l.maacha@managemgroup.com (L.M.)
[5] Faculté des Sciences, Université Moulay Ismaïl, BP 11201 Zitoune, Meknes 50 000, Morocco; badra_lakhlifi@yahoo.fr
[6] Laboratoire de Génie Civil et géo-Environnement–Lille Nord de France, EA 4515, Département des Sciences de la Terre, Université de Lille, Bât. SN5, 59655 Villeneuve d'Ascq, France; michel.dubois@univ-lille1.fr
* Correspondence: j.tuduri@brgm.fr; Tel.: +33-238-644-790

Received: 31 August 2018; Accepted: 10 December 2018; Published: 13 December 2018

Abstract: The Jbel Saghro is interpreted as part of a long-lived silicic large igneous province. The area comprises two lithostructural complexes. The Lower Complex consists of folded metagreywackes and N070–090°E dextral shear zones, which roughly results from a NW–SE to NNW–SSE shortening direction related to a D_1 transpressive tectonic stage. D_1 is also combined with syntectonic plutons emplaced between ca. 615 and 575 Ma. The Upper Complex is defined by ash-flow caldera emplacements, thick and widespread ignimbrites, lavas and volcaniclastic sedimentary rocks with related intrusives that were emplaced in three main magmatic flare ups at ca. 575, 565 and 555 Ma. It lies unconformably on the Lower Complex units and was affected by a D_2 transtensive tectonic stage. Between 550 and 540 Ma, the magmatic activity became slightly alkaline and of lower extent. Ore deposits show specific features, but remain controlled by the same structural setting: a NNW–SSE shortening direction related to both D_1 and D_2 stages. Porphyry Au(–Cu–Mo) and intrusion-related gold deposits were emplaced in an earlier stage between 580 and 565 Ma. Intermediate sulfidation epithermal deposits may have been emplaced during lull periods after the second and (or) the third flare-ups (560–550 Ma). Low sulfidation epithermal deposits were emplaced late during the felsic alkaline magmatic stage (550–520 Ma). The D_2 stage, therefore, provided extensional structures that enabled fluid circulations and magmatic-hydrothermal ore forming processes.

Keywords: structural control; silicic large igneous province; ignimbrite flare-ups; ash-flow caldera; epithermal; porphyry; IRGD; Anti-Atlas

1. Introduction

In northwest Africa, the Anti-Atlas, Ougarta and Hoggar domains consist of pericratonic terranes located at the margin of the West African Craton (WAC, Figure 1a) and that were mostly amalgamated from Palaeoproterozoic to Phanerozoic times [1–5]. These terranes were the site of recurring tectonic activity and periods of intense magmatic activity. The most important magmatic pulses occurred during the Mesoproterozoic [6–8] and at the end of Triassic when the Central Atlantic Magmatic Province (CAMP) was emplaced [9–11]. According to Ernst [12] and Ernst and Bleeker [13], both events have been related to large igneous provinces (LIPs). Indeed LIPs consist of large volumes of mainly mafic magma (>0.1 Mkm^3) in provinces whose areal extent might exceed 0.1 Mkm^2. LIPs are thought to emplace in a short duration pulse or multiple pulses (less than 1–10 Myr each) with a whole maximum duration of ca. 50 Myr. Intense felsic magmatism may also occur as silicic large igneous provinces (SLIPs) [12,14]. Dacite–rhyolite pyroclastic rocks (ignimbrites), along with transitional calc-alkaline I-type [15] to A-type granites, mainly characterise these SLIPs [16]. Further, LIPs are commonly related to a wide variety of metal deposits [17] including world-class deposits such as magmatic sulfide ore deposits associated with mafic and ultramafic magmatism (Ni, Cu, PGE, Cr, Ti, Fe [18,19]), with carbonatite and peralkaline complexes (Nb, Ti, REE, Zr [20,21]), or with diamondiferous kimberlites [22]. Iron oxide copper gold (IOCG) deposit types [23] and epithermal deposits of mostly low and intermediate sulfidation gold-based metal types [24–26] may be also related to more silicic LIPs. Another magmatic event described in peri-Gondwanan terranes of, e.g., Avalonian and Cadomian types, is mostly characterised by huge volumes of pyroclastic flows and was emplaced at the end of the Neoproterozoic era [27–30]. Conditions and geodynamical environment at the origin of such a silicic province remain insufficiently understood, although Moume et al. [31] recently proposed that this event might be related to the Central Iapetus Magmatic Province event (CIMP) of Ediacaran-Cambrian age [13,32]. In the Moroccan Anti Atlas, world-class deposits occur in an area mostly dominated by rhyolitic ignimbrites such as the giant Ag–Hg Imiter deposit, the Ag-Hg Zgounder deposit and the Co–Ni–Fe–As(–Au–Ag) Bou Azzer district (Figure 1b).

In fact, the Moroccan Anti-Atlas hosts several precious and base-metal deposits affected by at least four major tectonic phases: i.e., the Palaeoproterozoic, the Neoproterozoic, the Variscan and the Alpine cycles [29,33–35]. Until recently, most of the ore deposits from the Anti-Atlas were considered as Neoproterozoic in age due to their occurrence within Proterozoic inliers. One exception was the vein and stratabound copper deposits hosted within the early Palaeozoic cover, which were assumed to be syn-sedimentary or epigenetic and Variscan in age [36,37]. However, recent studies have reassessed the age and origin of numerous metal deposits, assigning younger ages than previously admitted. The arguments are essentially two-fold: (i) absolute dating and (ii) fluid chemistry by isotopic and fluid inclusion study methods. Indeed geochronological methods (e.g., Re/Os, Ar/Ar) frequently give younger ages than expected though possible resetting of dating materials and (or) ore remobilisation are rarely discussed. For instance, the Imiter deposit would coincide with the Permo-Triassic boundary [38]. Similarly, a late Carboniferous age has been proposed for ore enrichment at BouAzzer with a possible earlier pre-mineralising stage [39]. The strong association of mineral deposits with fluids of moderate to high salinities, suggests interpretations favouring the influence of basinal brines in the ore-forming processes [40,41]. Indeed, processes involving basin-related and(or) surface-related brines resulting from evaporation of seawater in Triassic basins in the formation of ore deposits, up to now interpreted as, deposits related to the late Neoproterozoic felsic magmatic event [42–48] have been defended by several recent works [40,41,49] although the debate is still open [50]. This controversy mainly concerns the Bou Azzer (Co–Ni–), Imiter (Ag–Hg) and Zgounder (Ag–Hg) mines that represent the main three world-class ore deposits of the Moroccan Anti-Atlas after the Akka gold mine was closed few years ago. The Imiter concentration has been proposed to be associated with the 550 ± 3 Ma rhyolitic magmatism [51], a hypothesis recently controverted by Ar-Ar geochronology [38] and palaeo-fluid geochemistry investigations [41]. The Zgounder deposit is supposed to be emplaced around 564 ± 5 Ma at the same time as rhyolitic intrusions [46,52] although fluid geochemistry would suggest a Triassic fluid contribution [41,49] sedimentary. The Bou Azzer Co–Ni–Fe–As(–Au–Ag) district hosts the only mine in the world where Co is produced as a primary commodity directly from Co- and As-bearing arsenide minerals [53]. It is interpreted as having experienced many ore remobilisations from Late Neoproterozoic, Variscan to Triassic magmatic-hydrothermal stages [39,40,44,54,55].

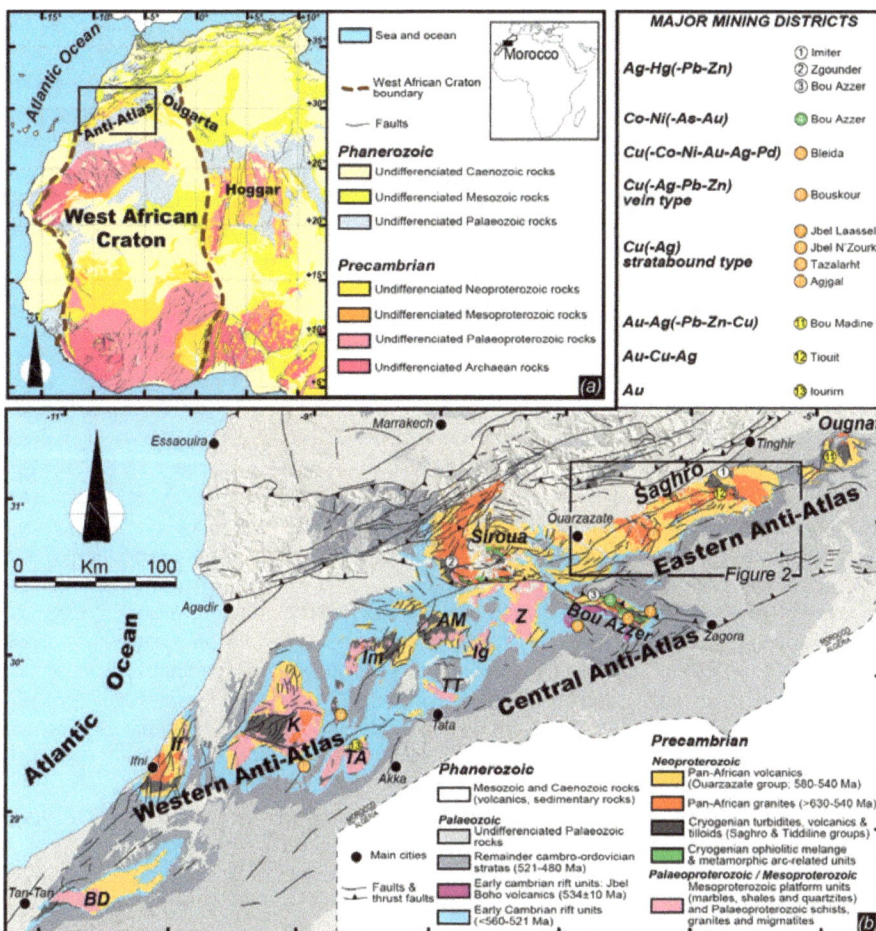

Figure 1. (**a**) Location of the Anti-Atlas belt at the northern limit of the West African Craton, after Thiéblemont et al. [56]. (**b**) Main geological units and major mining districts of the Moroccan Anti-Atlas [5,7,29,48,57–59]. Inliers–BD: Bas Drâa; If: Ifni; K: Kerdous; TA: Tagragra d'Akka; Im: Igherm; TT: Tagragra de Tata; Ig: Iguerda; AM: Agadir-Melloul; Z: Zenaga.

The debate as to whether the Anti-Atlas ore deposits are mostly Neoproterozoic in age and magmatic-related, or Phanerozoic and disconnected from any magmatic input, is similar to the one that exists between the orogenic and intrusion-related gold deposit models [60–65]. For instance, as applied to the Variscan gold ore deposits in the French Massif Central and beyond, this debate is focused on the involvement of magmatic fluids in the formation of mineralised systems. Arguments supporting the orogenic model are mostly based on fluid inclusion studies and isotopic data on quartz-bearing veins and highlight the meteoric and/or metamorphic signatures of the fluids [66,67]. In such cases, heat production from possible synchronous granitoids is supposed to generate only thermal convection cells. Arguments in favour of an intrusion-related model highlight the systematic spatial association between granite and hydrothermal systems, while fluid compositions and metal source are interpreted as showing a magmatic signature [68–70].

Therefore, it may be relevant to clarify whether most of the ore deposits of the Moroccan Anti-Atlas are linked to a Neoproterozoic magmatic input (the magmatic-related alternative), or to a more recent stage related to the penetration and circulation of sedimentary brines and (or) metamorphic fluids into the basement (the orogenic alternative). Indeed, as the Anti-Atlas domain may be considered as an important Neoproterozoic magmatic province with respect to the abundance of plutonic and volcanic rocks, evidences for a major Neoproterozoic metallogenetic province need to be deciphered. In this work, we present a review of the global geology and metallogeny of the Jbel Saghro in the Eastern Anti-Atlas with a specific emphasis on the formation, styles and tectonic controls of different ore deposit occurrences in order to assess and discuss a tectono-magmatic evolution of the studied area and a metallogenetic and geodynamic model. The world-class Bou Azzer deposit is not concerned in this study because of its location far from the Jbel Saghro. Its characterisation will be the subject of work currently in progress.

2. Geological Overview of the Anti-Atlas Mountains

The Moroccan Anti-Atlas belt, located in the northern part of the WAC (Figure 1a), constitutes an important segment of the Pan-African orogeny, also known as the Cadomian orogeny, that occurred from the Middle Neoproterozoic to Early Cambrian. This area is currently elevated at altitudes exceeding 1000 m in large areas due to a Caenozoic uplift [35,71,72] although they basically represent a Variscan intra-cratonic, thick-skinned basement inversion belt [34]. Consequently, the Anti-Atlas mountains present significant exposures of WSW–ENE trending inliers, (aka *boutonnières*), which consist of Proterozoic rocks in core of Phanerozoic, mainly Palaeozoic, sedimentary sequences (Figure 1b).

During Precambrian times, the Anti-Atlas domain has recorded two major orogenic cycles: the Eburnean, a Palaeoproterozoic cycle from about 2.1 to 2.0 Ga and the Pan-African/Cadomian cycle from about 885 to 540 Ma [5,33,73–75]. The Pan-African orogeny is characterised by three main tectono-magmatic events [29,73,75–77]. (i) The first event corresponds to an oceanic basin closure, oceanic subduction and arc-craton accretion, coeval with calc-alkaline magmatism and ophiolite obduction. This event is supposed to have occurred between 770 and 630 Ma and related formations are mainly observable in the Central Anti-Atlas [29,59,78–80]. (ii) Then, the development of an active margin along the amalgamated West African Craton was responsible for an intense high-K calc-alkaline magmatism and transpressive tectonics between ca. 615 and ca. 560 Ma [29,48,59,74,75,77,81,82]. (iii) Finally, late to post-orogenic granites with cogenetic volcanic and volcaniclastic cover are emplaced between ca. 560 and 550 Ma together with subsequent transtensional tectonics [5,29,47,74]. The transition from transpressive to transtensive tectonics remains the subject of debate but it likely occurred between 580 and 560 Ma [29,48,74]. There is still also an ongoing debate on the origin of the two late magmatic events, which are either described as arc-related [29] or as post-orogenic related to an asthenospheric rise beneath the West African Craton, without any active subduction [74,83]. Then, the Ediacaran–Cambrian transition is recognised as a carbonate-dominated succession (the Adoudou formation) that unconformably overlies the late Ediacaran plutonic and volcaniclastic rocks [84–86]. This sedimentation period shows dramatic and rapid thickness changes, consistent with an active extensional faulting related to incipient continental rifting [86,87]. Intercalated alkaline volcanic ash and flows dated between 550 and 520 Ma confirm that deposition of the Adoudou formation took place during the Early Cambrian and is in tectono-magmatic continuity with the Ediacaran volcano-tectonism [74,88–90].

In the Eastern Anti-Atlas of interest here, Jbel Saghro, Palaeoproterozoic terranes are not exposed (Figures 1b and 2), so, the oldest rocks outcropping consist of Middle Neoproterozoic (Cryogenian) metasedimentary rocks [91,92]. These are slightly deformed and unconformably overlain by a thick and widespread upper Neoproterozoic (Ediacaran) volcanic and volcaniclastic sequence (Figure 2). Both Cryogenian and Ediacaran units are intruded by Pan-African plutons (Figure 2).

Prior to the Variscan compression, the Palaeozoic sedimentary cover reaches an overall thickness of 8–10 km in the western Anti-Atlas and only 4 km in the eastern part of the Anti-Atlas, being mainly characterised by shallow marine sedimentary rocks [34,86,93–95]. Subsidence is assumed to be regular and characterised by two main steps [34,96]. The first step occurs during the Cambrian, from the Terreneuvian to the Miaolingian. It corresponds to a rifting episode [84,86,97,98]. The Adoudou formation and related alkaline volcanic rocks mentioned above belong to this step. The second step occurs during the Upper Devonian and is characterised by a multi-directional extension mainly controlled by NNW–SSE and ENE–WNW faults inherited from the Pan-African basement [93]. Fission track dating on zircon [99] yields a peak temperature affecting the basement around 328 ± 30 Ma (Carboniferous, Upper Mississipian). It is interpreted as the age of maximum burial. The subsequent uplift is attributed to the Variscan compression [34,100] during the Upper Carboniferous (early Pennsylvanian). It results in the inversion tectonics of these palaeofaults as thrusts and strike-slip–reverse faults. The main shortening directions are assumed to be NW–SE to N–S in the Western Anti-Atlas [34,100], and NE–SW in the Eastern Anti-Atlas [93]. The Precambrian basement is actually uplifted and folded into huge antiformal culminations probably at the origin of the present days inliers which characterise the Anti-Atlas fold belt. Thus, many authors e.g., [34,86,93,100] suggest that brittle deformations related to the Variscan compression occur in the whole Anti-Atlas and reactive structures from the Precambrian basement. Ductile deformation related to the Variscan tectonics has not yet been described in the Anti-Atlas. It might be suggested in the Western Anti-Atlas because the overall Palaeozoic rock thickness reaches 10 km making it possible a ductile–brittle transition. However, with a maximum thickness of only 4 km such ductile deformations seem precluded in the Eastern Anti-Atlas. Consequently, how and to what extent the Palaeozoic tectono-hydrothermal events may have an influence on the main ore deposits of the Eastern Anti-Atlas remains an important topic. For this reason, we focus our work below on the re-examination of the model of formation and structural control of different ore-bearing deposits of the Jbel Saghro. We highlight the relevance of a pluri-disciplinary approach involving analyses of regional tectonics, vein geometry, internal texture and mineralogy of ore deposits in order to decipher the mode of formation for each deposits. The ore-forming processes are specifically adressed with the view of their possible relationships with the Precambrian magmatism, in order to better assess whether they could have been mainly formed during the Neoproterozoic times, or mostly formed and remobilised during the Phanerozoic. We also stress that conventional approaches of ore deposits involving kinematic criteria on slip surfaces of faults and palaeostress reconstructions in the brittle regime are here excluded, since the whole Anti-Atlas is still an active seismic area [94]. So, the Precambrian or Palaeozoic features should be erased by recent tectonic reactivations.

The Jbel Saghro hosts numerous precious and base metal deposits (Figure 2). The most famous are: the giant Ag Imiter mine (8.5 Mt @ 700 g/t Ag; [41,45,51,101,102]) and the Cu Bouskour mine (21 Mt @ 1.3% Cu; [103]) and also the now closed Cu–Ag Tizi Mouddou (1.5 Mt @ 2% Cu, 250 g/t Ag; [43]) and Cu–Au–Ag Tiwit (1.06 Mt @ 8 g/t Au, 65 g/t Ag, ~0.4% Cu; [43]) mines. Mining activity in the Jbel Saghro area dates back a very long time as suggested for the Imiter and Bouskour mining sites [101,103]. Although no precise dating is available, some historical texts suggest that workings occurred since the beginning of medieval times. These sites were subsequently abandoned and forgotten and were rediscovered under the French Protectorate between 1912 to 1956, leading to subsequent intensive exploration and mining. Since the mid 1990s and after reinterpreting the regional geology of the Anti-Atlas area, the BRPM (Bureau de Recherches et de Participation Minière, currently ONHYM: Office National des HYdrocarbures et des Mines) and Reminex (a subsidiary of the Managem Group) conducted several exploration campaigns. Indeed, before the 1990s, the Anti-Atlas was mainly interpreted as a collisional belt and the formation of ore deposits was related to orogenic processes, e.g., [43,73,76]. Then, the geology and the ore forming processes were related to the complex evolution of active margin(s) in a subduction and accretion setting, e.g., [29,45,47,48,51,77,82]. This led to the discovery of new Au–Ag occurrences such as those of the Issarfane (0.5 Mt @ 1.8 g/t Au), Qal'at Mgouna and Thaghassa (Figure 2).

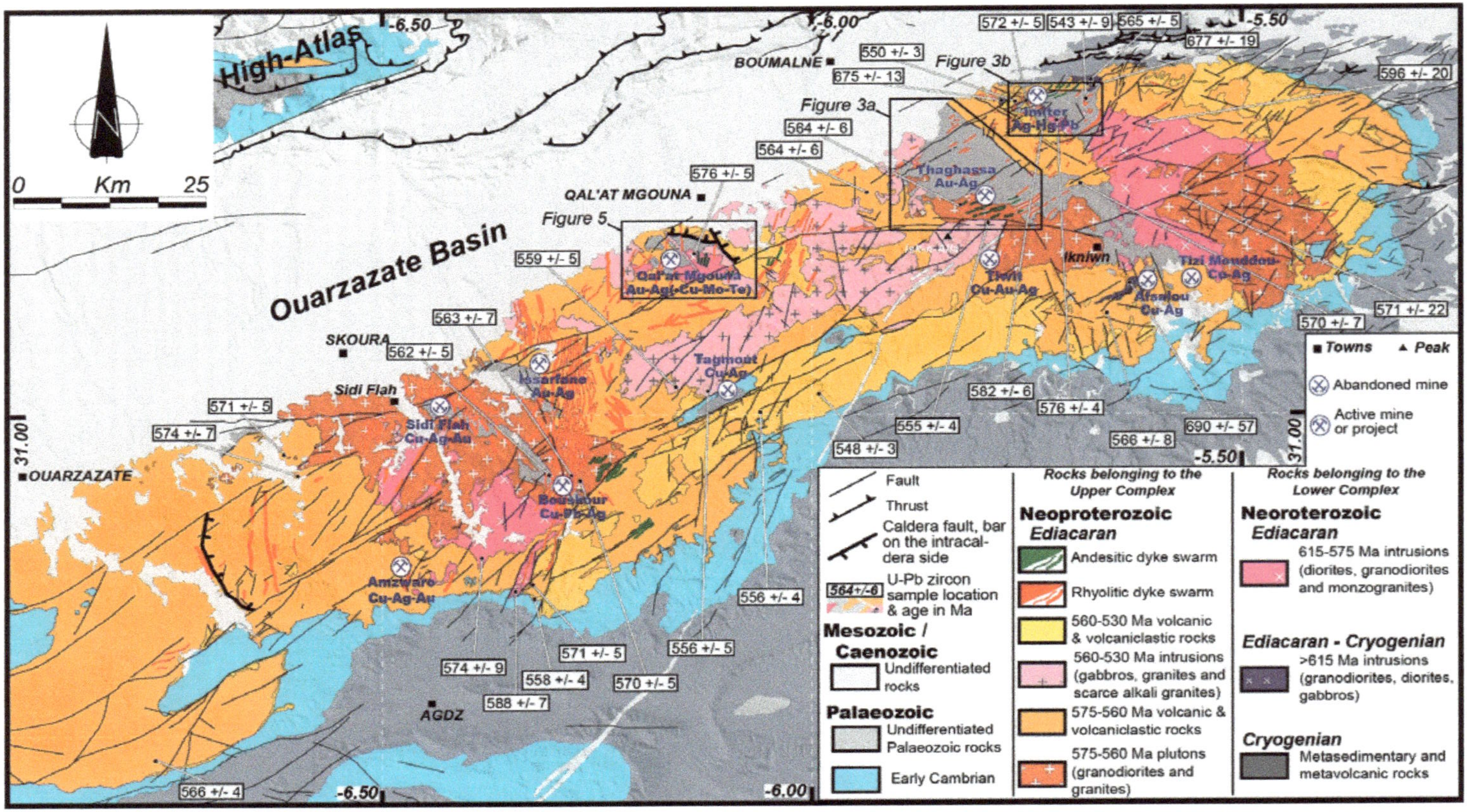

Figure 2. Simplified geologic map of the Jbel Saghro, after Hindermeyer et al. [104], Tuduri [47] and Tuduri et al. [48]. U-Pb radiometric ages obtained on zircon [29,48,51,78,105–108].

3. Tectono-Magmatic Evolution of the Jbel Saghro

Until the early 2000s, the Precambrian terranes in Morocco were traditionally subdivided into 3 main epochs: the PI for the Palaeoproterozoic era or Archaean aeon; PII for the Tonian and Cryogenian; and PIII for the Ediacaran periods [73,109–112]. These periods have been further subdivided into various units (e.g., P.II[1], P.II-III, P.IIsup, P.III3m-1a . . .) in order to take into account the extraordinary complexity of the different areas [77,109,113]. Geologists then used a lithostratigraphic approach in which rocks were correlated based on their lithological characteristics and grouped in Supergroups, Groups, Subgroups, Formations and Members for layered sedimentary and volcanic sequences, and in Suites for plutonic and metamorphic rocks [5,29,59,107]. However, these lithostratigraphic schemes confuse the geological and tectonic messages and the resulting legends on geological maps. Indeed, showing the cogenetic relationships in a volcano-plutonic setting where plutonic, hypabyssal and volcanic rocks coexist as in the Anti-Atlas, remains a challenge, especially when tectonic controls are combined. For the seek of consistency, we propose below a tectono-lithostratigraphic framework that groups the sedimentary, plutonic and volcanic rocks of the Saghro area in Lower and Upper Complexes with regard to the tectono-magmatic evolution of the eastern Anti-Atlas. Such a complex corresponds to the definition of the ICS (International Commission on Stratigraphy), that is, a lithostratigraphic unit composed of diverse types of any classes of rocks (sedimentary, igneous and metamorphic) and characterised by irregularly mixed lithology or by complicated structural relationships. The definition of the two complexes and rationale behind are given below.

3.1. The Lower Complex and the D_1 Transpressive Tectonics

The Lower Complex is composed of a thick succession of slightly deformed volcano-sedimentary rocks intruded by broadly coeval high-K calc-alkaline plutons (Figure 2). Volcano-sedimentary rocks are deformed and tightly upright folded (Figure 3a,b). They are also affected by a metamorphism below the amphibolite facies with hornfels around plutons [48,91,111]. Elsewhere, cleavage is well expressed in the vicinity of the large-scale, regional, strike-slip faults that mainly trend N070°E (Figure 2). The general structure is characterised by steeply dipping bedding (Figure 3a–c), large-scale folds (Figure 3a,b) and intense fracturing (see below).

3.1.1. The Earlier Arc-Related Metagreywackes and Metavolcanic Rocks

These oldest formations belong to the Anti-Atlas supergroup, Saghro Group or MGouna Group of Thomas et al. [5] or Habab Group of O'Connor et al. [107] and consist of metaturbidites with intercalated thin mafic metavolcanic layers, exposed, from west to east, near the cities of Sidi Flah, Qal'at Mgouna, Boumalne and Imiter (Figure 2). The metasedimentary sequences, hereafter, called metagreywackes, are dominated by sandstones and silty mudstone [91,92,114–118]. Although slightly deformed (Figures 3 and 4), they consist of 2000 to 6000 m of flysh-like turbiditic rocks [83,91,118] that locally alternate with basaltic flows, volcanic breccias, hyaloclastite and pillow structures (Figure 3a,b). Based on geochemical analyses, the interbedded mafic lava rocks may have a transitional character between tholeiites and alkali basalts and may be related to a back-arc environment formed in an extensional setting or correspond to the remnants of a passive-margin [91,115,118,119]. These rocks have a mean Nd depleted mantle model age (T_{DM}) of 650 ± 30 Ma, the ε_{Nd} range from +7.63 to +8.08, and the initial $^{87}Sr/^{86}Sr$ from 0.704 to 0.706 that have been interpreted as indicating a mantle origin without any old crust contribution [83,119]. These sedimentary sequences are considered to be Cryogenian [91,115,116]. However, Liégeois et al. [120] in Gasquet et al. [83], using U/Pb geochronology on detrital zircons from similar metagreywacke sequences in the Qal'at Mgouna area, have shown that zircon grains become younger towards the top of the metasedimentary sequences. They suggest that basins infilling were active until more recently than previously suspected, i.e., until 630–610 Ma. This corresponds to the onset of the Ediacaran Period and Pan-African magmatism that culminated during the next stages (see below).

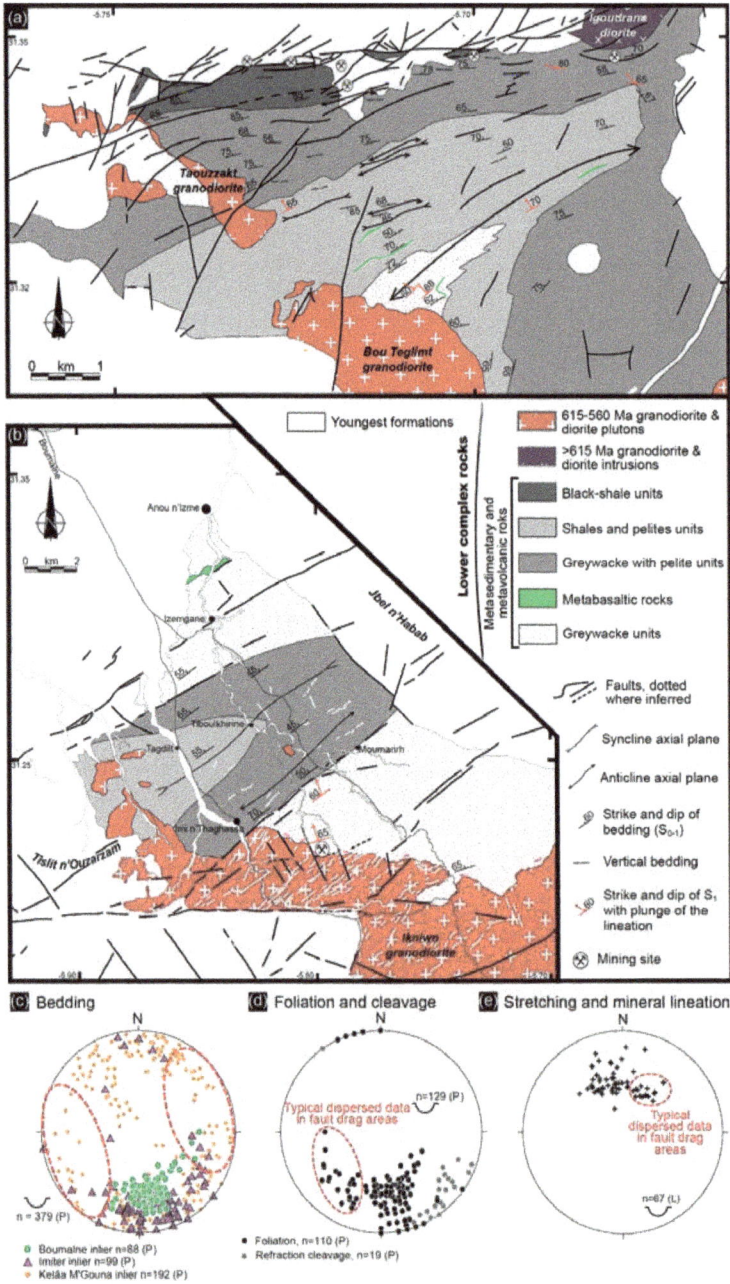

Figure 3. Structural maps of the (**a**) Imiter inlier modified from SMI (Société Métallurgique d'Imiter) data and Ighid et al. [121] and (**b**) Boumalne inlier (from Tuduri et al. [48]). Note that metabasaltic rocks with pilloid structures have never been observed in the Imiter inlier. Sills have been described and discussed as possibly syn-sedimentary in origin [100]. Stereoplots of structural orientation data: (**c**) bedding, (**d**) foliation and refraction cleavage, and (**e**) lineation. Bedding data are from the Qal'at

MGouna, Boumalne and Imiter Lower Complex inliers whereas foliation, refraction cleavage and lineation data are mainly from the Boumalne and Imiter Lower Complex inliers [47,48]. The dotted red lines in the bedding stereoplot highlight zones where data are locally reoriented because of drag fault zones and disturbance upon pluton emplacement.

3.1.2. The Intrusive Rocks

Diorites, granodiorites with minor gabbros and monzogranites [29,47,48,105,122–124], intrude these metaturbiditic and metavolcanic sequences (Figures 2 and 3a,b). The most representative are granodiorite and diorite intrusions that belong to the Ouarzazate Supergroup "precursor rocks" and Bardouz suite of Thomas et al. [5]. They are composed of plagioclase (An_{20-56}), biotite, amphibole (Mg-hornblende to Fe-tschermakite and pargasite to Fe-edenite), rare pyroxene and accessory minerals such as titanite, zircon, apatite and magnetite [47,119,123]. K-feldspar and quartz abundances vary depending on pluton compositions. At the contact with the country rocks, cordierite, andalusite and biotite related to contact metamorphism are observed [47,48,121,125].

Rock geochemistry shows that most plutons have moderate to high-K calc-alkaline affinities with trends that are consistent with formation in an active continental margin [29,119]. Available ages range from 677 ± 19 and 576 ± 4 Ma for this event, although most of the data are bracketed between 615 and 575 Ma in the Jbel Saghro [29,51,59,78,81,105–108]. With regard to the oldest intrusions, their age and emplacement conditions remain unclear and debated in the Jbel Saghro [48] and are only evidenced by poorly defined U-Pb zircon geochronology (i.e., >615 Ma) on calc-alkaline tonalite, diorite and granodiorite intrusions [107,108,126]. However, in the Central Anti-Atlas, similar plutons are widespread and were emplaced continuously between 660 and 615 Ma in a supra-subduction setting [29,78]. One can therefore question on a possible diachronism of magmatic activity between the Central Anti-Atlas and the Eastern Anti-Atlas during this period especially if we consider that some rare ages reported in the Eastern Anti-Atlas are inherited (677 ± 19, 675 ± 13, 645 ± 12 Ma, [108,126]). However, the following magmatic period spanning from 615 to 575 Ma is well expressed in both the Central and Eastern Anti-Atlas. This is mostly plutonic in the Eastern Anti-Atlas and both volcanic and plutonic in the Central Anti-Atlas.

3.1.3. The Main Pan-African D_1 Deformation Event

At map-scale (Figure 3a,b), the overall style of the deformation suggests that steep bedding and folds are the most conspicuous features of the metaturbiditic terranes [47,48,77,125–129]. Two types of folds are observed. (i) At the regional scale, first-order folds are easily discernible such as in the Imiter and Boumalne areas (Figure 3a,b). They are tight (Figure 3c) and the majority of the ENE–WSW axial surfaces are moderately to steeply inclined (50–60°), with a predominance of dips to the NNW suggesting that folds roughly verges toward the SE [125,127,128]. Their hinge lines gently plunge toward the ENE. (ii) At local scale, higher-order folds are metric and open to tight (Figure 4a). They are asymmetric and probably unrelated to lower-order folds as they always show a right-verging (dextral wrenching, [47]). Indeed, they are observed in shear zones where cleavage is well developed. These folds are interpreted as drag folds below. Their axial surfaces are upright (NE–SW to ENE–WSW) with a hinge line that steeply plunges toward the ENE or WSW [47]. In Figure 3c, most of the poles to bedding falling in the NNW and SSE quadrants are related to the regional scale, first-order folds. Gentle poles falling in the SW and NE quadrants are related to drag folds. However, these plots and more generally the ones comprised within the dotted red lines (Figure 3c) highlight areas where data are locally reoriented because of drag fault zones and plutonic interferences. Such interferences are abundant in the Qal'at MGouna area where folds may be also interpreted as open and upright (Figure 3c, [47]).

In detail, metaturbiditic rocks show finite strain markers including foliation and stretching and mineral lineation [47,48,77,81,128–130]. Foliation that usually consists in a slaty cleavage is evident throughout the entire inlier. It is noteworthy that both metamorphic planar fabrics and lineations have been only observed in the eastern part of the Jbel Saghro, in the Boumalne and Imiter areas. Indeed, these fabrics (Figures 3 and 4b) remain difficult to observe except in the hornfels zone surrounding the diorite and granodiorite intrusions described above (Figure 4b). Foliations show a constant ENE strike and dip toward the NW (N075°E 60°N, Figure 3d), except when reoriented in fault drag areas (i.e., NW–SE strikes) or when refracted (NE–SW strikes). Such foliation is axial planar and, thus, is generally parallel to the bedding of the metaturbiditic rocks when localised along the limbs of the first-order tight folds. It is called herein S_{0-1}. A S_2 cleavage refraction (Figure 4c) may also occur in specific areas, that define an obliquity with respect to the S_{0-1}. It is caused by localised shear developed in incompetent layers usually associated with the drag fold structures. Stretching and mineral lineations (L_1) are always carried by the S_{0-1} foliation. They are discrete and their orientation (Figure 3e) also appear fairly constant with a rather low dispersion from NW–SE to N–S trending directions (average: N170°E 55°N). According to mineralogic and micro-structural evidences, at least 2 distinct metamorphic assemblages are observed but appear to be related to the same tectonic event defined as a hornfels assemblage and a regional chlorite to amphibole assemblage. (i) The hornfels zone is observed in the contact metamorphic aureoles caused by the diorite and granodiorite intrusions within the Cryogenian greywackes [48,121,130]. It consists of a spotted phyllite zone in which foliation (phyllitic cleavage to gneissic foliation) and lineation are easily discernable. Rocks are mostly characterised by K-feldspar, muscovite, biotite and small spots of retrogressed andalusite and/or cordierite (Figure 4b). The latter frequently highlights the lineation and, therefore, suggests a probable relationship between pluton emplacement and the D_1 tectonics. Garnet-amphibole assemblage and spessartine occurrences are associated with skarns and skarnoids as reported by Benziane [130] and Tuduri [47]. (ii) When observable far from intrusions, foliation (slaty to phyllitic cleavage) is defined by a planar-linear fabric mainly formed by phyllosilicates such as chlorite, sericite to biotite [48]. In both assemblages, rolling structures and S-C fabrics (Figure 4d) define unambiguous south- to southeast-verging noncoaxial shear criteria, parallel to the average lineation trending direction. These structures indicate reverse sense top-to-the-south deformation in the Boumalne and Imiter areas [47,48,77,82,121,123].

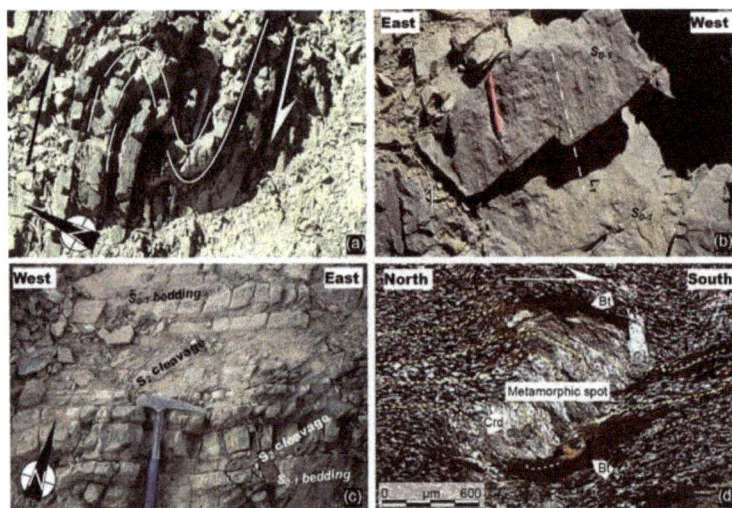

Figure 4. (**a**) Drag fold from the Qal'at Mgouna metagreywacke inlier. (**b**) Regional S_{0-1} metamorphic foliation and related lineation marked by elongate contact metamorphic minerals that affect the metagreywacke sequence. (**c**) S_2 cleavage refraction, from the Imiter metagreywacke inlier, consistent with a dextral shearing. (**d**) Microphotograph showing syn- to late kinematic metamorphic mineral, a probable cordierite (Crd) showing rolling structure consistent with a top-to-the-south shearing sense. Note the asymmetric tails composed of biotite (Bt), plane polar light.

The D_1 deformation is also well-expressed along the large-scale N070–090°E trending strike slip faults, about 100 to 150 km long that occur in the northern and central part of the Jbel Saghro (Figures 2 and 3a,b). On both sides of these shear zones, deformation appears more intense especially in the slaty to phyllitic cleavage domains where the S_2 refraction cleavage oriented N040–050°E 80°N is formed (Figure 4c) synchronously with drag-folds (Figure 4a). These structures are consistent with a dextral sense of shearing [47]. Where plutons are emplaced within shear zones, they develop both magmatic foliation and lineation (e.g., the Igoudrane pluton). There, the foliation is parallel to the fault (N075–090°E 80°N) and the lineation roughly horizontal. All structures linked with the D_1 events and the geometry of the associated intrusions indicate that this first Pan-African event developed in response to a NW–SE to WNW–ESE trending shortening.

3.2. The Upper Complex or the Inception of a Silicic Large Igneous Province

3.2.1. Generalities

Rocks of the Upper Complex cover nearly 80% of the surface of the Jbel Saghro (Figure 2). This complex, only affected by very low-grade metamorphism and weak deformation, consists of thick and regionally extensive felsic volcaniclastic sequences that are non-conformably above the more deformed and metamorphosed rocks of the Lower Complex [29,30,47,112,131–133]. Related dykes and plutons intrude both the rocks of the Lower and Upper Complexes. These rocks belong to the Ouarzazate Supergroup that includes the lower Mançour Group and the upper Imlas Group [5,107]. Plutonic suites are attributed to the Tanghourt Suite. These plutonic and volcanic rocks were emplaced between ca. 575 and 540 Ma [29,51,74,78,106–108,112,134,135]. Most of the volcanic rocks are ash-flow tuffs, felsic lavas, resedimented volcaniclastic deposits with some andesitic lavas and rare mafic intrusions [29,30,47,112,133] that cover an area of approximately 2000 km^2 and reach a maximum thickness of 1000–1500 m.

3.2.2. The Qal'at Mgouna Ash-Flow Caldera

In the vicinity of Qal'at Mgouna (Figures 2 and 5), the association of lava flows (rhyodacites, rhyolites and andesites), pyroclastic rocks (ash-flow tuffs and ash falls), re-deposited volcaniclastics and reworked volcanic rocks is consistent with an ash-flow caldera environment [47]. The pre-caldera formations mostly consist of metagreywacke basement rocks and diorite and granodiorite intrusion dated at 576 ± 5 Ma [29], all belonging to the Lower Complex. Related to the caldera formation stage, plutonic rocks are mostly represented by monzogranites and coeval porphyries (Figure 5). Volcaniclastic rocks are not deformed, but only tilted when localised inside the caldera (Figure 5). The structural limit of the proposed Qal'at Mgouna caldera can be traced over 5–6 km. Tuduri [47] interprets the near vertical arcuate lineament located at the margin of the caldera as a potential ring-fault that trends E–W to NW–SE (Figures 5 and 6a). Evidence is given because this boundary separates the near vertical Cryogenian metagreywacke bedding to the South from the NNW dipping (50° to 80°) monocline intra-caldera sequences, to the North (Figures 5 and 6b). Numerous rhyolite dykes, 10 to 20 meters wide are located along this lineament (Figures 5 and 6a,b). Dykes are vertical or near-vertical and intrude the Lower Complex rocks as well as the plutonic and volcaniclastic rocks of the Upper Complex. Because these rhyolitic dykes are emplaced along the proposed structural limit and generally present an arcuate shape, they are interpreted as ring dykes.

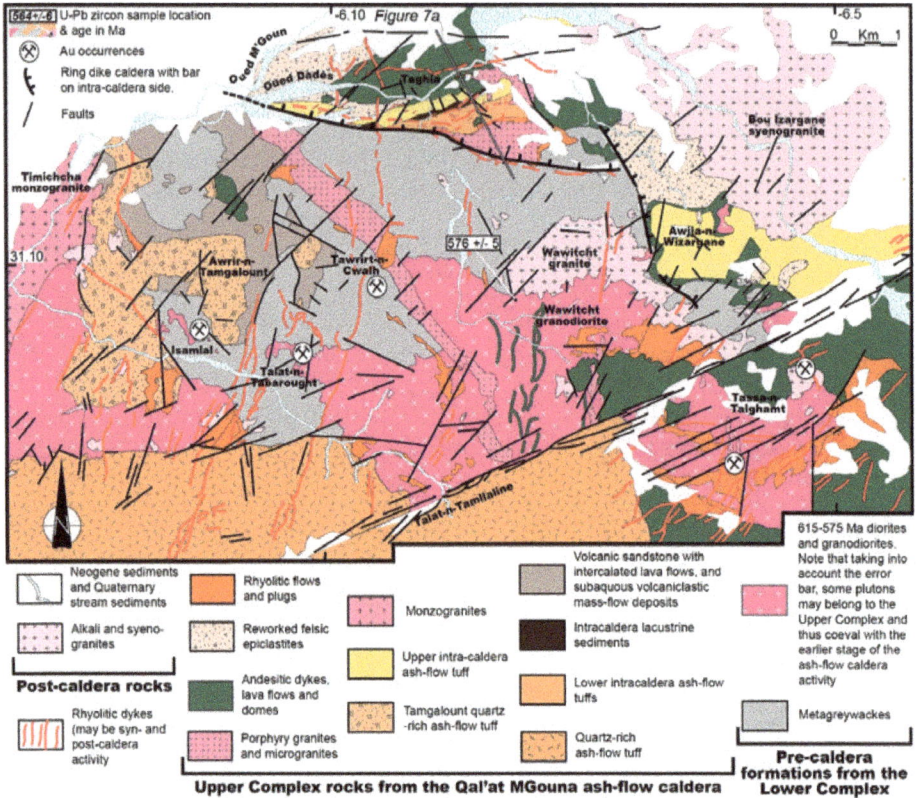

Figure 5. Detailed geologic map of the Qal'at MGouna district showing the extra- and intra-caldera rock units (from Benharref [132], Derré and Lécolle [113] and Tuduri [47]).

All the data presented above argue for the existence of a collapse caldera structure as shown on Figure 5 [136–139]. The peripheral structure of the ash-flow caldera where ring dykes occur is herein interpreted as having accommodated both subsidence (caldera collapse and intra-caldera sequences deposition) as well as subsequent uplift and tilting (magmatic resurgence and ring dykes injections). The geometry given by the NW–SE trending structural limit favours an elliptic shape rather than a sub-circular one for the caldera [41,140,141]. The consequence of such a shape will be discussed further. From bottom to top and according to Tuduri [47], the intra-caldera volcaniclastic sequence is detailed below (Figures 5–7):

(i). A lowermost pyroclastic layer consists of a 400–500 m thick, unwelded to slightly welded, moderately crystal-rich dacitic lapilli tuff (Figure 6c,d and Figure 7a,b) interpreted as an ash-flow deposit [142,143]. Internal stratification of the ash-flow tuff is crude, oriented N060–080°E 80°NW, as highlighted by discrete layers which are either pumice-richer, lithic-richer or entirely devitrified with spherulites. In a specific layer 40 m thick, greenish fibrous pumices displaying silicified tubular micro-vesicles and a silky/fibrous fabric can reach up to 5 cm long. Lithic clasts up to 20 cm in size are common throughout and consist mainly of basement greywackes, lavas and quartz-rich ignimbrite fragments. Phenocrysts are mostly broken plagioclase and K-feldspar, in various ratios with minor amounts of chloritised ferro-magnesian crystals (biotite and probable amphibole) and very scarce quartz.

Figure 6. (**a**) View of the structural limit of the caldera showing the contact between the intra-caldera sequence, rhyolite ring dykes and basement, Taghia area (see location on Figure 5). (**b**) Detailed view of the structural limit showing the ring dykes and the unconformity between the basement to the south and the intra-caldera sequences to the north. (**c**) Microphotograph of the lower intra-caldera ash-flow tuff. Slight compaction and welding of shards are characteristics of this tuff, Plane Polar Light. (**d**) View towards the east showing the relationships between the lower and upper intra-caldera tuffs and interbedded sedimentary ponded rock sequences Taghia area (see location on Figure 5). (**e**) Close-up view of the upper intra-caldera tuff characterised by compacted fiammes (flattened pumices) and lithic fragments of K-feldspar-rich granite. (**f**) In thin section, the upper ignimbrite shows strong compaction and welding of shards, Plane Polar Light. Bt: biotite, Kfs: K-feldspar, Pl: plagioclase, Qtz: quartz.

(ii). Above the ash-flow unit lies a ca. 200 m thick volcano-sedimentary (epiclastic) unit with very thin bedding (Figures 6d and 7a,b). The lower part (100 m thick) is made of layered tuffaceous breccias containing ignimbrites fragments. The upper part consists of laminated reddish and greenish mudstones and sandstone oriented N070° E 70° NW. When preserved from important silicification, the identifiable components are microscopic broken crystals and lithic fragments. Beds are broadly continuous laterally, being only sometimes disrupted by syn-sedimentary normal faults and slump-like structures. Faults are roughly oriented NW–SE. Fluid escape textures are common and allow assessment of the polarity of the intra-caldera sequence. All sedimentological features argue for a subaqueous emplacement, at least for the upper part of the epiclastic unit. In our model, and as no marine sediments have been hitherto recognised in the entire Jbel Saghro in the Ediacaran formations, such subaqueous environment may be reasonably related to a caldera lake.

(iii). Above the volcano-sedimentary unit lies a ca. 200–300 m thick crystal-rich rhyolitic ash-and-lapilli tuff (Figures 6d and 7a–c). Plastic deformation due to significant compaction is evidenced by reddish flattened pumices (Figure 6e). Glass shards and broken phenocrysts are visible under the microscope (Figure 6f). Phenocrysts are quartz, plagioclase, K-feldspar, and scarce amounts of chloritised Fe–Mg minerals (biotite and amphibole). Up to 2 m-sized lithic clasts are abundant, especially in the basal part, and consist of metagreywacke, ash-flow tuff fragments, jasperoids and monzogranite (Figure 6e). This voluminous quartz- and pumice-rich unit is readily interpreted here as a welded ignimbrite.

(iv). The top of the sequence is dominated by massive andesite lava flows and epiclastic polylithologic breccias (Figures 5, 6d and 7a,b). The bottom of this whole sequence is intruded by a monzogranite porphyry (Figures 5 and 7a,b). Plugs of similar porphyry facies also locally intrude the upper parts of the sequence. Such porphyry is here interpreted as a resurgent pluton [137,144] that tilted the intra-caldera sequence upon emplacement [47]. Because of the tilting, the thickness of the intra-caldera sequence is exposed over 1500–2000 m, of which 800–1000 m consists of ash-flow tuffs and epiclastic rocks. To the north, the intra-caldera sequence disappears beneath the young sedimentary rocks of the Dadès valley (Figure 5).

Volcanic and pyroclastic rocks also occur outside the caldera structure to the south and west. They form a broadly stratified pile up to 500 m thick with moderate dips toward the W–NW (Figure 5). In the vicinity of the Awrir-n-Tamgalount (Figure 5), the extra-caldera sequences are made up of two units overlain by the Tamgalount tuff (Figure 7d,e). The lower unit is dominated by hundreds of meters of well bedded, normally graded crystal-rich sandstones and siltstones (Figure 7e). Under the microscope some of the silt-sized layers are formed of formerly vitric, now devitrified, material and might be primary ash fall deposit. The stratified lower unit dips towards the west at variable angles (Figure 5), perhaps due to palaeo-topography effects and(or) syn-tectonic deposition. The upper unit is dominated by rhyo-dacitic lavas that display distinctive spherulitic devitrification microtexture and pilloïd texture in the field that suggests emplacement under water. The Tamgalount tuff (Figures 5 and 7d,e) is a porphyritic porphyritic rhyo-dacitic ash-flow tuff (ca. 25% phenocryst) with eutaxitic texture. Broken phenocrysts consist of quartz, plagioclase, K-feldspar and chloritised Fe-Mg minerals. Lithic fragments are abundant and composed of greywackes, ash-flow tuff fragments and jasperoids. all extra-caldera units lie unconformably on the Lower Complex and are pervasively affected to various degrees by hydrothermal alteration and/or silicification. The importance of this unconformity will be discussed later.

Numerous intrusions are related to the Upper Complex (Figures 2 and 5). They have been mapped according to their mineralogy, texture and geochemical features. Two types are distinguished: (i) gabbros and biotite- and amphibole-rich, pink-coloured coarse-grained granites (monzo- to syenogranites), which are calc-alkaline to highly potassic, abd coeval porphyries emplaced at shallower levels (e.g., the intra-caldera resurgent granite and its apophyses); and (ii) Si-rich alkali (K-rich, i.e., shoshonitic in composition) granites and related aplitic bodies (sills, dykes), which frequently appear as late magmatic events in the Upper Complex history. The pink mozogranites contain quartz, albite-oligoclase, Fe-edenite, annite, K-feldspar and accessory minerals such as thorite, zircon, allanite, apatite, magnetite, sulphides and W and Mo-rich minerals. By analogy with the Isk-n-Alla monzogranite in the central part of the Jbel Saghro (Figure 2), they may have been emplaced around 555 Ma [29,106]. The alkali granites are mainly composed of quartz, albite and K-feldspar displaying granophyric intergrowths. Accessory minerals include tourmaline (fluor-schorl) and metamict zircon [47]. They are associated with late N–S rhyolitic dykes that also display shoshonitic compositions (Figure 5). We assume they may have emplaced later, between 550 and 530–520 Ma [74,107,108].

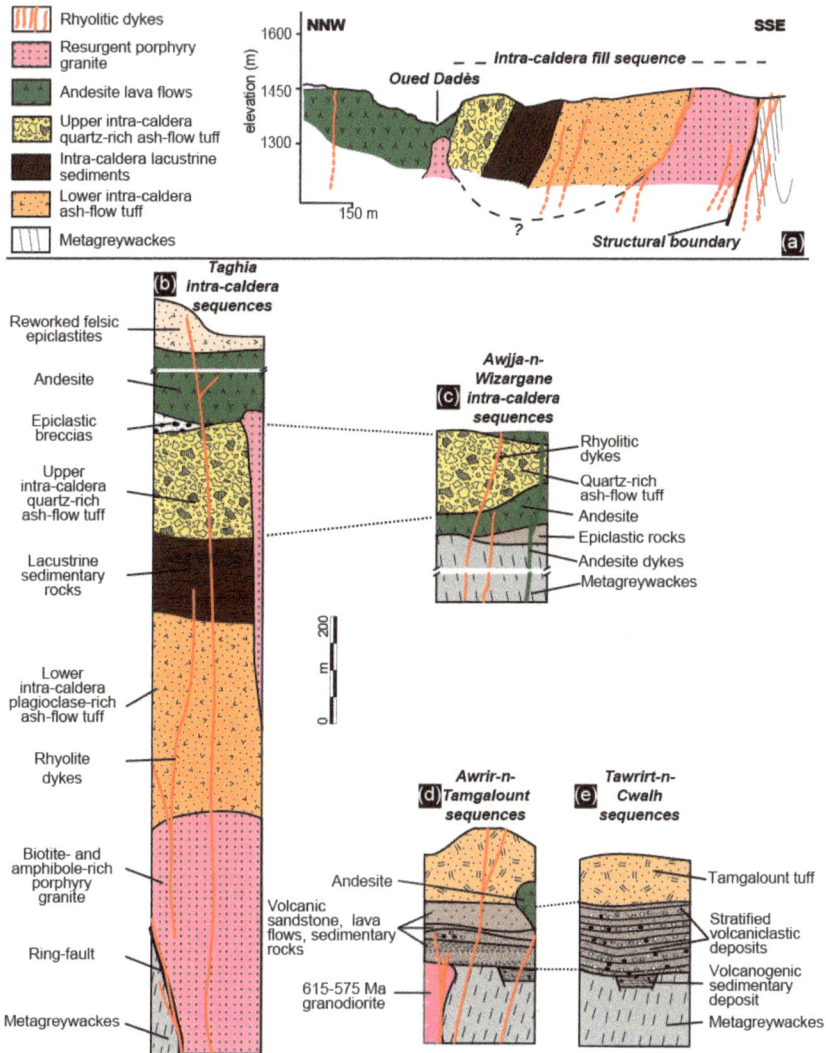

Figure 7. Simplified cross section and graphic logs of the Qal'at Mgouna volcanic sequences according to Tuduri [47]. (**a**) Schematic N–S cross section across the Qal'at MGouna ash-flow caldera. Summary stratigraphic sections for the (**b**) Taghia and (**c**) Awjja-n-Wizargane intra-caldera sequences, and extra-caldera sequences from (**d**) Awrir-n-Tamgalount and (**e**) Tawrirt-n-Cwalh (See location on Figure 5).

According to the relative and absolute chronology of both volcanic and plutonic rocks in both the Jbel Saghro and in the Qal'at MGouna area, three main ignimbrites flare-ups are herein evidenced. The earliest flare-up corresponds to the lower intra-caldera ash-flow tuff emplacement. It is mostly dacitic and may coeval with granodiorite plutons emplaced around 575 Ma [29,74]. Ignimbrites from the Oued Dar'a caldera described by Walsh et al. [29] are herein interpreted as belonging to this earliest flare-up. Then, the emplacement of the Tamgalount ash-flow tuff with rhyo-dacitic affinities, may correspond to a second high-volume magmatic event emplaced around 565 Ma [78,107]. Such an event would have produced similar tuffs that are coeval with the huge rhyo-dacitic dyke swarm [29] observed in the western and southern parts of the Jbel Saghro (Figure 2). The later ignimbrite flare-up corresponds to the upper intra-caldera rhyolitic ash-flow tuff and to the numerous rhyolitic lava flows and domes reported in the literature [29,74,106], coeval with pink monzogranite plutons around 555 Ma.

3.2.3. The D_2 Deformation Event

D_2 is interpreted as a strike-slip faulting event [29,47] that is also controlled by a WNW–ESE direction of shortening [47]. Tectonic features are faint, belong to the brittle regime, and affect both the Lower and Upper Complex units. Indeed, large-scale structures (except faults) and intense folding as observed in the Lower Complex are absent. It may be noted that some authors [29,78] described gentle folds that only affect the lower ignimbritic sequences (i.e., the 575 and 565 Ma ones) mainly in the western part of the Jbel Saghro and the Central and Western Anti-Atlas. Field works document: (i) ca. NNW–SSE normal faulting emplaced perpendicular to the extensional direction (i.e., NNE–SSW); and (ii) N070°E strike slip fault systems and associated veining. In the Qal'at Mgouna area, D_2-related normal faults control the emplacement and development of extra-caldera basins filled by the volcaniclastic rocks (Figures 5 and 7d,e). In contrast to the D_1-related ones, structures related to the D_2 event are characterised by dominant opening and extensional features [47]. Some specific locations (see below) demonstrate the re-activation of previously formed faults (i.e., the regional strike-slip faults trending N070°E) and fractures under the state of stress link with D_2. Others structures were formed, mainly in the Upper Complex. As fractures are favourable sites for fluid circulations and fluid trapping, intense veining occur within domains were intense fracturing is coeval of significant magmatic activity and related hydrothermal events. All these factors lead to the establishment of a general D_2-related extensional/transtensional setting that is particularly favourable for the emplacement of fluid-filled structures and, consequently, to the formation of the numerous ore deposits concerned by this study.

In the Qal'at Mgouna area, the structural map of the volcanic complex points to a control by a combination between E–W to NW–SE faults and NNE–SSW normal faults (Figure 5). Both structural direction controlled block-faulting, collapse and caldera formation. Because E–W to NW–SE faults are more developed and have a longer extent, they are assumed to represent the structural limit of the Qal'at Mgouna caldera (Figure 5). The orientation of these faults is consistent with a caldera formed under the control of a transtensional regime with a WNW–ESE shortening direction and a NNE–SSW extensional direction. Similar tectonic features have been described in the Oued Dar'a caldera which is localised 60 km to the WSW [29]. Few examples of ash-flow calderas developed in strike-slip and(or) extensional tectonic regimes have been documented [145–150].

4. Characteristics of Ore Deposits

Au–Ag showings and/or mines occur in several areas of the studied area (Figure 2). Four zones were selected. These showings are hosted in rocks of the Lower Complex (i.e., Thaghassa), the Upper Complex (i.e., Zone des Dykes area), or both (i.e., Imiter and Qal'at Mgouna) depending upon the depth of their formation. Mineralisation formed during the D_2 tectonic event within the four localities herein described in detail.

4.1. The Thaghassa Intrusion-Related Gold Deposit

The Thaghassa intrusion-related gold deposit (IRGD) is an exploration project with drill core data showing the presence of several intersects of 1–2 m at 5 g/t Au, and up to 400 g/t Ag. It is hosted in hornfelsed metagreywacke rocks (Figure 8a) that are adjacent to a large granodioritic pluton [47,48,151], while the metagreywackes belong to the Lower Complex, the intrusion may correspond to the earlier development of the Upper Complex. Two main tectono-magmatic stages control the formation of the deposit. (i) The first stage corresponds to the top-to-the-south asymmetry and the syn-kinematic Ikniwn pluton emplacement controlled by a transpressional strain regime. Zircon U-Pb dating yields a Concordia age of 564 ± 6 Ma for the intrusion [48]. (ii) The second stage (Figure 8b) is characterised, from older to younger and further away from the intrusion, by: metatexite with leucocratic stromatic bands, aplo-pegmatite sills (Figure 8c), intermediate veinlets composed of quartz, K-feldspar and muscovite (Figure 8d), and then gold-bearing striped foliation-veins (Figure 8e). All these are assumed to have been emplaced under large-scale ENE–WSW dextral shearing that results from an ESE–WNW shortening during transtensive tectonics (Figure 8b). Tuduri et al. [48] suggested that the progressive and continuous shearing was initiated at the aplo-pegmatite stage and achieved during the hydrothermal phase (Figure 8b–f). The existence of intermediate veins characterised by quartz-rich core and apatite-muscovite-feldspar-rich rims demonstrates a progressive evolution from a magmatic to a hydrothermal stage and the persistence of the magmatic character, at least until the onset of the hydrothermal process. The main Au-mineralization was concentrated at the end of such a magmatic-hydrothermal evolution. The ore paragenesis is characterised by arsenian pyrite with refractory gold (<5 μm) arsenopyrite, sphalerite and scarce grains of chalcopyrite, loellingite, pyrrhotite, tetraedrite, freibergite, argentite and cassiterite. Galena is abundant but always in the form of microscopic inclusions within pyrite. Fluid inclusion characterisation, based on the concept of fluid inclusion assemblages (FIA, [152]) as in all the cited references, combined with mineral geothermometry [48] suggests that the system evolved from hot fluids (~550 °C) dominated by N_2 and CH_4 to intermediate temperature (~300–450 °C) and low salinity aquo-carbonic fluids in the system (H_2O-NaCl-CO_2) + CH_4. Salinities are low to intermediate, being lower than 11.5 wt. % NaCl equiv. Gold precipitation is related to intermediate temperature mineralising fluids that have strongly interacted with the hornfelsed country rocks. According to Tuduri et al. [48], such a metallogenetic system is assumed to have developed due to migmatisation and partial melting of metagreywackes country rocks in response to heat transfer from the underlying Ikniwn intrusion. Fluid and metal sources may originate from magmatic processes (i.e., magmatic exsolution of incompatible elements from newly formed peraluminous melts and perhaps from the Ikniwn intrusion) and from the devolatilisation of the metamorphic host rocks. (iii) A third, later tectono-magmtic stage in the area developed a large volcanic dyke swarm and brittle faulting and is assumed to belong to the Upper Complex.

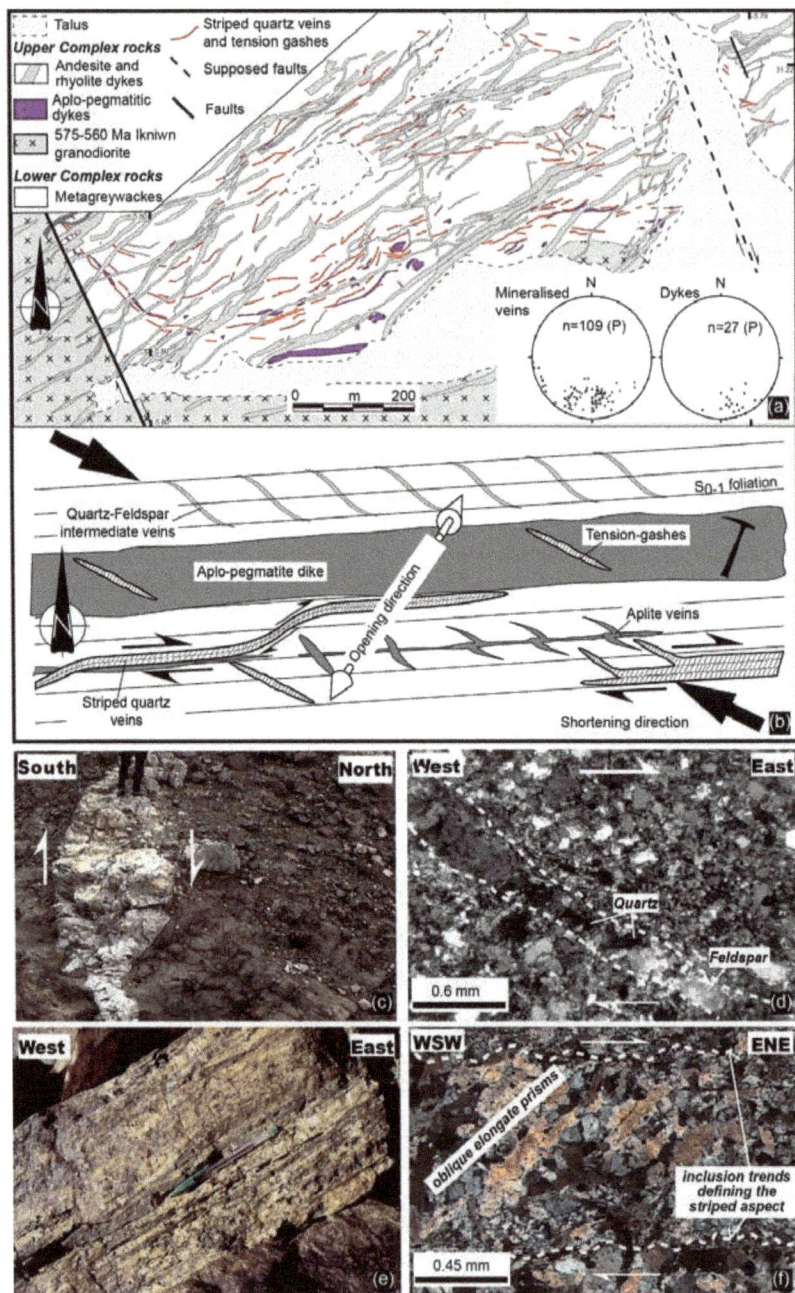

Figure 8. Main features of the Taghassa intrusion-related gold deposit (IRG) [47,48]. (**a**) General map and stereoplots of structural orientation data of the Au–Ag Thaghassa intrusion-related gold deposit. These reveal the high density of veins developed north of the 575–560 Ma Ikniwn granodiorite. (**b**) Interpretative sketch illustrating the magmatic-hydrothermal model that involved a progressive and continuous tectonic event including the aplo-pegmatitic dykes and sills emplacement, then the intermediate veins and the hydrothermal and gold-rich striped quartz veins. (**c**) Pegmatite dyke

showing a dextral pull-apart geometry. (**d**) N120°E trending intermediate veins filled by quartz, muscovite and feldspar assemblage, cross polar light. (**e**) Macrostructure illustrating the gold-bearing quartz vein stage. The layering texture defined the striped aspect of veins. (**f**) Microtextural characteristics of gold-bearing quartz veins. The internal texture shows elongate quartz grains with obliquity with respect to vein walls suggesting a dextral shearing, cross polar light.

4.2. The Qal'at Mgouna Au–Ag (Cu, Mo, Bi, Te) District

The Qal'at Mgouna district is composed of three main exploration projects: the Isamlal, Talat-n-Tabarought and Tawrirt-n-Cwalh districts, all of which located outside of the caldera structure (Figures 5 and 9). Based on mineralogical, chemical, textural and structural constraints, two distinct ore deposit types have been identified: an older porphyry ore deposit on which a younger epithermal system is superimposed [47,113,153,154].

4.2.1. The Isamlal Porphyry Au(–Cu–Mo) Deposit

The porphyry Au(–Cu–Mo) deposit type is related to high-temperature hydrothermal system observed in the Isamlal and Talat-n-Tabarought areas (Figure 9a–e). The Isamlal project appears as being the most promising [47,154–157]. Ore emplacement is assumed to be synchronous to slightly late with respect to the emplacement of diorite and granodiorite stocks [158]. Most of these intrusive stocks, that display porphyritic textures in drill cores (Figure 9e), present a preferential NW–SE orientation in map (Figure 9a,b). Their age of emplacement is still unknown. However, a large pluton, the Wawitcht granodiorite, displaying similar mineralogy and located ca. 4 km east of the Isamlal deposit, has been dated at 576 ± 5 Ma (U-Pb radiometric ages on zircon, [29]). Alterations are mostly observed within the Lower Complex and affect both the metagreywackes and diorite-granodiorite intrusions. The deposit is characterised by a Au mineralisation hosted by the metagreywackes that is also elongated along a ca. NW–SE trending direction (Figure 9b). Cu and Mo occurrences are additionally associated. In paragenetic order, the mineralisation includes magnetite, K-feldspar and late forming and less pervasive quartz veins that form a well developed stockwork. The stockwork occurs along a ca. N120°E preferred orientation, highlighting a strong structural control (Figure 9c). It is localised in the vicinity of the granodiorite intrusion or related apophyses but seems more developed within the metagreywackes at the hanging-wall (Figure 9d). Veins range in size from 0.5 to 30 cm in width and are ≤10 m in length except for the largest veins which are more than 30 m long. Such an important structural control on the stockwork may suggest that the main opening direction (i.e., NE–SW) is controlled by an ESE–WNW shortening direction. In drill cores, the central part of the mineralised zone is characterised by potassic-altered rocks. K-feldspar mostly occurs in highly reactive igneous rocks (Figure 9e). Biotite may coexist with K-feldspar but also occurs around the central zone whereas the propylitic alteration is more distal [158] In the Isamlal area, the quartz-rich stockwork is characterised by K-feldspar, magnetite, F-Cl-rich amphibole, Cl-F-rich biotite with scarce F-rich tourmaline, brannerite and rutile in the central part of the deposit, and muscovite with scarce iron oxides more externally. Note that such a stockwork has never been reported in rocks belonging to the Upper Compex [47]. Sulfides mostly occur in the central zone and mainly consist of pyrite, chalcopyrite, with scarce molybdenite, pyrrhotite, electrum, galena and tetradymite [47,155,157,158]. Fluid inclusions from the quartz-rich stockwork veinlets were used to constrain the palaeohydrothermal conditions [154,155,159]. Primary multiphase fluid inclusions are composed of liquid, vapour and halite cubes, as well as other salts such as sylvite, and $CaCl_2$. In addition, uncommon mineral inclusions that may be abundant have been identified, such as calcite, brookite–titanite, haematite, magnetite, and a solid phase with a very high refringence identified as andradite.

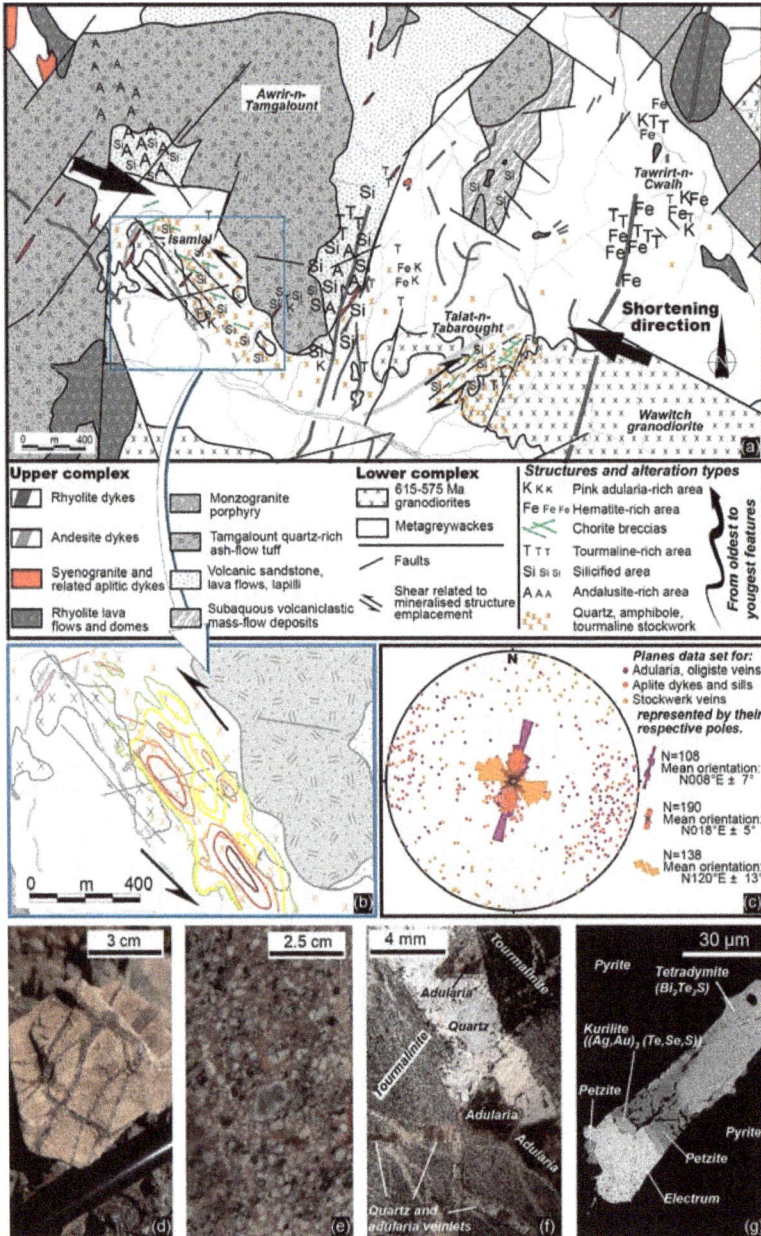

Figure 9. Main features of the Qal'at Mgouna deposit types [47,158]. (**a**) Detailed map of the Au(–Cu–Mo–Ag–Te–Bi) Qal'at Mgouna district (from Tuduri [47]). Note that the kinematics shown by shear zones are consistent with a WNW–ESE direction of shortening. (**b**) Kriging interpolation revealed that gold anomalies are correlated with both the quartz stockwork and the NW–SE faulted corridor in the Isamal porphyry deposit. The red colour is indicative of the highest Au grades [158]. (**c**) Stereoplots of structural orientations data for: the quartz stockwork related to the porphyry stage (mean orientation N120°E); the aplitic dykes and sills related to the alkali–syeno–granite stocks (mean

orientation N018°E); and the adularia-specularite-quartz veins from the epithermal stage (mean orientation N008°E). (**d**) Quartz stockwork from the Isamlal porphyry Au(–Cu–Mo) system. (**e**) Typical potassic alteration (pink coloured zones) of a porphyritic granodiorite (the Isamlal porphyry Au(–Cu–Mo) system). (**f**) Typical alteration and vein development of the epithermal stage: pervasive tourmalinites are cut by quartz and adularia-rich veins in the vicinity of the Tawrirt-n-Cwalh deposit, plane polar light. (**g**) Economic paragenesis of the epithermal stage characterised by Au-Ag tellurides, electrum and Bi-telluride veinlets within pyrite; Tawrirt-n-Cwalh deposit, SEM back scattered picture.

Chalcopyrite and gold were also observed in multiphase inclusions. Multiphase inclusions have a high though variable salinity (30 to 45 wt. % NaCl equiv.) and are characterised through homogenization by halite-disappearance. The large range of homogenization temperatures (160–460 °C) combined with a zoned potassic to propylitic alteration, stockwork structures and with an Au(–Cu–Mo) paragenesis is interpreted as characteristic of Au(–Cu–Mo) porphyry environments. This porphyry Au(–Cu–Mo) system is herein described as a vein-dominated deposit (stockwork) that is consistent with the emplacement of a porphyry stock, then exsolution and cooling of a magmatic-derived hydrothermal fluid. The overall system appears as mostly controlled by a ca. WNW–ESE trending direction. Indeed, this pattern suggests that the stockwork structure both reflects the magmatic stress associated with the porphyry emplacement and fluid exsolution, and also, a ca. NNE–SSW-oriented minimum principal stress (i.e., extensional direction) associated with a regional deformation that may be consistent with a WNW–ESE shortening direction although no clear tectonic regime has been proposed for the hydrothermal stage.

4.2.2. The Qal'at Mgouna Au–Ag(–Bi–Te) Epithermal System

In contrast to the porphyry deposit type, this low sulfidation epithermal deposit type appears more atypical (Figure 9a,c,f,g). From west (Timicha) to east (Isamlal then Tawrirt-n-Cwalh), the Qal'at Mgouna area (Figure 5) displays a progressive and continuous tectono-magmatic activity initiated as the plutonic stage (mostly observed to the west) and ending with a volcanic and hydrothermal stage observed from west to east [47]. The magmatic stage produced small pink-coloured Si-rich alkali granites, and sill and dyke intrusions with a typical fine-grained aplitic texture. Rhyolitic K-feldspar-phyric dykes are also assumed to belong to this stage. Such intrusive bodies may intrude rock units from both the Lower and Upper Complexes. The transition to hydrothermal stage is characterised by fluid exsolutions from the alkali granites and the formation of (i) quartz, Fe- and F-rich tourmaline with scarce F-rich muscovite miarolitic cavities; (ii) quartz, K-feldspar with scarce tourmaline stockscheider; (iii) tourmaline-rich quartz veins and NW–SE chlorite-rich transtensive cataclasites; and (iv) quartz, adularia, specularite veins with magnetite, fluorite, sulphides and gold (Figure 9f,g; [47,158–160]). The magmatic-hydrothermal processes strongly affected rocks especially the highly reactive volcaniclastic ones from the Upper Complex. These alterations correspond to a strong pervasive silicification. In addition to quartz, andalusite, diaspore pyrophyllite, Mg- F-rich tourmaline, F-rich phlogopite, F-rich muscovite, Cl-F-rich apatite occur along with rutile, haematite, monazite, xenotime, thorite and uranothorite as well as pyrite with inclusions of galena, coloradoite, hessite and altaite [47,113,153]. In some locations close to the Isamlal porphyry, the association of andalusite, pyrophyllite and diaspore with phlogopite and muscovite may be associated with the late epithermal or earlier porphyry stages [47,113,153]. Tourmalinisation is noteworthy and well developed along pervasive axes. Chlorite appears later in the paragenesis and mostly occurs in fault breccias. Ore concentration occurs during the hydrothermal stage with the quartz-adularia veins by crystallisation of As- and Co-rich pyrite, minor chalcopyrite and precious metal (Au–Ag telluride, electrum, Ag-telluride and Bi-telluride, Figure 9f) in the core of previously formed quartz–adularia–chlorite veins. Except for the chloritic breccias, which are strongly oriented along the NW–SE trending direction, all sills, dykes and veins related to the hydrothermal stage are roughly N–S with a maximum of 20° of dispersion toward the NNW and NNE (Figure 9c). The structural control of this event remains poorly evidenced as both extensional and transtensional features have been reported [47]. Paragenetic and microthermometric studies show the mineralising system is characterised by decreasing temperature of formation [154,160]. Indeed quartz from the miarolitic and stockscheider stages are characterised by high temperature of formation (400–600 °C), multiphase highly saline fluid inclusions (22–32 wt. % NaCl + CaCl$_2$ equiv.), while the intermediate stages related to massive tourmalinisation show temperatures of homogenization of 200–250 °C. The system then evolves toward lower salinity fluids probably belonging to the H$_2$O–NaCl–CaCl$_2$ (11–27 wt. % NaCl equiv.) system, with the absence of multiphase inclusions, and temperature around 180 ± 20 °C for the quartz-adularia veins. This hydrothermal stage corresponds to the formation of Au-Ag(–Te–Bi) epithermal showings in this part of the Jbel Saghro [47]. The issue whether this system belongs to an alkaline, low sulfidation or intermediate epithermal deposit remains open. The age of formation is unclear, although it may be coeval with the slightly alkaline N–S rhyolitic dykes dated between 550 and 530–520 Ma in the eastern Saghro [74,107,108].

4.3. The Zone des Dykes Intermediate Sulfidation Epithermal Au-Base Metal Deposit

The Zone des Dykes, also known as the Issarfane area, is located in the western part of the Jbel Saghro inlier (Figure 2) in the vicinity of a huge N–S rhyolitic dyke swarm emplaced around 565 Ma [29]. The Zone des Dykes ore district consists of quartz veins systems hosted by two ash-flow tuff units belonging to the Upper Complex (Figure 10). Therein, three mineralised systems, called the F1, F5 and Bou Issarfane structures, respectively (Figure10a), are identified [47]. The F1 structure consists of a 2 km long and 2 m width vein system that trends N180–160°E and dips about 50–60° to the east. The vein system shows several step-over zones showing a left lateral pull-apart geometry (Figure 10b) with a faint vertical component. The F5 structure also consists of a vein system, 1 km long and 2 m width, that is roughly oriented N080°E 60–70°S and is characterised by right-lateral shear structures (Figure 10c). The F1 and F5 structures are both interpreted as developed as conjugate pairs (Figure 10d). Because cross-cutting relationships are observed, we interpret the N080°E direction trend (i.e., F5) as the dominant direction. The F1 shear structures is herein interpreted as emplaced along a pre-existing NNW–SSE fracture analogous to the ones occuring between the Bouskour and Issarfane areas (Figure 2). The F5 structure is also emplaced along an important pre-existing fracture set that corresponds to a main ENE–WSW regional fault. This may explain why the F1 and F5 shear fractures are almost orthogonal yet conjugated, but also why the F1 pull-apart structures are always brecciated. The Bou Issarfane structure has a ca. N–S orientation like the F1 system, but unlikely lies at dip angles of 20–40° to the east. It consists of a 1.5 km long and 5 to 10 m thick silicified breccia system hosted by ignimbrites and rhyolitic tuff that is affected by E–W brittle faults (Figure 10a). While an unsilicified rhyodacitic lava flow occurs at the hanging wall, the footwall is made up of a 5–10 m thick anastomosed quartz stockwork. All veins are mostly filled in by quartz (Figure 10e,g–i) with scarce amount of adularia, sericite, chlorite, calcite and rare fluorite [47]. Sulphides are also common and mainly consist of arsenian pyrite (Figure 10f). Chalcopyrite, sphalerite, Pb–Cu–Bi assemblages (aikinite group) and electrum are accessory minerals and appear as inclusions within the As-rich pyrites (Figure 10f). Chlorite has a pycnochlorite composition and a Fe/Fe + Mg ratio close to 0.48 which is consistent with temperature of crystallisation bracketed between 210 and 280 °C [47]. All veins are characterised by internal textures typical of epithermal deposits according to Dong et al. [161]. The most representative quartz textures are those showing a partial replacement of a silica gel precursor characterised by colloform and moss texture (Figure 10g). Such textures are specific of siliceous sinters in active geothermal systems [161]. Ghost-bladed calcite textures [162] are also observed (Figure 10h,i). The occurrence of platy calcite (ghost-bladed calcite) demonstrates that boiling processes were active during vein formation [163–165]. Veins also show complex texture reflecting several stages of crystallisation, replacement and re-crystallisation occur. Within this complex process of vein formation, Tuduri [47] suggests that sulphides crystallised after the second stage of quartz replacement while electrum is mostly located within fissures of the pyrite. While the structural control of the F1 and F5 structures is clearly evidenced, the emplacement of the Bou Issarfane breccia remains unclear and needs to be discussed. Fragments from the Bou Issarfane breccia are silicified, sub-angular and of dimensions lower than 1 cm. Their probable volcanic origin suggest minor distance of transport. The matrix is highly silicified and seems composed by detritus of rock fragments (host rocks), and (or) by comminuted gangue minerals.

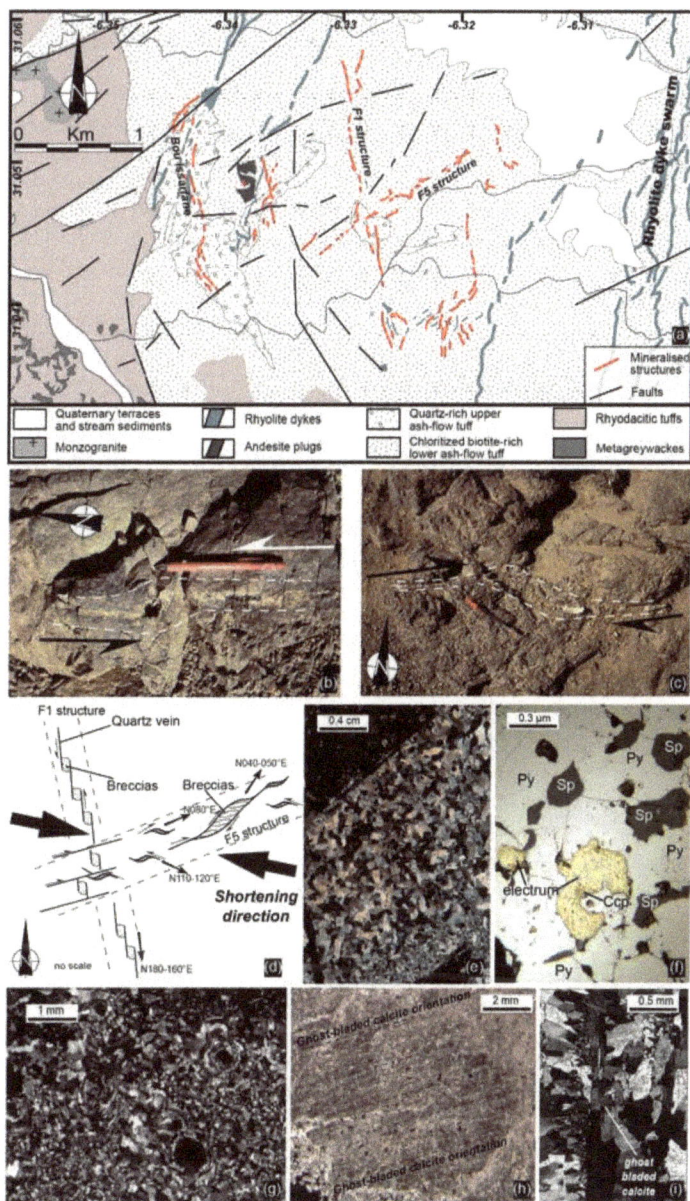

Figure 10. Main features of the Zone des Dykes deposits [47]. (**a**) General map of the Zone des Dykes Au-Ag deposit at the crossing between the F1 and F5 vein systems and Bou Issarfane area (after Tuduri [47]). (**b**) Pull-apart texture indicative of a left-lateral shearing movement along the F1 structure.

The filling is composed by quartz. (**c**) Pull-apart geometry of the F5 structure mainly filled by quartz and formed by dextral kinematics. (**d**) Global kinematic interpretation for the F1 and F5 structures integrating all the structural features observed in the field. (**e**) Microphotograph showing internal microtexture of the F1 structure characterized by a saccharoidal layout of quartz grains, cross polar light. (**f**) Typical gold-rich paragenesis of the F5 structure. Py: pyrite, Sp: sphalerite, Ccp: chalcopyrite. (**g**) Microphotograph of silica spheroid aggregates displaying moss texture, cross polar light. (**h**) Microphotograph of parallel ghost bladed calcites replaced by quartz, cross planar light and (**i**) plane polarized light microscopy.

The origin of this breccia may be therefore compared with phreatic or phreatomagmatic breccia pipes, although we cannot exclude a tectonic origin corresponding to a silicified cataclasite. According to salinity and homogenization temperatures, two type of fluids are assumed to be at the origin of the mineralised system [166]. The Bou Issarfane stockwork is characterised by primary multiphase fluid inclusions composed of liquid, vapour and halite cubes. Values obtained using the FIA concept indicate homogenization temperatures between 210 and 230 °C and moderate salinities (14–17 wt. % eq. NaCl). Secondary inclusions have lower homogenization temperatures between 130 and 180 °C. The F1–F5 veins consist of primary multiphase inclusions composed of liquid, vapour and halite cube with $CaCl_2$ assemblages.

Homogenization temperatures are bracketed between 160 and 180°C. Salinities are variable from 6% to 29% (wt. % NaCl + $CaCl_2$ equiv.) and may reflect the effects of boiling processes. Such textural, mineralogical and fluid characteristics suggest this hydrothermal Au–Ag(–Cu–Zn–Pb–Bi) system is comparable with intermediate sulfidation epithermal deposits as the ones reported in the Sierra Madre Occidental in Mexico [25,26,167]. The age of this ore deposit is still unknown. Considering the N–S trending direction of the F1 and Bou Issarfane structures, and the N–S trending direction of the rhyolitic dyke swarm, the mineralisation may have been formed coevally with this volcanic pulse around 565 Ma, i.e., after the emplacement of the ash-flow tuffs, that host the mineralisation dated, at 574 ± 7 and 571 ± 5, respectively [29]. However, we cannot exclude that the mineralisation is related to a later magmatic-hydrothermal period (e.g., the 550 or 530–520 Ma event) whilst no study has so far evidenced such later activity in the Bou Isserfane area.

4.4. The Giant Ag–Hg Imiter Deposit

The world-class Ag–Hg mining distirct of Imiter (Figures 2 and 11a) with 8.5 Mt of ore at a concentration of 700 g/t Ag consists of mineralised quartz–carbonate veins hosted by metagreywackes and more seldomly by the lower volcaniclastic units of the Upper Complex. The ore is located along a major E–W faulted corridor (Figure 11a) and results from a two-stage model of formation [47,102]. The main economic stage 1 developed within a N070–090°E 75–90°N trending dextral vein system filled by grey quartz, adularia and minor pink dolomite (Figure 11b,c, [102]). Satellite veins are common (e.g., F0 South, F0 North, R7 and R6, Figure 11b) and were formed synchronously with the main ones (e.g., F0 and B3, Figure 11b) within a global model of formation that involves double restraining bends along a strike-slip system. Indeed, such a push-up geometry results from irregular trends and stepovers developed along the N070–090°E dextral regional-scale faults (Figures 2 and 11a,b). The system also includes steeply-dipping, listric reverse faults and veins that flatten with depth (N065°E 50°SE, Figure 11b,d,e) and therefore that become flat-lying faults/veins in deeper parts of mine (N065°E 20–30°SE, Figure 11b). Reverse motions are clearly observed along all these structures (Figure 11d,e) and are assumed to have been formed during transpression. They serve as receptacles for the late emplacement of the high-grade silver deposit defined by Ag–Hg and Ag-rich minerals (Figure 11f, [102]). The main economic paragenesis is characterised by massive deposition of Ag–Hg amalgam (luanheite, eugenite), polybasite–pearceite, acanthite, stephanite, pyrargyrite-proustite and imiterite [47,168–170]. Sphalerite and galena are abundant while no silver was detected. Arsenopyrite is well represented but always as small euhedral crystals of a few hundred micrometers [47].

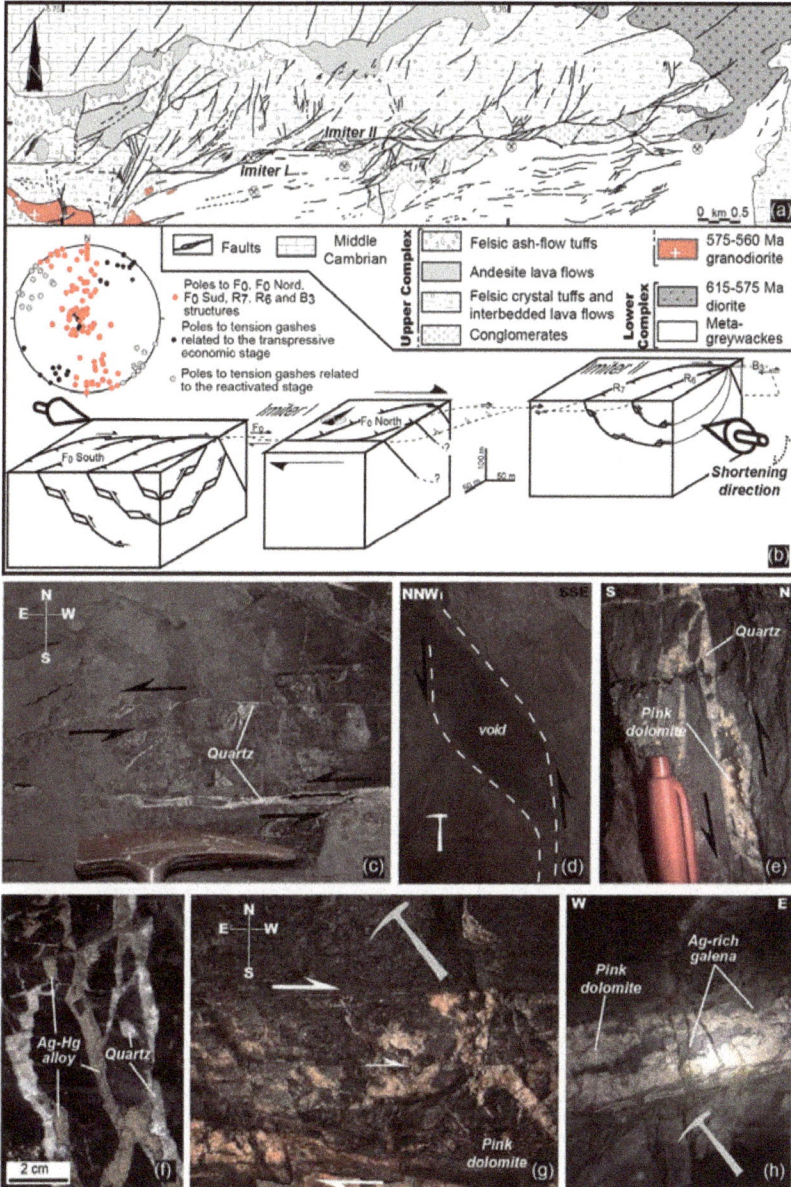

Figure 11. Main features of the world-class Ag–Hg Imiter mining district [47,102]. (**a**) General map of the giant Ag–Hg Imiter mine (after Leistel and Qadrouci [127] and Tuduri [47]) and (**b**) stereoplots of structural orientations and interpretative block diagram explaining the formation of the main ore-bearing vein system. Thrusts are formed in core of transpressive push-up structures. Note that the mineralised structures were everywhere controlled by a ESE–WNW direction of shortening. (**c**) Pull-apart and tension-gashes structures of the economic stage filled by geodic quartz and formed

during dextral kinematics, F_0 North vein systems, view realised towards the top of the mining gallery, Imiter I. (**d**) Pull-apart geometry of the economic stage indicative of a reverse shearing towards the north and showing void formations, R_6 structure, Imiter II. (**e**) Pull-apart texture of the F_0 South vein systems, thrusting towards the NW–NNW. The filling is composed by quartz (economic stage) and scarce pink dolomite in core of pull-apart texture. (**f**) Typical paragenesis of the economic stage 1 composed by quartz veins and huge concentration of Ag–Hg alloys, Imiter I, F_0 south. (**g**) Tension gashes and left lateral pull-apart structures filled with pink dolomite of the stage 2, F_0 structure, Imiter I. (**h**) Typical features of the stage 2 pink dolomite stage with large patches of Ag-rich galena, F_0 structure, Imiter I. Since photographs c and h were taken towards the top of exploration galleries, kinematics interpretation must be inverted.

Tuduri et al. [102] further demonstrated that a stage 2 reactivated the transpression-related structures in the opposite sense, and developed normal left-lateral motions associated with massive pink dolomite crystallisation, as well as prismatic quartz and variable amounts of Ag-rich galena and sphalerite, pyrite, chalcopyrite, arsenopyrite and freibergite (Figure 11g,h). Note that these two economic stages were preceded by a barren quartz vein network stage associated with sericite, illite-chlorite and base-metal sulphide minerals such as pyrite, galena, sphalerite with chalcopyrite exsolutions [45,47,51,127,169–171].

The main driving mechanism for silver ore deposition is assumed to be the dilution of ore-bearing fluids that were $CaCl_2$-dominated. Values obtained using the FIA concept [172,173] point to a general temperature decrease from stage 1 (280–100 °C) to stage 2 (110–60 °C). Note that the deepest levels of the mine workings (−220 m below the surface) record temperature in excess of 60 °C (i.e., lowest temperatures >160 °C) with respect to the shallow levels (−100 m) where the lowest temperatures are around 100°C. During stage 1, fluid salinities are moderate to high (8.4 to 26.1 wt. % NaCl + $CaCl_2$ equiv.), whereas they are very high when stage 2 dolomite precipitates (24.6 to 30 wt. % NaCl + $CaCl_2$ equiv.). Such value ranges are in agreement with data published by previous authors whether or not they used the FIA concept [41,45,101,171]. At shallower levels, additional supergene enrichment has been responsible for massive formation of native silver (1500 g/t Ag) associated with cerusite and mimetite [164].

Work in progress shows that Ag–Hg sulfohalides could also be related to the supergene processes [174]. Two opposing ore-forming models are strongly debated at Imiter. Some authors taking into account halogen composition of fluid inclusions, stable (C, O, S) and radiogenic (Pb, Re/O) isotope data together with noble gas (He) isotope compositions, suggest that the deposit is consistent with an epithermal model related to the felsic volcanic event at the Precambrian–Cambrian transition [45,47,51,101,127,170,175]. In that way, the huge Ag–Hg deposit would be comparable with the ones from the Mexican Sierra Madre Occidental Ag–Pb–Zn–Au belt, and should be considered as an intermediate sulfidation epithermal deposit [167,175]. On the other hand, a lithogene model [176] has been alternatively proposed in which fluids, according to laser ablation inductively coupled plasma-mass spectrometry (LA-ICP-MS) data on fluid inclusions, halogen signatures, and stable isotopes (H, C, O), are the products of diagenetic brine–evaporite interactions within a sedimentary basin [41]. The ore deposit might also be the result of basin inversion that expelled deep Ag-rich brines during, or at the end, of the Palaeozoic orogeny [41,175]. Recent $^{40}Ar/^{39}Ar$ age measured at 255 ± 3 Ma on adularia from stage 1 quartz vein supports the late Palaeozoic brine model [38].

5. Discussion

5.1. A Simplified Tectono-Magmatic Evolution Model of the Eastern Anti-Atlas

The evolution and transition between two main tectono-magmatic events is of importance in the Eastern Anti-Atlas and deserves to be discussed extensively and integrated with metallogenetic issues.

5.1.1. The Lower Complex and the D_1 Deformation

The D_1 deformation affects metagreywackes and metavolcanics and is associated with the syntectonic emplacement of calc-alkaline diorite and granodiorite plutons (Figure 12a). It is consistent with a NW–SE to WNW–ESE trending horizontal shortening. Structural studies have highlighted several important deformation structures related to that D_1 stage, i.e., upright tight folds with large-scale anticline and syncline, development of an axial planar S_{0-1} foliation, regional-scale N070–090°E trending dextral wrenching shear-zones, contact metamorphic minerals defining a stretching lineation (L_1) and rare S–SE-verging thrusts; Figures 2–4. Some plutons are inferred to be coeval with the D_1 deformation. Indeed, some of them are emplaced within the N070–090°E dextral shear-zones where they develop penetrative magmatic foliation and lineation. Elsewhere, other plutons seem to be emplaced in the core of folded sequences [47,121]. Furthermore, a tectonically-controlled fabric is best expressed in the thermal metamorphic aureoles developed around such plutons. There, foliations, lineations and related shear sense indicators show top-to-the-SSE–SE thrusting (Figure 4b,d) and dextral shearing depending on the distance from major shear-zones [47,121]. For those reasons, such diorite and granodiorite intrusions are herein interpreted as syntectonic and emplaced under the control of a NW–SE to WNW–ESE trending shortening [47,48,121]. This is consistent with the model of Saquaque et al. [77] who proposed that the main regional deformation results in top-to-the-SE thrusting, right-lateral wrenching along ca. E–W to ENE–WSW shear zones and syn-tectonic plutons. Since we herein combine strike-slip and shortening that is roughly perpendicular to the shear-zones, we infer a global transpressive tectonic regime as dominant in the D_1 stage. The characteristics of this tectonic stage that associates transpressive tectonics, pluton emplacement, strike-slip and rare thrusts explain the heterogeneity of the deformation observed within the entire Jbel Saghro.

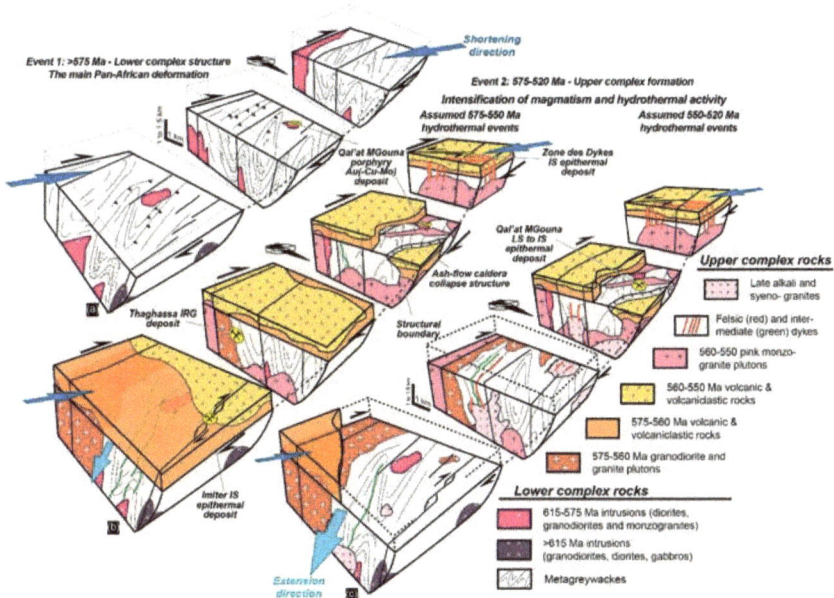

Figure 12. Interpretative three-phase model (**a–c**) explaining the tectono-magmatic evolution as well as the Lower and Upper Complexes definition of the Jbel Saghro and the formation of the ore-bearing vein systems of the Zone des Dykes, Qal'at MGouna, Thaghassa and Imiter districts. The size of the blue arrow relates with the inferred intensity of regional stress. See text for explanation. Note that the Qal'at MGouna porphyry deposit may belong to either the late stage of event 1 or the earlier stage of event 2 as the related porphyry stock emplaced at the transition between the two at 576 ± 5 Ma [29].

Previous interpretations tried to link each type of structures or pluton emplacement to one distinct tectonic event, increasing the complexity. From this study, it appears more appropriate to interpret all these features by the occurrence of a single and unique D_1 tectono-magmatic event. According to Gasquet et al. [83], as basins infilling with greywacke sequences was active until the onset of the Ediacaran period between 630 and 610 Ma; we suggest that the D_1 deformation probably occurred coevally with the syn-tectonic calc-alkaline magmatic occurrences, those being dated from ca. 615 Ma until ca. 575–565 Ma. This upper 575–565 Ma limit for the D_1 age is an important issue and will be further discussed below.

5.1.2. The Upper Complex and the D_2 Deformation

The Upper Complex is dominated by ash-flow tuffs emplaced in three main flare-ups and exposed inside or outside caldera structures, and by plutonic intrusions, mostly granitic, that were coeval with this D_2-related deformation (Figure 12b). Deformation was very weak and characterised by a brittle regime mainly represented by fault zones and vein formation throughout the Jbel Saghro area. Faults are mostly characterised by ca. N070–090°E orientations for the dextral strike-slip ones [47] whereas the ca. NW–SE trending direction may correspond to conjugate strike slip faults. However and according to Soulaimani et al. [86,87] and Azizi Samir et al. [177], the ca. NW–SE trending direction also corresponds to normal faulting yielding tilted blocks with syn-sedimentation processes. Such sedimentary deposits are volcaniclastic. The huge volumes of ash-flow tuffs associated with lava flows and related intrusions is considered as evidence for ash-flow caldera such as in the Qal'at Mgouna or Oued Dar'a areas (Figures 2 and 12b). We cannot exclude the presence of other caldera structures elsewhere in the Eastern Anti-Atlas. Because the structural limits of these caldera structures preferentially trend WNW–ESE to NW–SE [29,47], i.e., perpendicular to the direction of extension, the caldera yields an elliptic shape and is interpreted to be structurally controlled [141].

Three main ignimbrite flare-ups have been defined in the Upper Complex formation. The first one may have been emplaced at ca. 575 Ma and consists of dacitic ash-flow tuffs with related granodiorite and monzogranites [29]. The second flare-up occurred probably around 565 Ma and is related to monzogranite and granodiorite intrusions as well as the huge rhyo-dacitic dyke swarm in the western part of the Saghro [29,48,78,107]. The third ignimbrite flare-up is related to rhyolitic ash-flow tuff and cogenetic monzo-to syenogranite emplaced around 555 Ma [29,51,74,106,107]. Finally, a late magmatic stage is composed of the alkali-syenogranite plutons and related rhyolite dykes (Figure 12c). Such late alkali magmas highlight the persistence of the magmatism even after the emplacement of caldera-related rocks although they are less abundant. No radiometric ages are currently available on the alkali granites but related dykes provide ages bracketed between 550 and 520 Ma at the transition between the Proterozoic and Phanerozoic aeon [51,74,107,108]. Note that most of the intrusives such as aplitic and rhyolite dykes display a change in their orientation (i.e., ~N020°E, Figure 9c). Also, most of the hydrothermal veins related to those dykes and alkali plutons have the same direction pattern (Figure 9c). As no clear direction of shearing is evidenced in the horizontal plane, it is herein suggested that an extensional tectonic setting may control and assist the emplacement of both magmatic and hydrothermal structures in the ca. 550–520 Ma interval. The D_2 event is controlled by a ca. NW–SE to WNW–ESE direction of shortening and a ca. NE–SW to NNE–SSW direction of extension. Such control may evolve later toward a ca. N–S shortening direction and a ca. E–W extensional direction during the early Cambrian stage. This would be consistent with what is observed during the Adoudounian rifting period [86,87,106].

5.1.3. About the Transition between the Two Complexes and the D_1 and D_2 Tectonics

In the Eastern Anti-Atlas, the transition between the Lower and Upper Complexes is characterised by a significant increase of the magmatic addition rate [178] mostly evidenced by the Ediacaran volcanic flare-ups and by a change in the tectonic regime. However, such a transition appears as diachronous whether we consider the start of the volcanic activity or the change in tectonic regime.

The key evidence highlighting such a transition is the existence of an angular unconformity between the Lower and Upper Complex units [29,47,74,78]. However, the age of the transition remains elusive and debated. If we consider the available ages earlier volcanics of the Upper Complex that lie above the unconformity, the transition might occur between 580 and 570 Ma taking into account the error bars of the radiometric ages. This has been well described in the Western Anti-Atlas on the Tizgui geological map [29,179]. In the Central Anti-Atlas, Blein et al. [78] also suggest that most of the regional deformation observed there was completed by ca. 580 Ma and followed by an important erosional phase prior to the deposition of the Upper Complex volcaniclastic sequences. The 575 Ma age, therefore, represents a mean value.

Dating the deformations and thus the change in the tectonic regime from D_1 to D_2 is matter of more confusion. Indeed, the transition between the two complexes is estimated at ca. 575 Ma with the incipient volcanic activity and the occurrence of a strong angular uncorformity. By contrast, the transition between the two deformation events would be a little more recent and would have occurred around 565 Ma. The age assigned to the D_1 deformation appears thus mostly dependent on the radiometric ages obtained on the syntectonic plutons, and the issue of what plutons are syntectonic or not in the Eastern Anti-Atlas is clearly to be better assessed from field data. Currently, numerous plutons emplaced between ca. 575 and 565 Ma have been interpreted as syntectonic as they develop a coherent and homogenous ductile deformation in contact aureoles [48,105,106,121,123]. Given the error bars on the ages, we can actually question on their belonging to the earlier Upper Complex structuration, or to the later stage of the Lower Complex and D_1 deformation. As a matter of fact, discussions do exist about the possible intrusive character of some of these plutons in the lower part of the Upper Complex (e.g., The Igoudrane, Taouzzakt and Ikniwn plutons, Figures 2 and 3), while they may display an erosive roof on which lies Upper Complex volcaniclastics (at least the upper part of the Upper Complex). In addition, we have shown above that magmatic mushes at the origin of plutons emplaced around 575 and 565 Ma may be at the origin of the former ignimbritic flare-ups.

In any case, we assume that between 575 and 565 Ma, syn-D_1 plutons have developed in the Lower Complex metagreywackes, both contact metamorphism and ductile deformation that we interpret as the ongoing tectono-magmatic evolution between the Lower and Upper Complex structuration as suggested by Tuduri et al. [48]. Possibly, the S_2 cleavage refraction that developed after the main first-order folding event may be related to this transitional stage. It also remains difficult to assess whether some plutons were intruded within folds (e.g., the Bou Teglimt granodiorite in core of the Imiter inlier anticline, Figure 3a) and thus after the folding event (i.e., plutons would belong to the earlier stage of the Upper Complex) or if their emplacement played a part in the anticline structuring (i.e., plutons would belong to the Lower Complex tectono-magmatic history). Consequently, taking the important dioritic and granodioritic plutonism as systematically belonging to the Lower Complex might be misleading. As well, it is not straightforward to math the change in the tectonic regime (D_1-D_2 transition) with the transition between the Lower and Upper Complexes.

Therefore, we herein suggest that the reported tectonic transition must have been an ongoing process through the Lower-Upper Complex transition, given the deformation features (cleavage, weak upright folding) observed in the volcaniclastic rocks of the lower part of the Upper Complex in the western Saghro and Central Anti-Atlas [29,78]. Indeed, Blein et al. [78] recall that rapid variations in thickness of the volcanic and volcaniclastic rocks of the Upper Complex suggest they were deposited during active tectonics and on highly variable basement (i.e., Lower Complex units) topography. Such variations in the topography may be due to the earlier transpressive stage but also to later extensional-transtensional tectonics. In the rest of the Jbel Saghro no clear evidence of deformation (except cleavage) has been described affecting the lower part of the Upper Complex rocks. Note that the "Série molassique du Dadès" described by Walsh et al. [29] in the northern Qal'at Mgouna area, as being bedded, deformed and weakly metamorphosed in lower greenschist facies corresponds to the tilted intra-caldera volcanic sequence which underwent propylitic hydrothermal alteration. This formation herein belongs to the Upper Complex.

The preservation of the same shortening direction and tectonic style (i.e., strike slip dominated) further suggests that D_2 is a continuation of D_1 even though D_1 was associated with transpressive tectonics and D_2 linked with transtension to extension. Eventually, the transition between the D_1 and the D_2 deformation regimes occurred, while the shortening direction (parallel to Z strain axis) remained roughly constant (i.e., NWSE to WNW–ESE). However, one can note a shift in the two other strain directions showing a decrease of the vertical extension that becomes horizontal. This explains the numerous extensional features developed during the D2 stage (i.e., calderas formation and multiple ore concentrations within open structures). We note that the transition from D_1 to D_2 must be achieved prior the emplacement of the Thagassa IRGD around 565 Ma [48]. This transition is supposed to be progressive and the earliest mineralisation stages, such as the ones emplaced in the Thaghassa area, are the witnesses of this transition.

Therefore, we assume that the D_1 deformation might affect the lower units of the Upper Complex. Future works may focuse on testing this assumption. The contrasting structural levels that exist between the Lower Complex and Upper Complex units may argue against such a continuum. However, the combination of exhumation processes along strike-slip fault systems and of denudation history, integrating erosion processes [180–182] provides the first elements of an answer. Moreover, ignimbrite flare-ups and caldera formation are assumed to have been occurred rapidly as catastrophic events, contributing unconformities between the volcaniclastic rocks of the Upper Complex and the Lower Complex units.

5.2. The Mineralising Model

This study demonstrates the strong spatial link that exists between ore deposits and magmatism in the Eastern Anti-Atlas. A common hypothesis for the source of the mineralisation is to involve fluid exsolutions from intrusions. Even though we cannot demonstrate this assumption for all the ore deposits, a progressive and continuous vein formation model that began in the magmatic stage, developed an intermediate stage and finally ended under hydrothermal conditions is well constrained at Thaghassa and Qal'at Mgouna. Such a model of ore formation thus involves a three-stage model as stated in Table 1. About the economic stages, even though they were formed late during the veining stage, they benefited from the existence of previously formed veins. This model again highlights the significant role of the magmatism and related structures for ore deposition and concentration [183].

Table 1. Synthetic table summarizing the magmatic-hydrothermal features of main ore deposits of the Saghro area.

	The Zone des Dykes Deposit	The Qal'at MGouna District		The Thaghassa Deposit	The Imiter Deposit
Stage I (strong magmatic affinities)	Not expressed	Granodiorite stock emplacement and fluid exsolution	Alkali-granite emplacement with quartz, K-feldspar, F-tourmaline, F-muscovite in miarolitic cavities and stockscheider (400 < T°C < 600) [1]	Aplite dyke emplacement (500 < T°C < 600 °C) [1]	Not expressed
Stage II (intermediate stage between magmatic and hydrothermal conditions)	Not expressed	K-feldspar and magnetite alteration (disseminated and in veinlet stockwork)	Strong alteration with quartz, F-tourmaline, F-Cl-micas, andalusite, F-Cl apatite, monazite (250 < T°C < 400) [1]	Quartz, K-feldspar, muscovite, apatite, tourmaline and apatite (T°C < 500) [1]	Not expressed
Stage III (hydrothermal)	Quartz, adularia veins with calcite, chlorite, sulfides and gold (210 < T°C < 280) [1]	Quartz stockwork with Cl-biotite, Cl-amphibole, muscovite, sulfides (Cu-Mo) and gold (160 < T°C < 460) [1]	Chlorite-rich breccias, then Adularia, specularite, quartz veins with sulfides and "gold" (160 < T°C < 200) [1]	Striped quartz veins filled by quartz, sericite, sulfides and gold (300 < T°C < 450) [1]	Firstly, quartz-adularia veins with silver-rich sulfides and alloys (100 < T°C < 280) [1] and then dolomite-quartz veins with Ag-rich galena and sphalerite (60 < T°C < 110) [1]

[1] Temperatures have been estimated using selected geothermometers applied on micas [184], chlorites [185,186], arsenopyrites [187,188], activity diagrams generated for tourmaline, albite, andalusite, paragonite and kaolinite [189] and microthermometric data results [41,45,48,155,159,160,166,170–173]. See Tuduri [47] for details.

- The first stage shows strong magmatic influence. It is characterised by emplacement of porphyry stocks, aplite dykes and sills at high temperatures from 400 °C up to 600 °C. At Thaghassa, this stage was responsible for the partial melting of the metagreywackes in response to the Ikniwn granodiorite thermal effect and for the related genesis of leucocratic S-type haplogranitic sills. In the Qal'at Mgouna district, this stage was responsible for the formation of stockscheider and miarolitic cavities within sills and dykes.

- The second intermediate stage consists of magmatic-hydrothermal vein emplacement and associated pervasive alteration. The persistence of the magmatic character is shown by the occurrence of high-temperature alteration phases such as tourmaline, micas, andalusite, apatite, K-feldspar with quartz. In the Qal'at Mgouna district, at Isamlal, this stage can be compared with the classical potassic and magnetite alteration in some porphyry type systems [190]. It is also marked by the wide pervasive development of Al-silicate–Al-hydroxide–phosphate–muscovite–F-rich phlogopite and F-rich tourmaline alteration, related to the late Si-rich alkali granites. K-feldspar, apatite, white mica along vein rims are observed at the beginning of vein aperture at Thaghassa. Temperatures of formation are bracketed between 250 and 500 °C.

- The third stage is hydrothermal and formed at lower temperature (60 < T (°C) < 300) producing gangue minerals except for the Thaghassa and Isamlal deposits where high-temperature minerals were also formed (350–450 °C). This stage end with the emplacement of economic ore.

The conditions of vein formation vary depending on their location in the Lower or Upper Complex and are reflected by variations in mineralogy and internal texture. Such variations are basically due to different structural levels of formation. Indeed, the geometry of the mineralised structures is controlled by tectonics, hydrostatic pressure, effective vertical stress and volcano-related effects. The Thaghassa prospect, hosted by the Lower Complex metagreywakes, exhibits texture and mineralogy indicative of high-temperature conditions of formation. The Zone des Dykes vein system, entirely developed in Upper Complex ash-flow tuffs, shows internal textures consistent with low-temperature epithermal deposits. The Qal'at Mgouna deposits are developed at an intermediate structural level. Veins hosted by the Lower Complex rocks are related to high-temperature formation (i.e., the Isamlal porphyry Au(–Cu_Mo) deposit) whereas those formed at shallower levels reflect low-temperature conditions (Talat-n-Tabarought and Tawrirt-n-Cwalh districts). Tuduri et al. [154] suggest that high temperature systems emplaced at ca. 150–200 MPa, whereas the lower temperature systems were formed at a lower depth (20–50 MPa). The regional variations clearly document the transition from magmatic to hydrothermal conditions, i.e., from somewhat high temperature fluids (350 °C and higher) at Thaghassa and Isamlal, to lower temperature hydrothermal fluids (below 300 °C) at the Zone des Dykes and Imiter.

For all ore deposits described in this study, the Ag and Au economic concentrations (with Cu, Zn, Pb) correlate with ore-forming fluids with moderate to high salinity. This is consistent with transport by chloro-complexes and confirms the importance of brines in such ore formation. If brines are frequent in the formation of porphyry copper deposit [183,191] and probably IRGDs [64,68], their role in intermediate sulfidation epithermal systems remains a matter of debate. In large Mexican epithermal silver deposits of intermediate sulfidation state, Wilkinson et al. [26] suggested that the ore forming fluids, were injected into a geothermal circulation system in response to the ascent of a magmatic intrusion. Such hydrothermal diapirs would be sourced from a stratified hyper-saline brine reservoir, formed in response to incremental exsolution of magmatic fluid, and intense brine condensation at depth, with halite precipitation, being stored above the source magma reservoirs [26, 192,193]. By contrast, Essarraj et al. [40,41,49] suggested that the ore-forming fluids were related to deep-basinal sedimentary brines and that metals had no genetic relationship with Neoproterozoic magmatism, on the basis of numerous deposits they investigated in the Eastern and Central Anti-Atlas. They suggested that ore brines resulted from evaporatively concentrated seawater in Triassic basins producing hot basinal brines, comparable with conditions for Mississippi Valley-Type (MVT) deposits.

We here suggest that all ore deposits described above are related to the Ediacaran magmatic-hydrothermal complex emplaced from 575 to 530–520 Ma. The concept of stored hypersaline brines is emphasised following Scott et al. [193]. Nevertheless, while a magmatic origin of such stored brines is obvious, the origin for Ca-rich brines has yet to be defined. Though such Ca-rich brines may derive from magmatic processes, they are also widespread in the deeper parts of many sedimentary basins and involved in ore forming processes [194,195]. This is, pro parte, the reason why Essarraj et al. [40,41,49] support a deep-basinal sedimentary brine model in the Anti-Atlas. Surprisingly, they do not support an Ediacaran model for such a model but a more recent one that is Permo-Triasic in age. Nevertheless, prior and during the early stage of development of the SLIP that characterise the Upper Complex and the related ore deposits, periods that roughly belong to the Ediacaran Gaskiers glaciation on the West African Craton and around [196,197], would have occurred between 595 and 565 Ma in the Anti-Atlas [198]. Such evidence may also point to a possible contribution of natural cryogenic brines in the ore forming processes. Indeed, according to the Starinsky model of evolution of a marine-cryogenic basin [199], these brines usually are seawater-derived and also enriched in Mg, K, Na, Ca, Cl, SO$_4$ and Br, even if they may have been heavily modified by fluid–rock interactions and(or) dilution [200]. Future studies would have to address the possible role of Ediacaran glaciations on the ore-forming processes in the Anti-Atlas.

The age of the mineralisations in the Jbel Saghro remains poorly constrained due to the lack of absolute dating. Nevertheless, we can assume that the transtensional regime characteristic of the Upper Complex provided extensional structures for melt and fluid circulations that are favourable for magmatic-hydrothermal ore forming processes. This period of time (575–550 Ma and 550–520 Ma) was suitable for ore emplacement as suggested above. We thus assume that most of the mineralised systems may have been emplaced between ca. 575 and 520 Ma. Indeed, these ore deposits seem to have been emplaced when tectonic regime changes at the transition between the Lower and Upper Complexes and when the typical medium-K calc-alkaline arc magmatism became more abundant in between 580 and 570 Ma. Some porphyry Au(–Cu–Mo) deposit may have been emplaced earlier around 575 Ma. In the Sirwa mountains belonging to the Central Anti-Atlas, Cu–Mo ± Au mineralisations associated with high-K calc-alkaline intrusions are also assumed to have been formed between 575 and 560 Ma [59,201]. Similar ages are also suggested in the vicinity of Bouskour where Re/Os analyses on molybdenite related to a Cu-rich mineralised stage, yield a weighted average age of 574.9 ± 2.4 Ma [103]. Further, the Thaghassa deposit displays strong similarities with the earlier base metal assemblage observed at Imiter and Tiouit [43,45,47,48,51]. Indeed, in the Tiouit gold deposit, the ore body is closely associated with the Ikniwn granodiorite dated at 564 ± 6 [48]. At Imiter, ^{40}Ar/^{39}Ar dating on sericites related to the earlier base metal sulphide veins range from 577 ± 4 to 563 ± 5 Ma, in good agreement with the synchronous Taouzzakt granodiorite dated at 572 ± 5 Ma and the Thaghassa model of emplacement (U-Pb radiometric ages on zircon, [48,51]). The intermediate sulfidation epithermal Au(–Ag–Cu–Pb–Zn) deposit of the Zone des Dykes may have been emplaced during lull periods soon after the emplacement of the second ignimbrite flare-up and related huge dyke swarm in the western Saghro (around 560 Ma). Similarly, the huge intermediate sulfidation epithermal Ag(–Hg–Pb–Zn) deposit at Imiter may have been emplaced following the third ignimbrite flare-up around 550 Ma. This assumption adopts the age alreadyproposed for the Imiter mineralisation on the basis of absolute dating of felsic volcanism at 550 Ma [45,51,101]. Lastly, we suggest that the intermediate to low sulfidation epithermal Au–Ag–Te deposits were emplaced later when magmatism became less abundant, more silicic and alkaline between 550 and 530–520 Ma [74,107,108]. This is supported by the fact that some late rhyolitic dykes do not cross the Cambrian boundary while they cut across mineralised structures, at Qal'at Mgouna for example (Figure 9a)

5.3. Implication of the Tectonic Regime Changes for the Late Neoproterozoic—Early Cambrian Geodynamic Evolution and Ore Deposit Emplacement

In the light of our results, a geodynamic evolution model is proposed (Figure 13), showing the spatial and temporal distribution of metal deposits in the Eastern Anti-Atlas that may be reasonably extended to the whole Anti-Atlas. Considering the tectonic and metallogenetic framework at the scale of the Anti-Atlas regional scale, we suggest a first-order influence of subduction dynamics on the shift with time of metal concentrations and ore deposit types in the Anti-Atlas. In our scenario (Figure 13a,b) and according to Walsh et al. [29], the 615–575 Ma period is characterised by an Andean-type subduction of a wide continuous slab along the northern Gondwana margin (i.e., the currently northern side of South America and Africa). This long period of subduction (ca. 40 Myr) probably occurs because the large slab width (>2000 km) increases the viscous resistance of the mantle on the slab [202]. In that model, oceanic lithosphere *subducts* beneath the continental lithosphere that progressively becomes thicker due to the presence of the West African Craton. Suction between ocean and continent increases, favouring slab flattening and mantle wedge closure [203].

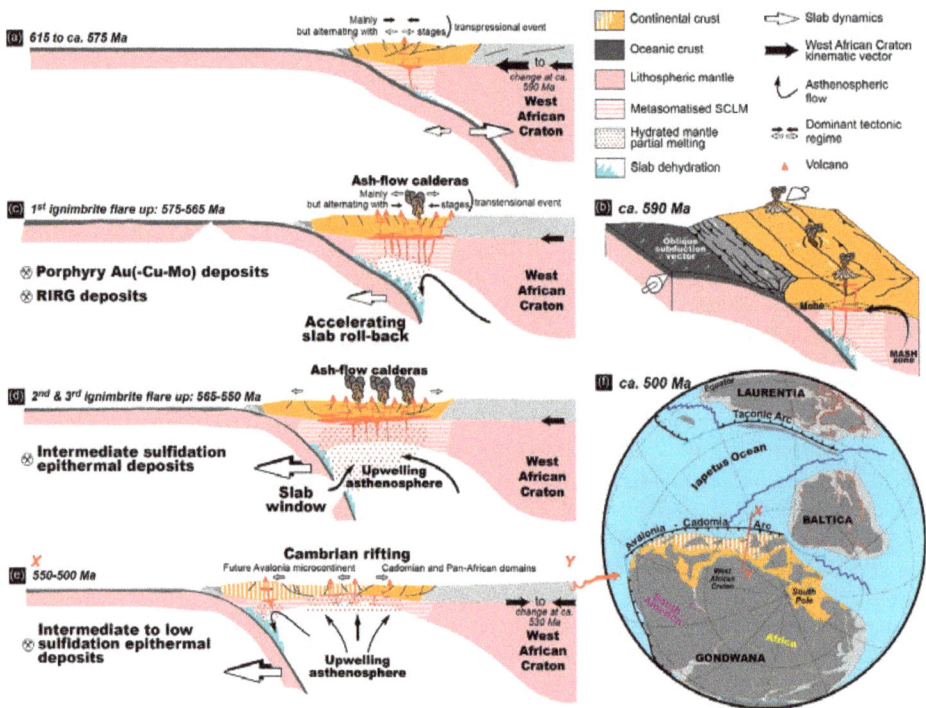

Figure 13. Compiled lithospheric-scale reconstructions (**a–e**) of the Pan-African/Cadomian belt systems from the West African Craton foreland to the Iapetus oceanic domain showing the possible progressive slab retreat since ca. 575 Ma to 500 Ma as suggested in text. Zones of partial melting in both the subcontinental lithospheric mantle (SCLM) and the lower crust, as well as the zones of storage and transfer of melts are shown in red. Kinematic vectors related to the West African Craton drifting are from Merdith et al. [204] using the GPlates software [205]. The palaeotectonic map reconstruction of Gondwana (**f**) has been realised using the GPlates software and the global plate models with kinematic continuity of Domeier [206].

Both slab flattening and probably relatively high convergence rate control the D_1 deformation in the overriding plate. A subduction vector (Figure 13b) oblique to the continental margin may explain the overall transpression regime that characterises the Lower Complex. Indeed, the lack of nappes, fold nappes, mylonitic fabrics and metamorphic gradient preclude a continental collision setting in that period. Possibly, a volcanic arc accretion event may locally emphasize this deformation. Melt generation remains limited given the restricted size of the mantle wedge while their ascent may occur when regional or local extensional regime occurred (Figure 13a,b). Ore forming processes are limited in this setting as they depend on the generation and ascent of fertile melts. However, melts produced and stored at depth in the MASH (melting, assimilation, storage and homogenization) zone may become more fertile in the following stages.

From 575 Ma, the ongoing evolution of the flattened subduction of oceanic beneath cratonic lithospheres causes a dynamic push on the slab surface [203]. This occurs as the rate of wedge closure increases, pushing the slab backward and initiating slab roll-back and high-volume magmatic production (Figure 13c), ultimately leading to the first ignimbrite flare-up. Ongoing slab roll-back along with tectonic regimes becoming more extensional initiate the second, and then the third ignimbrite flare-ups (Figure 13d). Slab tearing and (or) breakoff provide important asthenospheric flow that probably catalyse partial melting of both the asthenospheric mantle wedge and subcontinental lithosphere mantle (SCLM). The latest alkali magmas are assumed to have been emplaced at an extensional stage related to the Adoudounian rift in the Western Anti-Atlas [87,106] that corresponds to a back-arc setting. Such ongoing extensional tectonics [206,207] will ultimately result in the opening of the future Rheic Ocean (Figure 13e,f).

A collision then post-collision scenario cannot be applied to the Anti-Atlas region between 615 and 520 Ma, since subduction do not cease in our model but progressively migrate towards lower latitudes (Figure 13). Eventually, such subduction dynamics determine the dominant stress regime in the overriding plate, which influences the metals mobilization in the MASH zone and, their ascent through the crust, and thus controls the distribution of resulting metal occurrences [208–212].

According to Tosdal and Richards [211], most of the mineralised structures are suggested to have formed during the D_2-related transtensive regime caused by shortening in a WNW–ESE direction and extension along the NNE–SSW direction (Figures 12c and 13c–e). In our interpretation, porphyry Au(–Cu–Mo) and intrusion-related gold deposits are emplaced earlier than the first and(or) second ignimbrite flare-ups (i.e., 575 and(or) 565 Ma). Intermediate sulfidation epithermal Au, Ag deposits may be emplaced during lull periods after the second and (or) the third flare-ups (i.e., 560 and(or) 550 Ma). Compressive structures are indicated in the Zone des Dykes and Imiter districts, as the result of likely a specific structures geometry with respect to the shortening direction. Intermediate to low sulfidation epithermal Au–Ag–Te deposits are emplaced late and in relation with the felsic alkaline magmatic stage (550–520 Ma).

We have stressed here the existence of a long period of magmatism, i.e., over a 95 Myr duration, which is characterised by an increase in produced magmatic volumes, probably in response to geodynamical controls, marked by a late magmatic paroxysm in the form of several ignimbrite flare-ups over a shorter duration of ca. 25 Myr. Such long-lived magmatic activity is paralleled by a tectonic progressive evolution from beginning within transpression conditions to transtension then extension, allowing the mineralisations to take place.

At the regional scale, we may question a possible diachronism of the magmatic activity between the western Bou Azzer and Siroua and the eastern Saghro inliers (Figure 2). Indeed, the main volcanic event occurred between 580 and 560 Ma in Bou Azzer, Siroua and western Saghro inliers. In the central and eastern Jbel Saghro area, volcanic rocks as reported are somewhat younger and mostly dated between 570 and 550 Ma from west to east [51,74]. This suggests that the main volcanic stage progresses toward the east along with the D_2 tectonic regime. If we compare with the Sierra Madre Occidental as a model of large silicic volcanic province [14,147,148,213], the widespread ash-flow tuff deposits of the Anti-Atlas domain should be emplaced as ignimbrite flare-ups and are correlated with progressive and diachronic formation toward the east of caldera collapse structures, broadly oriented E–W to NW–SE. This volcano-structural framework developed coevally with a transtensive tectonic regime characterised by both NNE–SSW extension and WNW–ESE shortening. In terms of melting processes, the main controlling factor in the generation of such a SLIP is a crustal setting located along a long-lived active subduction zone that evolves into a post-subduction domain via slab-roll back processes. According to Bryan and Ferrari [14] and Ernst [12], a huge thermal pulse is at the origin of a large-scale crustal anataxis of fertile and hydrous lower-crustal materials as well as metasomatised subcontinental lithospheric mantle. Flare-up models are in part inherited from the late evolution of arc settings that underwent slab-roll back, slab-breakoff and slab-window [14,147,148,214], as propoded here and correlated with the D_1 and D_2 tectonic model (Figure 13). Because they represent transient events of high magmatic fluxes from the mantle [215], volcanic flare ups are considered here as highly potential for Ag(–Au) economic deposits emplacement.

6. Conclusions

We document in this paper a long-lived tectono-magmatic event that produced two main litho-structural units we refer to as the Lower and Uper Complexes, respectively. The Lower Complex is coeval with the main Pan-African D_1 deformation consisting of a transpressive regime responsible for folding, faulting and pluton emplacement under the effects of a NW–SE to WNW–ESE direction of shortening from 615 to 575 Ma. The Upper Complex is characterised by the emplacement of large volumes of ash-flow tuffs and volcanoclastic rocks and related intrusions. These were linked with the formation of ash-flow caldera structures which are uncommom examples of preserved Precambrian ash-flow calderas.

Ore deposits (porphyry, intermediate and low sulfidation epithermal deposit types, and IRGD) show strong spatial and temporal relationships with the emplacement of the widespread magmatic units belonging to the Upper Complex. For each ore deposit, fluid circulations associated with plutonic and/or volcanic systems can be invoked to be at the origin of the genesis of economic paragenesis. We suggest that magmatism of the Upper Complex and ore concentrations were both coeval with a D_2 deformation stage (575–540 Ma) and were controlled by the same transtensive tectonic regime under the effect of a nearly WNW–ESE shortening direction.

Despite an incomplete record by absolute datations, we infer that the Jbel Saghro was affected over a long period of time (i.e., 95 Myr) by successively magmatic, magmato-hydrothermal and hydrothermal events which formed a large mineralised province with substantial economic potential. Previous authors have already envisioned such a long period of magmatism and hydrothermalism in the area [29,47,74]. We further defend such a view and contend the Jbel Saghro province compares in this respect to numerous examples of large magmatic-related mineralised systems documented elsewere during the Archaean to Caenozoic times. It may be questioned whether the particular longevity of the magmatic activity in areas dominated by transpressive and transtensive tectonics could be related to the specific behaviour of ancient continental domains in which tectonics are dominated by vertical forces and are linked with especially huge magmatism [216–218]. This debate is currently open and our results illustrate one additional example of long-lived tectono-magmatic event that characterises the late Precambrian in this part or the African continent. Considering the large volume of ash-flow tuffs that crop out in the Western, Central and Eastern Anti-Atlas and their long-lived magmatic activity (575–550 Ma), we infer that the whole Anti-Atlas area (i.e., 700 km long) belongs to a continental silicic large igneous province as defined by Bryan and Ferrari [14], and Ernst [12], and emplaced in a subduction to post-subduction setting and that may be linked to a wide area including the Cadomian segments [28,206,207,219]. Anyway, our results offer further evidence that the majority of metal deposits in the Moroccan Anti-Atlas could be formed during the Neoproterozoic times coevally with the tectonic and magmatic evolutions in this period. They also demonstrate that structural geology can provide relevant constraints for debating the age and mode of formation of ore deposits, specifically in the context of a large silicic magmatic provinces.

Author Contributions: J.T., A.C., L.B. (Luc Barbanson), J.-L.B., M.L., A.E., L.B. (Lakhlifi Badra), C.E.-L., M.D. and S.S. took part in the field investigation; M.L., A.E. and L.M. supported the field investigation; J.T., A.C., L.B.

(Luc Barbanson), J.-L.B., C.E.-L., S.S., and M.D. interpreted the data and took part in the discussion; J.T., A.C., J.-L.B. and M.D. wrote the original draft; J.T. and A.C. reviewed and edited the paper.

Funding: This work has been undertaken with the help of the French-Moroccan programs "Action Intégrée No 222/STU/00".

Acknowledgments: The REMINEX exploration team and SMI mining company provided funds and logistics for field and laboratory studies. We particularly acknowledge El Hajj Bouiroukouten and the intern geologists of MANAGEM for their constant help and support. We are grateful to the Masters students from the BRGM Campus, University of Orléans, University of Lille and LaSalle Beauvais who were involved in the field for geological mapping exercises. Olivier Rouer and Gilles Drouet are warmly thanked for assistance and help with electronic microprobe analyses. The manuscript benefitted considerably from constructive reviews by four anonymous referees.

Conflicts of Interest: The authors declare no conflict of interest.

References

1. Dostal, J.; Caby, R.; Keppie, J.D.; Maza, M. Neoproterozoic magmatism in Southwestern Algeria (Sebkha el Melah inlier): A northerly extension of the Trans-Saharan orogen. *J. Afr. Earth Sci.* **2002**, *35*, 213–225. [CrossRef]

2. Ennih, N.; Liegeois, J.-P. The boundaries of the West African craton, with special reference to the basement of the Moroccan metacratonic Anti-Atlas belt. *Geol. Soc. Lond. Spec. Publ.* **2008**, *297*, 1–17. [CrossRef]

3. Liégeois, J.P.; Latouche, L.; Boughrara, M.; Navez, J.; Guiraud, M. The LATEA metacraton (Central Hoggar, Tuareg shield, Algeria): Behaviour of an old passive margin during the Pan-African orogeny. *J. Afr. Earth Sci.* **2003**, *37*, 161–190. [CrossRef]

4. Ouzegane, K.; Kienast, J.-R.; Bendaoud, A.; Drareni, A. A review of Archaean and Paleoproterozoic evolution of the In Ouzzal granulitic terrane (Western Hoggar, Algeria). *J. Afr. Earth Sci.* **2003**, *37*, 207–227. [CrossRef]

5. Thomas, R.J.; Fekkak, A.; Ennih, N.; Errami, E.; Loughlin, S.C.; Gresse, P.G.; Chevallier, L.P.; Liegeois, J.-P. A new lithostratigraphic framework for the Anti-Atlas Orogen, Morocco. *J. Afr. Earth Sci.* **2004**, *39*, 217–226. [CrossRef]

6. El Bahat, A.; Ikenne, M.; Söderlund, U.; Cousens, B.; Youbi, N.; Ernst, R.; Soulaimani, A.; El Janati, M.H.; Hafid, A. U–Pb baddeleyite ages and geochemistry of dolerite dykes in the Bas Drâa Inlier of the Anti-Atlas of Morocco: Newly identified 1380Ma event in the West African Craton. *Lithos* **2013**, *174*, 85–98. [CrossRef]

7. Ikenne, M.; Söderlund, U.; Ernst, R.E.; Pin, C.; Youbi, N.; El Aouli, E.H.; Hafid, A. A c. 1710 Ma mafic sill emplaced into a quartzite and calcareous series from Ighrem, Anti-Atlas—Morocco: Evidence that the Taghdout passive margin sedimentary group is nearly 1 Ga older than previously thought. *J. Afr. Earth Sci.* **2017**, *127*, 62–76. [CrossRef]

8. Youbi, N.; Kouyaté, D.; Söderlund, U.; Ernst, R.E.; Soulaimani, A.; Hafid, A.; Ikenne, M.; El Bahat, A.; Bertrand, H.; Rkha Chaham, K.; et al. The 1750Ma Magmatic Event of the West African Craton (Anti-Atlas, Morocco). *Precambrian Res.* **2013**, *236*, 106–123. [CrossRef]

9. Davies, J.H.F.L.; Marzoli, A.; Bertrand, H.; Youbi, N.; Ernesto, M.; Schaltegger, U. End-Triassic mass extinction started by intrusive CAMP activity. *Nat. Commun.* **2017**, *8*, 15596. [CrossRef]

10. Knight, K.B.; Nomade, S.; Renne, P.R.; Marzoli, A.; Bertrand, H.; Youbi, N. The Central Atlantic Magmatic Province at the Triassic-Jurassic boundary: Paleomagnetic and ^{40}Ar/^{39}Ar evidence from Morocco for brief, episodic volcanism. *Earth Planet. Sci. Lett.* **2004**, *228*, 143–160. [CrossRef]

11. Marzoli, A.; Callegaro, S.; Dal Corso, J.; Davies, J.H.F.L.; Chiaradia, M.; Youbi, N.; Bertrand, H.; Reisberg, L.; Merle, R.; Jourdan, F. The Central Atlantic Magmatic Province (CAMP): A Review. In *The Late Triassic World: Earth in a Time of Transition*; Tanner, L.H., Ed.; Springer International Publishing: Cham, Switzerland, 2018; pp. 91–125.

12. Ernst, R.E. *Large Igneous Provinces*; Cambridge University Press: Cambridge, UK, 2014.

13. Ernst, R.; Bleeker, W. Large igneous provinces (LIPs), giant dyke swarms, and mantle plumes: Significance for breakup events within Canada and adjacent regions from 2.5 Ga to the Present. *Can. J. Earth Sci.* **2010**, *47*, 695–739. [CrossRef]

14. Bryan, S.E.; Ferrari, L. Large igneous provinces and silicic large igneous provinces: Progress in our understanding over the last 25 years. *GSA Bull.* **2013**, *125*, 1053–1078. [CrossRef]

15. Chappell, B.W.; White, A.J.R. Two contrasting granite types: 25 years later. *Aust. J. Earth Sci.* **2001**, *48*, 489–499. [CrossRef]

16. Bonin, B. A-type granites and related rocks: Evolution of a concept, problems and prospects. *Lithos* **2007**, *97*, 1–29. [CrossRef]

17. Ernst, R.E.; Jowitt, S.M. Large Igneous Provinces (LIPs) and Metallogeny. In *Tectonics, Metallogeny, and Discovery: The North American Cordillera and Similar Accretionary Settings*; Colpron, M., Bissig, T., Rusk, B.G., Thompson, J.F.H., Eds.; Society of Economic Geologists: Littleton, CO, USA, 2013; Volume 17, pp. 17–51.

18. Arndt, N.T.; Lesher, C.M.; Czamanske, G.K. Mantle-derived magmas and magmatic Ni-Cu-(PGE) deposits. In *Economic Geology 100th Anniversary Volume*; Hedenquist, J.W., Thompson, J.F.H., Goldfarb, R.J., Richard, J.P., Eds.; Society of Economic Geologists: Littleton, CO, USA, 2005; pp. 5–23.

19. Barnes, S.J.; Holwell, D.A.; Le Vaillant, M. Magmatic Sulfide Ore Deposits. *Elements* **2017**, *13*, 89–95. [CrossRef]

20. Goodenough, K.M.; Schilling, J.; Jonsson, E.; Kalvig, P.; Charles, N.; Tuduri, J.; Deady, E.A.; Sadeghi, M.; Schiellerup, H.; Müller, A.; et al. Europe's rare earth element resource potential: An overview of REE metallogenetic provinces and their geodynamic setting. *Ore Geol. Rev.* **2016**, *72*, 838–856. [CrossRef]

21. Ernst, R.E.; Bell, K. Large igneous provinces (LIPs) and carbonatites. *Mineral. Petrol.* **2010**, *98*, 55–76. [CrossRef]

22. Rao, N.V.C.; Lehmann, B. Kimberlites, flood basalts and mantle plumes: New insights from the Deccan Large Igneous Province. *Earth-Sci. Rev.* **2011**, *107*, 315–324. [CrossRef]

23. Barton, M.D. 13.20—Iron Oxide(–Cu–Au–REE–P–Ag–U–Co) Systems A2—Holland, Heinrich, D. In *Treatise on Geochemistry (Second Edition)*; Turekian, K.K., Ed.; Elsevier: Oxford, UK, 2014; pp. 515–541.

24. Camprubí, A. Tectonic and Metallogenetic History of Mexico. In *Tectonics, Metallogeny, and Discovery: The North American Cordillera and Similar Accretionary Settings*; Colpron, M., Bissig, T., Rusk, B.G., Thompson, J.F.H., Eds.; Society of Economic Geologists: Littleton, CO, USA, 2013; Volume 17, pp. 201–243.

25. Camprubí, A.; Albinson, T. Epithermal deposits in México—Update of current knowledge, and an empirical reclassification. *Geol. Soc. Am. Spec. Pap.* **2007**, *422*, 377–415.

26. Wilkinson, J.J.; Simmons, S.F.; Stoffell, B. How metalliferous brines line Mexican epithermal veins with silver. *Sci. Rep.* **2013**, *3*, 2057. [CrossRef]

27. Ballèvre, M.; Le Goff, E.; Hébert, R. The tectonothermal evolution of the Cadomian belt of northern Brittany, France: A Neoproterozoic volcanic arc. *Tectonophysics* **2001**, *331*, 19–43. [CrossRef]

28. Linnemann, U.; Pereira, F.; Jeffries, T.E.; Drost, K.; Gerdes, A. The Cadomian Orogeny and the opening of the Rheic Ocean: The diacrony of geotectonic processes constrained by LA-ICP-MS U–Pb zircon dating (Ossa-Morena and Saxo-Thuringian Zones, Iberian and Bohemian Massifs). *Tectonophysics* **2008**, *461*, 21–43. [CrossRef]

29. Walsh, G.J.; Benziane, F.; Aleinikoff, J.N.; Harrison, R.W.; Yazidi, A.; Burton, W.C.; Quick, J.E.; Saadane, A. Neoproterozoic tectonic evolution of the Jebel Saghro and Bou Azzer—El Graara inliers, eastern and central Anti-Atlas, Morocco. *Precambrian Res.* **2012**, *216–219*, 23–62. [CrossRef]

30. Boyer, C.; Leblanc, M. Les appareils émissifs de la formation volcanique infracambriennes de Ouarzazate, Anti-Atlas (Maroc). *Comptes rendus hebdomadaires des séances de l'Académie des sciences* **1977**, *285*, 641–644.

31. Moume, W.; Youbi, N.; Marzoli, A.; Bertrand, H.; Gärtner, A.; Linnemann, U.; Gerdes, A.; Ernst, R.; Söderlund, U.; Hachimi Hind, E.; et al. The distribution of the Central Iapetus Magmatic Province (CIMP) into West African craton: U-Pb dating, geochemistry and petrology of Douar Eç-çour and Imiter mafic Dyke Swarms (High and Anti-Atlas, Morocco). In Proceedings of the 2nd Colloquium of the International Geoscience Programme (IGCP638), Casablanca, Morocco, 7–12 November 2017.

32. Puffer, J.H. A late Neoproterozoic eastern Laurentian superplume: Location, size, chemical composition, and environmental impact. *Am. J. Sci.* **2002**, *302*, 1–27. [CrossRef]

33. Barbey, P.; Oberli, F.; Burg, J.P.; Nachit, H.; Pons, J.; Meier, M. The Palaeoproterozoic in western Anti-Atlas (Morocco): A clarification. *J. Afr. Earth Sci.* **2004**, *39*, 239–245. [CrossRef]

34. Burkhard, M.; Caritg, S.; Helg, U.; Robert-Charrue, C.; Soulaimani, A. Tectonics of the Anti-Atlas of Morocco. *Comptes Rendus Geosci.* **2006**, *338*, 11–24. [CrossRef]

35. Missenard, Y.; Zeyen, H.; Frizon de Lamotte, D.; Leturmy, P.; Petit, C.; Sébrier, M.; Saddiqi, O. Crustal versus asthenospheric origin of relief of the Atlas Mountains of Morocco. *J. Geophys. Res. Solid Earth* **2006**, *111*, B03401. [CrossRef]

36. Bourque, H.; Barbanson, L.; Sizaret, S.; Branquet, Y.; Ramboz, C.; Ennaciri, A.; El Ghorfi, M.; Badra, L. A contribution to the synsedimentary versus epigenetic origin of the Cu mineralizations hosted by terminal Neoproterozoic to Cambrian formations of the Bou Azzer–El Graara inlier: New insights from the Jbel Laassel deposit (Anti Atlas, Morocco). *J. Afr. Earth Sci.* **2015**, *107*, 108–118. [CrossRef]

37. Pouit, G. Paléogéographie et répartition des minéralisations stratiformes de cuivre dans l'Anti-Atlas occidental (Maroc). *Chronique de la Recherche Minière* **1966**, *34*, 279–289.

38. Borisenko, A.S.; Lebedev, V.I.; Borovikov, A.A.; Pavlova, G.G.; Kalinin, Y.A.; Nevol'ko, P.A.; Maacha, L.; Kostin, A.V. Forming conditions and age of native silver deposits in Anti-Atlas (Morocco). *Dokl. Earth Sci.* **2014**, *456*, 663–666. [CrossRef]

39. Oberthur, T.; Melcher, F.; Henjes-Kunst, F.; Gerdes, A.; Stein, H.; Zimmerman, A.; El Ghorfi, M. Hercynian age of the cobalt-nickel-arsenide-(gold) ores, Bou Azzer, Anti-Atlas, Morocco: Re-Os, Sm-Nd, and U-Pb age determinations. *Econ. Geol.* **2009**, *104*, 1065–1079. [CrossRef]

40. Essarraj, S.; Boiron, M.-C.; Cathelineau, M.; Banks, D.A.; Benharref, M. Penetration of surface-evaporated brines into the Proterozoic basement and deposition of Co and Ag at Bou Azzer (Morocco): Evidence from fluid inclusions. *J. Afr. Earth Sci.* **2005**, *41*, 25–39. [CrossRef]

41. Essarraj, S.; Boiron, M.-C.; Cathelineau, M.; Tarantola, A.; Leisen, M.; Boulvais, P.; Maacha, L. Basinal Brines at the Origin of the Imiter Ag-Hg Deposit (Anti-Atlas, Morocco): Evidence from LA-ICP-MS Data on Fluid Inclusions, Halogen Signatures, and Stable Isotopes (H, C, O). *Econ. Geol.* **2016**, *111*, 1753–1781. [CrossRef]

42. Abia, E.H.; Nachit, H.; Marignac, C.; Ibhi, A.; Saadi, S.A. The polymetallic Au-Ag-bearing veins of Bou Madine (Jbel Ougnat, eastern Anti-Atlas, Morocco): Tectonic control and evolution of a Neoproterozoic epithermal deposit. *J. Afr. Earth Sci.* **2003**, *36*, 251–271. [CrossRef]

43. Al Ansari, A.E.; Sagon, J.P. Le gisement d'or de Tiouit (Jbel Saghro, Anti-Atlas, maroc). Un système mésothermal polyphasé à sulfures-or et hématite-or dans une granodiorite potassique d'âge Protérozoïque supérieur. *Chronique de la Recherche Minière* **1997**, *527*, 3–25.

44. Leblanc, M.; Lbouabi, M. Native silver mineralization along a rodingite tectonic contact between serpentinite and quartz diorite (Bou Azzer, Morocco). *Econ. Geol.* **1988**, *83*, 1379–1391. [CrossRef]

45. Levresse, G.; Cheilletz, A.; Gasquet, D.; Reisberg, L.; Deloule, E.; Marty, B.; Kyser, K. Osmium, sulphur, and helium isotopic results from the giant Neoproterozoic epithermal Imiter silver deposit, Morocco: Evidence for a mantle source. *Chem. Geol.* **2004**, *207*, 59–79. [CrossRef]

46. Marcoux, E.; Wadjinny, A. Le gisement Ag–Hg de Zgounder (Jebel Siroua, Anti-Atlas, Maroc): Un épithermal néoprotérozoïque de type Imiter. *Comptes Rendus Geosci.* **2005**, *337*, 1439–1446. [CrossRef]

47. Tuduri, J. Processus de formation et relations spatio-temporelles des minéralisations à or et argent en contexte volcanique Précambrien (Jbel Saghro, Anti-Atlas, Maroc). Implications sur les relations déformation-magmatisme-volcanisme-hydrothermalisme. Ph.D. Thesis, University of Orléans, Orléans, France, 2005.

48. Tuduri, J.; Chauvet, A.; Barbanson, L.; Labriki, M.; Dubois, M.; Trapy, P.-H.; Lahfid, A.; Poujol, M.; Melleton, J.; Badra, L.; et al. Structural control, magmatic-hydrothermal evolution and formation of hornfels-hosted, intrusion-related gold deposits: Insight from the Thaghassa deposit in Eastern Anti-Atlas, Morocco. *Ore Geol. Rev.* **2018**, *97*, 171–198. [CrossRef]

49. Essarraj, S.; Boiron, M.-C.; Cathelineau, M.; Banks, D.A.; El Boukhari, A.; Chouhaidi, M.Y. Brines related to Ag deposition in the Zgounder silver deposit (Anti-Atlas, Morocco). *Eur. J. Mineral.* **1998**, *10*, 1201–1214. [CrossRef]

50. Levresse, G.; Bouabdellah, M.; Gasquet, D.; Cheilletz, A. Basinal Brines at the Origin of the Imiter Ag-Hg Deposit (Anti-Atlas, Morocco): Evidence from LA-ICP-MS Data on Fluid Inclusions, Halogen Signatures, and Stable Isotopes (H, C, O)—A Discussion. *Econ. Geol.* **2017**, *112*, 1269–1272. [CrossRef]

51. Cheilletz, A.; Levresse, G.; Gasquet, D.; Azizi-Samir, M.R.; Zyadi, R.; Archibald, A.D.; Farrar, E. The giant Imiter silver deposit: Neoproterozoic epithermal mineralization in the Anti-Atlas, Morocco. *Miner. Depos.* **2002**, *37*, 772–781. [CrossRef]

52. Pelleter, E.; Cheilletz, A.; Gasquet, D.; Mouttaqi, A.; Annich, M.; Camus, Q.; Deloule, E.; Ouazzani, L.; Bounajma, H.; Ouchtouban, L. U/Pb Ages of Magmatism in the Zgounder Epithermal Ag–Hg Deposit, Sirwa Window, Anti-Atlas, Morocco. In *Mineral Deposits of North Africa*; Bouabdellah, M., Slack, J.F., Eds.; Springer International Publishing: Cham, Switzerland, 2016; pp. 143–165.

53. Ahmed, A.H.; Arai, S.; Ikenne, M. Mineralogy and Paragenesis of the Co-Ni Arsenide Ores of Bou Azzer, Anti-Atlas, Morocco. *Econ. Geol.* **2009**, *104*, 249–266. [CrossRef]

54. Ennaciri, A.; Barbanson, L.; Touray, J.C. Mineralized hydrothermal solution cavities in the Co-As Ait Ahmane mine (Bou Azzer, Morocco). *Miner. Depos.* **1995**, *30*, 75–77. [CrossRef]

55. Leblanc, M. Co-Ni arsenide deposits, with accessory gold, in ultramafic rocks from:Morocco. *Can. J. Earth Sci.* **1986**, *23*, 1592–1602. [CrossRef]

56. Thiéblemont, D.; Chêne, F.; Liégeois, J.-P.; Ouabadi, A.; Le Gall, B.; Maury, R.C.; Jalludin, M.; Ouattara Gbélé, C.; Tchaméni, R.; Fernandez-Alonso, M. *Geological Map of Africa at 1:10 Million Scale*, 35th International Geology Congress ed; CCGM-BRGM: Orléans, France, 2016.

57. Hollard, H.; Choubert, G.; Bronner, G.; Marchand, J.; Sougy, J. Carte géologique du Maroc, échelle: 1/1.000.000. *Notes et Mémoires du Service Géologique du Maroc* **1985**, *260*.

58. Mouttaqi, A.; Rjimati, E.; Maacha, A.; Michard, A.; Soulaimani, A.; Ibouh, H. Les principales mines du Maroc. *Notes et Mémoires du Service Géologique du Maroc* **2011**, *564*, 375.

59. Thomas, R.J.; Chevallier, L.P.; Gresse, P.G.; Harmer, R.E.; Eglington, B.M.; Armstrong, R.A.; de Beer, C.H.; Martini, J.E.J.; de Kock, G.S.; Macey, P.H.; et al. Precambrian evolution of the Sirwa Window, Anti-Atlas Orogen, Morocco. *Precambrian Res.* **2002**, *118*, 1–57. [CrossRef]

60. Goldfarb, R.J.; Groves, D.I. Orogenic gold: Common or evolving fluid and metal sources through time. *Lithos* **2015**, *233*, 2–26. [CrossRef]

61. Groves, D.I.; Santosh, M.; Goldfarb, R.J.; Zhang, L. Structural geometry of orogenic gold deposits: Implications for exploration of world-class and giant deposits. *Geosci. Front.* **2018**, *9*, 1163–1177. [CrossRef]

62. Hart, C.J. Reduced intrusion-related gold systems. In *Mineral Deposits of Canada: A Synthesis of Major Deposit Types, District Metallogeny, the Evolution of Geological Provinces, and Exploration Methods*; Special Publication; Geological Association of Canada, Mineral Deposits Division: St. John's, NL, Canada, 2007; pp. 95–112.

63. Kontak, D.; O'Reilly, G.; MacDonald, M.; Horne, R.; Smith, P. Gold in the Meguma Terrane, Southern Nova Scotia: Is There a Continuum between Mesothermal Lode Gold and Intrusion-related Gold Systems? In Proceedings of the 49th Annual Meeting of the GAC-MAC, St. Catharines, ON, Canada, 12–14 May2004; p. 128.

64. Lang, J.R.; Baker, T. Intrusion-related gold systems: The present level of understanding. *Miner. Depos.* **2001**, *36*, 477–489. [CrossRef]

65. Walshe, J.; Neumayr, P.; Cooke, D. Two boxes we don't need: Orogenic and intrusion-related gold systems. In Proceedings of the STOMP 2005: Structure, Tectonics and Ore Mineralisation Processes, Townsville, Australia, 29 August–2 September 2005; EGRU: Townsville, Australia, 2005; p. 143.

66. Boiron, M.-C.; Cathelineau, M.; Banks, D.A.; Fourcade, S.; Vallance, J. Mixing of metamorphic and surficial fluids during the uplift of the Hercynian upper crust: Consequences for gold deposition. *Chem. Geol.* **2003**, *194*, 119–141. [CrossRef]

67. Vallance, J.; Cathelineau, M.; Boiron, M.C.; Fourcade, S.; Shepherd, T.J.; Naden, J. Fluid-rock interactions and the role of late Hercynian aplite intrusion in the genesis of the Castromil gold deposit, northern Portugal. *Chem. Geol.* **2003**, *194*, 201–224. [CrossRef]

68. Baker, T.; Lang, J.R. Fluid inclusion characteristics of intrusion-related gold mineralization, Tombstone–Tungsten magmatic belt, Yukon Territory, Canada. *Mineral. Depos.* **2001**, *36*, 563–582. [CrossRef]

69. Chauvet, A.; Volland-Tuduri, N.; Lerouge, C.; Bouchot, V.; Monié, P.; Charonnat, X.; Faure, M. Geochronological and geochemical characterization of magmatic-hydrothermal events within the Southern Variscan external domain (Cévennes area, France). *Int. J. Earth Sci.* **2012**, *101*, 69–86. [CrossRef]

70. Mustard, R.; Ulrich, T.; Kamenetsky, V.S.; Mernagh, T. Gold and metal enrichment in natural granitic melts during fractional crystallization. *Geology* **2006**, *34*, 85–88. [CrossRef]

71. Gouiza, M.; Charton, R.; Bertotti, G.; Andriessen, P.; Storms, J.E.A. Post-Variscan evolution of the Anti-Atlas belt of Morocco constrained from low-temperature geochronology. *Int. J. Earth Sci.* **2017**, *106*, 593–616. [CrossRef]

72. Teixell, A.; Ayarza, P.; Zeyen, H.; Fernàndez, M.; Arboleya, M.-L. Effects of mantle upwelling in a compressional setting: The Atlas Mountains of Morocco. *Terra Nova* **2005**, *17*, 456–461. [CrossRef]

73. Choubert, G. Histoire géologique du Précambrien de l'Anti-Atlas de l'Archéen à l'aurore des temps primaires. *Notes et Mémoires du Service Géologique du Maroc* **1963**, *162*, 352.

74. Gasquet, D.; Levresse, G.; Cheilletz, A.; Azizi-Samir, M.R.; Mouttaqi, A. Contribution to a geodynamic reconstruction of the Anti-Atlas (Morocco) during Pan-African times with the emphasis on inversion tectonics and metallogenic activity at the Precambrian-Cambrian transition. *Precambrian Res.* **2005**, *140*, 157–182. [CrossRef]

75. Hefferan, K.; Soulaimani, A.; Samson, S.D.; Admou, H.; Inglis, J.; Saquaque, A.; Latifa, C.; Heywood, N. A reconsideration of Pan African orogenic cycle in the Anti-Atlas Mountains, Morocco. *J. Afr. Earth Sci.* **2014**, *98*, 34–46. [CrossRef]

76. Leblanc, M.; Lancelot, J.R. Interprétation géodynamique du domaine panafricain (Précambrien terminal) de l'Anti-Atlas (Maroc) à partir de données géologiques et géochronologiques. *Can. J. Earth Sci.* **1980**, *17*, 142–155. [CrossRef]

77. Saquaque, A.; Benharref, M.; Abia, H.; Mrini, Z.; Reuber, I.; Karson, J.A. Evidence for a Panafrican volcanic arc and wrench fault tectonics in Jbel Saghro, Morocco. *Geol. Rundsch.* **1992**, *81*, 1–13. [CrossRef]

78. Blein, O.; Baudin, T.; Chèvremont, P.; Soulaimani, A.; Admou, H.; Gasquet, P.; Cocherie, A.; Egal, E.; Youbi, N.; Razin, P.; et al. Geochronological constraints on the polycyclic magmatism in the Bou Azzer-El Graara inlier (Central Anti-Atlas Morocco). *J. Afr. Earth Sci.* **2014**, *99*, 287–306. [CrossRef]

79. El Hadi, H.; Simancas, J.F.; Martínez-Poyatos, D.; Azor, A.; Tahiri, A.; Montero, P.; Fanning, C.M.; Bea, F.; González-Lodeiro, F. Structural and geochronological constraints on the evolution of the Bou Azzer Neoproterozoic ophiolite (Anti-Atlas, Morocco). *Precambrian Res.* **2010**, *182*, 1–14. [CrossRef]

80. Inglis, J.D.; D'Lemos, R.S.; Samson, S.D.; Admou, H. Geochronological constraints on late Precambrian intrusions, metamorphism, and tectonism in the Anti-Atlas mountains. *J. Geol.* **2005**, *113*, 439–450. [CrossRef]

81. Inglis, J.D.; MacLean, J.S.; Samson, S.D.; D'Lemos, R.S.; Admou, H.; Hefferan, K. A precise U-Pb zircon age for the BleIda granodiorite, Anti-Atlas, Morocco: Implications for the timing of deformation and terrane assembly in the eastern Anti-Atlas. *J. Afr. Earth Sci.* **2004**, *39*, 277. [CrossRef]

82. Saquaque, A.; Admou, H.; Karson, J.; Hefferan, K.; Reuber, I. Precambrian accretionary tectonics in the Bou Azzer-El Graara region, Anti-Atlas, Morocco. *Geology* **1989**, *17*, 1107–1110. [CrossRef]

83. Gasquet, D.; Ennih, N.; Liégeois, J.-P.; Soulaimani, A.; Michard, A. The Pan-African Belt. In *Continental Evolution: The Geology of Morocco*; Michard, A., Saddiqi, O., Chalouan, A., Frizon de Lamotte, D., Eds.; Springer-Verlag: Berlin/Heidelberg, Germany, 2008; pp. 33–64.

84. Álvaro, J.J.; Benziane, F.; Thomas, R.; Walsh, G.J.; Yazidi, A. Neoproterozoic–Cambrian stratigraphic framework of the Anti-Atlas and Ouzellagh promontory (High Atlas), Morocco. *J. Afr. Earth Sci.* **2014**, *98*, 19–33. [CrossRef]

85. Choubert, G. In Essai d'application de la notion d'Infracambrien aux formations anciennes de l'Anti-Atlas (Maroc). In Proceedings of the 19th International Geological Congress, Alger, Algeria, 8–15 September 1952; pp. 33–71.

86. Soulaimani, A.; Michard, A.; Ouanaimi, H.; Baidder, L.; Raddi, Y.; Saddiqi, O.; Rjimati, E.C. Late Ediacaran–Cambrian structures and their reactivation during the Variscan and Alpine cycles in the Anti-Atlas (Morocco). *J. Afr. Earth Sci.* **2014**, *98*, 94–112. [CrossRef]

87. Soulaimani, A.; Bouabdelli, M.; Piqué, A. The Upper Neoproterozoic-Lower Cambrian continental extension in the Anti-Atlas (Morocco). *Bulletin de la Société Géologique de France* **2003**, *174*, 83–92. [CrossRef]

88. Chèvremont, P.; Blein, O.; Razin, P.; Baudin, T.; Barbanson, L.; Gasquet, D.; Soulaimani, A.; Admou, H.; Youbi, N.; Bouabdelli, M.; et al. Carte géologique du Maroc (1/50 000), feuille de Bou Azer. *Notes et Mémoires du Service Géologique du Maroc* **2013**, *535bis*, 153.

89. Ducrot, J.; Lancelot, J.R. Problème de la limite Précambrien–Cambrien: Étude radiochronologique par la méthode U–Pb sur zircons du volcan du Jbel Boho (Anti-Atlas marocain). *Can. J. Earth Sci.* **1977**, *14*, 2771–2777. [CrossRef]

90. Maloof, A.C.; Schrag, D.P.; Crowley, J.L.; Bowring, S.A. An expanded record of Early Cambrian carbon cycling from the Anti-Atlas Margin, Morocco. *Can. J. Earth Sci.* **2005**, *42*, 2195–2216. [CrossRef]

91. Fekkak, A.; Pouclet, A.; Ouguir, H.; Ouazzani, H.; Badra, L.; Gasquet, D. Géochimie et signification géotectonique des volcanites du Cryogénien inférieur du Saghro (Anti-Atlas oriental, Maroc). *Geodin. Acta* **2001**, *13*, 1–13.

92. Ouguir, H.; Macaudière, J.; Dagallier, G. Le Protérozoïque supérieur d'Imiter, Saghro oriental, Maroc: Un contexte géodynamique d'arrière arc. *J. Afr. Earth Sci.* **1996**, *22*, 173–189. [CrossRef]

93. Baidder, L.; Raddi, Y.; Tahiri, M.; Michard, A. Devonian extension of the Pan-African crust north of the West African craton, and its bearing on the Variscan foreland deformation: Evidence from eastern Anti-Atlas (Morocco). *Geol. Soc. Lond. Spec. Publ.* **2008**, *297*, 453–465. [CrossRef]

94. Malusà, M.G.; Polino, R.; Feroni, A.C.; Ellero, A.; Ottria, G.; Baidder, L.; Musumeci, G. Post-Variscan tectonics in eastern Anti-Atlas (Morocco). *Terra Nova* **2007**, *19*, 481–489. [CrossRef]

95. Michard, A.; Soulaimani, A.; Hoepffner, C.; Ouanaimi, H.; Baidder, L.; Rjimati, E.C.; Saddiqi, O. The South-Western Branch of the Variscan Belt: Evidence from Morocco. *Tectonophysics* **2010**, *492*, 1–24. [CrossRef]

96. Frizon de Lamotte, D.; Tavakoli-Shirazi, S.; Leturmy, P.; Averbuch, O.; Mouchot, N.; Raulin, C.; Leparmentier, F.; Blanpied, C.; Ringenbach, J.-C. Evidence for Late Devonian vertical movements and extensional deformation in northern Africa and Arabia: Integration in the geodynamics of the Devonian world. *Tectonics* **2013**, *32*, 107–122. [CrossRef]

97. Alvaro, J.J.; Macouin, M.; Ezzouhairi, H.; Charif, A.; Ayad, N.A.; Ribeiro, M.L.; Ader, M. Late Neoproterozoic carbonate productivity in a rifting context: The Adoudou Formation and its associated bimodal volcanism onlapping the western Saghro inlier, Morocco. *Geol. Soc. Lond. Spec. Publ.* **2008**, *297*, 285–302. [CrossRef]

98. Álvaro, J.J. Late Ediacaran syn-rift/post-rift transition and related fault-driven hydrothermal systems in the Anti-Atlas Mountains, Morocco. *Basin Res.* **2013**, *25*, 348–360. [CrossRef]

99. Sebti, S.; Saddiqi, O.; El Haimer, F.Z.; Michard, A.; Ruiz, G.; Bousquet, R.; Baidder, L.; Frizon de Lamotte, D. Vertical movements at the fringe of the West African Craton: First zircon fission track datings from the Anti-Atlas Precambrian basement, Morocco. *Comptes Rendus Geosci.* **2009**, *341*, 71–77. [CrossRef]

100. Caritg, S.; Burkhard, M.; Ducommun, R.; Helg, U.; Kopp, L.; Sue, C. Fold interference patterns in the Late Palaeozoic Anti-Atlas belt of Morocco. *Terra Nova* **2004**, *16*, 27–37.

101. Levresse, G.; Bouabdellah, M.; Cheilletz, A.; Gasquet, D.; Maacha, L.; Tritlla, J.; Banks, D.; Moulay Rachid, A.S. Degassing as the Main Ore-Forming Process at the Giant Imiter Ag–Hg Vein Deposit in the Anti-Atlas Mountains, Morocco. In *Mineral Deposits of North Africa*; Bouabdellah, M., Slack, J.F., Eds.; Springer International Publishing: Cham, Switzerland, 2016; pp. 85–106.

102. Tuduri, J.; Chauvet, A.; Ennaciri, A.; Barbanson, L. Modèle de formation du gisement d'argent d'Imiter (Anti-Atlas oriental, Maroc). Nouveaux apports de l'analyse structurale et minéralogique. *Comptes Rendus Geosci.* **2006**, *338*, 253–261. [CrossRef]

103. Bouabdellah, M.; Maacha, L.; Jébrak, M.; Zouhair, M. Re/Os Age Determination, Lead and Sulphur Isotope Constraints on the Origin of the Bouskour Cu–Pb–Zn Vein-Type Deposit (Eastern Anti-Atlas, Morocco) and Its Relationship to Neoproterozoic Granitic Magmatism. In *Mineral Deposits of North Africa*; Bouabdellah, M., Slack, F.J., Eds.; Springer International Publishing: Cham, Switzerland, 2016; pp. 277–290.

104. Hindermeyer, J.; Choubert, G.; Destombes, J.; Gauthier, H. Carte géologique de l'Anti-Atlas oriental: Feuille Dadès et Jbel Saghro 1/200 000. *Notes et Mémoires du Service Géologique du Maroc* **1977**, *161*.

105. Baidada, B.; Ikenne, M.; Barbey, P.; Soulaimani, A.; Cousens, B.; Haissen, F.; Ilmen, S.; Alansari, A. SHRIMP U-Pb zircon geochronology of the granitoids of the Imiter Inlier: Constraints on the Pan-African events in the Saghro massif, Anti-Atlas (Morocco). *J. Afr. Earth Sci.* **2018**. [CrossRef]

106. De Wall, H.; Kober, B.; Errami, E.; Ennih, N.; Greiling, R.O. Age de mise en place et contexte géologique des granitoïdes de la boutonnière d'Imiter (Saghro oriental, Anti-Atlas, Maroc). In Proceedings of the 2ème Colloque International 3MA (Magmatisme, Métamorphisme & Minéralisations Associées), Marrakech, Maroc, 10–12 May 2001; p. 19.

107. O'Connor, E.; Barnes, R.; Beddoe-Stephens, B.; Fletcher, T.; Gillespie, M.; Hawkins, M.; Loughlin, S.; Smith, M.; Smith, R.; Waters, C. *Geology of the Drâa, Kerdous, and Boumalne districts, Anti-Atlas, Morocco*; British Geological Survey: Nottingham, UK, 2010; p. 310.

108. Schiavo, A.; Taj Eddine, K.; Algouti, A.; Benvenuti, M.; Dal Piaz, G.V.; Eddebi, A.; El Boukhari, A.; Laftouhi, N.; Massironi, M.; Ounaimi, H.; et al. Carte géologique du Maroc au 1/50000, feuille Imtir. *Notes et Mémoires du Service Géologique du Maroc* **2007**, *518*.

109. Charlot, R.; Choubert, G.; Faure-Muret, A.; Tisserant, D. Etude géochronologique du Précambrien de l'Anti-Atlas (Maroc). *Notes et Mémoires du Service Géologique du Maroc* **1970**, *30*, 99–134.

110. Choubert, G. Sur le Précambrien marocain. *Comptes rendus hebdomadaires des séances de l'Académie des sciences* **1945**, *221*, 249–251.

111. Hindermeyer, J. Le Précambrien I et le Précambrien II du Saghro. *Comptes rendus hebdomadaires des séances de l'Académie des sciences* **1953**, *237*, 921–923.

112. Hindermeyer, J. Le Précambrien III du Saghro. *Comptes rendus hebdomadaires des séances de l'Académie des sciences* **1953**, *237*, 1024–1026.

113. Derré, C.; Lécolle, M. Altérations hydrothermales dans le Protérozoïque supérieur du Saghro (Anti-Atlas oriental). Relations avec les minéralisations. *Chronique de la Recherche Minière* **1999**, *536–537*, 39–61.

114. Fekkak, A.; Boualoul, M.; Badra, L.; Amenzou, M.; Saquaque, A.; El-Amrani, I.E. Origine et contexte géotectonique des dépôts détritiques du Groupe Néoprotérozoïque inférieur de Kelaat Mgouna (Anti-Atlas Oriental, Maroc). *J. Afr. Earth Sci.* **2000**, *30*, 295–311. [CrossRef]

115. Fekkak, A.; Pouclet, A.; Badra, L. The Pre-Panafrican rifting of Saghro (Anti-Atlas, Morocco): Exemple of the middle Neoproterozoic Basin of Boumalne. *Bulletin de la Société Géologique de France* **2002**, *173*, 25–35. [CrossRef]

116. Fekkak, A.; Pouclet, A.; Benharref, M. The Middle Neoproterozoic Sidi Flah Group (Anti-Atlas, Morocco): Synrift deposition in a Pan-African continent/ocean transition zone. *J. Afr. Earth Sci.* **2003**, *37*, 73–87. [CrossRef]

117. Fekkak, A.; Pouclet, A.; Ouguir, H.; Badra, L.; Gasquet, D. The Kelaat Mgouna early Neoproterozoic Group (Saghro, Anti-Atlas, Morocco): Witness of an initial stage of the pre-Pan-African extension. *Bulletin de la Société Géologique de France* **1999**, *170*, 789–797.

118. Marini, F.; Ouguir, H. Un nouveau jalon dans l'histoire de la distension pré-panafricaine au Maroc: Le Précambrien II des boutonnières du Jbel Saghro nord-oriental (Anti-Atlas, Maroc). *Comptes Rendus de l'Académie des Sciences Série II Mécanique-physique Chimie, Sciences de l'univers, Sciences de la Terre* **1990**, *310*, 577–582.

119. Errami, E.; Bonin, B.; Laduron, D.; Lasri, L. Petrology and geodynamic significance of the post-collisional Pan-African magmatism in the Eastern Saghro area (Anti-Atlas, Morocco). *J. Afr. Earth Sci.* **2009**, *55*, 105–124. [CrossRef]

120. Liégeois, J.-P.; Fekkak, A.; Bruguier, O.; Errami, E.; Ennih, N. The Lower Ediacaran (630–610 Ma) Saghro group: An orogenic transpressive basin development during the early metacratonic evolution of the Anti-Atlas (Morocco). In Proceedings of the IGCP485 4th Meeting, Algiers, Algeria, 2 September 2006; p. 57.

121. Ighid, L.; Saquaque, A.; Reuber, I. Plutons syn-cinématiques et la déformation panafricaine majeure dans le Saghro oriental (boutonnière d'Imiter, Anti-Atlas, Maroc). *Comptes Rendus de l'Académie des Sciences Série II Mécanique-physique Chimie, Sciences de l'univers Sciences de la Terre* **1989**, *309*, 615–620.

122. El Baghdadi, M.; El Boukhari, A.; Jouider, A.; Benyoucef, A.; Nadem, S. Calc-alkaline arc I-type granitoid associated with S-type granite in the Pan-African belt of eastern Anti-Atlas (Saghro and Ougnat, South Morocco). *Gondwana Res.* **2003**, *6*, 557–572. [CrossRef]

123. Errami, E.; Olivier, P. The Iknioun granodiorite, tectonic marker of Ediacaran SE-directed tangential movements in the Eastern Anti-Atlas, Morocco. *J. Afr. Earth Sci.* **2012**, *69*, 1–12. [CrossRef]

124. Karl, A.; de Wall, H.; Rieger, M.; Schmitt, T.; Errami, E.; Kober, B.; Greiling, R.O. Petrography and geochemistry of the Bou Teglimt, Taouzzakt and Igoudrane intrusions in the Eastern Saghro (Anti Atlas, Morocco). In *Magmatic evolution of a Neoproterozoic island-arc: Syn- to post-orogenic igneous activity in the Anti-Atlas (Morocco)*; de Wall, H., Greiling, R.O., Eds.; Forschungszentrum Jülich, International Cooperation, Scientific Series: Jülich, Germany, 2001; Volume 45, pp. 243–253.

125. Ouguir, H.; Macaudière, J.; Dagallier, G.; Qadrouci, A.; Leistel, J.-M. Cadre structural du gîte Ag-Hg d'Imiter (Anti-Atlas, Maroc); implication métallogénique. *Bulletin de la Société Géologique de France* **1994**, *165*, 233–248.

126. Massironi, M.; Moratti, G.; Algouti, A.; Benvenuti, M.; Dal Piaz, G.V.; Eddebi, A.; El Boukhari, A.; Laftouhi, N.; Ounaimi, H.; Schiavo, A.; et al. Carte géologique du Maroc au 1/50000, feuille Boumalne. *Notes et Mémoires du Service Géologique du Maroc* **2007**, *521*.

127. Leistel, J.-M.; Qadrouci, A. Le gisement argentifère d'Imiter (Protérozoïque supérieur de l'Anti-Atlas, Maroc). Contrôles des minéralisations, hypothèses génétiques et perspectives pour l'exploration. *Chronique de la Recherche Minière* **1991**, *502*, 5–22.

128. Benkirane, Y. Les minéralisations à W (Sn, Mo, Au, Bi, Ag, Cu, Pb, Zn) du granite de Taourirt-Tamellalt dans leur cadre géologique, la boutonnière protérozoïque du SE de Boumalne du Dadès (Saghro oriental, Anti-Atlas, Maroc). In *3ème Cycle*; Université de Paris VI: Paris, France, 1987.

129. Lécolle, M.; Derré, C.; Nerci, K. The Proterozoic sulphide alteration pipe of Sidi Flah and its host series. New data for the geotectonic evolution of the Pan-African Belt in the eastern Anti-Atlas (Morocco). *Ore Geol. Rev.* **1991**, *6*, 501–536. [CrossRef]

130. Benziane, F. Lithostratigraphie et évolution géodynamique de l'anti-Atlas (Maroc) du paléoprotérozoïque au néoprotérozoïque: Exemples de la boutonnière de Tagragra Tata et du Jebel Saghro. In *3ème Cycle*; Université de Chambéry: Chambéry, France, 2007.

131. Bajja, A. Volcanisme syn à post orogénique du Néoprotérozoïque de l'Anti-Atlas: Implications pétrogénétiques et géodynamiques. Ph.D. Thesis, Université Chouaib Doukkali, El Jadida, Maroc, 1998.

132. Benharref, M. Le Précambrien de la boutonnière d'El Kelaa des M'Gouna (Saghro, Anti-Atlas, Maroc). Pétrographie et structures de l'ensemble. Implications lithostratigraphiques et géodynamiques. In *3ème Cycle*; Université Cadi Ayyad: Marrakech, Maroc, 1991.

133. Bouladon, J.; Jouravsky, G. Les ignimbrites du Précambrien III de Tiouine et du sud marocain. *Notes et Mémoires du Service Géologique du Maroc* **1954**, *120*, 37–59.

134. Fauvelet, E.; Hindermeyer, J. Note préliminaire sur les granites associés à des coulées rhyolitiques au Sud de Ouarzazate (Anti-Atlas central) et dans le Sarho. *C. R. Hebd. Seances Acad. Sci.* **1952**, *234*, 2626–2628.

135. Mifdal, A.; Peucat, J. Datation U-Pb et Rb-Sr du volcanisme acide de l'Anti-Atlas marocain et du socle sous-jacent dans la région de Ouarzazate. Apport au problème de la limite Précambrien-Cambrien. *Sci. Géol. Bull.* **1985**, *38*, 185–200.

136. Acocella, V. Understanding caldera structure and development: An overview of analogue models compared to natural calderas. *Earth-Sci. Rev.* **2007**, *85*, 125–160. [CrossRef]

137. Lipman, P.W. The roots of ash flow calderas in western north america: Windows into the tops of granitic batholiths. *J. Geophys. Res. Solid Earth* **1984**, *89*, 8801–8841. [CrossRef]

138. Lipman, P.W. Subsidence of ash-flow calderas: Relation to caldera size and magma-chamber geometry. *Bull. Volcanol.* **1997**, *59*, 198–218. [CrossRef]

139. Williams, H. Calderas and their origin. University of California publications. *Bull. Dep. Geol. Sci.* **1941**, *25*, 239–346.

140. Acocella, V.; Korme, T.; Salvini, F.; Funiciello, R. Elliptic calderas in the Ethiopian Rift: Control of pre-existing structures. *J. Volcanol. Geotherm. Res.* **2003**, *119*, 189–203. [CrossRef]

141. Holohan, E.P.; Troll, V.R.; Walter, T.R.; Münn, S.; McDonnell, S.; Shipton, Z.K. Elliptical calderas in active tectonic settings: An experimental approach. *J. Volcanol. Geotherm. Res.* **2005**, *144*, 119–136. [CrossRef]

142. Ross, C.S.; Smith, R.L. Ash-flow tuffs: Their origin, geologic relations and identification. *Geol. Surv. Prof. Pap.* **1961**, *366*, 81.

143. Smith, R.L. Ash flows. *Geol. Soc. Am. Bull.* **1960**, *71*, 795–842. [CrossRef]

144. Smith, R.L.; Bailey, R.A. Resurgent cauldrons. *Geol. Soc. Am. Mem.* **1968**, *116*, 613–662.

145. Bellier, O.; Sébrier, M. Relationship between tectonism and volcanism along the Great Sumatran Fault Zone deduced by image analyses. *Tectonophysics* **1994**, *233*, 215–231. [CrossRef]

146. Chesner, C.A.; Rose, W.I. Stratigraphy of the Toba Tuffs and the evolution of the Toba Caldera Complex, Sumatra, Indonesia. *Bull. Volcanol.* **1991**, *53*, 343–356. [CrossRef]

147. Ferrari, L.; Valencia-Moreno, M.; Bryan, S. Magmatism and tectonics of the Sierra Madre Occidental and its relation with the evolution of the western margin of North America. *Geol. Soc. Am. Spec. Pap.* **2007**, *422*, 1–39.

148. Ferrari, L.; Lopez-Martinez, M.; Rosas-Elguera, J. Ignimbrite flare-up and deformation in the southern Sierra Madre Occidental, western Mexico: Implications for the late subduction history of the Farallon plate. *Tectonics* **2002**, *21*. [CrossRef]

149. Lécuyer, F.; Bellier, O.; Gourgaud, A.; Vincent, P.M. Tectonique active du Nord-Est de Sulawesi(Indonésie) et contrôle structural de la caldeira de Tondano. *Comptes Rendus de l'Academie des Sciences Ser. IIA Earth Planet. Sci.* **1997**, *325*, 607–613. [CrossRef]

150. Van Wyk de Vries, B.; Merle, O. Extension induced by volcanic loading in regional strike-slip zones. *Geology* **1998**, *26*, 983–986. [CrossRef]

151. Tuduri, J.; Chauvet, A.; Barbanson, L.; Labriki, M.; Badra, L. In Atypical gold mineralization within the Neoproterozoic of Morocco. Structural and mineralogical constraints from the Thaghassa prospect (Boumalne inlier, Jbel Saghro, Eastern Anti-Atlas). In Proceedings of the Mineral Exploration and Sustainable Development, Athens, Greece, 24–28 August 2003; Eliopoulos, D.G., Ed.; Millpress: Athens, Greece; pp. 537–540.

152. Goldstein, R.H.; Reynolds, T.J. *Systematics of Fluid Inclusions in Diagenetic Minerals*; Society for Sedimentary Geology: Broken Arrow, OK, USA, 1994; Volume 31, p. 199.

153. Lécolle, M.; Derré, C.; Rjimati, E.C.; Fonteilles, M.; Azza, A.; Benanni, A. Une altération hydrothermale peralumineuse à silicates, phosphates et rutile dans le Protérozoïque supérieur du Saghro (Anti-Atlas, Maroc). Genèse et implications métallogéniques. *Comptes Rendus de l'Académie des Sciences Série II Mécanique-physique Chimie Sciences de l'univers Sciences de la Terre* **1993**, *316*, 123–130.

154. Tuduri, J.; Dubois, M.; Try, E.; Chauvet, A.; Barbanson, L.; Ennaciri, A. The porphyry to epithermal transition in atypical late Neoproterozoic REE-Au-Ag-Te occurrences. *Acta Mineral.-Petrogr. Abstr. Ser.* **2010**, *6*, 288.

155. Delapierre, A. Etude de la minéralisation aurifère d'Isamlal (Jbel Saghro, Anti-Atlas, Maroc). In *Mem. Diplôme*; Université de Lausanne: Lausanne, Switzerland, 2000; p. 128.

156. Leloix, C. Etude des minéralisations aurifères épithermales d'Isamlal. District de Kelaat M'Gouna (Anti-Atlas, Maroc). In *Rapport Reminex*; Université d'Orléans: Orléans, France, 1999; p. 44.

157. Sizaret, S. Etude des minéralisations aurifères d'Isamlal (district de Kelâa M'Gouna, Anti-Atlas, Maroc). Master's Thesis, Université d'Orléans, Orléans, France, 1999.

158. Gaspard, E. Etude du prospect d'Isamlal (Anti-Atlas Oriental- Maroc): Caractérisation d'un porphyre à Au-Cu-Mo. Master's Thesis, Université d'Orléans—ENAG, Orléans, France, 2014.

159. Try, E.; Dubois, M.; Tuduri, J.; Ventalon, S.; Potdevin, J.-L.; Chauvet, A.; Barbanson, L. The transition between porphyric and epithermal styles: Insights from F.I. of the Kelâa M'Gouna prospect, Morocco. In Proceedings of the ECROFI-XX 20th Biennial Conferences, Granada, Spain, 21–27 September 2009; Volume 20, pp. 261–262.

160. Tomczyk, C. Âge de mise en place et modèle génétique du stockwerk du prospect à Au-Ag-Te de Kelâa M'Gouna (Maroc). Master's Thesis, University of Lille, Lille, France, 2010.

161. Dong, G.; Morrison, G.; Jaireth, S. Quartz textures in epithermal veins, Queensland; classification, origin and implication. *Econ. Geol.* **1995**, *90*, 1841–1856. [CrossRef]

162. Etoh, J.; Izawa, E.; Watanabe, K.; Taguchi, S.; Sekine, R. Bladed quartz and its relationship to gold mineralization in the Hishikari low-sulfidation epithermal gold deposit, Japan. *Econ. Geol.* **2002**, *97*, 1841–1851. [CrossRef]

163. André-Mayer, A.-S.; Leroy, J.L.; Bailly, L.; Chauvet, A.; Marcoux, E.; Grancea, L.; Llosa, F.; Rosas, J. Boiling and vertical mineralization zoning: A case study from the Apacheta low-sulfidation epithermal gold-silver deposit, southern Peru. *Mineral. Depos.* **2002**, *37*, 452–464. [CrossRef]

164. Chauvet, A.; Bailly, L.; André, A.-S.; Monié, P.; Cassard, D.; Tajada, F.; Vargas, J.; Tuduri, J. Internal vein texture and vein evolution of the epithermal Shila-Paula district, southern Peru. *Mineral. Depos.* **2006**, *41*, 387–410. [CrossRef]

165. Simmons, S.F.; Christenson, B.W. Origins of calcite in a boiling geothermal system. *Am. J. Sci.* **1994**, *294*, 361–400. [CrossRef]

166. Saule, A. La Zones des Dykes, Anti-Atlas Marocain: Caractérisation des fluides minéralisateurs et du gisement. Master's Thesis, Institut National Polytechnique de Lorraine, Nancy, France, 2012.

167. Albinson, T.; Norman, D.I.; Cole, D.; Chomiak, B. Controls on Formation of Low-Sulfidation Epithermal Deposits in Mexico: Constraints from Fluid Inclusion and Stable Isotope Data. In *New Mines and Discoveries in Mexico and Central America*; Society of Economic Geologists: Littleton, CO, USA, 2001; Volume 8, pp. 1–32.

168. Guillou, J.-J.; Monthel, J.; Picot, P.; Pillard, F.; Protas, J.; Samana, J.-C. L'imitérite, Ag_2HgS_2, nouvelle espèce minérale; propriétés et structure cristalline. *Bull. Mineral.* **1985**, *108*, 457–464.

169. Guillou, J.-J.; Monthel, J.; Samama, J.-C.; Tijani, A. Morphologie et chronologie relative des associations minérales du gisement mercuro-argentifère d'Imiter (Anti-Atlas—Maroc). *Notes et Mémoires du Service Géologique du Maroc* **1988**, *44*, 215–228.

170. Levresse, G. Contribution à l'établissement d'un modèle génétique des gisements d'Imiter (Ag-Hg), Bou Madine (Pb-Zn-Cu-Ag-Au), Bou Azzer (Co, Ni, As, Au, Ag) dans l'Anti-Atlas marocain. In *3ème Cycle*; Institut National Polytechnique de Lorraine: Nancy, France, 2001.

171. Baroudi, Z.; Beraaouz, E.H.; Rahimi, A.; Chouhaidi, M.Y. Minéralisations polymétalliques argentifères d'Imiter (Jbel Saghro, Maroc): Minéralogie, évolution des fluides minéralisateurs et mécanismes de dépôt. *Chronique de la Recherche Minière* **1999**, *536–537*, 91–111.

172. Hulin, C.; Dubois, M.; Tuduri, J.; Chauvet, A.; Boulvais, P.; Gaouzi, A.; Mouhajir, M.; Essalhi, M.; Outhounjite, S. New fluid inclusions and oxygen isotope data to constrain a formation model for the Imiter Ag world class deposit (Anti-Atlas, Morocco). In Proceedings of the ECROFI XXII 22nd Biennial Conferences, Antalya, Turkey, 4–9 June 2013; pp. 78–79.

173. Hulin, C.; Dubois, M.; Tuduri, J.; Chauvet, A.; Boulvais, P.; Gaouzi, A.; Mouhajir, M.; Essalhi, M.; Outhounjite, S. A fluid inclusion and stable isotope study of the world class Imiter silver deposit (Morocco). In Proceedings of the 24ème Réunion des Sciences de la Terre, Pau, France, 27–31 October 2014; p. 373.

174. Tuduri, J.; Pourret, O.; Chauvet, A.; Barbanson, L.; Gaouzi, A.; Ennaciri, A. Rare earth elements as proxies of supergene alteration processes from the giant Imiter silver deposit (Morocco). In *Let's Talk Ore Deposits, Proceeding of the Eleventh Biennial SGA Meeting*; Barra, F., Reich, M., Campos, E., Tornos, F., Eds.; Ediciones Universidad Católica del Norte: Antofagasta, Chile, 2011; Volume 2, pp. 826–828.

175. Tuduri, J.; Pourret, O.; Boulvais, P.; Chauvet, A.; Barbanson, L.; Gaouzzi, A.; Hulin, C.; Dubois, M. A reassessment of fluid-mineral relations in the world-class Imiter silver deposit (Anti-Atlas, Morocco). In Proceedings of the SEG 2012 Conference, Lima, Peru, 23–26 September 2012; Society of Economic Geologists: Lima, Peru, 2012.

176. Graybeal, F.T.; Vikre, P. A review of silver-rich mineral deposits and their metallogeny. In *SEG Special Publication: The Challenge of Finding New Mineral Resources: Global Metallogeny, Innovative Exploration, and New Discoveries*; Goldfarb, R.J., Marsh, E.E., Monecke, T., Eds.; Society of Economic Geologists: Littleton, CO, USA, 2010; Volume 15, pp. 85–117.

177. Azizi Samir, M.R.; Ferrandini, J.; Tane, J.L. Tectonique et volcanisme tardi-Pan Africains (580-560 M.a.) dans l'Anti-Atlas Central (Maroc): Interpretation geodynamique a l'echelle du NW de l'Afrique. *J. Afr. Earth Sci.* **1990**, *10*, 549–563. [CrossRef]

178. Ducea, M.N.; Paterson, S.R.; DeCelles, P.G. High-Volume Magmatic Events in Subduction Systems. *Elements* **2015**, *11*, 99–104. [CrossRef]

179. Harrison, R.W.; Yazidi, A.; Benziane, F.; Quick, J.E.; El Fahssi, A.; Stone, B.D.; Yazidi, M.; Saadane, A.; Walsh, G.J.; Aleinikoff, J.N.; et al. Carte géologique au 1/50 000, Feuille Tizgui. *Notes et Mémoires du Service Géologique du Maroc* **2008**, *470*, 131.

180. McQuarrie, N.; Barnes, J.B.; Ehlers, T.A. Geometric, kinematic, and erosional history of the central Andean Plateau, Bolivia (15–17° S). *Tectonics* **2008**, *27*, TC3007. [CrossRef]

181. Till, A.B.; Roeske, S.; Sample, J.C.; Foster, D.A. *Exhumation Associated with Continental Strike-Slip Fault Systems*; The Geological Society of America: Boulder, CO, USA, 2007; Volume 434, p. 264.

182. Willett, S.D.; Brandon, M.T. On steady states in mountain belts. *Geology* **2002**, *30*, 175–178. [CrossRef]

183. Hedenquist, J.W.; Lowenstern, J.B. The role of magmas in the formation of hydrothermal ore deposits. *Nature* **1994**, *370*, 519–527. [CrossRef]

184. Monier, G.; Robert, J.L. Muscovite solid solutions in the system K_2O, MgO, FeO, AL_2O_3, SiO_2, H_2O: An experimental study at 2 kbar P_{H2O} and comparison with natural Li-free white micas. *Mineral. Mag.* **1986**, *50*, 257–266. [CrossRef]

185. Cathelineau, M.; Nieva, D. A chlorite solid solution geothermometer: The Los Azufres (Mexico) geothermal system. *Contrib. Mineral. Petrol.* **1985**, *91*, 235–244. [CrossRef]

186. Kranidiotis, P.; MacLean, W.H. Systematics of chlorite alteration at the Phelps Dodge massive sulfide deposit, Matagami, Quebec. *Econ. Geol.* **1987**, *82*, 1898–1911. [CrossRef]

187. Kretschmar, U.; Scott, S.D. Phase relations involving arsenopyrite in the system Fe-As-S and their application. *Can. Mineral.* **1976**, *14*, 364–386.

188. Sundblad, K.; Zachrisson, E.; Smeds, S.A.; Berglund, S.; Aalinder, C. Sphalerite geobarometry and arsenopyrite geothermometry applied to metamorphosed sulfide ores in the Swedish Caledonides. *Econ. Geol.* **1984**, *79*, 1660–1668. [CrossRef]

189. Lynch, G.; Ortega, J. Hydrothermal alteration and tourmaline-albite equilibria at the Coxheat porphyry Cu-Mo-Au deposit, Nova Scotia. *Can. Mineral.* **1997**, *35*, 79–94.

190. Sillitoe, R.H. Porphyry Copper Systems. *Econ. Geol.* **2010**, *105*, 3–41. [CrossRef]

191. Kouzmanov, K.; Pokrovski, G.S. Hydrothermal controls on metal distribution in porphyry Cu (-Mo-Au) systems. In *Geology and Genesis of Major Copper Deposits and Districts of the World: A Tribute to Richard H. Sillitoe*; Hedenquist, J.W., Harris, M., Camus, F., Eds.; Special Publications of the Society of Economic Geologists: Littleton, CO, USA, 2012; Volume 16, pp. 573–618.

192. Rottier, B.; Kouzmanov, K.; Casanova, V.; Wälle, M.; Fontboté, L. Cyclic Dilution of Magmatic Metal-Rich Hypersaline Fluids by Magmatic Low-Salinity Fluid: A Major Process Generating the Giant Epithermal Polymetallic Deposit of Cerro de Pasco, Peru. *Econ. Geol.* **2018**, *113*, 825–856. [CrossRef]

193. Scott, S.; Driesner, T.; Weis, P. Boiling and condensation of saline geothermal fluids above magmatic intrusions. *Geophys. Res. Lett.* **2017**, *44*, 1696–1705. [CrossRef]

194. Letsch, D.; Large, S.J.E.; Buechi, M.W.; Winkler, W.; von Quadt, A. Ediacaran glaciations of the west African Craton—Evidence from Morocco. *Precambrian Res.* **2018**, *310*, 17–38. [CrossRef]

195. Pinneker, Y.V.; Lomonosov, I.S. Concentrated brines of Siberian Platform and their counterparts in Asia, Europe, Africa and America. *Int. Geol. Rev.* **1968**, *10*, 431–442. [CrossRef]

196. Richard, A.; Pettke, T.; Cathelineau, M.; Boiron, M.-C.; Mercadier, J.; Cuney, M.; Derome, D. Brine–rock interaction in the Athabasca basement (McArthur River U deposit, Canada): Consequences for fluid chemistry and uranium uptake. *Terra Nova* **2010**, *22*, 303–308. [CrossRef]

197. Linnemann, U.; Pidal, A.P.; Hofmann, M.; Drost, K.; Quesada, C.; Gerdes, A.; Marko, L.; Gärtner, A.; Zieger, J.; Ulrich, J.; et al. A ~565 Ma old glaciation in the Ediacaran of peri-Gondwanan West Africa. *Int. J. Earth Sci.* **2018**, *107*, 885–911. [CrossRef]

198. Vernhet, E.; Youbi, N.; Chellai, E.H.; Villeneuve, M.; El Archi, A. The Bou-Azzer glaciation: Evidence for an Ediacaran glaciation on the West African Craton (Anti-Atlas, Morocco). *Precambrian Res.* **2012**, *196–197*. [CrossRef]

199. Starinsky, A.; Katz, A. The formation of natural cryogenic brines. *Geochim. Cosmochim. Acta* **2003**, *67*, 1475–1484. [CrossRef]

200. Toner, J.D.; Catling, D.C.; Sletten, R.S. The geochemistry of Don Juan Pond: Evidence for a deep groundwater flow system in Wright Valley, Antarctica. *Earth Planet. Sci. Lett.* **2017**, *474*, 190–197. [CrossRef]

201. Belkacim, S.; Ikenne, M.; Souhassou, M.; Elbasbas, A.; Toummite, A. The Cu-Mo±Au mineralizations associated to the High-K calc-alkaline granitoids from Tifnoute valley (Siroua massif, anti-atlas, Morocco): An arc-Type porphyry in the late neoproterozoic series. *J. Environ. Earth Sci.* **2014**, *4*, 90–106.

202. Loiselet, C.; Husson, L.; Braun, J. From longitudinal slab curvature to slab rheology. *Geology* **2009**, *37*, 747–750. [CrossRef]

203. Manea, V.C.; Pérez-Gussinyé, M.; Manea, M. Chilean flat slab subduction controlled by overriding plate thickness and trench rollback. *Geology* **2012**, *40*, 35–38. [CrossRef]

204. Merdith, A.S.; Collins, A.S.; Williams, S.E.; Pisarevsky, S.; Foden, J.D.; Archibald, D.B.; Blades, M.L.; Alessio, B.L.; Armistead, S.; Plavsa, D.; et al. A full-plate global reconstruction of the Neoproterozoic. *Gondwana Res.* **2017**, *50*, 84–134. [CrossRef]

205. Boyden, J.A.; Müller, R.D.; Gurnis, M.; Torsvik, T.H.; Clark, J.A.; Turner, M.; Ivey-Law, H.; Watson, R.J.; Cannon, J.S. Next-generation plate-tectonic reconstructions using GPlates. In *Geoinformatics: Cyberinfrastructure for the Solid Earth Sciences*; Keller, G.R., Baru, C., Eds.; Cambridge University Press: Cambridge, UK, 2011; pp. 95–113.

206. Domeier, M. A plate tectonic scenario for the Iapetus and Rheic oceans. *Gondwana Res.* **2016**, *36*, 275–295. [CrossRef]

207. Torsvik, T.H.; Cocks, L.R.M. Gondwana from top to base in space and time. *Gondwana Res.* **2013**, *24*, 999–1030. [CrossRef]

208. Richards, J.P. Postsubduction porphyry Cu-Au and epithermal Au deposits: Products of remelting of subduction-modified lithosphere. *Geology* **2009**, *37*, 247–250. [CrossRef]

209. Richards, J.P. Magmatic to hydrothermal metal fluxes in convergent and collided margins. *Ore Geol. Rev.* **2011**, *40*, 1–26. [CrossRef]

210. Sillitoe, R.H.; Hedenquist, J.W. Linkages between volcanotectonic settings, ore-fluid compositions and epithermal precious metal deposits. In *Volcanic, Geothermal and Ore-Forming Fluids; Rulers and Witnesses of Processes within the Earth*; Simmons, S.F., Graham, I., Eds.; Society of Economic Geologist Special Publication: Littleton, CO, USA, 2003; Volume 10, pp. 315–343.

211. Tosdal, R.; Richards, J. Magmatic and structural controls on the development of porphyry Cu±Mo±Au deposits. *Rev. Econ. Geol.* **2001**, *14*, 157–181.

212. Menant, A.; Jolivet, L.; Tuduri, J.; Loiselet, C.; Bertrand, G.; Guillou-Frottier, L. 3D subduction dynamics: A first-order parameter of the transition from copper- to gold-rich deposits in the eastern Mediterranean region. *Ore Geol. Rev.* **2018**, *94*, 118–135. [CrossRef]

213. Bryan, S.E.; Orozco-Esquivel, T.; Ferrari, L.; López-Martínez, M. Pulling apart the Mid to Late Cenozoic magmatic record of the Gulf of California: Is there a Comondú Arc? *Geol. Soc. Lond. Spec. Publ.* **2013**, *385*. [CrossRef]

214. Thorkelson, D.J.; Breitsprecher, K. Partial melting of slab window margins: Genesis of adakitic and non-adakitic magmas. *Lithos* **2005**, *79*, 25–41. [CrossRef]

215. de Silva, S. Arc magmatism, calderas, and supervolcanoes. *Geology* **2008**, *36*, 671–672. [CrossRef]

216. Chauvet, A.; Alves Da Silva, F.C.; Faure, M.; Guerrot, C. Structural evolution of the Paleoproterozoic Rio Itapicuru granite-greenstone belt (Bahia, Brazil): The role of synkinematic plutons in the regional tectonics. *Precambrian Res.* **1997**, *84*, 139–162. [CrossRef]

217. Hickman, A.H. Two contrasting granite-greenstone terranes in the Pilbara Craton, Australia: Evidence for vertical and horizontal tectonic regimes prior to 2900 Ma. *Precambrian Res.* **2004**, *131*, 153–172. [CrossRef]

218. Van Kranendonk, M.J.; Collins, W.J.; Hickman, A.; Pawley, M.J. Critical tests of vertical vs. horizontal tectonic models for the Archaean East Pilbara Granite-Greenstone Terrane, Pilbara Craton, Western Australia. *Precambrian Res.* **2004**, *131*, 173–211. [CrossRef]

219. Nance, R.D.; Murphy, J.B.; Strachan, R.A.; Keppie, J.D.; Gutiérrez-Alonso, G.; Fernández-Suárez, J.; Quesada, C.; Linnemann, U.; D'lemos, R.; Pisarevsky, S.A. Neoproterozoic-early Palaeozoic tectonostratigraphy and palaeogeography of the peri-Gondwanan terranes: Amazonian v. West African connections. *Geol. Soc. Lond. Spec. Publ.* **2008**, *297*, 345–383. [CrossRef]

minerals

MDPI

Article

Fault Zone Evolution and Development of a Structural and Hydrological Barrier: The Quartz Breccia in the Kiggavik Area (Nunavut, Canada) and Its Control on Uranium Mineralization

Alexis Grare [1,*], Olivier Lacombe [1], Julien Mercadier [2], Antonio Benedicto [3], Marie Guilcher [2], Anna Trave [4] , Patrick Ledru [5] and John Robbins [5]

1   Sorbonne Université, CNRS-INSU, Institut des Sciences de la Terre de Paris, ISTeP UMR 7193, F-75005 Paris, France; olivier.lacombe@sorbonne-universite.fr
2   Université de Lorraine, CNRS, CREGU, GeoRessources lab, 54506 Vandoeuvre-lès-Nancy, France; julien.mercadier@univ-lorraine.fr (J.M.); marie.guilcher1@gmail.com (M.G.)
3   UMR Geops, Université Paris Sud, 91405 Orsay, France; antonio.benedicto@u-psud.fr
4   Departament de Mineralogia, Universitat de Barcelona (UB), Petrologia i Geologia Aplicada, Facultat de Ciències de la Terra, 08028 Barcelona, Spain; atrave@ub.edu
5   Orano Canada Inc., 817 45th Street, West Saskatoon, SK S7L 5X2, Canada; patrick.ledru@orano.group (P.L.); john.robbins@orano.group (J.R.)
*   Correspondence: alexisgrare@gmail.com

Received: 25 May 2018; Accepted: 24 July 2018; Published: 27 July 2018

Abstract: In the Kiggavik area (Nunavut, Canada), major fault zones along, or close to, where uranium deposits are found are often associated with occurrence of thick quartz breccia (QB) bodies. These bodies formed in an early stage (~1750 Ma) of the long-lasting tectonic history of the Archean basement, and of the Proterozoic Thelon basin. The main characteristics of the QB are addressed in this study; through field work, macro and microscopic observations, cathodoluminescence microscopy, trace elements, and oxygen isotopic signatures of the quartz forming the QB. Faults formed earlier during syn- to post-orogenic rifting (1850–1750 Ma) were subsequently reactivated, and underwent cycles of cataclasis, pervasive silicification, hydraulic brecciation, and quartz recrystallization. This was synchronous with the circulation of meteoric fluids mixing with Si-rich magmatic-derived fluids at depth, and were coeval with the emplacement of the Kivalliq igneous suite at 1750 Ma. These processes led to the emplacement of up to 30 m thick QB, which behaved as a mechanically strong, transverse hydraulic barrier that localized later fracturing, and compartmentalized/channelized vertical flow of uranium-bearing fluids after the deposition of the Thelon Basin (post 1750 Ma). The development and locations of QB control the location of uranium mineralization in the Kiggavik area.

Keywords: hydrothermal breccia; hydraulic breccia; uranium deposits; structural control; silicification; Kiggavik

1. Introduction

Fault zones are often associated with enhanced, focused, repeated fluid circulations in the earth's crust [1–7]. These fluids may have different origins: Meteoric, magmatic, metamorphic or basinal, and possibly transport metals to a favorable area of deposition [8,9]; that will ultimately allow for the formation of potential economic ore deposits. In many conceptual models of the formation of ore deposits, fault zones are important structural features acting as pathways [2,10] and/or as traps for fluids, and related metals [11]. In the uppermost crust, deformation is dominantly brittle and breccias are commonly observed in fault zones [12–15]. Among the different families of breccias, hydrothermal breccias are one sub-class that would develop early, in response to fracture propagation processes [13], through interaction between brecciated rocks and hydrothermal solutions. Hydrothermal breccias can be of various types depending on several parameters, such as pressure, temperature, depth of emplacement, and elements in the fluids [14]. Among them, quartz-cemented breccias can have an economic interest, being possibly associated with ore deposits such as epithermal (Au-Ag-Cu-Pb-Zn-Sb, [16,17]), orogenic gold (Au, [18]), and porphyric (Cu-Mo-Au-Ag, [19,20]). They display thickness from meter to several meters, thicker hydrothermal breccias being relatively rarely described. Quartz breccias in fault zones form progressively during several cycles of fluid pressure growth, seismogenic fault slip and quartz precipitation [21,22]. Unaltered, quartz-rich bodies have a lowered porosity and thus have an impact on later fluid circulation within the fault zone. Such silicification would be comparable to fluid-flow being constrained by horizontal barriers, such as sedimentary layers indurated through diagenesis (aquitards, [23,24]), or impermeable (clay-rich) layers in roll-front uranium deposits [25]. In addition, the likely hardening of the fault rocks in response to multiple cycles of quartz brecciation and healing may cause a significant rheological contrast between the "strong" fault zone and the expectedly "weaker" hosting terranes, possibly controlling localization of subsequent deformation.

In this contribution, we focus on one structural feature encountered in many fault zones within the Uranium (U)-rich district of the Kiggavik area (Nunavut, Canada): The so-called hydrothermal Quartz Breccia (QB). The importance of this breccia, only briefly described by previous authors [26–30] was recently highlighted by Grare et al. [31] who documented the control exerted by this breccia on later fracturing events, hydrothermal alterations and uranium mineralization at the Contact uranium prospect. However, despite observations in several locations of the Kiggavik area and its seemingly strong control on the current distribution of the uranium mineralization, the genetic model of the QB remains poorly characterized and explained to date. Grare et al. [31] showed that the QB emplaced along faults of inferred Archean age, and that this emplacement was a key event within a long-lasting (~1000 Ma) complex brittle tectonic history that led to uranium mineralization within or in the vicinity of the quartz breccia (Figure 1C). In order to better constrain the nature, emplacement, significance and role of the QB, we carried out a structural analysis combined with vein cement petrography using optical and cathodoluminescence observations, trace elements, and oxygen stable isotope analysis of quartz. Our study addresses the structural, mineralogical and geochemical characteristics of the QB. Combined with the reconstructed geochemical signature of the fluids, a model of formation of the QB is proposed and its role in controlling uranium mineralization in the Kiggavik area is highlighted.

2. Geological Setting

2.1. Regional Geological Setting

The Kiggavik area is located on the eastern border of the Proterozoic intracratonic Thelon Basin (ca. 1670–1540 Ma, [32,33]) in Nunavut, Canada, within the Churchill province. The Churchill province is known to host the Athabasca Basin (1740–1540 Ma, [34]); another Proterozoic basin, which itself hold the world-class Cigar Lake and McArthur River uranium deposits. The Thelon Basin is one analogue of the Athabasca Basin and the Kiggavik area displays several economically significant uranium orebodies: Four of the deposits yield calculated resources of 48,953 t of uranium at a grade of 0.47% U [35]. Exploration began in the 1980s by Urangesellschaft, and the property is now held by Orano Canada (formerly known as AREVA Resources Canada) in joint venture with JCU (Canada) Exploration Company Ltd. (Vancouver, BC, Canada).

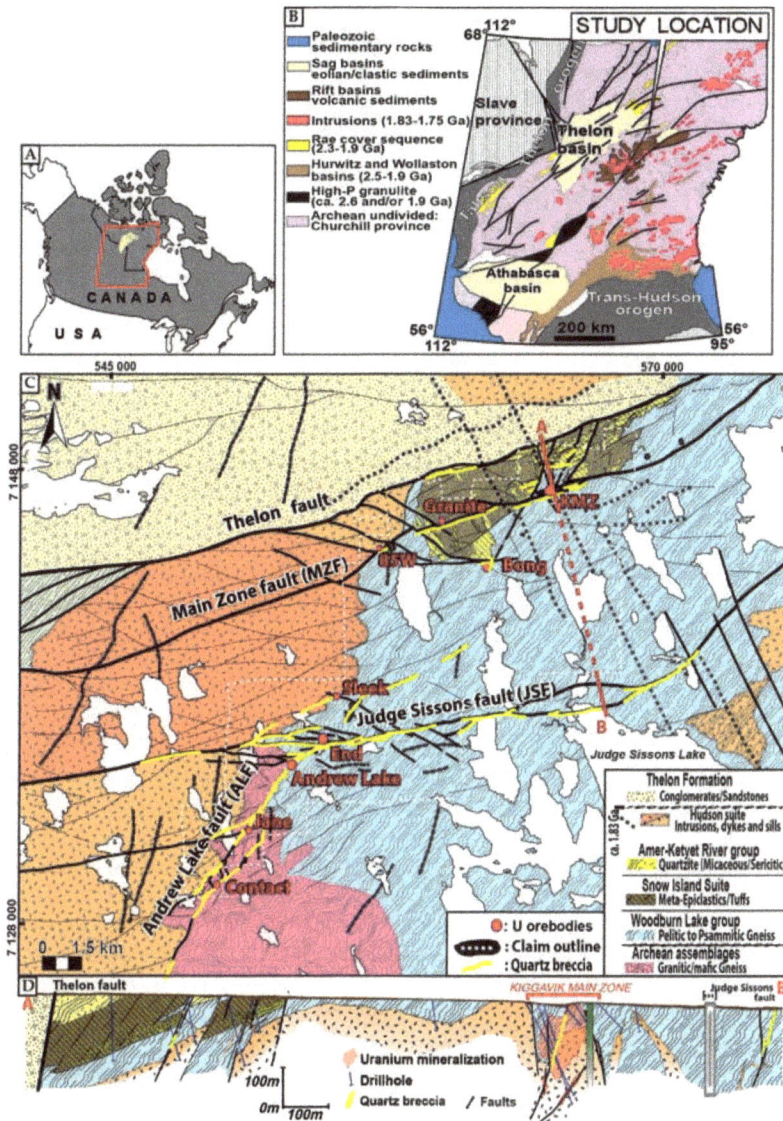

Figure 1. (**A**) Outline of Canada and location of the Thelon basin in yellow; (**B**) geological map of the Churchill-Wyoming craton showing the location of the Thelon basins and the Kiggavik area on its Eastern border; (**C**) simplified geological map of the Kiggavik area (Orano internal document) highlighting the occurrence of the QB (yellow) along the major faults; and (**D**) cross-section from the Thelon fault to the Judge Sisson fault. Deposits and prospects are indicated with red circles.

The Churchill province (Figure 1B) is bordered to the NW by the Thelon-Taltson (ca. 2020–1900 Ma), and to the SE by the Trans-Hudson orogenic belts (ca. 2070–1800 Ma). At the end of the Trans-Hudsonian orogeny, the Baker Lake Basin developed as a result of (retro-arc) extensional to transtensional rifting tectonics [36], and was filled with sedimentary and bi-modal volcanic-sedimentary rocks (Baker Lake and Wharton Grps, ca. 1850–1750 Ma, [37,38]). It was followed by uplift, extensive erosional peneplanation and regolith formation, over which deposited the eolian sandstones and conglomeratic red-beds of the Thelon formation (ca. 1670–1540 Ma [32,33]), linked to thermal subsidence in the sag, fault-controlled intracratonic Thelon basin [36,38,39]. This volcano-sedimentary pile unconformably overlies a metamorphosed basement consisting of Archean rocks that include Mesoarchean (ca. 2870 Ma) granitic gneisses, 2730–2680 Ma, supracrustal rocks of the Woodburn Lake Group [40], and a distinctive package of 2620–2580 Ma felsic volcanic and related hypabyssal rocks known as the Snow Island Suite [41–47].

Before emplacement of the Thelon formation, the Archean to Paleoproterozoic rocks of the Churchill province where intruded by three magmatic suites: (i) The late syn-orogenic (ca. 1830 Ma) Hudson Suite [48], (ii) the Dubawnt Minette Suite (contemporaneous of the Hudson Suite), with ultrapotassic intrusions, minette dikes and lamprophyres, and (iii) the anorogenic (ca. 1750 Ma) Kivalliq Igneous Suite (KIS) [46,49–51].

2.2. Local Geological Setting

A simplified geological map of the Kiggavik area is presented in Figure 1C. The local litho-structural pile consists of Mesoarchean granitic, granodioritic, and augen gneisses (2866 ± 6 Ma; [52]) tectonically overlain by a Neoarchean metavolcano- sedimentary package retromorphosed to greenschist facies: The Woodburn Lake Group. This package consists of quartzo-feldspathic wackes and minor quartzite with thin, interbedded banded iron formation layers, rare black shales, and locally komatiite and rhyolite (2710 ± 2.1 Ma). These rocks, together with overlying Paleoproterozoic (2300–2150 Ma) rocks of the Ketyet River Group [53], include a prominent unit of orthoquartzite [52]. These rocks are intruded by the Schultz Lake Intrusive Complex (SLIC, [51]). The SLIC comprises rocks from the two intrusive suites previously described [51]: (i) The "Hudson granite" consists of non-foliated granitoid sills, syenites and lamprophyre dikes of the late syn-orogenic Hudson Suite; and (ii) the "Nueltin granite" comprises anorogenic granite to rhyolite of the KIS [46].

The diabase dikes of the Mackenzie diabase swarm form prominent linear aeromagnetic features trending NNW-SSE [44,45] and cut across all previous lithologies. This intrusive event is dated at 1267 ± 2 Ma [54,55], and represents the last magmatic-tectonic event in the region.

The main structural features in the Kiggavik area are the ENE-trending Thelon fault (TF) and the Main Zone fault (MZF) in the northern part of the property, the ENE-trending Judge Sisson fault (JSF) in the central part, and the NE-trending Andrew Lake Fault (ALF) in the southwestern part of the study area (Figure 2). These faults date back to at least ~1920 Ma [56] and had a subsequent complex structural and kinematic evolution with several episodes of reactivation and fluid circulation during Proterozoic time [31,56]. These faults host several uranium orebodies; prospects and deposits, the main uranium mineralizing events being bracketed between 1540 and 1270 Ma [28,29,31]. The MZF hosts various deposits and prospects: 85 W, Granite Grid and Kiggavik (Main, Central and East Zones, Figure 1C). End is hosted by the JSF, while Andrew Lake, Jane and Contact occur along the ALF (Figure 1C).

Figure 2. (**A**) Outcrop view looking east on the N80-trending steeply dipping to the north Judge Sisson fault (JSF) underlain by at least 10 m of white quartz veins; (**B**) heterogeneous size, pervasively hematized clasts cemented by a white quartz matrix; (**C**) right lateral relay step, N80 trending main veins (outcrop on the JSF); (**D**) optical microphotograph picture (OM): Clasts bearing quartz veins in the Thelon sandstones; (**E**) oriented data of thick quartz veins for deposits and prospects; and (**F**) histogram of all measured quartz vein dips in the Kiggavik area.

3. Sampling and Methods

3.1. Drillhole Observations, Sampling Strategy and Collection of Oriented Data

The QB has been observed in the field but the scarcity of outcrops in the area is the reason why most observations and oriented measurements were taken from drill holes within the deposits and prospects in the Kiggavik area (location in Figure 1C). Porosity was measured systematically in the field following the fluid resaturation method. A clean and dried sample is weighted, saturated with a liquid of known density, and then reweighed. The weight change divided by the density of the fluid results in the pore volume. Many of the observations and samples come from the recently drilled Contact prospect (2014 and 2015 Orano exploration campaigns). More than 5000 m of drill core were reviewed, with several hundreds of meters dedicated to the characterization and study of the QB. Recent drilling (2014–2015) in the Kiggavik area was done with NQ^{TM} coring providing a 47.6 mm diameter sample. Oriented data measured on drill core were restored in their original position and plotted with Dips 6.0 software (Rocscience, Toronto, ON, Canada). Uncertainty on fault/fracture orientation measurements is estimated to be ±10°.

3.2. Quartz Microscopic Characterisation by Optical and Cathodoluminescence Microscopy

Quartz Fifty-five drill core samples (10 to 20 cm in length) displaying veins or breccias linked to the QB were collected, mainly from the Contact prospect, but also from End, Andrew Lake and Bong deposits. All samples were studied from the macro- to the micro-scale in order to characterize the macroscopic texture of the quartz breccia and its relationships with predating and postdating fracturing and faulting events. Thirty-five thin sections were prepared for petrographic and microstructural studies. Thin sections were observed through optical microscopy (plane polarized transmitted and reflected light microscope Motic BA310 POL Trinocular, equipped with a 5 M pixel Moticam camera) (Motic Instruments Inc., Richmond, BC, Canada), and cathodoluminescence microscopy (CITL Cold Cathodoluminescence device Model MK5-1, made at University of Barcelona (Barcelona, Spain), for deciphering quartz generations.

3.3. Fluid Characterization by Trace Elements and Oxygen Isotopes Analyses

Laser ablation ICP-MS analyses of quartz were conducted at GeoRessources, Université de Lorraine (Vandoeuvre-lès-Nancy, France), using a 7500e quadrupole ICP-MS (Agilent, Santa Clara, CA, USA) coupled with a nanosecond excimer laser (GEOLAS Pro; 193 nm wavelength). Zones free of fluid inclusions (FIs) were selected for analyses. Analyses were performed using a laser beam diameter of 60 (first session of analyses) and 90 (second session) μm, with a fluence of ~10 J/cm^2 and a repetition rate of 5 Hz. The laser beam was focused onto the sample with a Schwarztschild reflective objective (magnification $\times$25; numerical aperture = 0.4). Each analysis consisted of 20 s of background measurement during laser warm-up, 20 to 40 s of ablation (depending on the thickness of the quartz) and 15 s of washout before repeating the process on a nearby location. The external standards were NIST SRM610 and NIST SRM 612 [57], the external standards being analyses twice at the beginning and at the end of each set of samples, following a bracketing standardization procedure. LA-ICP-MS calibration was optimized for highest sensibility for the whole mass/charge range, while maintaining Th/U ~1 and ThO/Th < 0.5% as determined on NIST SRM 610 or 612. The following isotopes were measured: ^{7}Li, ^{11}B, ^{23}Na, ^{24}Mg, ^{27}Al, ^{28}Si, ^{39}K, ^{44}Ca, ^{48}Ti, ^{57}Fe, ^{74}Ge, ^{85}Rb, ^{88}Sr, ^{89}Y, ^{90}Zr, ^{133}Cs, ^{138}Ba and ^{153}Eu for the first session, and ^{7}Li, ^{11}B, ^{23}Na, ^{27}Al, ^{29}Si, ^{45}Sc, ^{47}Ti, ^{51}V, ^{53}Cr, ^{55}Mn, ^{59}Co, ^{60}Ni, ^{63}Cu, ^{66}Zn, ^{69}Ga, ^{72}Ge, ^{75}As, ^{85}Rb, ^{88}Sr, ^{90}Zr, ^{93}Nb, ^{95}Mo, ^{115}In, ^{118}Sn, ^{121}Sb, ^{133}Cs, ^{137}Ba, ^{181}Ta, ^{182}W, ^{197}Au, ^{208}Pb, and ^{209}Bi for the second session. ^{28}Si or ^{29}Si were used as internal standard, using a SiO$_2$ concentration of 100%. Data reduction was done using Iolite software [58].

In situ oxygen isotope analysis of the main quartz generations was performed by secondary ion mass spectrometry (SIMS, CAMECA, Gennevillier, France) using the Cameca IMS1270 at CRPG/CNRS in Vandoeuvre-les-Nancy, France, following the approach of Hervig et al. [59]. The isotopes ^{16}O and ^{18}O were measured, based on standard polished sections coated with gold. A ~4 nA defocused primary ion beam of Cs impact energy 10 keV was used, producing sub-circular ablation craters of ~10–20 μm diameter. A mass resolution (ΔM/M) of 5000 was used, to resolve potential interference of ^{17}O on ^{16}O. Two in-house standards were used (Brésil (δ^{18}O = 9.6‰) and Brésil-2 (δ^{18}O = 19.6‰)) to set-up the instrument and correct for drifts and fractionations using a standard bracketing approach. The internal precision for δ^{18}O was between 0.06 and 0.1‰ (measurements on the standards Brésil and Brésil-2 and on the different quartz generations of Kiggavik). δ^{18}O values are reported relative to the V-SMOW standard.

4. Results

4.1. Spatial Organisation and Macroscopic Characteristics of the QB

Occurrence of QB has been recognized along various segments of the major faults within the Kiggavik area (e.g., ALF, JSF, Figure 1C,D and Figure 2A). The QB consists of a up to 30 m thick complex network of mosaic quartz-sealed breccia and veins (Figure 2A–C), typically displaying angular fragments and jigsaw pattern (Figure 2B), and associated with a pervasive iron-oxidation of the host rock (Figure 2B). This kind of observation is common in drill holes. Lithologies within and around the QB display a pervasive red-purple hematization, as documented at the Contact prospect [31]. Clasts bearing veins of the QB are observed in the sandstones of the Thelon formation

(Figure 2D), indicating that QB predates formation of the Thelon Basin, as already suggested by several authors [29,31] and crosscuts, thus postdates, Hudsonian intrusions (ca. 1.83 Ga). Fault zones outlined by the QB are presumably better preserved in the field due to the silicification process that increases their resistance to erosion.

The outcrop shown in Figure 2B illustrates the complexity of the identification of the main structural trends on limited exposures. We considered that the most regionally significant structural trend of the breccia bodies is given by the thicker (>10 cm) veins and breccias, because where they are visible, minor quartz veins are more randomly oriented or give a mean statistical value that is different between two (2) nearby drill holes. By plotting the orientations of thick veins we infer the true orientation of the quartz breccia (Figure 2E), which was revealed to be consistent with the major fault trends in map view (Figure 1D). The QB usually displays a consistent high angle dip, reflecting the orientation of the main fault trend: N30, dip to the NW at Contact, N175, dip to the W at Bong, N50 and N90, dip to the NW and to the S, respectively, at End (Figure 2E). Even though the majority of minor quartz veins display throughout the Kiggavik area a steep dip (60–90°), a significant amount of veins (Figure 2F) shows relatively shallow dip angles (<30°).

Figure 3 summarizes the data collected on drill holes at the Contact prospect (Figure 3A). The QB bodies usually display two main distinct zones, an outer zone and an inner (core) zone. The outer zone (blue in Figure 3B) is represented by a dense to scarce network of millimeter to centimeter-thick quartz veins, while the inner (core) zone (red in Figure 3B) is represented by thick (>10 cm thick) quartz veins and a dense quartz vein network, where angular clasts of the fragmented host rock are barely observable. Several QB core zones were crosscut by drill holes (Cont-24, Cont-16, Cont-06). These core zones are discontinuous from the SW to the NE. They are tapering toward their ends (Figure 3C) both laterally (for example, between Cont-26 and Cont-25, Figure 3B), and vertically (for example, between Cont-10 and Cont-11, Figure 3B). This supports that they have elliptical shapes, connected by quartz vein networks. This observation explains the important changes in thickness of the QB between two nearby drill holes (e.g., Cont-06 and Cont-13).

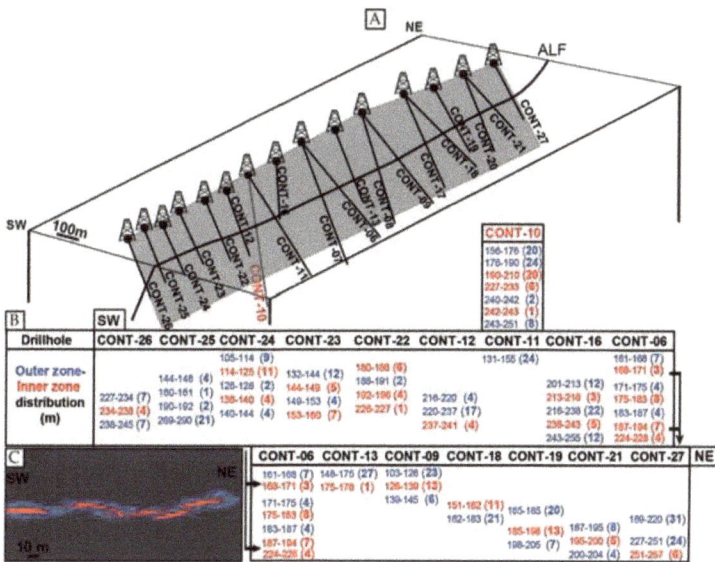

Figure 3. Organisation of inner (core) and outer zones of the quartz breccia (QB) crosscut in drillholes at Contact. (**A**) Plan view of the drill holes; (**B**) lateral variation in thickness of QB inner (core) and outer zones; and (**C**) simplified interpretative drawing of the QB intersected in drill holes (grey plane).

One observation not highlighted by previous studies in the Kiggavik area is the presence of a large (20–100 m) brittle fault zone predating emplacement of the QB but systematically spatially associated with it. Macroscopically, the QB consists of thin to massive quartz veins as described in Figure 2; however, our detailed observations document numerous quartz healing events crosscutting clay-altered cataclastic to ultra-cataclastic fault rocks that are now silicified and "preserved". Clasts are monomictic, sub-rounded, millimetric to centimetric in size and clay altered, embeded in a light red to brown matrix (Figure 4A,B).

Figure 4. (**A**) Pervasively silicified cataclastic fault rock; (**B**) same as (**A**), crosscut by a white quartz vein of the QB; (**C**) pervasively silicified fault zone crosscut by late fracturing and clay alteration event (End deposit); and (**D**) typical intersection of the QB displaying deep purple hematized rock, massive and minor white quartz veins. Jigsaw textures are locally observable (e.g., at 189 m, yellow arrow; Contact prospect).

The quartz veins of the QB were observed in several locations as cutting across the cataclasites (Figure 4B). These early cataclastic fault rocks therefore predate the QB; they could be related to extensional to trans-tensional faulting during formation of the Baker Lake Basin [31]. This early, now silicified fault zones and the QB are spatially associated, indicating that the pervasive silicification likely occurred at the onset of emplacement of the QB. However, even though the pervasive silicification of the fault zone is spatially and likely roughly temporally associated with the QB, we differentiate hereafter these two features: The silicified fault zone on one hand and the QB that results from brecciation sealed by quartz on the other hand.

Both features display different thicknesses: In Figure 4C, the pervasively silicified fault zone with its light reddish color is observable along 40 m of drill core and is cut by numerous small quartz veins and a 4 m thick core zone of the QB. A late faulting and white clay alteration pattern is observed at depth 389–395 m (Figure 4C, post ore faulting f7). In Figure 4D, the silicified fault zone is observed along 5 m of drill core and is cut by 23 m of QB.

The pre-QB silicified fault zone displays evidence of multiple events of tectonic brecciation and comminution. In the sample observed at micro-scale under transmitted light (Figure 5A), three generations of cataclastic fault rocks are observed, with each generation of cataclasis consuming the previous one. They are crosscut by at least three generations of quartz veins, building a complex pattern (Figure 5B,C). Minerals from the original host rock (psammo-pelitic gneiss with quartz, apatite, illite, muscovite, pyrite) are preserved in the first generation of clasts (pink, Figure 5B). A closer look at the cataclastic fault rocks reveals that the different cements are made of micro-crystalline quartz and white micas (Figure 5D,E). The superimposition of multiple generations of cataclasites indicates that the localized zone of deformation was repeatedly reactivated during progressive deformation.

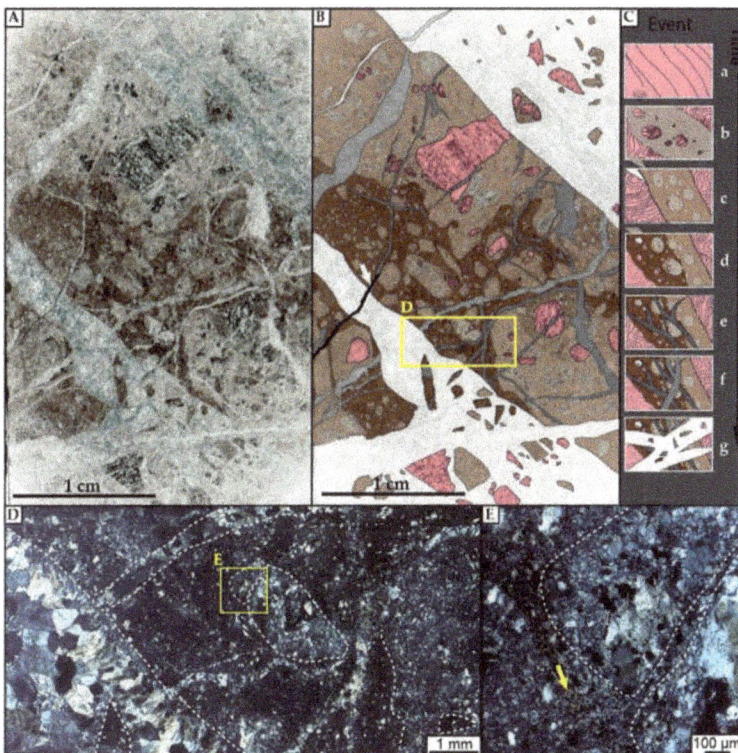

Figure 5. (**A,B**) Thin section of a polyphase cataclastic fault rock crosscut by several generations of quartz veins of the QB. White arrow indicates a late microcrystalline quartz veinlet; (**C**) simplified chronology of the events; (**D**) zoom on the different generations of clasts; and (**E**) matrix of the latest cataclastic event displaying white micas and micro-crystalline quartz.

In order to better understand and characterize the influence of silicification on fluid circulation, we selected porosity data measured in the field for four types of rocks: Fresh host rock (granitic gneiss, before fracturing and alteration), silicified type 1 (pervasively silicified fault zone), silicified type 2 (typical white QB), and clay-altered/fractured samples. Results are presented in Figure 6. Fresh granitic gneiss yields the lowest porosity values, <2%. Fault rock and samples displaying quartz brecciation and pervasive silicification yield values slightly higher but <5%. Fractured and clay altered fault rock display much higher values, up to 40%. Cataclastic fault rocks formed before the QB should have displayed a high porosity, but after pervasive silicification they have a porosity comparable to fresh rock (Figure 6), unlike strongly clay altered and fractured samples (Figure 6).

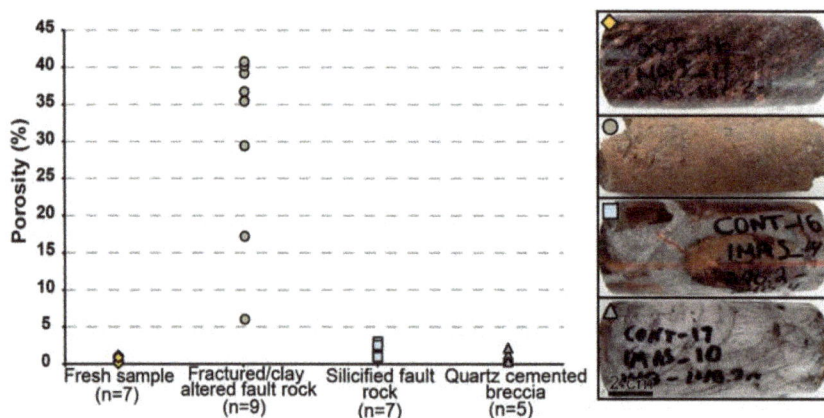

Figure 6. Porosity measured for fresh samples (granitic gneiss from the Contact prospect), pervasively silicified cataclastic fault rock, thick quartz veins within granitic gneiss, and clay-altered, fractured host rock (examples for each category are displayed on the right of the chart).

4.2. Microscopic-Scale Characteristics of the QB

Microscopic observations indicate that the pervasive hematization spatially associated with the QB is defined by disseminated micro-grains of hematite (aggregates of micrometric grains) and specular hematite (Figure 7A; specular hematite being less common in samples compared to hematite); possibly observed in banded veins synchronous with microcrystalline quartz (Figure 7B). Subhedral specular hematite (~100–200 μm) is observed filling quartz vugs and disseminated in the host rock. Where present, the specular hematite is responsible for the dark-red to purplish color of the oxidized host rock. Anhedral magnetite (50–100 μm) was locally observed as being mixed with (likely replaced by) hematite (Figure 7B) in banded quartz veins and likely represents changes in the oxidation state of the fluid. These observations, along with the spatial association of iron oxidation and quartz brecciation, support the overall synchronicity of the two phenomena. However, the precise timing of the oxidation, within the several episodes of silicification, remains unconstrained.

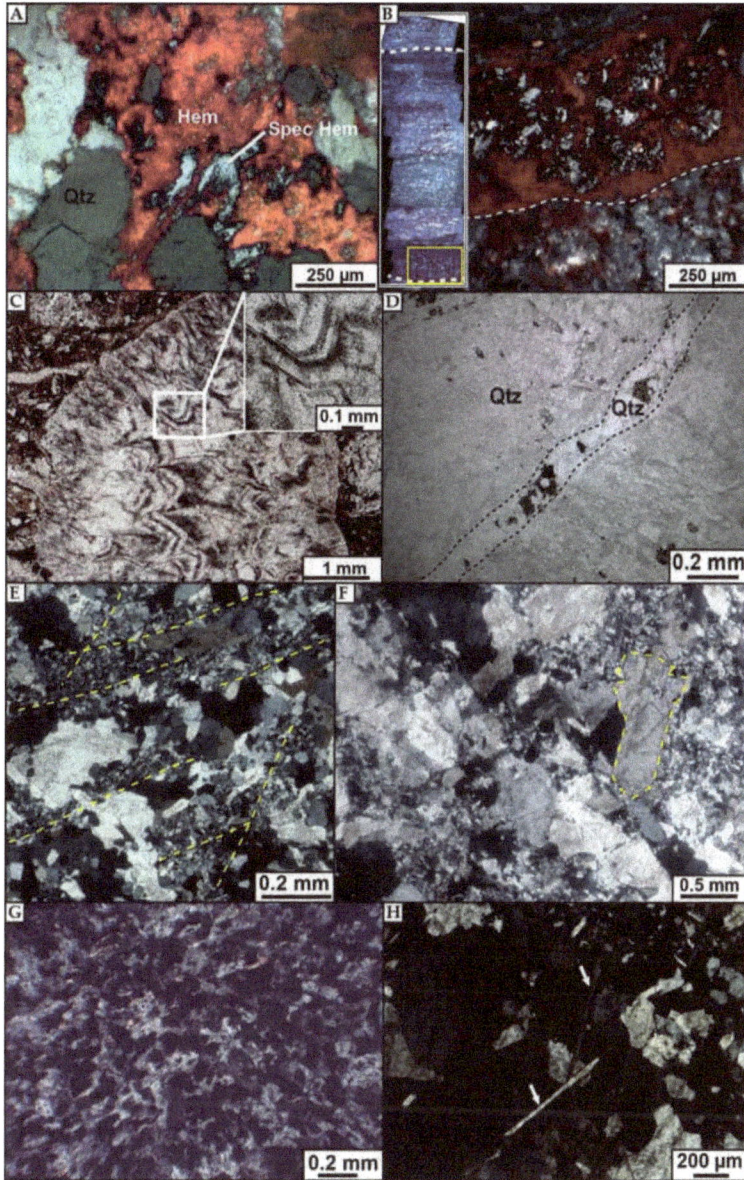

Figure 7. Optical microscope microphotograph (OM): (**A**) Disseminated hematite (Hem) and specular hematite (Spec Hem). Qtz: Quartz; (**B**) banded microcrystalline quartz (Qtz) with synchronous anhedral hematite and magnetite; (**C**) euhedral quartz crystals and arrays of dense monophase fluid inclusions (vapor rich); (**D**) euhedral clear quartz cement a fracture that crosscuts previous quartz generations; (**E**) trends of microcrystalline quartz (yellow); (**F**) comb quartz grains (example in yellow) engulfed in a fine-grained quartz matrix; (**G**) moss quartz texture; and (**H**) bladed lattice calcite (white arrow) replaced by quartz.

Microscopic observations also document a variety of quartz textures (Figure 7C–F for example), mutually crosscutting each other, and defining different conditions of quartz precipitation. The two most common types of quartz are: Euhedral quartz (comb quartz, ~200 µm in size) and microcrystalline quartz (~<50 µm). Two generations of euhedral white quartz can be distinguished: One (millimetric quartz) being characterized by dense arrays of monophase fluid inclusions (vapor rich), usually at the tip of the quartz crystal (Figure 7C); and the second, clearer, nearly fluid inclusions-free, usually observed as a late quartz generation (~100 µm, Figure 7D). In addition to the banded microcrystalline quartz-hematite texture, microcrystalline quartz is also observed filling vugs, and as conjugate "trends" (Figure 7E) in subhedral quartz veins. In other samples, subhedral quartz can be found as clasts in a microcrystalline quartz mass (Figure 7F). In term of quartz texture, comb quart, microcrystalline quartz and "moss" textures were observed (Figure 7G). Additionally, rare recrystallized bladed calcite were found (Figure 7H).

Quartz observed under cathodoluminescence display weak luminescence intensity, with a 20 s exposure time required in order to get enough signal for imaging. The most recurrent color observed under cathodoluminescence is a deep blue observed for microcrystalline quartz veins, sometimes synchronous with hematite (Figure 8A), and quartz cementing microbreccias. The fluid inclusion (FI)-rich euhedral quartz crystals exhibit alternating growth zones of brown and blue luminescence (oscillatory growth-zoning, Figure 8B). Clasts of euhedral quartz crystals are found within a blue luminescent quartz matrix (Figure 8C). The brown luminescence is also observed in breccias where the quartz has likely completely recrystallized, leaving the breccia texture only observable under cathodoluminescence; the "cement" of the breccia displays a brown luminescence (Figure 8D). These colors characterize the main generations of quartz in the QB.

The luminescence of the latest generation of quartz (i.e., euhedral quartz filling vugs and open fractures), is dark blue with rare concentric zoning. It also displays greenish luminescence associated with primary to pseudo-secondary fluid inclusions (Figure 8E). In terms of spatial occurrence of this quartz generation, it is more frequently observed in the vicinity of the QB than in its inner zone.

Quartz which was formerly in contact with uranium minerals displays a characteristic luminescence: Red/pink close to uranium-bearing minerals and yellow/greenish further from the uranium-bearing mineral (Figure 8F). This is especially well observed in quartz veins that were later microfractured as described by Grare et al. [31] and in quartz of the host rock (when not dissolved by circulation of the uranium-bearing fluid). This peculiar luminescence is brighter than the original luminescence of the quartz and displays a nearly uniform circular shape of 35–45 µm width (Figure 8F). The latest quartz generation, which fills vugs and open fractures, is characterized by dark blue luminescence with rare concentric zoning. This generation is observed more in the vicinity of the QB.

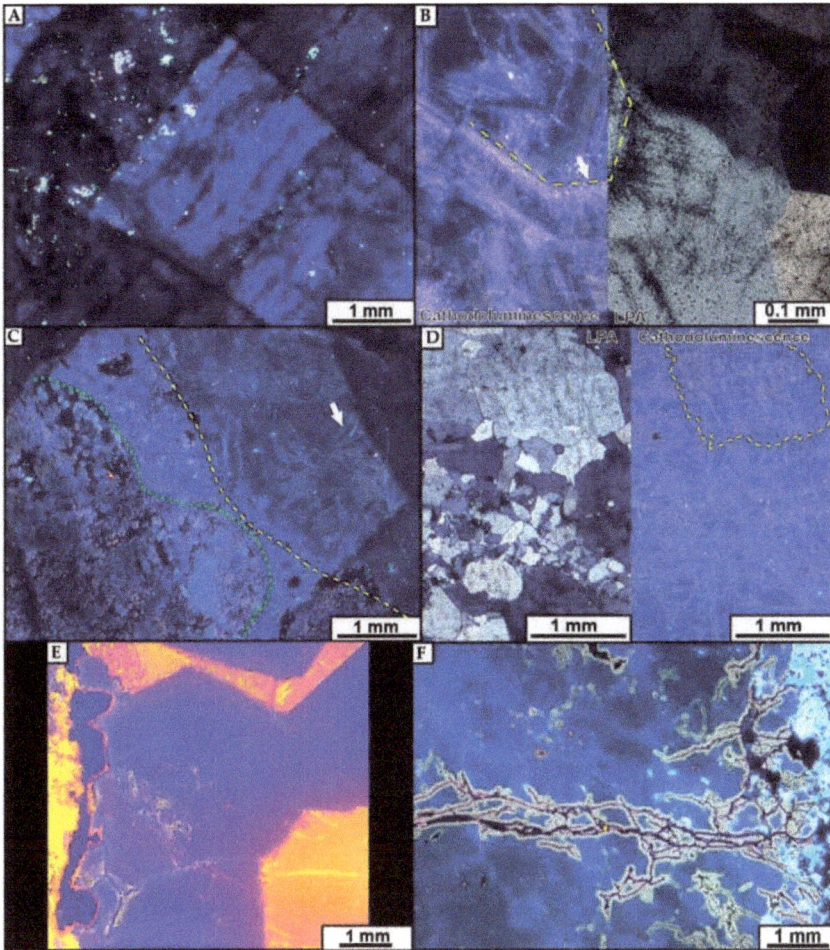

Figure 8. Cathodoluminescence microphotograph (CM): (**A**) Deep blue luminescent microcrystalline quartz and red luminescent hematite in banded vein; (**B**) fracture cemented with euhedral quartz (blue-brown luminescence in concentric zoning, yellow dotted line) and arrays of purplish FIs (white arrow); (**C**) quartz breccia (green) cemented by subhedral, sub-millimetric blue-purple quartz, crosscut by dark blue luminescent subhedral millimetric quartz vein (yellow), the white arrow highlight brown-luminescent overgrowth; (**D**) OM and CM of the same zone: Subhedral and microcrystalline quartz (blue luminescence) displaying microbrecciation/recrystallization (brown luminescence). Yellow dashed line outlines a quartz crystal; (**E**) dark blue luminescent euhedral syntaxial quartz. The boundary of the fracture on the left is lined with U-oxides (black and calcite (orange); and (**F**) quartz vein orthogonally crosscut by a microfracture and cemented with pitchblende. The boundaries of the microfractures display a "buffer area" where the luminescence of the quartz is modified, from pink to yellow.

4.3. Geochemical Signature of Quartz

4.3.1. Trace Element Concentrations

Concentrations of selected elements were measured with LA-ICP-MS for the three main quartz generations (identified through microscopic texture and luminescence color): Deep blue microcrystalline quartz, blue euhedral quartz with brown concentric zoning, late dark blue euhedral vuggy quartz. Results are presented in Figure 9A for Li, K, and Ba; in Figure 9B for Zr, Ti, and B; in Figure 9C for Mg, Na, Al, Fe, and Ca. Several elements we tried to measure show values below the limit of detection and thus are not displayed here.

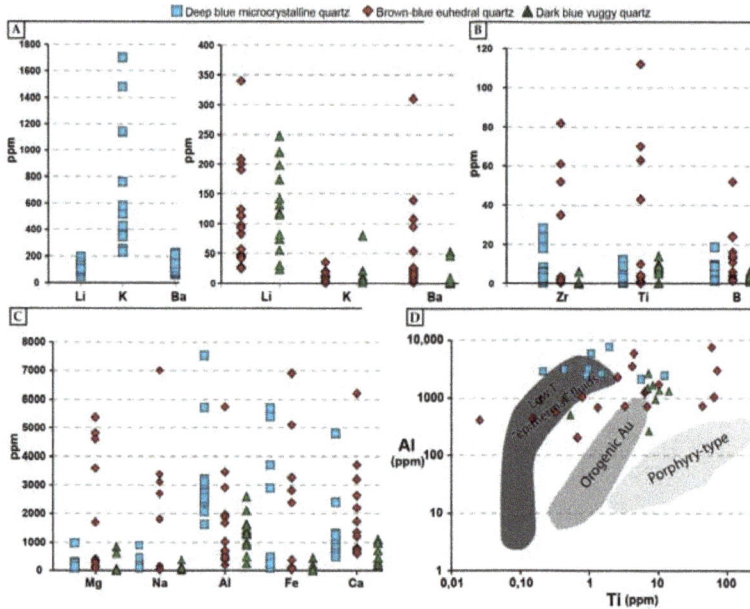

Figure 9. Concentration of trace elements measured in quartz through LA-ICP-MS, for main quartz generations: (**A**) Li, K, and Ba; (**B**) Zr, Ti, and B; (**C**) Mg, Na, Al, Fe, and Ca; and (**D**) Al vs. Ti concentrations of main quartz generation. Zones correspond to values of hydrothermal quartz from low T °C, orogenic Au, and porphyry-type deposits [60].

In some analyses, trace elements display extreme values above 10,000 ppm (e.g., Al or Mg, Figure 9) which likely represent analysis of undetected solid inclusions, hence are not displayed in Figure 9. The high Fe content in both the deep blue microcrystalline and brown-blue euhedral quartz could reflect the analysis of micro-inclusions of iron oxides, related to the pervasive hematization synchronous with the QB event. Microcrystalline quartz in banded veins associated with iron oxides shows a small range of values for all elements except for K and Fe. Brown-blue euhedral quartz displays bimodal concentrations for most of the elements consistent with observed concentric zoning. For all quartz generations, Li, K, and Na are positively correlated with Al. Dark blue vuggy quartz usually display a small range of values for most elements compared to other quartz generation, except for Li (23–248 ppm) and Al (255–2593 ppm).

Li contents are homogeneous between the three quartz generations and are below 250 ppm. Such values correlate positively with Al concentrations, Li balancing the replacement of Si by Al [61]. K is enriched in deep blue microcrystalline quartz with values up to 1700 ppm, compared to concentrations below 100 ppm in the case of the two other quartz generations. Fe yields high values (up to 7000 ppm) in the case of the two quartz generations of the QB (deep blue microcrystalline and brown-blue euhedral quartz). B displays concentrations below 20 ppm except for 2 analysis. Concentrations in B are lower in the case of post-QB dark blue vuggy quartz.

Deep blue microcrystalline and dark blue vuggy quartz yield low values of Ti (<20 ppm for most measurements). Bi-modal concentrations of Ti were measured for brown-blue euhedral quartz, with one group of values below 20 ppm and the other above 40 ppm.

4.3.2. Oxygen Isotope Values

Oxygen isotopes were measured in the main quartz generations while paying attention to variations in luminescence, for instance for quartz displaying concentric zoning. Results are shown in Table 1. $\delta^{18}O_{quartz}$ values are relatively homogeneous within one quartz generation (maximum variation of 5‰) and are independent of Cathodoluminescence (CL) color variations except for dark blue vuggy quartz: Zones of this quartz close to fluid inclusions display a green luminescence with higher $\delta^{18}O_{quartz}$.

The different quartz generation display different ranges of $\delta^{18}O_{quartz}$, with values ranging between +12‰ and +14‰ for microcrystalline quartz with hematite, between +7.5‰ and +9.3‰ for brown blue quartz and between +14.4‰ and +15.5‰ for druzy/vug-filling quartz. These ranges show two exceptions: Vug-filling euhedral quartz displaying green luminescence with higher isotopic values (from +16.2‰ to +21.9‰) in zones close to fluid inclusions (Figure 7C), and a microcrystalline quartz vein crosscutting the one cemented with microcrystalline quartz and iron-oxides exhibiting much higher isotopic values (from +18.8‰ to +23.9‰).

$\delta^{18}O_{quartz}$ values between the two main generations of quartz in the QB are different with an average of +12.9‰ for banded microcrystalline quartz and +8.4‰ for zoned euhedral quartz.

Table 1. $\delta^{18}O_{quartz}$ measured in main quartz generations and calculated values of $\delta^{18}O_{fluid}$. To calculate $\delta^{18}O_{fluid}$, an average temperature of 250 °C was used for quartz generations of the QB (lines 1–3 of the table), while an average temperature of 150 °C was used for late druzy quartz (lines 4–5).

Quartz Type	CL Luminescence Color	$\delta^{18}O_{quartz}$	Avg.	n	$\delta^{18}O_{fluid}$	Avg.
Banded microcrystalline quartz (alternated with iron oxides)	Deep blue	12.0–14.0	12.9	14	2.4–5.3	3.9
Euhedral quartz with concentric zoning	Alternating blue and brown luminescence	7.5–9.3	8.4	18	−1.6–0.3	−0.6
"Late" microcrystalline quartz	Deep blue	18.8–23.9	22.1	10	11.5–14.8	12.4
Vuggy quartz	Dark blue	14.4–15.5	14.9	29	−5.3–2.5	−3.3
Quartz alteration associated with fluid inclusions	Green	16.2–22.0	17.8	5	−2.5–3.8	−0.3

$\delta^{18}O_{quartz}$ and temperatures measured by fluid inclusion microthermometry in quartz veins in the area [26,29,62] were used to calculate the $\delta^{18}O_{fluid}$ following the equation of Clayton et al. [63], set for measuring oxygen isotope exchange between quartz and water (assuming that the fluid was in equilibrium with the quartz at the temperature of mineralisation). We considered homogenisation temperatures in the range of 200–300 °C, avg. 250 °C (i.e., the range of temperatures revealed by low salinity fluid inclusions), to be representative for the quartz generations of the QB and of 100–200 °C, avg. 150 °C (i.e., the range of temperature revealed by high salinity fluid inclusions), to be representative for late druzy quartz probably precipitating from basinal brines [29]. However, a microthermometric study on primary fluid inclusions for each quartz generation is missing actually and would give a more accurate calculation of fluid isotopic values. Results are displayed in Table 1. The $\delta^{18}O_{fluid}$ value range from 2.4‰ to 5.3‰ (+3.9‰ on average) for microcrystalline quartz associated with hematite. In contrast, late veinlets of micro-crystalline quartz display a much higher $\delta^{18}O_{fluid}$ value: Between +11.5‰ and +14.8‰ (+12.4‰ on average). Brown-blue quartz precipitated from a fluid with a lighter $\delta^{18}O_{fluid}$ value comprised between −1.6‰ and +0.3‰ (−0.6‰

on average). Late vug-filling euhedral quartz yield lighter isotopic values from −5.3‰ to −3.3‰ (−3.3‰ on average).

5. Interpretation of Results and Discussion

5.1. Origin and Nature of Silicifying Fluids

Although relations between most quartz trace elements and the conditions of quartz formation are not direct, recent studies suggest that the quartz trace element composition may be influenced by the rate of crystallization [64], pressure, temperature [65], and fluid composition [66,67]. Nevertheless, it has been shown that—like CL textures, colors and intensity—the concentrations of trace elements vary systematically among ore deposit types [67], and thus can be used to fingerprint the type of ore deposits. Combined with oxygen isotope signature, trace elements of quartz generations can be used to constrain the origin and nature of the fluids responsible for the formation of the QB.

Quartz analyses yield low but detectable (<20 ppm) values of Ti except for quartz with concentric zoning (>40 ppm). Additionally, all the quartz observed under cathodoluminescence display a weak luminescence. Several authors showed that Ti content in quartz is positively correlated with the fluid temperature and also with CL intensities [67–69], while Al is anti-correlated with CL intensity in most of low temperature deposits (e.g., [70]). Such Ti-T °C relationship can be used to directly estimate the temperature of the fluid in some specific case (i.e., the TitaniQ thermobarometer [71]).

Ti values for the QB indicate an overall temperature of formation below 350 °C [67], a value consistent with temperatures obtained from fluid inclusion microthermometric studies at the End and Andrew Lake deposit [26–29]. Concentric zoning with low and high Ti values (>40 ppm, characterizing higher temperature) could reflect cyclic episodes of circulation of >350 °C fluids, which would agree with concentric zoning and with the occurrence of dense trends of monophase vapor rich FIs at the tip of some growth zones (Figure 7B). Associated with the bi-modal concentration of Al [67], these observations support precipitation of quartz from a low temperature (<~350 °C) fluids, mixing cyclically with high temperature (>~350 °C) fluids.

The $\delta^{18}O_{fluid}$ value of +3.9‰ in average for microcrystalline quartz associated with hematite points toward a magmatic origin (+5.5 to +10‰, although this may vary for a particular intrusion) for the high temperature fluids [72,73]. This is consistent with the presence of the KIS at the presumable time of emplacement of the QB (ca. 1750 Ma [46]). Moreover, oxygen isotopic values for the main quartz generations of the QB are comparable to those obtained for quartz veins at the nearby Mallery lake epithermal (low-sulfidation) deposit by Turner et al. [62]. The Mallery lake deposit displays similar quartz brecciation formed at the time of emplacement of the KIS (ca. 1.75 Ga). In contrast, brown-blue quartz precipitated from a fluid with a lighter $\delta^{18}O_{fluid}$ value of −0.62‰ in average, which could represent mixing between the magmatic derived fluids and a meteoric fluid: The $\delta^{18}O_{fluid}$ value of meteoric fluids at the latitude of the Kiggavik area at this time would have been ~−13‰ [74]. Such interpretation is consistent with the bi-modal distribution of Ti concentrations for this quartz generation but also with the differences in concentrations for different trace elements between the two quartz within the QB. This phenomenon is more easily explained by a fluid mixing rather by changes in temperature for a same fluid (which could explained only the difference in $\delta^{18}O$ values). The two quartz generations have clearly different $\delta^{18}O$ values as the internal precision of the SIMS measurements is <0.1‰. The observation of hematite surrounding magnetite in some microcrystalline banded vein further supports an input of fluids with different redox signatures in the system.

Higher values of $\delta^{18}O_{fluid}$ (~12.9‰) for microcrystalline quartz (another generation within the same sample) could suggest a stronger input of magmatic fluids, which is however unsupported by Ti values for this quartz generation. This apparent discrepancy could be explained by cooling of the fluid before quartz precipitation. Contrasting $\delta^{18}O_{fluid}$ values between similar quartz generation in the same samples (2 generations of microcrystalline quartz, deciphered on the basis of $\delta^{18}O_{fluid}$) could be explained by a local effect of rock buffering within breccia cavity. As a result, veinlets would have formed from an isotopically isolated fluid reservoir, thus yielding higher $\delta^{18}O_{quartz}$ values.

Cathodoluminescence observations support the above interpretations, as blue-purple quartz luminescence is commonly found in quartz precipitated in magmatic/hydrothermal environments [75] while brown luminescence is rather observed in sedimentary-diagenetic (i.e., lower temperature) environments [76]).

The very low B content of the post-QB dark blue vuggy quartz could be explained by co-crystallization of other minerals enriched in B in the uranium deposits of the Kiggavik area; such as magnesiofoitite (dravite), which is commonly observed in environments seeing brine circulations [77]. Accordingly, the isotopic values for vug-filling euhedral quartz (post-QB) are consistent with those obtained for quartz precipitated from relatively low temperature (~150 °C) brines in the Proterozoic Athabasca [78] and Kombolgie basins [79,80] and linked to the formation of U deposits. Considering that the QB predates the formation of the Thelon Basin, from which brines are likely derived, this supports that brines circulated after the emplacement of the QB, in agreement with the findings of Grare et al. [31]. In the Kiggavik area, the fluid inclusion studies by Pagel [26] on the hydrothermal quartz, at Andrew Lake, and by Chi et al. [29] on the hydrothermal quartz, at End, are in agreement with our observations. Indeed, these authors documented low temperature (100–200 °C)-high salinity (25–38 wt % NaCl) fluids, low temperature (150–200 °C)-low salinity (<9 wt % NaCl) fluids, and high temperature (200–300 °C)-low salinity (<9 wt % NaCl) fluids within the QB. Our study indicates that low-T °C/high salinity fluids (brines) circulated after formation of the QB, while the high-T °C/low salinity fluids are more characteristic of the QB that formed earlier in the history of the Kiggavik area.

Using data from several deposits type, it has been shown that deposits linked to low (Mississippy Valley type, Carlin, Epithermal) and high (porphyry Cu-Au) fluid temperatures can be distinguished one from another based on Al and Ti concentrations in quartz associated with orebodies [60]. Figure 9D plots Al-Ti concentrations for the main quartz generations in our study. The epithermal domain of Figure 9D was built after data from low temperature (~100–350 °C) hydrothermal fluids. The distribution is scattered even within one quartz generation (e.g., deep blue microcrystalline). Primary (comb quartz) and secondary ("moss" quartz) textures [81,82] indicate primary quartz deposition and recrystallization. Bobis [83] also attributed the rounded shapes of the moss texture to recrystallisation of silica gel, which preserved the original structure and impurities of the silicate phase. These quartz textures, along with recrystallized bladed calcite, also characterize phases of silica precipitation by boiling and non-boiling hydrothermal fluids in a geothermal/epithermal system [84]. Some textures observed within the QB are typical of epithermal deposits, but they are rare. Commonly encountered precious metals (e.g., Au, Ag) are lacking within the breccia even though they were observed at the nearby Mallery Lake deposit [62]. However, such environment of formation is consistent with the geochemical signature of quartz in the QB and is much more plausible for the formation of the QB than orogenic Au and porphyric deposits.

To sum up, even though it is difficult to be truly conclusive with the measured trace elements concentrations only, the combination of these data with oxygen isotope values and quartz textures points toward a scenario in which high (magmatic-derived) and low (meteoric-derived) temperature fluids interacted and mixed during silicification of the fault zone that led to the formation of the QB. The important volumes of Si would have been provided by intrusive bodies of the KIS emplaced at depth and related to the rift-related extensional tectonics that occurred at ca. 1750 Ma (Figure 10).

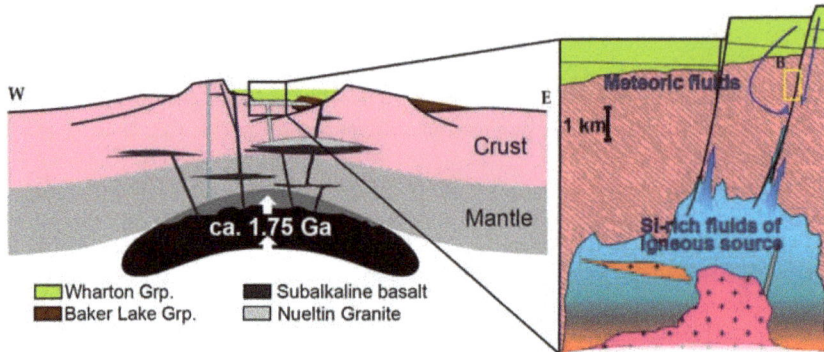

Figure 10. Cross-section after Peterson et al. [46] and zoom in the zone of formation of the QB.

5.2. Fault Zone Processes Leading to the Formation of the QB: Cataclasis, Silicification and Hydraulic Brecciation

Macroscopic observations of the QB and petrographic and textural observations on quartz, although lacking a simple and clear chronology of events, provide additional constraints on the processes behind its formation. Before emplacement of quartz cemented veins and breccias, the superimposition of multiple generations of cataclasites indicates that a localized zone of deformation was repeatedly reactivated during progressive deformation. The presence of microcrystalline quartz in clasts generated before emplacement of quartz veins show that the pervasive silicification of the fault zone was a syn-tectonic process. Regarding quartz-cemented fractures, the common macroscopic textural observation of quartz cemented breccia with angular fragments and jigsaw pattern indicates hydraulic brecciation [14] of the host rock. The pervasive silicification of the fault zone was a first step (Figure 11A,B) before emplacement of the quartz veins and breccias of the QB: It likely triggered fluid pressure build up in the fault zone leading to hydraulic brecciation of the host rock, hence to the "building" of the so-called QB.

The complex patchwork of quartz textures observed under optical microscope shows a still more complex pattern under cathodoluminescence, but highlights several events of quartz fracturing (reworked quartz clasts) and recrystallization. The white quartz mass which displays fine-grained subhedral quartz crystals also shows in some locations numerous fragments of earlier aggregates. The conjugate trends of microcrystalline quartz likely reflect shearing in the quartz mass and synchronous quartz recrystallization. A better evidence for such fracturing and synchronous quartz crystallization is provided by white quartz veins in which euhedral quartz grains are surrounded by microcrystalline quartz and other quartz of heterogeneous sizes (Figure 7F). We interpret this as cataclasis and tectonic comminution (i.e., fracture propagation and wear abrasion) of previously formed quartz mass and recrystallization of quartz (i.e., syn-tectonic). This process differs from the formation of sub-horizontal quartz veins and hydraulic breccias related to transient fluid overpressurization followed by fluid pressure drop and quartz precipitation [21].

To summarize, textures and crosscutting relationships of quartz cements reveal the following sequence of events: (1) Episodes of brittle faulting and cataclasis, before silicification and quartz-brecciation; (2) pervasive silicification of the fault zones (beginning of the QB event); and (3) episodes of brittle fracturing synchronous with the circulation of silica-rich fluids (QB event). During this last event, there were stages of hydrothermal hydraulic brecciation with slow and rapid silica precipitation in relation to boiling of magmatic and/or meteoric fluids (trace elements and $\delta^{18}O$ data inconclusive). This boiling process is supported by the monophase fluid inclusions within the quartz generations of the QB. The presence of 100% of monophase vapor inclusions can be only explained by a boiling process, affecting either magmatic fluids (with a spatial separation between vapor and brines) or meteoric fluids heated due to emplacement of a magmatic intrusion at low depth at ca. 1750 Ma. Boiling process can be marked in other geological environments by the presence of monophase vapor fluid inclusions spatially associated with multiphase and of relatively high-salinity brines due to demixion of the magmatic fluids. The absence of two-phase fluid inclusion in the observed samples of the QB could indicate that the vapor migrated farther than the magmatic brines. The hydraulic brecciation alternated with stages of fluid-assisted cataclasis and quartz recrystallization (Figure 11A). Arrays of monophase fluid inclusions (vapor-rich) also indicate abrupt pressure drops following rupture of the "seal" of the system [78]. The so-called QB therefore appears to be a composite structural feature much more complex than previously thought, which consists of a mass of quartz emplaced by alternating quartz healed hydraulic brecciation and tectonic-induced cataclasis with synchronous quartz recrystallization during fault zone reactivation.

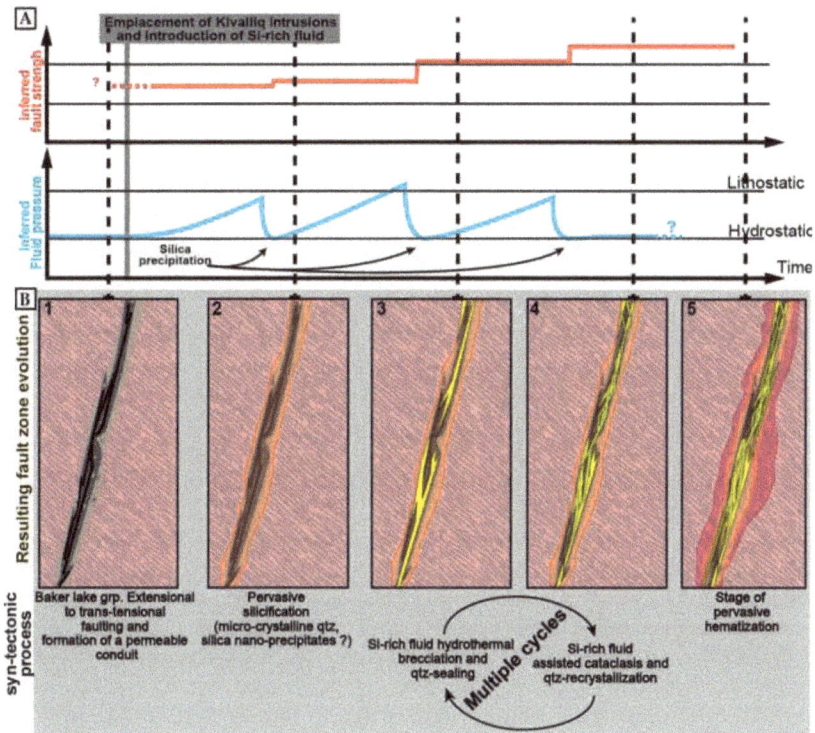

Figure 11. (**A**) Evolution of inferred fault strength (frictional shear resistance) and fluid pressure in the fault zone as a response of fluid pulse, fracturing and silica-precipitation. (**B**) Scheme depicting the succession of events that produced the QB.

Intense multi-episodic hydraulic brecciation of the early fault zone at the time of QB formation would have occurred during the interaction of two isotopically distinct fluids: Meteoric water, mixed with a magmatic-derived fluid. The processes of faulting/fracturing discussed in this section, that led to the formation of the QB, likely occurred at shallow depth (~2 km, [29]) and, looking at relative chronology and geochemical constraints, were likely initiated by the emplacement of the KIS (Figure 10). To a first glance, the fluid temperature of ~350 °C is not easy to reconcile with this shallow depth even if considering an abnormal thermal gradient related to the emplacement of the Kivalliq intrusions. We infer that hydrothermal fluids originated at a greater depth (about 5 km, which may indicate a 70°/km geothermal gradient), and flowed upward sufficiently fast to prevent any significant cooling before they mixed with downward-moving meteoric fluids and precipitated the quartz generations of the QB in thermal disequilibrium with the hosting basement rocks. Interestingly, such a quartz cemented breccia and its complex spatial organization are similar to the meter-thick hydrothermal quartz breccia related to the emplacement of an igneous intrusion described by Tanner et al. [85] in Scotland.

We therefore propose a conceptual tectono-hydrological model for the QB formation involving mixing of deep silica-rich fluids of igneous origin with downward-moving meteoric fluids. The possible mechanisms allowing for such meteoric fluid downward flow in fault zones are either active seismic pumping or passive meteoric infiltration throughout a permeable fault zone. We favor a mechanism of syn-tectonic seismic pumping because beside the formation of quartz-filled fractures, the intrinsic low permeability of the unaltered basement rock surrounding the fault zone and the impermeabilization of the fault zone—including its damage zone—through multiple silicification events presumably make a simple, gravity-driven downflow of meteoric fluids difficult, hence unlikely. Fluid pressure built up at depth through the input of meteoric fluids and magmatic-derived fluids together with likely pore cementation of basement rocks (that decreased porosity). Fluid mixed and flowed upward to higher crustal levels along the fault zones which served as conduits (Figures 10 and 11). This upward flow likely occurred cyclically as the fluid pressure evolved between hydrostatic and supralithostatic (Figure 11A), depending, among other factors, on the sealing effectivity of the reactivated fault zone by quartz precipitation [86]. In turn, fluid pressure increased during the QB event also likely favored multiple reactivations of the high angle fault zone under the regional stress field.

5.3. Evolution of the Fault Zone Properties though Time and Structural Control on Later Uranium Mineralization

The QB is found along many segments of the main fault (Figure 1C) trends, and uranium orebodies are systematically spatially associated with more or less thick bodies of QB along these fault zones. Even though a systematical study of QB thickness could not be undertaken throughout the area, the QB was observed as being usually thinner where it is not associated with uranium mineralization, which implies a possible control on later uranium mineralization by the thickness of the QB in fault zones.

Cataclastic fault rocks formed before the QB should have displayed an initial high porosity, but after pervasive silicification they likely ended with a low porosity comparable to that of the fresh basement rocks, unlike strongly clay altered and fractured samples (Figure 6). Since the evolution of the porosity can be to some extent directly linked to the evolution of permeability since it is controlled by fracturing and mineralogical destabilization/dissolution, we can safely infer that these multiple events of pervasive silicification, faulting/fracturing and quartz cementation caused the destruction of the porosity (hence of the permeability) and thus directly impacted the fluid circulation within the conduit.

At all deposits and prospects in Kiggavik, three main fracturing events postdate emplacement of the QB (two stages of faulting/fracturing and uranium mineralization and one stage of faulting and strong clay alteration [31]). The distribution of fractures and mineralization in some drill holes intersecting uranium orebodies in the vicinity of the QB is shown in Figure 12A for Contact, End, Bong and Andrew Lake. Post-QB fracturing and uranium mineralization are clearly restricted to the hanging wall of the QB in Contact, where the thickness of the breccia is far greater compared to the earlier silicified fault zone. At End, Andrew Lake and Bong, post-QB fracturing and uranium mineralization are observed in both the hanging wall and the footwall, but still not within the QB (inner zone). In the case of End, the QB displays lateral variations in thickness comparable—even less important—to what is observed at Contact. This distribution indicates that post-QB fracturing was preferentially localized in the hanging-wall and/or in the footwall of the QB, along its contact with the host rocks, while most of the QB (core zone) remained poorly fractured.

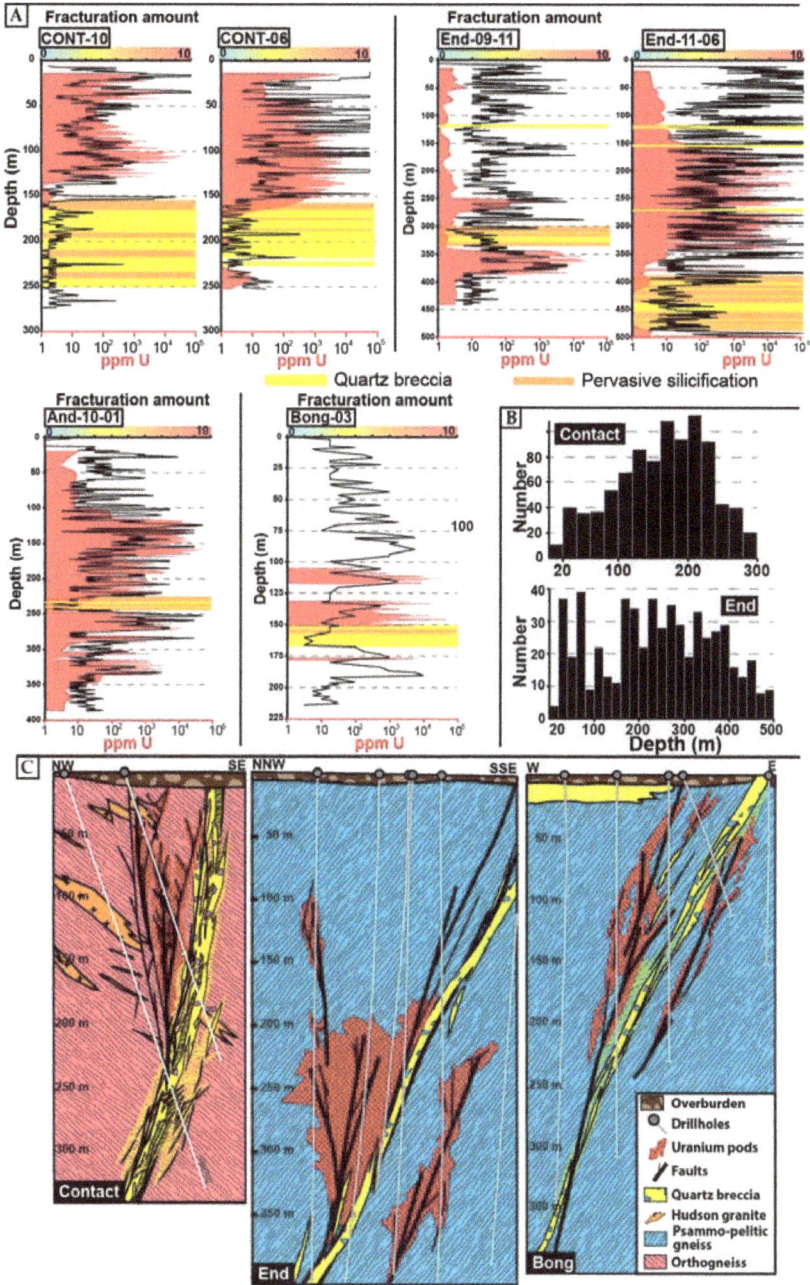

Figure 12. (**A**) Distribution of fracture density and uranium mineralization as a function of depth, for selected drillholes from Contact (Cont), End, Andrew Lake (And) and Bong. Fracture density as black lines; 0: Non-fractured drill-core, 10: Intensely fractured drill-core. Uranium mineralization in red: U in ppm measured by assays, logarithmic scale). (**B**) Number of quartz veins as a function of depth for Contact and End. (**C**) Simplified cross sections.

A significant amount of quartz veins (Figure 12B) were observed up to the top of drill holes, i.e., in the transition from the outer zone of the QB to the host rock. Such quartz veins are typically re-opened and were also used as pathways for uranium bearing fluids at the first stage of uranium mineralization [31]. Ore minerals are observed along the vein boundaries (Figure 13A) and cementing orthogonal microfractures (Figure 13B,C; see also Chi et al. [29]). Quartz with uranium-oxides in their vicinity display specific luminescence which has been described in many places worldwide [87–90]; it has been explained by the destabilization of the crystal lattice by radiation damages (due to liberation of alpha particles through U^{238} decay series).

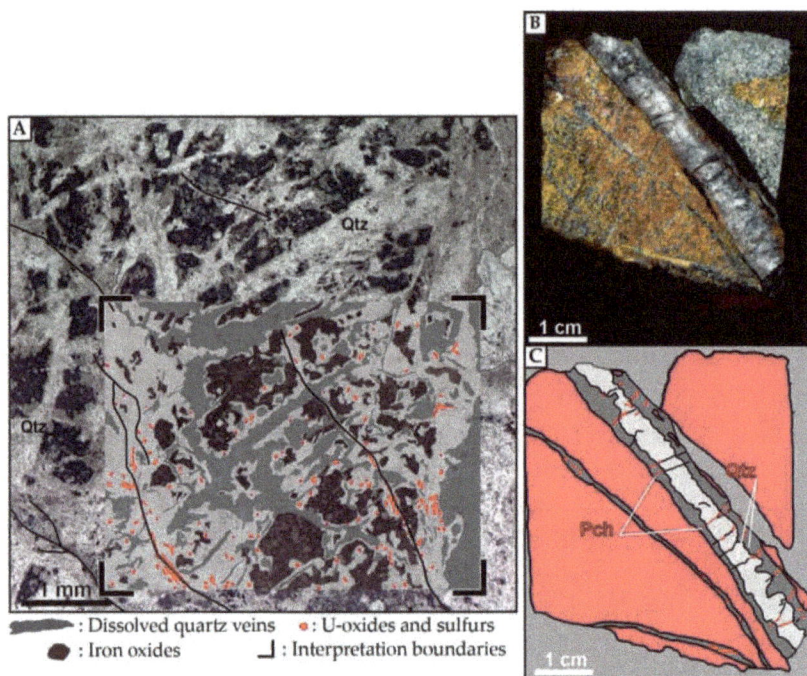

Figure 13. (**A**) Plane polarized light picture and interpretation drawing of a quartz vein network guiding the mineralizing fluid along its boundaries. Quartz and iron oxides display evidence for dissolution; (**B,C**) macroscopic drill core sample scan and interpretation drawing: Examples of a QB related quartz vein bearing pitchblende (Pch) along edges or in orthogonal microfractures.

This change of the luminescence, together with the fact that the micro-fractures cemented with uranium minerals crosscut several generations of quartz, show that formation of the QB and deposition of uranium bearing minerals are two distinctive events, supporting isotope data on dark blue vuggy quartz. Uranium minerals and associated specific luminescence are only observed in the vicinity of the QB. Vuggy quartz precipitated from basinal brines (potentially U-bearing; [29]) are observed mainly in the outer zone and in the vicinity of the QB and further demonstrate the barrier role played by the QB in fluid flow partitioning. These observations suggest that the QB behaved as a rigid and hard body compared to the weaker host rocks, so that later deformation preferentially localized in the host rocks along the contact with the strong QB body.

Quartz veins in the outer zone of the QB form a network that, when microfractured, helped focusing mineralizing fluid flow—thus creating local traps for uranium deposition. Post QB fractures located in its vicinity acted as preferential pathways parallel to the QB for later, uranium-rich brines, leading to deposition of uranium ore bodies at ca. 1500–1300 Ma for main stages [28–30]. To conclude, the silicification processes ultimately led to the building of a complex quartz-cemented breccia body, up to tens of meters thick, acting as a transverse hydraulic barrier depending on the vertical and lateral variations in thickness and the degree of quartz cementation. As a result, the distribution of mineralization in the Kiggavik area was heavily controlled—at different scales—by the mechanical and hydraulic properties of the reactivated pre-existing fault zones where the QB was emplaced. The QB behaved as a mechanically strong, transverse hydraulic barrier, that localized later fracturing and compartmentalized/channelized vertical flow of uranium-bearing fluids, hence orebodies (Figure 12C) in its hanging-wall and/or footwall during fault zone reactivation.

Even if the 3D architecture is not perfectly constrained, and would deserve proper 3D geometrical and kinematic modelling, we could expect that relays within the QB, vertical and horizontal variations in thickness, and overlap between QB bodies would likely influence fluid flow properties (e.g., fluid velocity)—hence would impact uranium deposition rate [91]. The quartz breccia in the Kiggavik area seems to be a good example of the "physical seal" developed by McCuaig and Hronsky [92] in their mineral system concept, in conjunction with other factors to generate ore in a considered area.

6. Conclusions

Based on a structural analysis combined with vein cement petrography, trace elements, and oxygen stable isotope analysis of quartz, we constrain the nature, emplacement and significance of the QB which strongly controlled the location of uranium deposits in the Kiggavik area (Nunavut, Canada). The formation of the breccia bodies appears to be linked to fluctuations in pressure, temperature and compositions of fluids during tectonic reactivation of the fault zones along which the QB was emplaced. Faults formed during syn- to post-orogenic rifting processes and formation of the Baker Lake basin (ca. 1850–1750 Ma) were subsequently reactivated, and cycles of underwent pervasive silicification, hydraulic brecciation, and quartz recrystallization linked to cataclasis. This was associated with the circulation of meteoric-derived fluids mixing with Si-rich magmatic-derived fluids at depth. This is interpreted to be linked to the emplacement of the KIS at ca. 1750 Ma.

Post-QB fracturing at 1500–1300 Ma was constrained in the hanging wall and footwall of the QB, with flow of basin-derived brines being channeled along the fault zones where QB emplaced. The network of quartz veins in the vicinity of the QB was a favorable pathway for circulation of these uranium-rich fluids and related uranium precipitation, as they were re-opened and micro-fractured. Thus, the QB bodies likely exerted a major structural and hydrological control on the formation of significant uranium orebodies in the Kiggavik area. Beyond regional implications, this study demonstrates how an unconventional trap was built in impermeable Archean basement rocks. It also emphasizes the importance of the spatial organization, and long-term evolution of fault zones in the location of uranium orebodies of economic interest.

Author Contributions: A.G. and O.L. conceptualized both the study and the final model and wrote the original draft. J.M., A.B., A.T., P.L., J.R. reviewed and edited the draft. A.G. performed macro to micro-scale petrographic and microstructural characterization; A.G. and A.T. performed cathodoluminescence microscopy; A.G. and J.M. performed trace elements and isotopes analyses and their interpretation. P.L. and J.R. gave their validation. Funding acquisition and project administration were performed by A.B. and P.L., and geochemical analyses were funded by O.L. and J.M.

Funding: This research was funded by Orano Canada and the laboratory ISTeP.

Acknowledgments: The authors thank ORANO and ORANO Canada for the full financial support and access to the Kiggavik camp and exploration data. Special thanks to geologists R. Zerff, R. Hutchinson, K. Martin, and D. Hrabok for their help and enriching discussions during field work. The authors also want to acknowledge the first exploration geologists (Cogema and/or Orano) that worked on, and developed preliminary concepts on the quartz breccia in the Kiggavik area: D. Baudemont, N. Flotte, J.-L. Feybesse, J.-L. Lescuyer. The authors thank The SIMS team of the CRPG (Vandoeuvre-lès-Nancy, France) for their assistance in measuring the O isotopic composition of the quartz by SIMS. Special thanks to A. Pêtre for his thoughtful comments.

Conflicts of Interest: The authors declare no conflict of interest. The founding sponsors had no role in the design of the study; in the collection, analyses, or interpretation of data; in the writing of the manuscript, and in the decision to publish the results".

References

1. Sibson, R.H.; Robert, F.; Poulsen, K.H. High-angle reverse faults, fluid-pressure cycling, and mesothermal gold-quartz deposits. *Geology* **1988**, *16*, 551–555. [CrossRef]
2. Blundell, D.J.; Karnkowski, P.H.; Alderton, D.H.M.; Oszczepalski, S.; Kucha, H. Copper mineralization of the polish Kupferschierfer: A proposed basement fault-fracture system of fluid flow. *Econ. Geol.* **2003**, *98*, 1487–1495. [CrossRef]
3. Micklethwaite, S.; Cox, S.F. Fault-segment rupture, aftershock-zone fluid flow, and mineralization. *Geology* **2004**, *32*, 813–816. [CrossRef]
4. Cox, S.F. Coupling between deformation, fluid pressures, and fluid flow in ore-producing hydrothermal systems at depth in the crust. *Econ. Geol.* **2005**, *100th Anniv. Vol*, 39–75. [CrossRef]
5. Muchez, P.; Heijlen, W.; Banks, D.; Blundell, D.; Boni, M.; Grandia, F. 7: Extensional tectonics and the timing and formation of basin-hosted deposits in Europe. *Ore Geol. Rev.* **2005**, *27*, 241–267. [CrossRef]
6. Micklethwaite, S.; Sheldon, H.A.; Baker, T. Active fault and shear processes and their implications for mineral deposit formation and discovery. *J. Struct. Geol.* **2010**, *32*, 151–165. [CrossRef]
7. Caine, J.S.; Evans, J.P.; Forster, C.B. Fault zone architecture and permeability structure. *Geology* **1996**, *24*, 1025–1028. [CrossRef]
8. McCuaig, T.; Kerrich, R. P-T-t-deformation-fluid characteristics of lode gold deposits: Evidence from alteration systematics. *Ore Geol. Rev.* **1998**, *12*, 381–453. [CrossRef]
9. Ridley, J.R.; Diamond, L. Fluid chemistry of orogenic lode gold deposits and implications for genetic models. *Rev. Econ. Geol.* **2000**, *13*, 141–162.
10. Kolb, J.; Rogers, A.; Meyer, F.M.; Vennemann, T.W. Development of fluid conduits in the auriferous shear zones of the Hutti Gold Mine, India: Evidence for spatially and temporally heterogeneous fluid flow. *Tectonophysics* **2004**, *378*, 65–84. [CrossRef]
11. Sibson, R. Earthquake rupturing as a mineralizing agent in hydrothermal systems. *Geology* **1987**, *15*, 701–704. [CrossRef]
12. Sibson, R.H. Fault rocks and fault mechanisms. *J. Geol. Soc. Lond.* **1977**, *133*, 191–213. [CrossRef]
13. Phillips, W.J. Hydraulic fracturing and mineralization. *J. Geol. Soc. Lond.* **1972**, *128*, 337–359. [CrossRef]
14. Jébrak, M. Hydrothermal breccias in vein-type ore deposits: A review of mechanisms, morphology and size distribution. *Ore Geol. Rev.* **1997**, *12*, 111–134. [CrossRef]
15. Laznicka, P. *Breccias and Coarse Fragmentites: Petrology, Environments, Associations, Ores*; Developments in Economic Geology Series; Elsevier Science & Technology Books: New Yrok, NY, USA, 1988; ISBN 9780444412508.
16. Simmons, S.F.; White, N.C.; John, D. Geological characteristics of epithermal precious and base metal deposits. *Econ. Geol.* **2005**, *100*, 485–522.
17. Zhong, J.; Pirajno, F.; Chen, Y.-J. Epithermal deposits in South China: Geology, geochemistry, geochronology and tectonic setting. *Gondwana Res.* **2017**, *42*, 193–219. [CrossRef]
18. Goldfarb, R.; Christie, A.; Bierlein, F. The orogenic gold deposit model and New Zealand: Consistencies and anomalies. In Proceedings of the 2005 New Zealand Minerals Conference: Realising New Zealand's Mineral Potential, Auckland, New Zealand, 13–16 November 2005; pp. 105–114.
19. Cannell, J.; Cooke, D.R.; Walshe, J.L.; Stein, H. Geology, mineralization, alteration, and structural evolution of the El Teniente Porphyry Cu-Mo Deposit. *Econ. Geol.* **2005**, *100*, 979–1003. [CrossRef]
20. Landtwing, M.R.; Furrer, C.; Redmond, P.B.; Pettke, T.; Guillong, M.; Heinrich, C.A. The Bingham Canyon Porphyry Cu-Mo-Au deposit. III. Zoned copper-gold ore deposition by magmatic vapor expansion. *Econ. Geol.* **2010**, *105*, 91–118. [CrossRef]
21. Henderson, I.H.C.; McCaig, A.M. Fluid pressure and salinity variations in shear zone-related veins, central Pyrenees, France: Implications for the fault-valve model. *Tectonophysics* **1996**, *262*, 321–348. [CrossRef]

22. Rusk, B.; Reed, M. Scanning electron microscope-cathodoluminescence analysis of quartz reveals complex growth histories in veins from the Butte porphyry copper deposit, Montana. *Geology* **2002**, *30*, 727–730. [CrossRef]

23. Tóth, J.; Corbet, T. Post-Palaeocene evolution of regional groundwater flow systems and their relation to petroleum accumulations, Taber Area, southern Alberta, Canada. *Geol. Soc. Lond. Spec. Publ.* **1987**, *34*, 45–77. [CrossRef]

24. Hiatt, E.E.; Palmer, S.E.; Kyser, K.; O'Connor, E.; Terrence, K.H. Basin evolution, diagenesis and uranium mineralization in the Paleoproterozoic Thelon Basin, Nunavut, Canada. *Basin Res.* **2010**, *22*, 302–323. [CrossRef]

25. Reynolds, R.L.; Goldhaber, M.B. Origin of a South Texas roll-type uranium deposit; I, Alteration of iron-titanium oxide minerals. *Econ. Geol.* **1978**, *73*, 1677–1689. [CrossRef]

26. Pagel, M.; Ahamdach, N. *Etude des Inclusions Fluides dans les Quartz des Gisements U de l'Athabasca et du Thelon*; Internal Report of Centre de Recherches sur la Géologie des Matières Premières Minérales et Energétiques (CREGU): Nancy, France, 1995.

27. Riegler, T.; Lescuyer, J.-L.; Wollenberg, P.; Quirt, D.; Beaufort, D. Alteration related to uranium deposits in the kiggavik-andrew lake structural trend, Nunavut, Canada: New insights from petrography and clay mineralogy. *Can. Mineral.* **2014**, *52*, 27–45. [CrossRef]

28. Sharpe, R.; Fayek, M.; Quirt, D.; Jefferson, C.W. Geochronology and genesis of the Bong Uranium deposit, Thelon Basin, Nunavut, Canada. *Econ. Geol.* **2015**, *110*, 1759–1777. [CrossRef]

29. Chi, G.; Haid, T.; Quirt, D.; Fayek, M.; Blamey, N.; Chu, H. Petrography, fluid inclusion analysis, and geochronology of the End uranium deposit, Kiggavik, Nunavut, Canada. *Miner. Depos.* **2017**, *52*, 211–232. [CrossRef]

30. Shabaga, B.M.; Fayek, M.; Quirt, D.; Jefferson, C.W.; Camacho, A. Mineralogy, geochronology, and genesis of the Andrew Lake uranium deposit, Thelon Basin, Nunavut, Canada. *Can. J. Earth Sci.* **2017**, *54*, 850–868. [CrossRef]

31. Grare, A.; Benedicto, A.; Lacombe, O.; Trave, A.; Ledru, P.; Blain, M.; Robbins, J. The Contact uranium prospect, Kiggavik project, Nunavut (Canada): Tectonic history, structural constraints and timing of mineralization. *Ore Geol. Rev.* **2018**, *93*, 141–167. [CrossRef]

32. Hiatt, E.E.; Kyser, K.; Dalrymple, R.W. Relationships among sedimentology, stratigraphy, and diagenesis in the Proterozoic Thelon Basin, Nunavut, Canada: Implications for paleoaquifers and sedimentary-hosted mineral deposits. *J. Geochem. Explor.* **2003**, *80*, 221–240. [CrossRef]

33. Davis, W.J.; Gall, Q.; Jefferson, C.W.; Rainbird, R.H. Fluorapatite in the Paleoproterozoic Thelon Basin: Structural-stratigraphic context, in situ ion microprobe U-Pb ages, and fluid-flow history. *GSA Bull.* **2011**, *123*, 1056–1073. [CrossRef]

34. Jefferson, C.; Thomas, D.J.; Gandhi, S.; Ramaekers, P.; Delauney, G.; Brisbin, D.; Cutts, C.; Portella, P.; Olson, R. Unconformity-associated uranium deposits of the Athabasca Basin, Saskatchewan and Alberta. *Bull. Geol. Surv. Can.* **2007**, *588*, 23–67.

35. AREVA (Areva, Paris, France). Internal Reference Document 2015. Available online: http://www.sa.areva.com/EN/finance-1176/regulated-financial-information.html (accessed on 24 July 2018).

36. Hadlari, T.; Rainbird, R.H. Retro-arc extension and continental rifting: A model for the Paleoproterozoic Baker Lake Basin, Nunavut1Geological Survey of Canada Contribution 2010 04 36. *Can. J. Earth Sci.* **2011**, *48*, 1232–1258. [CrossRef]

37. Rainbird, R.H.; Davis, W.J.; Stern, R.A.; Peterson, T.D.; Smith, S.R.; Parrish, R.R.; Hadlari, T. Ar-Ar and U-Pb geochronology of a Late Paleoproterozoic Rift Basin: Support for a Genetic Link with Hudsonian Orogenesis, Western Churchill Province, Nunavut, Canada. *J. Geol.* **2006**, *114*, 1–17. [CrossRef]

38. Rainbird, R.H.; Davis, W.J. U-Pb detrital zircon geochronology and provenance of the late Paleoproterozoic Dubawnt Supergroup: Linking sedimentation with tectonic reworking of the western Churchill Province, Canada. *GSA Bull.* **2007**, *119*, 314. [CrossRef]

39. Rainbird, R.H.; Hadlari, T.; Aspler, L.B.; Donaldson, J.A.; LeCheminant, A.N.; Peterson, T.D. Sequence stratigraphy and evolution of the paleoproterozoic intracontinental Baker Lake and Thelon basins, western Churchill Province, Nunavut, Canada. *Precambrian Res.* **2003**, *125*, 21–53. [CrossRef]

40. Pehrsson, S.J.; Berman, R.G.; Eglington, B.; Rainbird, R. Two Neoarchean supercontinents revisited: The case for a Rae family of cratons. *Precambrian Res.* **2013**, *232*, 27–43. [CrossRef]

41. Peterson, T.D. Geological setting and geochemistry of the ca. 2.6 Ga Snow island Suite in the central Rae Domain of the Western Churchill Province, Nunavut. *Geol. Surv. Can. Open File* **2015**, *7841*. [CrossRef]

42. Jefferson, C.; Pehrsson, S.; Peterson, T.; Chorlton, L.; Davis, B.; Keating, P.; Gandhi, S.; Fortin, R.; Buckle, J.; Miles, W.; et al. Northeast Thelon region geoscience framework–new maps and data for uranium in Nunavut. *Geol. Surv. Can.* **2011**, 288791. [CrossRef]

43. McEwan, B. Structural style and regional comparison of the Paleoproterozoic Ketyet River group in the region North-Northwest of Baker Lake, Nunavut. Master's Thesis, University of Regina, Regina, SK, Canada, 2012; p. 155.

44. Tschirhart, V.; Morris, W.A.; Jefferson, C.W. Framework geophysical modelling of granitoid vs. supracrustal basement to the northeast Thelon Basin around the Kiggavik uranium camp, Nunavut. *Can. J. Earth* **2013**, *50*, 667–677. [CrossRef]

45. Tschirhart, V.; Jefferson, C.W.; Morris, W.A. Basement geology beneath the northeast Thelon Basin, Nunavut: Insights from integrating new gravity, magnetic and geological data. *Geophys. Prospect.* **2017**, *65*, 617–636. [CrossRef]

46. Peterson, T.D.; Scott, J.M.J.; LeCheminant, A.N.; Jefferson, C.W.; Pehrsson, S.J. The Kivalliq Igneous Suite: Anorogenic bimodal magmatism at 1.75 Ga in the western Churchill Province, Canada. *Precambrian Res.* **2015**, *262*, 101–119. [CrossRef]

47. Johnstone, D.; Bethune, K.M.; Quirt, D. Lithostratigraphic and structural controls of uranium mineralization in the Kiggavik East, Centre and Main Zone deposits, Nunavut. In Proceedings of the Geological Association of Canada-Mineralogival Association of Canada, Joint Annual Meeting, WhiteHorse, YT, Canada, 1–3 June 2016.

48. Peterson, T.D.; Van Breemen, O.; Sandeman, H.; Cousens, B. Proterozoic (1.85–1.75 Ga) igneous suites of the Western Churchill Province: Granitoid and ultrapotassic magmatism in a reworked Archean hinterland. *Precambrian Res.* **2002**, *119*, 73–100. [CrossRef]

49. Hoffman, P.F. United Plates of America, The Birth of a Craton: Early Proterozoic Assembly and Growth of Laurentia. *Annu. Rev. Earth Planet. Sci.* **1988**, *16*, 543–603. [CrossRef]

50. Van Breemen, O.; Peterson, T.D.; Sandeman, H.A. U-Pb zircon geochronology and Nd isotope geochemistry of Proterozoic granitoids in the western Churchill Province: Intrusive age pattern and Archean source domains. *Can. J. Earth Sci.* **2005**, *42*, 339–377. [CrossRef]

51. Scott, J.M.J.; Peterson, T.D.; Davis, W.J.; Jefferson, C.W.; Cousens, B.L. Petrology and geochronology of Paleoproterozoic intrusive rocks, Kiggavik uranium camp, Nunavut. *Can. J. Earth Sci.* **2015**, *52*, 495–518. [CrossRef]

52. Zaleski, E.; Pehrsson, N.D.; Davis, W.J.; L'Heureux, R.; Greiner, E.; Kerswill, J.A. *Quartzite Sequences and Their Relationships, Woodburn Lake Group, Western Churchill Province, Nunavut*; West. Churchill NATMAP Project; Geological Survey of Canada: Ottawa, ON, Canada, 2000; pp. 1–10.

53. Rainbird, R.H.; Davis, W.J.; Pehrsson, S.J.; Wodicka, N.; Rayner, N.; Skulski, T. Early Paleoproterozoic supracrustal assemblages of the Rae domain, Nunavut, Canada: Intracratonic basin development during supercontinent break-up and assembly. *Precambrian Res.* **2010**, *181*, 167–186. [CrossRef]

54. LeCheminant, A.N.; Heaman, L.M. Mackenzie igneous events, Canada: Middle Proterozoic hotspot magmatism associated with ocean opening. *Earth Planet. Sci. Lett.* **1989**, *96*, 38–48. [CrossRef]

55. Heaman, L.M.; LeCheminant, A.N. Paragenesis and U-Pb systematics of baddeleyite (ZrO_2). *Chem. Geol.* **1993**, *110*, 95–126. [CrossRef]

56. Anand, A.; Jefferson, C.W. Reactivated fault systems and their effects on outcrop patterns of thin-skinned early thrust imbrications in the Kiggavik uranium camp, Nunavut. *Geol. Surv. Can.* **2017**. [CrossRef]

57. Peter, J.K.; Ulrike, W.; Brigitte, S.; Dmitry, K.; Qichao, Y.; Ingrid, R.; Andreas, S.; Karin, B.; Detlef, G.; Jacinta, E. Determination of Reference Values for NIST SRM 610–617 Glasses Following ISO Guidelines. *Geostand. Geoanal. Res.* **2010**, *35*, 397–429. [CrossRef]

58. Paton, C.; Hellstrom, J.; Paul, B.; Woodhead, J.; Hergt, J. Iolite: Freeware for the visualisation and processing of mass spectrometric data. *J. Anal. At. Spectrom.* **2011**, *26*, 2508–2518. [CrossRef]

59. Hervig, R.L.; Williams, L.B.; Kirkland, I.K.; Longstaffe, F.J. Oxygen isotope microanalyses of diagenetic quartz: Possible low temperature occlusion of pores. *Geochim. Cosmochim. Acta* **1995**, *59*, 2537–2543. [CrossRef]

60. Kempe, U.; Götze, J.; Dombon, E.; Monecke, T.; Poutivtsev, M. *Quartz: Deposits, Mineralogy and Analytics*; Springer: Berlin, Germany, 2012; pp. 331–355.

61. Dennen, W.H. Stoichiometric substitution in natural quartz. *Geochim. Cosmochim. Acta* **1966**, *30*, 1235–1241. [CrossRef]
62. Turner, W.; Richards, J.; Nesbitt, B.; Muehlenbachs, K.; Biczok, J. Proterozoic low-sulfidation epithermal Au-Ag mineralization in the Mallery Lake area, Nunavut, Canada. *Miner. Depos.* **2001**, *36*, 442–457. [CrossRef]
63. Clayton, R. Oxygen isotope exchange between quartz and water. *J. Geophys. Res.* **1972**, *77*, 3057–3067. [CrossRef]
64. Lowenstern, J.B.; Sinclair, W.D. Exsolved magmatic fluid and its role in the formation of comb-layered quartz at the Cretaceous Logtung W-Mo deposit, Yukon Territory, Canada. *Trans. R. Soc. Edinb. Earth Sci.* **1996**, *87*, 291–303. [CrossRef]
65. Thomas, J.B.; Bruce Watson, E.; Spear, F.S.; Shemella, P.T.; Nayak, S.K.; Lanzirotti, A. TitaniQ under pressure: The effect of pressure and temperature on the solubility of Ti in quartz. *Contrib. Mineral. Petrol.* **2010**, *160*, 743–759. [CrossRef]
66. Perny, B.; Eberhardt, P.; Ramseyer, K.; Pankrath, R. Microdistribution of Al, Li, and Na in α quartz: Possible causes and correlation with short-lived cathodoluminescence. *Am. Mineral.* **1992**, *77*, 534–544.
67. Rusk, B.G.; Lowers, H.A.; Reed, M.H. Trace elements in hydrothermal quartz: Relationships to cathodoluminescent textures and insights into vein formation. *Geology* **2008**, *36*, 547–550. [CrossRef]
68. Spear, F.; Wark, D. Cathodoluminescence imaging and titanium thermometry in metamorphic quartz. *J. Metamorph. Geol.* **2009**, *27*, 187–205. [CrossRef]
69. Leeman, W.P.; MacRae, C.M.; Wilson, N.C.; Torpy, A.; Lee, C.-T.; Student, J.J.; Thomas, J.B.; Vicenzi, E.P. A Study of cathodoluminescence and trace element compositional zoning in natural quartz from volcanic rocks: Mapping titanium content in quartz. *Microsc. Microanal.* **2012**, *18*, 1322–1341. [CrossRef] [PubMed]
70. Rusk, B.; Koenig, A.; Lowers, H. Visualizing trace element distribution in quartz using cathodoluminescence, electron microprobe, and laser ablation-inductively coupled plasma-mass spectrometry. *Am. Mineral.* **2011**, *96*, 703–708. [CrossRef]
71. Huang, R.; Audétat, A. The titanium-in-quartz (TitaniQ) thermobarometer: A critical examination and re-calibration. *Geochim. Cosmochim. Acta* **2012**, *84*, 75–89. [CrossRef]
72. Taylor, H.P. The Application of oxygen and hydrogen isotope studies to problems of hydrothermal alteration and ore deposition. *Econ. Geol.* **1974**, *69*, 843–883. [CrossRef]
73. Bettencourt, J.S.; Leite, W.B.; Goraieb, C.L.; Sparrenberger, I.; Bello, R.M.S.; Payolla, B.L. Sn-polymetallic greisen-type deposits associated with late-stage rapakivi granites, Brazil: Fluid inclusion and stable isotope characteristics. *Lithos* **2005**, *80*, 363–386. [CrossRef]
74. Bowen, G.J. Statistical and Geostatistical Mapping of Precipitation Water Isotope Ratios. In *Isoscapes: Understanding Movement, Pattern, and Process on Earth through Isotope Mapping*; West, J.B., Bowen, G.J., Dawson, T.E., Tu, K.P., Eds.; Springer: Dordrecht, The Netherlands, 2010; pp. 139–160, ISBN 978-90-481-3354-3.
75. Vollbrecht, A.; Oberthür, T.; Ruedrich, J.; Weber, K. Microfabric analyses applied to the Witwatersrand gold- and uranium-bearing conglomerates: Constraints on the provenance and post-depositional modification of rock and ore components. *Miner. Depos.* **2002**, *37*, 433–451. [CrossRef]
76. Kraishan, G.; Rezaee, R.; Worden, R. Significance of trace element composition of quartz cement as a key to reveal the origin of silica in sandstones: An example from the cretaceous of the Barrow Sub-Basin, Western Australia. *Quartz Cementation Sandstones* **2009**, *29*, 317–331.
77. Rosenberg, P.E.; Foit, F.F. Magnesiofoitite from the uranium deposits of the Athabasca Basin, Saskatchewan, Canada. *Can. Mineral.* **2006**, *44*, 959–965. [CrossRef]
78. Richard, A.; Boulvais, P.; Mercadier, J.; Boiron, M.-C.; Cathelineau, M.; Cuney, M.; France-Lanord, C. From evaporated seawater to uranium-mineralizing brines: Isotopic and trace element study of quartz–dolomite veins in the Athabasca system. *Geochim. Cosmochim. Acta* **2013**, *113*, 38–59. [CrossRef]
79. Polito, P.A.; Kyser, T.K.; Thomas, D.; Marlatt, J.; Drever, G. Re-evaluation of the petrogenesis of the Proterozoic Jabiluka unconformity-related uranium deposit, Northern Territory, Australia. *Miner. Depos.* **2005**, *40*, 257–288. [CrossRef]
80. Derome, D.; Cathelineau, M.; Fabre, C.; Boiron, M.-C.; Banks, D.; Lhomme, T.; Cuney, M. Reconstitution of paleo-fluid composition by Raman LIBS and crush-leach techniques: Application to mid-Proterozoic evaporitic brines (Kombolgie Formation basin, Northern Territory, Australia). *Chem. Geol.* **2007**, *237*, 240–254.

81. Bodnar, R.J.; Reynolds, T.J.; Kuehn, C.A. Fluid-Inclusion Systematics in Epithermal Systems. In *Geology and Geochemistry of Epithermal Systems*; Berger, B.R., Bethke, P.M., Eds.; Society of Economic Geologists: Littleton, CO, USA, 1985.

82. Dong, G.; Morrison, G.; Jaireth, S. Quartz textures in epithermal veins, Queensland; classification, origin and implication. *Econ. Geol.* **1995**, *90*, 1841–1856. [CrossRef]

83. Bobis, E.R. A review of the description, classification, and origin of quartz textures in low-sulphidation epithermal veins. *J. Geol. Soc. Philipp.* **1994**, *99*, 15–39.

84. Moncada, D.; Mutchler, S.; Nieto, A.; Reynolds, T.J.; Rimstidt, J.D.; Bodnar, R.J. Mineral textures and fluid inclusion petrography of the epithermal Ag–Au deposits at Guanajuato, Mexico: Application to exploration. *J. Geochem. Explor.* **2012**, *114*, 20–35. [CrossRef]

85. Tanner, P.W. The giant quartz-breccia veins of the Tyndrum-Dalmally area, Grampian Highlands, Scotland: Their geometry, origin and relationship to the Cononish gold-silver deposit. *Earth Environ. Sci. Trans. R. Soc. Edinb.* **2012**, *103*, 51–76. [CrossRef]

86. Sibson, R.H. Implications of fault-valve behaviour for rupture nucleation and recurrence. *Tectonophysics* **1992**, *211*, 283–293. [CrossRef]

87. Meunier, J.D.; Sellier, E.; Pagel, M. Radiation-damage rims in quartz from uranium-bearing sandstones. *J. Sediment. Res.* **1990**, *60*, 53–58. [CrossRef]

88. Hu, B.; Pan, Y.; Botis, S.; Rogers, B.; Kotzer, T.; Yeo, G. Radiation-induced defects in drusy quartz, Athabasca basin, Canada: A new aid to exploration of uranium deposits. *Econ. Geol.* **2008**, *103*, 1571–1580. [CrossRef]

89. MacRae, C.M.; Wilson, N.C.; Torpy, A. Hyperspectral cathodoluminescence. *Mineral. Petrol.* **2013**, *107*, 429–440. [CrossRef]

90. Cerin, D.; Götze, J.; Pan, Y. Radiation-Induced Damage In Quartz At the Arrow Uranium Deposit, Southwestern Athabasca Basin, Saskatchewan. *Can. Mineral.* **2017**, *55*, 457–472. [CrossRef]

91. Zhang, Y.; Robinson, J.; Schaubs, P.M. Numerical modelling of structural controls on fluid flow and mineralization. *Geosci. Front.* **2011**, *2*, 449–461. [CrossRef]

92. McCuaig, T.C.; Hronsky, J.M.A. The Mineral System Concept: The Key to Exploration Targeting. In *Building Exploration Capability for the 21st Century*; Kelley, K.D., Golden, H.C., Eds.; Society of Economic Geologists: Littleton, CO, USA, 2014; ISBN 9781629491424.

minerals

MDPI

Article

Structural Control on Clay Mineral Authigenesis in Faulted Arkosic Sandstone of the Rio do Peixe Basin, Brazil

**Ingrid B. Maciel [1], Angela Dettori [2], Fabrizio Balsamo [2,*] , Francisco H.R. Bezerra [1,3],
Marcela M. Vieira [3], Francisco C.C. Nogueira [4], Emma Salvioli-Mariani [2] and
Jorge André B. Sousa [5]**

[1]   Post-Graduation Program on Geodynamics and Geophysics, Universidade Federal do Rio Grande do Norte,
      Natal, RN 59078-970, Brazil; ingridbmaciel@gmail.com (I.B.M.); bezerrafh@geologia.ufrn.br (F.H.R.B.)
[2]   Department of Chemistry, Life Sciences and Environmental Sustainability, University of Parma,
      I-43124 Parma, Italy; angela.dettori@studenti.unipr.it (A.D.); emma.salviolimariani@unipr.it (E.S.-M.)
[3]   Department of Geology, Federal University of Rio Grande do Norte, Natal, RN 59078-970, Brazil;
      marcela@geologia.ufrn.br
[4]   Department of Petroleum Engineering, Federal University of Campina Grande,
      Campina Grande, PB 58100-000, Brazil; aulascezar@gmail.com
[5]   Petrobras Research Center—CENPES, Rio de Janeiro, RJ 21941-915, Brazil; jorgeabs@petrobras.com.br
*   Correspondence: fabrizio.balsamo@unipr.it; Tel.: +39-0521-905365

Received: 3 July 2018; Accepted: 13 September 2018; Published: 14 September 2018

Abstract: Clay minerals in structurally complex settings influence fault zone behavior and characteristics such as permeability and frictional properties. This work aims to understand the role of fault zones on clay authigenesis in arkosic, high-porosity sandstones of the Cretaceous Rio do Peixe basin, northeast Brazil. We integrated field, petrographic and scanning electron microscopy (SEM) observations with X-ray diffraction data (bulk and clay-size fractions). Fault zones in the field are characterized by low-porosity deformation bands, typical secondary structures developed in high-porosity sandstones. Laboratory results indicate that in the host rock far from faults, smectite, illite and subordinately kaolinite, are present within the pores of the Rio do Peixe sandstones. Such clay minerals formed after sediment deposition, most likely during shallow diagenetic processes (feldspar dissolution) associated with meteoric water circulation. Surprisingly, within fault zones the same clay minerals are absent or are present in amounts which are significantly lower than those in the undeformed sandstone. This occurs because fault activity obliterates porosity and reduces permeability by cataclasis, thus: (1) destroying the space in which clay minerals can form; and (2) providing a generally impermeable tight fabric in which external meteoric fluid flow is inhibited. We conclude that the development of fault zones in high-porosity arkosic sandstones, contrary to other low-porosity lithologies, inhibits clay mineral authigenesis.

Keywords: fault zones; deformation bands; clay authigenesis; shallow diagenesis

1. Introduction

Clay minerals have important economic applications in industry—e.g., [1]. Additionally, in most geological settings clay minerals can occur in faults, thus influencing their permeability, frictional properties [2–8] and subsurface fluid flow [4,9–14]. Therefore, the understanding of the feedback between faulting and clay mineral authigenesis has important implications for seismicity, the migration and accumulation of oil and gas in the subsurface, and contaminant transport in aquifers.

Faults in high-porosity sandstones are generally considered as barriers to fluid flow, due to the combined effect of grain size and porosity reduction within fault cores and associated deformation bands in damage zones [15–22]. In this context, clay minerals are commonly described as mechanically weak minerals, and because of this weakness their presence in faults commonly contributes to stable sliding failures [23,24]. Furthermore, the origin and distribution of clays in sandstone are also important in oil industry, because these minerals contribute to increases in the sealing potential of faults and can determine reservoir compartmentalization [3,6,8].

Several studies have described the clay mineralogy of fault zones [24–26], however little attention has been paid to the role of faults in determining the type and amount of clay mineral transformation in faulted, arkosic sandstones. The goal of this study is to investigate how fault zones in arkosic sandstones (composed of a fault core surrounded by deformation bands) modify grain-scale fabric and control clay mineral authigenesis at shallow burial depths. We selected the Cretaceous Rio do Peixe basin in northeast Brazil (Figure 1) as a case study, due to its excellent exposures of undeformed sandstones and well-preserved fault zones. By integrating field analysis with laboratory data, we conclude that deformation-band faulting in arkosic, high-porosity sandstones inhibits clay mineral authigenesis, rather than promoting alteration and clay mineral formation.

2. Geological Background of the Rio do Peixe Basin

The Rio do Peixe basin (RPB) is a pull-apart Early Cretaceous basin situated in northeastern Brazil. The basin was generated during the reactivation of Precambrian basement shear zones during the opening of the South Atlantic Ocean [27–30]. The basin's deeper depocenters were established based on gravity data, and reach depths of ~2420 m [30]. These depocenters are filled by continental siliciclastic sedimentary units which were deposited in fluvial and lacustrine depositional systems. These deposits are divided into three main stratigraphic units, namely, from the base to the top: (1) the Antenor Navarro Formation, represented by conglomerates and mudstones; (2) the Sousa Formation, composed of mudstones; and (3) the Rio Piranhas Formation, composed of conglomerates and coarse sandstones [31–33].

The Antenor Navarro Formation is the basal unit. It represents the main fill of the basin and contains typical syn-rift deposits. The formation consists of siliciclastic fluvial deposits that are exposed in large and continuous outcrops in different sectors of the basin (Figure 1). The sandstones and conglomerates are composed of quartz, feldspars, rock fragments and biotite. Their matrix consists of silt and dark brown clay (approximately 1–1.5%) [34]. The Souza Formation is the intermediate unit and consists mostly of mudstones and a few occurrences of sandstones and marls. These units were deposited in floodplains or shallow lakes on meandering rivers. The Rio Piranhas Formation is the top unit and consists of conglomerates and coarse sandstones interfingered with sandy mudstones [32].

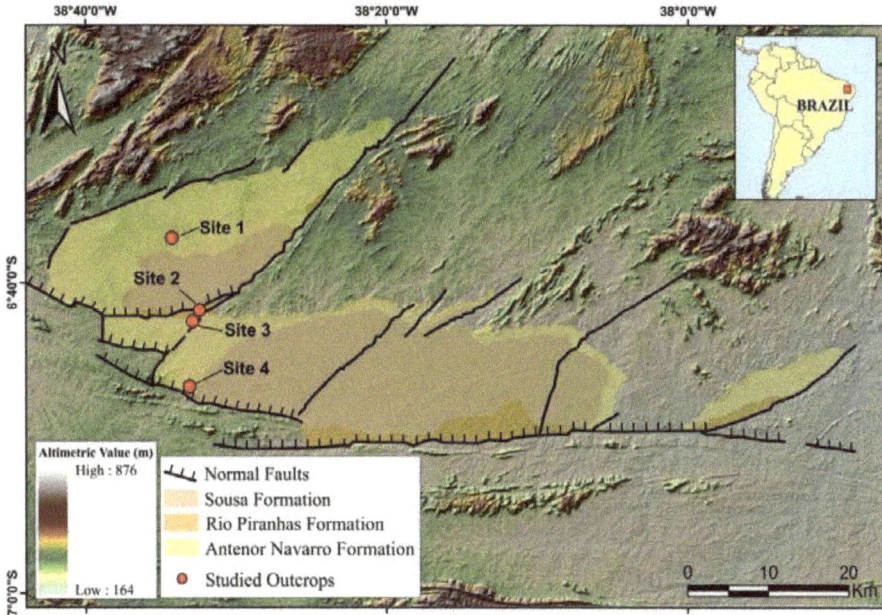

Figure 1. Simplified geological map of the Rio do Peixe Basin, showing major faults and the lithostratigraphics units. The location of the four selected outcrops is indicated. Modified from [27,29,35].

The sedimentary rocks of the Rio do Peixe basin were affected by two main tectonic phases: an Early to Late Cretaceous NW–SE oriented extension [29] followed by a basin inversion in a strike–slip regime from the Late Cretaceous to Cenozoic [35]. The extensional faults that developed during the first phase are dominated by deformation bands, often associated with slickensided surfaces [34]. The deformation bands occur as cm-thick tabular structures developed in the fault damage zone, and are arranged as single elements or in clusters. Within deformation bands, a cataclastic foliation was formed by preferential grain alignment and the selective fragmentation of feldspar grains [34].

3. Methods and Materials

This study focused on the western side of the Rio do Peixe basin, both in its central part far from major faults and in the proximity of the basin-boundary faults (Figure 1). Field analysis and sampling were performed at four main outcrops: one represents undeformed host rocks (Site 1) and the other three are deformed sites near major faults (Sites 2, 3 and 4). Site 1 is located in the undeformed part of the basin, where the basal Antenor Navarro sandstones are not affected by faults and fractures. In this site, we constructed a vertical sedimentary log to characterize the undeformed rocks. Sites 2 and 3 are located along the major intra-basinal fault zones in the Rio Piranhas sandstones and conglomerates, and are characterized by abundant fault zones with meter-scale offsets. Site 4 is located in the hanging wall block of the major basin-boundary fault in the Antenor Navarro sandstones, and has a displacement of ~170 m [36].

In the studied field sites, we collected a total of 95 samples of undeformed and faulted rocks, from which we made 34 thin sections. In the lab, we examined the following materials: (1) undeformed sandstones and conglomerates (i.e., host rocks); (2) deformed rocks collected within the fault zones, consisting of both deformation bands (single or clustered) and foliated cataclasites (see Section 4.1 for a description of fault zone structure). Thin section analysis was performed using an optical microscope, and scanning electron microscopy (SEM) combined with energy dispersive spectroscopy (EDS). The thin sections were impregnated with stained blue epoxy to highlight porosity. For observations using the light petrographic microscope, we focused on grain size and roundness, sorting, packing, porosity, mineral composition and amount of clay. We also described depositional and diagenetic features in undeformed and faulted samples. Small representative samples were analyzed using a scanning electron microscope (SUPERSCAN SSX-550, Shimadzu Corporation, Kyoto, Japan) to improve clay mineral identification and textural analysis. EDS was used to identify the main chemical elements and mineral composition of the samples. X-ray diffraction (XRD) analyses were performed using a Bruker (Billerica, MA, USA) D2 Phaser powder diffractometer (CuKα radiation, voltage of 30 kV, current of 10 mA, step size of 0.018, interval of 1 s per step) on powdered bulk samples (n = 10) and fraction samples <2 µm in size (n = 6) for clay mineral identification in undeformed and faulted rocks. The oriented samples of the clay fractions were analyzed under three different conditions: air dried; ethylene glycol saturated; and heated to a temperature of 550 °C. The powdered bulk samples were measured in the range 2–80° 2θ, and the clay fraction samples were measured in the range 2–20° 2θ. Mineral phase identification and semi-quantitative estimations were performed using the DIFFRAC.EVA suite software provided by Bruker Corporation (Billerica, MA, USA). The results of the XRD analyses, together with sample description and location, are listed in Table 1.

4. Results

4.1. Fault Zone Structure

The studied fault zones exhibit three major structural domains (Figure 2): (1) the host rock, i.e., the undeformed sandstone and conglomerates without any significant deformational features; (2) the fault core, where most of the fault slip is accommodated; and (3) the surrounding footwall and hanging wall damage zones, interposed between the host rock and fault core, and composed of deformation bands.

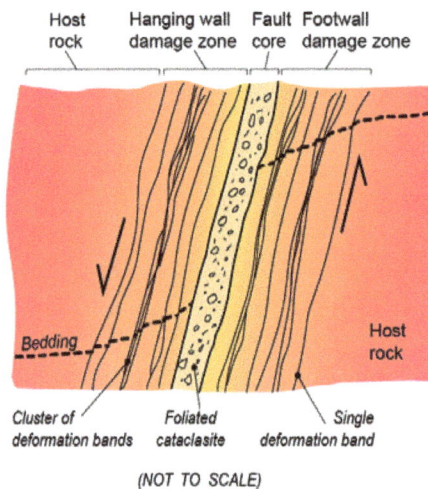

Figure 2. Conceptual sketch showing the typical architecture of fault zones in the Rio do Peixe basin. The host rock represents the sandstones and conglomerates with pristine textures and sedimentary structures not affected by faults. The fault core is the most deformed part of the fault zone, where slip surfaces were frequently developed and where several movements (different slicken lines) are observed. The damage zone is the deformed rock volume next to the fault core that has single or clusters of deformation bands (Sites 2, 3 and 4 in this study). Colors are indicative of the amount of weathering observed in the field.

Concerning the host rock, the original undeformed fluvial facies of the Rio do Peixe basin exhibit a massive laminated structure with trough–festoon crossbedding stratification (Figure 3). These units vary from silty sandstones to fine conglomerates. In a few cases, thin silt lenses are also observed (Figure 3). The sandstones are generally clast-supported with a granular texture, and grains are locally fractured. The grain sizes vary between silt and gravel.

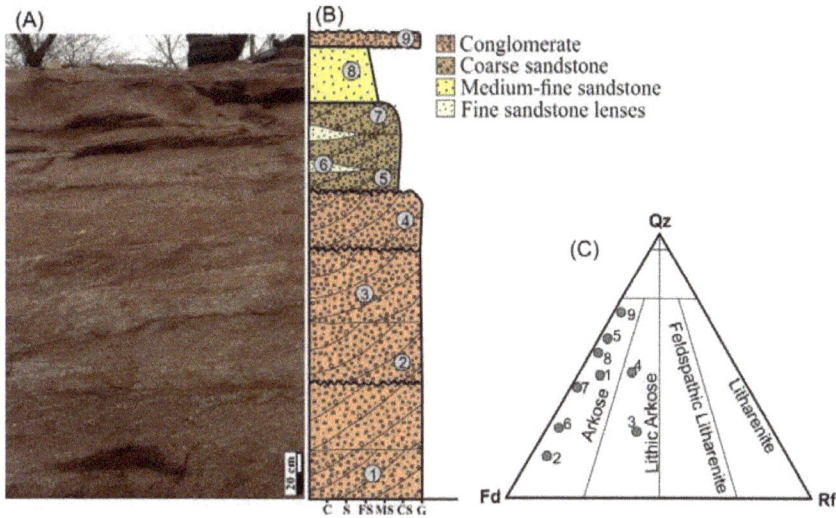

Figure 3. Schematic profile and compositional classification of the host rock (Antenor Navarro Formation) in the Rio do Peixe basin, Site 1. (**A**) Outcrop photograph showing fluvial sedimentary structures with tabular and lenticular shapes of fine sandstones and trough–festoon conglomerates. Note the intense red-orange coloration of undeformed rocks. (**B**) Vertical sedimentary log showing sampling position (Samples 1 to 9) in the fluvial succession. (**C**) Compositional classification of analyzed Samples 1 to 9, based on [37]. Key: c—clay; s—sand; fs—fine sand; ms—medium sand; cs—coarse sand; g—gravel; Qz—quartz; Fd—feldspar; Rf—rock fragments.

The fault cores range from 0.1 m to 0.3 m in thickness, whereas the width of the damage zones broadly range from ~5 m to 10 m (in small faults of Sites 2 and 3) up to ~200 m (in the hanging wall damage zone of Site 4). The fault cores and the inner damage zone generally form topographic relief up to 1 m in height with respect to the surrounding undeformed rock (Figure 4A). The sandstones in the fault core show a strong decrease in grain size and a preferential grain alignment, which forms a tectonic foliation visible at the hand scale (Figure 4B). Most offsets are extensional or slightly oblique. The footwall and hanging wall damage zones consist of clusters of anastomosing deformation bands (Figure 4C) and isolated single deformation bands (Figure 4D). Deformation bands also form a small positive relief. The fault cores and deformation bands exhibit lighter colors than surrounding host rocks (Figure 4B–D) and in some cases an orange to red coloration is also observed.

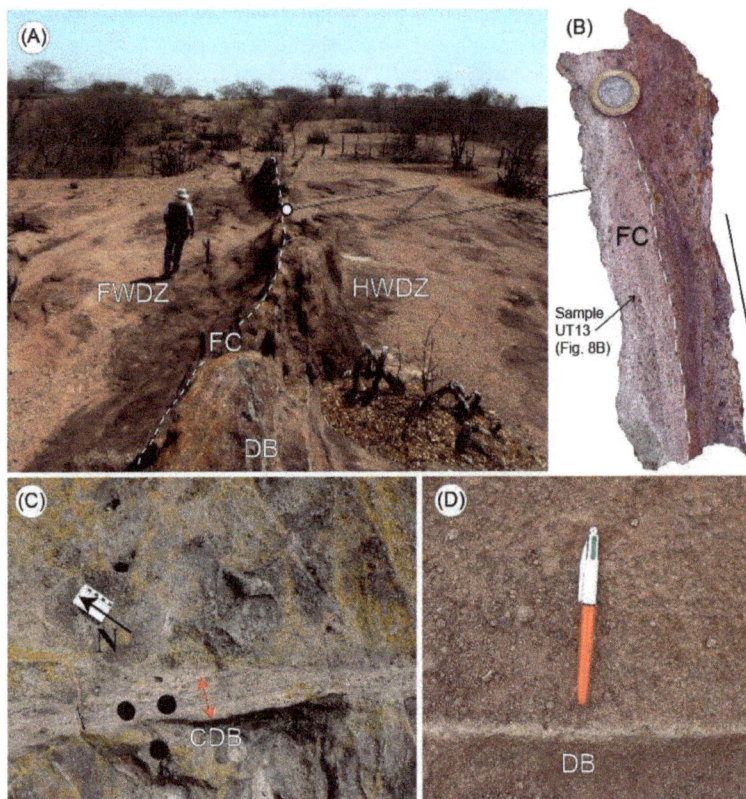

Figure 4. Field photographs showing the main structural features of studied fault zones in the Rio do Peixe basin. (**A**) Example of an extensional fault zone with m-scale offset showing positive relief with respect to the host sandstone, Site 2. The dotted line indicates the approximate position of the fault core. (**B**) Foliated fault core rock (sample UT13 in Table 1) showing light grey to red colors, Site 2. The diameter of the coin is ~2.5 cm. (**C**) Example of a 12.0 cm-thick cluster of deformation bands, in positive relief, developed in the fault damage zone, Site 4. The length of the white scale is ~8.0 cm. (**D**) Whitish single deformation band in positive relief developed in the damage zone, Site 2. The length of the pen is ~14.0 cm. Key: FWDZ—footwall damage zone; HWDZ—hanging wall damage zone; FC—fault core; DB—deformation band; CDB—cluster of deformation bands.

4.2. Petrography

4.2.1. Host Rock

The host rock comprises: (1) fine-grained sandstones with moderate sorting and angular grains (Figure 5A); and (2) coarse-grained, generally poorly sorted sandstones with sub-rounded grains (Figure 5B,C). Point, line and concave–convex grain contacts predominate, while floating and sutured grains are rare. These grain contacts indicate that the sandstones have moderate packing and shallow burial conditions. The sandstones are composed of feldspar, quartz, chert, metamorphic lithoclasts and opaque minerals, although the percentage of these minerals varies in each sample. In thin sections, feldspar is usually the most abundant constituent (~60%), followed by quartz (~40%), lithoclasts (up to 10%) and opaque minerals (~1%). Feldspar grains generally show microfractures and are often partially to completely dissolved, while quartz grains are generally intact. The sandstones vary from arkose to lithic arkose according to the Folk (1968) classification (Figure 3C). All analyzed samples commonly exhibit primary intergranular porosity and, subordinately, secondary moldic porosity associated with the selective dissolution of feldspar grains (Figure 5E,F). Fracture porosity is also observed (Figure 5C), although fractures are mostly filled by clay minerals. In thin sections, visual porosity was observed to be 17% and 28% in fine and coarse sandstones, respectively. Generally, grains are coated by thin layers of clay minerals; these are even more abundant within intergranular pores and microfractures. In some samples, pores show shrinkage (Figure 5B,C). In rare cases, the porosity is almost entirely filled by clay minerals (Figure 5A).

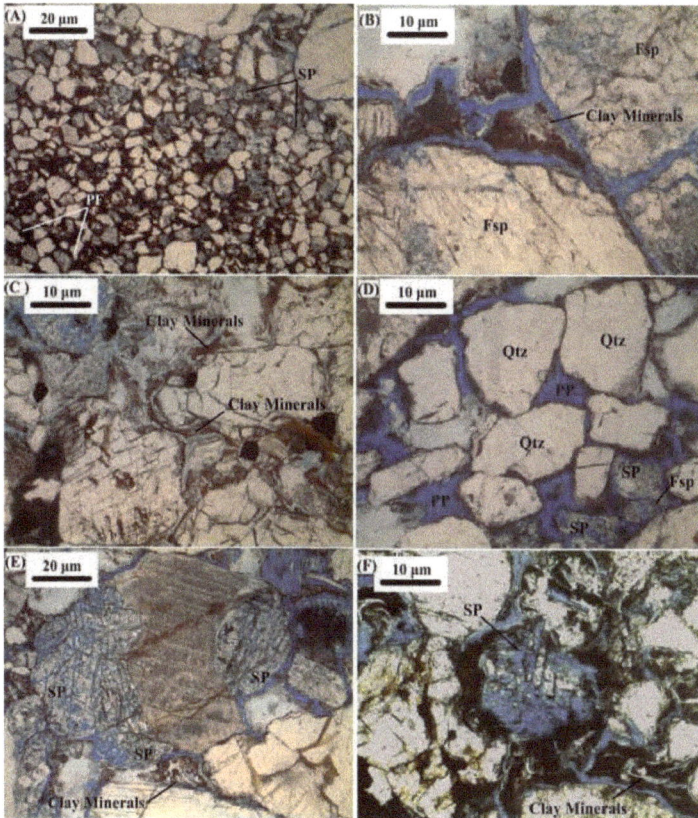

Figure 5. Optical microscopic images of the undeformed Rio do Peixe basin sandstones at Site 1. (**A**) Host rock, with very fine grains and abundant small pores which are frequently filled by clay minerals (pore filling, PF). The porosity is mainly secondary porosity (SP). (**B**) Coarse-grained sandstone showing sub-rounded grains and primary porosity, partially filled by smectite and illite. (**C**) Example of a feldspar clast dissolved and replaced by abundant clays. (**D–F**) show intergranular primary porosity (PP) between quartz grains (Qtz) and SP resulting from the dissolution of feldspar grains (Fsp).

4.2.2. Fault Rocks

The deformed sandstones show different degrees of deformation, most likely due to the amount of offset accommodated within the fault zones (Figure 6). The samples from single deformation bands exhibit a tight fabric and limited intergranular and secondary moldic (feldspar dissolution) porosity compared to the host rock (Figure 6A,B). The visual porosity inside the deformation bands is around 11%, i.e., lower than host rock samples. Intragranular fractures are common in quartz and feldspar grains within the deformation bands (Figure 6C). In places, fractures are open and filled by fine-grained angular cataclastic material (Figure 6D).

Samples from fault cores exhibit a strong reduction in grain size with abundant fractions ranging from fine sand to silt (Figure 6E,F). In strong contrast to the undeformed host rock, fault core samples are clearly matrix-supported and very poorly sorted (Figure 6E,F). The brown colored, fine-grained matrix consists of crushed feldspar grains, in agreement with recent observations [34]. In the fault cores, the visual porosity determined using the optical microscope is practically zero due to the presence of the cataclastic matrix (Figure 6F).

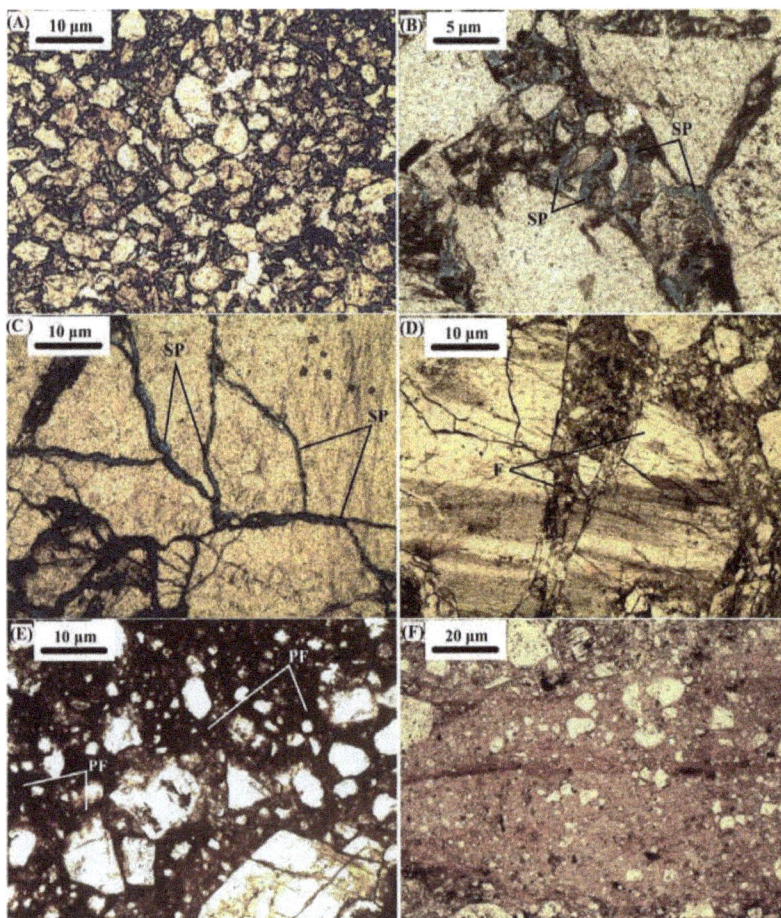

Figure 6. Fault rocks viewed under optical microscope. (**A**) Example of tight fabric in a deformation band within the damage zone, Site 2. (**B**) Detail of primary porosity filled by small angular clasts generated by the cataclasis process in a deformation band, Site 2. Note that feldspar grain dissolution is limited. (**C**) Secondary porosity (SP) developed by intragranular microfractures in a deformation band. (**D**) Detail of intragranular fractures filled by fine-grained angular cataclastic material in a small fault core. (**E**) Fine-grained matrix in the fault core resulting from high grain comminution, in which the pore space was completely destroyed, Site 3. (**F**) A high degree of cataclasis within the fault core, showing a dramatic reduction in grain size and porosity, Site 4. The brown crushed material in (**E**) and (**F**) mostly consist of very small feldspar grains developed during a cataclastic process cf. [34].

4.3. Clay Minerals

The clay minerals observed in the undeformed host hock are smectite, illite and subordinately kaolinite (Figure 7). These generally form a coating around clasts. The feldspars have a pore lining with an arrangement similar to the surrounding clay minerals, indicating growth into an open void (Figure 7A). The smectite and kaolinite exhibit a pore-filling texture (Figure 7B), while the illite shows a pore-lining geometry on quartz (Figure 7C). The smectite is marked by contraction fractures, likely due to sample drying (Figure 7D).

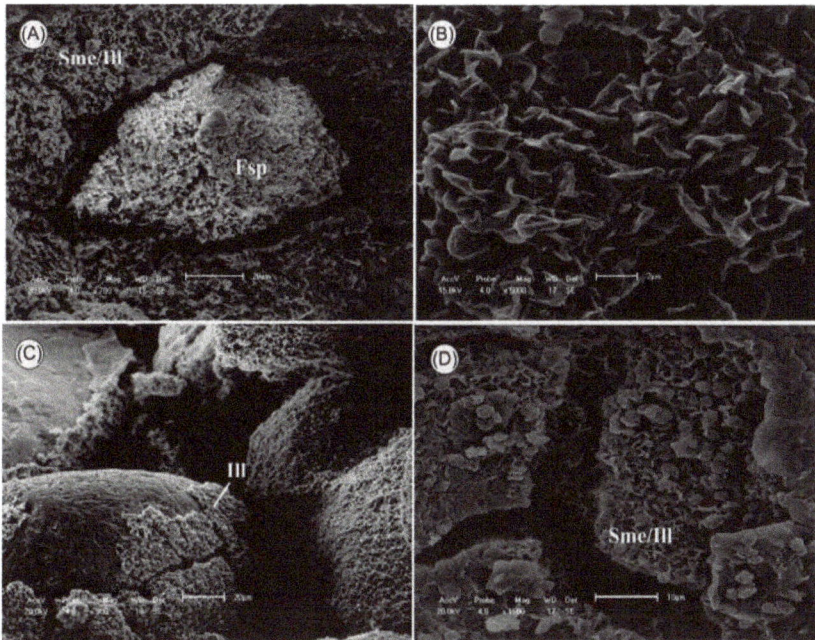

Figure 7. SEM images of clay minerals in undeformed sandstones. (**A**) Feldspar grain surrounded by fractured smectite. (**B**) Well developed smectite flakes inside a pore. (**C**) Pore-line illite coatings capping quartz grains. (**D**) Mixed layers of smectite and illite. Key: Sme—smectite; Fsp—feldspar; Ill—illite.

4.4. XRD Data

XRD bulk analysis was carried out on 10 samples as listed in Table 1. The results show that the main mineral phases identified in both undeformed and faulted sandstones are very similar (Figure 8A–C). Undeformed and faulted sandstones contain mostly quartz and feldspar (orthoclase, microcline and albite) and subordinately micas (Table 1). Hematite was only detected in three out of five undeformed sandstones, and was not detected in faulted samples. At low 2θ values, when clay minerals can be identified, undeformed sandstones show small spectral peaks (see enlargement in Figure 8A) which have a very low intensity (or are absent) in faulted sandstones (see enlargements in Figure 8B,C). Overall, semi-quantitative estimates based on XRD peak intensity, combined with thin section observations, indicate that the amount of clay minerals is systematically <1–2%. In three faulted samples (one of fault core rock and two of deformation bands), either no clay minerals were detected or insufficient material was available for clay fraction analysis (Table 1).

Table 1. Results of XRD analyses (bulk and clay fractions) performed in undeformed and faulted rocks from Sites 1, 2 and 4. (I-S: illite-smectite; DB: deformation bands). Sample labels are the same as Figures 8 and 9.

Sample #	Site	Formation	Structural Domain	Bulk Mineralogy	Clay Mineralogy
SCP01	Site 1	Antenor Navarro	Host rock (fine conglomerate)	Quartz, feldspars, muscovite, hematite	Illite, smectite
SCP05	Site 1	Antenor Navarro	Host rock (fine sandstone)	Quartz, feldspars, muscovite, hematite	Illite, smectite
UT01	Site 2	Rio Piranhas	Host rock (fine sand)	Quartz, feldspars, muscovite	Illite, smectite, I-S mixed layers
UT11	Site 2	Rio Piranhas	Host rock (Fine sand)	Quartz, feldspars, muscovite	Illite, smectite, I-S mixed layers
UT13	Site 2	Rio Piranhas	Fault core (foliated cataclasite)	Quartz, feldspars, muscovite	Not analyzed (no enough clay)
UT14	Site 2	Rio Piranhas	Fault core (foliated cataclasite)	Quartz, feldspars	Illite, chlorite
SVEM1	Site 4	Antenor Navarro	Host rock (fine sandstone)	Quartz, feldspars, muscovite, hematite	Illite, smectite, I-S mixed layers
SVEF3	Site 4	Antenor Navarro	Damage zone (deformation band)	Quartz, feldspars, muscovite	No clay phase
SVEA1	Site 4	Antenor Navarro	Damage zone (cluster of DB)	Quartz, feldspars	Illite, smectite
SVEB2	Site 4	Antenor Navarro	Damage zone (cluster of DB)	Quartz, feldspars	No clay phase

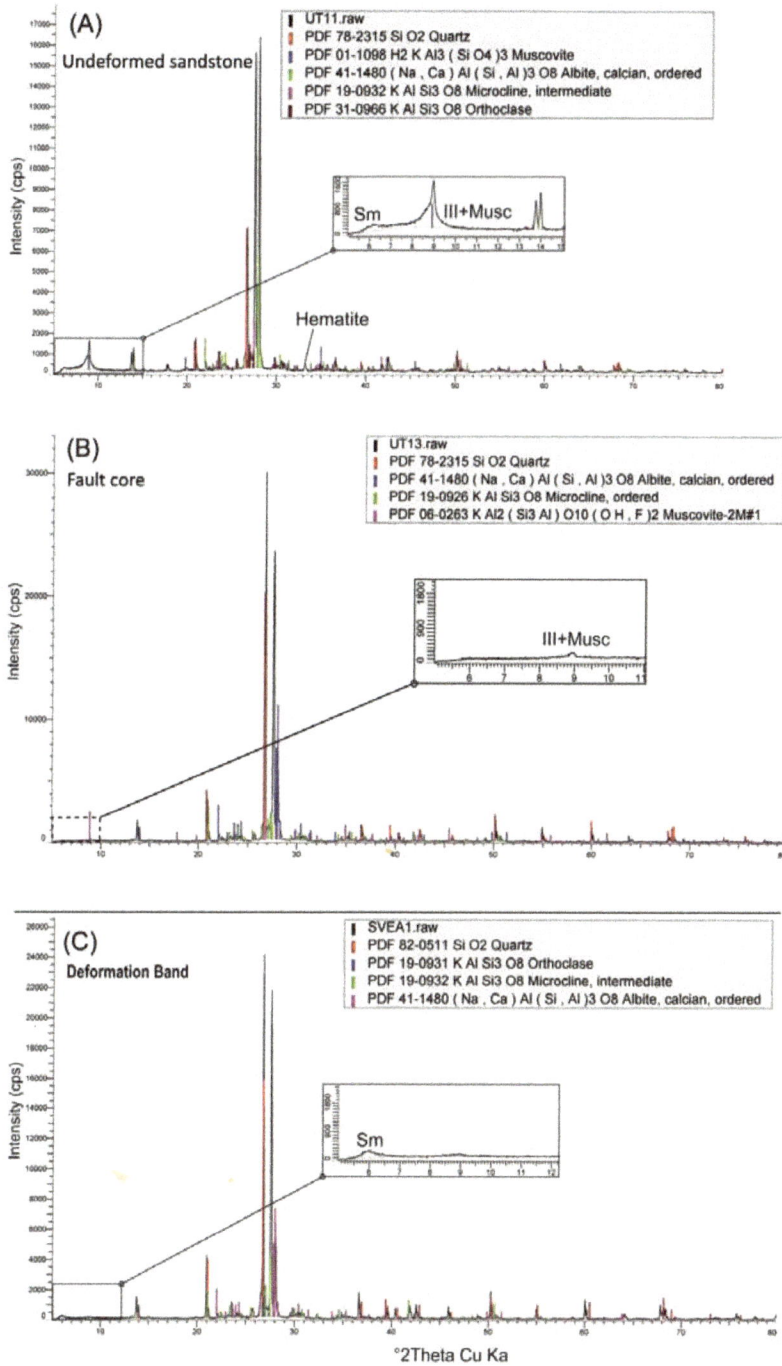

Figure 8. Examples of XRD analyses on bulk samples representative of undeformed sandstone (**A**), foliated cataclasite in fault core rock (**B**) and deformation band (**C**).

XRD analyses of clay fractions, performed on the six samples which had sufficient clay minerals in bulk analyses, indicate that the main types of clay minerals in both undeformed and faulted sandstones are smectite and subordinately illite (Figure 9), as indicated by the comparison between the XRD spectra of air-dried, glycolated and heated samples. Based on spectral peak intensities, in all the spectra the amount of illite was found to be less than that of smectite. Both illite and smectite occur as both distinct phases and mixed layers (Figure 9). In undeformed samples (Figure 9A–D) the spectral peaks of illite and smectite have greater intensities than in faulted samples (Figure 9E,F). Smectite is absent in the foliated fault-core rock (Figure 9E), which also shows the lowest amount of illite of all the analyzed samples; this is consistent with the non-weathered, whitish foliated cataclasites that are often observed in the field (e.g., Figure 4B) and the observed lack of clay minerals in thin sections (Figure 7E,F). A very small amount of chlorite is also observed in the fault core rock sample (Figure 9E).

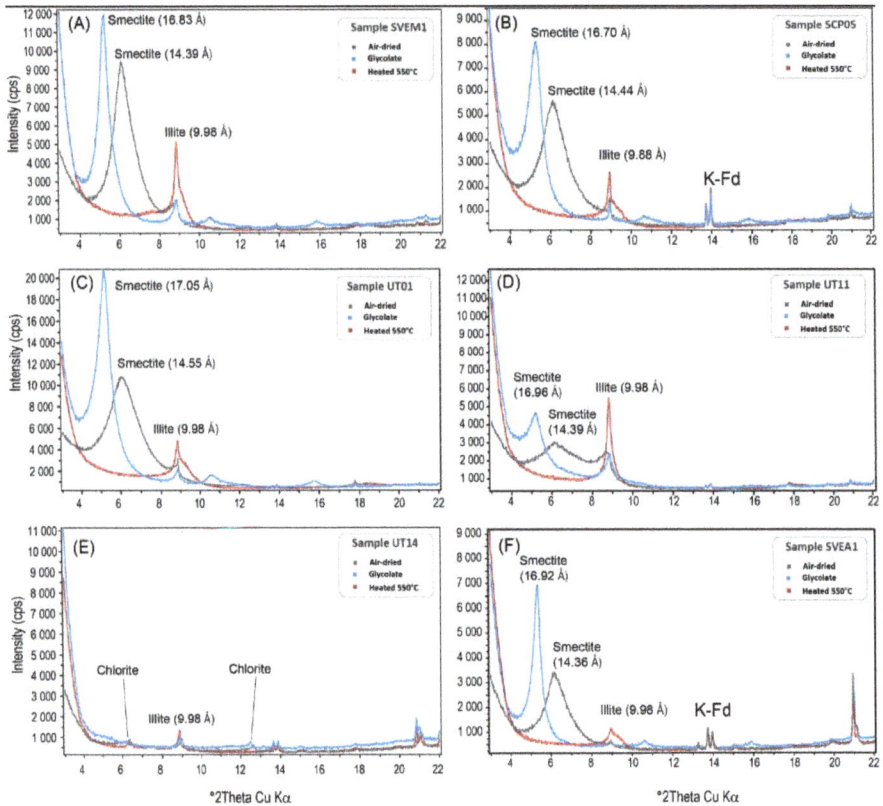

Figure 9. XRD diffractograms of aggregates of clay-size fractions of undeformed (**A–D**) and fault core (**E**) rocks, and a deformation band in a fault damage zone (**F**). Air-dried (in grey), heated at 550 °C (in red), and treated with ethylene glycol (in blue) conditions are shown.

5. Discussion

It is well known that the development of fault zones in sandstones can significantly modify fluid circulation pathways e.g., [12], thus influencing a variety of shallow diagenetic processes [21,24,38]. In this study, fault zones that developed in high-porosity arkosic sandstones are found to have the typical architecture described in other settings [9,24,38], being organized in a foliated fault-core surrounded by a damage zone hosting cataclastic deformation bands [34].

5.1. Origin of Clay Minerals

In the analyzed samples, illite and smectite were found to be the most abundant clay minerals (Figures 9 and 10) that typically occupy the spaces of the intergranular porosity of undeformed sandstones (Figures 5 and 7B). These phases are not easily distinguishable from one another under optical and electronic microscopy, however the results of XRD analysis on clay fractions indicate that illite and smectite occur both in mixed layers and as independent phases (Figure 9). Illite and smectite are among the most common clay minerals in sedimentary rocks. These minerals may form under diagenetic conditions at low pressure and temperature at near-surface conditions [5,8,24], typical of the shallow crust of the studied Rio do Peixe basin. The shallow burial depth during faulting (<1–2 km) is attested by the following conditions: (1) the syn-rift nature of extensional faults; (2) poorly lithified lithology; (3) the high-porosity framework in thin sections of undeformed sandstones; (4) the absence of high temperature mineral phases; and (5) the absence of quartz overgrowth in thin sections. Therefore, we interpret the occurrence of illite and smectite as the result of the partial weathering of K-feldspar and plagioclase grains during shallow early diagenesis under meteoric conditions [14]. This interpretation is strongly supported by the intense and selective dissolution of feldspar grains observed in thin sections (Figure 5). A detrital origin of the clay minerals in the studied samples is excluded since such clays are much more abundant in undeformed (porous) sandstones than in their faulted (non-porous) counterparts (Figure 10A–C), as discussed below.

5.2. Timing Between Clay Authigenesis and Faulting

The greater presence of clay minerals in the host rocks, and their scarcity or absence in the fault zone domains (both in thin sections and in X-ray diffractograms) indicates that, in the Rio do Peixe basin, clay authigenesis mostly occurred after the formation of faults. This interpretation implies that the studied fault zones acted as barriers to weathering meteoric fluids rather than preferential conduits, which is consistent with previously published data in similar lithologies. This hydraulic behavior is in agreement with pervasive grain fragmentation and cataclasis, as documented in thin sections (Figure 6), which provides a more compact, tight and impermeable fabric within the fault zones cf. [34]. Accordingly, when the studied extensional faults and deformation bands were formed at shallow burial depth, the rock volume incorporated into the fault zones could not provide an effective pathway for meteoric influx, thus compromising the process of clay authigenesis and limiting the development of clay phases. This is also consistent with: (1) the presence of Fe-oxides in the host rocks, and not in their faulted counterparts (Table 1); and (2) the intense reddish coloration of host rocks and the orange to whitish coloration of faulted rocks observed in the field (as shown schematically in Figure 3).

Therefore, we believe that no significant amount of detrital clays was present in the sandstone before deformation, due to their absence in faulted samples. If clay minerals were present in the host sandstones at the time of faulting, they would have certainly been incorporated into the fault zones, probably reducing the friction between grains and preventing (or at least hindering) both cataclasis [23,24] and the development of deformation bands. In summary, we suggest that when the diagenetic process of clay authigenesis occurred in the high-porosity undeformed sandstones, there was no significant porosity and sufficient permeability within the fault zones for meteoric fluid circulation and intense alteration of feldspars.

5.3. Evolutionary Model

Based on field observations and laboratory results, we propose the following evolutionary model for the generation of clay minerals after sediment deposition in the following sequence: (1) the syn- to post-sedimentary formation of extensional faults and deformation bands in poorly lithified sandstones, inducing localized grain compaction and early comminution within the fault zones; (2) the further evolution of failure and the generation of low-permeability cataclastic fault zones; (3) the beginning of the weathering process during shallow diagenesis and meteoric fluid circulation; (4) the selective weathering and dissolution of feldspar grains in a semi-arid environment (as shown by moldic porosity in thin sections) and clay mineral authigenesis in the high-porosity undeformed sandstones and conglomerates; (5) the exhumation of faults during regional basin inversion and the formation of positive reliefs of fault zones (due to differential surface erosion) caused by a tight cataclastic fabric.

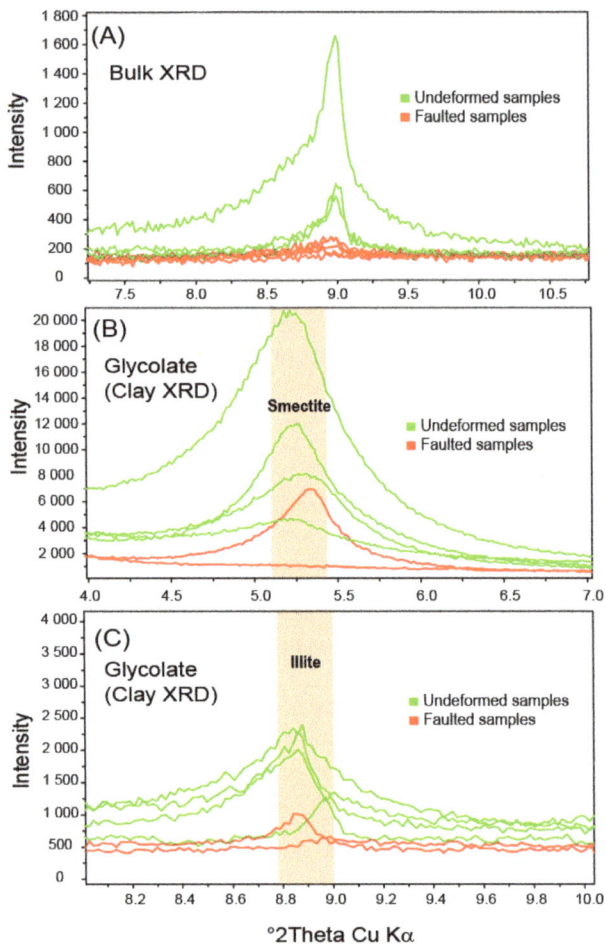

Figure 10. Summary diagrams showing a comparison between diffractograms of undeformed host rocks (green lines) and faulted rocks (red lines). (**A**) Bulk XRD analysis indicating that that the amount of clay minerals in the undeformed samples is higher than that in faulted samples. (**B**) The relative abundance of smectite in undeformed and faulted samples. (**C**) The relative abundance of illite in undeformed and faulted samples.

6. Conclusions

We studied clay mineral assemblages in faulted, high-porosity arkosic sandstone of the Rio do Peixe basin (northeast Brazil) to understand the role of faults in clay mineral authigenesis. We integrated field observations with analysis of microstructures, optical and scanning electron microscopy and XRD (bulk and clay-fraction) mineralogy. The results obtained in this study indicate the following conclusions:

(1) The bulk mineralogy of the Rio do Peixe sandstone does not change significantly between the undeformed and faulted domains, consisting of lithic arkose with feldspar grains generally comprising >50%.

(2) In both undeformed and faulted domains, clay minerals are <1–2% and consist of smectite and illite, and subordinately illite–smectite mixed layers. Despite the similar mineralogy, the amount of clay is systematically less in the faulted domain than in pristine rocks and in some cases is not observed at all.

(3) Clay minerals in the studied arkosic sandstones most likely developed during feldspar weathering processes in a shallow meteoric environment. A detrital origin of clay is excluded in the analyzed sandstones and conglomerates.

(4) Contrary to the results of other fault rock studies in similar lithologies, clay is found to be less abundant in the faulted domains (fault core and damage zone) than in the host rocks. We conclude that this is due to the tight fabric that developed in the faulted porous sandstone, which inhibited meteoric fluid circulation and clay mineral authigenesis.

We conclude that, contrary to several other faulted settings which have a high abundance of authigenic clays, the development of fault zones in high-porosity arkosic sandstone in semi-arid regions prevents the authigenesis of clay minerals. Consequently, clay authigenesis is more efficient in undeformed sandstones than faulted domains, which has important implications for oil and water reservoir quality in siliciclastic rocks and fault behavior in structurally complex settings.

Author Contributions: I.B.M. participated to the fieldwork and sampling, performed petrographic and SEM analysis, contributed to manuscript writing. A.D. participated to the fieldwork, performed XRD analyses and thin section observations. F.B. conceived the research, participated to the fieldwork and sampling, contributed to manuscript writing; F.H.R.B. participated to fieldwork, contributed to manuscript writing; M.M.V. supervised SEM and petrographic analyses; F.C.C.N. organized the overall fieldwork and participated to sampling; E.S.-M. supervised and interpreted the XRD analysis; J.A.B.S. participated to the fieldwork and supported the project.

Funding: This research was funded by Petrobras/Federal University of Campina Grande project (TC 0050.0096065.15.9 grant to Francisco C. C. Nogueira); Fieldwork of Angela Dettori and Fabrizio Balsamo was funded by University of Parma, Italy (Overworld progam 2016–2017 grant to Fabrizio Balsamo); Ingrid Maciel was supported by a Brazilian CAPES grant.

Acknowledgments: We kindly thank two anonymous reviewers which significantly improve the original early version of this manuscript. We thank Luca Aldega and Luciana Mantovani for helful discussion on XRD data. We also thank the Brazilian Agency of Oil, Gas, and Biofuels (*Agência Nacional do Petróleo, ANP*) for sharing data on the Rio do Peixe basin. Fabrizio Balsamo wishes to dedicate this work in faulted sandstones to his mother Isabella Bellina, died in Albano Laziale (Rome, Italy) the 14 June 2018.

Conflicts of Interest: The authors declare no conflicts of interest.

References

1. Xi, K.; Cao, Y.; Liu, K.; Jahren, J.; Zhu, R.; Yuan, G.; Hellevang, H. Authigenic minerals related to wettability and their impacts on oil accumulation in tight sandstone reservoirs: An example from the Lower Cretaceous Quantou Formation in the southern Songliao Basin, China. *J. Asian Earth Sci.* **2018**. [CrossRef]

2. Rice, J.R. Fault stress states, pore pressure distributions, and the weakness of the San Andreas Fault. In *International Geophysics*; Academic Press: Cambridge, MA, USA, 1992; pp. 475–503.

3. Fisher, Q.J.; Knipe, R.J. Fault sealing processes in siliciclastic sediments. In Faulting, Fault Sealing and Fluid Flow in Hydrocarbon Reservoirs. *Geol. Soc. Spec. Publ.* **1998**, *147*, 117–134. [CrossRef]

4. Haines, S.H.; Van der Pluijm, B.A.; Ikari, M.J.; Saffer, D.M.; Marone, C. Clay fabric intensity in natural and artificial fault gouges: Implications for brittle fault zone processes and sedimentary basin clay fabric evolution. *J. Geophys. Res.* **2009**, *114*, B05406. [CrossRef]

5. Lander, R.H.; Bonnell, L.M. A model for fibrous illite nucleation and growth in sandstones. *AAPG Bull.* **2010**, *94*, 1161–1187. [CrossRef]

6. Faulkner, D.R.; Jackson, C.A.L.; Lunn, R.J.; Schlische, R.W.; Shipton, Z.K.; Wibberley, C.A.J.; Withjack, M.O. A review of recent developments concerning the structure, mechanics and fluid flow properties of fault zones. *J. Struct. Geol.* **2010**, *32*, 1557–1575. [CrossRef]

7. Balsamo, F.; Aldega, L.; De Paola, N.; Faoro, I.; Storti, F. The signature and mechanics of earthquake ruptures along shallow creeping faults in sediments. *Geology* **2014**, *42*, 435–438. [CrossRef]

8. Buatier, M.D.; Cavailhes, T.; Charpentier, D.; Lerat, J.; Sizun, J.P.; Labaume, P.; Gout, C. Evidence of multi-stage faulting by clay mineral analysis: Example in a normal fault zone affecting arkosic sandstones (Annot sandstones). *J. Struct. Geol.* **2015**, *75*, 101–117. [CrossRef]

9. Antonellini, M.; Aydin, A. Effect of Faulting on Fluid Flow in Porous Sandstones: Petrophysical Properties. *AAPG Bull.* **1994**, *78*, 355–377.

10. Rawling, G.C.; Goodwin, L.B.; Wilson, J.L. Internal architecture, permeability structure, and hydrologic significance of contrasting fault-zone types. *Geology* **2001**, *29*, 43–46. [CrossRef]

11. Eichhbl, P.; Taylor, W.L.; Pollard, D.D.; Aydin, A. Paleo-fluid flow and deformation in the Aztec Sandstone at the Valley of Fire, Nevada—Evidence for the coupling of hidrogeologic, diagenetic, and tectonic process. *GSA Bull.* **2004**, *116*, 1120–1136. [CrossRef]

12. Fossen, H.; Schultz, R.A.; Shipton, Z.K.; Mair, K. Deformation bands in sandstone: A review. *J. Geol. Soc.* **2007**, *164*, 755–769. [CrossRef]

13. Caine, J.S.; Minor, S.A. Structural and geochemical characteristics of faulted sediments and inferences on the role of water in deformation, Rio Grande Rift, New Mexico. *GSA Bull.* **2009**, *121*, 1325–1340. [CrossRef]

14. Vrolijk, P.; Van der Pluijm, B. Clay gouge. *J. Struct. Geol.* **1999**, *21*, 1039–1048. [CrossRef]

15. Gibson, R. Physical character and fluid-flow properties of sandstonederived fault gouge, in Structural Geology in Reservoir Characterization. *Geol. Soc. Spec. Publ.* **1998**, *127*, 87–93. [CrossRef]

16. Fisher, Q.; Knipe, R.J. The permeability of faults within siliclastic petroleum reservoirs of the North Sea and Norwegian Continental Shelf. *Mar. Pet. Geol.* **2001**, *18*, 1063–1081. [CrossRef]

17. Fossen, H.; Bale, A. Deformation bands and their influence on fluid flow. *AAPG Bull.* **2007**, *91*, 1685–1700. [CrossRef]

18. Rotevatn, A.; Torabi, A.; Fossen, H.; Braathen, A. Slipped deformation bands: A new type of cataclastic deformation bands in Western Sinai, Suez rift, Egypt. *J. Struct. Geol.* **2008**, *30*, 1317–1331. [CrossRef]

19. Eichhubl, P.; Davatzes, N.C.; Becker, S.P. Structural and diagenetic control of fluid migration and cementation along the Moab Fault, Utah. *AAPG Bull.* **2009**, *93*, 653–681. [CrossRef]

20. Balsamo, F.; Storti, F.; Salvini, F.; Lima, C.C. Structural and petrophysical evolution of extensional fault zones in low-porosity, poorly lithified sandstones of the Barreiras Formation NE Brazil. *J. Struct. Geol.* **2010**, *32*, 1806–1826. [CrossRef]

21. Balsamo, F.; Bezerra, F.H.; Vieira, M.; Storti, F. Structural control on the formation of iron oxide concretions and Liesegang bands in faulted, poorly lithified Cenozoic sandstones of the Paraiba basin, Brazil. *Bulletin* **2013**, *125*, 913–931. [CrossRef]

22. Williams, J.N.; Toy, V.G.; Massiot, C.; McNamara, D.D.; Wang, T. Damaged beyond repair? Characterising the damage zone of a fault late in its interseismic cycle, the Alpine Fault, New Zealand. *J. Struct. Geol.* **2016**, *90*, 76–94. [CrossRef]

23. Hoffman, U.; Endell, K.; Wilm, M.D. Kristallstruktur und Quellung von Montmorillonit. *Z. Kristallogr. Cryst. Mater.* **1933**, *86*, 340–348. [CrossRef]

24. Solum, J.G.; Davatzes, N.C.; Lockner, D.A. Fault-related clay authigenesis along the Moab Fault: Implications for calculations of fault rock composition and mechanical and hydrologic fault zone properties. *J. Struct. Geol.* **2010**, *32*, 1899–1911. [CrossRef]

25. Van der Pluijm, R. Out-of-Plane Bending of Masonry: Behaviour and Strength Eindhoven. Ph.D. Thesis, Technische Universiteit Eindhoven, Eindhoven, The Neitherlands, 1999.

26. Solum, J.G.; Van der Pluijm, B.A.; Peacor, D.R. Neocrystallization, fabrics and age of clay minerals from an exposure of the Moab Fault, Utah. *J. Struct. Geol.* **2005**, *27*, 1563–1576. [CrossRef]

27. Sénant, J.; Popoff, M. Early Cretaceous extension in northeast Brazil related to the South Atlantic opening. *Tectonophysics* **1991**, *198*, 35–46. [CrossRef]

28. Matos, R.M.D. The Northeast Brazilian Rift System. *Tectonics* **1992**, *11*, 766–791. [CrossRef]

29. Françolin, J.B.L.; Cobbold, P.R.; Szatmari, P. Faulting in the early Cretaceous Rio do Peixe basin (NE Brazil) and its significance for the opening of the Atlantic. *J. Struct. Geol.* **1994**, *16*, 647–661. [CrossRef]

30. De Castro, D.L.; De Oliveira, D.C.; Gomes Castelo Branco, R.M. On the tectonics of the Neocomian Rio do Peixe Rift Basin, NE Brazil: Lessons from gravity, magnetics, and radiometric data. *J. South. Am. Earth Sci.* **2007**, *24*, 184–202. [CrossRef]

31. Albuquerque, J.P.T. *Inventário Hidrogeológico do Nordeste*; Folha 15; Sudene, Divisão de Documentação: Recife, Brazil, 1970; p. 187.

32. Lima, M.R.; Coelho, M.P.C.A. Estudo palinológico da sondagem de Lagoa do Forno Bacia do Rio do Peixe Cretáceo do Nordeste do Brasil. São Paulo. *Bol. IG-USP Sci.* **1987**, *18*, 67–83.

33. Córdoba, V.C.; Antunes, A.F.; Jardim de Sá, E.F.; Nunes da Silva, A.; Sousa, D.C.; Lins, F.A.P.L. Análise estratigráfica e estrutural da Bacia do Rio do Peixe Nordeste do Brasil: Integração de dados a partir do levantamento sísmico pioneiro 0295_rio_do_peixe_2d. *Bol. Geoci. Petrobras* **2008**, *16*, 53–68.

34. Nicchio, M.A.; Nogueira, F.C.C.; Balsamo, F.; Souza, J.A.B.; Carvalho, B.R.B.; Bezerra, F.H.R. Development of cataclastic foliation in deformation bands in feldspar-rich conglomerates of the Rio do Peixe Basin, NE Brazil. *J. Struct. Geol.* **2018**, *107*, 132–141. [CrossRef]

35. Nogueira, F.C.C.; Marques, F.O.; Bezerra, F.H.R.; de Castro, D.L.; Fuck, R.A. Cretaceous intracontinental rifting and post-rift inversion in NE Brazil: Insights from the Rio do Peixe Basin. *Tectonophysics* **2015**, *644*, 92–107. [CrossRef]

36. Araujo, R.E.B.; Bezerra, F.H.R.; Nogueira, F.C.C.; Balsamo, F.; Carvalho, B.R.B.M.; Souza, J.A.B.; Sanglard, J.C.D.; de Castro, D.L.; Melo, A.C.C. Basement control on fault formation and deformation band damage zone evolution in the Rio do Peixe Basin, Brazil. *Tectonophysics* **2018**, *745*, 117–131. [CrossRef]

37. Folk, R.L. *Petrology of Sedimentary Rocks*; Hemphill Publishing Company: Austin, TX, USA, 1968; p. 182.

38. Balsamo, F.; Storti, F.; Grocke, D. Fault-related fluid flow history in shallow marine sediments from carbonate concretions, Crotone Basin, south Italy. *J. Geol. Soc.* **2012**, *169*, 613–626. [CrossRef]

![minerals logo] *minerals*

MDPI

Article

Structural Controls on Copper Mineralization in the Tongling Ore District, Eastern China: Evidence from Spatial Analysis

Tao Sun [1,2,*] [ID], Ying Xu [3], Xuhui Yu [4], Weiming Liu [1], Ruixue Li [1], Zijuan Hu [1] and Yun Wang [5]

[1] School of Resources and Environmental Engineering, Jiangxi University of Science and Technology, Ganzhou 341000, China; wm_liu@sina.com (W.L.); liruixue0911@163.com (R.L.); m18370956200@163.com (Z.H.)

[2] Jiangxi Key Laboratory of Mining Engineering, Jiangxi University of Science and Technology, Ganzhou 341000, China

[3] Institute of Multipurpose Utilization of Mineral Resources, CAGS, Chengdu 610041, China; yingxuhui@foxmail.com

[4] College of Earth Sciences, Chengdu University of Technology, Chengdu 61005, China; yuxuhui@foxmail.com

[5] School of Water Resource and Environment, China University of Geosciences, Beijing 341515, China; yunwang_1123@163.com

* Correspondence: suntao@jxust.edu.cn; Tel.: +86-0797-831-2751

Received: 4 May 2018; Accepted: 14 June 2018; Published: 15 June 2018

Abstract: Structures exert significant controls on hydrothermal mineralization, although such controls commonly have cryptic expression in geological datasets dominated by 2D maps. Analysis of spatial patterns of mineral deposits and quantification of their correlation with detailed structural features are beneficial to understand the plausible structural controls on mineralization. In this paper, a series of GIS-based spatial methods, including fractal, Fry, distance distribution and weights-of-evidence analyses, were employed to reveal structural controls on copper mineralization in the Tongling ore district, eastern China. The results indicate that Yanshanian intrusions exert the most significant control on copper mineralization, followed by EW-trending faults, intersections of basement faults and folds. The scale-variable distribution patterns of copper occurrences are attributed to the different structural controls operating in the basement and sedimentary cover. In the basement, EW-trending faults serve as pathways for channeling Yanshanian magma from a deep magma chamber to structurally controlled trap zones in the caprocks, imposing an important regional control on the spatial distribution of Cretaceous magmatic-hydrothermal system genetically related to copper mineralization. In the sedimentary cover, bedding-parallel shear zones, formed during the progressive folding and shearing in Indosinian and overprinted by tensional deformation in Yanshanian, act as favorable sites for hosting, focusing and depositing the ore-bearing fluids, playing a vital role in the localization of stratabound deposits at fine scale.

Keywords: structural control; spatial analysis; fractal; buffer-based analysis; data-driven model; Tongling

1. Introduction

Structural controls on hydrothermal mineralization at various scales have been widely recognized [1–5]. At a global scale, hydrothermal systems usually form in specific tectonic settings, e.g., porphyry systems mostly occur in magmatic arc settings [1–3]. At a regional scale, hydrothermal deposits show close proximity to regional faults system or shear zones, which sever as pathways for transporting ore-forming fluids from deep-seated sources to shallow depositing spots [6,7]. At a deposit scale, hydrothermal replacement disseminations, breccias and veins, which are related to subsidiary fracture zones of regional structures, serve as favorable sites for focusing and depositing the ore-bearing fluids and are interpreted to be responsible for localization of orebodies [7]. However, such controls may usually have cryptic expression in various sources of geological records, because (i) structures, especially large-scale structures, may have variable expressions from depth to surface (e.g., mylonite zone at depth and fault zone near the surface) [7]; (ii) spatial associations between map-generalized structures and surface-projected deposits in 2D maps may lead to an inaccurate view or even misunderstanding with respect to controls of mineralization; and (iii) structural features together with structurally controlled mineralization may be formed through successive deformation and polyphase tectonics [8]. Thus, it is a challenge to identify ore-related structural features and elucidate structural controls, as well as measure their contributions to the formation of mineral deposits.

GIS-based spatial analysis has been well-established and developed in the last three decades, assisting in identification of inherent patterns of ore-related geological features and delineation of interplay of the processes that constrain the formation of mineral deposits [9–12]. More specifically, with the help of quantitative methods and easy-to-use GIS software, delineating the spatial patterns of known occurrences of mineral deposits and their associations with geological features (e.g., structural, lithological and geochemical features) can, in addition to field observations, geochemical and mineralogical laboratory methods, provide insights into the controlling mechanisms operating at different scales [10,13–15], especially in the brownfield areas where a relatively large number of mineral deposits have been well-explored [15]. Furthermore, recognition of geological features controlling the mineralization is critical for defining exploration criteria in future prospecting [16].

Since mineral occurrences are simplified to be represented as points on large-scale maps in various applications of spatial analyses, methods of spatial analysis for point patterns have been increasingly employed in studying spatial distribution and geological controls of mineral deposits, mostly involving fractal geometry [17,18] and Fry analysis [19,20]. Through statistical calculation, fractal and Fry analyses are able to highlight the distribution pattern of mineral deposits that may be difficult to be recognized by exclusively relying on visual interpretation [5]. Moreover, distance distribution [14,21] and weights-of-evidence (WofE) analyses [22,23] can further quantify the strength of the spatial association between mineral deposits and geological features believed to be favorable in predicting the location of the mineralization. A joint application of these methods is necessary as individual methods only characterizes a particular aspect, such as non-random clustering of deposits or preferential direction of deposits distribution, of complex spatial features of mineralization systems [10,24].

The Tongling ore district (TOD) is one of the most important Cu producers in China, with totally estimated reserves of over 5 Mt copper [25]. Large stratabound copper deposits constitute the majority of the copper reserves in this area, e.g., the Donggushan deposit with 1 Mt Cu @ 1.01% [26] and the Xinqiao deposit with 0.5 Mt Cu @ 0.71% [27], which have attracted many studies focusing on their genesis [28–36]. These stratabound deposits were firstly considered to be of SEDEX origin by many researchers because of their stratiform orebodies and massive sulfide ores, and the major orebodies occurring in the Carboniferous strata were thought to be products of Late Paleozoic (Hercynian) sedimentary exhalative system [28–30]. Some researchers further proposed an exhalative origin overprinted by Yanshanian magmatic-hydrothermal processes, based on the restricted occurrences of the stratabound orebodies in areas where Yanshanian intrusions are particularly extensive [31–33]. In contrast, some authors advocated that the stratabound mineralization is of epigenetic origin and genetically associated with the Jurassic-Cretaceous tectono-thermal events [34–36]. The precise geochronological data derived from recent studies confirmed that the massive sulfide and skarn orebodies were coeval with the Yanshanian intrusions [25,27,37], supporting the magmatic-hydrothermal origin of these stratabound deposits.

Although the genesis of stratabound deposits is still disputable, most of the researchers tend to agree with the epigenetic origin or at least the dominant contribution of Yanshanian magmatic-hydrothermal activities in the superimposed ore-forming processes [38,39]. In the magmatic-related genetic model, the stratiform orebodies were formed as a result of progressive fluid-rock interaction along the bedding-parallel structurally controlled conduits and were integral but distal parts of a large hydrothermal system that produced the proximal skarn orebodies at the contact zones and porphyry orebodies in the Yanshanian intrusions [35]. Such hydrothermal system and stratabound deposits are similar to their counterparts elsewhere [35], among which the porphyry-skarn polymetallic deposits in the Ertsberg district of Indonesia and manto-type copper deposits in Chile are two representative examples. In the Ertsberg district, the Ertsberg East skarn orebody, one of the largest orebodies, is hosted by a bedding-parallel fault-bounded zone between the limestone of the Faumai Formation and dolomitic carbonate of the Waripi Formation [40]. In the Punta del Cobre district in Chile, the stratabound tabular orebodies occur in the andesite breccia horizons between underlying massive andesite and overlying shale, while the sites of economic copper concentration appear to be controlled by faults [41]. Since structure is an important controlling factor of these stratabound deposits, some relevant studies have been conducted in the TOD, including the spatial patterns [34], deformation model [42] and formative process [43] of ore-controlling structures. However, these studies mostly focus on theoretical deduction and qualitative analysis, and lack quantitative analysis concerning detailed structural features. Hence, this paper attempts to delineate the structural controls by both qualitative and quantitative analytical methods, focusing on the structural controlling mechanisms operating at different scales, which can facilitate the understanding of the formation of copper deposits and provide criteria for future exploration in the TOD.

2. Materials and Methods

2.1. Study Area

The TOD is situated in the central part of the Middle-Lower Yangtze Cu–Au–Fe metallogenic belt (MLYMB) along the Northern margin of the Yangtze craton, bordered by the Qingling-Dabieshan orogenic belt and the North China craton to the North (Figure 1a).

The Northern Yangtze craton is underlain by tonalitic-trondjhemitic-granitic (TTG) gneisses aged from 3.45 to 2.87 Ga [44]. The TTG gneisses and unconformably overlying Archean to Paleoproterozoic metamorphic rocks constituted the basement of the TOD [44]. From Cambrian to Middle Triassic, the TOD represented a stable trough filled with carbonate and clastic rocks of shallow marine facies [36]. Two sedimentary sequences developed in this period, including the Lower Silurian to Upper Devonian regressive bathyal to littoral clastic rocks and overlying Upper Carboniferous to Middle Triassic littoral to neritic carbonates interbedded with bathyal and alternative marine-continental clastic rocks [45]. The sedimentary strata were folded during the Indosinian movement which is initiated at the end of Middle Triassic due to the collision between the Yangtze craton and North China craton [35]. From Jurassic to Cretaceous, this region experienced an event that has long been interpreted as an intracontinental deformation stage with abundant magmatism [35,38,45], and developed thick terrestrial sedimentary and volcanic sequences which unconformably overlie the Silurian to Triassic strata. The detailed lithological description and contact relationship of sedimentary strata in the TOD are listed in Table 1.

The regional structures in the TOD are dominated by several folds with NE-striking axial surfaces and sigmoidal-shaped axes (Figure 1b). Secondary structures include NE-trending thrust faults, NW- and NNW-trending strike-slip faults [46]. The regional gravity anomalies [47] and deep seismic reflection profiles [48] indicate the presence of EW- and NS-trending basement faults. The Yanshanian magmatism resulted in more than 70 intrusions that are mainly composed of granodiorite, quartz monzonite, gabbro monzonite and their hypabyssal equivalents [26,49]. High-precision zircon U-Pb dating results have shown that the intrusions of this region were formed in the Early Cretaceous (mainly 145–129 Ma) [50,51]. The copper-polymetallic deposits discovered in the TOD are dominated by skarn-type, with minor porphyry-type copper deposits occurring in the deeper parts of some skarn deposits [45,50,51]. More than 60 copper-polymetallic skarn deposits and prospects have been discovered in the TOD, mainly clustering in four ore fields designated as Tongguanshan, Shizishan, Fenghuangshan and Shatanjiao ore field from west to east (Figure 1b).

Table 1. Stratigraphy and tectonic events in the Tongling ore district

Epoch	Lithostratigraphic Unit	Code	Lithological Description	Tectonic Activity
Upper Cretaceous	Xuannan Formation	K_2x	Conglomerate and sandstone	Yanshanian movement (ca. 135 Ma)
Middle Jurassic	Luoling Formation	J_2l	Feldspar sandstone, siltstone and shale	
Lower Jurassic	Moshan Formation	J_1m	Feldspar sandstone with interlays of silty shale and coal, conglomerate at bottom	Indosinian movement (ca. 195 Ma)
Middle Triassic	Tongtoujian Formation	T_2t	Siltstone with interlays of sandy shale	
	Yueshan Formation	T_2y	Limestone, dolomite in upper and siltstone in lower	
Lower Triassic	Nanlinghu Formation	T_1n	Limestone	
	Helongshan Formation	T_1h	Limestone	
	Yingkeng Formation	T_1y	Limestone with interlays of silt shale	
Upper Permian	Dalong Formation	P_2d	Siliceous shale with interlays of limestone	
	Longtan Formation	P_2l	Fine sandstone and silt shale with interlays of coal	
Lower Permian	Gufeng Formation	P_1g	Siliceous slate and siliceous shale	
	Qixia Formation	P_1q	Bioclastic limestone in upper and carbonaceous shale in lower	
Upper Carboniferous	Chuanshan Formation	C_2c	Orbicular limestone and bioclastic limestone	
	Huanglong Formation	C_2h	Bioclastic limestone and dolomite	
Upper Devonian	Wutong Formation	D_3w	Quartz sandstone and silty shale	
Middle Silurian	Fentou Formation	S_2f	Sandstone, siltstone and sandy shale	
Lower Silurian	Gaojiabian Formation	S_1g	Black shale	
Upper Ordovician	Wufeng Formation	O_3w	Black siliceous shale	
	Tangtou Formation	O_3t	Calcareous shale with interlayers of limestone	
Middle Ordovician	Tangshan Formation	O_2t	Limestone with interlayers of thin slate	
Lower Ordovician	Lunshan Formation	O_1l	Limestone in upper and dolomite in lower	
Cambrian	Huangjiabang Formation	ϵ	Limestone	
Precambrian	Dongling Group	Pt_3d	Biotite quartz schist and gneiss	Jinning movement (ca. 850–800 Ma)

Modified from [26]; dashed line represents disconformity; and double line represents angular unconformity.

The copper occurrences (including known deposits and prospects) and structural features employed in this study were derived from Geological Database of Bureau of Geological and Mineral Resources of Anhui Province based on 1:50,000 geological survey and complemented by the literature available for the study area concerning regional geological settings [26,42,45,49,51]. The raw data were examined before being inputted into a spatial database. Only those copper and copper-dominated polymetallic deposits were included in the analysis, since the other types of copper-related polymetallic deposits may be products of different structurally controlled processes when compared with copper mineralization. The structural features were reclassified into three categories including the basement faults, cover faults and folds. All the examined data were compiled to vector formats and imported into the ArcGIS 10 platform (Environmental Systems Research Institute, Redlands, CA, USA) for the subsequent spatial analyses.

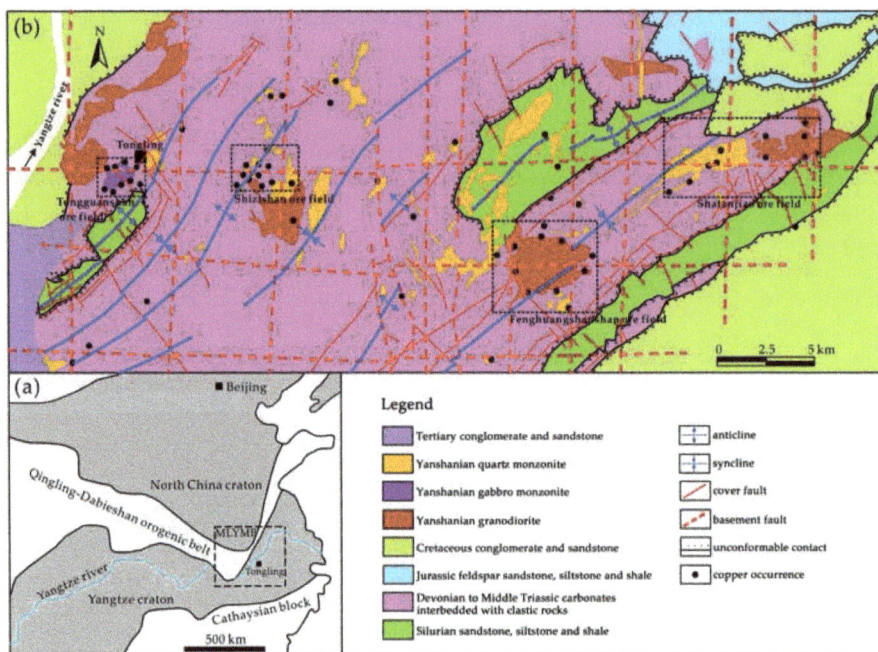

Figure 1. Geological map of the study area: (**a**) simplified tectonic map showing the location of the TOD; and (**b**) geological map of the TOD showing the locations of copper occurrences, modified from a 1:50,000 scale geological map [52] and [37,42,43,53].

2.2. Fractal Analysis

Fractals are entities that have similar geometrical patterns when observed in ranges of scales [17]. This scale-invariance can be depicted by a power-law proportional relationship between a measurement of the target pattern and its scale [17]. Various methods have been proposed for estimating the fractal dimension of a given pattern, each of which reveals an aspect of geometrical complexity of the target pattern [10,54]. The box-counting and radial-density methods, which are the most commonly used methods in analyzing geological point patterns, were employed to estimate corresponding fractal dimensions in this study. The fractal analysis can reveal the statistical scale-related laws of the distribution of copper occurrences, which may be products of structurally controlled processes.

In the box-counting method, the study area involving geological features of interest (e.g., mineral deposits) is overlain by a grid that comprises square cells or boxes with side length δ, and then the number $N(\delta)$ of those boxes containing parts of target features is counted (Figure 2a). The above process is repeated using different box size δ to obtain corresponding box number $N(\delta)$ (Figure 2b,c). If the pattern under analysis pertains to fractal pattern, the relationship between $N(\delta)$ and δ should follow a power-law function as below [17]:

$$N(\delta) \propto A\delta^{-D_B} \tag{1}$$

where D_B is the box-counting fractal dimension, and A is a constant. Practically, a graph of $\log(N(\delta))$ versus $\log(\delta)$ is plotted and then a best-fit regression line is drawn by the least square method, while the slope of the regression line represents the box-counting fractal dimension (Figure 2d).

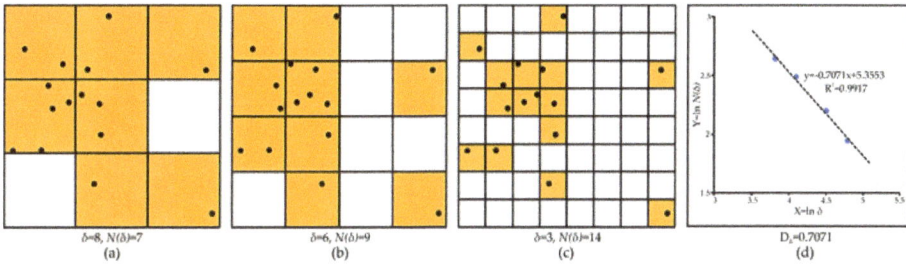

Figure 2. Schematic diagram of box-counting analysis: (**a**) 7 boxes containing target points with box size $\delta = 8$; (**b**) 9 boxes counted with box size $\delta = 6$; (**c**) 14 boxes counted with box size $\delta = 3$; and (**d**) log-log plot revealing the power-law relationship of counted box number $N(\delta)$ and box size δ, obtaining box-counting fractal dimension $D_B = 0.7071$.

In the radial-density method, fractal points, also called fractal dusts, have been demonstrated to satisfy a radial-density relationship, which can be described as [55]:

$$d \propto Br^{D_R-2} \tag{2}$$

where, d is the average point density of the circles with a radius r that center in every point, and B is a constant, while D_R is the radial-density fractal dimension. Likewise, D_R is usually obtained by calculating the slope of a regression line that presents the linear relationship of d and r in a log-log plot.

2.3. Fry Analysis

Fry analysis is a geometrical method of spatial autocorrelation for analyzing point patterns [22], which is implemented by the construction of an autocorrelation diagram called Fry plot. Figure 3 shows a basic procedure for creating Fry plot [10,24]: (i) two analogue sheets including an original sheet recording raw points (Figure 3a) and a blank tracing sheet are prepared; (ii) the origin of the original sheet O is placed on one of the raw points, thus preserving the orientations and distances of all the other points (Figure 3b); (iii) the points in the original sheet are then translated to the tracing sheet with O coinciding with the origin of the tracing sheet O' (Figure 3c); (iv) the origin O moves to another raw point (Figure 3d), and the new distribution pattern of raw points is copied to the tracing sheet following step (iii) (Figure 3f). This procedure is repeated until every point in the original sheet is used as the origin O (Figure 3e,f), resulting in (n^2-n) points in the tracing sheet for n raw points (Figure 3g). The final tracing sheet is termed as Fry plot, and the points in this sheet are called Fry points.

Fry plot records the distances and orientations of each raw point relative to every other point, thus enhancing subtle patterns of target point features [24], based on which rose diagrams are usually constructed to analyze preferential orientations of point pairs within specific distances that reveal the directional controls on mineralization at different map scales.

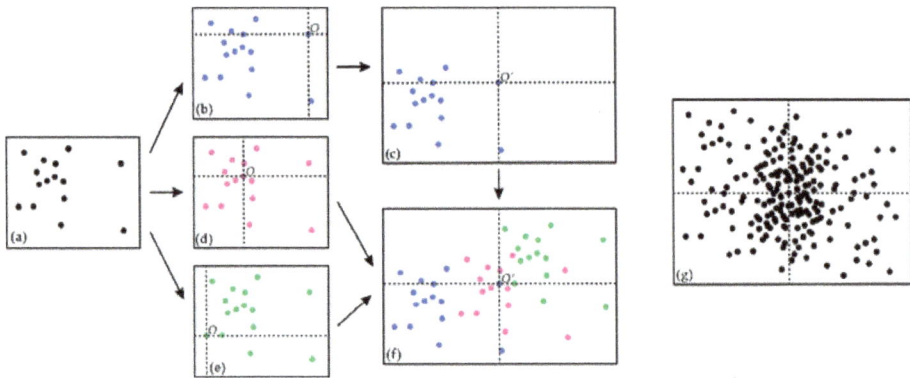

Figure 3. Schematic diagram for constructing a Fry plot: (**a**) the original sheet records raw points; (**b**) the origin O is placed on one of the raw points; (**c**) distribution pattern of raw points according to the origin O is transferred to the tracing sheet; (**d**,**e**) the origin O is re-placed on every raw point; (**f**) the tracing sheet records all distribution patterns of raw points with respect to different origins; and (**g**) Fry plot is constructed.

2.4. Distance Distribution Analysis

Distance distribution analysis is a spatially buffer-based method for quantifying spatial association between a set of points (i.e., mineral occurrences) and another set of spatial features [24,56]. This analysis involves calculating and comparing the cumulative relative frequency according to given distances from a certain set of geological features to (i) mineral occurrence locations (denoted as D_M) and (ii) non-occurrence locations (denoted as D_N). D_N indicates naturally random probability density distribution of regular patterns within a given buffer distance, while D_M presents a non-random probability density distribution of mineralized patterns that is characterized by unevenly clustering of mineral occurrences within the corresponding buffer. The difference D, which is calculated by $(D_M - D_N)$, represents how much the cumulative frequency of mineral occurrences (i.e., D_M) is higher or lower than that expected due to chance (i.e., D_N), measuring the intensity of spatial association between the analyzed geological feature and mineralization.

In order to manifest statistically if D_M is significantly greater than D_N, an upper confidence band for the curve of D_N (denoted as uc) can be given by [57]:

$$uc = D_N + \sqrt{9.21(M+N)/4MN} \tag{3}$$

where M is the number of mineral occurrences that used to estimate D_M, while N is the number of non-occurrence locations using for calculating D_N, and 9.21 is a constant for significance level $\alpha = 0.01$ [24].

2.5. Weights of Evidence (WofE) Analysis

The WofE analysis is a data-driven Bayesian statistical method that offers a quantitative measurement of spatial association between a set of given geological features and the target occurrences (e.g., mineral deposits, prospects or geological anomalies) [58,59].

A detailed mathematical explanation of the WofE method is available in Bonham-Carter (1994) [9]. In the GIS-based application of a mineral occurrence-related analysis, the WofE analysis is implemented on the basis of several binary predictor maps of geological features [9]. Firstly, the study area is subdivided into T square cells with an equal size, among which D cells are occupied by mineral occurrences. The prior probability can be defined as:

$$P_{prior} = P(D) = \frac{D}{T} \tag{4}$$

and the relative importance of spatial association between the geological feature B_i and mineralization is estimated by a pair of weights, namely positive weight W^+ and negative weight W^-, which can be given by:

$$W^+ = \ln\left\{\frac{P(B|D)}{P(B|\overline{D})}\right\}, \quad W^- = \ln\left\{\frac{P(\overline{B}|D)}{P(\overline{B}|\overline{D})}\right\} \tag{5}$$

where P denotes the corresponding probability; B and $\overline{B}$ are the presence and absence of geological features; D and $\overline{D}$ are the presence and absence of mineral occurrences. $P(B|D)$, for example, represents the probability of B occurring given the presence of D. The contrast C is defined as an overall measurement of spatial correlation, which is given by:

$$C = W^+ - W^- \tag{6}$$

In order to evaluate the significance of the contrast C, the confidence of the contrast (denoted as C_S), obtained from a Student t-test, is employed here and defined as:

$$C_S = \frac{C}{S(C)} = \frac{C}{\sqrt{S^2(W^+) + S^2(W^-)}} \tag{7}$$

where S denotes the standard deviation of the corresponding parameter.

3. Results and Discussion

3.1. Spatial Patterns of Copper Occurrences

The box-counting log-log graph shows that the distribution of copper occurrences in the study area exhibits a bifractal pattern, i.e., the log-log plot of box number $N(\delta)$ versus box size δ can be fitted with two regression lines (Figure 4a), resulting in two fractal dimensions of 0.2468 ($\delta \leq 1.6$ km) and 0.75 ($\delta > 1.6$ km). In contrast, the radial-density analysis yields a trifractal pattern, as indicated by three regression lines that represent three fractal dimensions varying from 0.796 ($r \leq 1.4$ km), 1.1722 (between 1.5 and 4.5 km) to 0.8092 ($r > 4.5$ km) (Figure 4b). A single regression line represents a power-law (fractal) relationship between the measurements and their scales, implying a scale-invariance pattern resulted from a nonlinear process. In this study, the different fractal patterns of copper occurrences indicated by the multi-line fractal model could be ascribed to the different ore-controlling processes operating at different scales. It is noteworthy that the log-log graph, especially for radial-density analysis, seems to be alternatively fitted with one single regression line. However, the two-lines and three-lines fractal models shown in Figure 4 are assumed to be optimum because they reach the maximum regression coefficients (R^2) of the fitted lines, meaning that the reduction of any regression line would lower the regression coefficients.

Although apparent differences in the fractal dimensions are noted in the above analyses, there is a general agreement between the result of box-counting method and that of radial-density analysis, such that variations in fractal dimensions, indicated by intersection of the neighbor regressing lines, both occur at around 1.5 km, suggesting that the different fractal structures exist within identical ranges (within 1.5 km and beyond 1.5 km) in both box-counting and radial-density fractal relationship. It is also noted that there is an intersection at 4.5 km in the radial-density fractal plot; however, it is not clear whether there also exists another fractal dimension in the box-counting analysis when taking box size greater than 4.5 km, since box number under such situation is not large enough to be statistically counted.

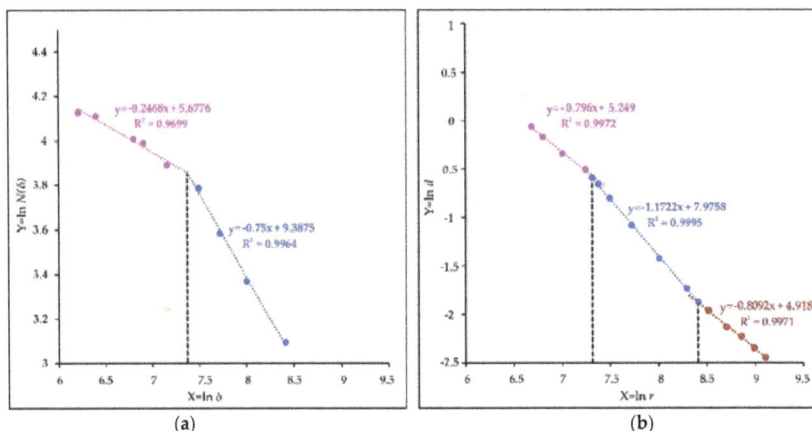

Figure 4. Log-log plot defining the fractal dimensions of spatial pattern of copper occurrences in the TOD: (**a**) box-counting linear relationship; and (**b**) radial-density linear relationship.

The results of fractal analyses in this study, including the multi-fractal dimension model and fractal structures occurring within identical ranges, are consistent with those of some previous studies [10,15,16,24]. It is considered that discrepancies in fractal dimensions are plausibly linked to different geological controls operating at diverse scales, e.g., regional-, local- and prospect-scale [10]. Nevertheless, such scale-variable geological controls are still cryptic and need to be delineated by further analysis.

Fry analysis has been performed to investigate the orientations of plausible controls on copper mineralization. 3906 Fry points were delivered from 63 copper occurrences in the TOD (Figure 5a), based on which rose diagrams were constructed. The rose diagram for all Fry points illustrates a simply dominant EW trend (Figure 5b), suggesting a fundamental EW-trending control on copper mineralization at regional scale. Since fractal analyses indicate variations in fractal dimensions around 1.5 and 4.5 km, we also analyzed the characteristics of Fry points within these ranges. The rose diagram for Fry points within 4.5 km of each other indicates a preferential NNE trend, with subordinate NE and EW trends (Figure 5c). The rose diagram for Fry points within 1.5 km of each other exhibits a main NE-NEE trend, with subsidiary trends in EW and NS directions (Figure 5d).

The results of Fry analysis infer different directional controls at regional- (>4.5 km) and fine- (<4.5 km) scales, which could be correlated to detailed structural features in the TOD. However, such correlation is not specific. For example, the NE-trending control at fine scale may be related to the NE-trending faults or be linked to the folds with NE-striking axes. Further analysis is necessary so as to delineate the one-to-one correspondence between the scale-variable controls and detailed structural features.

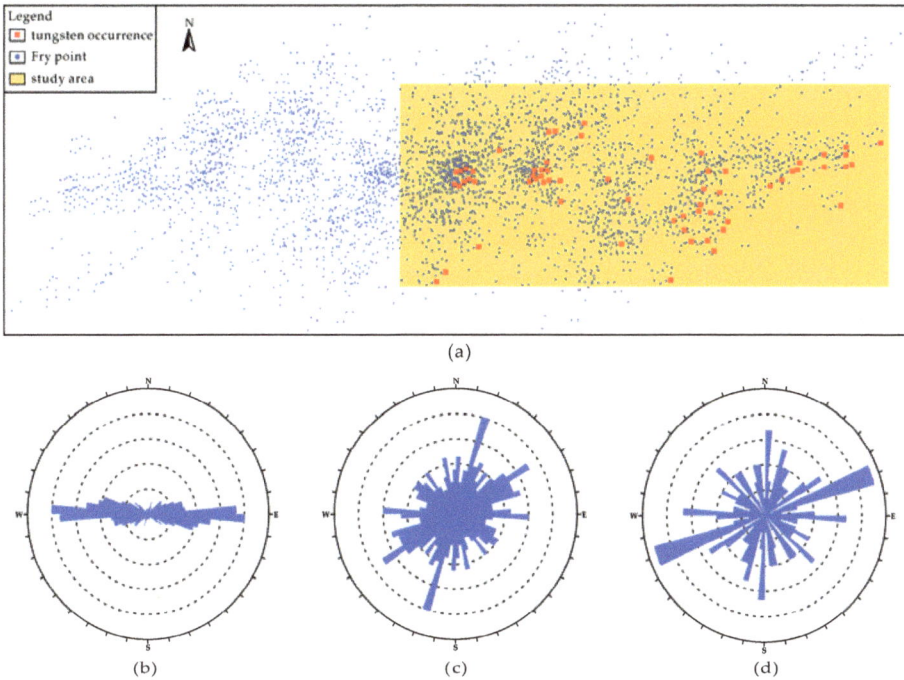

Figure 5. (**a**) Fry plot showing spatial distribution of Fry points derived from 63 copper occurrences; and rose diagram for (**b**) all Fry points; (**c**) Fry points within 4.5 km; (**d**) Fry points within 1.5 km.

3.2. Spatial Correlation of Structural Features with Copper Mineralization

Structural features are the outcomes of diverse geological processes, only a few of which are associated with ore-forming processes and sever as structural controls on mineralization [10,24]. In order to reveal the subtle structural controls in the TOD, the structural features were grouped in terms of their orientations, and the spatial associations of these features with copper mineralization were quantitatively assessed by distance distribution analysis. The study area was subdivided into 20,250 square cells with side length of 200 m, among which 63 cells containing copper occurrences represent occurrence samples and the other cells are taken as non-occurrence samples.

The EW-trending basement faults exhibit a positive correlation with copper occurrences according to the curve of D (Figure 6). Within an optimal buffer distance of 1.5 km, there is at most 21% higher frequency of copper occurrences than what would be expected due to chance. Such correlation is verified to be statistically significant (at $\alpha = 0.01$) since the curve of D_M is plotted above the upper confidence band of D_N within a 1.5 km buffer (Figure 6b).

The NS-trending basement faults have a positive but weak association with copper occurrences beyond the buffer distance of 1 km, reaching only 2% higher frequency than what would be expected (Figure 7). However, the curve of D_M is plotted below the upper confidence band of D_N in the whole range of buffer analysis (Figure 7b), indicating that the weak association between NS-trending faults and copper mineralization is not of statistical significance.

The intersections of the basements faults have a statistically significantly positive correlation with copper occurrences between the buffer distances of 2 and 3 km (Figure 8). At an optimal buffer distance of 2.5 km, there is 23% higher frequency of copper occurrences than what would be expected (Figure 8b).

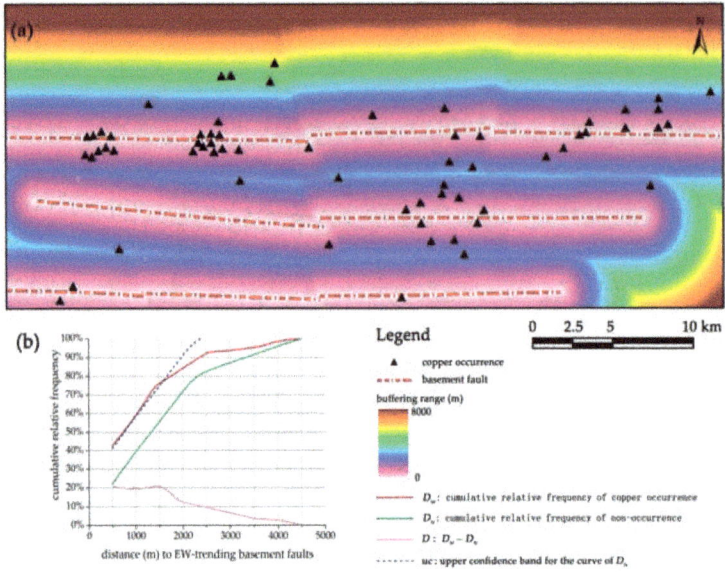

Figure 6. (**a**) Buffer analysis and (**b**) graph of cumulative relative frequency concerning distance to EW-trending faults.

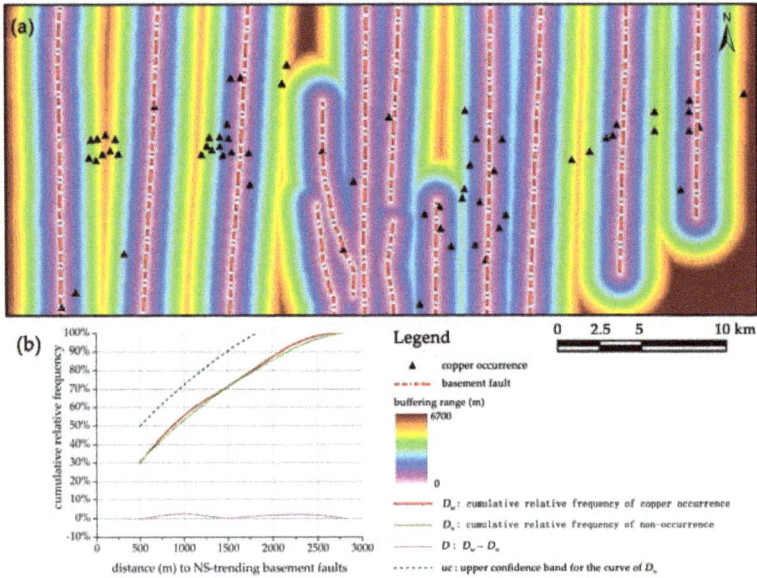

Figure 7. (**a**) Buffer analysis and (**b**) graph of cumulative relative frequency concerning distance to NS-trending faults.

The folds exhibit a statistically significantly positive correlation with copper occurrences in the buffers ranging from 1.5 to 3 km (Figure 9). There is 22% higher frequency of copper occurrences than what would be expected at a 2.5 km buffer (Figure 9b).

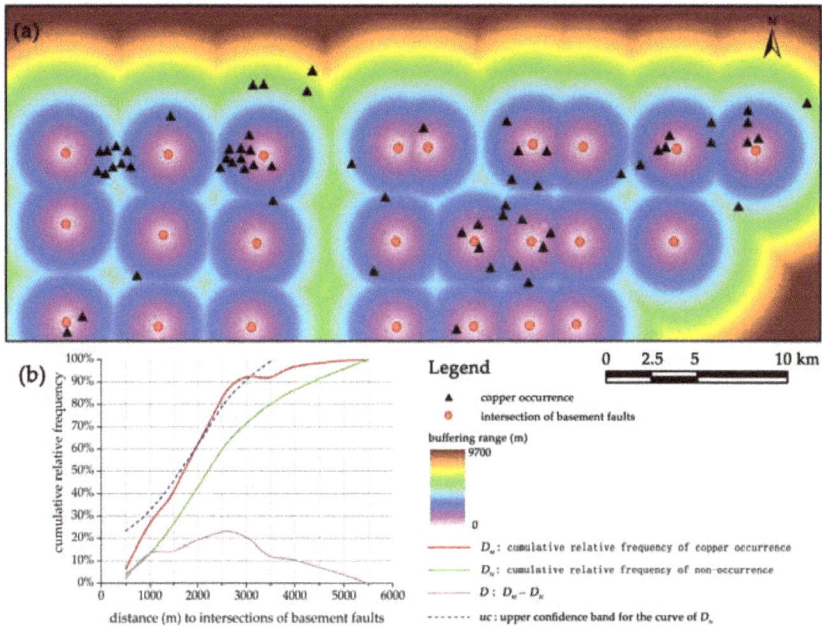

Figure 8. (**a**) Buffer analysis and (**b**) graph of cumulative relative frequency concerning distance to intersections of basement faults.

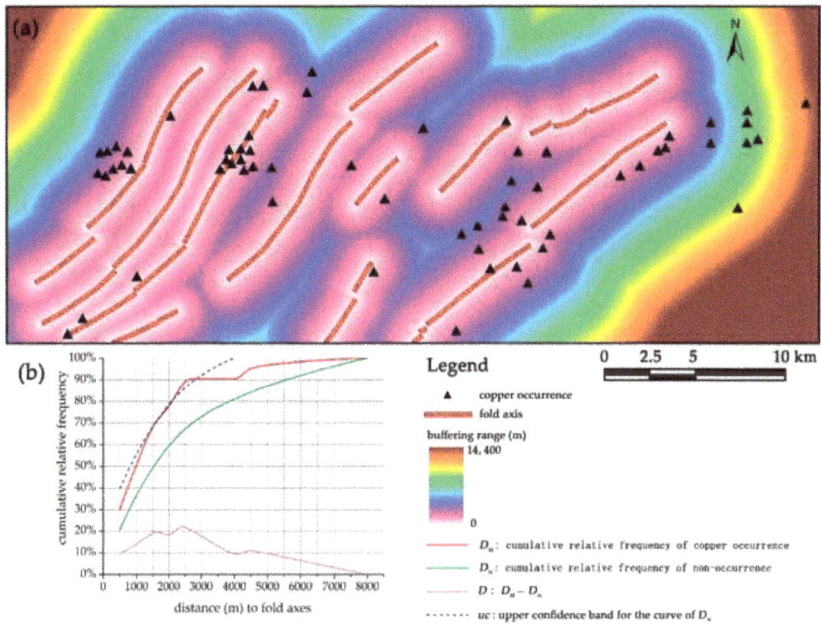

Figure 9. (**a**) Buffer analysis and (**b**) graph of cumulative relative frequency concerning distance to folds.

The cover faults consisting of NE- and NW-trending faults as well as the intersections of these faults show positive spatial association with copper occurrences. There are at most 11%, 10% and 9% higher frequencies of occurrences than what would be expected within the optimal buffers of NE-, NW-trending faults and their intersections, respectively (Figures 10–12). Nevertheless, none of these structural features have a statistically significantly associated with copper occurrence at any buffer distance (Figures 10b, 11b and 12b).

WofE analysis was also implemented to investigate the association of structural features with copper occurrences. At corresponding optimal buffer distances, the contrast values and confidences of contrast were calculated. As depicted in Figure 13 and Table 2, the EW-trending faults, intersections of basement faults and folds have top three highest values of both contrast and confidence of contrast, which are remarkably greater than those of the other structural features. The contrasts and confidences of contrast, which can assist in evaluating the intensity of spatial association, show exactly the same variations as the results derived from distance distribution analysis, implying that the EW-trending faults, intersections of basement faults and folds are plausibly major structural controls on copper mineralization in the TOD.

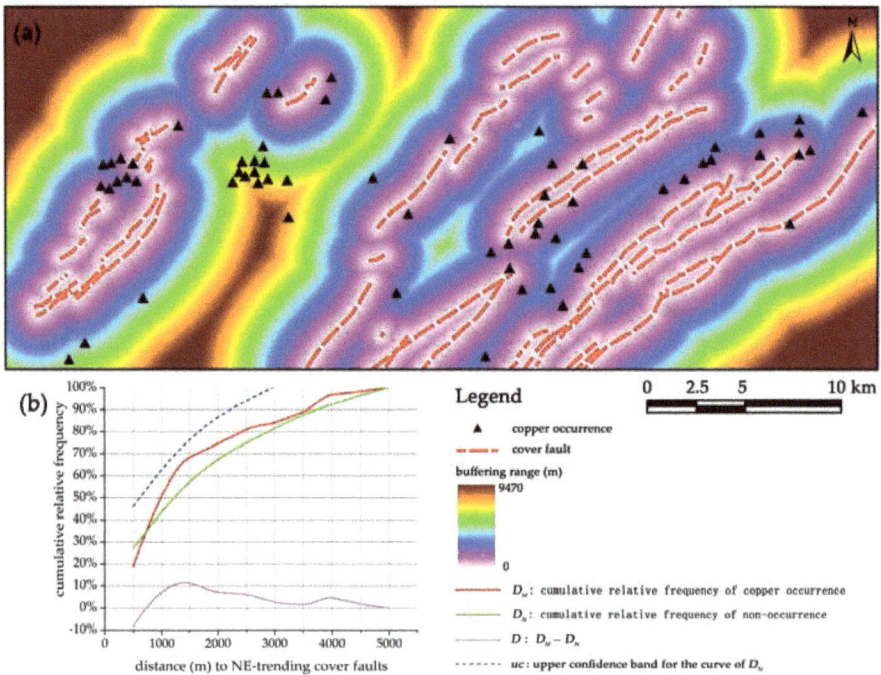

Figure 10. (a) Buffer analysis and (b) graph of cumulative relative frequency concerning distance to NE-trending faults.

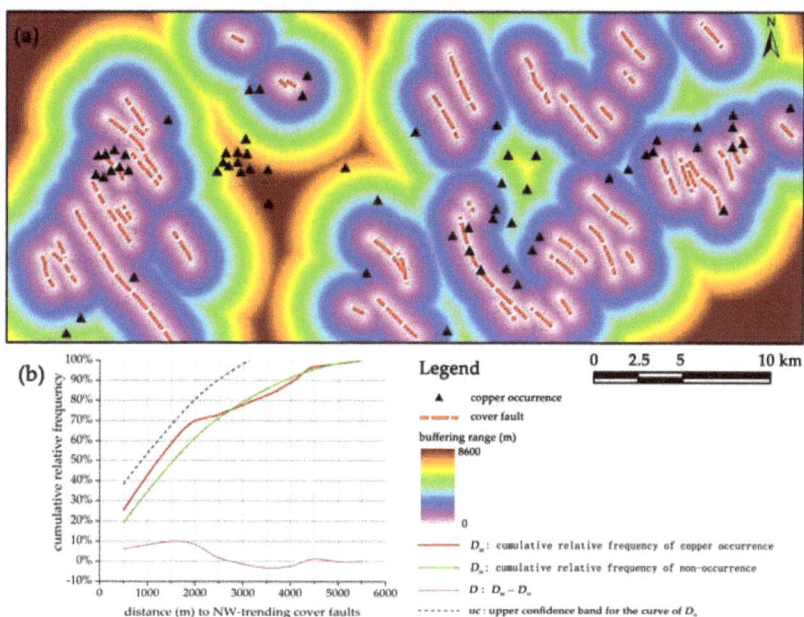

Figure 11. (**a**) Buffer analysis and (**b**) graph of cumulative relative frequency concerning distance to NW-trending faults.

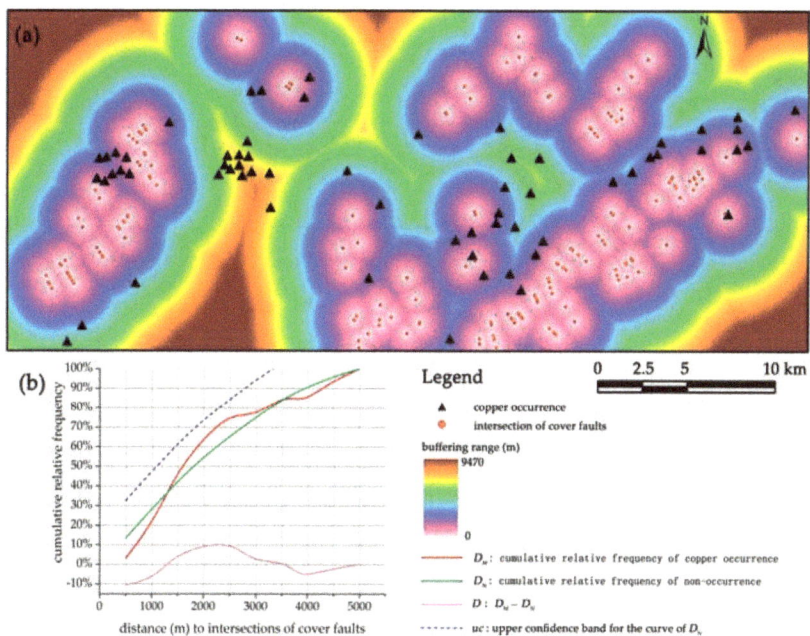

Figure 12. (**a**) Buffer analysis and (**b**) graph of cumulative relative frequency concerning distance to intersections of cover faults.

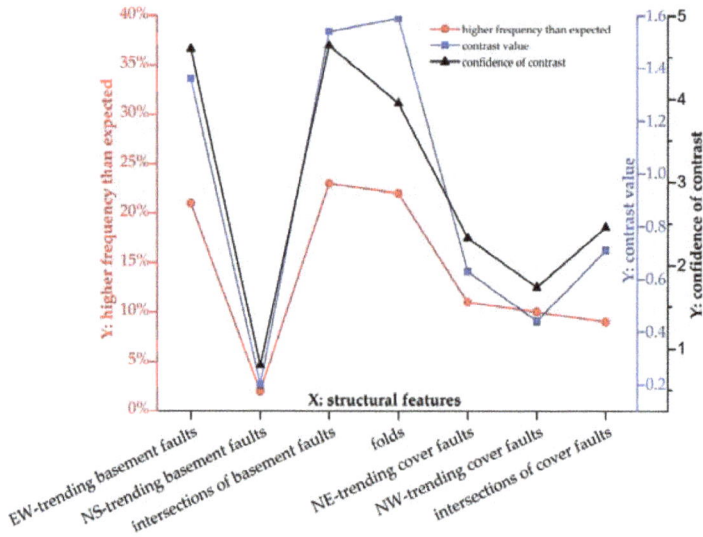

Figure 13. Graph showing variations of higher frequencies than expected, contrasts and confidences of contrast of detailed structural features in the TOD.

Table 2. Results of distance distribution and WofE analyses.

Structural Features	Optimal Buffer Distance (m)	Distance Distribution Analysis				WofE Analysis	
		D_M	D_N	D	uc	C	C_s
EW-trending faults	1500	76%	55%	21%	75%	1.36	4.6
NS-trending faults	1000	55%	53%	2%	72%	0.2	0.8
intersections of basement faults	2500	83%	60%	23%	79%	1.54	4.64
folds	2500	89%	67%	22%	86%	1.59	3.95
NE-trending faults	1500	68%	57%	11%	76%	0.63	2.33
NW-trending faults	1500	59%	49%	10%	68%	0.44	1.74
intersections of cover faults	2500	74%	65%	9%	84%	0.71	2.46
contact of Yanshanian intrusion	350	87%	21%	55%	52%	3.04	8.03

D_M: cumulative relative frequency of copper occurrence; D_N: cumulative relative frequency of non-occurrence; D: $D_N - D_N$; uc: upper confidence band for the curve D_N; C: contrast value; and C_s: confidence of contrast.

3.3. Spatial Correlation of Faults with Intrusions

Considering that copper deposits in the TOD are dominated by skarn-type, Yanshanian intrusion is a key ore-controlling factor and its contact with wall rock can be considered as a special structure. The result of distance distribution analysis shows the strongest association of the contact with copper occurrences. Within a 1.7 km buffer of the contact, there is at most 55% higher frequency of copper occurrences than what would be expected, and such strong association is manifested to be statistically significant (Figure 14). The WofE analysis yields a contrast value of 3.04 and a confidence value of 8.03 which are markedly higher than the corresponding values of the other structural features (Table 2), supporting the most significant association of the contact with copper mineralization.

Since regional faults are considered to control the emplacement of intrusions according to many previous studies [45,48,50], we also performed distance distribution analysis to investigate the correlation of intrusion with the faults of various orientations. The results show that the EW-, NS-trending faults and their intersections have statistically significantly positive correlations with intrusion regions at most of the buffer distances (Figure 15a–c). There are 26% and 17% higher frequencies of intrusion regions than what would be expected at the optimal buffers of the intersections of basement faults and EW-trending faults, respectively (Figure 15a,c), suggesting strong associations of these structural features with intrusions. The NS-trending basement faults have a moderate correlation with intrusion, delineated by 11% higher frequency of intrusion regions than what would be expected (Figure 15b). In contrast, cover faults and their intersections show negative correlations with intrusion regions within a 1.5 km buffer (Figure 15d–f). Beyond the buffer distance of 1.5 km, they show positive but weak associations with intrusions. There are 6%, 9% and 5% higher frequencies of intrusion regions than what would be expected at the optimal buffers of NE-, NW-trending faults and their intersections, respectively (Figure 15d–f).

It is noteworthy that the EW-trending faults and intersections of basement faults, which show the strongest correlations with intrusions, also exhibit significant associations with copper occurrences in the previous distance distribution analysis. It is necessary to evaluate what extent of these structural controls on intrusion determine their strong correlations with copper mineralization. The EW-trending faults and Yanshanian intrusions were buffered with their optimal distances, and the copper occurrences located within the corresponding buffered zones were counted. It appears that 98% (47 out of 48) of the copper occurrences distributed within the buffers of EW-trending faults are located in the overlapping zones of the buffered EW-trending faults and intrusions which account for 33.58% of total area. Only one occurrence is included in the buffered zones where intrusions are absent (occupying 66.42% of total area) (Figure 16). Likewise, 96% (49 out of 51) of copper occurrences located within the buffers of intersections of basement faults are included in the overlapping zones of buffered intersections of faults and intrusions that occupy 37.11% of total area (Figure 17). It is inferred that the significantly strong associations of EW-trending faults and intersections of basement faults with copper mineralization are attributed to the controls of these structural features on intrusions.

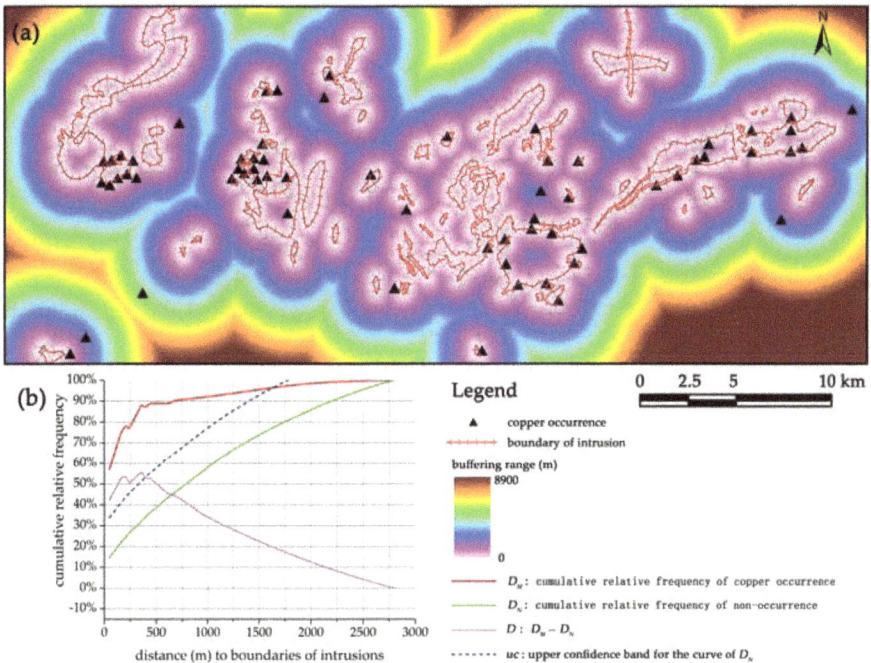

Figure 14. (a) Buffer analysis and (b) graph of cumulative relative frequency concerning distance to boundaries of intrusions.

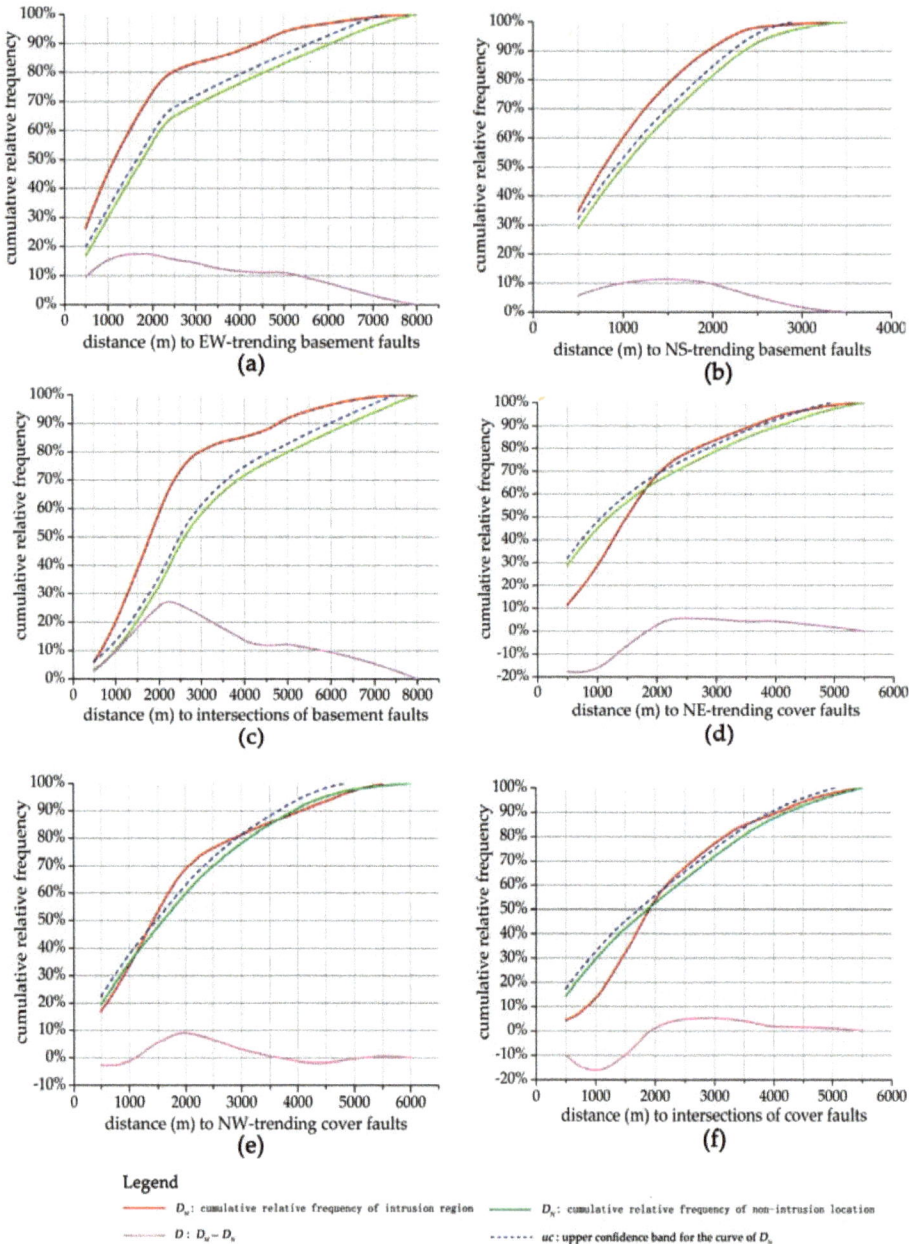

Figure 15. Graph of cumulative relative frequency concerning distance to (**a**) EW-trending faults; (**b**) NS-trending faults; (**c**) intersections of basement faults; (**d**) NE-trending faults; (**e**) NW-trending faults and (**f**) intersections of cover faults.

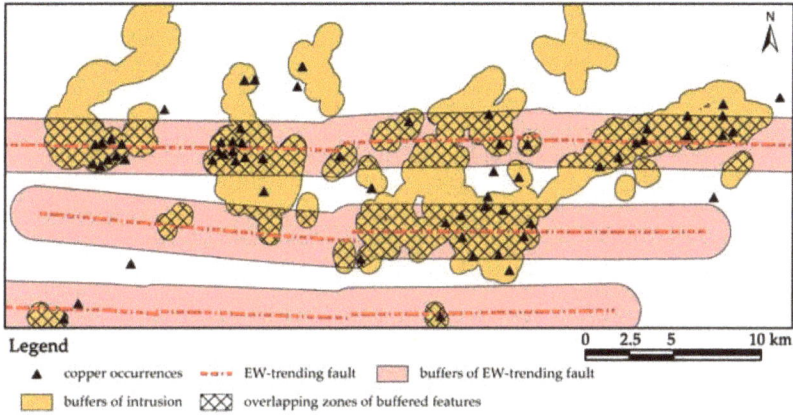

Figure 16. Buffer analysis showing the distribution of copper occurrences in buffered EW-trending faults and intrusions.

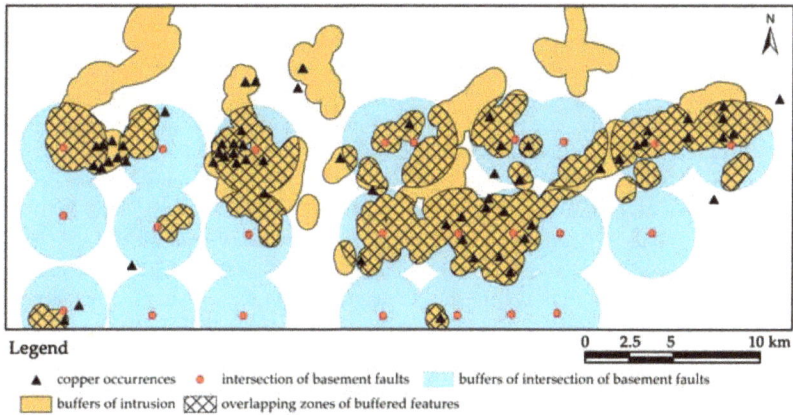

Figure 17. Buffer analysis showing the distribution of copper occurrences in buffered intersections of basement faults and intrusions.

3.4. Interpretation of Structural Controls on Copper Mineralization

The tectonic evolution of the TOD can be divided into four stages. The first stage is the formation and development of the basement of the Lower Yangtze Terrain (LYT) before Jinning movement (ca. 850–800 Ma) when the TOD was still an integral part of the LYT [38]. Secondly, after Jinning movement and before Indosinian movement (ca. 195 Ma), the LYT gradually developed into an archipelagic ocean stage, and the major sedimentary cover in this region formed. Contemporarily, the Cathaysian block and LYT drifted gradually to the North China Craton, leading to several soft collisions [60]. The vertical movement induced by the opening-closing effect related to soft collision was dominated in this stage, resulting in some disconformities (Table 1). Thirdly, the convergence of the Yangtze craton and North China craton (referred to as the Indosinian orogeny) commenced at the end of Triassic, which induced the formation of a series of significant structural features including angular unconformity between the Triassic and Jurassic strata (Table 1), folds and faults [35,60]. It is considered that the Indosinian movement has produced the present structural framework in the TOD, and even in the South China [39]. Eventually, the TOD experienced Yanshanian movement (ca. 135 Ma) characterized by transformation from contraction to extension since early Cretaceous [35], which induced the formation of widespread intermediate-felsic intrusions and associated mineralization. The multi-stage tectonic evolution is responsible for the structural features in both basement and sedimentary cover that are related to epigenetic copper mineralization.

The basement structures are dominated by EW- and NS-trending faults. These faults, totally overlain by Mesozoic strata, are considered to be formed before Indosinian period and reactivated in the Mesozoic [45], although the detailed geometrical and kinematical characteristics of these faults are still not clear. In previous studies, faults have been proven to act as favorable pathways for transporting ore-related magma and ore-forming fluids from deep sources to shallow trap zones, resulting in a strong association of these faults with hydrothermal mineral deposits [10,16,19]. In the study area, the petrological data and geophysical profiles evidence that a magma chamber was developed in the Mesozoic at about −10 km from the surface [42]. The EW-trending basement faults are interpreted to play a vital role of channeling the magma from the magma chamber to the shallow trap zones during the Yanshanian period. This significant control of the EW-trending faults on Yanshanian intrusions is supported by distance distribution and WofE analyses, which is fully responsible for the strong correlation between the EW-trending faults and copper mineralization. This interpretation can explain the result of Fry analysis that exhibits a predominant EW trend at regional scale.

The known copper deposits are situated in the sedimentary cover where the folds with sigmoidal-shaped axes are the dominant structures, thus delineating the formative process of the folds is crucial for understanding the structural framework and copper mineralization in the cover level. Since the youngest stratum involved in the folds is Middle Triassic, it is deduced that the folds were formed during the Indosinian movement which resulted in the angular unconformity between Middle Triassic and Lower Jurassic (Table 1). A classic model of dextral simple-shear deformation in a strike-slip fault zone is introduced to illustrate the formation of folds and faults under the deformation regime of Indosinian movement dominated by NW-SE compression and dextral shear (Figures 18 and 19). As the fault zone initiates, a structural system forms consisting of (i) conjugate strike-slip faults, (ii) folds, (iii) reverse faults, and (iv) normal faults (Figure 19a) [61,62]. The initially formed folds and reverse faults trend perpendicular to the direction of the greatest shortening, while the normal faults trend parallel to the direction of the greatest shortening. Subsequently, the continued strike-slip shearing can lead to a rotation of the elements in this system [62]. The axes of previously formed folds turn to sigmoidal shape. The earlier formed normal faults accommodate sinistral strike-slip motion, and the reverse faults accommodate dextral strike-slip motion (Figure 19b). The NE-trending thrust faults observed in the field [63] and sinistral strike-slip motion of NW-trending faults identified in the geological map (Figure 1) support the rationality of this model.

In the Mesozoic strata, there existed several interfaces between two adjacent strata which have distinct mechanical properties, some of which also represented the interfaces of disconformity, e.g., the interface between the quartz sandstone of Upper Devonian and limestone of Upper Carboniferous. During the formative process of the folds in Indosinian period, the abovementioned interfaces were subjected to the progressive deformation of folding and shearing, leading to extensive bedding-parallel shear zones [43] (Figures 20 and 21). In particular, the bedding detachments occur in the cores of the folds due to the layer-parallel slippage in the formative process of folds. These shear zones were overprinted by tensional deformation in the Cretaceous when the tectonic regime in this region changed from compression to extension, thus being favorable for trapping and localizing mineralized fluids. This inference is supported by (i) the clearly discordant boundaries between stratiform orebodies and wall rocks which suggest that the ores were deposited in mechanical dilation spaces (Figure 22) [27,35,64], and (ii) the result of a numerical modeling on the Dongguashan deposit which demonstrates that the stratiform high dilation zones induced by extensional stress are favorable for fluids focusing and consistent with those positions where orebodies actually occur [64]. In addition, the bedding-parallel trap zones are located near the contacts of intrusions where sufficient sources of heat and fluid are available, and hosted in a set of carbonate strata suitable for forming skarn (Figure 20).

Therefore, the bedding-parallel structures in the folded strata are favorable for hosting, focusing and depositing ore-bearing fluids, assisting in the formation of the stratabound orebodies in this area. The thickening of orebodies in the cores of folds is attributed to the detachments occurring there (e.g., major orebody within C_2 in Figure 20). This interpretation is supported by distance distribution and WofE analyses, which both exhibit strong spatial association of the folds with copper mineralization. It is also inferred that the dominance of NE, NNE and NEE trends in the rose diagrams of Fry points at fine scales (<4.5 km) is attributed to the control of the folds with NE-striking axes, rather than those of the NE-trending faults which show poor correlation with copper mineralization through spatial analyses. Moreover, neither cover faults of various orientations nor the intersections of these faults show statistically significant correlation with copper mineralization, suggesting that they may only play a role in migrating the ore-bearing fluids towards the favorable host structures (i.e., multi-layered bedding-parallel shear zones) where fluid concentration and mineral deposition actually occurred, therefore leading to a lesser significant association of these cover faults with copper occurrences.

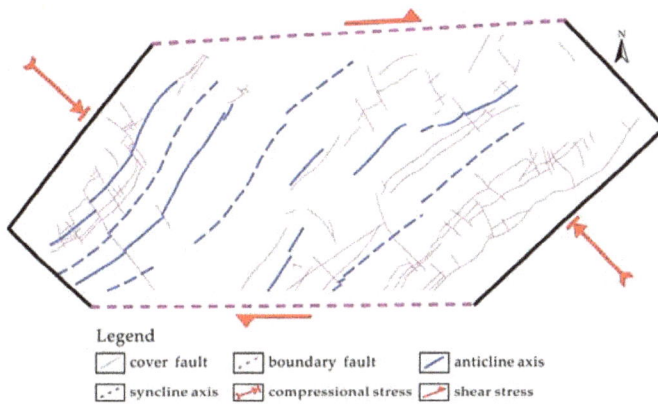

Figure 18. Stress regime during the formative process of the folds with sigmoidal axes, modified from [42].

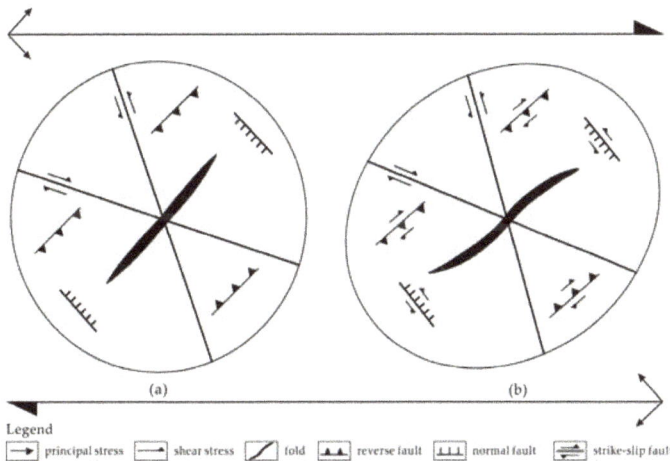

Figure 19. Deformation model of dextral shearing in a strike-slip fault zone, modified from [61]. (**a**) a structural system formed in initial stage of deformation; and (**b**) a rotation of structural elements during continued strike-slip shearing.

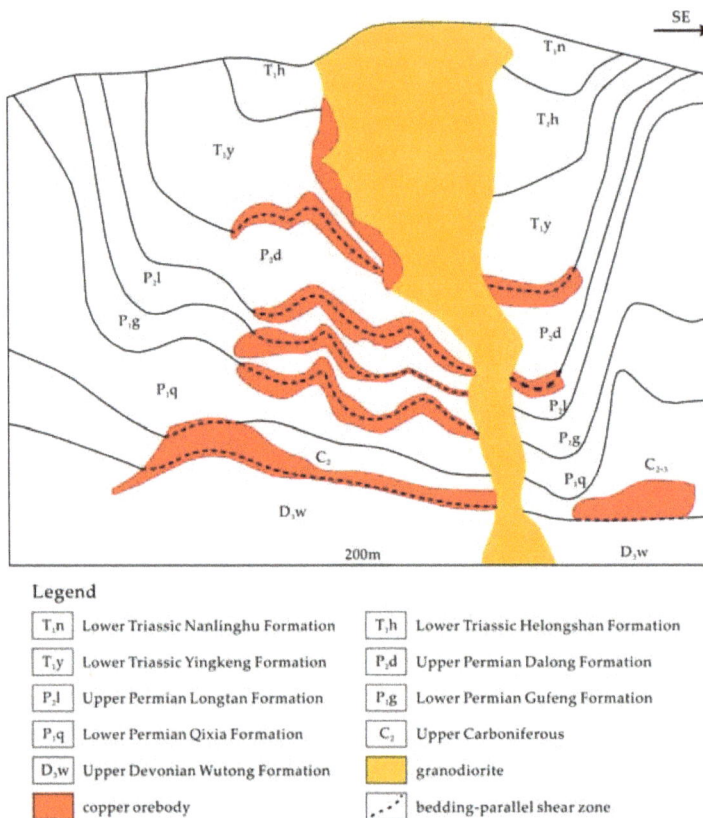

Figure 20. Typical cross-section of Shizishan ore field showing the characteristic stratabound skarn orebodies hosted in the folds, modified from [43].

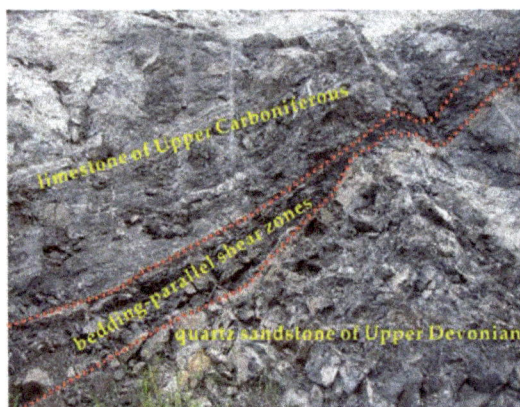

Figure 21. Field photograph of outcropped bedding-parallel shear zone between limestone of Upper Carboniferous and quartz sandstone of Upper Devonian in the Xinqiao deposit.

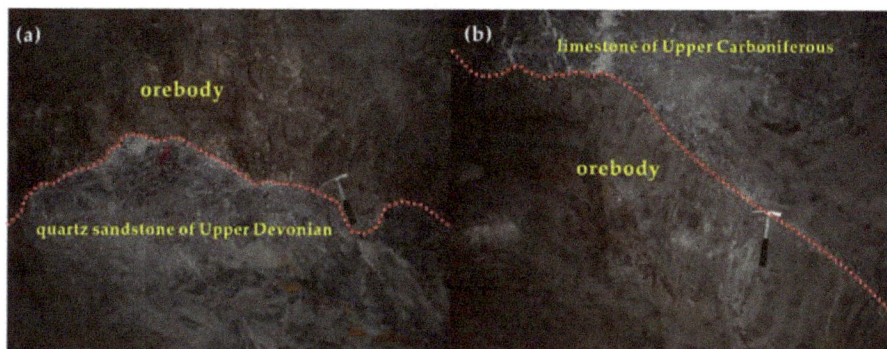

Figure 22. Photographs showing the discordant boundaries between stratiform orebodies and wall rocks in the Xinqiao deposit. (**a**) the boundary between orebody and underlying Upper Devonian quartz sandstone; and (**b**) the boundary between orebody and overlying Upper Carboniferous limestone.

4. Conclusions

(i) Fractal dimensions obtained from box-counting and radial-density analyses suggest that different structural controls operate at diverse scales of <1.5 km, 1.5–4.5 km and >4.5 km. This scale-variable controlling behavior is supported and explored by the results of Fry analysis, which illustrates a dominant EW trend at regional scale (>4.5 km) and preferential NE-NNE-NEE trends at fine scale (<4.5 km).

(ii) The spatial associations of detailed structural features with copper mineralization are further investigated by quantitative spatial analyses. The Yanshanian intrusions, EW-trending faults, intersections of basement faults, and folds have significant associations with copper mineralization, indicated by their high values of quantitative parameters in both distance distribution and WofE analyses.

(iii) The interpretation of structural controls on copper mineralization is made in combination of foregoing analytical results. The scale-variable patterns of mineral occurrences are attributed to the different structural controls operating in the basement and sedimentary cover. In the basement, the EW-trending faults serve as pathways for channeling magma from a magma chamber into trap zones in the caprocks during Yanshanian period. The significant control of the EW-trending faults on Yanshanian intrusion is fully responsible for the strong correlation between the EW-trending faults and copper mineralization. This inference is supported by the result of Fry analysis which shows a dominant EW trend at regional scale (>4.5 km). In the sedimentary cover, the bedding-parallel shear zones formed during Indosinian folding and shearing and overprinted by tensional deformation in Yanshanian period act as favorable sites for hosting, focusing and depositing the ore-bearing fluids, which is responsible for the dominance of NE-NNE-NEE trends at fine scale (<4.5 m) in the results of Fry analysis. Such bedding-parallel structures, together with the contact zones of intrusion, exert an important control on the formation of characteristic stratabound skarn deposits in the TOD.

Author Contributions: T.S. conducted the GIS-based computational experiments, analyzed the results and wrote the draft paper; Y.X. and X.Y. participated in the analysis of experimental results; W.L. and R.L. revised the calculation scheme; Z.H. and Y.W. collected the original data.

Acknowledgments: The research leading to this paper was jointly supported by National Natural Science Foundation of China (Grant No. 41602335), Natural Science Foundation of Jiangxi Province (Grant No. 20161BAB213084), Science and Technology Project of Jiangxi Provincial Department of Education (Grants No. GJJ150625 and No. GJJ170537), Program of Qingjiang Excellent Young Talents (Grant No. JXUSTQJYX2017001) and Doctoral Scientific Research Foundation of Jiangxi University of Science and Technology (Grant No. jxxjbs15002).

We would like to express our gratitude to two anonymous *Minerals* reviewers for their constructive comments and suggestions that greatly improved the manuscript. Thanks are also given to Zhongfa Liu from Central South University for assistance with field evidences.

Conflicts of Interest: The authors declare no conflict of interest.

References

1. Sillitoe, R.H. A plate tectonic model for the origin of porphyry copper deposits. *Econ. Geol.* **1972**, *67*, 184–197. [CrossRef]

2. Tosdal, R.M.; Richards, J.P. Magmatic and structural controls on the developments of porphyry Cu ± Mo ± Au deposits. *Rev. Econ. Geol.* **2001**, *14*, 157–181.

3. Kwelwa, S.D.; Dirks, P.H.G.M.; Sanislav, I.V.; Blenkinsop, T.; Kolling, S.L. Archaean gold mineralization in an extensional setting: The structural history of the Kukuluma and Matandani Deposits, Geita Greenstone Belt, Tanzania. *Minerals* **2018**, *8*, 171. [CrossRef]

4. Cox, S.F.; Knackstedt, M.A.; Braun, J. Principles of structural control on permeability and fluid flow in hydrothermal systems. *Rev. Econ. Geol.* **2001**, *14*, 1–24.

5. Austin, J.R.; Blenkinsop, T.G. Local to regional scale structural controls on mineralisation and the importance of a major lineament in the eastern Mount Isa Inlier, Australia: Review and analysis with autocorrelation and weights of evidence. *Ore Geol. Rev.* **2009**, *35*, 298–316. [CrossRef]

6. Sillito, R.H. Iron oxide-copper-gold deposits: An Andean view. *Miner. Deposita* **2003**, *38*, 787–812. [CrossRef]

7. Zeng, M.; Zhang, D.; Zhang, Z.; Liu, T.; Li, C.; Wei, C. Structural controls on the Lala iron-copper deposit of the Kangdian metallogenic province, Southwestern China: Tectonic and metallogenic implications. *Ore Geol. Rev.* **2018**, *97*, 35–54. [CrossRef]

8. Chauvet, A.; Piantone, P.; Barbanson, L.; Nehlig, P.; Pedroletti, I. Gold deposit formation during collapse tectonics: Structural, mineralogical, geochronological, and fluid inclusion constraints in the Ouro Preto Gold Mines, Quadrilátero Ferrífero, Brazil. *Econ. Geol.* **2001**, *96*, 25–48. [CrossRef]

9. Bonham-Carter, G.F. *Geographic Information System for Geoscientists, Modeling with GIS*; Pergamon: Elmsford, NY, USA, 1994; pp. 238–333.

10. Haddad-Martim, P.M.; Filho, C.R.D.S.; Carranza, E.J.M. Spatial analysis of mineral deposit distribution: A review of methods and implications for structural controls on iron oxide-copper-gold mineralization in Carajás, Brazil. *Ore Geol. Rev.* **2017**, *81*, 230–244. [CrossRef]

11. Schetselaar, E.; Ames, D.; Grunsky, E. Integrated 3D geological modeling to gain insight in the effects of hydrothermal alteration on post-ore deformation style and strain localization in the Flin Flon Volcanogenic Massive Sulfide Ore System. *Minerals* **2018**, *8*, 3. [CrossRef]

12. Sun, T.; Wu, K.X.; Chen, L.K.; Liu, W.M.; Wang, Y.; Zhang, C.S. Joint application of fractal analysis and weights-of-evidence method for revealing the geological controls on regional-scale tungsten mineralization in Southern Jiangxi Province, China. *Minerals* **2017**, *7*, 243. [CrossRef]

13. Li, X.H.; Yuan, F.; Zhang, M.M.; Jia, C.; Jowitt, S.M.; Ord, A.; Zheng, T.K.; Hu, X.Y.; Li, Y. Three-dimensional mineral prospectivity modeling for targeting of concealed mineralization within the Zhonggu iron orefield, Ningwu Basin, China. *Ore Geol. Rev.* **2015**, *71*, 633–654. [CrossRef]

14. Xie, J.Y.; Wang, G.W.; Sha, Y.Z.; Liu, J.J.; Wen, B.T.; Nie, M.; Zhang, S. GIS prospectivity mapping and 3D modeling validation for potential uranium deposit targets in Shangnan district, China. *J. Afr. Earth Sci.* **2017**, *128*, 161–175. [CrossRef]

15. Carranza, E.J.M. Developments in GIS-based mineral prospectivity mapping: An overview. In Proceedings of the Mineral Prospectivity, Current Approaches and Future Innovations, Orléans, France, 24–26 October 2017.

16. Parsa, M.; Maghsoudi, A.; Yousefi, M. Spatial analyses of exploration evidence data to model skarn-type copper prospectivity in the Varzaghan district, NW Iran. *Ore Geol. Rev.* **2018**, *92*, 97–112. [CrossRef]

17. Mandelbrot, B.B. *Fractals: Form, Chances and Dimension*; W.H. Freeman: New York, NY, USA, 1977; pp. 1–23.

18. Roberts, S.; Sanderson, D.J.; Gumiel, P. Fractal analysis of Sn-W mineralization from central Iberia; insights into the role of fracture connectivity in the formation of an ore deposit. *Econ. Geol.* **1998**, *93*, 360–365. [CrossRef]

19. Carranza, E.J.M.; Owusu, E.A.; Hale, M. Mapping of prospectivity and estimation of number of undiscovered prospects for lode gold, Southwestern Ashanti Belt, Ghana. *Miner. Deposita.* **2009**, *44*, 915–938. [CrossRef]

20. Mehrabi, B.; Ghasemi, S.M.; Tale, F.E. Structural control on epithermal mineralization in the Troud-Chah Shirin belt using point pattern and Fry analyses, North of Iran. *Geotectonics* **2015**, *49*, 320–331. [CrossRef]

21. Agterberg, F.P.; Bonham-Carter, G.F.; Wrigh, D.F. Statistical pattern integration for mineral exploration. In *Computer Application in Resource Estimation Prediction and Assessment for Metals and Petroleum*; Gaal, G., Merriam, D.F., Eds.; Pergamon: Elmsford, NY, USA, 1990; pp. 1–21.

22. Cheng, Q.M.; Agterberg, F.P. Fuzzy weights of evidence method and its application in mineral potential mapping. *Nat. Resour. Res.* **1999**, *8*, 27–35. [CrossRef]

23. Yuan, F.; Li, X.H.; Zhang, M.M.; Jowitt, S.M.; Jia, C.; Zheng, T.K.; Zhou, T.F. Three-dimensional weights of evidence-based prospectivity modeling: A case study of the Baixiangshan mining area, Ningwu Basin, Middle and Lower Yangtze Metallogenic Belt, China. *J. Geochem. Explor.* **2014**, *145*, 82–97. [CrossRef]

24. Carranza, E.J.M. Controls on mineral deposit occurrence inferred from analysis of their spatial pattern and spatial association with geological features. *Ore Geol. Rev.* **2009**, *35*, 383–400. [CrossRef]

25. Cao, Y.; Zheng, Z.; Du, Y.; Gao, F.; Qin, X.; Yang, H.; Lu, Y.; Du, Y. Ore geology and fluid inclusions of the Hucunnan deposit, Tongling, Eastern China: Implications for the separation of copper and molybdenum in skarn deposits. *Ore Geol. Rev.* **2017**, *81*, 925–939. [CrossRef]

26. Liu, L.M.; Zhao, Y.L.; Zhao, C.B. Coupled geodynamics in the formation of Cu skarn deposits in the Tongling–Anqing district, China: Computational modeling and implications for exploration. *J. Geochem. Explor.* **2010**, *106*, 146–155. [CrossRef]

27. Zhang, Y.; Shao, Y.J.; Li, H.B.; Liu, Z.F. Genesis of the Xinqiao Cu–S–Fe–Au deposit in the Middle-Lower Yangtze River Valley metallogenic belt, Eastern China: Constraints from U–Pb–Hf, Rb–Sr, S, and Pb isotopes. *Ore Geol. Rev.* **2017**, *86*, 100–116. [CrossRef]

28. Fu, S.G.; Yan, X.Y.; Yuan, C.X. Geologic feature of submarine volcanic eruption-sedimentary pyrite type deposit in Carboniferous in the Middle-Lower Yangtze River Valley metallogenic belt, Eastern China. *J. Nanjing Univ. Nat. Sci. Ed.* **1977**, *4*, 43–67. (In Chinese)

29. Gu, L.X.; Xu, K.Q. On the carboniferous submarine massive sulfide deposit in the lower reaches of the Yangtze River. *Acta Geol. Sin.* **1986**, *60*, 176–188. (In Chinese)

30. Gu, L.X.; Hu, W.X.; He, J.X. Regional variations in ore composition and fluid features of massive sulfide deposits in South China: Implications for genetic modeling. *Episodes* **2000**, *23*, 110–118.

31. Yang, D.F.; Fu, D.X.; Wu, N.X. Genesis of pyrite type copper in Xinqiao and its neighboring region according to ore composition and structure. *Issue Nanjing Inst. Geol. Miner. Resour. Chin. Acad. Geol. Sci.* **1982**, *3*, 59–68. (In Chinese)

32. Xie, H.G.; Wang, W.B.; Li, W.D. The genesis and metallogenetic of Xinqiao Cu–S–Fe deposit, Anhui Province. *Volcanol. Miner. Resour.* **1995**, *16*, 101–107. (In Chinese)

33. Zhou, T.F.; Zhang, L.J.; Yuan, F.; Fang, Y.; Cooke, D.R. LA-ICP-MS in situ trace element analysis of pyrite from the Xinqiao Cu–Au–S Deposit in Tongling, Anhui, and its constrains on the ore genesis. *Earth Sci. Front.* **2010**, *17*, 306–319. (In Chinese)

34. Chang, Y.F.; Liu, X.G. Layer control type skarn type deposit—Some deposits in the Middle-Lower Yangtze Depression in Anhui Province as an example. *Miner. Depos.* **1983**, *2*, 11–20. (In Chinese)

35. Pan, Y.; Dong, P. The lower Changjiang (Yangtzi/Yangtze River) metallogenic belt, East-center China: Intrusion and wall rock hosted Cu–Fe–Au, Mo, Zn, Pb, Ag deposits. *Ore. Geol. Rev.* **1999**, *15*, 177–242. [CrossRef]

36. Mao, J.W.; Shao, Y.J.; Xie, G.Q.; Zhang, J.D.; Chen, Y.C. Mineral deposit model for porphyry-skarn polymetallic copper deposits in Tongling ore dense district of Middle-Lower Yangtze Valley metallogenic belt. *Miner. Depos.* **2009**, *28*, 109–119. (In Chinese)

37. Zhang, Y.; Shao, Y.; Zhang, R.; Li, D.; Liu, Z.; Chen, H. Dating ore deposit using garnet U–Pb geochronology: Example from the Xinqiao Cu–S–Fe–Au deposit, Eastern China. *Minerals* **2018**, *8*, 31. [CrossRef]

38. Zhou, T.; Wang, S.; Fan, Y.; Yuan, F.; Zhang, D.; White, N.C. A review of the intracontinental porphyry deposits in the Middle-Lower Yangtze River Valley metallogenic belt, Eastern China. *Ore Geol. Rev.* **2015**, *65*, 433–456. [CrossRef]

39. Hu, R.Z.; Chen, W.T.; Xu, D.R.; Zhou, M.F. Reviews and new metallogenic models of mineral deposits in South China: An introduction. *J. Asian Earth Sci.* **2017**, *137*, 1–8. [CrossRef]

40. Mertig, H.J.; Rubin, J.N.; Kyle, J.R. Skarn Cu–Au orebodies of the Gunung Bijih (Ertsberg) district, Irian Jaya, Indonesia. *J. Geochem. Explor.* **1994**, *50*, 179–202. [CrossRef]

41. Sato, T. Manto type copper deposit in Chile—A review. *Bull. Geo. Surv. Japan* **1984**, *35*, 565–582.

42. Wang, Q.F.; Deng, J.; Huang, D.H.; Xiao, C.H.; Yang, L.Q.; Wang, Y.R. Deformation model for the Tongling ore cluster region, East-Central China. *Int. Geol. Rev.* **2011**, *53*, 562–579. [CrossRef]

43. Wu, G.G.; Zhang, D.; Zang, W.S. Study of tectonic layering motion and layering mineralization in the Tongling metallogenic cluster. *Sci. China Ser. D Earth Sci.* **2003**, *46*, 852–863. [CrossRef]

44. Li, Y.; Li, J.W.; Li, X.H.; Selby, D.; Huang, G.H.; Chen, L.J.; Zheng, K. A carbonate replacement origin for the Xinqiao stratabound massive sulfide deposit, middle-lower Yangtze Metallogenic Belt, China. *Ore Geol. Rev.* **2017**, *80*, 985–1003. [CrossRef]

45. Chang, Y.F.; Liu, X.P.; Wu, Y.C. *The Copper–Iron Belt of the Low and Middle Reaches of the Changjiang River*; Geological Publish House: Beijing, China, 1991; pp. 1–359. (In Chinese)

46. Liu, L.M.; Yang, G.Y.; Peng, S.L.; Zhao, C.B. Numerical modeling of coupled geodynamical processes and its role in facilitating predictive ore discovery: An example from Tongling, China. *Resour. Geol.* **2005**, *55*, 21–31. [CrossRef]

47. Liu, W.C.; Li, D.X.; Gao, D.Z. Analysis on the time sequence of compounding of structural deformation systems and resulting effects in Tongling area. *J. Geomech.* **1996**, *2*, 42–48. (In Chinese)

48. Lü, Q.T.; Hou, Z.Q.; Zhao, J.H.; Shi, D.N.; Wu, X.Z.; Chang, Y.F.; Pei, R.F.; Huang, D.D.; Kuang, C.Y. Complex crustal structure of Tongling ore district: Insights from deep seismic reflection profiling. *Sci. China Ser. D* **2003**, *33*, 442–449. (In Chinese)

49. Liu, Z.F.; Shao, Y.J.; Wei, H.T.; Wang, C. Rock-forming mechanism of Qingshanjiao intrusion in Dongguashan copper (gold) deposit, Tongling area, Anhui province, China. *Trans. Nonferr. Met. Soc. China* **2016**, *26*, 2449–2461. [CrossRef]

50. Xie, J.C.; Yang, X.Y.; Sun, W.D.; Du, J.G. Early Cretaceous dioritic rocks in the Tongling region, Eastern China: Implications for the tectonic settings. *Lithos* **2012**, *150*, 49–61. [CrossRef]

51. Liu, L.M.; Peng, S.L. Prediction of hidden ore bodies by synthesis of geological, geophysical and geochemical information based on dynamic model in Fenghuangshan ore field, Tongling district, China. *J. Geochem. Explor.* **2004**, *81*, 81–98. [CrossRef]

52. 321 Geological Team. *Structural Maps of Tongling Area*; Bureau of Geological and Mineral Resources of Anhui Province: Hefei, China, 1989; pp. 1–33.

53. Du, Y.L. Ore-Controlling Factors and Metallogenic Model of Stratabound Skarn Deposits in Tongling Area, Anhui Province. Ph.D. Thesis, China University of Geosciences, Beijing, China, 2013. (In Chinese)

54. Zuo, R.G.; Wang, J. Fractal/multifractal modeling of geochemical data: A review. *J. Geochem. Explor.* **2016**, *164*, 33–41. [CrossRef]

55. Mandelbrot, B.B. *The Fractal Geometry of Nature: Updated and Augmented*; W.H. Freeman: New York, NY, USA, 1983; pp. 1–31.

56. Berman, M. Distance distributions associated with poisson processes of geometric figures. *J. Appl. Probab.* **1977**, *14*, 195–199. [CrossRef]

57. Berman, M. Testing for spatial association between a point process and another stochastic process. *J. R. Stat. Soc. C Appl.* **1986**, *35*, 54–62. [CrossRef]

58. Allek, K.; Boubaya, D.; Bouguern, A.; Hamoudi, M. Spatial association analysis between hydrocarbon fields and sedimentary residual magnetic anomalies using weights of evidence: An example from the Triassic Province of Algeria. *J. Appl. Geophys.* **2016**, *135*, 100–110. [CrossRef]

59. Sang, X.J.; Xue, L.F.; Liu, J.W.; Zhan, L. A novel Workflow for geothermal prospectively mapping weights-of-evidence in Liaoning Province, Northeast China. *Energies* **2017**, *10*, 1069. [CrossRef]

60. Deng, J.; Huang, D.H.; Wang, Q.F.; Hou, Z.Q.; Lü, Q.T.; Yao, L.Q.; Xin, H.B.; Zhang, Q.; Wei, Y.G. Formation mechanism of "drag depressions" and irregular boundaries in intraplate deformation. *Acta Geol. Sin.* **2004**, *78*, 267–272.

61. Waldron, J.W.F. Extensional fault arrays in strike-slip and transtension. *J. Struct. Geol.* **2005**, *27*, 23–34. [CrossRef]

62. David, G.H.; Reynolds, S.J.; Kluth, C.F. *Structural Geology of Rocks and Regions*, 3rd ed.; JohnWiley & Sons, Inc.: Westwood, MA, USA, 2011; pp. 336–338.

63. Wang, Q.F. Model study of the tectonic-magmatic-metallogenical system in Tongling ore cluster area. Ph.D. Thesis, China University of Geosciences, Beijing, China, 2005. (In Chinese)

64. Liu, L.M.; Sun, T.; Zhou, R.C. Epigenetic genesis and magmatic intrusion's control on the Dongguashan stratabound Cu-Au deposit, Tongling, China: Evidence from field geology and numerical modeling. *J. Geochem. Explor.* **2014**, *144*, 97–114. [CrossRef]

![minerals logo] *minerals*

MDPI

Article

The Hajjar Regional Transpressive Shear Zone (Guemassa Massif, Morocco): Consequences on the Deformation of the Base-Metal Massive Sulfide Ore

Safouane Admou [1,2,*], Yannick Branquet [2,3], Lakhlifi Badra [1], Luc Barbanson [2], Mohamed Outhounjite [4], Abdelali Khalifa [4], Mohamed Zouhair [4] and Lhou Maacha [4]

1   Département des Sciences de la Terre, Faculté des Sciences, Université Moulay Ismaïl de Meknès, B.P. 11201 Zitoune Meknès, Morocco; badra_lakhlifi@yahoo.fr (L.B.)
2   Institut des Sciences de la Terre d'Orléans (ISTO), Université Orléans, CNRS BRGM UMR7327, Campus Géosciences 1A, rue de la Férollerie, 45071 Orléans, CEDEX 2, France; yannick.branquet@univ-orleans.fr (Y.B.); luc.barbanson@univ-orleans.fr (L.B.)
3   Géosciences Rennes (GR), Université de Rennes 1, CNRS UMR6118, Campus de Beaulieu, CS 74205, 35042 Rennes CEDEX, France
4   Groupe MANAGEM, Twin center, Tour A, BP 5199, Casablanca, Morocco; M.OUTHOUNJITE@managemgroup.com (M.O.); A.KHALIFA@managemgroup.com (A.K.); M.ZOUHAIR@managemgroup.com (M.Z.); L.MAACHA@managemgroup.com (L.M.)
*   Correspondence: admou.safouane@gmail.com

Received: 30 June 2018; Accepted: 2 October 2018; Published: 7 October 2018

Abstract: The genesis of the base-metal massive sulfide deposits hosted within the Moroccan Hercynian Jebilet and Guemassa Massifs is still under debate. No consensus currently exists between the two models that have been proposed to explain the deposits, i.e., (1) syngenetic volcanogenic massive sulfide mineralization, and (2) synmetamorphic tectonic fluid-assisted epigenetic mineralization. Conversely, researchers agree that all Hercynian massive sulfide deposits in Morocco are deformed, even though 3D structural mapping at the deposit scale is still lacking. Therefore, while avoiding the use of a model-driven approach, the main aim of this contribution is to establish a first-order structural pattern and the controls of the Hajjar base metal deposit. We used a classical structural geology toolbox in surface and subsurface mining work to image finite strain at different levels. Our data demonstrate that: i) the Hajjar area is affected by a single foliation plane (not two) which developed during a single tectonic event encompassing a HT metamorphism. This syn-metamorphic deformation is not restricted to the Hajjar area, as it is widespread at the western Meseta scale, and it occurred during Late Carboniferous times; ii) the Hajjar ore deposit is hosted within a regional transpressive right-lateral NE-trending shear zone in which syn- to post-metamorphic ductile to brittle shear planes are responsible for significant inflexion (or virgation) of the foliation yielding an anastomosing pattern within the Hajjar shear zone. Again, this feature is not an exception, as various Late Carboniferous-Permian regional scale wrenching shear zones are recognized throughout the Hercynian Meseta orogenic segment. Finally, we present several lines of evidence emphasizing the role of deformation in terms of mechanical and fluid-assisted ore concentrations.

Keywords: Hajjar; shear zone; base metal massive sulfide deposits; structural control; remobilization

---

1. Introduction

Most Volcanogenic Massive Sulfide Deposits (VMSDs) are assumed to form within extensional and subsiding basins during both divergent and convergent plate tectonic settings (e.g., [1]). As a result, in convergent settings leading to continental collision for instance, many VMSDs underwent deformation, burial, and metamorphism. During these transformations, syngenetic massive sulfide bodies (e.g., stratoid lenses, chimneys and stockwerks) were reworked, and primary metallic bearing mineral assemblages may have been remobilized (e.g., either depleted or enriched). For this reason, the deformation and (re)mobilization of the primary sulfide concentration is a fundamental and economic matter which has been recognized and studied for a long time (e.g., [2–6]).

However, in spite of recent advances in modern textural (e.g., electron backscatter diffraction coupled to chemistry) and opaque mineral strain characterization (e.g., [7–10]), it still remains difficult for economic geologists dealing with deformed VMSD to decipher the respective parts of primary syngenetic vs. epigenetic mineralizing processes. As a result, metallogenic models of very large base metal concentrations all over the world are still ambiguous and under debate.

Currently, the genesis of polymetallic base-metal massive sulfide deposits (MSD) from the western Meseta domain in Morocco are currently under debate. This debate is particularly relevant for MSD from the Central Jebilet unit (Figure 1), e.g., the Kettara, Draa Sfar, Koudiat Aïcha, and Lachach deposits. Many authors consider these MSD as metamorphosed and deformed primary VMS and/or sedimentary exhalative (SEDEX) deposits [11–17]; however, other authors argue for a fluid-assisted syn-metamorphic origin during the major Hercynian deformation event [18–22]. In contrast, the Hajjar MSD located in the Hercynian Guemassa Massif (Figure 1) is considered as a metamorphosed and deformed syngenetic VMS/SEDEX deposit [12,23–25]. Although Hajjar shares many similar geological and mineralogical features (e.g., predominance of pyrrhotite) with the Central Jebilet MSD to the north, the hypothesis of either an epigenetic or a syn-metamorphic origin has not yet been put forward.

Since the pioneering works of Hibti (1993) [23] on the Hajjar MSD, very few studies dealing with the structural controls of this ore deposit have been carried out and published in the international literature. However, on a larger scale, much thermal and geochronological data dealing with the tectono-magmatic evolution of the western segment of the Hercynian Meseta have been published [26–28]. Therefore, using data collected from new outcrops, the aim of this work is to complete the Hajjar MSD structural dataset and to re-evaluate the structural context and controls; this is a prerequisite to being able to have a potential syngenetic vs. syn-metamorphic debate, if required. Our approach is to perform structural mapping at each subsurface exploitation level, yielding a 3D view of the deformation pattern. This pattern is then compared to the structural map of the surface outcrops in the Guemassa Massif.

2. The Hajjar Geological and Ore Deposit Framework

The Hajjar MSD is located in the southern part of the Hercynian Occidental Meseta in Morocco, within the Guemassa Massif south of Marrakech (Figure 1). The Guemassa Massif is composed of metasediments, metavolcanites, and intrusions, all of which are Carboniferous in age (see [29–31]) for a detailed description of the volcano-sedimentary series). Massive sulfide lenses (and scarce magnetite bodies) are found and exploited within this volcano-sedimentary sequence [12,23] and references therein). The Hajjar mineralization corresponds to sub-lenticular bodies of various sizes containing 50–75% vol. pyrrhotite, with sphalerite, galena, chalcopyrite, pyrite, and arsenopyrite as the related major ore minerals. The tonnage is about 20 MT of ore with grades of 8% Zn, 2.3% Pb and 04–0.6% Cu [32]. The Hajjar MSD have been classified as an intermediate type between SEDEX and VMS deposits such as the Iberian Pyrite Belt giant deposits, within the "Guemassa-Jebilet" sub-type owing to its high content in pyrrhotite [25]. In the Hajjar MSD, the primary economic massive mineralization is assumed to form in a Visean basin in which an intense syn-sedimentary volcanism occurred [12,23,25]. Like the other MSD of the Occidental Meseta, the Hajjar MSD is strongly folded, faulted, and metamorphosed, which makes it difficult to recognize syngenetic/diagenetic structures and textures.

Based on the literature, the Guemassa rocks were deformed and metamorphosed during several tectonic/thermal events which affected the Guemassa Massif area [23]: i) a D0 syn-sedimentary event at the Visean-Namurian with slumps, intraformational breccias attesting to slope instabilities in the basin.

These syn-sedimentary structures are encountered both within host rocks and sulfide mineralized bodies; ii) a D1 event corresponding to the incipient Hercynian deformation and responsible for a steep NW-SE foliation (S1) in the Oriental Guemassa associated with folding under regional greenschist facies metamorphic conditions. It should be noted that S1 cannot be observed clearly within the Hajjar MSD; iii) a D2 Hercynian tectono-thermal event with P2 folds and associated S2 planar cleavage oriented NE-SW under low-grade metamorphism with sericite. S2 is the predominant foliation observable in the Hajjar mine; and iv) finally, a post-kinematic thermal event, likely related to "hidden plutons", responsible for the crystallization of static biotite porphyroblasts with cordierite and andalousite locally described at Hajjar. In this ore deposit, this thermal event has been dated using "hydrothermal" biotite at ca. 301 Ma [33]. Moreover, for Carboniferous times, the Guemassa Massif is affected by intense multiscale ductile to brittle faulting [34,35], with probable components of Atlasic reactivation during the Tertiary High Atlas orogen (the Guemassa Massif is 15 km to the north of the Atlasic thrusting front, Figure 1B). On a structural map (Figure 1B), these faults and shear zones cross-cut and delineated several blocks within the Guemassa Massif. In the Oriental Guemassa, in which the Hajjar mine is located, the N'Fis block appears to present a peculiar "anarchic" foliation orientation with respect to the bulk NNE-trend of the main Hercynian foliation in the western Meseta domain. These "anarchic" foliation orientations have been explained by deflection or virgations (here defined as a bulk inflexion of foliation plane trajectories) induced by conjugate shear zones during or shortly after a broad E-W-oriented D1 shortening [34,35]: the dominant and earlier shear zones are dextral and trend ENE–WSW (e.g., the Imi-In-Tanout Fault, the eastern branch of the Amizmiz Fault, and the Guemassa Fault, Figure 1B), whereas WNW-ESE-trending shear zones are sinistral, such as the Lalla Takerkoust Fault (Figure 1B). This "virgation model" is compatible with a W–E horizontal shortening, in contrast to Hibti's hypothesis (1993) [23], which argued for a NE–SW horizontal shortening during the D1 event (cf. supra).

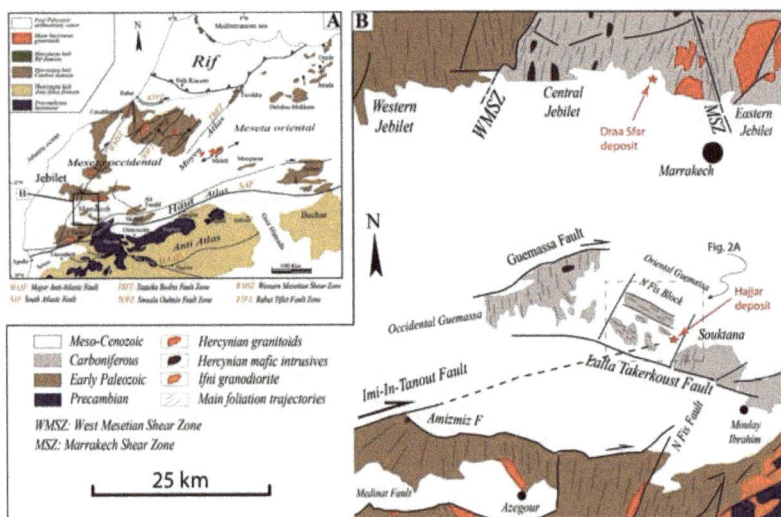

Figure 1. (**A**) Structural map of Morocco showing the major bounding-fault domains. The arrows indicate the sense of shear for the late Variscan structures (modified from Hoepffner et al., 2005 [36]); (**B**) Geological and structural map of the central domain of the Hercynian belt (from [35,37]. The main foliation trajectories in the Jebilet are reported from Essaifi, 1995 [18]). Within the Guemassa Massif, the Hajjar base metal deposit is located in the N'Fis block which presents an "anarchic" foliation orientation with respect to the bulk N to NNE trend reported in the Jebilet, Occidental Guemassa and western High Atlas Variscan Massifs.

The geology of the Guemassa Massif is similar to the Central Jebilet domain (Figure 1B). Both Massifs host the major MSD of the Occidental Meseta. Thus, recent advances in the tectono-metamorphic and magmatic history of the Jebilet [26,28] may help better constrain the Guemassa Massif evolution. Based on petro-structural data, new absolute dating and thermal investigations, these authors improve the time constraints and the succession of the deformational events as follows: i) from 370 to 325 Ma (D0 of Delchini, 2018 [26]), the Jebilet area was a basin filled with syn- to post-rift sediments (the Sarhlef and Teksmin formations, respectively) intruded by shallow sills and dykes and deeper plutonic laccoliths originating from a tholeitic bimodal magmatism (e.g., the mafic/ultramafic Kettara and Sarhlef intrusions) and from a calc-alkaline magmatic suite (e.g., the Oulad-Ouaslam granodiorite) respectively; ii) from 325 to around 310 Ma, a first Hercynian event (D1) is marked by the emplacement of shallow thin skinned nappes with syn-sedimentary breccias. The internal strain is very low and no regional foliation/cleavage (S1) is reported; iii) from ca. 310 Ma to 280 Ma, the main Hercynian deformation (D2), which is polyphased and characterized by a first regional metamorphism (M2a), locally reaches the amphibolite facies (Grt-St) and a second HT/BP "contact" metamorphism in the syn-to post tectonic hornfels facies (M2b, biot + Crd + And) is associated with the leucogranite emplacement around 295 Ma. The successive foliations (S2a and S2b), sub-vertical and oriented N0/30, marked a homoaxial progressive and continual strain regime from a coaxial to a non-coaxial transpression with a broad horizontal NW-SE-trending shortening axis. Last, the D2 increments correspond to a right-lateral transpression accommodated and located along the vertical and conjugate ductile shear zones as the sinistral MSZ (Figure 1B). Therefore, the tectonic scenario proposed by Hibti (1993) [23] for the Guemassa which implies strain axis rotation between D1 and D2 and post-tectonic HT/LP metamorphism diverges from the one proposed by Delchini (2018) [26] for the Jebilet domain.

3. Surface Structural Data

The surface outcrops of the N'Fis block and Souktana Massif have been mapped and studied in terms of the strain analysis and micro-tectonics (Figures 2 and 3). The lithologies encountered are pelites alternating with sandy- to pure limestones intruded by felsic and basic sills and dykes (Figure 2). Major volcanic rock bodies correspond to rhyolitic domes and plugs.

Many outcrops of the Imarine Massif present soft-sediment deformation as slumps and convolutes (Figure 3A,B), suggesting slope instabilities within syn-rift sediments. These soft sediment structures are cross-cut by a sub-vertical foliation (S1) which is often oblique with respect to the axial plane of isoclinal folds (Figure 3B). This suggests that most isoclinal and disharmonic folds are slumps, and therefore, that they pre-date the development of the planar axial foliation. This S1 foliation is well-developed though the N'Fis block, with a sub-vertical dip and a NW-SE orientation (Figure 2). This widespread S1 planar fabric corresponds to a P1 axial planar cleavage (Figure 3C), and locally transposes the bedding planes (S0//S1, e.g., Figure 3D).

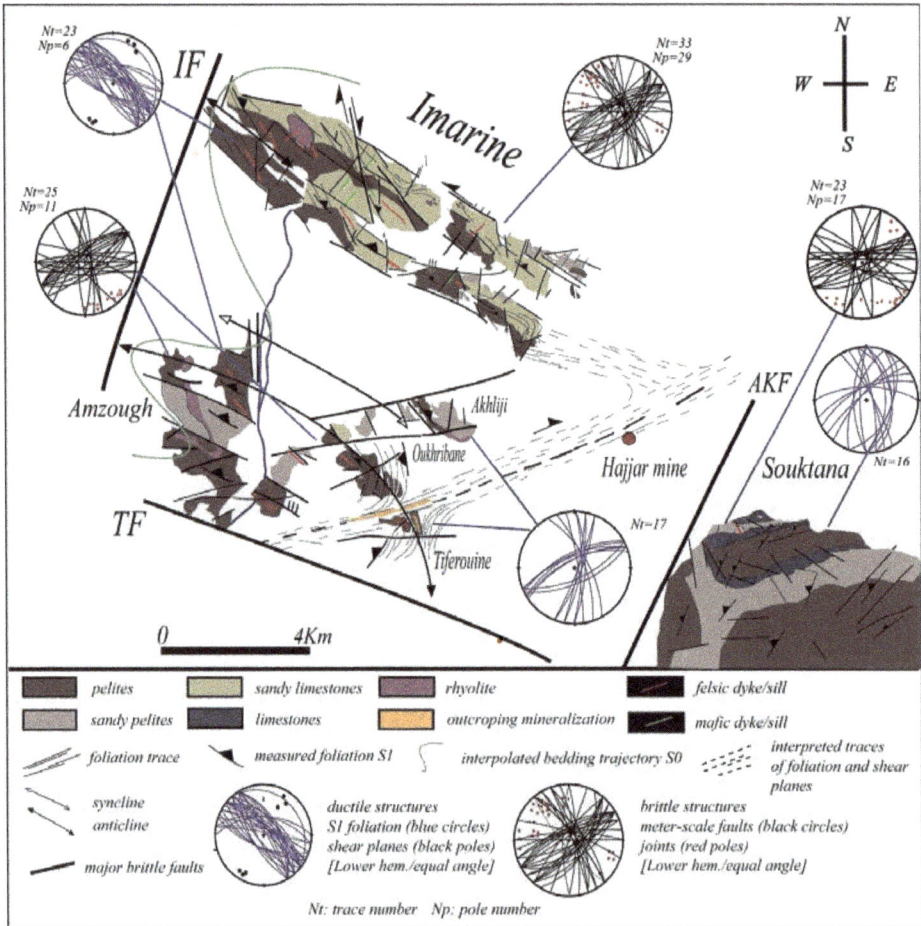

Figure 2. Geological map and surface structural data of the N'Fis block and the Souktana Massif. All of the sedimentary formations are Carboniferous in age and are affected by both metamorphism and deformation. IF: Imarine Fault; TF: Lala Takerkoust Fault; AKF: Ait Khaled Fault. The interpolation of the foliation/shear planes is also supported by sub-surface structural data from underground mine works (cf. infra). Note that the S1 trajectories depict a dextral drag fold against the ENE-trending Tiferouine mineralized body.

Locally, the NW-trending S1 is marked by elongated and aligned biotite porphyroblasts, parallel to the stretching of pyrrhotite grains (Figure 3D), suggesting a syn-tectonic growth of biotite. No stretching lineation has been observed in the N'Fis block. Decimeter-scale sinistral WNW to NW-trending vertical ductile shear planes, occurring sparsely and slightly oblique to S1, are responsible for the local deflection of the S1 planes in the Imarine outcrops (six observations plotted on the stereogram, Figure 2). Brittle faults and joints show a predominant NE-trending orientation with a sub-vertical dip (Figure 2). Due to unfavorable rock materials, the precise kinematics of brittle faults are difficult to establish, which enable the reconstruction of the paleo-stress using the right dihedral method, for instance. However, when it can be observed, the apparent map offsets of the NE-trending decimeter-scale faults indicate a dominant dextral sense of shear.

Finally, the Tiferouine outcrop (Figures 2 and 3E) shows a N70-trending gossan which corresponds to the weathered part of a magnetite-bearing body recognized at depth [12]. The supergene alteration appears to overprint an early cataclasite. Along and within the cataclased mineralized body, the S1 foliation orientation is strongly disturbed (Figure 3E), suggesting drag folding along a right-lateral N70-trending wrench fault (also, see Figure 2 for a map view of the drag folding in the Tiferouine area).

Figure 3. Structures observed in the outcrops. (**A**) syn-sedimentary and soft sediment deformation occurring as slumps and convolutes are widespread in the sandy limestones of the N'Fis block; (**B**) obliquity between the S1 foliation plane and recumbent fold axial plane suggests that some isoclinal folds are former slumps rather than P1 folds; (**C**) NW-trending S1 foliation plane developed within the P1 hinge zone; (**D**) thin sections (cross polars normal to foliation) of metapelite with sulfide ribbons (Po: pyrrhotite) from the N'Fis block. The bedding plane is transposed by the S1 foliation plane, the sulfide ribbon and patches disseminated in the matrix are flattened. Biotite porphyroblasts are elongated broadly parallel to the foliation plane; (**E**) mineralized Tiferouine body (see location in Figure 2) with an associated gossan inside an ENE-trending dextral shear zone evidenced by cataclasites and the re-orientation of S1.

4. Sub-Surface Structural Data from the Hajjar Deposit

Five mine levels have been mapped in the Hajjar deposit (Figure 4). Moreover, we selected two peculiar cross-sections along the galleries to present the meso-scale structures (Figures 5 and 6). The micro scale structure and texture data are summarized in Figures 7 and 8.

4.1. Strain Pattern and Meso-Scale Structures

Bedding, foliation, and shear planes have been reported at each exploitation level within either host rocks or mineralized bodies (Figure 5). Due to the exploitation, the mineralized bodies are not all accessible yet, and foliation data from Hibti (1993) [23] were used to interpolate the S1 trajectories. The micro petro-structural description of the foliation and shear planes are presented below in the following sections within both host rocks and ore.

The resulting maps show that in the Hajjar MSD, the S1 foliation is near vertical and trends from N0 to N45. With respect to the surface data from the N'Fis block (Figures 2 and 3), NW-oriented foliation has not been measured. The interpolation of the bedding trace emphasizes large-scale tight folding which affects both ore bodies and host rocks. The mineralized bodies have been mapped considering historical grade cut-offs for the exploitation. Geologically, the margins of the ore bodies are much less sharp than those shown on the maps provided in Figure 4. Despite this, the ore body morphologies are distributed among various shapes from a group of lensoid decametric bodies to multi-lobate and "dendritic" shapes (e.g., CP in Figure 4C). It is noteworthy that most of the lensoid decametric bodies, often distributed in clusters, are elongated parallel to the local foliation (e.g., CEWD, CWD, CP in Figure 4D,E).

The brittle deformation marked by fault offsets and joints makes it difficult to locally follow the ductile foliation and shear planes. This brittle deformation is marked by large cataclasite zones (thick lines in Figure 4) with different apparent kinematics that indicate complex and likely diachronous activities. Even though these polyphased cataclasite zones are very important for the continuation, exploration, and production of ore bodies, they require a specific structural study which is outside the scope of this paper which focuses on Variscan ductile strain.

Ductile strain is marked by foliated zones that are heterogeneously distributed, suggesting strain localization in corridors between less deformed areas (Figures 4, 5A and 6A). The strain pattern presents two types of high strain corridors (indicated in Figure 4B): N to NNE-trending and NE to ENE-trending strain corridors.

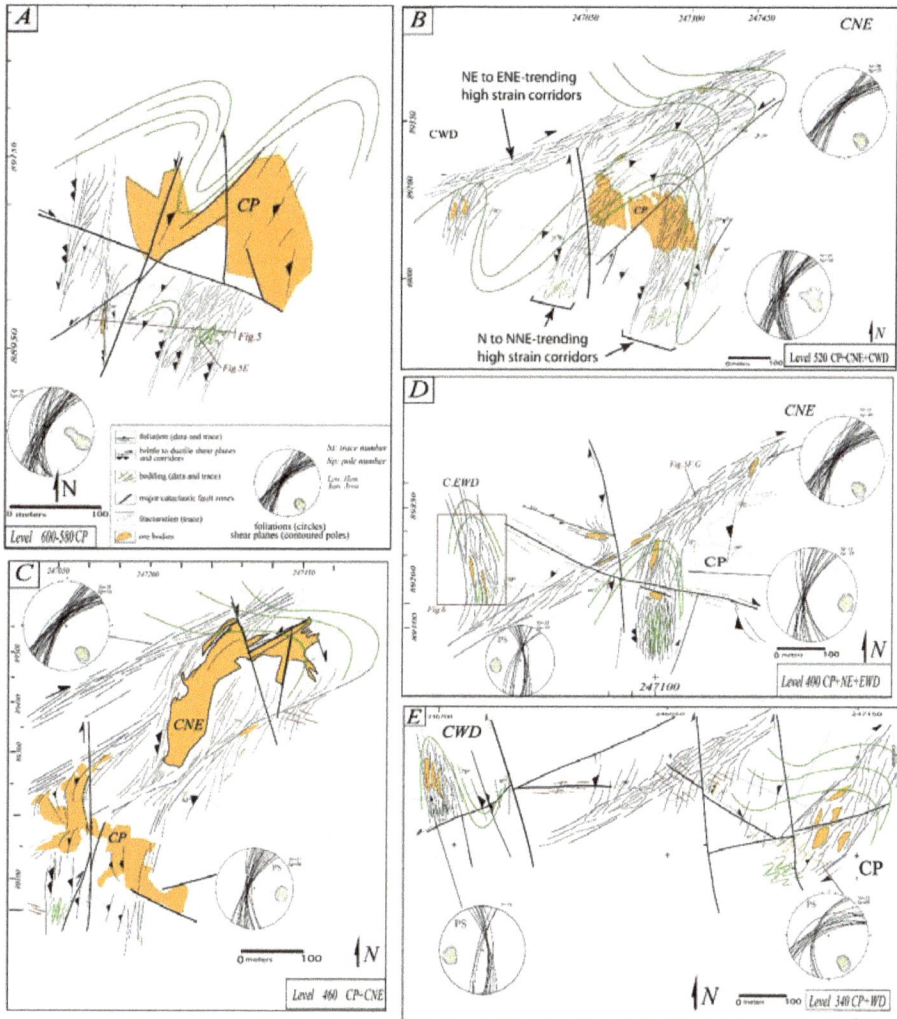

Figure 4. Structural maps of the five main exploitation levels (decreasing altitude from **A** to **E**) in the Hajjar mine. High strain corridors are marked by the development of dense foliation and shear planes. The light blue traces are galleries. The coordinates are taken from the mine's own system. Note that the scale is slightly different for each level. The following acronyms are used for the ore bodies (translated from French): CP = main body; CNE = north-eastern body; CWD = western body; CEWD = extreme western body. The CP ore body has been intensively exploited and some zones are no longer accessible, structural data from Hibti (1993) [23] were then added and carefully projected in these areas (see text for explanation).

4.2. The N to NNE-Trending High Strain Corridors

These corridors have been almost fully mapped at all exploitation levels (Figure 4). Two types of strain corridors can be distinguished: the first is characterized by dominant reverse shear planes and folds with an axial planar cleavage S1 (called a "reverse corridor" below, Figure 5A–E), whereas the second corresponds to the development of a strong and penetrative S1 foliation with horizontal stretching lineation without the occurrence of reverse shear planes (called a "flattening corridor" below, Figure 6).

Within reverse corridors, shear planes present dominant reverse rather than strike-slip kinematics. In the map view (Figure 4), the obliquity between S1 and the shear planes, which seems to indicate a sinistral sense of shear, is an artifact as the strike-slip component which is low and dextral when it is observed. The noticeable meso-scale structures are: i) eastward verging thrusts and decollements, most of time using a weak pyrrhotite-rich layer/body as the sole, which is near-parallel to the bedding in the foot-wall (Figure 5A–C). The associated folds in the hanging-walls developed an axial-planar cleavage S1. Typical meter-scale detachment folds, with thickening of the sulfide-rich decollement level, are frequent (Figure 5C), which might explain the "corrugation" observed along the decollement plane (Figure 5B); ii) the high strain corridors are characterized by the development of an intense foliation associated with similar upright NS-oriented folds (Figure 5A and D) which are frequently in association with reverse shear bands responsible for "pop ups" (Figure 5E). Local evidence of the oblique-slip component is provided by oblique stria, the "pop ups" then corresponding to dextral positive flower structures (Figure 5E).

Within the flattening corridors where thrusting is not observed, bedding marked by sulfide-rich ribbons is fully overprinted by the S1 foliation which bears a horizontal NS stretching lineation (Figure 6A). With increasing strain, the rock color changes to a very dark and black tint. To the west, a massive sulfide body is exploited (CEWD). This body is not continuous as it is instead composed of several distinct massive sulfide lenses aligned parallel to S1. The termination of the sulfide lenses is wavy due to the occurrence of small-scale folds of sulfide ribbons or host rocks. This sulfide lens morphology is frequently observed throughout the mine (e.g., Figure 8D). Near the termination, these lenses integrate clasts of host rocks (Figure 6). Cm- to dm-thick veins are abundant along the high strain corridor (Figure 6A). Locally, tips of massive sulfide lenses present triangular veins (or "saddle reef") at a "triple junction" position with respect to the foliation (Figure 6).

4.3. The NE to ENE-Trending High Strain Corridors

They are typical dextral shear zones, as indicated by drag folds in map view and obliquity between the near vertical S1 and the ductile shear planes (Figures 4 and 5F). The lineation is horizontal along the shear planes which often presents a graphitic/silvery mirror surface. The angle between S1 and the shear planes may be very low to null thus defining a mylonitic foliation locally (e.g., Figure 4C). These dextral shear planes, steeply dipping and trending NE to ENE (Figure 4), may present brittle characteristics as a gouge zone at the outcrop scale (Figure 5G). In the gallery, this type of high strain corridor is generally responsible for slope/wall instabilities, which makes access, oriented sampling, and structural data collection difficult, particularly where the strain corridors intersect large ore bodies (e.g., northern border of the CNE mineralized body, Figure 4C).

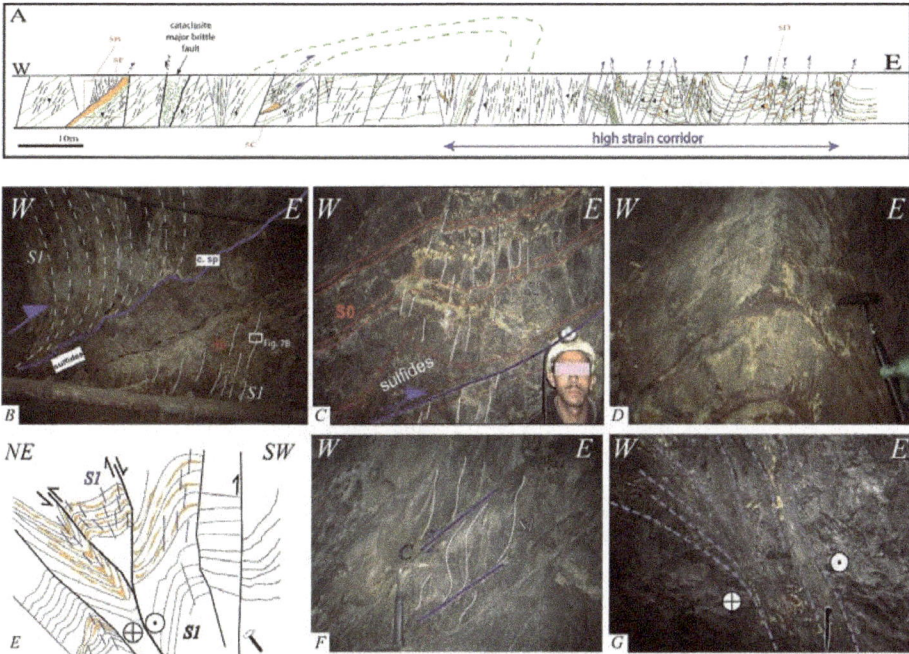

Figure 5. Structures and deformation of the Hajjar ore deposit. A to E are from the N to NNE-trending high strain "reverse"corridors; F and G are from the NE to ENE-trending high strain corridors. (**A**) Cross-section along the gallery from level 600-580 (see location in Figure 4A). S0 is shown in green, S1 is in black, the brittle to ductile shear planes are given in blue, the main massive sulfide bodies are shown in orange. The section is located within the footwall of the CP and is mainly composed of stratified greso-pelites and tuffs with mm- to cm-thick sulfide ribbons (containing mostly pyrrhotite and pyrite with a small amount of chalcopyrite) with no economic interest. The intensity/spacing of the foliation and high frequency of the shear planes can be used to depict the high strain corridors. Most of the brittle to ductile shear planes have an apparent reverse component: (**B**) an east-verging thrust developed within a pyrrhotite-rich massive sulfide deposit acting as a decollement layer. In the hanging-wall, the bedding is not observed whereas the S1 cleavage is curved by top-to-the-east drag folding. Both massive sulfide wallrocks are corrugated (c. sp: corrugated shear plane); (**C**) Detachment fold above a pyrrhotite-rich sulfide layer thickened within the core of a disharmonic fold hinge The S1 axial planar foliation is well-developed in the hanging-wall; (**D**) Upright similar fold with associated axial planar S1 cleavage. The pyrrhotite-rich red ribbons are extremely thinned in the limbs and thickened within the hinge zone; (**E**) Positive flower structure associated with similar drag folds and S1 cleavage (line drawing from level 600-580, Figure 4A). Along the N15E-trending faults, high dipping stria show that the reverse component is dominant relatively to the dextral strike-slip one; (**F**) Ductile dextral NE-trending near the vertical shear planes (C) and associated S1 foliation within a NE-trending right-lateral high strain corridor in meta-siltstones (location CNE area, Figure 4D); (**G**) ENE-trending steep dextral shear zones marked by foliated gouges and various branches (sense of shear is determined in the gallery roof, location in Figure 4D).

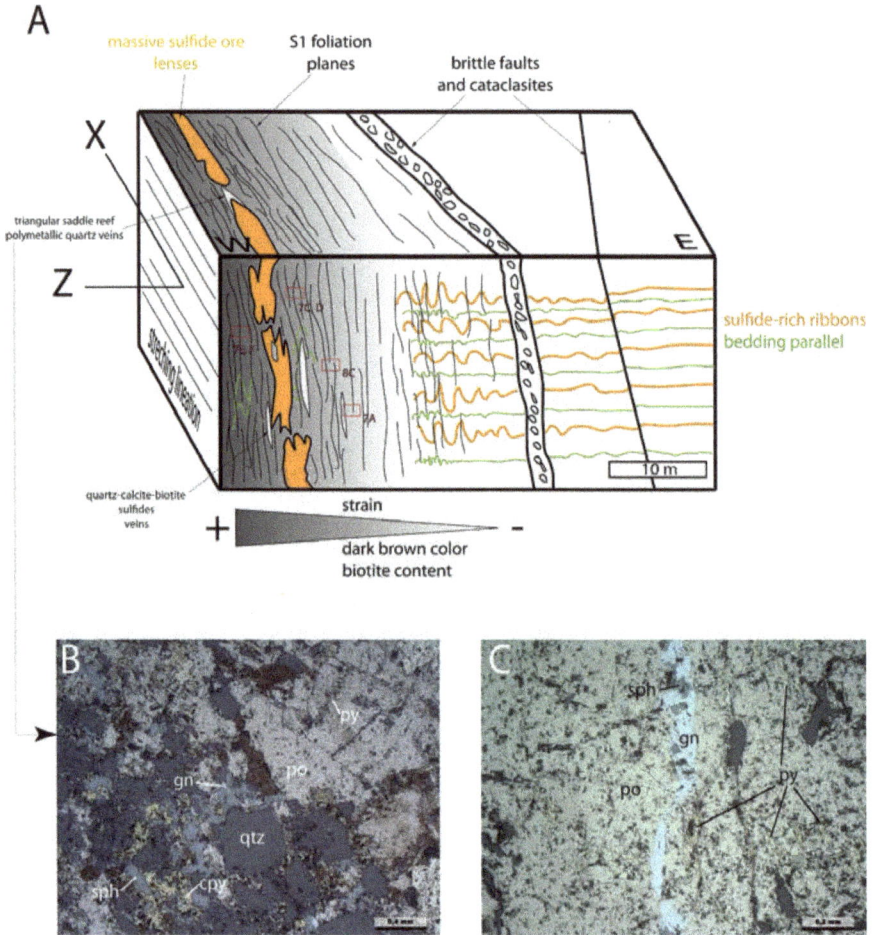

Figure 6. Outcrops of the extreme western body (CEWD) gallery, a typical N-trending "flattening" corridor. (**A**) 3D man-made sketch of the CEWD cross-section located in Figure 6A. The exploited massive ore bodies correspond to meter-scale lenses aligned within a high strain zone marked by an intense foliation in dark host rocks with high biotite and sulfide content. X and Z are the long and short axis of the strain ellipsoid respectively; (**B,C**) Thin-section photographs of the triangular veins developed at the massive ore lenses termination (RL). The vein is mainly filled with quartz associated with a polymetallic assemblage. Pyrrhotite is replaced by pyrite along cracks (**B,C**) and sphalerite/galena (± chalcopyrite) veinlets crosscut the former pyrrhotite and arsenopyrite grains (not shown).

4.4. Microstructures and Textures in the Host Rocks

Oriented thin sections were taken from the Hajjar MSD, especially from the N to NNE-trending high strain corridors described above.

The primary lithologies and associated syn- or diagenetic hydrothermal halos have been metamorphosed and/or altered. This metamorphism appears to be better expressed in high strain corridors, i.e., when the S1 foliation planes are densely represented (e.g., Figure 6A). Two metamorphic assemblages can be distinguished in the metapelites. The first one is comprised of quartz + biotite + andalousite (± calcite). (Figure 7A,C,D). Cordierite has not been observed, but the shape of some porphyroblasts replaced by white mica aggregates indicate the presence of this mineral (e.g., Figure 8E). The second assemblage is made of quartz + chlorite + muscovite (± carbonate). This last assemblage can also be observed in sandy-pelites with sulfide-rich ribbons, where it post-dates and locally replaces the biotites (Figure 7B). Foliation-parallel veinlets are filled with quartz and large biotite crystals associated with calcite in the geodic cavities (Figure 7A).

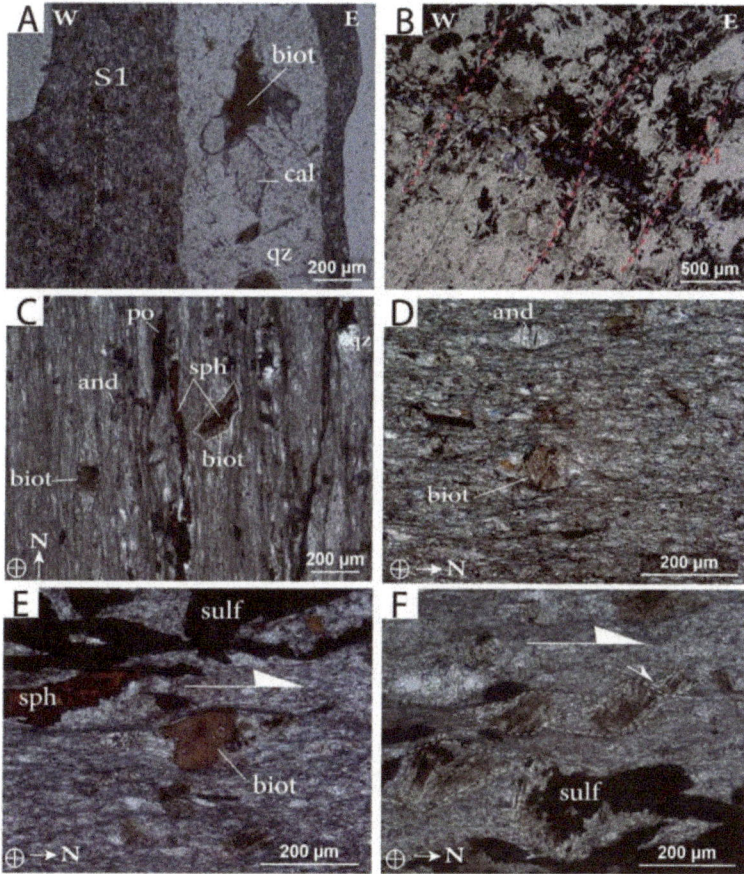

Figure 7. Thin section microphotographs of the S1 foliation and associated porphyroblasts within the Hajjar host rocks. A, C, D, E and F are from the flattening corridors; B is from the reverse corridors. (**A**) Quartz (qz), calcite (cal) and biotite (biot) vein parallel to the incipient S1 foliation, vertical section, see location in Figure 6, NAPL. The host rock presents a fine-grained granoblastic texture composed of biotite and andalousite grains with a local preferred orientation defining an incipient foliation plane; (**B**) Footwall of the thrust (see location in Figure 5) with the So plane marked by sulfide-rich ribbons (in blue) and discrete S1 planes (in red) characterized by muscovite (white laths) crystallization (NAPL). Please note that the non-oriented biotite (i.e., "static") porphyroblasts are replaced by chlorite (pale green); (**C–F**) Horizontal thin sections parallel to the stretching lineation showing the main foliation plane S1 marked by elongated sulfides (sulf) and particularly pyrrhotite (po) and sphalerite (sph); see location in Figure 6. Like the fine-grained foliation, the pressure shadows and caps around the andalousite (and) and biotite (biot) grains are composed of quartz, white micas, chlorite and local carbonates. In the high strain area, asymmetric pressure shadows around the biotite indicate a non-coaxial regime with a dextral sense of shear (**E,F**).

In areas where the foliation is weakly developed, biotite and andalousite porphyroblasts show a granoblastic "static" texture with a very subtle preferred orientation locally (Figure 7A). With increasing strain, biotite porphyroblasts are generally coarser and present a preferred orientation parallel to the fine-grained S1 foliation, a planar axial surface with micro-folds (Figure 8C). In high strain zones, pressure shadows and strain caps are found around some biotite and andalousite crystals (Figure 7C to F), whereas other biotite crystals remain nearly free of foliation deflection (e.g., a biotite crystal growing around a sphalerite core in Figure 7C). The pressure shadows are generally composed of quartz, muscovite, and chlorite, i.e., the same assemblage constituting the fined-grained foliation (Figure 7D,E). Asymmetric pressure shadows around biotite are common in flattening corridors attesting to a non-coaxial regime, at least locally (e.g., dextral in the CEWD outcrop, Figure 6, Figure 7E,F). Therefore, in high strain and non-coaxial zones, biotite crystals appear as pre-tectonic prophyroblasts, suggesting severe non-coaxial strain increments after the HT/LP metamorphism peak.

4.5. Microstructures and Textures in Sulfides

The internal ductile/plastic strain of the sulfide ribbons and bodies is widespread and high in the Hajjar MSD. This is particularly due to the high content of pyrrhotite within the ore.

Associated with folding, the S1 axial planar cleavage is marked by flattened sulfide ribbons within the XY plane with refractions and hinge thickening (Figure 8A,B). The preferred interpretation is that a refraction mechanism is responsible for this rather than a "bed to bed" flexural slip, because no shear planes have been observed in quartz (Figure 8B). Normal thin sections with XY planes cannot be used to characterize a preferred stretching X direction with respect to the 3D sulfide micro-lens dimensions. Therefore, the strain ellipsoid is mainly oblate and the strain regime is close to pure flattening. In greater detail, different rheological behaviors of the sulfides are expressed along S1, with pyrrhotite behaving much more plastically than chalcopyrite and sphalerite (Figure 8B), which present both a "ductile" and brittle response to the stress. Frequently, the sulfide micro-lenses present an "X-shape" or "chromosome-like" morphology (Figure 8A,C). The elongated sphalerite grains surrounded by pyrrhotite within a foliated silicate matrix (Figure 8C) suggest that the conjugate effects of folding, recrystallization, and rheological contrasts explain this "X" morphology. As mentioned above, this peculiar "X-shape" morphology is also encountered at a larger scale in the edge and tip areas of massive ore bodies (Figures 6 and 8D).

In many places at the Hajjar MSD massive sulfide bodies present an internal planar fabric. Three types of fabrics can be distinguished: i) a planar fabric parallel to the S1 foliation within the host rocks (Figure 8D). In this case, the fabric corresponds to flattened pyrrhotite grains (with various chalcopyrite, sphalerite and galena contents), separated by elongated metamorphic silicate slices/lenses (Figure 8E). The metamorphic assemblage is represented by chlorite and white micas replacing former biotite/andalousite (/cordierite?) blasts (Figure 8E). These silicate slices can be very thin or even absent in the most enriched ore. It is noteworthy that the mechanical twinning of pyrrhotite is regularly distributed in a direction normal to the planar fabric, providing evidence for the tectonic origin of this foliation; ii) mylonitic zones affecting weakly deformed massive sulfides (Figure 8G). These mylonites can be observed where high strain corridors intersect or encompass mineralized bodies. C/S-type structures are common within the sulfide mylonites (Figure 8G). Flattened pyrrhotite grains define the S planes, whereas C planes present finely cataclased sphalerite and chalcopyrite in a foliated silicate gouge; iii) the third planar fabric corresponds to a mineralogical and textural banding marked by alternations of sphalerite-rich and sphalerite-poor ribbons (Figure 8F). For instance, this banding is either parallel to the S1 foliation in the wall rocks or parallel to the bedding planes in the footwall of decollement layers (Figures 5 and 8F). Pyrrhotite grains are elongated with no systematic mechanical twinning. Sphalerite does not show systematic elongation, and the quartz grains are elongated with undulose extinction (Figure 8F). Therefore, the respective part of the syngenetic and diagenetic vs. tectonic processes are still unclear, and cannot be used to explain this banding.

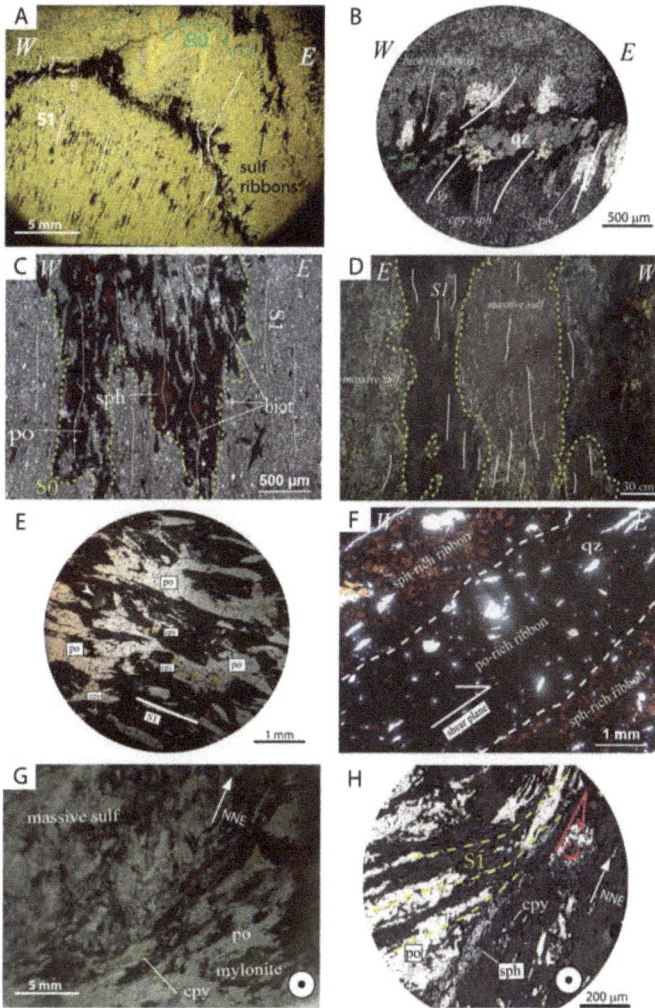

Figure 8. Deformation and textures of the sulfides in the Hajjar deposit. (**A**) The folding and associated S1 foliation of fine-grained sediments containing early sulfide-rich ribbons parallel to the bedding (S0). Note the cleavage refraction and thickening of the hinge zone due to the plastic behavior of pyrrhotite; (**B**) Details of A with pyrrhotite flowing along the stretching direction whereas the behavior of chalcopyrite and sphalerite is less plastic. A metamorphic assemblage mainly composed of muscovite and chlorite (± biotite) grows parallel to S1; (**C**) Micro-fold affecting a sphalerite and pyrrhotite-rich thin ribbon (CEWD, location in Figure 6A). The axial planar cleavage S1 is marked by the stretching of sulfides and elongated biotite blasts; (**D**) Massive sulfide lenses separated by strongly foliated host rock slices (south of CP, altitude 500 m). The ore bodies are internally banded parallel to the S1 foliation; (**E**) Texture of deformed pyrrhotite-rich massive sulfide (RL) parallel to the S1 foliation. The dark grey areas correspond to a muscovite/chlorite (replacing biotite locally) assemblage. Andalousite or cordierite porphyroblast ghosts are replaced by white micas (arrow); (**F**) textural and mineralogical banding within a massive sulfide body in the sole thrust (see location Figure 5A,B). Note the elongation of the quartz grains; (**G**) Massive sulfide sample affected by ductile shearing and mylonitization (SE part of the CP, level 400, the local name is "la bande Sud-Est"); (**H**) Details of G, thin section, RL. The sulfide mylonites present typical C/S structures. It should be noted that sphalerite appears to be "localized" in the C planes. The sample view from the bottom shows a dextral sense of shear.

5. Interpretation

5.1. Hajjar Mine and N'Fis Block: One Single Foliation (Not Two)

The rocks of the Hajjar mine are affected by one single flattening XY plane which is near vertical and trends from N0 to N45. The maps of the S1 trajectories (Figure 4) show that the deformation is not homogenous at the mine scale. In the high strain corridors, this XY plane corresponds to a S1 penetrative foliation overprinting the entire rock, whereas in less deformed areas, S1 is a slaty cleavage that is axial planar in similar folds. Host rocks and sulfide bodies present the same silicate metamorphic assemblages (Figures 7 and 8). With respect to this foliation, the qtz + biot and assemblage presents either a "static" granoblastic texture when the strain is low (i.e., weakly developed foliation, Figure 7A) or pre- to syn-tectonic features when the foliation is strongly expressed (Figures 7C to E, 8C). The texture, shapes, and aggregates of the biotite and andalousite (± suspected cordierite) are typical of HT/LP "contact" metamorphism in the hornfels facies. The syn-tectonic assemblage is composed of quartz + chlorite + white micas (± calcite) and partially replaced the former biotite and andalousite blasts (Figures 7B and 8E).

Similarly, surface data from the N'Fis block (Figures 2 and 3) show the occurrence of a single sub-vertical XY plane oriented N130. This flattening plane is a penetrative foliation secant to slumps (Figure 3B) and axial-planar to P1 folds (Figure 3C). Contact metamorphic biotite blasts are elongated parallel to the foliation and appear as flattened sulfide grains (Figure 3D).

Therefore, these data imply that the Hajjar MSD and the N'Fis block are affected by a single foliation which encompasses a HT/LP contact metamorphism. Although a single Variscan foliation was similarly recognized by Dias et al. (2011) [35] at the regional scale, our results disagree with the previously published works on the Hajjar mine/N'Fis area: i) first, two foliations were identified and consequently two successive tectonic events with sub-normal horizontal shortening directions were invoked [12,23]. In particular, the N20-30 dry joints affecting the N'Fis block at the surface (Figure 2) cannot be related to the N0-30 penetrative and ductile foliation observed in the Hajjar mine. Moreover, there has been no direct observation of an early foliation/cleavage in the Hajjar galleries during our study; ii) second, the biotite blasts were interpreted as post-tectonic with respect to the last deformation event [25].

5.2. The Hajjar Mine is Located within a Regional-Scale Shear Zone

The direct consequence of the previous result is the occurrence of a large foliation virgation from the Imarine outcrops to the Hajjar MSD (Figure 2). This virgation in the orogeny is typically caused by wrenching along regional shear zones. Our structural data from the surface (e.g., Tiferouine outcrops, Figures 2 and 3E) and from sub-surface structural maps (Figures 4 and 5) in the Hajjar underground mine fully support the occurrence of a major right-lateral ENE-trending transpressive shear zone at Hajjar.

The strain pattern on the maps show that shear planes are heterogeneously distributed as they are clustered within the shear corridors. Along a broad ENE direction, we identified various types of high strain corridors (see the 2D pattern in Figure 9): i) the N to NNE-trending corridors correspond to either reverse corridors characterized by thrusting and associated folding with a low amount of dextral oblique-slip (Figures 4 and 5A–E) or to flattening corridors characterized by a tight and penetrative foliation with horizontal stretching lineation and a local dextral sense of shear (Figures 6 and 7); ii) The NE to ENE-trending corridors correspond to unequivocal vertical dextral shear zones. The orientation of both strain corridors are connected and form an asymmetric 3D anastomosed pattern that is compatible with a bulk dextral sense of shear along a N60-70 direction (Figure 9). Reverse corridors with vertical thickening indicate that the Hajjar shear zone is transpressive. This result is fully in coherence with previous works dealing with the Western Meseta, in which dominant dextral strike-slip tectonics were clearly identified during the Variscan orogen [35]. However, in the location near the Hajjar mine, a regional scale shear zone of this type has not been previously recognized and constitutes a key structural feature of the Guemassa Hercynian orogenic segment.

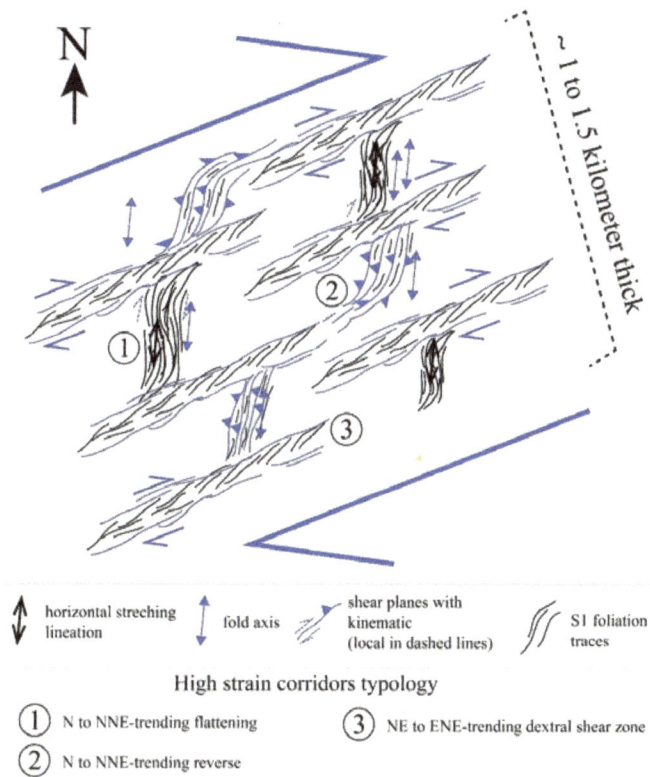

Figure 9. Simplified and conceptual map view model of the internal strain pattern within the Hajjar transpressive right-lateral shear zone (see text for explanation).

The shear planes of the Hajjar MSD present both ductile and brittle features (Figures 5 and 8G, H). The last brittle increments cross-cut and offset the former S1 foliation along the gouge zones (Figure 5G). Asymmetric biotite blasts with pressure shadows filled with the chlorite and white mica assemblage (Figure 7E,F) argue for simple shearing after the thermal peak of the HT/LP contact metamorphism. Lower or retrograde metamorphic conditions during simple shearing are also indicated via the cataclasis of sphalerite and chalcopyrite along the shear planes within mylonitic zones affecting massive sulfides (Figure 8H). Therefore, the Hajjar shear zone records simple shearing increments during and after the development of the widespread S1 foliation.

Last, the Atlasic brittle reactivation of this Hercynian shear zone cannot be ruled out, however it is still difficult to precisely depict this.

5.3. Ore Deformation and Remobilization

As recognized in previous studies [25], our data indicate that the Hajjar mineralization is strongly deformed and metamorphosed. It is affected by folding, foliation and mylonitic bands within a regional scale shear zone. Structures such as pyrrhotite-rich ribbons clearly pre-date the deformation and the HT/LP contact metamorphism (e.g., Figures 6 and 8A). The primary syn- or digenetic mineralization is then strongly reworked by deformation. In particular, at the meter scale, we present clear evidence of tectonic thickening within the fold hinge zone. The wavy termination of the metric-scale massive sulfide lenses parallel to S1 suggests that these lenses were likely thickened by folding before they were flattened within the XY plane of S1 (Figures 6 and 8D). This mechanism is enhanced by the high "plasticity" of pyrrhotite, which is by far the dominant sulfide at Hajjar. The pre-to syntectonic HT/LP metamorphism greatly favor the ductile behavior and recrystallization of sulfides including chalcopyrite and sphalerite. This is observable at the thin section scale, where the tectonic thickening induced the stress-oriented recrystallization of sphalerite, leading to an incipient "banding" of sphalerite-rich/sphalerite-poor slices parallel to S1 (Figure 8C). We suggest that, in Hajjar MSD, this solid-state thickening and remobilization are effective at a larger scale, but further modern textural and mineralogical studies are required in order to be able to investigate this point.

Remobilization of the primary metal stock by fluids (e.g., the fluid state processes and chemical remobilization described by Gilligan and Marshal (1987) [3] is also expressed in the Hajjar MSD. Even though the metal mass balance quantification is outside the scope of this study, the polymetallic veins argue for hydrothermal fluid-assisted remobilization during deformation. In particular, the polymetallic triangular veins at the tips of the massive sulfide lenses indicate such remobilization. This type of vein with a polymetallic assemblage associated with quartz, newly formed sphalerite and galena veinlets, and pyrrhotite replacement by vermicular pyrite (Figure 6B,C), is similar to the so-called "piercement veins" described by authors working on deformed MSD (e.g., [3,38–40]). It has been hypothesized that the metamorphic fluids liberated during the prograde HT/LP contact metamorphism (e.g., quartz veins with biotite in Figure 7A), combined with potential advective hot magmatic fluids exsolved from deeper granitic bodies, are able to chemically rework the primary sulfides and concentrate metals into dilatant sites as triple junction veins during the last increments of deformation [3]. Due to high reactive chemistry, the fluid-assisted chemical reworking of primary VMSD is common in many metamorphic contexts other than HT/LP metamorphic conditions (e.g., [41] and references therein).

6. Discussion: Toward an Integrated Tectono-Metamorphic Model for the MSD-Bearing Jebilet and Guemassa Massifs

These interpretations must be discussed in terms of the ages and tectono-metamorphic evolutions established for the Guemassa and Jebilet Massifs; both of these Massifs bear the major MSD in Morocco.

First, syn-sedimentary structures and soft sediment deformation have been identified in the N'Fis block (Figure 3A,B) and in the Hajjar mineralization [23]. These structures are well known within the Visean Sarhlef syn-rift formation in the Jebilet Massif, and correspond to slope instabilities during the opening of the Jebilet basin from 370 to 325 Ma (the D0 transtensive event described by Delchini (2018) [26]). Coeval with this sedimentation, the basin underwent significant bimodal and calc-alkaline magmatism, leading to many intrusions within the sediments and the basement. Consequently, the thermal gradient is very high [26], and primary syn- to diagenetic massive sulfide mineralization occurred within the volcano-sedimentary sequences (Figure 10A). The initial morphology (e.g., normal fault locations and trends, depocenters, etc.) of these basins and sub-basins is not constrained in the Guemassa Massif contrary to the Jebilet Massif where the basins are interpreted as pull-apart systems with NNE-trending normal faults and associated N70E-trending left-lateral strike-slip faults [42]. Therefore, it is likely that the Guemassa Massif and the Hajjar shear zones acted as sinistral strike-slip faults during this Early Carboniferous period (Figure 10A), lateral N to NNE trending normal faults accommodating the formation of local subsiding basins such as the Hajjar one (Figure 10A). However, because it may control the initial MSD distribution, further detailed work is required to specify the Early Carboniferous basin geometries in the Guemassa Massif.

Second, absolute dating of the Hajjar biotites related to thermal aureole metamorphism has been performed by Watanabe (2002) [33] using ^{40}Ar/^{39}Ar dating and yields an approximate age of ca. 301 Ma. This age and the associated HT/LP metamorphic assemblage are compatible with the M2b metamorphism reported in the Jebilet Massif [26,28]. In other words, the S1 foliation and HT/LP metamorphism that we document in the Hajjar MSD and in the N'Fis block within the Guemassa

Massif are structurally and temporally similar to the D2b tectono-metamorphic event described in the Jebilet Massif to the north (see the section on geological settings above and [26]). It is noteworthy that this thermal event is not restricted to the Hajjar mine, as it has been traced by Raman Spectroscopy of Carbonaceous Materials geothermometry method (RSCM) throughout the whole N'Fis block [27]. In the Jebilet as in the Guemassa Massifs, it has been reported that this thermal event is the consequence of hidden plutonic intrusions. Our data suggests the presence of fluid-assisted HT/LP "contact" metamorphism (Figure 7A). Therefore, the vigorous advection of hot fluids exsolved from melts and/or which come from metamorphic devolatilization may also partly explain the large extent of this HP/LP metamorphism observed in the Guemassa and Jebilet Massifs close to 300 Ma. This regional thermal anomaly is represented in Figure 10B. No former foliation/cleavage has been observed in either the N'Fis block or in the Hajjar mine, suggesting that the D2a/M2a event identified by Delchini (2018) [26] in the Jebilet Massif is not expressed in the Guemassa Massif. This is in agreement with the fact that the D2a/M2a event, which reaches the garnet-staurolite amphibolite facies, is poorly represented in the Jebilet Massif, and better expressed northward in the Rehamna Massif. Thus, the S1 foliation/cleavage characterized in this study matches the S2b foliation identified in the Jebilet Massif to the north by Delchini et al. (2016) [28]. Biotites related to the HT/LP metamorphism are not post-kinematic, as proposed by Hibti (1993) [23]. They are pre- to syn-kinematic, which implies that deformation occurred during the peak of the HT/LP "contact" metamorphism (Figure 9B). Based on the biotite blasts vs. strain relationship, we suggest that during the HT peak, the deformation was predominantly coaxial before shifting to a bulk non-coaxial regime.

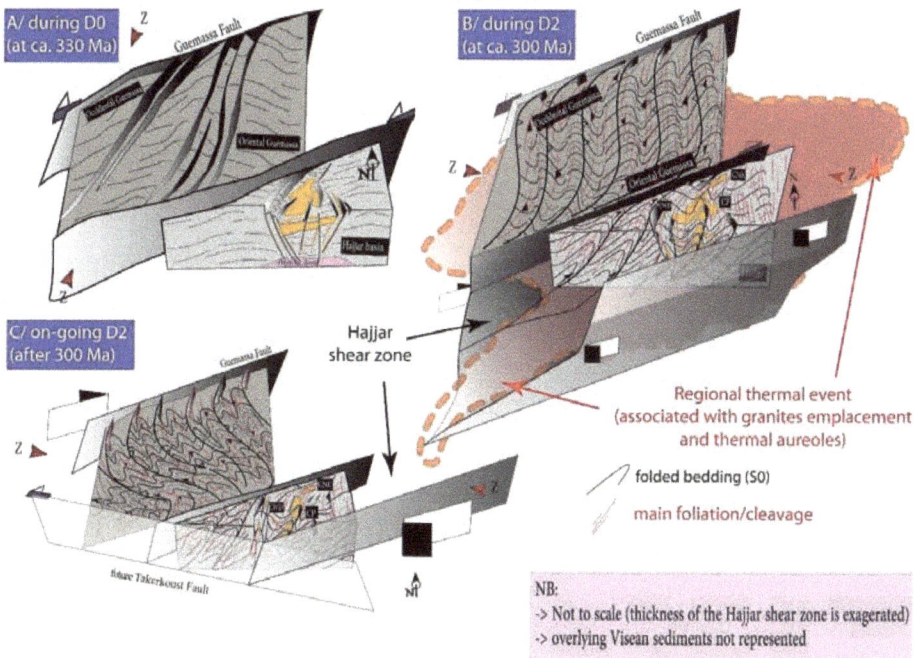

Figure 10. Tectono-metamorphic model of the Hajjar shear zone and associated MSD. The name of the tectonic events (D0, D2) corresponds to the tectonic events that have recently been established for the Jebilet Massif by Delchini (2018) [26]. D1 has not been identified in this study. See text for explanations.

Third, during the Early Permian, the D2 event identified in the Jebilet Massif ended with transpressive conjugate regional shear zones, oriented NE/ENE and SE/SSE with a dextral and sinistral (e.g., the MSZ, Figure 1) sense of shear respectively (i.e., the D2c event described by Delchini,

2018) [26]. This led to the development of a regional scale "flower structuration" of the Jebilet Massif. This strain localization along the shear zones appears to post-date the HT/LP contact metamorphism. Our data from the Guemassa Massif are fully compatible with this scenario (Figure 10C): the Hajjar regional shear zone we recognized in this study appears to be one of the dextral shear zones responsible for the large virgation of the main foliation planes. As observed in the Jebilet Massif, this shear zone corresponds to a progressive strain localization during the retrograde metamorphism when the D2 event ended. Last, as proposed by Dias et al. (2011) [35], conjugate WNW-ESE trending sinistral shear zones activated as the Lalla Takerkoust fault (Figure 10 C). This sinistral wrench zone accentuated and is responsible for the virgation of the S1 foliation, resulting in the "anarchic" WNW-orientation of the foliation observed through the N'Fis block.

7. Conclusion

The Guemassa Massif and the Hajjar base-metal massive sulfide deposit have been affected by a single foliation during a major Late Carboniferous-Early Permian Hercynian tectonic event. This foliation is strongly affected and deflected by regional scale shear zones such as the Hajjar N70-trending and right-lateral shear zone. Structural mapping in the Hajjar mine demonstrates that the Hajjar shear zone is complex with anastomosing shear plane patterns combined with thrusting and folding. This deformation is partially coeval, with a large thermal anomaly responsible for the HT/LP metamorphism. The tectono-metamorphic evolution of the Oriental Guemassa Hercynian segment is highly compatible with the evolution depicted for the Jebilet Massif. Strain under a high heat flux favored the deformation of the massive sulfides bodies which partly underwent fluid-assisted remobilization in the Hajjar mine. The tectonic thickening of the mineralization is observed at the meter scale, and must be re-examined at a larger scale.

Author Contributions: S.A. and Y.B. conceptualized both the study and the final model and wrote the original draft. L.B (Lakhlifi Badra) and L.B. (Luc Barbanson) reviewed and edited the draft. M.O., A.K., M.Z., L.M. gave their validation, Funding acquisition and project administration.

Funding: The PhD thesis of S. Admou has been partly funded by the "Office Mediterannéen de la Jeunesse" through a partnership between Orleans University (France) and Moulay Ismael University (Meknès, Morocco).

Acknowledgments: We are grateful to S. Janiec from ISTO and X. Le Coz from Geosciences Rennes who performed high quality thin sections. Our discussion with S. Delchini was greatly appreciated. We thank the reviewers and specially R. Dias for very fruitful and constructive review. The Guest Editor A. Chauvet is also thanked for inviting us to submit our work.

Conflicts of Interest: The authors declare no conflict of interest.

References

1. Cawood, P.A.; Hawkesworth, C.J. Temporal relations between mineral deposits and global tectonic cycles. In *Ore Deposits in an Evolving Earth*; Jenkin, G.R.T., Lusty, P.A.J., Mcdonald, I., Smith, M.P., Boyce, A.J., Wilkinson, J.J., Eds.; Geological Society of London: London, UK, 2013; pp. 9–21.
2. Graf, J.; Skinner, B. Strength and deformation of pyrite and pyrrhotite. *Econ. Geol.* **1970**, *65*, 206–215. [CrossRef]
3. Marshall, B.; Gilligan, L.B. An introduction to remobilisation: information from ore-body geometry and experimental considerations. *Ore Geol. Rev.* **1987**, *2*, 87–131.
4. Marshall, B.; Spry, P.G. Discriminating between regional metamorphic remobilization and syntectonic emplacement in the genesis of massive sulfide ores. *Rev. Econ. Geol.* **1998**, *11*, 39–80.
5. Marignac, C.; Diagana, B.; Cathelineau, M.; Boiron, M.-C.; Banks, D.; Fourcade, S.; Vallance, J. Remobilisation of base metals and gold by Variscan metamorphic fluids in the south Iberian pyrite belt: evidence from the Tharsis VMS deposit. *Chem. Geol.* **2003**, *194*, 143–165. [CrossRef]
6. Chauvet, A.; Onézime, J.; Charvet, J.; Barbanson, L.; Faure, M. Syn- to late-tectonic stockwork emplacement within the spanish section of the iberian pyrite belt: Structural, textural, and mineralogical constraints in the tharsis and la zarza areas. *Econ. Geol.* **2004**, *99*, 1781–1792. [CrossRef]

7. Barrie, C.D.; Boyle, A.P.; Prior, D.J. An analysis of the microstructures developed in experimentally deformed polycrystalline pyrite and minor sulphide phases using electron backscatter diffraction. *J. Struct. Geol.* **2007**, *29*, 1494–1511. [CrossRef]

8. Barrie, C.D.; Boyle, A.P.; Cook, N.J.; Prior, D.J. Pyrite deformation textures in the massive sulfide ore deposits of the Norwegian Caledonides. *Tectonophysics* **2010**, *483*, 269–286. [CrossRef]

9. Barrie, C.D.; Peare, M.A.; Boyle, A.P. Reconstructing the pyrite deformation mechanism map. *Ore Geol. Rev.* **2011**, *39*, 265–276. [CrossRef]

10. Reddy, S.M.; Hough, R.M. Microstructural evolution and trace element mobility in Witwatersrand pyrite. *Contrib. Mineral. Petrol.* **2013**, *166*, 1269–1284. [CrossRef]

11. Bernard, A.J.; Maier, O.W. Aperçus sur les amas sulfurés Massifs des hercynides Marocaines. *Miner. Depos.* **1988**, *23*, 104–114. [CrossRef]

12. Hibti, M. Les amas Sulfurés des Guemassa et des Jebilet (Meseta Sud-Occidentale, Maroc): Temoins de L'hydrothermalisme Précoce dans le Bassin Mesetien. Ph.D Thesis, University Cadi Ayyad, Marrakech, Morocco, 2001.

13. Belkabir, A.; Gibson, H.L.; Marcoux, E.; Lentz, D.; Rziki, S. Geology and wall rock alteration at the Hercynian Draa Sfar Zn–Pb–Cu massive sulphide deposit, Morocco. *Ore Geol. Rev.* **2008**, *33*, 280–306. [CrossRef]

14. Marcoux, E.; Belkabir, A.; Gibson, H.L.; Lentz, D.; Ruffet, G. Draa Sfar, Morocco: A Visean (331 Ma) pyrrhotite-rich, polymetallic volcanogenic massive sulphide deposit in a Hercynian sedimentdominant terrane. *Ore Geol. Rev.* **2008**, *33*, 307–328. [CrossRef]

15. Moreno, C.; Sáez, R.; González, F.; Almodóvar, G.; Toscano, M.; Playford, G.; Alansari, A.; Rziki, S.; Bajddi, A. Age and depositional environment of the Draa Sfar massive sulfide deposit, Morocco. *Miner. Depos.* **2008**, *43*, 891–911. [CrossRef]

16. Ben aissi, l. Contribution à L'étude Gîtologique des Amas Sulfurés Polymétalliques de Draa Sfar et de Koudiat Aïcha: Comparaison avec les Gisements de Ben Slimane et de Kettara (Jebilet Centrales, Maroc Hercynien). Ph.D Thesis, University Cadi Ayyad, Marrakech, Morocco, 2008.

17. Lotfi, F.; Belkabir, A.; Brown, A.C.; Marcoux, E.; Brunet, S.; Maacha, L. Geology and Mineralogy of the Hercynian Koudiat Aïcha Polymetallic (Zn-Pb-Cu) Massive Sulfide Deposit, Central Jebilet, Morocco. *Explor. Min. Geol.* **2008**, *17*, 145–162. [CrossRef]

18. Essaifi, A. Relations entre Magmatisme-Déformation et al.tération Hydrothermale: L'exemple des Jebilet Centrales (Hercynien, Maroc). Ph.D Thesis, Unversity of RennesI, Rennes, France, 1995.

19. Essaifi, A.; Hibti, M. The hydrothermal system of Central Jebilet (Variscan Belt, Morocco): A genetic association between bimodal plutonism and massive sulphide deposits? *J. Afr. Earth Sci.* **2008**, *50*, 188–203. [CrossRef]

20. Essaifi, A.; Goodenough, K.M.; Lusty, P.A.J.; Outigua, A. Microstructural and Textural Evidence for Protracted Polymetallic Sulphide Mineralization in the Jebilet Massif (Variscan Belt of Morocco). *Min. Resour. Sustain. World* **2015**, *1–5*, 1603–1606.

21. Lusty, P.A.J.; Goodenough, K.M.; Essaifi, A.; Maacha, L. Developing the lithotectonic framework and model for sulfide mineralization in the Jebilet Massif, Morocco: implications for regional exploration. In *Mineral Resources in a Sustainable World, Proceedings of the 13th Biennial SGA Meeting, Nancy, France, 24–27 August 2015*; André-Mayer, A.S., Cathelineau, M., Muchez, P.h., Pirard, E., Sindern, S., Eds.; Society for Geology Applied to Mineral Deposits (SGA): Genéve, Switzerland, 2015; pp. 1635–1638.

22. N'Diaye, I.; Essaifi, A.; Dubois, M.; Lacroix, B.; Goodenough, K.M.; Maacha, L. Fluid flow and polymetallic sulfide mineralization in the Kettara shear zone (Jebilet Massif, Variscan Belt, Morocco). *J. Afr. Earth Sci.* **2016**, *119*, 17–37. [CrossRef]

23. Hibti, M. L'amas Sulfuré de Hajjar, Contexte Géologique de mie en Place et Déformations Superposées (Haouz de Marrakech, Méseta Sudoccidentale, Maroc). Ph.D Thesis, University Cadi Ayyad, Marrakech, Morocco, 1993.

24. Zouhry, S. Étude Métallogénique D'un amas Sulfuré Viséen à Zn Pb Cu: cas de Hajar, Guemassa, Maroc. Ph.D Thesis, Ecole polytechnique de Montréal, Montréal, Canada, 1999.

25. Hibti, M.; Marignac, C. The Hajjar deposit of Guemassa (SW Meseta, Morocco): A metamorphosed syn-sedimentary massive sulfide ore body of the Iberian type of volcano-sedimentary massive sulfide deposits. In *Mineral Deposits at the Beginning of the 21st Century, Proceedings of the Joint Sixth Biennial SGA-SEG Meeting, Krakow, Poland, 26–29 August 2001*; A.A. Balkema: Lisse, The Netherlands, 2001; pp. 281–284.

26. Delchini, S. Etude Tectono-Thermique D'un Segment Orogénique Varisque à Histoire Géologique Complexe: Analyse Structurale, Géochronologique et Thermique du Massif des Jebilet, de L'extension à la Compression. Ph.D Thesis, University of Orléans, Orléans, France, 2018.

27. Delchini, S.; Lahfid, A.; Ramboz, C.; Branquet, Y.; Maacha, L. New Peak Temperature Constraints using RSCM Geothermometry on the Hajjar Zn-Pb-Cu Mine and its Surroundings (Guemassa Massif, Morocco). In Proceedings of the 13th SGA Biennial Meeting, Nancy, France, 24–27 August 2015.

28. Delchini, S.; Lahfid, A.; Plunder, A.; Michard, A. Applicability of the RSCM geothermometry approach in a complex tectono-metamorphic context: The Jebilet Massif case study (Variscan Belt, Morocco). *Lithos* **2016**, *256*, 1–12. [CrossRef]

29. Haimmeur, J. Contribution à L'étude de L'environnement Volcano-Sédimentaire et du Minerai de Douar Lahjar (Guemassa, Maroc), Lithologie, Paléo-Volcanisme, Géochimie et Métallogénie. Ph.D Thesis, École Nationale Supérieure de Géologie, Nancy, France, 1988.

30. Raqiq, H. Le bassin Carbonifère des Guemassa (Meseta Sud occidentale, Maroc): Lithostratigraphie, sédimentologie et évolution structurale. Ph.D Thesis, University Cadi Ayyad, Marrakech, Morocco, 1997.

31. Ouadjou, A. Pétrographie, Géochimie et Structure des Roches Magmatiques Antéschisteuses des Massifs Hercyniens des Guemassa et Souktana. Ph.D Thesis, University Cadi Ayyad, Marrakech, Morocco, 1997.

32. Ed Debi, A.; Saquaque, A.; Kersit, M.; Chbiti, A. L'amas sulfuré de Hajar (Guemassa, Maroc). *Chronique de la Recherche Minière* **1998**, *531–532*, 45–54.

33. Watanabe, Y. 40Ar/39Ar geochronologic constraints on the timing of massive sulfide and vein-Type Pb-Zn mineralization in the Western Meseta of Morocco. *Econ. Geol.* **2002**, *97*, 147–157. [CrossRef]

34. Soulaimani, A. L'évolution structurale des Massifs hercyniens du Haouz de Marrakech: Guemassa- N'fis (Maroc). Ph.D Thesis, University Cadi Ayyad, Marrakech, Morocco, 1991.

35. Dias, R.; Hadani, M.; Leal Machado, I.; Adnane, N.; Hendaq, Y.; Madih, K.; Matos, C. Variscan structural evolution of the western High Atlas and the Haouz plain (Morocco). *J. Afr. Earth Sci.* **2011**, *61*, 331–342. [CrossRef]

36. Hoepffner, C.; Soulaimani, A.; Piqué, A. The Moroccan Hercynides. *J. Afr. Earth Sci.* **2005**, *43*, 144–165. [CrossRef]

37. Saadi, M.; Hilali, E.A.; Bensaîd, M.; Boudda, A.; Dahmani, M. Carte géologique du Maroc, échelle 1:1 000 000. *Notes Mém. Serv. Géol. Maroc*. 1985. Available online: https://geodata.mit.edu/catalog/mit-gfcc2renabn5c (accessed on 6 October 2018).

38. Pedersen, F.D. Remobilization of the massive sulfide ore of the Black Angel Mine, central West Greenland. *Econ. Geol.* **1980**, *75*, 1022–1041. [CrossRef]

39. Maiden, K.J.; Chimimba, L.R.; Smalley, T.J. Cuspate ore-wall rock interfaces, piercement structures and the localization of some sulfide ores in deformed sulfide deposits. *Econ. Geol.* **1986**, *81*, 1464–1472. [CrossRef]

40. Plimer, I.R. Remobilization in high-grade metamorphic environments. *Ore Geol. Rev.* **1987**, *2*, 231–245. [CrossRef]

41. Gu, L.; Zheng, Y.; Tang, X.; Zaw, K.; Della-Pasque, F.; Wu, C.; Tian, Z.; Lu, J.; Li, X.; Yang, F.; et al. Copper, gold and silver enrichment in ore mylonites within massive sulphide orebodies at Hongtoushan VHMS deposit, NE China. *Ore Geol. Rev.* **2007**, *30*, 1–29. [CrossRef]

42. Aarab, E.M.; Beauchamp, J. Le magmatisme carbonifère pré-orogénique des Jebilet centrales (Maroc). Précisions pétrographiques et sédimentaires. Implications géodynamiques. *CR Acad. Sci. Paris* **1987**, *304*, 169–174.

MDPI

St. Alban-Anlage 66

4052 Basel

Switzerland

Tel. +41 61 683 77 34

Fax +41 61 302 89 18

www.mdpi.com

Minerals Editorial Office

E-mail: minerals@mdpi.com

www.mdpi.com/journal/minerals

www.ingramcontent.com/pod-product-compliance
Lightning Source LLC
Chambersburg PA
CBHW051727210326
41597CB00032B/5632